Nutrition *for* Foodservice *and* Culinary Professionals

Nutrition *for* Foodservice *and* Culinary Professionals

SEVENTH EDITION

Karen Eich Drummond,
Ed.D., R.D., L.D.N., F.A.D.A, F.M.P.

Lisa M. Brefere,
C.E.C., A.A.C.

WILEY

John Wiley & Sons, Inc.

Library of Congress Cataloging-in-Publication Data:

Drummond, Karen Eich.
    Nutrition for foodservice and culinary professionals / Karen Eich Drummond, Lisa M. Brefere. — 7th ed.
        p. cm.
    Includes bibliographical references.
    ISBN 978-0-470-05242-6 (cloth)
1. Nutrition.   2. Food service.   I. Brefere, Lisa M.   II. Title.

    TX353.D78 2010
    613.2—dc22
                                                                              2008044153

Printed in the United States of America
10 9 8 7 6 5 4 3 2

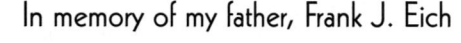

In memory of my father, Frank J. Eich

CONTENTS

PREFACE

Nutrition for Foodservice and Culinary Professionals, Seventh Edition, is written for students in culinary programs, as well as those in hotel, restaurant, and onsite management programs. Practicing culinary and management professionals will find it useful as well. As with previous editions, this is meant to be a practical how-to book tailored to the needs of students and professionals. It is written for those who need to use nutritional principles to evaluate and modify menus and recipes, as well as to respond knowledgeably to customers' questions and needs. As in the sixth edition, co-author Lisa Brefere, C.E.C., A.A.C., lends her firsthand experiences applying nutrition to selecting, cooking, and menuing healthy foods in restaurants and foodservices. After all, we eat foods not nutrients!

What's New for the Seventh Edition

Many important changes and additions have been made to *Nutrition for Foodservice and Culinary Professionals* to make this text even more user-friendly and up-to-date. Among the most significant changes are the following.

- The seventh edition's new, larger size allows for additional photos, drawings, menus, and tables to more effectively convey nutrition concepts and applications.
- Along with every book comes a copy of *Culinary Nutrition Manager,* an interactive diet and activity software program. *Culinary Nutrition Manager* allows students to keep a food journal, compare his or her nutrient intake to MyPyramid and Dietary Reference Intake values, and much more. Culinary students will also find *Culinary Nutrition Manager* to be particularly handy, as the comprehensive database of foods and ingredients allows them to create recipes and then view the caloric and nutritional content of each dish.
- Six new "Food Facts" and "Hot Topic" sections discuss timely issues such as trans fats, the glycemic index, functional foods and superfoods, herbs and botanicals, childhood obesity, and the impact of the American diet on the environment. The remaining "Food Facts" and "Hot Topic" sections are completely updated.
- Chapter 9, Healthy Menus and Recipes, now includes over a dozen examples of favorite recipes that have been reworked into healthier, yet just as tasty dishes. These recipes include not only main dishes but also sauces, dressings, desserts, and many others.
- Chapters 3 through 7, on carbohydrates, lipids, proteins, vitamins, and minerals, have been updated to include a greater focus on nutrition science. A "Nutrition Science Focus" feature in each chapter provides more in-depth information on the science and chemistry of nutrition.

Organization

The seventh edition of **Nutrition for Foodservice and Culinary Professionals** is organized into three major parts, beginning with an introduction to nutrition and foods, continuing on to provide advice on healthy recipes and menus, and finally relating nutrition to human health and lifespan.

Part I: Fundamentals of Nutrition and Foods (Chapters 1–7) consists of two introductory chapters, followed by five chapters concerned with specific nutrients, vitamins, and minerals. The first two chapters introduce basic nutrition concepts and explain how to use the Dietary Guidelines, MyPyramid, and food labels when planning menus. The next chapters focus on particular nutrients: carbohydrates, lipids, proteins, vitamins, water, and minerals. In-depth science information related to nutrients appears in each of these chapters.

Part II: Developing and Marketing Healthy Recipes and Menus (Chapters 8–10) begins with a chapter dedicated to the foundations of balanced cooking, including descriptions of how to use ingredients, flavoring principles, and cooking techniques to create healthy and delicious dishes. Chapter 9 introduces healthy menus, how to modify recipes, and hundreds of examples of healthy menu items for meals and snacks. Chapter 10 is concerned with the marketing of healthy foods in restaurants, foodservices, and beverage operations, with a new emphasis on responding to guest requests for options such as vegetarian, low-kcalorie, low-lactose, and gluten-free options.

Part III: Nutrition's Relationship to Health and Life Span (Chapters 11–13) looks at nutrition and health issues such as the relationship between nutrition and heart disease, cancer, diabetes, and obesity. Weight loss plans are also discussed in detail. Chapter 13 focuses on nutrition over the human life span, from pregnancy to the infant, child, adolescent, and older adult.

Learning Tools

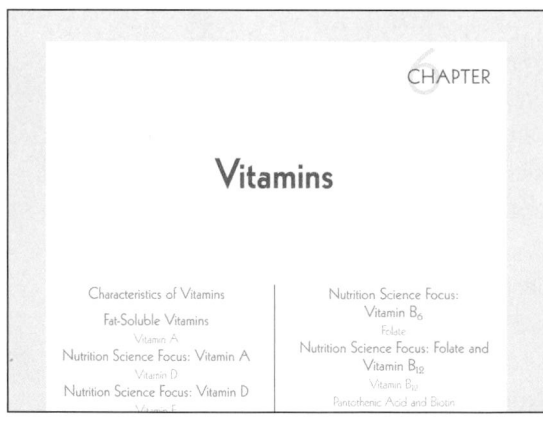

Nutrition for Foodservice and Culinary Professionals contains many special features that enable students to better understand concepts and extend and test their knowledge. These pedagogical tools include tables, charts, and illustrations, as well as the following:

Chapter Outline Each chapter begins with a brief overview of that chapter's content, allowing students to visualize the chapter as a whole.

Learning Objectives A bulleted list of learning objectives at the beginning of each chapter provides students with key points and a sequential organization of the chapter.

Key Terms and Concepts Whenever key terms and concepts are first introduced, their definitions can be found in these sidebars, located right next to the bolded term.

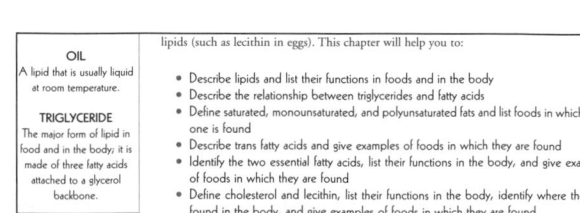

Mini-Summary Designed to help students focus on the important concepts within each chapter, a mini-summary is given after each section within a chapter.

Nutrition Science Focus Found in Chapters 3–7, this feature gives more in-depth scientific content on macronutrients and micronutrients.

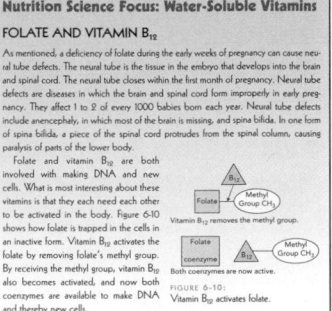

Nutrition Science Focus: Water-Soluble Vitamins

FOLATE AND VITAMIN B₁₂

CHEF'S TIPS

- See Figure 3-15 for information on flavor, uses, and cooking times for many grains.
- Figures 3-16 to 3-18 showcase many popular grains.
- Figure 3-19 and 3-20 highlight types of rice.
- Grains work very well as main dishes when mixed with each other or with lentils. For example, couscous and wheat berries are attractive, as is barley with quinoa. To either dish you could add lentils, vegetables, and seasonings. Keep in mind that each of the grains must be cooked separately with some bay leaf, onion, and thyme leaf and then strained, cooled, and mixed together with other grains and beans, depending on its application.

(*Text continues on page 108*)

Chef's Tips Chef's Tips provide an experienced chef's advice on all aspects of cooking, including which foods go together, how to use foods' natural colors to create an attractive dish, and how to use culinary techniques to create healthy and delicious dishes.

Food Facts These facts present in-depth information on relevant food-related topics, such as different types of oils and margarines, low-carbohydrate foods, caffeine, sports drinks, and fat substitutes.

FOOD FACTS: OILS AND MARGARINES

There is an ever-widening variety of oils and margarines on the market. They can differ markedly in their color, flavor, uses, and nutrient makeup (Figure 4-16).

When choosing vegetable oils, pick those high in monounsaturated fats, such as olive oil, canola oil, and peanut oil, or high in polyunsaturated fats, such as corn oil, safflower oil, sunflower oil, and soy-

on the olives, a more expensive process than using heat and chemicals.
- Olive oil, also called pure olive oil, is golden and has a mild, classic flavor. It is an ideal, all-purpose product that is great for sautéing, stir-frying, salad dressings, pasta sauces, and marinades.
- Light olive oil refers only to color or

War II, when it was introduced as a low-cost replacement for butter. Margarine must contain vegetable oil and water and/or milk or milk solids. Flavorings, coloring, salt, emulsifiers, preservatives, and vitamins are usually added. The mixture is heated and blended, then firmed by exposure to hydrogen gas at very high temperatures (see information about

HOT TOPIC: BIOTECHNOLOGY

Background

Potatoes with built-in insecticide. Rice with extra vitamin A. Decaf coffee beans fresh off the tree. What do these foods have in common? They have all been created using biotechnology and genetic engineering. *Biotechnology* is a collection of scientific techniques, including genetic engineering, that are used to create, improve, or modify plants, animals, and

Some traits are produced from the code contained in one gene; more complex traits depend on several genes. However, not all genes are switched on in every cell. The genes active in a liver cell are different from the genes active in a brain cell because the cells have different functions.

The language of DNA is common to all organisms. Humans share 7000 genes

3. Plant foods with desirable nutritional characteristics

One of the first GM foods to appear in the supermarket was the fresh tomato, called Flav Savr. If picked when ripe, tomatoes rot quickly, and so they are usually picked when green. The Flav Savr tomato was engineered to remain on the vine longer to ripen to full flavor before harvest. Once harvested, it did

Hot Topic The Hot Topic sections promote critical thinking and discussion forums on current issues related to nutrition, including fad diets, trans fats, botanicals and herbs, and childhood obesity.

Check-Out Quiz At the end of each chapter, a Check-Out Quiz allows students to check their comprehension of the chapter's concepts. Answers to these quizzes are found in Appendix E.

CHECK-OUT QUIZ

1. Match the nutrients with their functions/qualities. The functions/qualities may be used more than once.

Nutrients	Functions
Carbohydrate	Provides energy
Lipid	Promotes growth and maintenance

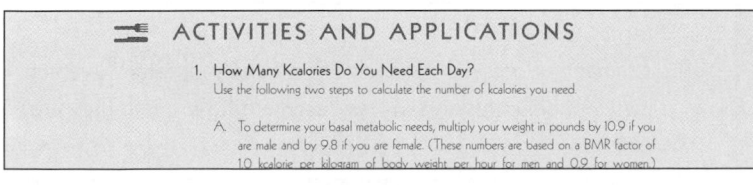

ACTIVITIES AND APPLICATIONS

1. How Many Kcalories Do You Need Each Day?
   Use the following two steps to calculate the number of kcalories you need.

   A. To determine your basal metabolic needs, multiply your weight in pounds by 10.9 if you are male and by 9.8 if you are female. (These numbers are based on a BMR factor of 1.0 kcalorie per kilogram of body weight per hour for men and 0.9 for women.)

Activities and Applications This section encourages students to extend their grasp of chapter concepts through analysis, problem-solving, and evaluation of nutrition-related questions and activities.

Glossary All key terms and definitions are listed in the glossary, easily found in the back of the book.

Appendices A very useful reference for readers, the appendices includes a variety of useful information, including a listing of nutritive values of foods commonly consumed in the United States, Dietary Reference Intakes for individuals by age, growth charts for children from birth to 20 years, and answers to the Check-Out Quizzes.

Culinary Nutrition Manager

Culinary Nutrition Manager software now accompanies every book and can be used in a variety of ways.

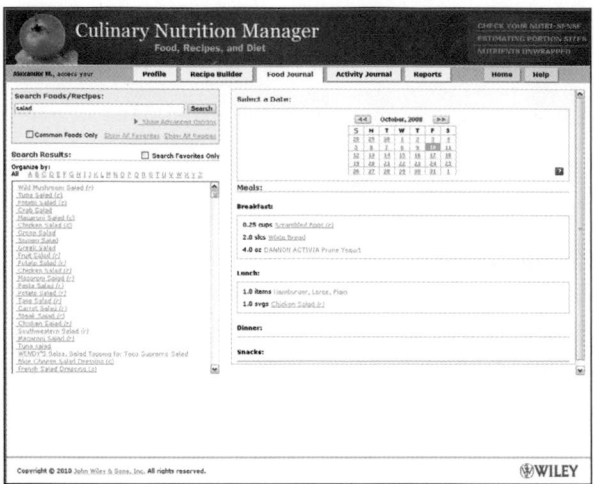

1. The most popular use of *Culinary Nutrition Manager* will be for students to enter what they eat in a Food Journal and then use a variety of reports to see how nutritious their diet is. The Food Journal is easy to use and the database is complete. There are four reports that students can use to analyze their diets: Macronutrient Distribution, Intake Spreadsheet, Intake Compared to DRI, and MyPyramid. The Intake Spreadsheet shows exactly how much of each nutrient was taken in. The other reports use excellent graphics and percentages to show how the student's intake compares to the DRI or MyPyramid, as examples.

2. Students can also use the Activity Journal to see how many kcalories they burn each day. By completing both the Food Journal and Activity Journal, students can use the "Energy Balance" report to see if they ate more kcalories than burned or the opposite.

3. Using the "Recipe Builder" feature, students can put their own recipes into the database and view the nutritional analysis.

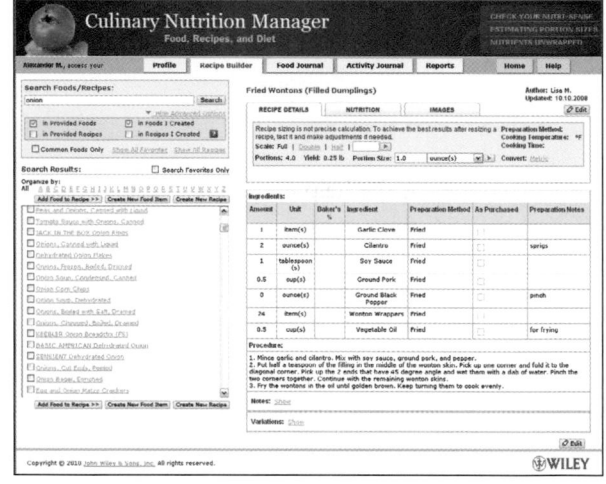

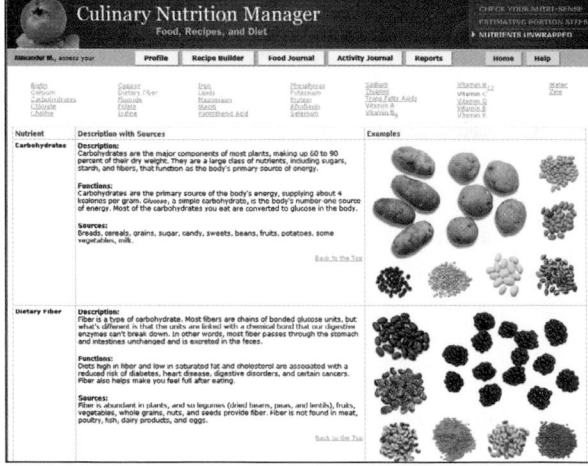

Culinary Nutrition Manager also includes these three useful features.

1. *Check Your Nutri-Sense* includes nine quizzes based on topics in the book.
2. *Estimating Portion Sizes* includes a variety of interactive activities that help students visually estimate how much they are eating.
3. *Nutrients Unwrapped* is a summary of important information on all nutrients, including carbohydrate, lipid, protein, vitamins, minerals, and water. It is an excellent review of nutrient functions and sources.

Supplementary Materials

A *Study Guide* (ISBN 978-0-470-28547-3) for students is available to help reinforce nutrition concepts and allow students to make nutrition applications.

A **Companion Website** (www.wiley.com/college/drummond) provides links to both the Student and Instructor Websites. The *Student Website* includes Powerpoints and Student Worksheets for each chapter. Powerpoints and Student Worksheets have been completely updated and expanded. The Student Website also includes Supplementary Recipes.

From the *Instructor Website*, instructors can download the Instructor's Manual and Test Bank as well as Powerpoint slides and Student Worksheets for each chapter. A selection of recipes are also available here.

The **Instructor's Manual** (ISBN 978-0-470-25728-9) includes class outlines, student worksheets, visual aids, and test questions and answers. Please contact your Wiley representative for a copy or go to www.wiley.com/college/drummond.

The **Test Bank** for this text has been specifically formatted for Respondus, an easy-to-use software for creating and managing exams that can be printed to paper or published directly to Blackboard, WebCT, Desire2Learn, eCollege, ANGEL, and other eLearning systems. Instructors who adopt **Nutrition for Foodservice and Culinary Professionals** can download the **Test Bank** for free. Additional Wiley resources can also be uploaded into your LMS course at no charge. To view and access these resources and the **Test Bank**, visit www.wiley.com/college/drummond, click on the "Instructor Companion Website" link, then click on "LMS Course Student Resources."

Acknowledgements

We are grateful for the help of all the educators who have contributed to this and previous editions through their constructive comments.

Chuck Becker, Pueblo Community College, CO
Marian Benz, Milwaukee Area Technical College, WI
Alex Bladowski, North Georgia Technical College, Currahee Campus
William J. Easter, Des Moines Area Community College, IA
Collen Engle, Sullivan University, KY
Dona Greenwood, Florida International University
Keith E. Gardiner, Guilford Technical Community College, NC
Julienne M. Guyette, Atlantic Culinary/McIntosh College, NH
Chef Herve Le Biavant, California Culinary Academy
Marjorie Livingston, Culinary Institute of America, NY
Debra Macchia, College of DuPage, IL
Aminta Martinez Hermosilla, Le Cordon Bleu College of Culinary Arts—Las Vegas, NV
Kevin Monti, Western Culinary Institute, OR
Renee Reagan-Moreno, California Culinary Academy
Mary L. Rhiner, Kirkwood Community College, IA
Richard Roberts, Wake Technical Community College, NC
Vickie S. Schwartz, Drexel University, PA
Joan Vogt, Kendall College, IL
Donna Wamsley, Hocking Technical College, OH
Jane Ziegler, Cedar Crest College, PA

PART ONE

Fundamentals of Nutrition & Foods

Introduction to Nutrition

Americans are fascinated with food: choosing foods, reading newspaper articles on food, perusing cookbooks, preparing and cooking foods, checking out new restaurants, and, of course, eating foods. Why are we so interested in food? Of course, eating is fun, enjoyable, and satisfying, especially when we are eating with other people whose company we like.

Beyond the physical and emotional satisfaction of eating, we often are concerned about how food choices affect our health. Our choice of diet strongly influences whether we will get certain diseases, such as heart disease, cancer, and stroke—the three biggest killers in the United States. Indeed, high costs are associated with poor eating patterns. In 2007, the Centers for Disease Control and Prevention estimated the annual cost of heart disease and stroke in the United States at $431 billion, including health-care expenditures and lost productivity from death and disability.

A 2005 survey by the International Food Information Council found that at least 89 percent of American adults sampled indicated that they believe diet, exercise, and physical activity influence health. These beliefs are reflected in the popularity of books, magazines, and weight-loss programs offering dietary and health advice. Recent consumption statistics, however, show that many of us are still choosing diets that are out of sync with dietary guidance. Many Americans eat too much sodium, saturated fat, and added sugar, and too few fruits, vegetables, and whole grains. And the prevalence of obesity and diet-related illnesses continues to rise. Although we may intend to have a healthy diet, other preferences often beguile us into food choices that may eventually harm our health.

Eating a healthy diet and exercising is not just a concern for adults but for children and teenagers as well. Overweight is a serious health concern for children and adolescents. Data from 1976 to 2003–2004 show that the prevalence of overweight is increasing. For children aged 6–11 years, prevalence increased from 6.5 to 18.8 percent; and for those aged 12–19 years, prevalence increased from 5.0 to 17.4 percent. Overweight children and adolescents are at risk for health problems during their youth and as adults. For example, during their youth, overweight children and adolescents are more likely to have risk factors associated with cardiovascular disease and type 2 diabetes than are other children and adolescents.

Young adults who go to college also face the challenge of not gaining what is called the "Freshman 15." As teenagers leave home and become more responsible for themselves and their eating habits, they often gain weight, although not always 15 pounds. Causes for freshmen gaining weight (particularly females) include eating unhealthy foods in the cafeteria, keeping unhealthy foods and snacks in the dorm room, drinking too much alcohol, and exercising too infrequently. College can be stressful and lead to poor eating choices.

This introductory chapter explores why we choose the foods we eat and then explains important nutrition concepts that build a foundation for the remaining chapters. It will help you to:

- Identify factors that influence food selection
- Define *nutrition, kilocalorie, nutrient,* and *nutrient density*
- Identify the classes of nutrients and their characteristics
- Describe four characteristics of a nutritious diet
- Define Dietary Reference Intakes and explain their function
- Compare the EAR, RDA, AI, and UL
- Describe the processes of digestion, absorption, and metabolism
- Explain how the digestive system works
- Distinguish between whole, processed, and organic foods
- Compare how a meat-based or a plant-based diet impacts the environment

FACTORS INFLUENCING FOOD SELECTION

Why do people choose the foods they do? This is a very complex question. As you can see from this list, many factors influence what you eat.

- Flavor
- Other aspects of food (such as cost, convenience, nutrition)
- Demographics
- Culture and religion
- Health
- Social and emotional influences
- Food industry and the media
- Environmental concerns

Now we will look at these factors in depth.

FLAVOR

The most important consideration when choosing something to eat is the taste of the food (Figure 1-1). You may think that taste and flavor are the same thing, but taste is actually a component of flavor. **Flavor** is an attribute of a food that includes its taste, smell, feel in the mouth, texture, temperature, and even the sounds made when it is chewed. Flavor is a combination of all five senses: taste, smell, touch, sight, and sound.

> **FLAVOR**
> An attribute of a food that includes its appearance, smell, taste, feel in the mouth, texture, temperature, and even the sounds made when it is chewed.

FIGURE 1-1:
The most important consideration when choosing something to eat is taste.
Courtesy of Digital Image.

Taste comes from 10,000 *taste buds*—clusters of cells that resemble the sections of an orange. Taste buds, found on the tongue, cheeks, throat, and roof of the mouth, house 60 to 100 receptor cells each. The body regenerates taste buds about every three days.

These taste cells bind food molecules dissolved in saliva and alert the brain to interpret them. Although the tongue is often depicted as having regions that specialize in particular taste sensations—for example, the tip is said to detect sweetness—researchers know that taste buds for each sensation (sweet, salty, sour, bitter, and umami) are actually scattered around the tongue. In fact, a single taste bud can have receptors for all five sensations. We also know that the back of the tongue is more sensitive to bitter, and that food temperature influences taste.

Taste buds are most numerous in children under age six, and this may explain why youngsters are such picky eaters. Children generally prefer higher levels of sweetness and saltiness in their food than adults do. This will change in adolescence, when their taste preferences become more like those of adults. Children will also develop food preferences that reflect their culture. For instance, in many Asian cultures, combining sweet and umami is common, whereas this would not be common in the United States. Cultural food preferences often adapt when people relocate into another culture.

Umami, the fifth basic taste, differs from the traditional sweet, sour, salty, and bitter tastes by providing a savory, sometimes meaty, sensation. Umami is a Japanese word and the taste is evident in many Japanese ingredients and flavorings, such as seaweed, dashi stock, and mushrooms, as well as other foods. The umami taste receptor is very sensitive to glutamate, which occurs naturally in foods such as meat, fish, and milk, and it is often added to processed foods in the form of the flavor enhancer monosodium glutamate (MSG). Despite the frequent description of umami as meaty, many foods, including mushrooms, tomatoes, and Parmesan cheese, have a higher level of glutamate than an equal amount of beef or pork. This explains why foods that are cooked with mushrooms or tomatoes seem to have a fuller, rounder taste than when cooked alone.

If you could taste only sweet, salty, sour, bitter, and umami, how could you taste the flavor of cinnamon, chicken, or any other food? This is where smell comes in. Your ability to identify the flavors of specific foods requires smell.

The ability to detect the strong scent of a fish market, the antiseptic odor of a hospital, the aroma of a ripe melon, and thousands of other smells is possible thanks to a yellowish patch of tissue the size of a quarter high up in your nose. This patch is actually a layer of 12 million specialized cells, each sporting 10 to 20 hairlike growths called cilia that bind with the smell and send a message to the brain. Our sense of smell may not be as refined as that of dogs, which have billions of olfactory cells, but we can distinguish among about 10,000 scents.

Of course, if you have a bad cold and mucus clogs up your nose, you lose some sense of smell and taste. With a cold, you can still taste salty and sweet, but you will have a hard time distinguishing the difference between flavors, such as beef from lamb.

You can smell foods in two ways. If you smell coffee brewing while you are getting dressed, you smell it directly through your nose. But if you are drinking coffee, the smell of the coffee goes to the back of your mouth and then up into your nose. To some extent, what you smell (or taste) is determined by your genetics and also your age.

All foods have texture, a natural texture granted by Mother Nature. It may be coarse or fine, rough or smooth, tender or tough. Whichever the texture, it influences whether you like the food. The natural texture of a food may not be the most desirable texture for a finished dish, and so a cook may create different texture. For example, a fresh apple may be too crunchy to serve at dinner, and so it is baked or sautéed for a softer texture. Or a cream soup may be too thin, and so a thickening agent is used to increase the viscosity of the soup or, simply stated, make it harder to pour.

Food appearance or presentation strongly influences which foods you choose to eat. Eye appeal is the purpose of food presentation, whether the food is hot or cold. It is especially important for cold foods because they lack the come-on of an appetizing aroma. Just the sight of something delicious to eat can start your digestive juices flowing.

OTHER ASPECTS OF FOOD

Food cost is a major consideration. For example, breakfast cereals were inexpensive for many years. Then their prices jumped, and it seemed that most boxes of cereal cost over $3.00. Some consumers switched from cereal to bacon and eggs because the bacon and eggs became less expensive. Cost is a factor in many purchasing decisions at the supermarket, whether one is buying dry beans at $0.69 per pound or fresh salmon at $13.99 per pound.

Convenience is more of a concern now than at any time in the past because of the lack of time to prepare meals. Just think about the variety of foods you can purchase today that are already cooked and can simply be microwaved. Even if you desire fresh fruits and vegetables, supermarkets offer them already cut up and ready to eat. Of course, convenience foods are more expensive than their raw counterparts, and not every budget can afford them. Take-out meals are also more expensive, but common in certain households.

Everyone's food choices are affected by availability and familiarity. Whether it is a wide choice of foods at an upscale supermarket or a choice of only two restaurants within walking distance of where you work, you can eat only what is available. The availability of foods is very much influenced by the way food is produced and distributed. For example, the increasing number of soft drink vending machines, particularly in schools and workplaces, has contributed to increasing soft drink consumption year-round. Fresh fruits and vegetables are perfect examples of foods that are most available (and at their lowest prices) when in season. Of course, you are more likely to eat fruits and vegetables, or any food for that matter, with which you are familiar and which you have eaten before.

The nutritional content of a food can be an important factor in deciding what to eat. You probably have watched people reading nutritional labels on a food package, or perhaps you have read nutritional labels yourself. Current estimates show that about 75 percent of Americans use nutrition information labels. Older people tend to read labels more often than younger people do.

DEMOGRAPHICS

Demographic factors that influence food choices include age, gender, educational level, income, and cultural background (discussed next). Women and older adults tend to consider nutrition more often than do men or young adults when choosing what to eat. Older adults are probably more nutrition-minded because they have more health problems, such as heart disease and high blood pressure, and are more likely to have to change their diet for health reasons. Older adults also have more concerns with poor dental health, swallowing problems, and digestive problems. People with higher incomes and educational levels tend to think about nutrition more often when choosing what to eat.

CULTURE AND RELIGION

Culture can be defined as the behaviors and beliefs of a certain social, ethnic, or age group. A culture strongly influences the eating habits of its members. Each culture has norms about which foods are edible, which foods have high or low status, how often foods are consumed, what foods are eaten together, when foods are eaten, and what foods are served at special events and celebrations (such as weddings).

In short, your culture influences your attitudes toward and beliefs about food. For example, some French people eat horsemeat, but Americans do not consider horsemeat acceptable to eat. Likewise, many common American practices seem strange or illogical to persons from

> **CULTURE**
> The behaviors and beliefs of a certain social, ethnic, or age group.

other cultures. For example, what could be more unusual than boiling water to make tea and adding ice to make it cold again, sugar to sweeten it, and then lemon to make it tart? When immigrants come to live in the United States, their eating habits gradually change, but they are among the last habits to adapt to the new culture.

For many people, religion affects their day-to-day food choices. For example, many Jewish people abide by the Jewish dietary laws, called the Kashrut. They do not eat pork, nor do they eat meat and dairy products together. Muslims also have their own dietary laws. Like Jews, they will not eat pork. Their religion also prohibits drinking alcoholic beverages. For other people, religion influences what they eat mostly during religious holidays and celebrations. Religious holidays such as Passover are observed with appropriate foods. Figure 1-2 explains the food practices of different religions.

HEALTH

Have you ever dieted to lose weight? Most Americans are trying to lose weight or keep from gaining it. You probably know that obesity and overweight can increase your risk of cancer, heart disease, diabetes, and other health problems. What you eat influences your health.

FIGURE 1-2: Food Practices of World Religions	
Religion	**Dietary Practices**
Judaism	Kashrut: Jewish dietary law of keeping kosher.
	1. Meat and poultry. Permitted: Meat of animals with a split hoof that chew their cud (includes cattle, sheep, goats, deer); a specific list of birds (includes chicken, turkey, goose, pheasant, duck). Not permitted: Pig and pork products, mammals that don't have split hooves and chew their cud (such as rabbit), birds not specified (such as ostrich). All animals require ritual slaughtering. All meat and poultry foods must be free of blood, which is done by soaking and salting the food or by broiling it. Forequarter cuts of mammals are also not eaten.
	2. Fish. Permitted: Fish with fins and scales. Not permitted: Shellfish (scallops, oysters, clams), crustaceans (crab, shrimp, lobster), fishlike mammals (dolphin, whale), frog, shark, eel. Do not cook fish with meat or poultry.
	3. Meat and dairy are not eaten or prepared together. Meals are dairy or meat, not both. It is also necessary to have two sets of cooking equipment, dishes, and silverware for dairy and meat.
	4. All fruits, vegetables, grains, and eggs can be served with dairy or meat meals.
	5. A processed food is considered kosher only if the package has a rabbinical authority's name or insignia.
Roman Catholicism	1. Abstain from eating meat on Fridays during Lent (the 40 days before Easter).
	2. Fast (one meal is allowed) and abstain from meat on Ash Wednesday (beginning of Lent) and Good Friday (the Friday before Easter).
Eastern Orthodox Christianity	Numerous feast days and fast days. On fast days, no fish, meat, or other animal products (including dairy products) are allowed. They also abstain from wine and oil, except for certain feast days that may fall during a fasting period. Shellfish are allowed. Wednesdays and Fridays are also fast days throughout the year.
Protestantism	1. Food on religious holidays is largely determined by a family's cultural background and preferences.
	2. Fasting is uncommon.
	(continued)

Mormonism	1. Prohibit tea, coffee, and alcohol. Some Mormons abstain from anything containing caffeine.
	2. Eat only small amounts of meat and base diet on grains.
	3. Some Mormons fast once a month.
Seventh-Day Adventist Church	1. Many members are lacto-ovo vegetarians (eat dairy products and eggs but no meat or poultry).
	2. Avoid pork and shellfish.
	3. Prohibit coffee, tea, and alcohol.
	4. Drink water before and after meals, not during.
	5. Avoid highly seasoned foods and eating between meals.
Islam	1. All foods are permitted (halal) except for swine (pigs), four-legged animals that catch prey with the mouth, birds of prey that grab prey with their claws, animals (except fish and seafood) that have not been slaughtered according to ritual, and alcoholic beverages. Use of coffee and tea is discouraged.
	2. Celebrate many feast and fast days. On fast days, they do not eat or drink from sunup to sundown.
Hinduism	1. Encourages eating in moderation.
	2. Meat is allowed, but the cow is sacred and is not eaten. Also avoided are pork and certain fish. Many Hindus are vegetarian.
	3. Many Hindus avoid garlic, onions, mushrooms, and red foods such as tomatoes.
	4. Water is taken with meals.
	5. Some Hindus abstain from alcohol.
	6. Hindus have a number of feast and fast days.
Buddhism	1. Dietary laws vary depending on the country and the sect. Many Buddhists do not believe in taking life, and so they are lacto-ovo vegetarians (eat dairy products and eggs but no meat or poultry).
	2. Celebrate feast and fast days.

Even if you are healthy, you may base food choices on a desire to prevent health problems and/or improve your appearance.

A knowledge of nutrition and a positive attitude toward nutrition may translate into nutritious eating practices. Just knowing that eating lots of fruits and vegetables may prevent heart disease does not mean that someone will automatically start eating more of those foods. For some people, knowledge is enough to stimulate new eating behaviors, but for most people, knowledge is not enough and change is difficult. Many circumstances and beliefs prevent change, such as a lack of time or money to eat right. But some people manage to change their eating habits, especially if they feel that the advantages (such as losing weight or preventing cancer) outweigh the disadvantages.

SOCIAL AND EMOTIONAL INFLUENCES

People have historically eaten meals together, making meals important social occasions. Our food choices are influenced by the social situations we find ourselves in, whether in the comfort of our own home or eating out in a restaurant. For example, social influences are involved when several members of a group of college friends are vegetarian. Peer pressure no doubt influences many food choices among children and young adults. Even as adults, we tend to eat the same foods that our friends and neighbors eat. This is due to cultural influences as well.

Food is often used to convey social status. For example, in a trendy, upscale New York City restaurant, you will find prime cuts of beef and high-priced wine.

Emotions are closely tied to some of our food selections. As a child, you may have been given something sweet to eat, such as cake or candy, whenever you were unhappy or upset. As an adult, you may gravitate to those kinds of foods, called comfort foods, when under stress. Carbohydrates, such as in cake or candy, tend have calming effects. Eating in response to emotions can lead to overeating and overweight.

FOOD INDUSTRY AND THE MEDIA

The food industry very much influences what you choose to eat. After all, the food companies decide what foods to produce and where to sell them. They also use advertising, product labeling and displays, information provided by their consumer services departments, and websites to sell their products.

On a daily basis, the media (television, newspapers, magazines, radio, and the like) portray food in many ways: paid advertisements, articles on food in magazines and newspapers, and foods eaten on television shows. Much research has been done on the impact of television food commercials on children. Quite often the commercials succeed in getting children to eat foods such as cookies, candies, and fast food. Television commercials probably are contributing to higher calorie and fat intakes.

The media also report frequently on new studies related to food, nutrition, and health topics. It is hard to avoid hearing sound bites such as "more fruits and vegetables lower blood pressure." Media reports can certainly influence which foods people eat.

ENVIRONMENTAL CONCERNS

Some people have environmental concerns, such as the use of chemical pesticides, and so they often, or always, choose organically grown foods (which are grown without such chemicals—see Food Facts on page 26 for more information). Many vegetarians won't eat meat or chicken because livestock and poultry require so much land, energy, water, and plant food, which they consider wasteful. See Hot Topic on page 28 for more information on some of the environmental concerns of commercial food production.

Now that you have a better understanding of why we eat the foods we do, we can look at some basic nutrition concepts and terms.

Figure 1-3 summarizes factors that influence what we eat.

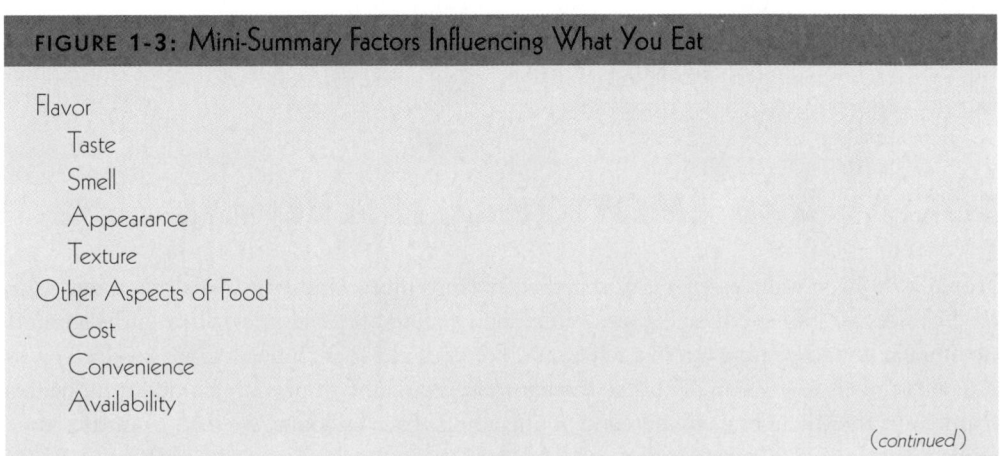

FIGURE 1-3: Mini-Summary Factors Influencing What You Eat

Flavor
    Taste
    Smell
    Appearance
    Texture
Other Aspects of Food
    Cost
    Convenience
    Availability

(continued)

Familiarity
   Nutrition
Demographics
   Age
   Gender
   Educational level
   Income
Culture and Religion
   Traditional foods and food habits
   Attitudes and beliefs
   Special events and celebrations
   Religious foods and food practices
Health
   Health status and desire to improve health
   Desire to improve appearance
   Nutrition knowledge and attitudes
Social and Emotional Influences
   Social status
   Peer pressure
   Emotional status
   Food associations
Food Industry and the Media
   Food industry
   Food advertising
   Food portrayal in media
   Reporting of nutrition/health studies
Environmental Concerns
   Use of synthetic fertilizers and pesticides
   Wastefulness of fattening up livestock/poultry

BASIC NUTRITION CONCEPTS

NUTRITION

Nutrition is a science. Compared with some other sciences, such as chemistry, that have been studied for thousands of years, nutrition is a young science. Many nutritional facts revolve around nutrients, such as carbohydrates. *Nutrients* are the nourishing substances in food that provide energy and promote the growth and maintenance of the body. In addition, nutrients aid in regulating body processes such as heart rate and digestion and in supporting the body's optimum health.

Nutrition researchers look at how nutrients and other substances in food relate to health and disease. Almost daily we are bombarded with news reports that something in the food we eat, such as fat, is not good for us—that it may indeed cause or complicate conditions

NUTRITION
A science that studies nutrients and other substances in foods and in the body and the way those nutrients relate to health and disease. Nutrition also explores why you choose particular foods and the type of diet you eat.

NUTRIENTS
The nourishing substances in food that provide energy and promote the growth and maintenance of your body.

such as heart disease and certain cancers. Researchers look closely at the relationships between nutrients and disease, as well as the processes by which you choose what to eat and the balance of foods and nutrients in your diet.

In summary, nutrition is a science that studies nutrients and other substances in foods, and how they affect the body, especially in terms of health and disease. Nutrition also explores why you choose the foods you do and the type of *diet* you eat. Diet is a word that has several meanings. Anyone who has tried to lose weight has no doubt been on a diet. In this sense, diet means weight-reducing diet and is often thought of in a negative way. But a more general definition of diet is the foods and beverages you normally eat and drink.

> **DIET**
> The food and beverages you normally eat and drink.

KILOCALORIES

Food energy, as well as the energy needs of the body, is measured in units of energy called *kilocalories*. The number of kilocalories in a particular food can be determined by burning a weighed portion of that food and measuring the amount of heat (or kilocalories) it produces. A kilocalorie raises the temperature of 1 kilogram of water 1 degree Celsius. Just as 1 kilogram contains 1000 grams, 1 kilocalorie contains 1000 calories.

When you read in a magazine that a cheeseburger has 350 calories, understand that it is actually 350 kilocalories. The American public has been told for years that an apple has 80 calories, a glass of regular milk has 150 calories, and so on, when the correct term is not calories but kilocalories. This has been done in part to make the numbers easier to read and to ease calculations. Imagine adding up your calories for the day, and having most numbers be 6 digits long, such as 350,000 calories for a cheeseburger. This book uses the term *kilocalorie* and its abbreviations, kcalorie and kcal, throughout each chapter.

The number of kcalories you need is based on three factors: your energy needs when your body is at rest and awake (referred to as *basal metabolism*), your level of physical activity, and the energy you need to digest and absorb food (referred to as the *thermic effect of food*). Basal metabolic needs include energy needed for vital bodily functions when the body is at rest but awake. For example, your heart is pumping blood to all parts of your body, your cells are making proteins, and so on. Your basal metabolic rate (BMR) depends on the following factors:

> **KILOCALORIE**
> A measure of the energy in food, specifically the energy-yielding nutrients.
>
> **BASAL METABOLISM**
> The minimum energy needed by the body for vital functions when at rest and awake.
>
> **THERMIC EFFECT OF FOOD**
> The energy needed to digest and absorb food.

1. Gender. Men have a higher BMR than women do because men have a higher proportion of muscle tissue (muscle requires more energy for metabolism than fat does).
2. Age. As people age, they generally gain fat tissue and lose muscle tissue. BMR declines about 2 percent per decade after age 30.
3. Growth. Children, pregnant women, and lactating women have higher BMRs.
4. Height. Tall people have more body surface than shorter people do and lose body heat faster. Their BMR is therefore higher.
5. Temperature. BMR increases in both hot and cold environments, to keep the temperature inside the body constant.
6. Fever and stress. Both of these increase BMR. Fever raises BMR by 7 percent for each 1 degree Fahrenheit above normal. The body reacts to stress by secreting hormones that speed up metabolism so that the body can respond quickly and efficiently.
7. Exercise. Exercise increases BMR for several hours afterward.
8. Smoking and caffeine. Both cause increased energy expenditure.
9. Sleep. Your BMR is at its lowest when you are sleeping.

The basal metabolic rate also decreases when you diet or eat fewer kcalories than normal. The BMR accounts for the largest percentage of energy expended—about two-thirds for individuals who are not very active.

FIGURE 1-4: Kcalories per Hour Expended in Common Physical Activities

Moderate Physical Activity	Kcals/Hour for a 154-pound Person
Hiking	367
Light gardening/yard work	331
Dancing	331
Golf (walking and carrying clubs)	331
Bicycling (less than 10 mph)	294
Walking (3.5 mph)	279
Weight lifting (general light workout)	220
Stretching	184

Vigorous Physical Activity	Kcals/Hour for a 154-pound Person
Running/jogging (5 mph)	588
Bicycling (over 10 mph)	588
Swimming (slow freestyle laps)	514
Aerobics	478
Walking (4.5 mph)	464
Heavy yard work (chopping wood)	441
Weight lifting (vigorous effort)	441
Basketball (vigorous)	441

Source: 2005 Report of the Dietary Guidelines Advisory Committee.

Your level of physical activity strongly influences how many kcalories you need. Figure 1-4 shows the kcalories burned per hour for a variety of activities. The number of kcalories burned depends on the type of activity, how long and how hard it is performed, and the individual's size. The larger your body is, the more energy you use in physical activity. Aerobic activities such as walking, jogging, cycling, and swimming are excellent ways to burn calories if they are brisk enough to raise heart and breathing rates. Physical activity accounts for 25 to 40 percent of total energy needs.

The thermic effect of food is the smallest contributor to your energy needs: from 5 to 10 percent of the total. In other words, for every 100 kcalories you eat, 5 to 10 are used for digestion, absorption, and metabolism of nutrients, our next topic.

NUTRIENTS

As stated, nutrients provide energy or kcalories, promote the growth and maintenance of the body, and/or regulate body processes. There are about 50 nutrients that can be arranged into six classes, as follows:

1. Carbohydrates
2. Fats (the proper name is lipids)
3. Protein
4. Vitamins
5. Minerals
6. Water

Each nutrient class performs different functions in the body, as shown in Figure 1-5.

FIGURE 1-5: Functions of Nutrients

Nutrients	Provide Energy	Promote Growth and Maintenance	Regulate Body Processes
Carbohydrates	X		
Lipids	X	X	X
Protein	X	X	X
Vitamins		X	X
Minerals		X	X
Water		X	X

ENERGY-YIELDING NUTRIENTS

Nutrients that can be burned as fuel to provide energy for the body, including carbohydrates, fats, and proteins.

MICRONUTRIENTS

Nutrients needed by the body in small amounts, including vitamins and minerals.

MACRONUTRIENTS

Nutrients needed by the body in large amounts, including carbohydrates, lipids, and proteins.

ORGANIC

In chemistry, any compound that contains carbon.

INORGANIC

In chemistry, any compound that does not contain carbon.

CARBOHYDRATES

A large class of nutrients, including sugars, starch, and fibers, that function as the body's primary source of energy.

LIPIDS

A group of fatty substances, including triglycerides and cholesterol, that are soluble in fat, not water, and that provide a rich source of energy and structure to cells.

Foods rarely contain just one nutrient. Most foods provide a mix of nutrients. For example, bread often is thought of as providing primarily carbohydrates, but it is also an important source of certain vitamins and minerals. Food contains more than just nutrients. Depending on the food, it may contain colorings, flavorings, caffeine, phytochemicals (minute substances in plants that are biologically active in the body and may protect health), and other substances.

Carbohydrates, lipids, and protein are called *energy-yielding nutrients* because they can be burned as fuel to provide energy for the body. They provide kcalories as follows:

Carbohydrates:	4 kcalories per gram
Lipids:	9 kcalories per gram
Protein:	4 kcalories per gram

(A gram is a unit of weight in the metric system; there are about 28 grams in 1 ounce.) Vitamins, minerals, and water do not provide energy or calories. Alcohol, although not considered a nutrient because it does not promote growth or maintenance of the body, does yield energy: Seven kcalories per gram.

The body needs vitamins and minerals in small amounts, and so these nutrients are called *micronutrients* (micro means small). In contrast, the body needs large amounts of carbohydrates, lipids, and protein, and so they are called *macronutrients* (macro means large).

Another way to group the classes of nutrients is to look at them from a chemical point of view. In chemistry, any compound that contains carbon is called *organic*. If a compound does not contain carbon, it is called *inorganic*. Carbohydrates, lipids, proteins, and vitamins are all organic. Minerals and water are inorganic.

Carbohydrates are a large class of nutrients, including sugars, starches, and fibers, that function as the body's primary source of energy. Sugar is most familiar in its refined forms, such as table sugar and high-fructose corn syrup, which are used in soft drinks, cookies, cakes, pies, candies, jams, jellies, and other sweetened foods. Sugar is also present naturally in fruits and milk (even though milk does not taste sweet). Starch is found in breads, breakfast cereals, pastas, potatoes, and beans. Both sugar and starch are important sources of energy for the body.

Fiber can't be broken down or digested in the body, and so it is excreted. It therefore does not provide energy for the body. Fiber does a number of good things in the body, such as improve the health of the digestive tract. Good sources of fiber include legumes (dried beans and peas), fruits, vegetables, whole-grain foods such as whole-wheat bread and cereal, nuts, and seeds.

Lipids are a group of fatty substances, including triglycerides and cholesterol, that are soluble in fat, not water, and that provide a rich source of energy and structure to cells. The most familiar lipids are fats and oils, which are found in butter, margarine, vegetable oils, mayonnaise, and salad dressings. Lipids are also found in the fatty streaks in meat, the fat under the skin of poultry, the fat in milk and cheese (except fat-free milk and products made with it), baked goods such as cakes, fried foods, nuts, and many processed foods, such as canned soups and frozen dinners. Most breads, cereals, pasta, fruits, and vegetables have little or no fat. Triglycerides are the major form of lipids. They provide energy for the body as well as a way to store energy as fat.

Most of the kcalories we eat come from carbohydrates or fats. Only about 15 percent of total kcalories come from *protein*. This doesn't mean that protein is less important. On the contrary, protein is the main structural component of all the body's cells. It is made of units called amino acids, which are unique in that they contain nitrogen. Besides its role as an important part of cells, protein regulates body processes and can be burned to provide energy (although the body prefers to burn carbohydrates and lipids so protein can be used to build new cells). Protein is present in significant amounts in foods from animal sources, such as beef, pork, chicken, fish, eggs, milk, and cheese. Protein appears in plant foods, such as grains, beans, and vegetables, in smaller quantities. Fruits contain only very small amounts of protein.

There are 13 different vitamins in food. *Vitamins* are noncaloric, organic nutrients found in a wide variety of foods. They are essential in small quantities to regulate body processes, maintain the body, and allow growth and reproduction. Instead of being burned to provide energy for the body, vitamins work as helpers. They assist in the processes of the body that keep you healthy. For example, vitamin A is needed by the eyes for vision in dim light. Vitamins are found in fruits, vegetables, grains, meat, dairy products, and other foods. Unlike other nutrients, many vitamins are susceptible to being destroyed by heat, light, and other agents.

Minerals are also required by the body in small amounts and do not provide energy. Like vitamins, they work as helpers in the body and are found in a variety of foods. Some minerals, such as calcium and phosphorus, become part of the body's structure by building bones and teeth. Unlike vitamins, minerals are indestructible and inorganic.

Although deficiencies of energy or nutrients can be sustained for months or even years, a person can survive only a few days without water. Experts rank water second only to oxygen as essential to life. Water plays a vital role in all bodily processes and makes up just over half the body's weight. It supplies the medium in which various chemical changes of the body occur and aids digestion and absorption, circulation, and lubrication of body joints. For example, as a major component of blood, water helps deliver nutrients to body cells and removes waste to the kidneys for excretion.

It's been said many times, "You are what you eat." This is certainly true; the nutrients you eat can be found in your body. As mentioned, water is the most plentiful nutrient in the body, accounting for about 60 percent of your weight. Protein accounts for about 15 percent of your weight, fat for 20 to 25 percent, and carbohydrates for only 0.5 percent. The remainder of your weight includes minerals, such as calcium in bones, and traces of vitamins (Figure 1-6).

Most, but not all, nutrients are considered *essential nutrients*. Essential nutrients either cannot be made in the body or cannot be made in the quantities needed by the body; therefore, we must obtain them from food. Carbohydrates (in the form of glucose), vitamins, minerals, water, some lipids, and some parts of protein are considered essential.

NUTRIENT DENSITY

All foods were not created equal in terms of the kcalories and nutrients they provide. Some foods, such as milk, contribute much calcium to your diet, especially when you compare them with other beverages, such as soft drinks. The typical can of cola (12 fluid ounces) contributes large amounts of sugar (40 grams, or about 10 teaspoons), no vitamins, and virtually no minerals. When you compare calories, you will find that skim milk (at 86 kcalories per cup along with many vitamins and minerals) packs fewer calories than does cola (at 97 kcalories per cup). Therefore, we can say that milk is more "nutrient-dense" than cola, meaning that milk contains more nutrients per kcalorie than colas do.

PROTEIN
Major structural component of the body's cells that is made of nitrogen-containing amino acids assembled in chains, particularly rich in animal foods.

VITAMINS
Noncaloric, organic nutrients found in a wide variety of foods that are essential in small quantities to regulate body processes, maintain the body, and allow growth and reproduction.

MINERALS
Noncaloric, inorganic chemical substances found in a wide variety of foods; needed to regulate body processes, maintain the body, and allow growth and reproduction.

ESSENTIAL NUTRIENTS
Nutrients that either cannot be made in the body or cannot be made in the quantities needed by the body; therefore, we must obtain them from food.

FIGURE 1-6:
Body Composition.

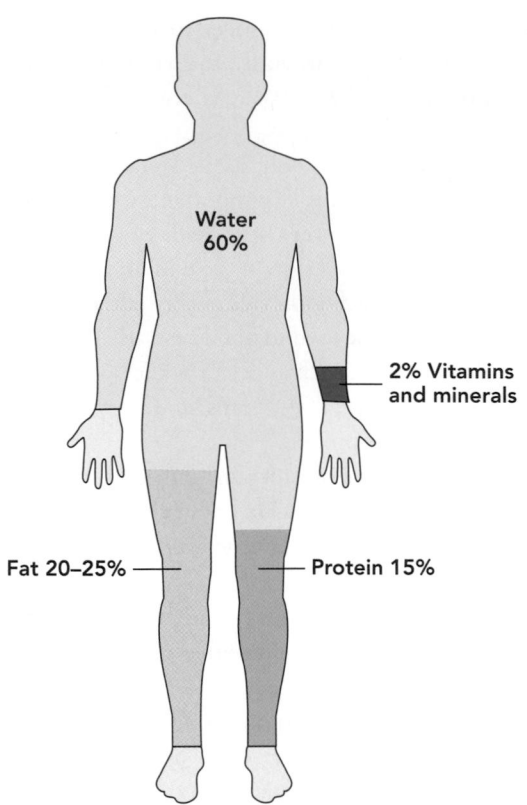

Water
60%

2% Vitamins
and minerals

Fat 20–25%

Protein 15%

NUTRIENT DENSITY

A measure of the nutrients provided in a food per kcalorie of that food.

EMPTY-KCALORIE FOODS

Foods that provide few nutrients for the number of kcalories they contain.

The *nutrient density* of a food depends on the amount of nutrients it contains and the comparison of that to its caloric content. In other words, nutrient density is a measure of the nutrients provided per kcalorie of a food. As Figure 1-7 shows, broccoli offers many nutrients for its few calories. Broccoli is considered to have a high nutrient density because it is high in nutrients relative to its caloric value. Vegetables and fruits are examples of nutrient-dense foods. In comparison, a cupcake contains many more kcalories and few nutrients. By now, you no doubt recognize that some foods, such as candy bars, have a low nutrient density, meaning that they are low in nutrients and high in kcalories. These foods are called *empty-kcalorie foods* because the kcalories they provide are "empty" (that is, they deliver few nutrients). The next section will tell you more about what a nutritious diet is.

FIGURE 1-7:
Nutrition density comparison.*

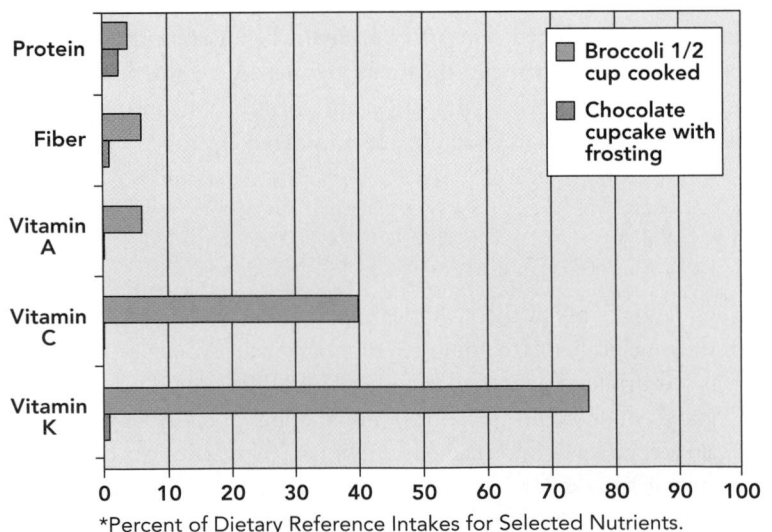

Protein

Fiber

Vitamin
A

Vitamin
C

Vitamin
K

Broccoli 1/2
cup cooked

Chocolate
cupcake with
frosting

0   10   20   30   40   50   60   70   80   90   100

*Percent of Dietary Reference Intakes for Selected Nutrients.

1. Nutrition is a science that studies nutrients and other substances in foods, and how they affect the body, especially in terms of health and disease. Nutrition also explores why you choose the foods you do and the type of diet you eat.

2. The number of kcalories (a measure of the energy in food) you need is based on three factors: your energy needs when your body is at rest and awake (basal metabolism), your level of physical activity, and the energy you need to digest and absorb food (thermic effect of food).

3. Nutrients are the nourishing substances in food, providing energy and promoting the growth and maintenance of the body. In addition, nutrients regulate the many body processes and support the body's optimum health and growth.

4. The six classes of nutrients are carbohydrates, fats (properly called lipids), protein, vitamins, minerals, and water. Carbohydrates, fats, and proteins, are macronutrients, while vitamins and minerals are micronutrients. Their characteristics are summarized in Figure 1-8.

5. Nutrient density is a measure of the nutrients provided per kcalorie of a food.

Carbohydrates – A large class of nutrients including sugar, starches, and fibers that are the body's primary source of energy.

Lipids (fats) – A group of fatty substances including triglycerides and cholesterol that are not soluble in water and that provide a rich source of energy and structure to the body's cells.

Proteins – Major structural part of body's cells composed of nitrogen-containing amino acids, particularly rich in animal foods.

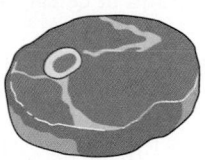

Vitamins – 13 noncaloric nutrients found in a wide variety of foods (especially fruits and vegetables)

Minerals – Noncaloric, inorganic chemical substances found in a wide variety of foods

Both vitamins and minerals are essential in small amounts to maintain the body, regulate body processes, and for growth and reproduction.

Water – Inorganic nutrient that plays a vital role in all bodily processes and makes up just over half of the body's weight.

FIGURE 1-8:
Six classes of nutrients.

CHARACTERISTICS OF A NUTRITIOUS DIET

A nutritious diet has four characteristics. It is:

1. Adequate
2. Balanced
3. Moderate
4. Varied

Your diet must provide enough nutrients, but not too many. This is where adequate and moderate diets fit in. An ***adequate diet*** provides enough kcalories, essential nutrients, and fiber to keep you healthy, whereas a ***moderate diet*** avoids taking in excessive amounts of kcalories or eating more of one food or food group than is recommended. In the case of kcalories, for example, consuming too many leads to obesity. The concept of moderation allows you to choose appropriate portion sizes of any food as well as to indulge occasionally in high-kcalorie, high-fat foods such as french fries and premium ice cream.

Although it may sound simple to eat enough, but not too much, of the necessary nutrients, surveys show that most adult Americans find this hard to do. One of the best ways to overcome this problem is to select nutrient-dense foods. As stated earlier, nutrient-dense foods contain many nutrients for the kcalories they provide.

Next, you need a ***balanced diet***. Eating a balanced diet means eating more servings of nutrient-dense foods such as whole grains, fruits, and vegetables and fewer servings of foods such as cakes, cookies, and chips, which supply few nutrients. For example, if you drink a lot of soft drinks, you will be getting too much sugar and possibly not enough calcium, a mineral found in milk. This is a particular concern for children, whose bones are growing and who are more likely than ever before to be obese. The typical American diet is unbalanced. We eat more fried foods and fatty meats than we need, and we drink too much soda. At the same time we eat too few fruits, vegetables, and whole grains. A balanced diet is also likely to be adequate and moderate.

Last, you need a ***varied diet***—in other words, you need to eat a wide selection of foods to get the necessary nutrients. If you imagine everything you eat for one week piled in a grocery cart, how much variety is in that cart from week to week? Do you eat the same bread, the same brand of cereal, the same types of fresh fruit, and so on, every week? Do you constantly eat favorite foods? Do you try new foods? A varied diet is important because it makes it more likely that you will get the essential nutrients in the right amounts. Our next topic, the Dietary Reference Intakes, gets specific about the amounts we need of most nutrients.

MINI-SUMMARY

A nutritious diet is adequate, moderate, balanced, varied, and packed with nutrient-dense foods.

NUTRIENT RECOMMENDATIONS: DIETARY REFERENCE INTAKES

The ***Dietary Reference Intakes (DRIs)*** expand and replace what you may have known as the Recommended Dietary Allowances (RDAs) in the United States and the Recommended Nutrient Intakes in Canada. The DRIs are developed by a committee of scientists within the National Academy of Scientists.

DRIs are a set of values that serve as standards for nutrient intakes for healthy persons in the United States and Canada. The DRIs are greatly expanded from the original RDAs and include the original RDAs as well as three new values. These values are:

1. Estimated Average Requirement (EAR). The dietary intake value that is estimated to meet the requirement of half the healthy individuals in a group. At this level of intake, the remaining 50 percent would not have its needs met. An EAR is set only when there is conclusive scientific research. The EAR is used to assess the nutritional adequacy of intakes of groups or populations and in nutrition research.

2. Recommended Dietary Allowance (RDA). The dietary intake value that is sufficient to meet the nutrient requirements of 97 to 98 percent of all healthy individuals in a group. The RDA is based on the Estimated Average Requirement (EAR), but is set higher so that the needs of most healthy people will be met. The RDAs may be used as nutrient goals for individuals. If there is not enough scientific evidence to justify setting an EAR, an RDA can't be established, so an Adequate Intake (discussed next) is given.

3. Adequate Intake (AI). The dietary intake value that is used when an RDA cannot be based on an EAR. An AI is given when there is insufficient scientific research to support an RDA. It is based on observed intakes of a nutrient by a group of healthy persons. For example, there is no EAR or RDA for calcium, only an AI. Like the RDA, the AI may be used as a goal for individual intake or to assess individual intake. Unlike the RDA, the AI is more tentative in part because it is based more on scientific judgment, rather than scientific evidence.

4. Tolerable Upper Intake Level (UL). The maximum intake level above which the risk of toxicity increases. Intakes below the UL are unlikely to pose a risk of adverse health effects in healthy people. For most nutrients, this figure refers to total intakes from food, fortified food, and nutrient supplements. UL cannot be established for some nutrients, due to inadequate research.

The DRIs vary depending on age and gender, and there are DRIs for pregnant and lactating women. The DRIs are meant to help healthy people maintain health and prevent disease. They are not designed for seriously ill people, whose nutrient needs may be much different.

The 2002 Dietary Reference Intake report established an *Estimated Energy Requirement (EER)* for healthy individuals. EER is the average energy intake (measured in kcalories) that is needed to maintain energy balance in a healthy adult so that he or she does not gain or lose weight. Your actual EER depends on your age, gender, weight, height, and level of physical activity. There is no RDA or UL for kcalories because these concepts do not apply to energy and would lead to weight gain.

The 2002 Dietary Reference Intake report also established *Acceptable Macronutrient Distribution Ranges (AMDR)* for carbohydrate, fat, and protein (Figure 1-9). AMDR is defined as the percent of total kilocalories coming from carbohydrate, fat, or protein that is associated with a reduced risk of chronic disease while providing adequate intake and

ESTIMATED AVERAGE REQUIREMENT (EAR)
The dietary intake value that is estimated to meet the requirement of half the healthy individuals in a group.

RECOMMENDED DIETARY ALLOWANCE (RDA)
The dietary intake value that is sufficient to meet the nutrient requirements of 97 to 98 percent of all healthy individuals in a group.

ADEQUATE INTAKE (AI)
The dietary intake that is used when there is not enough scientific research to support an RDA.

TOLERABLE UPPER INTAKE LEVEL (UL)
The maximum intake level above which the risk of toxicity would increase.

ACCEPTABLE MACRONUTRIENT DISTRIBUTION RANGE (AMDR)
The percent of total kilocalories coming from carbohydrate, fat, or protein that is associated with a reduced risk of chronic disease while providing adequate intake.

FIGURE 1-9: Acceptable Macronutrient Distribution Ranges			
Age	AMDR for Carbohydrate	AMDR for Fat	AMDR for Protein
1 to 3 years old	45–65%	30–40%	5–20%
4 to 18 years old	45–65%	25–35%	10–30%
Over 18 years old	45–65%	20–35%	10–35%

nutrients. For example, adults (and children over 1 year old) should obtain 45 to 65 percent of their total kcalories from carbohydrates. The AMDR for adults is 20 to 35 percent of total kcalories from fat and 10 to 35 percent of total kcalories from protein. The wide range allows for more flexibility in dietary planning for healthy people.

MINI-SUMMARY

1. The DRI includes four dietary intake values: EAR (value estimated to meet the requirements of half the healthy individuals in a group), RDA (value estimated to meet the requirements of 97 to 98 percent of healthy individuals in a group), AI (the dietary intake used when there is not enough scientific basis for an EAR or RDA), and UL (maximum intake).
2. The DRIs also include Estimated Energy Requirements and Acceptable Macronutrient Distribution Ranges for carbohydrate, fat, and protein.
3. The DRIs are used to assess dietary intakes as well as to plan diets. The RDA and AI are useful in planning diets for individuals. The EAR can be used to plan diets for groups.

WHAT HAPPENS WHEN YOU EAT

DIGESTION, ABSORPTION, AND METABOLISM

DIGESTION

The process by which food is broken down into its components in the mouth, stomach, and small intestine with the help of digestive enzymes.

ENZYMES

Compounds that speed up the breaking down of food so that nutrients can be absorbed. Also perform other functions in the body.

ABSORPTION

The passage of digested nutrients through the walls of the intestines or stomach into the body's cells. Nutrients are then transported through the body via the blood or lymph system.

METABOLISM

All the chemical processes by which nutrients are used to support life.

ANABOLISM

The metabolic process by which body tissues and substances are built.

CATABOLISM

The metabolic processes by which large, complex molecules are converted to simpler ones.

To become part of the body, food must be digested and absorbed. *Digestion* is the process by which food is broken down into its components in the mouth, stomach, and small intestine with the help of digestive *enzymes*.

Protein is digested, or broken down, into its building blocks, called amino acids; complex carbohydrates are reduced to simple sugars such as glucose; and fat molecules are broken down into fatty acids.

Before the body can use any nutrients that are present in food, the nutrients must pass through the walls of the stomach or intestines into the body's tissues, a process called *absorption*. Nutrients are absorbed into either the blood or the lymph, two fluids that circulate throughout the body, delivering needed products to the cells and picking up wastes. Blood is composed mostly of:

- Water
- Red blood cells (which carry and deliver oxygen to the cells)
- White blood cells (which are important in resistance to disease, called immunity)
- Nutrients
- Other components

Lymph is similar to blood but has no red blood cells. It goes into areas where there are no blood vessels to feed the cells.

Within each cell, *metabolism* takes place. Metabolism refers to all the chemical processes by which nutrients are used to support life. Metabolism has two parts: the building up of substances (called *anabolism*) and the breaking down of substances (called *catabolism*). Within each cell, nutrients such as glucose are split into smaller units in a catabolic reaction that releases energy. The energy is either converted to heat to maintain body temperature or used to perform work within the cell. During anabolism, substances such as proteins are built from their amino-acid building blocks.

GASTROINTESTINAL TRACT

Once we have smelled and tasted food, our meal goes on a journey through the **gastrointestinal tract** (also called the digestive tract), a hollow tube that runs down the middle of your body (Figure 1-10). The top of the tube is your mouth, which is connected in turn to your pharynx, esophagus, stomach, small intestine, large intestine, rectum, and anus, where solid wastes leave the body. The gastrointestinal tract is such a busy place that the cells lining it are replaced every few days.

The digestive system starts with the mouth, also called the **oral cavity**. Your tongue and teeth help with chewing. The tongue, which extends across the floor of the mouth, moves food around the mouth during chewing. Your 32 permanent teeth grind and break down food. Chewing is important because it breaks up the food into smaller pieces so that it can be swallowed. **Saliva**, a fluid secreted into the mouth from the salivary glands, contains important digestive enzymes and lubricates the food so that it may pass readily down the esophagus. Digestive enzymes help break down food into forms of nutrients that can be used by the body. Enzymes in the saliva start the digestion of carbohydrate. The tongue rolls the chewed food into a **bolus** (ball) to be swallowed.

The **pharynx** is a passageway about 5 inches long that connects the oral and nasal cavities to the esophagus and the air tubes to the lungs. When swallowing occurs, a flap of tissue, the **epiglottis**, covers the air tubes so that food does not get into the lungs. Food now enters the **esophagus**, a muscular tube that leads to the stomach. Food is propelled down the esophagus by **peristalsis**, rhythmic contractions of muscles in the wall of the esophagus. You might think of this involuntary contraction that forces food through the entire digestive system as squeezing a marble (the bolus) through a rubber tube. Peristalsis also helps break up food into smaller and smaller particles.

Food passes from the esophagus through the **lower esophageal (cardiac) sphincter**, a muscle that relaxes and contracts (in other words, opens and closes) to move food from the esophagus into the stomach. The **stomach**, a J-shaped muscular sac that holds about 4 cups

Mouth:	Tastes food.
	Chews food.
	Makes saliva.
Pharynx:	Directs food from mouth to esophagus.
Esophagus:	Passes food to stomach.
Stomach:	Makes enzyme that breaks down protein.
	Makes hydrochloric acid.
	Churns and mixes food.
	Acts like holding tank.
Small intestine:	Makes enzymes.
	Digests most of food.
	Absorbs nutrients across villi into blood and lymph.
Large intestine:	Passes waste to be excreted.
	Reabsorbs water and some minerals.
	Absorbs vitamins made by bacteria.
Rectum:	Stores feces.
Anus:	Keeps rectum closed.
	Opens for elimination.

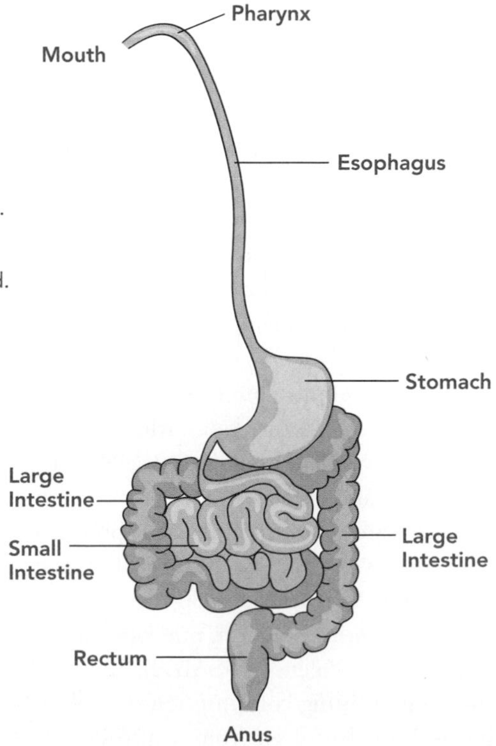

Pharynx
Mouth
Esophagus
Stomach
Large Intestine
Small Intestine
Large Intestine
Rectum
Anus

FIGURE 1-10:
Human digestive tract.

GASTROINTESTINAL TRACT
A hollow tube running down the middle of the body in which digestion of food and absorption of nutrients take place.

ORAL CAVITY
The mouth.

SALIVA
A fluid secreted into the mouth from the salivary glands that contains important digestive enzymes and lubricates the food so that it may readily pass down the esophagus.

BOLUS
A ball of chewed food that travels from the mouth through the esophagus to the stomach.

PHARYNX
A passageway that connects the oral and nasal cavities to the esophagus and air tubes to the lungs.

EPIGLOTTIS
The flap that covers the air tubes to the lungs so that food does not enter the lungs during swallowing.

ESOPHAGUS
The muscular tube that connects the pharynx to the stomach.

PERISTALSIS
Involuntary muscular contraction that forces food through the entire digestive system.

(or 1 liter) of food when full, is lined with a mucous membrane. Within the folds of the mucous membrane are digestive glands that make **hydrochloric acid** and an enzyme to break down proteins. Hydrochloric acid aids in protein digestion, destroys harmful bacteria, and increases the ability of calcium and iron to be absorbed. Because hydrochloric acid can damage the stomach, the stomach protects itself with a thick lining of mucus. Also, acid is produced only when we eat or think about eating.

It is the hydrochloric acid in the stomach that contributes to heartburn. Heartburn is a painful burning sensation in the esophagus, just below or behind the breastbone. Heartburn occurs when the lower esophageal sphincter fails to close tightly enough, allowing the stomach contents to back up (also called reflux) into the esophagus. This partially digested material is usually acidic and can irritate the esophagus. Frequent, ongoing heartburn may be a sign of gastroesophageal reflux disease (GERD). Ways to treat heartburn and GERD include eating small meals, avoiding foods and beverages that aggravate heartburn, losing weight (if overweight), and possibly medications.

From the top part of the stomach, food is slowly moved to the lower part, where the stomach churns it with the hydrochloric acid and digestive enzymes. The stomach has the strongest muscles and thickest walls of all the organs in the gastrointestinal tract. The food is now called **chyme** and has a semiliquid consistency. Chyme is next passed into the first part of the small intestine in small amounts (the small intestine can't process too much food at one time) through the **pyloric sphincter**, which operates like the lower esophageal sphincter. Liquids leave the stomach faster than solids do, and carbohydrate or protein foods leave faster than fatty foods do. The stomach absorbs few nutrients, but it does absorb alcohol. It takes 1.5 to 4 hours after you have eaten for the stomach to empty.

The **small intestine**, about 15 to 20 feet long, has three parts: the **duodenum**, the **jejunum**, and the **ileum**. The small intestine was so named because its diameter is smaller (about 1 inch) than that of the large intestine (about 2 1/2 inches), not because it is shorter. Actually, the small intestine is longer.

The duodenum, about 1 foot long, receives the digested food from the stomach as well as enzymes from other organs in the body, such as the pancreas and liver. The liver provides **bile**, a substance that is necessary for fat digestion. Bile is stored in the gallbladder and released into the duodenum when fat is present. The pancreas provides bicarbonate, a substance that neutralizes stomach acid. The small intestine itself produces digestive enzymes.

On the folds of the duodenal wall (and throughout the entire small intestine) are tiny, fingerlike projections called **villi**. Under a microscope you will see hairlike structures on the villi. These are called **microvilli** or the **brush border**. The villi and microvilli increase the surface area of the small intestine and therefore allow for more absorption of nutrients into the body. The muscular walls of the small intestine mix the chyme with the digestive juices and bring the nutrients into contact with the villi. Most nutrients pass through the villi of the duodenum and jejunum into either the blood or the lymph vessels, where they are transported to the liver and to the body cells.

The duodenum connects with the second section of the small intestine, the jejunum, which connects to the ileum. Most digestion is completed in the first half of the small intestine; whatever is left goes into the large intestine. Food is in the small intestine for about 7 to 8 hours and spends about 18 to 24 hours in the large intestine.

Ulcers are a common digestive problem that can affect the duodenum or the stomach. A peptic ulcer is a sore on the lining of the stomach or duodenum. Peptic ulcers are common: One in ten Americans develops an ulcer at some time in his or her life. One cause of peptic ulcer is bacterial infection, but some ulcers are caused by long-term use of nonsteroidal anti-inflammatory agents (NSAIDs), like aspirin and ibuprofen. Taking antibiotics, quitting smoking, limiting consumption of caffeine and alcohol, and reducing stress can speed healing and prevent ulcers from recurring.

The *large intestine* (also called the *colon*) is about 5 feet long and extends from the end of the ileum to a cavity called the rectum. One of the functions of the large intestine is to receive the waste products of digestion and pass them on to the rectum. Waste products are the materials that were not absorbed into the body. The large intestine does absorb water, some minerals (such as sodium and potassium), and a few vitamins made by bacteria residing there. Bacteria are normally found in the large intestine and are necessary for a healthy intestine. Intestinal bacteria make some important substances, such as vitamin K. They also can digest some components of food that we don't digest, such as fiber.

The *rectum* stores the waste products until they are released as solid feces through the *anus*, which opens to allow elimination.

MINI-SUMMARY

1. Before the body can use the nutrients in food, the food must be digested and the nutrients absorbed through the walls of the stomach and/or intestine into either the blood or the lymph system.
2. Within each cell, metabolism (all the chemical processes by which nutrients are used to support life) takes place. Metabolism has two parts: anabolism (building up) and catabolism (breaking down).
3. Figure 1-10 summarizes food digestion and absorption.

ILEUM
The final segment of the small intestine.

BILE
A substance made by the liver that is stored in the gallbladder and released when fat enters the small intestine to help digest fat.

VILLI
Tiny fingerlike projections in the wall of the small intestines that are involved in absorption.

MICROVILLI (BRUSH BORDER)
Hair-like projections on the villi that increase the surface area for absorbing nutrients.

LARGE INTESTINE (COLON)
The part of the gastrointestinal tract between the small intestine and the rectum.

RECTUM
The last section of the large intestine, in which feces, the waste products of digestion, is stored until elimination.

ANUS
The opening of the digestive tract through which feces travels out of the body.

⫷ CHECK-OUT QUIZ

1. Match the nutrients with their functions/qualities. The functions/qualities may be used more than once.

Nutrients	Functions
Carbohydrate	Provides energy
Lipid	Promotes growth and maintenance
Protein	Supplies the medium in which chemical changes of the body occur
Vitamins	Works as main structure of cells
Minerals	Regulates body processes
Water	

2. Match the Dietary Reference Intake values with their definition.

DRI Value	Definition
RDA	Value for kcalories
AI	Maximum safe intake level
UL	Value that meets requirements of 50 percent of individuals in a group
EAR	Value that meets requirements of 97 to 98 percent of individuals
EER	Value used when there is not enough scientific data to support an RDA

3. Match the terms on the left with their definitions on the right.

Term	Definition
Absorption	Process of building substances
Enzyme	Involuntary muscular contraction
Anabolism	Substance that speeds up chemical reactions
Peristalsis	Process of breaking down substances
Catabolism	Process of nutrients entering the tissues from the gastrointestinal tract

4. Which digestive organ passes waste to be excreted and reabsorbs water and minerals?

   a. stomach
   b. small intestine
   c. large intestine
   d. liver

5. Which nutrient supplies the highest number of calories per gram?

   a. carbohydrate
   b. fat
   c. protein
   d. vitamin pills

6. Flavor is a combination of all five senses.

   a. True
   b. False

7. Women have a higher basal metabolic rate than men do.

    a. True
    b. False

8. Hydrochloric acid aids in protein digestion, destroys harmful bacteria, and increases the ability of calcium and iron to be absorbed.

    a. True
    b. False

9. The nutrient density of a food depends on the amount of nutrients it contains and the comparison of that value to its caloric content.

    a. True
    b. False

10. The DRIs are designed for both healthy and sick people.

    a. True
    b. False

ACTIVITIES AND APPLICATIONS

1. How Many Kcalories Do You Need Each Day?

   Use the following two steps to calculate the number of kcalories you need.

   A. To determine your basal metabolic needs, multiply your weight in pounds by 10.9 if you are male and by 9.8 if you are female. (These numbers are based on a BMR factor of 1.0 kcalorie per kilogram of body weight per hour for men and 0.9 for women.) Example: 150-pound woman × 9.8 = 1470 kcalories

   B. To determine how much you use each day for physical activity, first determine your level of activity.

      Very light activity: You spend most of your day seated or standing.
      Light activity: You spend part of your day up and about, such as in teaching or cleaning house.
      Moderate activity: You engage in exercise for an hour or so at least every other day, or your job requires some physical work.
      Heavy activity: You engage in manual labor, such as construction.

      Once you have picked your activity level, you need to multiply your answer in A by one of the following numbers.

      Very light (men and women): Multiply by 1.3
      Light (men): Multiply by 1.6

Light (women): Multiply by 1.5
Moderate (men): Multiply by 1.7
Moderate (women): Multiply by 1.6
Heavy (men): Multiply by 2.1
Heavy (women): Multiply by 1.9
Example: A woman with light activity.
1470 kcalories × 1.5 = 2205 kcalories needed daily

Compare the number of kcalories you need with your Estimated Energy Requirement, using Appendix B. The results should be similar.

2. **Factors Influencing What You Eat**

Answer the following questions to try to understand the factors influencing what you eat. Compare your answers with a friend or classmate.

A. How many meals and snacks do you eat each day, and when are they eaten?
B. What are your favorite foods?
C. What foods do you avoid eating and why?
D. Rate the importance of each of these factors when selecting foods (1 = very important, 3 = somewhat important, 5 = not important)

    Cost
    Convenience
    Availability
    Familiarity
    Nutrition

E. Are you usually willing to try a new food?
F. What holidays do you and your family celebrate? What foods are served?
G. Do your food habits differ from those of your family? Your friends? Your coworkers? If yes, describe how your food habits are different and why you think this is so.
H. What foods, if any, do you eat to stay healthy or improve your appearance?
I. How much do you know about nutrition? How important is good nutrition to you?
J. Do you eat differently when you are with others than you do when alone?
K. Which foods do you eat when you are under stress?
L. Which foods do you eat when you are sick?
M. Do you think that food advertising affects what you eat? Describe.
N. Do you prefer organic fruits and vegetables? Why or why not?
O. Are you a vegetarian, and if so, why did you choose this eating style?

3. **Taste and Smell**

Pick one of your favorite foods, eat it normally, and then take a bite of it while holding your nose. How does it taste when you can't smell very well? What influence does smell have on taste?

4. **Nutrient-Dense Foods**

Pick one food that you ate yesterday that could be considered nutrient-dense. Also pick one food that would not be considered nutrient-dense. Compare the nutrition labels, or compare their kcalorie and nutrient content by going to this website: www.nal.usda.gov/fnic/foodcomp/search/

Type in a food next to "Keyword," and select the specific food and then the portion size. The next screen will give a nutrient analysis that you can print and compare.

NUTRITION WEB EXPLORER

U.S. Government Healthfinder   www.healthfinder.gov

This government site can help you find information on virtually any health topic. On the home page, click on "H" under "Health A to Z." Next, click on heart disease. Using the links, find 5 ways to reduce your risk of heart disease.

Nutrition.gov   www.nutrition.gov

From this government site, you can access many nutrition topics right from the home page. Click on "In the News." Then click on a nutrition article and summarize this article in one paragraph.

National Organic Program   www.ams.usda.gov/nop

Visit the website for the National Organic Program to find out if a "natural" food can also be labeled as "organic." Click on "Labeling."

Center for Science in the Public Interest: Eating Green   www.cspinet.org/EatingGreen

Click on "Eating Green Calculator" and fill in how much in the way of animal products you eat each week. Then click on "Calculate Impact" and find out the environmental impact of your eating habits. Also use the "Score Your Diet" tool to show how your diet scores on nutrition, the environment, and animal welfare.

Center for Young Womens' Health   www.youngwomenshealth.org/college101.html

Read "College Eating and Fitness 101." List 5 suggestions they make to help you not gain the Freshman 15.

Alcohol Calorie Calculator   www.collegedrinkingprevention.gov/ CollegeStudents/calculator/alcoholcalc.aspx

Fill in the "Average Drinks per Week" column and then press "Compute." You will see how many calories you take in each month and in one year from alcoholic beverages.

FOOD FACTS: HOW TO RECOGNIZE WHOLE FOODS, PROCESSED FOODS, AND ORGANIC FOODS

When people talk about food, you may hear some terms with which you are not familiar or are unsure of. *Whole foods* (besides being the name of a chain of stores), are foods pretty much as we get them from nature. Examples include eggs, fresh fruits and vegetables, beans and peas, whole grains, and fish. Whole foods are often not processed, but some are minimally processed. Milk, for example, is minimally processed to make it safe to drink. Fresh meat is also minimally processed so that consumers can buy just what they want.

Processed foods have been prepared using a certain procedure: milling (white flour), cooking and freezing (such as frozen pancakes or dinners), canning (canned vegetables), dehydrating (dried fruits), or culturing with bacteria (yogurt). In some cases, processing removes nutrients, such as when whole wheat is milled to make white flour. In other cases, processing helps retain nutrients, such as when freshly picked vegetables are frozen.

Whereas the food supply once contained mostly whole farm-grown foods, today's supermarket shelves are stocked primarily with processed foods. Many processed foods contain parts of whole foods and often have added ingredients such as sugars, or sugar or fat substitutes. For instance, cookies are made with eggs and flour. Then sugar and fat are added. Highly processed foods, such as many breakfast cereals, cookies, crackers, sauces, soups, baking mixes, frozen entrees, pasta, snack foods, and condiments, are staples nowadays.

When processing adds nutrients, the resulting food is either an enriched or a fortified food. For example, white flour must be enriched with several vitamins and iron to make up for some of the nutrients lost during milling. A food is considered *enriched* when nutrients are added to it to replace the same nutrients that are lost in processing.

Milk is often fortified with vitamin D because there are few good food sources of this vitamin. A food is considered *fortified* when nutrients are added that were not present originally or nutrients are added that increase the amount already present. For example, orange juice does not contain calcium, and so when calcium is added to orange juice, the product is called calcium-fortified orange juice. Probably the most notable fortified food is iodized salt. Iodized salt was introduced in 1924 to combat iodine deficiencies.

Organic foods are becoming more and more popular in supermarkets and restaurants. Common organic foods include fruits, vegetables, and cereals. Meat, poultry, and eggs can also be organic.

Organic food is produced by farmers who emphasize the use of renewable resources and the conservation of soil and water to enhance environmental quality for future generations. Organic meat, poultry, eggs, and dairy products come from animals that are given no antibiotics or growth hormones. Organic food is produced without using most conventional pesticides; fertilizers made with synthetic ingredients or sewage sludge; genetic engineering or irradiation. Before a product can be labeled "organic," a government-approved certifier inspects the farm where the food is grown to make sure the farmer is following all the rules necessary to meet USDA organic standards. Companies that handle or process organic food before it gets to your local supermarket or restaurant must be certified too.

So what makes organic fruits and vegetables different from nonorganic fruits and vegetables? The organic crop production standards state:

A. The land will have no prohibited substances applied to it for at least three years before the harvesting of an organic crop.
B. The use of genetic engineering, ionizing radiation, and sewage sludge is prohibited.
C. Soil fertility and crop nutrients will be managed through tillage and cultivation practices, crop rotations, and cover crops, supplemented with animal and crop waste materials and allowed synthetic materials.
D. Preference will be given to the use of organic seeds and other planting stock, but a farmer may use nonorganic seeding and planting stock under specified conditions.
E. Crop pests, weeds, and diseases will be controlled primarily through management practices, including physical, mechanical, and biological controls.

Livestock must be fed 100 percent organic feed. Organically raised animals may not be given hormones to promote growth or antibiotics for any reason (unless an animal is sick or injured, in which case the animal can't be sold as organic). Preventive management practices, including the use of vaccines, are used to keep animals healthy. Also, livestock may be given allowed vitamin and mineral supplements. All organically raised animals must have access to the outdoors, including access to pasture. They may be temporarily confined only for reasons such as health and safety.

Many consumers, as well as chefs, feel that organic foods taste better than their conventional counterparts. Whether organic foods taste better is to some extent a matter of personal taste. Also, taste will vary among any fresh produce, depending on their freshness, the seeds used, where they were grown, and so on.

As for nutrition, some studies show that organic foods may be higher in vitamins (especially vitamin C), minerals, and polyphenols (substances in plants that have antioxidant activity) compared with conventionally grown foods. However, there is no solid body of research yet. The nutrient composition of any food grown in soil will vary due to many factors, such as variations in the soil quality, the amount of sunshine, and the amount of rain. Vitamins in plants are created by the plants themselves as long as they get adequate sunshine, water, carbon dioxide, and fertilizer. Minerals must come from the soil.

Figure 1-11 shows how organic foods are labeled. There are four categories of organic labeling.

1. Foods labeled "100 percent organic" must contain only organically produced ingredients (excluding water and salt).
2. Foods labeled "organic" must consist of at least 95 percent organically produced ingredients (excluding water and salt). Any remaining ingredients must consist of nonagricultural products approved on the national list or agricultural products that are not commercially available in organic form.
3. Processed foods labeled "made with organic ingredients" must contain at least 70 percent organic ingredients, and they can list up to three of the organic ingredients or food groups on the principal display panel. For example, soup made with at least 70 percent organic ingredients and only organic vegetables may be labeled either "soup made with organic peas, potatoes, and carrots" or "soup made with organic vegetables."
4. Processed foods that contain less than 70 percent organic ingredients cannot use the term "organic" anywhere on the principal display panel. However, they may identify the specific ingredients that are organically produced on the ingredients statement.

Foods that are 100 or 95 percent organic may display the USDA organic seal, shown in Figure 1-12. Use of the seal is voluntary.

HOT TOPIC: HOW THE AMERICAN DIET IMPACTS THE ENVIRONMENT AND HOW RESTAURANTS ARE GOING GREEN

As you drive by many farms across America, you might be inclined to think that we eat a lot of corn, soybeans, and grains. However, it is not Americans who are eating most of these foods: it is livestock, such as beef cattle, dairy cattle, hogs, chickens, and turkeys. Eventually these livestock (except dairy cattle, which give us milk) will be slaughtered to produce meat and poultry. The typical American eats about 8 ounces of meat a day (including beef, poultry, and fish). This amount is about double the global average.

Producing large quantities of meat in America uses many resources and has serious environmental consequences, such as the following.

1. According to the Food and Agriculture Organization (FAO) of the United Nations, livestock now use 30 percent of the earth's entire land surface, which includes pastures as well as land used to produce feed for livestock. In Latin America, some 70 percent of former forests in the Amazon have been turned over to grazing. Forests have a huge impact on the environment. The trees help balance the oxygen-carbon dioxide balance of the earth by absorbing carbon dioxide from the environment and releasing oxygen. Increased deforestation has led to the accumulation of greenhouse gases in the atmosphere, such as carbon dioxide and methane. The accumulation of these greenhouse gases has enhanced the earth's natural greenhouse effect, by which the temperature of the earth is maintained, leading to global warming. Trees also absorb rainfall by soaking up moisture through their roots, thus preventing runoff and the accompanying soil erosion and flooding. Leaves that fall on the forest ground act as a nutrient source and increase soil fertility. Forests provide essential ecological services by absorbing carbon dioxide, preventing runoff and flooding, and providing homes to diverse forms of wildlife.

2. Livestock farms are major air and water polluters. Cattle naturally produce methane, a potent greenhouse gas that can contribute to global climate change. Livestock production systems also produce other greenhouse gases such as nitrous oxide and carbon dioxide. People who live near or work at these farms breathe in hundreds of gases, which are formed as manure decomposes. The stench can be unbearable. And, of course, there is the problem of waste output: U.S. livestock produce about 900 million tons of manure each year, or about 3 tons for each American. On most factory farms, animals are crowded into relatively small areas; their wastes are funneled into massive cesspools called lagoons. These cesspools often break, leak, or overflow, sending dangerous microbes, nitrate pollution, and drug-resistant bacteria into water supplies.

3. Enormous quantities of water, fuel, fertilizers, and pesticides are required to grow the feed for livestock, utilizing many acres of farmland. In drier climates, huge amounts of irrigation water are used to produce feed grains such as corn. To produce 100 kcalories of plant foods only requires about 50 kcalories from fossil fuels, but to get the equivalent amount of kcalories from beef requires almost 1,600 kcalories. Fertilizers also require a lot of energy to produce, and along with pesticides, they often wind up polluting waterways and drinking water.

Our current method of producing meat is costly and unsustainable, and its harmful environmental consequences will become more troublesome as the world's population grows and demand for meat increases. New government programs and policies will be necessary to reduce damage to the environment while ensuring adequate nutrition for the world's population. One answer to this dilemma can be found in sustainable agriculture.

Sustainable agriculture produces abundant food without depleting the earth's resources or polluting its environment. It is agriculture that follows the principles of nature to develop systems for raising crops and livestock that are, like nature, self-sustaining (see Figure 1-13). In recent decades, sustainable farmers and researchers around the world have used a variety of techniques to farm with nature. Sustainable practices lend themselves to smaller, family-scale farms. These farms, in turn, tend to find their best niches in local markets, within local food systems, often selling directly to consumers in farmers' markets or to local restaurants.

FIGURE 1-13:
Sustainable agriculture produces abundant food without depleting the earth's resources or polluting its environment. Courtesy of Digital Vision.

New government programs and policies will also need to encourage Americans to eat less animal protein and eat more fruits, vegetables, beans, grains, nuts, and seeds. Even a meat-free diet that includes dairy and eggs is still much less harmful to the environment than a meat-based diet. Choosing to eat lower down on the food chain creates a lower environmental impact with fewer negative ecological consequences.

Many restaurants are also trying to do their part to purchase sustainable foods including locally grown and organic foods. The mission of Green Foodservice Alliance in Georgia is to "incorporate environmental conservation and sustainability practices into the daily operations of Georgia foodservice establishments." More and more restaurants across the United States are implementing environmentally responsible practices that also include:

1. Saving energy through more energy-efficient equipment and lighting—

for example, ENERGY STAR refrigeration models have better insulation and use almost half as much energy as older models

2. Buying tableware and cups made of recycled and renewable materials

3. Buying nontoxic cleaning and sanitation supplies

4. Installing flow restrictors on faucets and using low-flush toilets

5. Recycling

6. Using an energy management program

ENERGY STAR, a joint program of the U.S. Environmental Protection Agency and the U.S. Department of Energy, identifies and promotes energy-efficient products to reduce greenhouse gas emissions. The ENERGY STAR label can be found on restaurant equipment and lighting.

According to the Environmental Protection Agency, approximately 64 billion paper cups and plates, 73 billion Styrofoam and plastic plates and 190 billion plastic

containers and bottles are thrown away every year in the United States. Since paper made from virgin wood contributes to forest depletion, many environment-conscious groups recommend using products made from recycled materials. Currently there is an environment-friendly substitute for just about every disposable item in foodservice, including napkins, paper towels, and facial tissues; trays and tray liners; cold and hot cups; lids; straws; forks, knives, spoons; to-go packaging; salad containers; cleaning products; plates and bowls.

Many factors are driving the green movement in restaurants, from meeting consumer demand for eco-friendly products to conserving resources and dollars to joining a global effort to protect and preserve our natural environment. The changes won't be accomplished overnight, but will take time. Sustainable food programs, waste reduction, recycling and energy-conservation are major endeavors that take time, planning and money.

CHAPTER 2

Using Dietary Recommendations, Food Guides, and Food Labels to Plan Menus

Dietary Recommendations
and Food Guides
Dietary Guidelines for Americans
MyPyramid

Food Labels
Nutrition Facts

Nutrient Claims
Health Claims

Portion Size Comparisons

Food Facts: Nutrient Analysis of Recipes

Hot Topic: Quack! Quack!

Dietary recommendations have been published for the healthy American public for almost 100 years. Early recommendations centered on encouraging intake of certain foods to prevent deficiencies, fight disease, and enhance growth. Although deficiency diseases have been virtually eliminated, they have been replaced by diseases of dietary excess and imbalance—problems that now rank among the leading causes of illness and death in the United States. Diseases such as heart disease, cancer, and diabetes touch the lives of most Americans and generate substantial health-care costs. More recent dietary guidelines therefore have centered on modifying the diet, in most cases cutting back on certain foods, to reduce risk factors for chronic disease such as heart disease and obesity.

This chapter looks at dietary recommendations, food guides, and food labels. It will help you to:

- Discuss the Dietary Guidelines for Americans with regard to adequate nutrients within kcalorie needs, weight management, physical activity, food groups to encourage, fat, carbohydrates, sodium and potassium, alcoholic beverages, and food safety
- Recommend ways to implement each Dietary Guideline
- Describe each food group in MyPyramid, including subgroups as appropriate
- Explain the concept of discretionary kcalories
- Gives examples of portion sizes from each food group
- Describe how MyPyramid illustrates variety, proportionality, and moderation
- Plan menus using MyPyramid
- List the information required on a food label
- Read and interpret information from the Nutrition Facts label
- Distinguish between a nutrient claim and a health claim
- Explain how an "A" health claim differs from those ranked "B," "C," or "D"
- Discuss the relationship between portion size on food labels and portions in MyPyramid

DIETARY RECOMMENDATIONS AND FOOD GUIDES

Whereas dietary recommendations discuss specific foods to eat for optimum health, *food guides* tell us the amounts of foods we need to eat to have a nutritionally adequate diet. Their primary role, whether in the United States or around the world, is to communicate an optimum diet for the overall health of the population. Food guides are based on current dietary recommendations, the nutrient content of foods, and the eating habits of the targeted population. MyPyramid, a food guide developed by the U.S. Department of Agriculture, is based on the Dietary Guidelines for Americans and nutrient recommendations. To better understand MyPyramid, let's first look at the Dietary Guidelines for Americans.

DIETARY GUIDELINES FOR AMERICANS

The most recent set of U.S. dietary recommendations, the *Dietary Guidelines for Americans*, was published in 2005. Updated every five years by a joint advisory committee of the U.S. Department of Agriculture and the Department of Health and Human Services, the Dietary Guidelines for Americans form the basis of federal food, nutrition education, and information programs. They provide sound advice to help people make food choices for a healthy, active life and reflect a consensus of the most current scientific and medical knowledge available. The recommendations contained in the Dietary Guidelines are targeted to members of the general public over two years of age who are living in the United States.

Dietary recommendations are quite different from Dietary Reference Intakes (DRIs). Whereas DRIs deal with specific nutrients, dietary recommendations discuss specific foods and food groups that will help individuals meet the DRIs. DRIs also tend to be written in a technical style. Dietary recommendations are generally written in easy-to-understand terms.

The 2005 Dietary Guidelines are very important in preventing disease. Good nutrition is vital to good health and absolutely essential for the healthy growth and development of children and adolescents. Poor diet is linked to diseases and conditions such as heart disease, high blood pressure, type 2 diabetes, overweight and obesity, iron deficiency anemia, and some cancers. Lack of physical activity has been associated with heart disease, high blood pressure, overweight and obesity, osteoporosis, diabetes, and certain cancers. Together with physical activity, a high-quality diet (without too many kcalories) should enhance the health of most individuals.

A basic premise of the Dietary Guidelines is that nutrient needs should be met primarily through consuming foods. Foods provide an array of nutrients and other compounds that may have beneficial effects on health. Supplements may be useful when they fill a specific identified nutrient gap that cannot be met by an individual's intake of food.

The key recommendations of the Dietary Guidelines are presented in nine categories: adequate nutrients within kcalorie needs, weight management, physical activity, food groups to encourage, fats, carbohydrates, sodium and potassium, alcoholic beverages, and food safety. A discussion of each category follows. Also see Figure 2-1 for key recommendations for each category.

1. Adequate Nutrients within Kcalorie Needs. Many Americans consume more kcalories than they need without meeting recommended intakes for a number of nutrients. This means that most people need to choose meals and snacks that are high in nutrients but low to moderate in energy content. Doing so offers important benefits: normal growth and development of children, health promotion for people of all ages, and less risk of chronic diseases. In particular, intake levels of the following nutrients may be of concern for these groups:
   - Adults: calcium, potassium, fiber, magnesium, and vitamins A (as carotenoids), C, and E
   - Children and adolescents: calcium, potassium, fiber, magnesium, and vitamin E
     At the same time, Americans often consume too many kcalories and too much saturated and trans fats, cholesterol, added sugars, and salt.

2. Weight Management. The prevalence of obesity in the United States has doubled in the last two decades. Nearly one-third of adults are obese, meaning their body mass index (BMI) is 30 or greater. (Figure 2-2 on page 36 will help you locate your BMI.) Also, as much as 16 percent of children and adolescents are overweight. That represents a doubling of the rate of overweight among children and a tripling among adolescents. Overweight and obesity are of concern because excess body fat leads to a higher risk for premature death, type 2 diabetes, heart disease, stroke, high blood pressure, certain kinds of cancers, and other problems.

   Ideally, the goal for adults is to achieve and maintain a body weight that optimizes their health. However, for obese adults, even modest weight loss (such as 10 pounds) has health benefits, and the prevention of further weigh gain is very important. For overweight children and adolescents, the goal is to slow the rate of weight gain while achieving normal growth and development. The keys to losing body weight are to eat fewer kcalories and increase physical activity.

FIGURE 2-1: Dietary Guidelines (2005)

1. Adequate Nutrients within Kcalorie Needs

Key Recommendations

- Meet recommended intakes within energy needs by adopting a balanced eating pattern such as that in MyPyramid (page 39). This food guide is designed to integrate dietary recommendations into a healthy way to eat. MyPyramid differs in important ways from common food consumption patterns in the United States. In general, MyPyramid recommends:
  - More dark green vegetables, orange vegetables, legumes, fruits, whole grains, and low-fat milk and milk products
  - Less refined grains, total fats (especially cholesterol, and saturated and trans fats), added sugar, and kcalories.
- Consume a variety of nutrient-dense foods and beverages within and among the basic food groups while choosing foods that limit the intake of saturated and trans fats, cholesterol, added sugars, salt, and alcohol.

2. Weight Management

Key Recommendations

- To maintain body weight in a healthy range, balance kcalories from foods and beverages with kcalories expended.
- To prevent gradual weight gain over time, make small decreases in food and beverage kcalories and increase physical activity.
- If you need to lose weight, aim for a slow, steady weight loss by decreasing kcalorie intake while maintaining an adequate nutrient intake and increasing physical activity. A reduction of 500 kcalories per day is often needed and will result in weight loss of about 1 pound a week. When it comes to losing weight, it is kcalories that count—not the proportions of fat, carbohydrates, and protein in the diet.

3. Physical Activity

Key Recommendations

- Engage in regular physical activity and reduce sedentary activities to promote health, psychological well-being, and a healthy body weight.
  - To reduce the risk of chronic disease in adulthood, engage in at least 30 minutes of moderate-intensity physical activity, above usual activity, on most days of the week.
  - For most people, greater health benefits can be obtained by engaging in physical activity of more vigorous intensity or longer duration.
  - To help manage body weight and prevent gradual, unhealthy body weight gain in adulthood, engage in approximately 60 minutes of moderate- to vigorous-intensity activity on most days of the week while not exceeding caloric intake requirements.
  - To sustain weight loss in adulthood: participate in at least 60 to 90 minutes of daily moderate-intensity physical activity while not exceeding caloric intake requirements. Some people may need to consult with a health-care provider before participating in this level of activity.
- Achieve physical fitness by including cardiovascular conditioning, stretching exercises for flexibility, and resistance exercises or calisthenics for muscle strength and endurance.

4. Food Groups to Encourage

Key Recommendations

- Consume a sufficient amount of fruits and vegetables while staying within energy needs. Two cups of fruit and 2½ cups of vegetables per day are recommended for a reference 2000 kcalorie intake.
- Choose a variety of fruits and vegetables each day. In particular, select from all five vegetable subgroups (dark green, orange, legumes, starchy vegetables, and other vegetables) several times a week.
- Consume 3 or more ounce-equivalents of whole-grain products per day, with the rest of the recommended grains coming from enriched or whole-grain products. In general, at least half the grains should be whole grains.
- Consume 3 cups per day of fat-free or low-fat milk or equivalent milk products.

5. Fats

Key Recommendations

- Consume less than 10 percent of kcalories from saturated fatty acids and less than 300 mg/day of cholesterol, and keep trans fatty acid consumption as low as possible.
- Keep total fat intake between 20 to 35 percent of kcalories, with most fats coming from sources of polyunsaturated and monounsaturated fatty acids, such as fish, nuts, and vegetable oils.
- When selecting and preparing meat, poultry, dry beans, and milk or milk products, make choices that are lean, low-fat, or fat-free.
- Limit intake of fats and oils in high in saturated and/or trans fatty acids, and choose products low in such fats and oils.

6. Carbohydrates

Key Recommendations

- Choose fiber-rich fruits, vegetables, and whole grains often.
- Choose and prepare foods and beverages with little added sugars or caloric sweeteners, such as amounts suggested by MyPyramid.
- Reduce the incidence of dental caries by practicing good oral hygiene and consuming sugar- and starch-containing foods and beverages less frequently.

7. Sodium and Potassium

Key Recommendations

- Consume less than 2,300 mg (about 1 teaspoon of salt) of sodium per day.
- Choose and prepare foods with little salt.
- Eat potassium-rich foods such as fruits and vegetables.

8. Alcoholic Beverages

Key Recommendations

- Those who choose to drink alcoholic beverages should do so sensibly and in moderation—defined as the consumption of up to one drink per day for women and up to two drinks per day for men. One drink is 12 ounces of beer, 5 ounces of wine, or 1-½ ounces of hard liquor such as 80 proof gin or whiskey.
- Alcoholic beverages should not be consumed by some individuals, including those who cannot restrict their alcohol intake, women of childbearing age who may become pregnant, pregnant and lactating women, children and adolescents, individuals taking medications that can interact with alcohol, and those with specific medical conditions.
- Alcoholic beverages should be avoided by individuals engaging in activities that require attention, skill, or coordination, such as driving or operating machinery.

9. Food Safety

Key Recommendations

- To avoid microbial food-borne illness:
  - Clean hands, food contact surfaces, and fruit and vegetables. Meat and poultry should not be washed or rinsed.
  - Separate raw, cooked, and ready-to-eat foods while shopping, preparing, or storing foods.
  - Cook foods to a safe temperature to kill microorganisms.
  - Chill (refrigerate) perishable food promptly and defrost foods properly.
  - Avoid raw (unpasteurized) milk or any products made from raw milk, raw or partially cooked eggs or foods containing raw eggs, raw or undercooked meat and poultry, unpasteurized juices, and raw sprouts.

Source: U.S. Department of Health and Human Services, U.S. Department of Agriculture.

FIGURE 2-2: Adult Body Mass Index Chart

Locate the height of interest in the left column and read across the row for that height to the weight of interest. Follow the column of the weight up to the top row that lists the BMI. BMI of 19 to 24 is the healthy weight range, BMI of 25 to 29 is the overweight range, and BMI of 30 and above is in the obese range.

BMI	19	20	21	22	23	24	25	26	27	28	29	30	31	32	33	34	35
Height							**Weight in Pounds**										
4'10"	91	96	100	110	105	115	119	124	129	134	138	143	148	153	158	162	167
4'11"	94	99	104	109	114	119	124	128	133	138	143	148	153	158	163	168	173
5'	97	102	107	112	118	123	128	133	138	143	148	153	158	163	168	174	179
5'1"	100	106	111	116	122	127	132	137	143	148	153	158	164	169	174	180	185
5'2"	104	109	115	120	126	131	136	142	147	153	158	164	169	175	180	186	191
5'3"	107	113	118	124	130	135	141	146	152	158	163	169	175	180	186	191	197
5'4"	110	116	122	128	134	140	145	151	157	163	169	174	180	186	192	197	204
5'5"	114	120	126	132	138	144	150	156	162	168	174	180	186	192	198	204	210
5'6"	118	124	130	136	142	148	155	161	167	173	179	186	192	198	204	210	216
5'7"	121	127	134	140	146	153	159	166	172	178	185	191	198	204	211	217	223
5'8"	125	131	138	144	151	158	164	171	177	184	190	197	203	210	216	223	230
5'9"	128	135	142	149	155	162	169	176	182	189	196	203	209	216	223	230	236
5'10"	132	139	146	153	160	167	174	181	188	195	202	209	216	222	229	236	243
5'11"	136	143	150	157	165	172	179	186	193	200	208	215	222	229	236	243	250
6'	140	147	154	162	169	177	184	191	199	206	213	221	228	235	242	250	258
6'1"	144	151	159	166	174	182	189	197	204	212	219	227	235	242	250	257	265
6'2"	148	155	163	171	179	186	194	202	210	218	225	253	241	249	256	264	272
6'3"	152	160	168	176	184	192	200	208	216	224	232	240	248	256	264	272	279
	Healthy Weight						Overweight				Obese						

Source: Evidence Report of Clinical Guidelines on the Identification, Evaluation and Treatment of Overweight and Obesity in Adults, 1998, NIH/National Heart, Lung and Blood Institute (NHLBI).

3. **Physical Activity.** Americans tend to be relatively inactive. In 2005, less than half of U.S. adults met the recommendation for 30 minutes of moderate-intensity physical activity five days a week. Regular physical activity and physical fitness make important contributions to one's health, sense of well-being, and maintenance of a healthy body weight. Conversely, a sedentary lifestyle increases the risk for overweight and obesity and many chronic diseases, including heart disease, high blood pressure, type 2 diabetes, certain types of cancer, and osteoporosis. Overall, mortality rates from all causes of death are lower in physically active people than in sedentary people. Also, physical activity can aid in managing mild to moderate depression and anxiety. Figure 2-3 lists the kcalories per hour expended in common physical activities.

4. **Food Groups to Encourage.** Increased intakes of fruits, vegetables, whole grains, and fat-free or low-fat milk and milk products are likely to have important health benefits for most Americans (Figure 2-4). Compared with the many people who consume only small amounts of fruits and vegetables, those who eat more generous amounts as part of a healthful diet are likely to have a reduced risk of chronic diseases, including stroke and perhaps other cardiovascular diseases, type 2 diabetes, and cancer in certain sites (such as the lungs, stomach, and colon). Diets rich in foods containing fiber, such as fruits, vegetables, and whole grains, may reduce the risk of coronary heart disease.

FIGURE 2-3: Kcalories per Hour Expended in Common Physical Activities

Moderate Physical Activity	Approximate Kcalories/Hour for a 154-Pound Person
Hiking	370
Light gardening/yard work	330
Dancing	330
Golf (walking and carrying clubs)	330
Bicycling (<10 mph)	290
Walking (3.5 mph)	280
Weight lifting (general light workout)	220
Stretching	180

Vigorous Physical Activity	Approximate Kcalories/Hour for a 154-Pound Person
Running/jogging (5 mph)	590
Bicycling (>10 mph)	590
Swimming (slow freestyle laps)	510
Aerobics	480
Walking (4.5 mph)	460
Heavy yard work (chopping wood)	440
Weight lifting (vigorous effort)	440
Basketball (vigorous)	440

Source: Adapted from the 2005 Dietary Guidelines Advisory Committee Report.

Diets rich in milk and milk products can reduce the risk of low bone mass throughout the life cycle. The consumption of milk products is especially important for children and adolescents who are building peak bone mass and developing lifelong habits.

5. **Fats.** Fats and oils are part of a healthful diet, but the type of fat makes a difference for heart health, and the total amount of fat consumed is also important. High intake of saturated fats, trans fats, and cholesterol increases the risk of unhealthy blood lipid levels, which in turn increases the risk of coronary heart disease. The biggest sources of saturated fat in the American diet are animal foods: cheese, beef (more than half comes from ground beef), whole milk, fats in baked goods, butter, and margarine. Saturated fat is also found in eggs, poultry skin, and other full-fat dairy products, such as ice cream.

A high intake of fat (more than 35 percent of kcalories) generally increases saturated fat intake and makes it more difficult to avoid consuming excess kcalories. A low intake of fats and oils (less than 20 percent of kcalories) increases the risk of inadequate intakes of vitamin E and essential fatty acids and may contribute to unfavorable changes in high-density lipoprotein (HDL) blood cholesterol and triglycerides.

Most dietary fats should come from sources of polyunsaturated and monounsaturated fatty acids. Sources of polyunsaturated fatty acids are liquid vegetable oils, including soybean oil, corn oil, and safflower oil. Plants sources that are rich in monounsaturated fatty acids include vegetable oils (such as canola, olive, high-oleic safflower, and sunflower oils) that are liquid at room temperature and nuts.

6. **Carbohydrates.** Carbohydrates are part of a healthful diet. The Acceptable Macronutrient Distribution Range for carbohydrates is 45 to 65 percent of total kcalories. Sugars and starches supply energy to the body in the form of glucose.

Mix up your choices within each food group.

Focus on fruits. Eat a variety of fruits—whether fresh, frozen, canned, or dried—rather than fruit juice for most of your fruit choices. For a 2,000-calorie diet, you will need 2 cups of fruit each day (for example, 1 small banana, 1 large orange, and ¼ cup of dried apricots or peaches).

Vary your veggies. Eat more dark green veggies, such as broccoli, kale, and other dark leafy greens; orange veggies, such as carrots, sweetpotatoes, pumpkin, and winter squash; and beans and peas, such as pinto beans, kidney beans, black beans, garbanzo beans, split peas, and lentils.

Get your calcium-rich foods. Get 3 cups of low-fat or fat-free milk—or an equivalent amount of low-fat yogurt and/or low-fat cheese (1½ ounces of cheese equals 1 cup of milk)—every day. For kids aged 2 to 8, it's 2 cups of milk. If you don't or can't consume milk, choose lactose-free milk products and/or calcium-fortified foods and beverages.

Make half your grains whole. Eat at least 3 ounces of whole-grain cereals, breads, crackers, rice, or pasta every day. One ounce is about 1 slice of bread, 1 cup of breakfast cereal, or ½ cup of cooked rice or pasta. Look to see that grains such as wheat, rice, oats, or corn are referred to as "whole" in the list of ingredients.

Go lean with protein. Choose lean meats and poultry. Bake it, broil it, or grill it. And vary your protein choices—with more fish, beans, peas, nuts, and seeds.

Know the limits on fats, salt, and sugars. Read the Nutrition Facts label on foods. Look for foods low in saturated fats and *trans* fats. Choose and prepare foods and beverages with little salt (sodium) and/or added sugars (caloric sweeteners).

FIGURE 2-4:
Food Groups to Encourage (Dietary Guidelines for Americans).
Courtesy of U.S. Department of Agriculture.

Sugars can be naturally present in foods, such as fructose in fruit, or added to foods. Added sugars are sugars and syrups that are added to foods such as soda and cookies during processing. The more you eat of foods containing large amounts of added sugars, the more difficult it is to eat enough nutrients without gaining weight. Consumption of added sugars provides kcalories while providing little, if any, of the essential nutrients. Both sugars and starches contribute to dental caries.

Diets rich in dietary fiber have been shown to have a number of beneficial effects, including a decreased risk of coronary heart disease and less constipation. There is also interest in the potential relationship between diets containing fiber-rich foods and a lower risk of type 2 diabetes.

7. **Sodium and Potassium.** On average, the higher your salt (sodium chloride) intake, the higher your blood pressure. Nearly all Americans consume substantially more salt than they need. Decreasing salt intake is advisable to reduce the risk of elevated blood pressure. Keeping blood pressure in the normal range reduces an individual's risk of coronary heart disease, stroke, congestive heart failure, and kidney disease.

Many American adults will develop hypertension (high blood pressure) during their lifetimes. Lifestyle change can prevent or delay the onset of high blood pressure and lower elevated blood pressure. These changes include reducing salt intake, increasing potassium intake, losing excess body weight, increasing physical activity, and eating an overall healthful diet. Fruits and vegetables are excellent sources of potassium, especially winter squash, potatoes, oranges, grapefruits, and bananas.

8. **Alcoholic Beverages.** Alcoholic beverages supply kcalories but few essential nutrients. As a result, excessive alcohol consumption makes it hard to consume enough nutrients within an individual's daily kcalorie allotment and to maintain a healthy weight. Although the consumption of one or two alcoholic beverages per day is not associated with nutritional deficiencies or with overall dietary quality, heavy drinkers may be at risk of malnutrition if the kcalories from alcohol are substituted for those in nutritious foods.

Alcohol may have beneficial effects when consumed in moderation. For example, compared with nondrinkers, adults who consume one to two drinks a day seem to have a lower risk of heart disease.

9. **Food Safety.** Avoiding foods that are contaminated with harmful bacteria, viruses, parasites, toxins, and chemical and physical contaminants is vital for healthful eating. The signs and symptoms of foodborne illness range from upset stomach, diarrhea,

fever, vomiting, abdominal cramps, and dehydration to more serious ones, such as paralysis and meningitis. You can take simple measures to reduce the risk of food-borne illness, especially at home.

MYPYRAMID

The original Food Guide Pyramid was developed in 1992 as an educational tool to help Americans select healthful diets. The new MyPyramid (Figure 2-5), which was developed by the U.S. Department of Agriculture, has replaced the original Food Guide Pyramid. MyPyramid

FIGURE 2-5:
MyPyramid.
Courtesy of U.S.
Department of
Agriculture.

GRAINS Make half your grains whole	VEGETABLES Vary your veggies	FRUITS Focus on fruits	MILK Get your calcium-rich foods	MEAT & BEANS Go lean with protein
Eat at least 3 oz. of whole-grain cereals, breads, crackers, rice, or pasta every day 1 oz. is about 1 slice of bread, about 1 cup of breakfast cereal, or 1/2 cup of cooked rice, cereal, or pasta	Eat more dark-green veggies like broccoli, spinach, and other dark leafy greens Eat more orange vegetables like carrots and sweet potatoes Eat more dry beans and peas like pinto beans, kidney beans, and lentils	Eat a variety of fruit Choose fresh, frozen, canned, or dried fruit Go easy on fruit juices	Go low-fat or fat-free when you choose milk, yogurt, and other milk products If you don't or can't consume milk, choose lactose-free products or other calcium sources such as fortified foods and beverages	Choose low-fat or lean meats and poultry Bake it, broil it, or grill it Vary your protein routine – choose more fish, beans, peas, nuts, and seeds

**For a 2,000-calorie diet, you need the amounts below from each food group.
To find the amounts that are right for you, go to MyPyramid.gov.**

Eat 6 oz. every day	Eat 2 1/2 cups every day	Eat 2 cups every day	Get 3 cups every day; for kids aged 2 to 8, it's 2	Eat 5 1/2 oz. every day

Find your balance between food and physical activity
- Be sure to stay within your daily calorie needs.
- Be physically active for at least 30 minutes most days of the week.
- About 60 minutes a day of physical activity may be needed to prevent weight gain.
- For sustaining weight loss, at least 60 to 90 minutes a day of physical activity may be required.
- Children and teenagers should be physically active for 60 minutes every day, or most days.

Know the limits on fats, sugars, and salt (sodium)
- Make most of your fat sources from fish, nuts, and vegetable oils.
- Limit solid fats like butter, margarine, shortening, and lard, as well as foods that contain these.
- Check the Nutrition Facts label to keep saturated fats, *trans* fats, and sodium low.
- Choose food and beverages low in added sugars. Added sugars contribute calories with few, if any, nutrients.

MyPyramid.gov
STEPS TO A HEALTHIER YOU

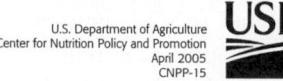 

U.S. Department of Agriculture
Center for Nutrition Policy and Promotion
April 2005
CNPP-15

USDA is an equal opportunity provider and employer.

translates the principles of the 2005 Dietary Guidelines and other nutritional standards to assist consumers in making healthier food and physical activity choices. It was developed to carry the messages of the dietary guidelines and to make Americans aware of the vital health benefits of simple and modest improvements in nutrition, physical activity, and lifestyle behavior.

The MyPyramid symbol represents the recommended proportion of foods from each food group and focuses on the importance of making smart food choices in every food group, every day. Physical activity is a new element in the symbol.

MyPyramid illustrates a number of principles.

1. One size doesn't fit all. MyPyramid symbolizes a personalized approach to healthy eating and physical activity. There are actually twelve MyPyramids that range from 1000 to 3200 kcalories. By using Figure 2-6 or MyPyramid Plan at MyPyramid.gov, you can get an estimate of what kcalorie level is best for you on the basis of your age, gender, and activity level. Once you know your kcalorie level, you can see how much you need to eat from each food group (Figure 2-7).

2. Activity. Activity is represented by the steps and the person climbing them, as a reminder of the importance of daily physical activity.

3. Moderation. Moderation is represented by the narrowing of each food group from bottom to top. The wider base stands for foods with little or no solid fats or added sugars. These foods should be selected more often. The narrower top area stands for foods containing added sugars and solid fats. The more active a person is, the more of these foods they can fit into their diet.

4. Proportionality. Proportionality is shown by the different widths of the food group bands. The widths suggest how much food a person should choose from each group. The widths are just a general guide, not exact proportions.

5. Variety. Variety is symbolized by the six color bands representing the five food groups of the Pyramid plus oils. This illustrates that foods from all groups are needed each day for good health.

6. Gradual improvement. Gradual improvement is encouraged by the slogan "Steps to a Healthier You." It suggests that you can benefit from taking small steps to improve your diet and lifestyle each day.

For example, if you can eat 2000 kcalories a day and maintain a healthy weight, you can use MyPyramid to see how many servings you need each day from each food group. At 2000 kcalories you can eat:

- 6 ounces or the equivalent of grains
- 2.5 cups of vegetables (five servings)
- 2 cups of fruit (or four servings)
- 3 cups of milk or the equivalent
- 5.5 ounces or the equivalent of lean meat and beans
- 27 grams of oils
- 267 discretionary kcalories

The exact amounts of foods in these plans do not need to be achieved every day, but on average, over time.

Now, let's take a look at each food group (grains, vegetables, fruits, milk, and meat and beans) and then see how to use your allowance of oils and discretionary kcalories to complete your understanding of MyPyramid. The nutrient contribution of each food group is summarized in Figure 2-8. Some serving sizes are given in Figure 2-9; additional serving sizes are in Appendix C.

FIGURE 2-6: How Many Kcalories Do You Need?*

	MALES Activity Level				FEMALES Activity Level		
AGE	Sedentary	Mod. Active	Active	**AGE**	Sedentary	Mod. Active	Active
2	1000	1000	1000	2	1000	1000	1000
3	1000	1400	1400	3	1000	1200	1400
4	1200	1400	1600	4	1200	1400	1400
5	1200	1400	1600	5	1200	1400	1600
6	1400	1600	1800	6	1200	1400	1600
7	1400	1600	1800	7	1200	1600	1800
8	1400	1600	2000	8	1400	1600	1800
9	1600	1800	2000	9	1400	1600	1800
10	1600	1800	2200	10	1400	1800	2000
11	1800	2000	2200	11	1600	1800	2000
12	1800	2200	2400	12	1600	2000	2200
13	2000	2200	2600	13	1600	2000	2200
14	2000	2400	2800	14	1800	2000	2400
15	2200	2600	3000	15	1800	2000	2400
16	2400	2800	3200	16	1800	2000	2400
17	2400	2800	3200	17	1800	2000	2400
18	2400	2800	3200	18	1800	2000	2400
19–20	2600	2800	3000	19–20	2000	2200	2400
21–25	2400	2800	3000	21–25	2000	2200	2400
26–30	2400	2600	3000	26–30	1800	2000	2400
31–35	2400	2600	3000	31–35	1800	2000	2200
36–40	2400	2600	2800	36–40	1800	2000	2200
41–45	2200	2600	2800	41–45	1800	2000	2200
46–50	2200	2400	2800	46–50	1800	2000	2200
51–55	2200	2400	2800	51–55	1600	1800	2200
56–60	2200	2400	2600	56–60	1600	1800	2200
61–65	2000	2400	2600	61–65	1600	1800	2000
66–70	2000	2200	2600	66–70	1600	1800	2000
71–75	2000	2200	2600	71–75	1600	1800	2000
76 and up	2000	2200	2400	76 and up	1600	1800	2000

*Kcalorie levels are based on the Estimated Energy Requirements (EER) and activity levels from the Institute of Medicine Dietary Reference Intakes Macronutrients Report, 2002. Sedentary means less than 30 minutes a day of moderate physical activity in addition to daily activities. Moderately active means at least 30 to 60 minutes a day of moderate physical activity in addition to daily activities. Active means 60 or more minutes a day of moderate physical activity in addition to daily activities. Source: United States Department of Agriculture, Center for Nutrition Policy and Promotion (2005).

Grain group

In the MyPyramid graphic, the grain group is represented by the orange band on the left. Any food made from wheat, rice, oats, cornmeal, barley, or another cereal grain is a grain product. Bread, pasta, oatmeal, breakfast cereals, tortillas, and grits are examples of grain products.

FIGURE 2-7: MyPyramid by Kcalorie Level

The suggested amounts of food to consume from the basic food groups, subgroups, and oils to meet recommended nutrient intakes at 12 different calorie levels. Nutrient and energy contributions from each group are calculated according to the nutrient-dense forms of foods in each group (e.g., lean meats and fat-free milk). The table also shows the discretionary kcalorie allowance that can be accommodated within each kcalorie level, in addition to the suggested amounts of nutrient-dense forms of foods in each group.

DAILY AMOUNT OF FOOD FROM EACH GROUP (VEGETABLE SUBGROUP AMOUNTS ARE PER WEEK)

Kcalorie Level	1,000	1,200	1,400	1,600	1,800	2,000	2,200	2,400	2,600	2,800	3,000	3,200
Food Group	Food group amounts shown in cup (c) or ounce-equivalents (oz-eq), with number of servings (srv) in parentheses when it differs from the other units. See note for quantity equivalents for foods in each group.[1] Oils are shown in grams (g).											
Fruits	1 c (2 srv)	1 c (2 srv)	1.5 c (3 srv)	1.5 c (3 srv)	1.5 c (3 srv)	2 c (4 srv)	2 c (4 srv)	2 c (4 srv)	2 c (4 srv)	2.5 c (5 srv)	2.5 c (5 srv)	2.5 c (5 srv)
Vegetables	1 c (2 srv)	1.5 c (3 srv)	1.5 c (3 srv)	2 c (4 srv)	2.5 c (5 srv)	2.5 c (5 srv)	3 c (6 srv)	3 c (6 srv)	3.5 c (7 srv)	3.5 c (7 srv)	4 cc (8 srv)	4 c (8 srv)
Dark green veg	1 c/wk	1.5 c/wk	1.5 c/wk	2 c/wk	3 c/wk	3 c/wk	3 c/wk	3 c/wk	3 c/wk	3 c/wk	3 c/wk	3 c/wk
Orange veg.	.5 c/wk	1 c/wk	1 c/wk	1.5 c/wk	2 c/wk	2 c/wk	2 c/wk	2 c/wk	2.5 c/wk	2.5 c/wk	2.5 c/wk	2.5 c/wk
Legumes	.5 c/wk	1 c/wk	1 c/wk	2.5 c/wk	3 c/wk	3 c/wk	3 c/wk	3 c/wk	3.5 c/wk	3.5 c/wk	3.5 c/wk	3.5 c/wk
Starchy veg.	1.5 c/wk	2.5 c/wk	2.5 c/wk	2.5 c/wk	3 c/wk	3 c/wk	6 c/wk	6 c/wk	7 c/wk	7 c/wk	9 c/wk	9 c/wk
Other veg.	4 c/wk	4.5 c/wk	4.5 c/wk	5.5 c/wk	6.5 c/wk	6.5 c/wk	7 c/wk	7 c/wk	8.5 c/wk	8.5 c/wk	10 c/wk	10 c/wk
Grains	3 oz-eq	4 oz-eq	5 oz-eq	5 oz-eq	6 oz-eq	6 oz-eq	7 oz-eq	8 oz-eq	9 oz-eq	10 oz-eq	10 oz-eq	10 oz-q
Whole grains	1.5	2	2.5	3	3	3	3.5	4	4.5	5	5	5
Other grains	1.5	2	2.5	2	3	3	3.5	4	4.5	5	5	5
Lean meat and beans	2 oz-eq	3 oz-eq	4 oz-eq	5 oz-eq	5 oz-eq	5.5 oz-eq	6 oz-eq	6.5 oz-eq	6.5 oz-eq	7 oz-eq	7 oz-eq	7 oz-eq
Milk	2 c	2 c	2 c	3 c	3 c	3 c	3 c	3 c	3 c	3 c	3 c	3 c

Oils	15 g	17 g	17 g	22 g	24 g	27 g	29 g	31 g	34 g	36 g	44 g	51 g
Discretionary calorie allowance[2]	165	171	171	132	195	267	290	362	410	426	512	648

[1]Quantity equivalents for each food group:

Grains The following each count as 1 ounce-equivalent (1 serving) of grains: 1/2 cup cooked rice, pasta, or cooked cereal; 1 ounce dry pasta or rice; 1 slice bread; 1 small muffin (1 oz); 1 cup ready-to-eat cereal flakes.

Fruits and vegetables The following each count as 1 cup (2 servings) of fruits or vegetables: 1 cup cut-up raw or cooked fruit or vegetable, 1 cup fruit or vegetable juice, 2 cups leafy salad greens.

Meat and beans The following each count as 1 ounce-equivalent: 1 ounce lean meat, poultry, or fish, 1 egg, 1/4 cup cooked dry beans or tofu, 1 Tbsp peanut butter, 1/2 ounce nuts or seeds.

Milk The following each count as 1 cup (1 serving) of milk: 1 cup milk or yogurt, 1 1/2 ounces natural cheese such as cheddar cheese or 2 ounces processed cheese. Discretionary calories must be counted for all choices, except fat-free milk.

[2]Explanation of discretionary calorie allowance: The discretionary calorie allowance is the remaining amount of calories in each food pattern after selecting the specified number of nutrient-dense forms of foods in each food group. The number of discretionary calories assumes that food items in each food group are selected in nutrient-dense forms (that is, forms that are fat-free or low-fat and that contain no added sugars). Solid fat and sugar calories always need to be counted as discretionary calories, as in the following examples:

- The fat in low-fat, reduced-fat, or whole milk or milk products or cheese and the sugar and fat in chocolate milk, ice cream, pudding, and the like
- The fat in higher-fat meats (e.g., ground beef with more than 5 percent fat by weight, poultry with skin, higher-fat luncheon meats, sausages)
- The sugars added to fruits and fruit juices with added sugars to fruits canned in syrup
- The added fats and/or sugars in vegetables prepared with added fat or sugars.
- The added fats and/or sugars in grain products containing higher levels of fats and/or sugars (e.g., sweetened cereals, higher-fat crackers, pies and other pastries, cakes, cookies)

Total discretionary calories should be limited to the amounts shown in the table at each calorie level. The number of discretionary calories is lower in the 1,600-calorie pattern than in the 1,000-, 1,200-, and 1,400-calorie patterns. These lower calorie patterns are designed to meet the nutrient needs of children 2 to 8 years old. The nutrient goals for the 1,600-calorie pattern are set to meet the needs of adult women, which are higher and require that more calories be used in selections from the basic food groups.

FIGURE 2-8: Nutrient Contributions of Each Food Group

Food Group	Major Contribution(s)*
Fruit Group	Vitamin C
Vegetable Group	Vitamin A
Vegetable Subgroups	Vitamin A
Dark Green Vegetables	
Orange Vegetables	
Legumes	
Starchy Vegetables	
Other Vegetables	
Grain Group	Thiamin
	Folate
	Magnesium
	Iron
	Copper
	Carbohydrate
	Fiber
Grain Subgroups	
Whole Grains	Folate
	Magnesium
	Iron
	Copper
	Carbohydrate
	Fiber
Enriched Grains	Folate
	Thiamin
	Carbohydrate
Meat, Poultry, Fish, Dry Beans, Eggs, and Nuts Group	Niacin
	Vitamin B6
	Zinc
	Protein
Milk Group	Riboflavin
	Vitamin B_{12}
	Calcium
	Phosphorus
Oils and soft margarines	Vitamin E
	Linoleic acid
	Alpha-linolenic acid

*Major contribution means that the food group or subgroup provides more of the nutrient than does any other single food group, averaged over all calorie levels.

Source: 2005 Report of the Dietary Guidelines Advisory Committee.

    Grains are divided into two subgroups: whole grains and refined grains. Whole grains contain the entire grain kernel: the fiber-rich bran, the germ, and the endosperm. Examples include whole-wheat flour, bulgur, oatmeal, brown rice, and whole cornmeal.

    Refined grains have been milled, a process that removes the bran and the germ. This is done to give grains a finer texture and improve their shelf life, but it also removes dietary

FIGURE 2-9: Serving Sizes for MyPyramid

Food Group

Grains

1 ounce-equivalent =
1 slice bread
1 small muffin (1 ounce)
1 cup ready-to-eat cereal flakes
½ cup cooked rice, pasta, or cooked cereal
1 ounce dry pasta or rice

Vegetables

1 cup vegetables =
1 cup cut-up raw or cooked vegetables
1 cup vegetable juice
2 cups leafy salad greens (raw)

Fruits

1 cup fruit =
1 cup cut-up raw or cooked fruit
1 cup 100 percent fruit juice

Milk

1 cup milk =
1 cup milk or yogurt
1½ ounces natural cheese such as cheddar cheese
2 ounces processed cheese

Meats and beans

1 ounce-equivalent =
1 ounce lean meat, poultry, or fish
1 egg
¼ cup cooked dry beans or tofu
1 tablespoon peanut butter
½ ounce nuts or seeds

*See Appendix C for an expanded list.

fiber, iron, and many B vitamins. Some examples of refined grain products are white flour, white bread, white rice, and degermed cornmeal. Most refined grains are enriched. This means that certain B vitamins and iron are added back after processing. Fiber is not added back to enriched grains. Some food products, such as breads, can be made from mixtures of whole grains and refined grains.

At the 2000-kcalorie level, you need to eat 6 ounce-equivalents daily, and at least 3 of those ounce-equivalents should be whole grain. At all kcalorie levels, all age groups should eat at least half the grains as whole grains to achieve the fiber recommendation. In general, 1 slice of bread, 1 cup of ready-to-eat cereal, 1 small muffin, or ½ cup of cooked rice, cooked pasta, or cooked cereal can be considered as 1 ounce-equivalent from the grains group. Appendix C gives additional information on serving sizes.

Grains are important sources of many nutrients, including several B vitamins (thiamin, riboflavin, niacin, and folate) and minerals (iron and copper). Thiamin, riboflavin, and niacin play a key role in metabolism—they help the body release energy from protein, fat, and carbohydrates. B vitamins are also essential for a healthy nervous system. Folate helps the body form red blood cells. Eating grains fortified with folate before and during pregnancy helps prevent certain types of birth defects.

Whole grains are also good sources of dietary fiber, magnesium, and selenium. Dietary fiber from whole grains, as part of an overall healthy diet, helps reduce blood cholesterol levels and may lower the risk of heart disease. Fiber is also important for proper bowel function.

It helps reduce constipation and diverticulosis. Fiber-containing foods such as whole grains also help provide a feeling of fullness with fewer kcalories. Magnesium is used in building bones and releasing energy from muscles. Selenium protects cells from oxidation and is also important for a healthy immune system.

Figure 2-10 gives tips to help you eat whole grains.

Vegetable group

Any vegetable or 100 percent vegetable juice counts as a member of the vegetable group. Vegetables are organized into five subgroups, based on their nutrient content. Some commonly eaten vegetables in each subgroup are:

1. Dark green vegetables: dark green leafy lettuce, romaine lettuce, spinach, bok choy, collard greens, turnip greens
2. Orange vegetables: carrots, sweet potatoes, acorn squash, butternut squash
3. Dry beans and peas: kidney beans, pinto beans, navy beans, split peas, lentils, soybeans, black-eyed peas
4. Starchy vegetables: potatoes, corn, peas, lima beans
5. Other vegetables: onions, tomatoes, celery, cucumbers, green or red peppers, iceberg lettuce, mushrooms, vegetable juice, wax beans

A weekly intake of specific amounts from each of the five vegetable subgroups is recommended for adequate nutrient intake. Each subgroup provides somewhat different nutrients. At the 2000-kcalorie level, 2.5 cups of vegetables is recommended each day. The following weekly amounts are suggested from each subgroup.

Dark green vegetables	3 cups/week
Orange vegetables	2 cups/week
Beans	3 cups/week
Starchy vegetables	3 cups/week
Other vegetables	6½ cups/week

The recommendations from the vegetable group are given in cups. In general, 1 cup of raw or cooked vegetables or vegetable juice can be considered as 1 cup from the vegetable group. Two cups of raw leafy greens are considered as 1 cup from the vegetable group. Appendix C gives additional information on serving sizes.

Most vegetables are naturally low in kcalories and fat. None have cholesterol, although sauces or seasonings may add fat, kcalories, and/or cholesterol. Vegetables are important sources of many nutrients, including dietary fiber, vitamin A, vitamin C, vitamin E, folate, magnesium, and potassium. Vitamin A keeps the eyes and skin healthy and helps protect against infections. Vitamin C helps heal cuts and wounds and keep teeth and gums healthy. Vitamin C also helps iron get absorbed. Vitamin E is needed by the immune system and nervous system and also helps prevent vitamin A from being oxidized. Magnesium is used in building bones and releasing energy from muscles. Diets rich in potassium may help maintain healthy blood pressure.

Figure 2-11 gives you tips to help you eat more vegetables.

Fruit group

Any fruit or 100 percent fruit juice counts as part of the fruit group. Fruits may be fresh, canned, frozen, or dried and may be whole, cut up, or puréed. Consumption of whole fruits

FIGURE 2-10: Tips to Help You Eat Whole Grains

IN GENERAL

- To eat more whole grains, substitute a whole-grain product for a refined product, such as eating whole-wheat bread instead of white bread or brown rice instead of white rice. It's important to *substitute* the whole-grain product for the refined one, rather than *adding* the whole-grain product.
- For a change, try brown rice or whole-wheat pasta. Use brown rice stuffing in baked green peppers or tomatoes and whole-wheat macaroni in macaroni and cheese.
- Use whole grains in mixed dishes, such as barley in vegetable soup or stews and bulgur wheat in casseroles or stir-fries.
- Create a whole-grain pilaf with a mixture of barley, wild rice, brown rice, broth, and spices. For a special touch, stir in toasted nuts or chopped dried fruit.
- Experiment by substituting whole-wheat or oat flour for up to half the flour in pancake, waffle, muffin, or other flour-based recipes. They may need a bit more leavening.
- Use whole-grain bread or cracker crumbs in meatloaf.
- Try rolled oats or a crushed, unsweetened whole-grain cereal as breading for baked chicken, fish, veal cutlets, or eggplant parmesan.
- Try an unsweetened, whole-grain, ready-to-eat cereal as croutons in salad or in place of crackers with soup.
- Freeze leftover cooked brown rice, bulgur, or barley. Heat and serve it later as a quick side dish.

AS SNACKS

- Snack on ready-to-eat, whole-grain cereals such as toasted oat cereal.
- Add whole-grain oatmeal when making cookies or other baked treats.
- Try a whole-grain snack chip, such as baked whole-grain tortilla chips.
- Popcorn, a whole grain, can be a healthy snack with little or no added salt or butter.

WHAT TO LOOK FOR ON THE FOOD LABEL

- Choose foods that name one of the following whole-grain ingredients first on the label's ingredient list:
  Brown rice
  Bulgur
  Oatmeal
  Whole oats
  Whole rye
  Whole wheat
  Whole-grain corn
  Wild rice
- Color is not an indication of a whole grain. Bread can be brown because of molasses or other added ingredients. Read the ingredient list to see if it is a whole grain.

Source: U.S. Department of Agriculture.

(fresh, canned, or dried) rather than fruit juice for the majority of the total daily amount is suggested to ensure adequate fiber intake.

At the 2000-kcalorie level, 2 cups of fruits are recommended daily. In general, 1 cup of fruit or 100 percent fruit juice or ½ cup of dried fruit can be considered as 1 cup from the fruit group. You can also count the following as 1 cup: 1 small apple, 1 large banana, 32 seedless

FIGURE 2-11: Tips to Help You Eat Vegetables

IN GENERAL

- Buy fresh vegetables in season. They cost less and are likely to be at their peak flavor.
- Stock up on frozen vegetables for quick and easy cooking in the microwave.
- Buy vegetables that are easy to prepare. Pick up prewashed bags of salad greens and add baby carrots or grape tomatoes for a salad in minutes. Buy packages of baby carrots or celery sticks for quick snacks.
- Use a microwave to quickly "zap" vegetables. White or sweet potatoes can be baked quickly this way.
- Vary your veggie choices to keep meals interesting.
- Try crunchy vegetables, raw or lightly steamed.

FOR THE BEST NUTRITIONAL VALUE

- Select vegetables with more potassium often, such as sweet potatoes, white potatoes, white beans, tomato products (paste, sauce, and juice), beet greens, soybeans, lima beans, winter squash, spinach, lentils, kidney beans, and split peas.
- Sauces or seasonings can add calories, fat, and sodium to vegetables. Use the Nutrition Facts label to compare the calories and Percent Daily Value for fat and sodium in plain and seasoned vegetables.
- Prepare more foods from fresh ingredients to lower sodium intake. Most sodium in the food supply comes from packaged or processed foods.
- Buy canned vegetables labeled "no salt added." If you want to add a little salt, it will likely be less than the amount in the regular canned product.

AT MEALS

- Plan some meals around a vegetable main dish, such as a vegetable stir-fry or soup. Then add other foods to complement it.
- Try a main dish salad for lunch. Go light on the salad dressing.
- Include a green salad with your dinner every night.
- Shred carrots or zucchini into meatloaf, casseroles, quick breads, and muffins.
- Include chopped vegetables in pasta sauce or lasagna.
- Order a veggie pizza with toppings such as mushrooms, green peppers, and onions and ask for extra veggies.
- Use puréed, cooked vegetables such as potatoes to thicken stews, soups, and gravies. They add flavor, nutrients, and texture.
- Grill vegetable kabobs as part of a barbecue meal. Try tomatoes, mushrooms, green peppers, and onions.

MAKE VEGETABLES MORE APPEALING

- Many vegetables taste great with a dip or dressing. Try a low-fat salad dressing with raw broccoli, red and green peppers, celery sticks, or cauliflower.
- Add color to salads by adding baby carrots, shredded red cabbage, or spinach leaves. Include in-season vegetables for variety through the year.
- Include cooked dry beans or peas in flavorful mixed dishes, such as chili or minestrone soup.
- Garnish plates or serving dishes with vegetable slices.
- Keep a bowl of cut-up vegetables in a see-through container in the refrigerator. Carrot and celery sticks are traditional, but consider broccoli florets, cucumber slices, and red or green pepper strips.

Source: U.S. Department of Agriculture.

grapes, 1 medium pear, 2 large plums, or 1 large orange. Appendix C gives information on additional serving sizes.

Most fruits are naturally low in kcalories, fat, and sodium. None have cholesterol. Fruits are important sources of many nutrients, including vitamin C, potassium, folate, and dietary fiber.

Eating foods such as fruits or vegetables that are low in kcalories instead of a higher-kcalorie food may be useful in helping to lower kcalorie intake. Eating a diet rich in fruits and vegetables as part of an overall healthy diet may reduce the risk for heart disease, stroke, type 2 diabetes, and certain types of cancer, such as stomach and colon cancer.

Figure 2-12 gives you tips to help you eat more fruits.

Milk group

All fluid milk products and many foods made from milk are included in the milk group. Foods made from milk that retain their calcium content are part of this group, while foods made from milk that have little to no calcium, such as cream cheese, cream, and butter, are not. Most choices in the milk group should be fat-free or low-fat. If you choose milk or yogurt that is not fat-free or cheese that is not low-fat, the fat in the product counts as part of your discretionary kcalorie allowance (to be discussed in a moment). If sweetened milk products are chosen, such as chocolate milk and drinkable yogurt, the added sugars also count as part of the discretionary kcalorie allowance.

The recommendations for the milk group are given in cups. At the 2000-kcalorie level, 3 cups of milk or equivalents are suggested. In general, 1 cup of milk or yogurt, 1½ ounces of natural cheese, or 2 ounces of processed cheese can be considered as 1 cup from the milk group. Appendix C gives additional information on serving sizes.

Foods in the milk group provide nutrients vital to health, such as calcium, vitamin D, potassium, and protein.

- Calcium is used for building bones and teeth and maintaining bone mass. Milk products are the primary source of calcium in American diets. Diets that provide 3 cups or the equivalent of milk products daily can improve bone mass.
- Vitamin D functions in the body to maintain proper levels of calcium and phosphorus in the blood, thus helping to build and maintain bones. Vitamin D is added to milk, some milk products, and some ready-to-eat breakfast cereals.
- Diets rich in potassium may help maintain healthy blood pressure.

Diets rich in milk and milk products help build and maintain bone mass throughout the life cycle. The intake of milk products is especially crucial for bone growth during childhood and adolescence, when bone mass is being built.

Figure 2-13 gives you tips for making wise choices in the milk group.

Meat and beans group

All foods made from meat, poultry, fish, dry beans or peas, eggs, nuts, and seeds are considered part of this group. Most meat and poultry choices should be lean or low-fat. Fish, nuts, and seeds contain healthy oils, and so you can often choose these foods instead of meat or poultry. If higher-fat choices are made, such as regular ground beef (75 to 80 percent lean) or chicken with skin, the fat in the product counts as part of the discretionary kcalorie allowance. If solid fat is added in cooking, such as frying chicken in shortening or frying eggs in butter or stick margarine, this also counts as part of the discretionary kcalorie allowance.

Dry beans and peas are the mature forms of legumes such as kidney beans, pinto beans, lima beans, black-eyed peas, and lentils. These foods are excellent sources of plant protein and also provide other nutrients, such as iron and zinc. They are similar to meats, poultry, and fish in their contribution of these nutrients and are low in fat. Many people consider dry beans and peas as vegetarian alternatives for meat. However, they are also excellent sources

FIGURE 2-12: Tips to Help You Eat Fruits

IN GENERAL

- Keep a bowl of whole fruit on the table, on the counter, or in the refrigerator.
- Refrigerate cut-up fruit to store for later.
- Buy fresh fruits in season, when they may be less expensive and at their peak flavor.
- Buy fruits that are dried, frozen, and canned (in water or juice) as well as fresh so that you always have a supply on hand.
- Consider convenience when shopping. Buy precut packages of fruit (such as melon or pineapple chunks) for a healthy snack in seconds. Choose packaged fruits that do not have added sugars.

FOR THE BEST NUTRITIONAL VALUE

- Make most of your choices with whole or cut-up fruit rather than juice for the benefits that dietary fiber provides.
- Select fruits with more potassium often, such as bananas, prunes and prune juice, dried peaches and apricots, cantaloupe, honeydew melon, and orange juice.
- When choosing canned fruits, select fruit canned in 100 percent fruit juice or water rather than syrup.
- Vary your fruit choices. Fruits differ in nutrient content.

AT MEALS

- At breakfast, top your cereal with bananas or peaches, add blueberries to pancakes, and drink 100 percent orange or grapefruit juice. Or try a fruit mixed with low-fat or fat-free yogurt.
- At lunch, pack a tangerine, banana, or grapes to eat or choose fruits from a salad bar. Individual containers of fruits such as peaches or applesauce are easy and convenient.
- At dinner, add crushed pineapple to coleslaw or include mandarin oranges or grapes in a tossed salad.
- Make a Waldorf salad with apples, celery, walnuts, and dressing.
- Try meat dishes that incorporate fruit, such as chicken with apricots or mango chutney.
- Add fruit such as pineapple or peaches to kabobs as part of a barbecue meal.
- For dessert, have baked apples, pears, or a fruit salad.

AS SNACKS

- Cut-up fruit makes a great snack. Either cut them yourself or buy precut packages of fruit pieces such as pineapples or melons. Or try whole fresh berries or grapes.
- Dried fruits also make a great snack. They are easy to carry and store well. Because they are dried, $\frac{1}{4}$ cup is equivalent to $\frac{1}{2}$ cup of other fruits.
- Keep a package of dried fruit in your desk or bag. Some fruits that are available dried include apricots, apples, pineapple, bananas, cherries, figs, dates, cranberries, blueberries, prunes (dried plums), and raisins (dried grapes).
- As a snack, spread peanut butter on apple slices or top frozen yogurt with berries or slices of kiwi fruit.
- Frozen juice bars (100 percent juice) make healthy alternatives to high-fat snacks.

MAKE FRUIT MORE APPEALING

- Many fruits taste great with a dip or dressing. Try low-fat yogurt or pudding as a dip for fruits such as strawberries or melons.
- Make a fruit smoothie by blending fat-free or low-fat milk or yogurt with fresh or frozen fruit. Try bananas, peaches, strawberries, or other berries.
- Try applesauce as a fat-free substitute for some of the oil when baking cakes.
- Try different textures of fruits. For example, apples are crunchy, bananas are smooth and creamy, and oranges are juicy.
- For fresh fruit salads, mix apples, bananas, or pears with acidic fruits such as oranges, pineapple, or lemon juice to keep them from turning brown.

Source: U.S. Department of Agriculture.

FIGURE 2-13: Tips for Making Wise Choices from the Milk Group

FROM THE MILK GROUP IN GENERAL

- Include milk as a beverage at meals. Choose fat-free or low-fat milk.
- If you usually drink whole milk, switch gradually to fat-free milk to lower saturated fat and calories. Try reduced-fat (2 percent), then low-fat (1 percent), and finally fat-free (skim) milk.
- If you drink cappuccinos or lattes, ask for them with fat-free (skim) milk.
- Add fat-free or low-fat milk instead of water to oatmeal and hot cereals.
- Use fat-free or low-fat milk when making condensed cream soups (such as cream of tomato).
- Have fat-free or low-fat yogurt as a snack.
- Make a dip for fruits or vegetables from yogurt.
- Make fruit-yogurt smoothies in the blender.
- For dessert, make chocolate or butterscotch pudding with fat-free or low-fat milk.
- Top cut-up fruit with flavored yogurt for a quick dessert.
- Top casseroles, soups, stews, or vegetables with shredded low-fat cheese.
- Top a baked potato with fat-free or low-fat yogurt.

FOR THOSE WHO CHOOSE NOT TO CONSUME MILK PRODUCTS

- If you avoid milk because of lactose intolerance, the most reliable way to get the health benefits of milk is to choose lactose-free alternatives within the milk group, such as cheese, yogurt, and lactose-free milk, or to consume the enzyme lactase before consuming milk products.
- Calcium choices for those who do not consume milk products include:
- Calcium-fortified juices, cereals, breads, soy beverages, or rice beverages.
- Canned fish (sardines, salmon with bones) soybeans and other soy products (soy-based beverages, soy yogurt, tempeh), some other dried beans, and some leafy greens (collard and turnip greens, kale, bok choy). The amount of calcium that can be absorbed from these foods varies.

Source: U.S. Department of Agriculture.

of dietary fiber and nutrients such as folate that are low in diets of many Americans. These nutrients are found in plant foods such as vegetables. Because of their high nutrient content, the consumption of dry beans and peas is recommended for everyone, including people who also eat meat, poultry, and fish regularly.

Dry beans and peas can be counted either as vegetables (dry beans and peas subgroup) or in the meat, poultry, fish, dry beans, eggs, and nuts (meat and beans) group. Generally, individuals who regularly eat meat, poultry, and fish would count dry beans and peas in the vegetable group. Individuals who seldom eat meat, poultry, or fish (vegetarians) would count some of the dry beans and peas they eat in the meat, poultry, fish, dry beans, eggs, and nuts group.

The recommendations for the meat and beans group are given in ounce-equivalents. At the 2000-kcalorie level, you are allowed 5.5 ounce-equivalents. In general, 1 ounce-equivalent is equal to 1 ounce of meat, poultry, or fish; $\frac{1}{4}$ cup cooked dry beans; 1 egg; 1 tablespoon of peanut butter; or $\frac{1}{2}$ ounce of nuts or seeds. Appendix C gives additional information on serving sizes.

The meat and bean group supplies many nutrients, including protein, B vitamins (niacin, thiamin, riboflavin, and B_6), vitamin E, iron, zinc, and magnesium. Protein has many functions in the body, where it is part of most body structures; builds and maintains the body; is a part of many enzymes, hormones, and antibodies; transports substances; maintains fluid and acid-base balance; provides energy for the body; and helps in blood clotting.

Some food choices in this group are high in saturated fat. Diets that are high in saturated fats raise "bad" cholesterol levels in the blood. The "bad" cholesterol is called LDL (low-density lipoprotein) cholesterol. High levels of LDL cholesterol, in turn, increase the risk for heart disease. Food choices that are high in saturated fat include fatty cuts of beef, pork, and lamb; regular ground beef (75 to 85 percent lean); regular sausages, hot dogs, and bacon; some luncheon meats, such as regular bologna and salami; and duck. To help keep blood cholesterol levels healthy, limit the amount of these foods you eat.

Some foods in this group are high in cholesterol, which can raise LDL cholesterol levels in the blood. Cholesterol is found only in foods from animal sources. Egg yolks and organ meats such as liver are high in cholesterol. (Egg whites are cholesterol-free.) To help keep blood cholesterol levels healthy, limit the amount of these foods you eat.

Figure 2-14 gives you tips for making wise choices in the meat and bean group.

Oils

The thin yellow column in MyPyramid represents oils. Oils are fats that are liquid at room temperature, such as the vegetable oils used in cooking. Oils come from many different plants and from fish. Some common vegetable oils include canola oil, corn oil, olive oil, safflower oil, and soybean oil. Some oils are used mainly as flavorings, such as walnut oil and sesame oil. A number of foods are naturally high in oils, such as nuts, olives, some fish, and avocados. Foods that are mainly oil include certain salad dressings, mayonnaise, and soft (tub or squeeze) margarine.

Most of the fats you eat should be polyunsaturated fatty acids or monounsaturated fatty acids. Oils are the major sources of both of these types of fatty acids in the diet. Polyunsaturated fatty acids are found in the greatest amounts in safflower, corn, soybean, sesame, and sunflower oils. Good examples of oils high in monounsaturated fatty acids include olive oil, canola oil, and peanut oil. Polyunsaturated and monounsaturated fatty acids are also found in nuts and seeds. They do not raise LDL ("bad") choelsterol levels in the blood.

MyPyramid shows suggested amounts of "Oils" to consume (Figure 2-15) because oils are also major contributors of vitamin E and essential fatty acids. Oils include vegetable oils and soft vegetable oil table spreads that have no trans fats. Oils may also come from mayonnaise, salad dressings, and olives (see Figure 2-16).

While consuming some oil is needed for health, oils still contain kcalories. In fact, oils and solid fats both contain about 120 kcalories per tablespoon. Therefore, the amount of oil consumed needs to be limited to balance total kcalorie intake.

Discretionary kcalories

MyPyramid counts most solid fats (such as butter, stick margarine, and shortening) and all added sugars (sugars and syrups that are added to foods or beverages during processing or preparation) as *discretionary kcalories* even if they are part of your selections from the grains group, lean meat group, or another food group. When you choose foods from the food groups, it is assumed that your selections of vegetables, fruits, grains, milk, and meat and beans are fat-free or low-fat and contain no added sugars. For example, if you have milk with fat, such as reduced-fat or whole milk, the extra fat (and therefore kcalories) in these choices will come from your discretionary kcalorie allowance. Here are some more examples of solid fat and sugar in foods that count toward your discretionary kcalorie allowance.

- The fat in cheese
- The fat and sugar in chocolate milk, ice cream, and pudding
- The fat in well-marbled cuts of meat or ground beef with more than 5 percent fat by weight

DISCRETIONARY KCALORIES

The balance of kcalories you have after meeting the recommended nutrient intakes by eating foods in low-fat or no added sugar forms. Your discretionary kcalorie allowance may be used to select forms of foods that are not the most nutrient-dense (such as whole milk rather than fat-free milk) or may be additions to foods, such as sugar or butter.

GO LEAN WITH PROTEIN

- Start with a lean choice:
  - The leanest beef cuts include round steaks and rounds (round eye, top round, bottom round, round tip), top loin, top sirloin, and chuck shoulder and arm roasts.
  - The leanest pork choices include pork loin, tenderloin, and center loin.
  - Choose extra-lean ground beef. The label should say at least "90 percent lean." You may be able to find ground beef that is 93 percent or 95 percent lean.
  - Buy skinless chicken parts or take off the skin before cooking.
  - Boneless skinless chicken breasts and turkey cutlets are the leanest poultry choices.
  - Choose lean turkey, roast beef, ham, or low-fat luncheon meats for sandwiches instead of luncheon meats with more fat, such as regular bologna or salami.
- Keep it lean:
  - Trim away all the visible fat from meats and poultry before cooking.
  - Broil, grill, roast, poach, or boil meat, poultry, or fish instead of frying.
  - Drain off any fat that appears during cooking.
  - Skip or limit the breading on meat, poultry, or fish. Breading adds fat and calories. It will also cause the food to soak up more fat during frying.
  - Prepare dry beans and peas without added fats.
  - Choose and prepare foods without high-fat sauces or gravies.

VARY YOUR PROTEIN CHOICES

- Choose fish more often for lunch or dinner. Look for fish rich in omega-3 fatty acids, such as salmon, trout, and herring. Some ideas are:
  - Salmon steak or filet
  - Salmon loaf
  - Grilled or baked trout
- Choose dry beans or peas as a main dish or part of a meal often. Some choices are:
  - Chili with kidney or pinto beans
  - Stir-fried tofu
  - Split-pea, lentil, minestrone, or white bean soup
  - Baked beans
  - Black bean enchiladas
  - Garbanzo or kidney beans on a chef's salad
  - Rice and beans
  - Salsa with beans
  - Veggie burgers or garden burgers
  - Hummus (chickpeas) spread on pita bread
- Choose nuts as a snack, on salads, or in main dishes. Use nuts to replace meat or poultry, not in addition to these items:
  - Use pine nuts in pesto sauce for pasta.
  - Add slivered almonds to steamed vegetables.
  - Add toasted peanuts or cashews to a vegetable stir-fry instead of meat.
  - Sprinkle a few nuts on top of low-fat ice cream or frozen yogurt.
  - Add walnuts or pecans to a green salad instead of cheese or meat.

WHAT TO LOOK FOR ON THE FOOD LABEL

- Check the Nutrition Facts label for the saturated fat, trans fat, cholesterol, and sodium content of packaged foods.
- Lower-fat versions of many processed meats are available. Look on the Nutrition Facts label to choose products with less fat and saturated fat.

Source: U.S. Department of Agriculture.

FIGURE 2-15: Daily Allowance for Oils*

Group	Age	Amount of Oil
Children	2–3 years old	3 teaspoons
	4–8 years old	4 teaspoons
Girls	9–13 years old	5 teaspoons
	14–18 years old	5 teaspoons
Boys	9–13 years old	5 teaspoons
	14–18 years old	6 teaspoons
Women	19–30 years old	6 teaspoons
	31–50 years old	5 teaspoons
	51+ years old	5 teaspoons
Men	19–30 years old	7 teaspoons
	31–50 years old	6 teaspoons
	51+ years old	6 teaspoons

*These amounts are appropriate for individuals who get less than 30 minutes per day of moderate physical activity beyond their normal daily activities. Those who are more physically active may be able to consume more while staying within kcalorie needs.

Source: U.S. Department of Agriculture.

- The fat in poultry skin
- The fat in higher-fat luncheon meats, bacon, and sausage
- The sugar added to fruits and fruit juices
- The added fat and/or sugars in vegetables prepared with added fat or sugars
- The added fats and/or sugars in grain products containing higher levels of fats and/or sugars, such as sweetened cereals, higher-fat crackers, pies and other pastries, cakes, and cookies

Discretionary kcalories therefore represent the balance of kcalories you have after meeting the recommended nutrient intakes by eating foods in low-fat or no added sugar forms.

For example, assume your kcalorie budget is 2000 kcalories per day. Of these kcalories, you need to spend about 1735 kcalories for essential nutrients if you choose foods without added fat and sugar. Then you have 265 discretionary kcalories left. You may use these on

FIGURE 2-16: How to Count the Oils You Eat

Oils/Foods	Amount of Food	Amount of Oil	Kcalories From Fat	Total Kcalories
Vegetable oils (such as canola, corn, olive, soybean, safflower, peanut)	1 tablespoon	3 teaspoons	120	120
Margarine (trans fat-free)	1 tablespoon	2½ teaspoons	100	100
Mayonnaise	1 tablespoon	2½ teaspoons	100	100
Mayonnaise-type salad dressing	1 tablespoon	1 teaspoon	45	55
Italian dressing	2 tablespoons	2 teaspoons	75	85
Thousand Island dressing	2 tablespoons	2½ teaspoons	100	120
Olives, ripe, canned	4 large	½ teaspoon	15	20

Source: U.S. Department of Agriculture.

"luxury" versions of the foods in each group, such as higher-fat meat, sweetened cereal, and cookies. Or, you can spend them on sweets, soda, wine, or beer or by adding fats (such as butter) or sweeteners (such as syrup or table sugar) to foods. Figure 2-17 shows you how to count the discretionary kcalories you eat.

Most solid fats are high in saturated fats and/or trans fats, both of which raise LDL cholesterol levels and increase the risk of heart disease. A few plant oils, including coconut oil and palm kernel oil, are high in saturated fats and for nutritional purposes should be considered solid fats.

Total discretionary kcalories should be limited to the amounts shown at the bottom of the MyPyramid chart (Figure 2-7). The best way to increase the number of discretionary kcalories is to increase physical activity. The greater the amount of physical activity, the more discretionary kcalories will be available.

FIGURE 2-17: How to Count Discretionary Kcalories

Food	Amount	Estimated Total Calories	Estimated Discretionary Calories
MILK GROUP			
Fat-free milk	1 cup	85	0
1 percent milk	1 cup	100	20
2 percent milk (reduced-fat)	1 cup	125	40
Whole milk	1 cup	145	65
Low-fat chocolate milk	1 cup	160	75
Cheddar cheese	1½ ounces	170	90
Nonfat mozzarella	1½ ounces	65	0
Whole-milk mozzarella	1½ ounces	130	45
Fruit-flavored low-fat yogurt	1 cup (8 fl oz.)	240 to 250	100 to 115
Frozen yogurt	1 cup	220	140
Ice cream, vanilla	1 cup	290	205
Cheese sauce	¼ cup	120	75
MEAT AND BEANS GROUP			
Extra-lean ground beef, 95 percent lean	3 oz., cooked	165	0
Regular ground beef, 80 percent lean	3 oz., cooked	230	65
Turkey roll, light meat	3 slices (1 oz. each)	125	0
Roasted chicken breast (skinless)	3 oz.	140	0
Roasted chicken thigh with skin	3 oz.	210	70
Fried chicken with skin and batter	3 wings	475	335
Beef sausage, precooked	3 oz., cooked	345	180
Pork sausage	3 oz., cooked	290	125
Beef bologna	3 slices (1 oz. each)	265	100
GRAINS			
Whole-wheat bread	1 slice (1 oz.)	70	0
White bread	1 slice (1 oz.)	70	0
English muffin	1 muffin	135	0
Blueberry muffin	1 small (2 oz.)	185	45

(continued)

FIGURE 2-17: How to Count Discretionary Kcalories (*Continued*)

Food	Amount	Estimated Total Calories	Estimated Discretionary Calories
Croissant	1 med. (2 oz.)	230	95
Biscuit, plain	1 2½ inches diameter	130	60
Cornbread	1 piece (2½ by 2½ by 1¼ inches)	190	50
Graham crackers	2 large pieces	120	50
Whole-wheat crackers	5 crackers	90	20
Round snack crackers	7 crackers	105	35
Chocolate chip cookies	2 large	135	70
Cake-type doughnuts, plain	2 mini doughnuts, 1½ inches diameter	120	50
Glazed doughnut, yeast type	1 medium, 3¾ inches in diameter	240	165
Cinnamon sweet roll	1 3 oz. roll	310	100
VEGETABLES			
French fries	1 medium order	460	325
Onion rings	1 order (8 to 9 rings)	275	160
EXTRAS*			
Regular soda	1 can (12 fluid ounces)	155	155
Regular soda	1 20 ounce bottle	260	260
Diet soda	1 can (12 fluid ounces)	5	5
Fruit punch	1 cup	115	115
Table wine	5 fluid ounces	115	115
Beer (regular)	12 fluid ounces	145	145
Beer (light)	12 fluid ounces	110	110
Distilled spirits (80 proof)	1½ fluid ounces	95	95
Butter	1 teaspoon	35	35
Stick margarine	1 teaspoon	35	35
Cream cheese	1 tablespoon	50	50
Heavy (whipping) cream	1 tablespoon	50	50
Dessert topping, frozen, semisolid	1 tablespoon	15	15
Gravy, canned	¼ cup	30	30

*All the calories in candy, sodas, alcoholic beverages, and solid fats are discretionary calories. The calories per serving are listed on the Nutrition Facts label on food packages. Be sure to compare the stated serving size to the amount actually eaten. If you eat twice the stated serving size, you will get twice the calories.

Source: U.S. Department of Agriculture.

Physical activity

Physical activity simply means movement of the body that uses energy. Walking, gardening, briskly pushing a baby stroller, climbing the stairs, playing soccer, and dancing the night away are all good examples of being active. For health benefits, physical activity should be moderate or vigorous and add up to at least 30 minutes a day. Moderate physical activities include:

- Walking briskly (about 3½ miles per hour)
- Hiking

- Gardening/yard work
- Dancing
- Golf (walking and carrying clubs)
- Bicycling (less than 10 miles per hour)
- Weight training (general light workout)

Vigorous physical activities include:

- Running or jogging (5 miles per hour)
- Bicycling (more than 10 miles per hour)
- Swimming (freestyle laps)
- Aerobics
- Walking very fast (4½ miles per hour)
- Heavy yard work, such as chopping wood
- Weight lifting (vigorous effort)
- Basketball (competitive)

Some physical activities are not intense enough to help you meet the recommendations. Although you are moving, these activities do not increase your heart rate, and so you should not count them toward the 30 or more minutes a day for which you should strive. These include walking at a casual pace, such as while grocery shopping, and doing light household chores.

Some types of physical activity are especially beneficial:

- Aerobic activities: Speed heart rate and breathing and improves heart and lung fitness. Examples are brisk walking, jogging, and swimming.
- Resistance, strength building, and weight-bearing activities: Help build and maintain bones and muscles by working them against gravity. Examples are carrying a child, lifting weights, and walking.
- Balance and stretching activities: Enhance physical stability and flexibility, which reduces the risk of injuries. Examples are gentle stretching, dancing, yoga, martial arts, and t'ai chi.

At a minimum, do moderate-intensity activity for 30 minutes most days or preferably every day. This is in addition to your usual daily activities. Increasing the intensity or the amount of time of this activity can have additional health benefits and may be needed to control body weight.

About 60 minutes a day of moderate physical activity may be needed to prevent weight gain. For those who have lost weight, at least 60 to 90 minutes a day may be needed to maintain the weight loss. At the same time, kcalorie needs should not be exceeded. Children and teenagers should be physically active for at least 60 minutes every day or most days.

While 30 minutes a day of moderate-intensity physical activities provides health benefits, being active for longer or doing more vigorous activities can provide even greater health benefits. It also uses up more kcalories per hour. No matter what activity you choose, it can be done all at once or divided into two or three parts during the day. Even 10-minutes bouts of activity count toward your total.

Most adults do not need to see their health-care provider before starting to exercise at a moderate level. However, men over age 40 and women over age 50 planning to start vigorous physical activity should consult a health-care provider.

Planning menus using MyPyramid

Planning menus gives you the opportunity to include a variety of foods from each food group, especially foods from subgroups that provide nutrients that often are low in American diets. It also provides a chance to balance fat and sodium to maintain healthful levels over time.

Use the following questions to ensure that your menu follows both the Dietary Guidelines for Americans and MyPyramid.

1. Does a day's menu on the average provide at least the number of servings required from each of the major food groups for a 2000-kcalorie diet?
2. Are most of the menu items nutrient dense (without solid fat or sugars added)?
3. Does the menu have whole-grain breads and grains available at each meal?
4. Are most meat and poultry items lean?
5. Are fish, beans, and other meat alternates available?
6. Does the menu include servings from each of the vegetable subgroups: dark green, orange, beans, starchy, and other?
7. Do most vegetables and fruits have their skins and seeds (baked potatoes with skin, berries, and apples or pears with peels)?
8. Are there more choices for fresh, canned, or dried fruit than for fruit juices?
9. Are low-fat or fat-free milk and other dairy choices available?
10. Are the fruit juices 100 percent juice?
11. Are foods (especially desserts) high in fat, sugar, and/or sodium balanced with choices lower in these nutrients?
12. Is a soft margarine available that does not contain trans fat?
13. Are unsweetened beverages available?

Figures 2-18 to 2-20 show the Mediterranean, Asian, and Latin American Diet Pyramids. These pyramids were developed by the Oldways Preservation and Exchange Trust to illustrate traditional food patterns that are healthy. These pyramids illustrate proportions rather than specific amounts of foods to eat. Like MyPyramid, they emphasize grains, fruits, and vegetables.

FIGURE 2-18:
The Traditional Healthy Mediterranean Diet Pyramid.
Courtesy of Oldways Preservation & Exchange Trust, Cambridge, MA.

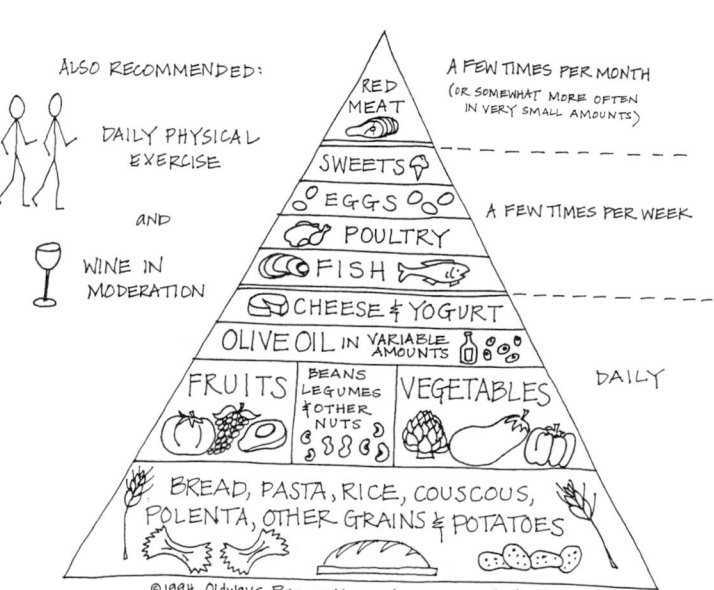

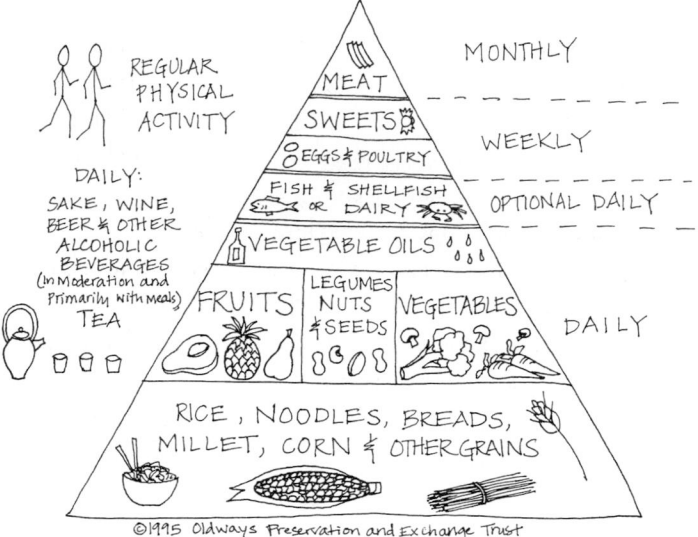

FIGURE 2-19:

The Traditional Healthy Asian Diet Pyramid.
Courtesy of Oldways Preservation & Exchange Trust,
Cambridge, MA.

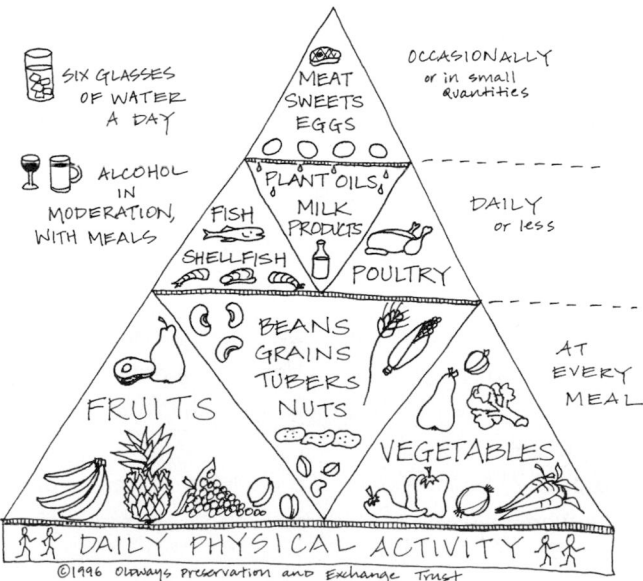

FIGURE 2-20:

The Traditional Healthy Latin American Diet Pyramid.
Courtesy of Oldways Preservation & Exchange Trust,
Cambridge, MA.

1. Dietary recommendations have been published for the healthy American public for almost 100 years.
2. The Dietary Guidelines include key recommendations as detailed on pages 34 to 35 for each of these areas: adequate nutrients within kcalorie needs, weight management, physical activity, foods groups to encourage, fat, carbohydrates, sodium and potassium, alcoholic beverages, and food safety.
3. MyPyramid has replaced the original Food Guide Pyramid. It translates the principles of the 2005 Dietary Guidelines and other nutritional standards to assist consumers in making healthier food and physical activity choices. Figures 2-5 and 2-7 describe MyPyramid.
4. Discretionary kcalories represent the balance of kcalories you have after meeting the recommended nutrient intakes by eating foods in low-fat or no added sugar forms. Your discretionary kcalorie allowance may be used to select forms of foods that are not the most nutrient-dense (such as whole milk rather than fat-free milk) or may be additions to foods, such as sugar and butter. Discretionary kcalorie allowances are generally small and represent solid fats and added sugars. Figure 2-17 shows how to count discretionary kcalories.
5. Figure 2-8 shows the major nutrient contributions of each food group.
6. Figure 2-9 lists serving sizes for MyPyramid.
7. At a minimum, do moderate-intensity activity for 30 minutes most days or preferably every day. This is in addition to your usual daily activities. Increasing the intensity or the amount of time of activity can have additional health benefits and may be needed to control body weight. About 60 minutes a day of moderate physical activity may be needed to prevent weight gain. For those who have lost weight, at least 60 to 90 minutes a day may be needed to maintain the weight loss.
8. Guidelines are given for using MyPyramid and the Dietary Guidelines to plan menus.

FOOD LABELS

Since 1938, the federal government has required basic information on food labels (Figure 2-21). The U.S. Food and Drug Administration (FDA) regulates labels on all packaged foods except for meat, poultry, and egg products—foods regulated by the U.S. Department of Agriculture (USDA). The amount of information on food labels varies, but all food labels must contain at least:

- The name of the food.
- A list of ingredients. The ingredient that is present in the largest amount, by weight, must be listed first. Other ingredients follow in descending order of weight.
- The net contents or net weight—the quantity of the food without the packaging (in English and metric units).
- The name and place of business of the manufacturer, packer, or distributor.
- Nutrition information is also required for most foods, our next topic.

For most foods, all ingredients must be listed on the label and identified by their common names so that consumers can identify the presence of any of eight major food allergens. Food manufacturers must use plain English words such as milk or wheat rather than less familiar words such as casein or semolina to identify the most common food allergens on ingredients lists. While more than 160 foods can cause allergic reactions in people with food

FIGURE 2-21:
Food label.

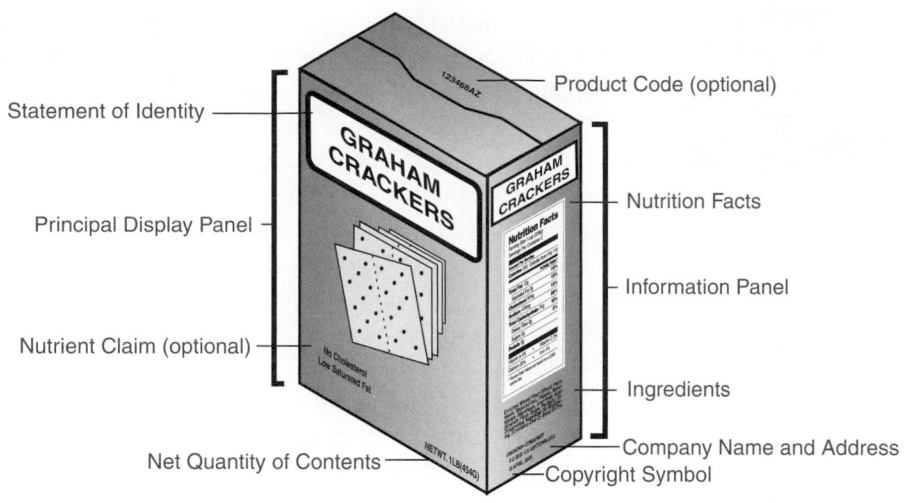

allergies, the law identifies the eight most common allergenic foods. These foods account for 90 percent of food allergic reactions.

1. Milk
2. Eggs
3. Fish
4. Crustacean shellfish (such as crab, lobster, shrimp)
5. Tree nuts (such as almonds, walnuts, pecans)
6. Peanuts
7. Wheat
8. Soybeans

Food labels must also indicate the presence of any of these allergens used in spices, flavorings, additives, and colorings.

Symptoms of food allergies usually appear within a few minutes to two hours after a person has eaten the food to which he or she is allergic. Symptoms might include hives, skin rash, vomiting and/or diarrhea, coughing, wheezing, swelling of the throat and vocal cords, and difficulty breathing. Following ingestion of a food allergen(s), a person with food allergies can experience a severe, life-threatening allergic reaction called anaphylaxis. This can lead to:

- Constricted airways in the lungs
- Severe lowering of blood pressure and shock (anaphylactic shock)
- Suffocation by swelling of the throat

NUTRITION FACTS

Figure 2-22 shows a sample Nutrition Facts panel from a package of macaroni and cheese.

Serving size

Serving size and number of servings in the package are the first stop when you read the Nutrition Facts. Just how big is a serving? Serving sizes are provided in familiar units, such as cups or pieces, followed by the metric amount (the number of grams). A serving of applesauce would read "½ cup (114 g)." The household measure is easier to understand, but the metric measure gives a more precise idea of the amount. For example, "114 g" means

Nutrition Facts

Serving Size 1 cup (228 g)
Servings Per Container 2

Amount Per Serving

Calories 250	Calories from Fat 110

	% Daily Value*
Total Fat 12g	18%
Saturated Fat 3g	15%
Trans Fat 3g	
Cholesterol 30mg	10%
Sodium 470mg	20%
Total Carbohydrates 31g	10%
Dietary Fiber 0g	0%
Sugars 5g	
Protein 5g	

Vitamin A 4%	-	Vitamin C 2%
Calcium 20%	-	Iron 4%

*Percent Daily Values are based on a 2,000 calorie diet. Your daily values may be higher or lower depending on your calorie needs.

		Calories	2,000	2,500
Total Fat	less than		65g	80g
Sat. Fat	less than		20g	25g
Cholesterol	less than		300 mg	300 mg
Sodium	less than		2,400 mg	2,400 mg
Total Carbohydrate			300 g	375 g
Dietary Fiber			25g	30g

FIGURE 2-22:

Nutrition Facts label.
Courtesy of U.S. Department of Agriculture.

DAILY VALUE

A set of nutrient-intake values developed by the Food and Drug Administration that are used as a reference for expressing nutrient content on nutrition labels.

---

114 grams, a measure of weight. There are approximately 28 grams in 1 ounce. The label helps you get familiar with metrics, too.

Serving sizes are designed to reflect the amounts people actually eat. Compare the serving size, including how many servings there are in the food package, to how much you actually eat. The size of the serving on the food package influences all the nutrient amounts listed on the top part of the label. In the sample label, one serving of macaroni and cheese equals 1 cup. If you ate the whole package, you would eat 2 cups, and that doubles the kcalories and other nutrient numbers.

If you check the serving size of different brands of macaroni and cheese, you'll see that the sizes are similar. That means you don't need to be a math whiz to compare two foods. Likewise, it's easy to see the kcalorie and nutrient differences between similar servings of canned fruit packed in syrup and the same fruit in natural juices.

Kcalories

The next stop on the Nutrition Facts panel is the kcalories per serving category, which lists the total kcalories in one serving as well as the kcalories from fat. In the example, there are 250 kcalories in one serving of this macaroni and cheese. How many kcalories from fat are there in one serving? Answer: 110 kcalories, which means almost half the food's kcalories come from fat. If you ate the whole package, you would consume 500 kcalories, and 220 would come from fat.

Nutrients

Nutrients are listed next. Information about some nutrients is required: total fat, saturated fat, trans fat, cholesterol, sodium, total carbohydrates, dietary fiber, sugars, protein, vitamin A, vitamin C, calcium, and iron. Information about additional nutrients may be given voluntarily, but it is required in two cases: if a claim is made about the nutrients on the label or if the nutrients are added to the food. For example, fortified breakfast cereals must give Nutrition Facts for any added vitamins and minerals.

The nutrients listed first (total fat, saturated fat, trans fat, cholesterol, sodium) are the ones Americans generally eat in adequate amounts or eat too much of. Eating too much fat, saturated fat, trans fat, cholesterol, or sodium may increase your risk for certain chronic diseases, such as heart disease, some cancers, and high blood pressure. Americans often don't get enough of some of the other nutrients listed: fiber, vitamins A and C, calcium, and iron. Eating enough of these nutrients can improve your health and help reduce the risk of some diseases and conditions. For example, a diet high in dietary fiber promotes healthy bowel function. You can use the food label to help limit those nutrients you want to cut back on and also increase those nutrients you want to consume in greater amounts.

Nutrient amounts are listed in two ways: in metric amounts (in grams) and as a percentage of the **Daily Value**. Developed by the Food and Drug Administration, Daily Values are recommended levels of intake specially developed for food labels (Figure 2-23). DRIs can't be used on nutrition labels because they are set for specific age and gender categories, and you can't have several nutrition labels on one food for males and females in various age groups.

The table at the bottom of the food label shows the Daily Values for certain nutrients at both the 2000- and 2500-calorie levels. For example, if you eat a 2000-calorie diet, you should get less than 65 grams of fat from all the foods you eat in a day. If you consume 2500 kcalories per

FIGURE 2-23: Selected Daily Values

Nutrient	Daily Value
Carbohydrate	300 grams
Fiber	25 grams
Fat	65 grams
Saturated fat	20 grams
Cholesterol	300 milligrams
Sodium	2400 milligrams
Potassium	3500 milligrams
Vitamin A	1500 micrograms RAE
Vitamin C	60 milligrams
Folate	400 micrograms
Calcium	1000 milligrams
Iron	18 milligrams

Source: Food and Drug Administration.

day, the amounts of cholesterol and sodium you eat in a day are not different from those eaten by others eating 2000 kcalories per day. This table is found only on larger packages and does not change from product to product.

The percentage of the Daily Value (%DV) is based on a 2000-kcalorie diet. Therefore, the Daily Value may be a little high, a little low, or right on target for you. The percentage of the Daily Value shows you how much of the Daily Value is in one serving of food. For example, in Figure 2-22, the %DV for total fat is 18 percent and for dietary fiber it is 0 percent. When one serving of macaroni and cheese contains 18 percent of the DV for Total Fat, that means you have 82 percent of your fat allowance left for all the other foods you eat that day.

When you are looking at the %DV on a food label, use this guide.

- Foods that contain 5 percent or less of the Daily Value for a nutrient are generally considered low in that nutrient.
- Foods that contain 10 to 19 percent of the Daily Value for a nutrient are generally considered good sources of that nutrient.
- Foods that contain 20 percent or more of the Daily Value for a nutrient are generally considered high in that nutrient. For example, one cup of macaroni and cheese contains 18 percent of the Daily Value for fat, which is just below 20 percent. Therefore, the macaroni and cheese is pretty high in fat, particularly if you eat 1½ to 2 cups.

You can use the %DV to help you make dietary trade-offs with other foods throughout the day. You don't have to give up a favorite food to eat a healthy diet. When a food you like is high in fat, balance it by eating foods that are low in fat at other times of the day.

The values listed for total carbohydrate include all carbohydrates, including dietary fiber and sugars listed below it. The sugar values include naturally present sugars, such as lactose in milk and fructose in fruits, as well as those added to the food, such as table sugar and corn syrup. The label can claim no sugar added but still include naturally occurring sugars. An example is fruit juice.

The values listed for total fat refer to all the fat in the food. Only total fat, saturated fat, and trans fat information is required on the label, because high intakes are linked to high blood cholesterol, which is linked to an increased risk of coronary heart disease. Trans fat is a specific type of fat that forms when liquid oils are made into solid fats such as margarine and shortening.

Trans fat behaves like saturated fat in raising low-density lipoprotein ("bad") cholesterol. Listing the amount of polyunsaturated and monounsaturated fats in the food is voluntary.

Note that trans fat, sugars, and protein do not list a %DV on the Nutrition Facts label.

- Experts could not provide a reference value for trans fat or any other information that the FDA believes is enough to establish a Daily Value or %DV. Health experts do recommend that you keep your intake of trans fat (and saturated fat and cholesterol) as low as possible.
- A %DV is required to be listed if a claim is made for protein, such as "high in protein," or if the food is meant for use by infants and children under 4 years old. Current scientific evidence indicates that protein intake is not a public health concern for adults and children over 4 years of age.
- No daily reference value has been established for sugars because no recommendations have been made for the total amount to eat in a day.

To limit nutrients that have no %DV, such as trans fats and sugars, compare the labels of similar products and choose the food with the lowest amount.

The Daily Value for calcium is 1000 milligrams (mg). Experts advise adolescents, especially girls, to consume 1300 mg and postmenopausal women to consume 1200 mg of calcium daily. The daily target for teenagers should therefore be 130%DV, and the daily target for postmenopausal women should be 120%DV.

NUTRIENT CLAIMS

NUTRIENT CONTENT CLAIMS
Claims on food labels about the nutrient composition of a food, regulated by the Food and Drug Administration.

Nutrient content claims, such as "good source of calcium" and "fat-free," can appear on food packages only if they follow legal definitions (Figure 2-24). For example, a food that is a good source of calcium must provide 10 to 19 percent of the Daily Value for calcium in one serving. Phrases such as "sugar-free" describe the amount of a nutrient in a food but don't indicate exactly how much. These nutrient content claims differ from Nutrition Facts, which do list specific nutrient amounts.

If a food label contains a descriptor for a certain nutrient but the food contains other nutrients at levels known to be less healthy, the label has to bring that to consumers' attention. For example, if a food making a low-sodium claim is also high in fat, the label must state "see back panel for information about fat and other nutrients."

HEALTH CLAIMS

HEALTH CLAIMS
Claims on food labels that state that certain foods or food substances—as part of an overall healthy diet— may reduce the risk of certain diseases.

Health claims state that certain foods or components of foods (such as calcium) may reduce the risk of a disease or health-related condition. Examples include calcium and osteoporosis and dietary saturated fat and the risk of coronary heart disease. Although food manufacturers may use health claims approved by the FDA to market their products, the intended purpose of health claims is to benefit consumers by providing information on healthful eating patterns that may help reduce the risk of heart disease, cancer, osteoporosis, high blood pressure, dental cavities, and certain birth defects.

The FDA reviews and approves all health claims before they can be used on food labels. Until recently, the only permitted health claims were those supported by evidence that met "significant scientific agreement." In other words, solid proof was necessary to establish a clear link between a food or its component and a disease. For example, decreasing sodium intake is clearly advisable to reduce the risk of high blood pressure. Therefore, a claim such as "Diets low in sodium may reduce the risk of high blood pressure" can be used on the labels of foods low in sodium, such as unsalted tuna fish. Figure 2-25 lists food label health

FIGURE 2-24: Nutrient Content Claims: A Dictionary

Nutrient (Content Claim)	Definition (per Serving)
CALORIES	
Calorie-free	Less than 5 kcalories
Low-calorie	40 kcalories or less
Reduced or fewer calories	At least 25 percent fewer kcalories than the comparison food
Light	One-third fewer kcalories than the comparison food
SUGAR	
Sugar-free	Less than 0.5 gram sugars
Reduced sugar or less sugar	At least 25 percent less sugars than the comparison food
No added sugar	No sugars added during processing or packing, including ingredients that contain sugars, such as juice and dry fruit
FIBER	
High-fiber	5 grams or more
Good source of fiber	2.5 to 4.9 grams
More or added fiber	At least 2.5 grams more than the comparison food
FAT AND CHOLESTEROL	
Fat-free (nonfat or no-fat)	Less than 0.5 gram fat
Percent fat-free	The amount of fat in 100 grams; may be used only if the product meets the definition of low-fat or fat-free
Low-fat	3 grams or less of fat
Light	50 percent or less of the fat than in the comparison food
Reduced or less fat	At least 25 percent less fat than the comparison food
Saturated fat-free	Less than 0.5 gram saturated fat and less than 0.5 gram trans fat
Low saturated fat	1 gram or less saturated fat and less than 0.5 gram trans fat
Reduced or less saturated fat	At least 25 percent less saturated fat than the comparison food
Trans fat-free	Less than 0.5 gram of trans fat and less than 0.5 gram of saturated fat
Cholesterol-free	Less than 2 milligrams cholesterol and 2 grams or less of saturated fat and trans fat combined
Low-cholesterol	20 milligrams cholesterol or less and 2 grams or less saturated fat
Reduced or less cholesterol	25 percent or less cholesterol than the comparison food and 2 grams or less saturated fat
Lean	Less than 10 grams of fat, 4.5 grams of saturated fat and trans fat combined, and 95 milligrams of cholesterol per 100 grams of meat, poultry, and seafood
Extra-lean	Less than 5 grams of fat, 2 grams of saturated fat and trans fat combined, and 95 milligrams of cholesterol per 100 grams of meat, poultry, and seafood
SODIUM	
Sodium-free, salt-free	Less than 5 milligrams sodium
Very low sodium	35 milligrams or less sodium
Low-sodium	140 milligrams or less sodium
Reduced or less sodium	At least 25 percent less sodium than the comparison food
Light in sodium	50 percent less sodium than the comparison food
Light (for sodium-reduced products)	If food is "low-calorie" and "low-fat" and sodium is reduced by at least 50 percent
GENERAL CLAIMS	
High, rich in, excellent	Provides 20 percent or more of the Daily Value for a given nutrient source of
Good source	Provides 10 to 19 percent of the Daily Value for a given nutrient

(continued)

FIGURE 2-24: Nutrient Content Claims: A Dictionary (*Continued*)

Nutrient (Content Claim)	Definition (per Serving)
More	Provides at least 10 percent or more of the Daily Value for a given nutrient than the comparison food
Fresh	Raw, unprocessed, or minimally processed, with no added preservatives
Healthy	Low in fat, saturated fat, cholesterol, and sodium and contains at least 10 percent of the Daily Value for one of the following: vitamin A, vitamin C, calcium, iron, protein, fiber

Source: Food and Drug Administration.

FIGURE 2-25: Food Label Health Claims: The "A" List

Calcium and reduced risk of osteoporosis	"Regular exercise and a healthy diet with enough calcium helps teen and young adult white and Asian women maintain good bone health and may reduce their high risk of osteoporosis later in life."
Dietary fat and reduced risk of cancer	"Development of cancer depends on many factors. A diet low in total fat may reduce the risk of some cancers."
Dietary saturated fat and cholesterol and reduced risk of coronary heart disease	"While many factors affect heart disease, diets low in saturated fat and cholesterol may reduce the risk of this disease."
Polyols (sugar alcohols) and reduced risk of dental caries	"Frequent eating of foods high in sugars and starches as between-meal snacks can promote tooth decay. The sugar alcohol used to sweeten this food may reduce the risk of dental caries."
Fiber-containing grain products, fruits, and vegetables and reduced risk of cancer	"Low-fat diets rich in fiber-containing grain and products, fruits, and vegetables may reduce the risk of cancer, a disease associated with many factors."
Folic acid and reduced risk of neural tube defects	"Healthful diets with adequate folate may reduce a woman's risk of having a child with a brain or spinal cord birth defect."
Fruits and vegetables and reduced risk of cancer	"Low-fat diets rich in fruits and vegetables (foods that are low in fat and may contain dietary fiber, vitamin A, and vitamin C) may reduce the risk of some types of cancer, a disease associated with many factors. Broccoli is high in vitamins A and C, and it is a good source of dietary fiber."
Fruits, vegetables, and grain products that contain fiber, particularly soluble fiber, and reduced risk of coronary heart disease	"Diets low in saturated fat and cholesterol and rich in fruits, vegetables, and grain products that contain some types of dietary fiber, particularly soluble fiber, may reduce the risk of heart disease, a disease associated with many factors."
Sodium and reduced risk of hypertension	"Diets low in sodium may reduce the risk of high blood pressure, a disease associated with many factors."
Soluble fiber from whole oats and psyllium seed husk and reduced risk of coronary heart disease	"Three grams of soluble fiber from oatmeal daily in a diet low in saturated fat and cholesterol may reduce the risk of heart disease. This cereal has 2 grams per serving."
Soy protein and reduced risk of coronary heart disease	"25 grams of soy protein a day, as part of a diet low in saturated fat and cholesterol, may reduce the risk of heart disease. A serving of this food supplies _____ grams of soy protein."
Plant stanols/sterol esters and reduced risk of coronary heart disease	"Diets low in saturated fat and cholesterol that include two servings of foods that provide a daily total of at least 1.3 g of vegetable oil sterol esters in two meals may reduce the risk of heart disease. A serving of this food supplies _____ grams of vegetable oil sterol esters."
Potassium and reduced risk of hypertension and stroke	"Diets containing foods that are good sources of potassium and low in sodium may reduce the risk of high blood pressure and stroke."
Whole grains and reduced risk of heart disease	"Diets rich in whole-grain foods and other plant foods and low in saturated fat and cholesterol may help reduce the risk of heart disease."

Source: Food and Drug Administration.

claims on which there is significant scientific agreement. Using the FDA's new ranking system for health claims, these types of claims are given the highest grade, which is grade A.

The new ranking system, from A to D, categorizes the quality and strength of the scientific evidence for every proposed health claim (Figure 2-26). Under this system, the grade of B is assigned to claims for which there is good supporting scientific evidence but the evidence is not entirely conclusive. Grades of C apply to claims for which the evidence is limited and not conclusive. The fourth level, D, is given to claims with little scientific evidence to support them. Health claims graded B, C, or D are referred to as *qualified health claims* because they require a disclaimer or other qualifying language to ensure that they do not mislead consumers. For example, supplements containing selenium (a mineral thought to possibly reduce the risk of some cancers) must include this disclaimer: "Selenium may reduce the risk of certain cancers. Some scientific evidence suggests that consumption of selenium may reduce the risk of certain forms of cancer. However, the FDA has determined this evidence is limited and not conclusive."

Whereas health claims describe a relationship between a specific food (or a component) and a disease, you may see claims on labels that refer to a broad class of foods and a disease, such as "diets rich in fruits and vegetables may reduce the risk of some types of cancer." This is a statement using current dietary guidance. Truthful, non-misleading dietary guidance statements may be used on food labels and do not have to be reviewed by the FDA. However, once the food is marketed with the statement, the FDA can consider whether the statement meets the requirement to be truthful and not misleading.

Health Claims Report Card

A	**High** Significant scientific agreement	1
B	**Moderate** Evidence is not conclusive	2
C	**Low** Evidence is limited and not conclusive	3
D	**Extremely Low** Little scientific evidence supporting this claim	4

FIGURE 2-26:
Ranking system for health claims
Source: U.S. Department of Agriculture.

> **QUALIFIED HEALTH CLAIMS**
> Health claims graded B, C, or D that require a disclaimer or other qualifying language to ensure that they do not mislead consumers.

MINI-SUMMARY

1. All food labels must contain the name of the product; the net contents or net weight; the name and place of business of the manufacturer, packer, or distributor; a list of ingredients using common names in order of predominance by weight; and nutrition information (Figure 2-21).
2. Daily Values are nutrient standards used on food labels to allow nutrient comparisons among foods.
3. Any nutrient claim on food labels must comply with Food and Drug Administration regulations and definitions as outlined in this chapter.
4. Health claims state that certain foods or components of foods (such as calcium) may reduce the risk of a disease or health-related condition. Health claims may be ranked A, B, C, or D (Figure 2-26). They must be approved by the FDA.
5. Qualified health claims (always ranked B, C, or D) require a disclaimer or other qualifying language to ensure that they do not mislead consumers.

PORTION SIZE COMPARISONS

Portion size is an important concept for anyone involved in preparing, serving, and consuming foods. Serving sizes vary from kitchen to kitchen, but American serving sizes have been increasing steadily. In comparison with MyPyramid portion sizes as well as those

1 Serving Looks Like . . .

GRAIN PRODUCTS

1 cup of cereal flakes = fist

1 pancake = compact disc

½ cup of cooked rice, pasta, or potato = ½ baseball

1 slice of bread = cassette tape

1 piece of cornbread = bar of soap

1 Serving Looks Like . . .

VEGETABLES AND FRUIT

1 cup of salad greens = baseball

1 baked potato = fist

1 med. fruit = baseball

½ cup of fresh fruit = ½ baseball

¼ cup of raisins = large egg

1 Serving Looks Like . . .

DAIRY AND CHEESE

1½ oz. cheese = 4 stacked dice or 2 cheese slices

½ cup of ice cream = ½ baseball

FATS

1 tsp. margarine or spreads = 1 die

1 Serving Looks Like . . .

MEAT AND ALTERNATIVES

3 oz. meat, fish, and poultry = deck of cards

3 oz. grilled/baked fish = checkbook

2 Tbsp. peanut butter = ping-pong ball

FIGURE 2-27:
Portion sizes.

served in many European countries, our portion sizes are huge. It wasn't that long ago that a "large" soft drink was typically 16 fluid ounces. Now that's often the "small" size.

What you may consider one serving of the bread group may actually be three or four servings. For example, MyPyramid considers 1 ounce of bread, about one slice, to be one serving. A typical New York–style bagel is about four ounces, or about four servings.

You may have noticed that the portion sizes in MyPyramid do not always match the serving sizes found on food labels. This is the case because the purpose of MyPyramid is not the same as that of nutrition labeling. MyPyramid was designed to be very simple to use. Therefore, the USDA specified only a few serving sizes for each food group so that those sizes could be remembered easily. Food labels have a different purpose: to allow the consumer to compare the nutrients in equal amounts of foods. To compare the nutrient amounts in equal amounts of pasta, the portion size on the label is 2 ounces of uncooked pasta (about 56 grams), which will cook up to about 1 cup of spaghetti or as much as 2 cups of a large shaped pasta such as ziti. Using MyPyramid, the portion size for cooked pasta is only half a cup.

In many cases, the portion sizes are similar on labels and in the food guide, especially when expressed as household measures. For foods falling into only one major group, such as fruit juices, the household measures provided on the label (1 cup or 8 fluid ounces) can help you relate the label serving size to MyPyramid serving size. For mixed dishes, MyPyramid serving sizes may be used to visually estimate the food item's contribution to each food group as the food is eaten—for example, the amounts of bread, vegetables, and cheese contributed by a portion of pizza.

Use Figure 2-27 to help you visualize appropriate portion sizes. It may also help to compare serving sizes to everyday objects. For example, ¼ cup of raisins is about the size of a large egg. Three ounces of meat or poultry is about the size of a deck of cards. See other serving size comparisons. (Keep in mind that these size comparisons are approximations.)

CHECK-OUT QUIZ

1. Draw a line from the name of the item to the appropriate recommendation found in the Dietary Guidelines for Americans.

Item	Dietary Guidelines for Americans
Energy	20 to 35 percent of kcalories
Physical activity	300 mg or less
Whole grains	2300 mg or less
Total fat	Eat 3 ounce-equivalents per day
Saturated fat	Don't exceed kcaloric needs
Trans fat	Less than 10 percent of kcalories
Sodium	Keep consumption as low as possible
Cholesterol	At least 30 minutes most days of moderate intensity

2. What are the serving sizes for MyPyramid? Fill in the blanks with the correct number.

1 ounce-equivalent of grains = _____ slice(s) bread

_____ cup(s) ready-to-eat cereal

_____ cup(s) cooked rice, pasta, or cooked cereal

1 cup vegetables = _____ cup(s) leafy salad greens (raw)

_____ cup(s) vegetable juice

_____ cup(s) cooked vegetables

1 cup milk = _____ cup(s) yogurt

1 ounce-equivalent meat = _____ egg(s)

_____ cup(s) dry beans

_____ tablespoon(s) peanut butter

3. Which food group(s) provides the most protein?

   a. bread, cereal, rice, and pasta
   b. vegetable
   c. milk, yogurt, and cheese
   d. meat, poultry, fish, dry beans, eggs, and nuts

4. Which food group(s) provides the most vitamin A?

   a. bread, cereal, rice, and pasta
   b. vegetable
   c. milk, yogurt, and cheese
   d. fruit

5. Which food group(s) provides the most vitamin C?

   a. bread, cereal, rice, and pasta
   b. vegetable
   c. milk, yogurt, and cheese
   d. fruit

6. Which food group provides the most calcium?

   a. bread, cereal, rice, and pasta
   b. vegetable
   c. milk, yogurt, and cheese
   d. meat, poultry, fish, dry beans, eggs, and nuts

7. Daily Values are based on an 1800-calorie diet.

   a. True
   b. False

8. Claims such as "good source of calcium" and "fat-free" on food labels are examples of health claims.

   a. True
   b. False

9. Qualified health claims must include a disclaimer on the food label.

   a. True
   b. False

10. All ingredients must be listed on the label and identified by their common names so that consumers can identify the presence of any of eight major food allergens.

    a. True
    b. False

ACTIVITIES AND APPLICATIONS

1. Checking Out Nutrient and Health Claims
   Look at food labels from two of the following sections of the supermarket. Write down nutrient claims (such as "low-fat") given on at least two different foods from each section. Don't forget: Fresh fruits and vegetables, meat, poultry, and seafood don't have labels—look for nutrition information nearby. Also look at the label to see which nutrition facts support this claim.

   Produce
   Frozen foods
   Fresh meats, poultry, and fish
   Dairy
   Cereals
   Cookies

   During your search, also find one food item with a health claim and write it down. Use Figure 2-25 to determine if it is an A claim or an unqualified health claim.

2. **Label Reading at Breakfast**

   Look closely at the Nutrition Facts for each food you normally eat for breakfast, such as cereal, milk, and juice. Add up the %DVs for fat, saturated fat, cholesterol, sodium, total carbohydrates, vitamin A, vitamin C, calcium, and iron. How nutritious is your breakfast?

3. **Label Comparison**

   Below is the nutrition label information from regular mayonnaise and low-fat mayonnaise dressing. Examine the labels and then answer these questions

   A. Which label is the regular mayonnaise? How do you know that?
   B. Does either mayonnaise contain significant amounts of vitamins and minerals?
   C. What is the percent of total kcalories coming from fat in both products? (For example, in product A, divide 10 by 25 and then multiply by 100 to get 40 percent.)
   D. Which product contains sugar?
   E. Which product contains more saturated fat and cholesterol? Why do you think that is so?

   PRODUCT A NUTRITION FACTS
   Serving size 1 tablespoon
   AMOUNT PER SERVING
   Calories 25
   Calories from fat 10
   Total Fat 1 g
   Saturated 0 g
   Trans Fat 0 g

   PRODUCT B NUTRITION FACTS
   Serving size 1 tablespoon
   AMOUNT PER SERVING
   Calories 100
   Calories from fat 99
   Total Fat 11 g
   Saturated 2 g
   Trans Fat 0 g

   PRODUCT A NUTRITION FACTS
   Serving size 1 tablespoon
   AMOUNT PER SERVING
   Polyunsaturated 0.5 g
   Monounsaturated 0 g
   Cholesterol 0 mg
   Sodium 140 mg
   Total Carbohydrate 4 g
   Sugars 3 g
   Protein 0 g
   Not a significant source of dietary
   fiber, vitamin A, vitamin C, calcium, or iron

   PRODUCT B NUTRITION FACTS
   Serving size 1 tablespoon
   AMOUNT PER SERVING
   Polyunsaturated 6 g
   Monounsaturated 3 g
   Cholesterol 5 mg
   Sodium 180 mg
   Total Carbohydrate 0 g
   Sugars 0 g
   Protein 0 g
   Not a significant source of dietary fiber,
   vitamin A, vitamin C, calcium, or iron

4. **Menu/Diet Evaluation**

   Obtain a cycle menu (a menu rotated at specific time intervals, such as two or four weeks) from a college dining hall, school foodservice, or other foodservice. Evaluate the menu using MyPyramid and the questions on pages 58.

5. **Fat in Snacks**

   Obtain a single serving of one of your favorite snacks. Write down the percentage of kcalories from fat the snack contains and write it on an index card. In class, you are to line up from the lowest percentage of kcalories from fat to the highest percentage. Once the class is in the correct order, each of you will identify your food and the percentage of kcalories from fat.

6. **Mystery Food**

   Bring to class the Nutrition Facts panel from a food product. Exchange your Nutrition Facts panel with a partner and examine the panel from your partner carefully. One at a time, you are each to guess what food category it falls into. Once you have the correct food category (your partner can tell you if you are right), guess what food it is (again with feedback from your partner).

7. **Using iProfile, click on "Estimating Portion Sizes" at the top of the page.**

   Do the exercise "Visualizing Serving Sizes" and check out all the foods listed on the left.

✕ NUTRITION WEB EXPLORER

MyPyramid   www.mypyramid.gov

At the MyPyramid website, perform each of the following:

1. Under "MyPyramid Plan," enter your age, gender, and activity level to see how many total kcalories, servings from each food group, and discretionary kcalories you are allowed.
2. Next, click on "Meal Tracking Worksheet" to make a copy of your MyPyramid worksheet. Write down everything you eat in one day, and compare it to the recommendations.
3. To get an idea of portion sizes, click on "Inside the Pyramid," and then on a food group such as Meat & Beans. Then click on "What's in the Meat & Beans Group?" Next click on, "View Meat & Beans Food Gallery." Do this for each food group.
4. To get a nutrient analysis of your one day intake, click on "MyPyramid Tracker," and register as a new user. Then click on "Assess Your Food Intake." Enter in the foods and serving sizes that you ate yesterday. Once you have done that, click on "Save and Analyze" and obtain each of these reports: Meeting 2005 Dietary Guidelines, Nutrient Intake, and MyPyramid Recommendation.
5. You can also plan menus use the MyPyramid Menu Planner. The Menu Planner shows whether your food choices are balanced for the day, or on average over a week. You can also use it to help plan upcoming meals to meet MyPyramid goals.

American Dietetic Association   www.eatright.org

Visit the ADA website and get its "Tip of the Day." Also use "Find a Nutrition Professional" to find a list of dietitians in your area.

Dietitians of Canada   www.dietitians.ca

Visit the Dietitians of Canada website, and click on "Find a Nutrition Professional" to find a list of dietitians in your area. On the home page, click on "Eat Well, Live Well," then click on "Let's Make a Meal." This interactive program let you choose various menu items for breakfast, lunch, dinner, and snacks and compares your choices with the Canadian food guide.

Quackwatch   www.quackwatch.com

Visit this website and click on "25 Ways to Spot It" under "Quackery." What are 10 ways to spot quackery?

International Dietary Guidelines

http://fnic.nal.usda.gov/nal_display/index.php?
info_center=4&tax_level=2&tax_subject=
270&topic_id=1339

On this page, click on "Dietary Guidelines from Around the World" (at the bottom of the page) and read about the diets of people in another country.

Food Label Quiz

www.cfsan.fda.gov/~dms/flquiz1.html

Take the "Test Your Food Label Knowledge Quiz" at this government website.

University of Florida Libraries
Tips on Web Search

http://web.uflib.ufl.edu/admin/wwwtips.pdf

This article presents eight questions to keep in mind when searching for reliable information on the Web.

FOOD FACTS: NUTRIENT ANALYSIS OF RECIPES

Computer software is the standard for analyzing the amount of nutrients in a recipe. Little wonder, when a task that used to take a half hour or more is done in a matter of minutes. Before computers, it was necessary to look up the nutritional content of each recipe ingredient and record it on a piece of paper. If the amount of the ingredient was not the same as the amount listed in the reference book, you would have to do some multiplication or division on all the nutrient values to come up with the right numbers. Then, after looking up all the ingredients, you would add up all your columns to get totals. In the final step, you would divide the totals by the yield of the recipe to get the amount of nutrients per serving. Sounds complicated! It sure is, and very time-consuming, too.

Nutrient-analysis software contains nutrient information from many different resources. When the name of an ingredient is typed in, the computer lists similar ingredients so that you can choose exactly which one is appropriate. Then you click on or type in the amount of the ingredient you want to be used in the analysis, such as 1 cup. After inputting all the ingredients, you can ask the computer to divide the results by the yield, such as 12 portions. Then the computer will tell you exactly how much of each nutrient (and what percentage of the RDA or AI) is contained in one portion. Most nutrient-analysis programs can also give you a percentage breakdown of kcalories from protein, fat, carbohydrate, and alcohol. Of course, these figures can be printed out and/or stored in the computer's memory. A sample recipe analysis is shown in Figure 2-28.

When considering a nutrient-analysis program, consider the following:

- What do you need the program to do? What different functions can the program perform, and how many of those functions do you need?
- What kind of computer system do you have to run this software? Be sure you have enough hard-disk space and random access memory (RAM) and an appropriate-speed microprocessor.
- How large is the nutrient database? What is the source of the data? The database may contain from several thousand to more than 30,000 foods. Most databases use the U.S. Department of Agriculture's Handbook #8 and manufacturers' information.
- Can foods be added to the database? It's also a good idea to check how many foods can be added.
- How many nutrients are provided for each food?
- How is output presented (graphs, tables, pie charts, etc.), and how easily can it be printed?
- How easy is it to use this program?
- What's the price?
- What service and support are available once you purchase the program? Is online help available? Is there an annual update fee? How much does it cost?
- The United States Department of Agriculture offers its National Nutrient Database for Standard Reference (Release 21) at its website (www.nal.usda.gov/fnic/foodcomp/). Release 21 is also available at this website to download onto a handheld personal digital assistant (PDA). Many companies offer demonstration software at no cost. This is a real benefit, because you can try out the program before buying it. Following are a number of companies that offer software for nutrient analysis.

Axxya Systems: www.nutritionistpro.com
DINE Systems: www.dinesystems.com
ESHA Research: www.esha.com
Nutribase: www.nutribase.com

FIGURE 2-28: Sample Computerized Nutrient Analysis Output

BREAKFAST: 3 PANCAKES, 1 OZ. SAUSAGE, 2 T SYRUP, 1 CUP ORANGE JUICE

Recipe Nutrient Analysis Source: Custom
Recipe Food ID: 29

Yield: 1.00 (1.00 SERVING)
Goal: DAILY VALUES/RDI–ADULT/CHILD

Nutrient	Value	Goal	% Goal	Nutrient	Value	Goal	% Goal
Weight (gm)	430.350			Zinc (mg)	0.778	15.000	5%
Kilocalories (kcal)	566.180	2000.000	28%	Copper (mg)	0.250	2.000	13%
Protein (gm)	12.503	50.000	25%	Manganese (mg)	0.298	2.000	15%
Carbohydrate (gm)	88.334	300.000	29%	Selenium (mg)	0.017	0.070	24%
Fat, total (gm)	19.654	65.000	30%	Fluoride (mg)			
Alcohol (gm)	0.000			Chromium (mg)		0.120	
Cholesterol (mg)	84.617	300.000	28%	Molybdenum (mg)		75.000	
Saturated Fat (gm)	5.372	20.000	27%	Dietary Fiber, total (gm)	1.624	25.000	6%
Monounsaturated Fat (gm)	4.888			Soluble Fiber (gm)	0.000		
Polyunsaturated Fat (gm)	6.702			Insoluble Fiber (gm)	0.000		
MFA 18:1, Oleic (gm)	2.786			Crude Fiber (gm)	0.250		
PFA 18:2, Linoleic (gm)	4.533			Sugar, total (gm)	52.544		
PFA 18:3, Linolenic (gm)	0.608			Glucose (gm)	14.826		
PFA 20:5, EPA (gm)	0.000			Galactose (gm)	0.000		
PFA 22:6, DHA (gm)	0.006			Fructose (gm)	9.440		
Sodium (mg)	686.569	2400.000	29%	Sucrose (gm)	15.002		
Potassium (mg)	647.280	3500.000	18%	Lactose (gm)	0.000		
Vitamin A (RE)	116.946	1000.000	12%	Maltose (gm)	4.522		
Vitamin A (IU)	738.724	5000.000	15%	Tryptophan (mg)	96.160		
Beta-carotene (mg)	0.000			Threonine (mg)	290.020		
Vitamin C (mg)	124.689	60.000	208%	Isoleucine (mg)	358.420		
Calcium (mg)	283.126	1000.000	28%	Leucine (mg)	617.060		
Iron (mg)	2.792	18.000	16%	Lysine (mg)	388.260		
Vitamin D (mg)	0.000	10.000	0%	Methionine (mg)	175.020		

1 SERVING

% of Kcals	
Protein	9%
Carbohydrate	62%
Fat, total	31%
Alcohol	0%

(continued)

FIGURE 2-28: Sample Computerized Nutrient Analysis Output (*Continued*)

BREAKFAST: 3 PANCAKES, 1 OZ. SAUSAGE, 2 T SYRUP, 1 CUP ORANGE JUICE

Recipe Nutrient Analysis
Recipe Food ID: 29   Source: Custom

Yield: 1.00 (1.00 SERVING)
Goal: DAILY VALUES/RDI-ADULT/CHILD

Nutrient	Value	Goal	% Goal	Nutrient	Value	Goal	% Goal
Vitamin D (IU)	0.000	400.000	0%	Cystine (mg)	144.640		
Vitamin E (ATE)	0.223	20.000	1%	Phenylalanine (mg)	385.980		
Vitamin E (IU)		30.000		Tyrosine (mg)	283.520		
Alpha-tocopherol (mg)	0.099			Valine (mg)	409.180		
Thiamin (mg)	0.456	1.500	30%	Arginine (mg)	435.760		
Riboflavin (mg)	0.398	1.700	23%	Histidine (mg)	180.720		
Niacin (mg)	2.786	20.000	14%	Alanine (mg)	316.500		
Pyridoxine/Vit B$_6$ (mg)	0.151	2.000	8%	Aspartic Acid (mg)	664.800		
Folate (mg)	118.464	400.000	30%	Glutamic Acid (mg)	1871.640		
Cobalamin/Vit B$_{12}$ (mg)	0.251	6.000	4%	Glycine (mg)	248.040		
Biotin (mg)	0.800	300.000	0%	Proline (mg)	765.760		
Pantothenic Acid (mg)	0.942	10.000	9%	Serine (mg)	446.060		
Vitamin K (mg)	0.248	80.000	0%	Moisture (gm)	288.930		
Phosphorus (mg)	227.020	1000.000	23%	Ash (gm)	4.036		
Iodine (mg)		150.000		Caffeine (mg)	0.000		
Magnesium (mg)	46.320	400.000	12%				

Source: © 2000 First Data Bank, Inc., a wholly owned subsidiary of the Hearst Corporation. Reprinted with permission.

HOT TOPIC: QUACK! QUACK!

Food quackery has been defined by the U.S. Surgeon General as "the promotion for profit of special foods, products, processes, or appliances with false or misleading health or therapeutic claims." Have you ever seen advertisements for supplements that are guaranteed to help you lose weight or herbal remedies to prevent serious disease? If a product's claim seems just too good to be true, it probably is too good to be true. The problem with quackery is not just loss of money—you can be harmed as well. Unproven remedies sometimes give patients false hope while important medical care is delayed.

Nutrition is brimming with quackery, in part because nutrition is such a young science. Questions on many fundamental nutrition issues, such as the relationship between sugar and obesity, are far from being resolved, yet the media publicize the results of research studies long before those results can be said to really prove a scientific theory. Unfortunately, because much research is only in its early stages, the public has been bombarded with conflicting ideas about issues that relate directly to two very important parts of their lives: people's health and their eating habits. This conflict leaves the public confused about the truth and vulnerable to dubious health products (most often nutrition products) and practices—on which people spend $10 billion to $30 billion annually.

Much misinformation proliferates also because in some states anyone can call himself or herself a dietitian or nutritionist. In addition, one may even buy mail-order B.S., M.S., or Ph.D. degrees in nutrition from "schools" in the United States. In all states, nutrition books that are entirely bogus can be published and sold in bookstores, dressed up to look like legitimate health books.

A quack is someone who makes excessive promises and guarantees for a nutrition product or practice that is said to enhance your physical and mental health by, for example, preventing or curing a disease, extending your life, or improving some facet of performance. Health schemes and misinformation proliferate because they thrive on wishful thinking. Many people want easy answers to their medical concerns, such as a quick and easy way to lose weight. Often, claims appear to be grounded in science. Here's how to recognize quacks.

1. Their products make claims such as:
   - Quick, painless, and/or effortless
   - Contains special, secret, foreign, ancient, or natural ingredients
   - Effective cure-all for a wide variety of conditions
   - Exclusive product not available through any other source
2. They use dubious diagnostic tests, such as hair analysis, to detect supposed nutritional deficiencies and illnesses. Then they offer you a variety of nutritional supplements, such as bee pollen or coenzymes, as remedies against deficiencies and disease.
3. They rely on personal stories of success (testimonials) rather than on scientific data for proof of effectiveness.
4. They use food essentially as medicine.
5. They often lack any valid medical or health-care credentials.
6. They come across more as salespeople than as medical professionals.
7. They offer simple answers to complex problems.
8. They claim that the medical community or government agencies refuse to acknowledge the effectiveness of their products or treatments.
9. They make dramatic statements that are refuted by reputable scientific organizations.
10. Their theories and promises are not written in medical journals using a peer-review process but appear in books written only for the lay public.

Keep in mind that there are few, if any, sudden scientific breakthroughs. Science is evolutionary—even downright slow—not revolutionary.

So, where can you find accurate nutritional information? In the United States, over 50,000 registered dietitians (R.D.s) represent the largest and most visible group of professionals in the nutrition field. Registered dietitians are recognized by the medical profession as legitimate providers of nutrition care. They have specialized education in human anatomy and physiology, chemistry, medical nutrition therapy, foods and food science, the behavioral sciences, and foodservice management. Registered dietitians must complete at least a bachelor's degree from an accredited college or university, a program of college-level dietetics courses, a supervised practice experience, and a qualifying examination. Continuing education is required to maintain R.D. status. Registered dietitians work in private practice, hospitals, nursing homes, wellness centers, business and industry, and many other settings. Most are members of the American Dietetic Association, and most are licensed or certified by the state in which they live. Over 40 states have licensure or certification laws that regulate dietitians/nutritionists.

In addition to using the expertise of an R.D., you can ask some simple questions that will help you judge the validity of nutrition information seen in the media or heard from friends.

1. What are the credentials of the source? Does the person have academic degrees in a scientific or nutrition-related field?
2. Does the source rely on emotions rather than scientific evidence or use sensationalism to get a message across?
3. Are the promises of results for a certain dietary program reasonable or exaggerated? Is the program based on hard scientific information?
4. Is the nutrition information presented in a reliable magazine or newspaper, or is it published in an advertisement or a publication of questionable reputation?
5. Is the information someone's opinion or the result of years of valid scientific studies with possible practical nutrition implications?

Much nutrition information that we see or read is based on scientific research. It is helpful to understand how research studies are designed, as well as the pitfalls in each design, so that you can evaluate a study's results. The following three types of studies are used commonly in research.

Laboratory studies use animals such as mice or guinea pigs or tissue samples in test tubes to find out more about a process that occurs in people to determine if a substance might be beneficial or hazardous in humans or to test the effect of a treatment. A major advantage of using laboratory animals is that researchers can control many factors that they can't control in human studies. For instance, researchers can make sure that comparison groups are genetically identical and that the conditions to which they are exposed are the same. However,

mice and other animals are not the same as humans, and so the results from these studies can't automatically be generalized to humans. For example, laboratory studies have indicated that the artificial sweetener saccharin causes cancer in mice, but this has never been proved for humans.

Another type of research, called epidemiological research, looks at how disease rates vary among different populations and also examines factors associated with disease. Epidemiological studies rely on observational data from human populations, and so they can only suggest a relationship between two factors; they cannot establish that a particular factor causes a disease. This type of observational study may compare factors found among people with a disease, such as cancer, with factors among a comparable group without that disease or may try to identify factors associated with diseases that develop over time within a population group. Researchers may find, for example, fewer cases of osteoporosis in women who take estrogen after menopause.

A third type of research goes beyond using animals or observational data and uses humans as subjects. Clinical trials are studies that assign similar participants randomly to two groups. One group receives the experimental treatment; the other does not. Neither the researchers nor the participants know who is in which group. For example, a clinical trial to test the effects of estrogen after menopause would randomly assign each participant to one of two groups. Both groups would take a pill, but for one group this would be a dummy pill, called a placebo. Clinical studies are used to assess the effects of nutrition-education programs and medical nutrition therapy. Unlike epidemiological studies, clinical studies can observe cause-and-effect relationships.

When reading or listening to a news account of a particular study, it is helpful to have a few key questions in the back of your mind not only to help evaluate the merits of the study but also to determine whether it is applicable to you. Look to news reports to address the following:

1. How does this work fit with the body of existing research on the subject? Even the most well-written article does not have enough space to discuss all relevant research on an issue. Yet it is extremely important for the article to address whether a study is confirming previous research and therefore adding more weight to scientific beliefs or whether the study's results and conclusions take a wild departure from current thinking on the subject.
2. Could the study be interpreted to say something else? Scientists often reach different conclusions when commenting on the same or similar data. Look for varying conclusions from experts, because certain issues they address may be important in putting the findings into context.
3. Are there any flaws in how the study was undertaken that should be considered when making conclusions? The more experts are quoted or provide background in a news story, the more likely that potential flaws will be described.
4. Are the study's results generalizable to other groups? Not all research incorporates all types of people: men, women, older adults, and people of various ethnicities. Also, a study may have been conducted on animals and not humans. If study results are applicable only to a narrow group of people, that should be reported as such.

FIGURE 2-29: Websites with Reliable Nutrition Information

www.healthfinder.gov

This is a gateway to reliable consumer health and human services information developed by the U.S. Department of Health and Human Services.

www.eatright.org

This is the website for the American Dietetic Association. It contains information on many nutrition topics and issues.

http://vm.cfsan.fda.gov

The Center for Food Safety and Applied Nutrition contains much information on food safety issues, nutrition labels, and other topics.

www.usda.gov/cnpp

The Center for Nutrition Policy and Promotion of the USDA has the Dietary Guidelines for Americans and lots more nutrition information.

www.nal.usda.gov/fnic

FNIC's (Food and Nutrition Information Center) website provides a directory to credible, accurate, and practical resources. FNIC is part of the USDA.

www.foodandhealth.com

Food and Health Communications provides reliable nutrition information, as well as clip art, in many areas.

www.mayoclinic.com

This is the website of the Mayo Clinic, which has much information on many health topics.

http://nccam.nih.gov

This is the website for the National Center for Complementary and Alternative Medicine.

www.acsm.org

This is the website for the American College of Sports Medicine.

www.nutrition.gov

This government website has many links and much information.

Here are some websites that will help you separate fact from fiction:
www.quackwatch.com (Quack Watch)
www.ncahf.org (National Council Against Health Fraud)

Figure 2-29 lists websites with reliable nutrition information.

Source: With permission, this Hot Topic used sections of "If It Sounds Too Good to Be True . . . It Probably Needs a Second Look" from *Food Insight,* published by the International Food Information Council Foundation, March/April 1999.

Carbohydrates

Carbohydrate literally means hydrate (water) of carbon. The name was created by early chemists who found that heating sugars for a long period in an open test tube produced droplets of water on the sides of the tube and a black substance, carbon. Later chemical analysis of sugars and other carbohydrates indicated that they all contain at least carbon, hydrogen, and oxygen atoms.

Carbohydrates are the major components of most plants, making up 60 to 90 percent of their dry weight. In contrast, animals and humans contain only a small amount of carbohydrates. Plants are able to make their own carbohydrates from carbon dioxide in the air and water taken from the soil in a process known as *photosynthesis*. Photosynthesis converts energy from sunlight into energy stored in carbohydrates. The plant uses the carbohydrates to grow and be healthy. Animals are incapable of photosynthesis and therefore depend on plants as a source of carbohydrates. Plants, such as wheat and broccoli, supply most of the carbohydrates in our diets. Milk also contains some carbohydrate.

Carbohydrates are separated into two categories: simple and complex. Also called sugars, *simple carbohydrates* include sugars that occur naturally in foods, such as fructose in fruits and glucose in honey, as well as sugars that are added to foods, such as white or brown sugar in a chocolate chip cookie.

Carbohydrates are much more than just sugars, though, and include the *complex carbohydrates* starch and fiber. Another name for complex carbohydrates is *polysaccharides* (poly- means many), a good name for starch and most fibers because both consist of long chains of many sugars.

After completing this chapter, you should be able to:

- Identify the functions of carbohydrates
- List important monosaccharides and disaccharides and give examples of foods in which each is found
- Identify foods high in natural sugars, added sugars, and fiber
- List the potential health risks of consuming too much added sugar
- Identify food sources of starch and list the uses of starch in cooking
- Distinguish between the two types of dietary fiber and list examples of food containing each one
- Describe the health benefits of a high-fiber diet
- Describe how carbohydrates are digested, absorbed, and metabolized by the body
- State the dietary recommendations for carbohydrates
- Identify foods as being made from whole grains or refined grains
- Discuss the nutritional value and use of grains and legumes on a menu
- Examine the usefulness of the glycemic index
- Recognize alternatives to sugar in foods

 # FUNCTIONS OF CARBOHYDRATES

Carbohydrates are the primary source of the body's energy, supplying about 4 kcalories per gram. *Glucose*, a simple carbohydrate, is the body's number-one source of energy. Most of the carbohydrates you eat are converted to glucose in the body.

Our cells can burn protein and fat for energy, but the body uses glucose first, in part because glucose is the most efficient energy source. The brain and other nerve cells are picky about their food, and under most circumstances, they will use only glucose for energy.

Some glucose is stored in your body in a form called **glycogen**. This way the body has a constant, available glucose source. Glycogen is stored in two places in the body: the liver and the muscles. An active 150-pound man has about 400 kcalories stored in his liver glycogen and about 1400 kcalories stored in his muscle glycogen. When the blood sugar level starts to dip and more energy is needed, the liver converts glycogen into glucose, which then is delivered by the bloodstream. Muscle glycogen does not supply glucose to the bloodstream but is used strictly to supply energy for exercise.

If you run out of glycogen and do not eat any carbohydrates, the body will break down protein in muscles to some extent. Protein can be converted to glucose to maintain glucose levels in the blood and supply glucose to the central nervous system. Carbohydrates spare protein from being burned for energy so that protein can be used to build and repair the body.

An inadequate supply of carbohydrates can also cause the body to convert some fat to glucose, but this is also not desirable. When fat is burned for energy without any carbohydrates present, the process is incomplete and results in the production of **ketone bodies**. Ketone bodies can be used by the brain for energy, but too many can cause the blood to become too acidic (called **ketosis**), a condition that interferes with the transport of oxygen in the blood. Ketosis can cause dehydration and may even lead to a fatal coma. Carbohydrates are important to help the body use fat efficiently.

You need at least 130 grams of carbohydrates daily to prevent protein (and fat) from being burned for fuel and to provide glucose to the central nervous system (brain and spinal cord). This amount represents what you minimally need, not what is desirable (about two times more). We obtain 50 to 60 percent of our energy intake from carbohydrates. Therefore, if you eat 2000 kcalories per day, you take in 1000 to 1200 calories of carbohydrates, which represents 250 to 300 grams.

Carbohydrates are part of various materials found in the body, such as connective tissues, some hormones and enzymes, and genetic material.

Fiber, a complex carbohydrate, promotes the normal functioning of the intestinal tract, lowers blood cholesterol, and is associated with a reduced risk of developing type 2 diabetes (a disease characterized by high blood glucose levels).

GLYCOGEN
The storage form of glucose in the body; it is stored in the liver and muscles.

KETONE BODIES
A group of organic compounds that cause the blood to become too acidic as a result of fat being burned for energy without any carbohydrates present.

KETOSIS
Excessive level of ketone bodies in the blood and urine.

MINI-SUMMARY

1. Carbohydrates are the primary source of the body's energy. The brain and other neuron cells rely almost exclusively on glucose for energy.

2. Glycogen is a storage form of glucose in the liver and muscles. When your blood sugar level drops, the liver converts glycogen to glucose. Muscle glycogen is used strictly to provide energy to muscles during exercise.

3. Carbohydrate spares protein from being burned for energy.

4. Carbohydrates are also important to help the body use fat efficiently. When fat is burned for energy without any carbohydrates present, the process is incomplete and can result in the production of ketone bodies and ketosis.

5. You need at least 100 to 150 grams of carbohydrate daily to spare protein (and fat) from being burned for fuel and to provide glucose to the central nervous system and red blood cells.

6. Carbohydrates are part of various materials found in the body, such as connective tissues, some hormones and enzymes, and genetic material.

7. Fiber, a complex carbohydrate, promotes the normal functioning of the intestinal tract, lowers blood cholesterol, and is associated with a reduced risk of developing type 2 diabetes.

SIMPLE CARBOHYDRATES (SUGARS)

Simple carbohydrates include monosaccharides and disaccharides. The chemical names of the six sugars to be discussed all end in "-ose," which means sugar.

MONOSACCHARIDES

Monosaccharides include these simple sugars:

1. Glucose
2. Fructose
3. Galactose

The prefix mono- means "one"; these sugars consist of a single ring of atoms. **Monosaccharides** are the building blocks of other carbohydrates, such as disaccharides and starch.

Glucose is the most abundant sugar found in nature. In photosynthesis, plants make glucose, which provides energy for growth and other activities. Also called dextrose, glucose is our primary energy source as well. As already mentioned, most of the carbohydrates you eat are converted to glucose in the body. The concentration of glucose in the blood, referred to as the **blood glucose level**, is vital to the proper functioning of the human body. Glucose is found in fruits such as grapes, in honey, and in small amounts in many plant foods.

Fructose, the sweetest natural sugar, is also found in honey as well as in fruits. Although it is a natural sugar, honey (made by bees) is primarily fructose and glucose, the two components of white sugar. Fructose is about 1.3 times as sweet as white sugar. Fructose and glucose are the most common monosaccharides in nature.

The last single sugar, **galactose**, is almost always linked to glucose to make milk sugar, a disaccharide.

DISACCHARIDES

Most naturally occurring carbohydrates contain two or more monosaccharide units linked together. **Disaccharides**, the double sugars, include sucrose, maltose, and lactose. They each contain glucose (Figure 3-1). **Sucrose** is the chemical name for what is commonly called white sugar, table sugar, granulated sugar, or simply **sugar**. Sugar cane and sugar beets both contain much sucrose but have to be refined to extract the sucrose from them. As Figure 3-1 indicates, sucrose is simply two common single sugars—glucose and fructose—linked together. Although the primary source of sucrose in the American diet is refined sugar, sucrose occurs naturally in small amounts in many fruits and vegetables. Table sugar is more than 99 percent pure sugar and provides virtually no nutrients for its 16 kcalories per teaspoon.

Maltose, which consists of two bonded glucose units, does not occur in nature to any appreciable extent. It is fairly abundant in germinating (sprouting) seeds and is produced in the manufacture of beer.

The last disaccharide, **lactose**, is commonly called milk sugar because it occurs in milk. Although milk is not a food you think of as sweet, there is some sugar there. However, if you look at Figure 3-2, you will see that lactose is one of the lowest-ranking sugars in terms of sweetness. Lactose, or milk sugar, is present in milk and products made from milk. Unlike most carbohydrates, which are in plant products, lactose is one of the few carbohydrates associated exclusively with animal products.

MONOSACCHARIDES
Simple sugars, including glucose, fructose, and galactose, which consist of a single ring of atoms and are the building blocks for other carbohydrates, such as disaccharides and starch.

BLOOD GLUCOSE LEVEL (BLOOD SUGAR LEVEL)
The amount of glucose found in the blood; glucose is vital to the proper functioning of the body.

FRUCTOSE
A monosaccharide found in fruits and honey.

GALACTOSE
A monosaccharide found linked to glucose to form lactose, or milk sugar.

DISACCHARIDES
Double sugars such as sucrose.

SUCROSE (SUGAR)
A disaccharide commonly called table sugar, granulated sugar, or simply sugar.

MALTOSE
A disaccharide made of two glucose units bonded together.

LACTOSE
A disaccharide found in milk and milk products that is made of glucose and galactose.

Glucose

Fructose

Galactose

FIGURE 3-1:
Monosaccharides and disaccharides.

Sucrose (glucose + fructose)

Maltose (glucose + glucose)

Lactose (glucose + galactose)

ADDED SUGARS

Added sugars (Figure 3-3) include white sugar, high-fructose corn syrup, and other sweeteners added to foods in processing, as well as sugars added to foods at the table. Added sugars are the major source of simple sugars in most diets.

High-fructose corn syrup is corn syrup that has been treated with an enzyme to convert part of the glucose it contains to fructose. The reason for changing the glucose to fructose lies in the fact that fructose is twice as sweet as glucose. High-fructose corn syrup is therefore sweeter, ounce for ounce, than corn syrup, and so smaller amounts can be used (making it

ADDED SUGARS
Sugars added to a food for sweetening or other purposes; they do not include the naturally occurring sugars in foods such as fruit and milk.

HIGH-FRUCTOSE CORN SYRUP
Corn syrup that has been treated with an enzyme that converts part of the glucose it contains to fructose; found in most regular sodas as well as other foods.

FIGURE 3-2: Relative Sweetness of Sugars and Alternative Sweeteners	
Sweetener	Rating
SUGARS	
Lactose	20
Glucose	70–80
Sucrose	100
High-fructose corn syrup	120–160
Fructose	140
ARTIFICIAL SWEETENERS	
Aspartame (Nutrasweet, Equal)	160–220
Acesulfame-K (Sunette)	200
Saccharin (Sweet 'N Low)	200–700
Sucralose (Splenda)	600
Neotame	7,000–13,000

FIGURE 3-3: Examples of Added Sugars

Name	Description
White sugar	Made from beet sugar or cane sugar. Refining removes the yellow-brown pigments of unrefined sugar. Also called table sugar, granulated sugar, or sugar.
High-fructose corn syrup	Corn syrup treated with an enzyme that converts some of the glucose to fructose. Sweeter than corn syrup and used extensively in soft drinks, baked goods, jelly, fruit drinks, and other foods.
Corn syrup	A thick, sweet syrup made from cornstarch. Mostly glucose with some maltose. Only 75 percent as sweet as sucrose. Less expensive than sucrose. Used in baked goods and other foods.
Invert sugar	White sugar that has been heated with water and acid, resulting in some of the sucrose being broken down to glucose and fructose. Sweeter than sucrose. Available only as a liquid.
Brown sugar	Sugar crystals contained in a molasses syrup with natural flavor and color—91 to 96 percent sucrose.
Molasses	Thick syrup left over after making sugar from sugar cane. Brown in color with a high sugar concentration.
Honey	Sweet syrupy fluid made by bees from the nectar collected from flowers and stored in nests or hives as food. Made of glucose, fructose, and sucrose.
Maple sugar	A sugar made from the sap of sugar maple trees
Powdered or confectioners' sugar	Granulated sugar that has been crushed into a fine powder, and sugar combined with a small amount of cornstarch to prevent clumping. Available in several degrees of fineness, designated by the number of Xs following the name. 6X is the standard confectioners' sugar and is used in icing and toppings.
Raw sugar	Produced in the initial stages of white sugar's manufacturing process. Because it contains dirt, molds, and bacteria, it is not allowed to be sold in the United States. Products that are sold as "raw sugar" have gone through many more processing steps than real raw sugar but not as many steps as white sugar. These products may be sold as "demerara," "turbinado," or "muscavado" sugar.
Demerara sugar	A light brown sugar with large crystals that is named after the region in South American from where it originally came. Not as processed as white sugar, it is popular for tea and coffee in Canada, England, and Australia.
Turbinado sugar	Made from sugar cane but not processed as much as white sugar is. Retains some of the natural molasses, and so it is generally light golden in color. It has large crystals and a mild flavor.

cheaper). It is used to sweeten almost all regular soft drinks and is frequently used in fruit drinks, sweetened teas, cookies, jams and jellies, syrups, and sweet pickles.

Added sugars perform several functions in foods besides sweetening. They prevent spoilage in jams and jellies and perform several functions in baking, such as browning the crust and retaining moisture in baked goods so that they stay fresh. Sugar also acts as a food for yeast in breads and other baked goods that use yeast for leavening.

The major sources of added sugars in the diet are soft drinks, candy, tabletop sugars, baked goods, fruit drinks, and dairy desserts (such as ice cream) and sweetened milk (Figure 3-4). The added sugar content of various foods is listed in Figure 3-5. Foods high in added sugars often contain few nutrients for the number of kcalories they provide. In other words, these foods are not nutrient-dense, as discussed in Chapter 1.

When you look at "Sugars" on the Nutrition Facts panel, keep in mind that the number of grams given includes naturally occurring sugars and added sugars. Your body processes natural and added sugars in the same way. In fact, your body does not see any difference between a natural sugar, such as fructose in an apple, and the

FIGURE 3-4:
Foods high in added sugars. Clockwise from top: regular soda, candy, sweetened cereal, table sugar, cookies, candy bars, cupcakes.
Photo by Frank Pronesti.

FIGURE 3-5: Amount of Added Sugars in Selected Foods

Food	Portion Size	Amount of Sugar*
FOODS HIGH IN SUGAR (4 OR MORE TEASPOONS/SERVING)		
Beverage, cola	12 fl. oz.	10 teaspoons
Ginger ale	12 fl. oz.	8
Milk chocolate	1½ oz.	6
Reese's Peanut Butter Cup	2	5.5
Jellybeans	10 large	5
Sherbet, orange	½ cup	4.5
Honey	1 Tbsp.	4.3
Chocolate cupcake, crème-filled	1 cupcake	4.3
Ice cream, vanilla	½ cup	4
FOODS MODERATE IN SUGAR (1 TO 3.9 TEASPOONS/SERVING)		
Angel food cake	1 piece	3.8
Kellogg's Corn Pops	1 cup	3.5
Pancake syrup	1 Tbsp.	3
Frosted Flakes	1 cup	3.0
Toaster pastries, fruit	1 pastry	2.8
Syrups, chocolate, fudge-type	1 Tbsp.	2.8
Doughnuts, cake-type, plain	1 medium	2.6
Jellies	1 Tbsp.	1.9
Barbecue sauce	2 Tbsp.	2.1
Catsup	2 Tbsp.	1.7
French dressing	2 Tbsp.	1.2

*4 grams sugar = 1 teaspoon

Source: U.S. Department of Agriculture, Agricultural Research Service, 2004. USDA Nutrient Database for Standard Reference, Release 17. Nutrient Data Laboratory Home Page, www.nal.usda.gov/fnic/foodcomp.

high-fructose corn syrup in a soda. However, many natural sugars occur in foods such as fruits that contain fiber. Fiber helps slow down glucose absorption, to be discussed in a moment.

To find out whether a food contains added sugar and how much, you need to look at the ingredient list. For example, here are the ingredients for an apple pie: "Apples, corn syrup, sugar, wheat flour, water, modified corn starch, dextrose, brown sugar, sodium alginate, spices, citric acid, salt, lecithin, dicalcium phosphate." The ingredients are listed in order by weight. The ingredient in the greatest amount by weight is listed first, and the one in the least amount is listed last. In this example, corn syrup is the second ingredient listed and sugar is the third, which means that combined, these two sugars are the main ingredients in the apple pie.

HEALTH ISSUES

Added sugars

Is all this sugar good for you? Let's take a look at how it affects health.

Dental caries

> **DENTAL CARIES**
> Tooth decay.
>
> **PLAQUE**
> Deposits of bacteria, protein, and polysaccharides found on teeth that contribute to tooth decay.

Sugar (and starches, too) contribute to the development of *dental caries*, or cavities. The more often sugars and starches—even small amounts—are eaten and the longer they are in the mouth before the teeth are brushed, the greater is the risk for tooth decay. Dental caries are a major cause of tooth loss. This is so because every time you eat something sweet, the bacteria living on your teeth ferment, or digest, the carbohydrates, and this produces acid. This acid eats away at the teeth for 20 to 30 minutes, and cavities eventually develop. The deposit of bacteria, protein, and polysaccharides that forms on the teeth in the absence of tooth-brushing during a period of 12 to 24 hours is called *plaque*. Without good tooth-brushing habits, plaque may cover all surfaces of the teeth.

Other factors influence how much impact foods will have on the development of cavities: the foods' form (liquid or solid and sticky) and which foods are eaten together. At meals, if an unsweetened food such as cheese is eaten after a sugared food, the plaque will be less acidic, and so less acid eats away at the teeth. Cheese also stimulates more saliva, which helps wash away acids. This is why eating sugary or starchy foods as frequent between-meals snacks is more harmful to the teeth than having them at meals. Sticky carbohydrate foods such as raisins and caramels cause more problems than do liquid carbohydrate foods because they stick to the teeth and provide a constant source of fermentable carbohydrates for the bacteria until washed away. Liquids containing sugars, such as sodas, have been considered less harmful to teeth than solid sweets because they clear the mouth quickly.

Food such as dried fruits, breads, cereals, cookies, crackers, and potato chips increase the chances of dental caries when eaten frequently. Foods that do not seem to cause cavities include cheese, peanuts, sugar-free gum, some vegetables, meats, and fish. To prevent dental caries, brush your teeth twice a day, floss your teeth every day, try to limit sweets to mealtimes, and see your dentist regularly.

Sweeteners can add to the pleasure of eating and can help you improve the quality of your diet if selected in appropriate quantities and in forms that are high in vitamins and minerals.

Obesity

Although there is no research stating that added sugars cause obesity, they are undoubtedly a factor in rising obesity rates among adults and children. Individuals who consume food or beverages high in added sugars tend to consume more kcalories than do those who consume low amounts of added sugars, and they also tend to consume lower amounts of vitamins and minerals. Add just one 12-ounce soft drink to your diet every day for a year, and you will gain 15 pounds!

Other high-sugar foods, such as candy and cookies, are almost always teamed up with fat and high in kcalories. These foods are also typically low in nutrients and are therefore are referred to as *empty kcalorie foods*. For example, a cupcake supplies 170 kcalories with virtually no nutrients. If you look at foods with natural sugars, such as fruits and milk, you will notice that they also contain many essential nutrients. Fruits provide sugars, usually at a relatively low kcalorie cost, and are important sources of fiber and at least eight additional nutrients. Other nutrient dense foods besides fruits include vegetables, no- and low-fat milk and milk products, and whole grains. The foods highest in added sugars are, unfortunately, not nearly as rich in other nutrients.

Eating too many foods and beverages rich in added sugars contributes to overweight and obesity if more kcalories are consumed each day than are used. Being overweight or obese increases your risk for high blood pressure, high blood cholesterol, heart disease, stroke, diabetes, and other diseases. High consumption of added sugars also makes it difficult to get in all the required micronutrients (vitamins and minerals) that are necessary, which is a real concern for children. To grow properly, children need to eat a varied and adequate diet.

Diabetes

There is no evidence that total sugar intake is associated with the development of diabetes. *Diabetes* is a disorder in which the body does not metabolize carbohydrates properly. It results from having inadequate or ineffective insulin. *Insulin* is a hormone that increases the movement of glucose from the bloodstream into the body's cells, where it is used to produce energy. People with untreated diabetes have high blood sugar levels. Treatment for diabetes is individualized to the patient and includes a balanced diet that supports a healthy weight and physical activity, as well as insulin if needed. Being overweight and inactive are risk factors. Diabetes is discussed in more depth in Chapter 11.

Heart disease

A moderate intake of sugars does not increase the risk for heart disease. However, diets high in fructose and sucrose seem to increase blood triglyceride (fat) levels and low-density lipoprotein (LDL) cholesterol levels, which then increase the risk of heart disease.

Hypoglycemia

Hypoglycemia is the term used to describe an abnormally low blood glucose level. It occurs most often in people who have diabetes and take insulin. If they take too much insulin, eat too little food, and/or exercise a lot, their blood sugar level may drop to the point where they are hypoglycemic. Symptoms include quickened heartbeat, shakiness, weakness, anxiety, sweating, and dizziness, mimicking anxiety or stress symptoms.

Hypoglycemia is not seen often in healthy people. Regular, well-balanced meals with reasonable amounts of refined sugars and sweets, as well as protein, fiber, and fat, can moderate swings in blood glucose levels.

Hyperactivity in children

Extensive research has failed to show that high sugar intake causes hyperactivity or *attention deficit hyperactivity disorder (ADHD)*, a developmental disorder of children characterized by impulsiveness, distractibility, and hyperactivity.

Lactose intolerance

Lactose (milk sugar) is a problem for certain people who have a deficiency of the enzyme *lactase*. Lactase is needed to split lactose into its components in the small intestine. If lactose

---

EMPTY KCALORIE FOODS
Foods that provide few nutrients for the number of kcalories they contain.

DIABETES
A disorder in which the body does not metabolize carbohydrate properly due to inadequate or ineffective insulin.

INSULIN
A hormone that increases the movement of glucose from the bloodstream into the body's cells.

HYPOGLYCEMIA
A symptom in which blood sugar levels are low.

ATTENTION DEFICIT HYPERACTIVITY DISORDER (ADHD)
A developmental disorder of children characterized by impulsiveness, distractibility, and hyperactivity.

LACTASE
An enzyme needed to split lactose into its components in the intestines.

is not split, it travels to the colon, where it attracts water and causes bloating and diarrhea. Intestinal bacteria also ferment the lactose and produce fatty acids and gas, making the stomach discomfort worse and contributing still more to the diarrhea. Symptoms usually occur within about 30 minutes to 2 hours after the ingestion of milk products and clear up within 2 to 5 hours. This problem, called ***lactose intolerance***, seems to be an inherited problem that is especially prevalent among Asian Americans, Native Americans, African Americans, and Latinos, as well as some other population groups. For most people, the body begins to produce less lactase after about age two. However, many people may not experience symptoms until they are much older.

Treatment for lactose intolerance includes a diet that is limited in lactose, which is present in large amounts in milk, ice cream, ice milk, sherbet, cottage cheese, eggnog, and cream. Most individuals can drink small amounts of milk without any symptoms, especially if it is taken with food. Lactose-reduced milk and some other lactose-reduced dairy products are available in supermarkets. Eight fluid ounces of lactose-reduced milk contain only 3 grams of lactose, compared with 12 grams in regular milk. Reducing the lactose content of milk by 50 percent is often adequate to prevent the symptoms of lactose intolerance. Although lactose-reduced milk and other lactose-digestive aids are available, they may not be necessary when lactose intake is limited to 1 cup of milk (or the equivalent) or less a day.

Yogurt is usually well tolerated because it is cultured with live bacteria that digest lactose. This is not always the case with frozen yogurt, because most brands do not contain nearly the number of bacteria found in fresh yogurt (there are no federal standards for frozen yogurt at this time). Also, some yogurts have milk solids added to them that can cause problems. Many hard cheeses contain very little lactose and usually do not cause symptoms because most of the lactose is removed during processing or digested by the bacteria used in making cheese.

People who have difficulty digesting lactose report tremendous variation in which lactose-containing foods they can eat and even the time of day they can eat them. For example, one individual may not tolerate milk at all, whereas another can tolerate milk as part of a big meal. The ability to tolerate lactose is not an all-or-nothing phenomenon. As people with lactase deficiency usually decrease their intake of dairy products and thus their calcium intake, they should try different dairy products to see what they can tolerate. Soymilk and rice milk are good milk substitutes for people with lactose intolerance.

MINI-SUMMARY

1. Simple carbohydrates (also called simple sugars) include monosaccharides and disaccharides.

2. Monosaccharides include glucose, fructose, and galactose.

3. Most of the carbohydrates you eat are converted to glucose in the body. Your blood glucose level is vital to the proper functioning of the body. Glucose is found in fruits, honey, and many plant foods.

4. Fructose, the sweetest natural sugar, is found in fruits and honey.

5. Galactose is almost always linked to glucose to make lactose (milk sugar).

6. Disaccharides include sucrose (white sugar), maltose, and lactose (milk sugar). (See Figure 3-1.)

7. Added sugars include white sugar, high-fructose corn syrup, and other sweeteners added to foods in processing, as well as sugars added at the table. Added sugars are the major source of simple sugars in most diets.

8. The major sources of added sugars in the diet are soft drinks, candy and sugars, baked goods, fruit drinks, and dairy desserts and sweetened milk.

9. When you look at "Sugars" on the Nutrition Facts panel, the number of grams given includes naturally occurring sugars and added sugars.

10. Sugar and starches contribute to the development of dental caries, or cavities.

11. Individuals who consume food or beverages high in added sugars tend to consume more kcalories than do those who consume low amounts of added sugars, and they also tend to consume lower amounts of vitamins and minerals.

12. Cakes and cookies are examples of empty kcalories because they are high in fat and sugar and supply few nutrients for the number of kcalories they provide.

13. There is no evidence that total sugar intake is associated with the development of diabetes. However, obesity and being inactive are risk factors for type 2 diabetes.

14. A moderate intake of sugars does not increase the risk for heart disease. However, diets high in fructose and sucrose seem to increase blood fat levels and low-density lipoprotein cholesterol levels, which increase the risk of heart disease.

15. Hypoglycemia is not seen often in healthy people. Regular, well-balanced meals with reasonable amounts of refined sugars and sweets, as well as protein, fiber, and fat, can moderate swings in blood glucose levels.

16. Extensive research has failed to show that high sugar intake causes hyperactivity or attention deficit hyperactivity disorder in children.

17. Symptoms of lactose intolerance include stomach discomfort, bloating, and diarrhea. They usually occur within 30 minutes to 2 hours after the ingestion of milk products and clear up within 2 to 5 hours. Lactose intolerance seems to be an inherited problem that is especially prevalent among Asian Americans, Native Americans, African Americans, and Latinos. People with lactose intolerance report tremendous variation in which lactose-containing foods they can eat. Yogurt, hard cheeses, and lactose-reduced milk are generally tolerated.

COMPLEX CARBOHYDRATES (STARCHES AND FIBER)

STARCHES

Plants such as peas store glucose in the form of **_starch_**. Starch is made of many chains of hundreds to thousands of glucoses linked together. The chains may be straight or have treelike branches (Figure 3-6). The straight form of starch is called amylose, and the branched form is called amylopectin.

> **STARCH**
> A complex carbohydrate made up of a long chain of glucoses linked together; found in grains, legumes, vegetables, and some fruits; the straight form is called amylose, and the branched form is called amylopectin.

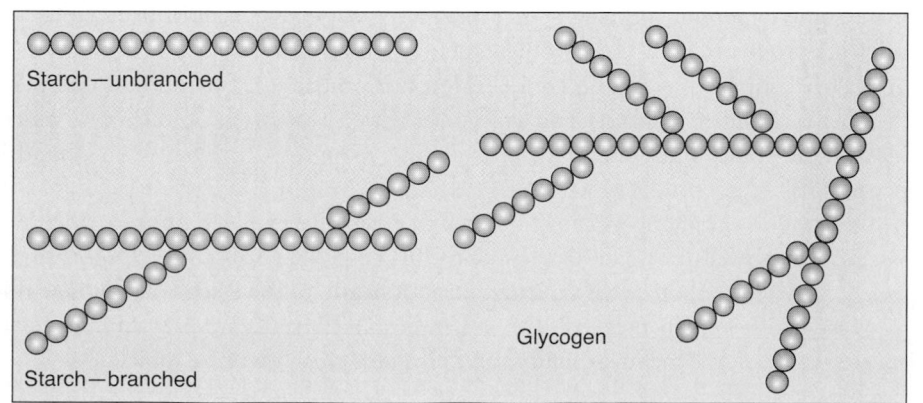

Starch—unbranched

Starch—branched

Glycogen

FIGURE 3-6:
The structures of starch and glycogen.

Just as plants store glucose in the form of starch, your body stores glucose as glycogen. Like starch, glycogen is a polysaccharide. It is a chain of glucose units, but the chains are longer and have more branches than is the case with starch (Figure 3-6).

Starch is found only in plant foods. Cereal grains, the fruits or seeds of cultivated grasses, are rich sources of starch and include wheat, corn, rice, rye, barley, and oats. Cereal grains are used to make flours, breads, baked goods, breakfast cereals, and pastas. Starches are also found in root vegetables such as potatoes and in dried beans and peas such as navy beans. Starchy vegetables, like all vegetables, are important sources of fiber and over 15 or more nutrients.

HEALTH EFFECTS OF STARCHES

Starch creates the same problem as sugar in the mouth and therefore contributes to tooth decay. Starch from whole-grain sources is preferable to starch found in refined grains such as white flour. This is discussed in detail in the following section.

MINI-SUMMARY

1. Starch is a storage form of glucose found in plants. Like starch, glycogen is a polysaccharide.
2. Starches are found in cereal grains, breads, baked goods, cereals, pastas, root vegetables, and dried beans and peas.
3. Starchy foods must generally be cooked to make them better tasting and more digestible. They are commonly used as thickeners because they gelatinize.
4. Like sugar, starch contributes to dental caries.

FIBERS

DIETARY FIBER
Polysaccharides and lignin (a nonpolysaccharide) that are not digested and absorbed.

SOLUBLE FIBER
A classification of fiber that includes gums, mucilages, pectin, and some hemicelluloses; they are generally found around and inside plant cells.

INSOLUBLE FIBER
A classification of fiber that includes cellulose, lignin, resistant starch, and the remaining hemicelluloses; they generally form the structural parts of plants.

The Food and Nutrition Board defines *dietary fiber* as the polysaccharides found in plant foods that are not digested and absorbed. That definition also includes lignin, a part of plant cells that is not technically a polysaccharide. Fiber is defined differently around the world and may include other nonpolysaccharides.

Like starch, most fibers are chains of bonded glucose units, but what's different is that the units are linked with a chemical bond that our digestive enzymes can't break down. In other words, most fiber passes through the stomach and intestines unchanged and is excreted in the feces.

However, some fiber is digested by bacteria in the large intestine. Bacteria are normally found in the large intestine and are necessary for a healthy intestine. Intestinal bacteria also make some important substances, such as vitamin K. When bacteria digest fiber, some fatty acids are produced, absorbed by our bodies, and used for energy. Determining the amount of kcalories supplied by fiber is complex, but it is probably 1.5 to 2.5 kcalories per gram.

Fiber is abundant in plants, and so legumes (dried beans, peas, and lentils), fruits, vegetables, whole grains, nuts, and seeds provide fiber (Figure 3-7). Fiber is not found in meat, poultry, fish, dairy products, and eggs.

One way to classify fibers is by how well they dissolve in water. *Soluble fiber* (also called viscous fiber) swells in water, like a sponge, into a gel-like substance. *Insoluble fiber* (also called nonviscous fiber) swells in water, but not nearly to the extent that soluble fiber does.

Most foods contain both soluble and insoluble fiber. Figure 3-8 summarizes the food sources and health benefits of soluble and insoluble fiber, our next topic.

FIGURE 3-7: Fiber Content of Selected Foods

Food	Portion Size	Amount of Fiber (Grams)
FOODS HIGH IN DIETARY FIBER (5 GRAMS OR MORE/SERVING)		
Kellogg's All-Bran	½ cup	9
Lentils, cooked	½ cup	8
Refried beans	½ cup	7
Beans, kidney, cooked	½ cup	7
Raspberries, frozen, red	½ cup	6
Cornmeal	½ cup	5
FOODS MODERATELY HIGH IN DIETARY FIBER (2.5 TO 4.9 GRAMS/SERVING)		
Bulgur, cooked	½ cup	4
Vegetables, mixed, cooked	½ cup	4
FOODS MODERATELY HIGH IN DIETARY FIBER (2.5 TO 4.9 GRAMS/SERVING)		
Prunes, stewed	½ cup	4
Kellogg's Raisin Bran	½ cup	4
Pear, raw	1 medium	4
Potato, baked	1 potato	4
Squash, winter, cooked	½ cup	3
Broccoli, cooked	½ cup	3
Raisins	½ cup	3
Blueberries, frozen	½ cup	3
Carrots, boiled	½ cup	3
Seeds, sunflower	¼ cup	3
Oatmeal, instant	1 pack	3
Peanuts	1 oz.	3
Apples, raw with skin	1 apple	3
Banana, raw	1 banana	3
FOODS MODERATE IN DIETARY FIBER (1 TO 2.4 GRAMS/SERVING)		
Oats, cooked	½ cup	2
Pears, canned	½ cup	2
Corn, sweet	½ cup	2
Rice, brown, cooked	½ cup	2
Mushrooms, cooked	½ cup	2
Bread, whole wheat	1 slice	2
Bread, rye	1 slice	2
Potatoes, mashed	½ cup	1.5
Carrots, raw	½ cup	1.5
Wild rice, cooked	½ cup	1.5
Applesauce, canned	½ cup	1.5
Squash, summer, cooked	½ cup	1.5

(continued)

FIGURE 3-7: Fiber Content of Selected Foods (*Continued*)

Food	Portion Size	Amount of Fiber (Grams)
FOODS LOW IN DIETARY FIBER (LESS THAN 1 GRAM/SERVING)		
Macaroni, cooked	½ cup	<1
Doughnuts, cake-type	1 medium	0.7
Rice, white, cooked	½ cup	0.4
Cookies, oatmeal	1 cookie	0.7
Bread, white	1 slice	0.6
Orange juice	½ cup	0.25

Source: U.S. Department of Agriculture, Agricultural Research Service, 2004. USDA Nutrient Database for Standard Reference, Release 17. Nutrient Data Laboratory Home Page, www.nal.usda.gov/fnic/foodcomp.

HEALTH EFFECTS OF FIBERS

What can fiber do for you? Numerous studies have found that diets high in fiber and low in saturated fat and cholesterol are associated with a reduced risk of diabetes, heart disease, digestive disorders, and certain cancers. However, this doesn't mean that fiber reduces the risk. Since high-fiber foods also often contain antioxidant vitamins, phytochemicals (substances in plants that may reduce the risk of cancer and heart disease), and other substances that may offer protection against these diseases, fiber alone is not responsible for the reduced health risks. Findings on the health effects of fiber show that it may play a role in the following:

- **Diabetes.** Dietary fiber reduces the risk of type 2 diabetes and also helps control diabetes. Type 2 diabetes is seen mostly in adults, many of whom are overweight. Soluble

FIGURE 3-8: Soluble and Insoluble Fiber: Foods and Health Benefits*

Type of Fiber	Food Sources	Health Benefits
Soluble (viscous)	Beans and peas Some cereal grains, such as barley, oats, rye Many fruits, such as citrus fruits, pears, apples, grapes Many vegetables, such as brussels sprouts and carrots	Traps carbohydrates to slow digestion and absorption of glucose. Binds to cholesterol in gastrointestinal tract. Reduces risk of diabetes and heart disease.
Insoluble (nonviscous)	Wheat bran Whole grains, such as whole wheat and brown rice Many vegetables Many fruits Beans and peas Seeds	Increases fecal weight so that feces travels quickly through the colon. Provides feeling of fullness. Helps prevent and treat constipation, diverticulosis, and hemorrhoids. Helpful in weight management. May help reduce risk of heart disease.

*Most foods contain both soluble and insoluble fiber.

fiber traps carbohydrates to slow their digestion and absorption. The slower absorption of glucose helps prevent wide swings in blood sugar levels throughout the day.

- **Heart disease.** Clinical studies show that a diet high in fruits, vegetables, beans, and grain products that contain soluble fiber can lower blood cholesterol levels and therefore lower the risk of heart disease. As it passes through the gastrointestinal tract, soluble fiber binds to dietary cholesterol, helping the body eliminate the cholesterol. This reduces blood cholesterol levels, a major risk factor for heart disease. Although research is not conclusive yet, insoluble fiber may protect the heart by reducing blood pressure or the risk of blood clots.

- **Digestive disorders.** Because insoluble fiber increases feces weight and bulk, it promotes normal bowel movements in adults and children. In general, the greater the weight of the feces is, the more quickly it passes through the large intestine. Insoluble fiber is therefore helpful in preventing **constipation** (infrequent passage of feces). Fiber also helps reduce the risk of **diverticulosis**, a condition in which small pouches form in the colon wall, usually from the pressure created within the colon by the small bulk and/or from straining during bowel movements. A diet high in insoluble fiber is also used to prevent or treat **hemorrhoids**, enlarged veins in the lower rectum. Larger, soft feces are easier to eliminate, and so there is less pressure and the rectal veins are less likely to swell.

- **Colon cancer.** Studies have generally noted an association between low total fat intake, high fiber intake, and a reduced incidence of colon cancer. The exact mechanism for reducing the risk is not known, but one possibility is that insoluble fiber adds bulk to stool, which in turn dilutes carcinogens (cancer-causing substances) and speeds their transit through the intestines and out of the body. More research needs to be done before we know for certain that fiber decreases the risk of colon cancer, but until then, health-care professionals strongly suggest eating plenty of fruits, vegetables, beans and peas, and whole grains.

CONSTIPATION Infrequent passage of feces. **DIVERTICULOSIS** A disease of the large intestine in which the intestinal walls become weakened, bulge out into pockets, and at times become inflamed. **HEMORRHOIDS** Enlarged veins in the lower rectum.

Lastly, fiber can be helpful in weight management. Because fiber-containing foods tend to be low in fat and sugar, they often contain fewer kcalories. And since fiber-containing foods slow digestion and provide bulk, you tend to feel full longer and eat less. If you want to increase the fiber in your diet, look at Figure 3-9 for tips. It is important to add fiber gradually to your diet and to drink more fluids as well. If too much fiber is added too rapidly you can experience intestinal discomfort, gas, and diarrhea.

FIGURE 3-9: How to Increase Fiber in Your Diet

Instead of:	Choose:
White bread	Whole-grain bread
Cereals with 1 gram or less of fiber/serving	Whole-grain cereals with at least 3 grams of fiber/serving
White pasta	Whole-wheat pasta
Baked goods made with white flour	Baked goods made with whole-wheat flour
Vegetable and fruit juices	Fresh, canned, or frozen fruits and vegetables
Meat stews and chili	Meatless stews and chilis using beans and whole grains
Meat sandwiches	Whole-grain bread or crackers with peanut butter or tahini
White rice	Brown rice, bulgur, barley
Creamy soups	Soups with vegetables, beans, and lentils
Cookies, chips, candy	Fresh fruit, vegetables, popcorn, nuts, or seeds

1. Dietary fiber includes the polysaccharides (and one nonpolysaccharide called lignin) found in plant foods that are not digested and absorbed.

2. Most fibers are chains of bonded glucose units, but our digestive enzymes can't split the glucoses. Most fiber simply passes through the gastrointestinal tract. Some fiber is digested by bacteria in the large intestine, producing some fatty acids, which are absorbed. Determining the amount of kcalories supplied by fiber is complex, but it is probably 1.5 to 2.5 kcalories/gram.

3. Figure 3-8 summarizes the food sources and health benefits of soluble and insoluble fiber.

4. When adding fiber to the diet, do so gradually to give the intestinal tract time to adapt, and drink lots of fluid.

5. It is recommended to eat three servings of whole grains per day. By eating three servings, you can reduce your risk of diabetes, coronary heart disease, and certain cancers.

Nutrition Science Focus: Carbohydrates

REGULATION OF BLOOD GLUCOSE

The level of glucose in your blood is very closely regulated because glucose is the preferred fuel, especially for the brain and nervous system. When your blood glucose level is within the normal range, most of your body's cells will use glucose for their energy source.

Carbohydrates you eat are absorbed mostly as glucose into the bloodstream. As blood glucose levels increase after eating, glucose is picked up by your cells to use for energy. Excess glucose is stored as glycogen in a process called glycogenesis. Glycogen stores are limited, so any additional excess glucose is converted to fat and stored in the body's fat cells. The process by which glucose is converted to fat is called lipogenesis (Figure 3-10).

If blood glucose concentrations decrease, as happens at night when you sleep or during strenuous exercise, the body converts glycogen back to glucose through a process called glycogenolysis.

Two hormones, insulin and glucagon, are responsible for hour-to-hour regulation of your blood sugar. The organ called the pancreas secretes both of these hormones. The alpha cells of the pancreas secrete glucagon, while the beta cells produce insulin. Insulin and glucagons act like opposites to keep blood glucose within an acceptable range. Both hormones are present in the blood most of the time. Depending on the ratio of insulin to glucagon, one of them will dominate.

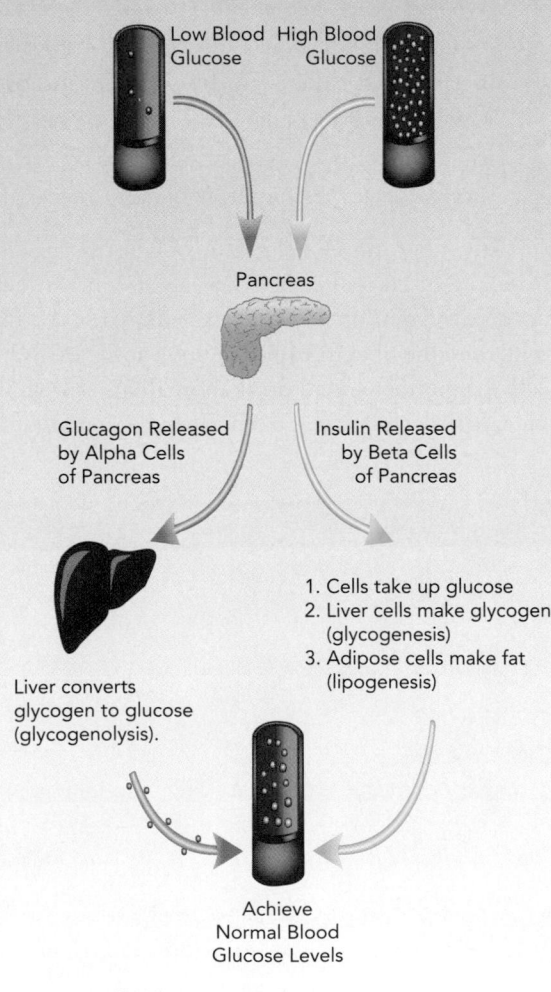

FIGURE 3-10:

How insulin and glucagon maintain normal blood glucose levels.

As you are absorbing nutrients from food, insulin is the dominant hormone. Once your blood glucose levels start increasing, insulin is released. Insulin is necessary for glucose to get into most of your cells. Insulin receptors on cells bind to a specific region of the insulin molecule, and this interaction results in glucose crossing into the cell. Insulin also activates enzymes so glucose in the cell can be made into glycogen or fat.

When blood glucose levels fall between meals and at night, insulin secretion slows and glucagon becomes more dominant. The goal of glucagon is to prevent low blood sugar levels. When blood glucose levels falls below a certain point, glucagon secretion therefore increases. The primary target for glucagon is the liver. Glucagon stimulates the liver to convert glycogen to glucose and release the glucose into the bloodstream.

The most common disease involving the pancreatic hormones is diabetes mellitus In people with diabetes, the pancreas either produces little or no insulin, or the cells do not respond appropriately to the insulin that is produced. Glucose does not get absorbed by the cells so it builds up in the blood and overflows into the urine. Thus, the body loses its main source of fuel even though the blood contains large amounts of glucose. Diabetes is covered in more detail on pages 411–415 in Chapter 11.

SOLUBLE AND INSOLUBLE FIBER

Fiber can be classified by how well it dissolves in water. Most foods contain both types of fiber: soluble and insoluble. The soluble fibers include gums, mucilages, pectin, and some hemicelluloses. They are generally found around and inside plant cells. The insoluble fibers include cellulose, lignin, resistant starch, and the remaining hemicelluloses. They generally form the structural parts of plants. The amount of fiber in a plant varies among plants and may vary within a species or variety, depending on growing conditions and the plant's maturity at harvest. Lignin is an insoluble fiber but is not technically a carbohydrate.

Foods containing soluble fiber are as follows (see Figure 3-11 on page 98).

- Beans and peas, such as kidney beans, pinto beans, chickpeas, split peas, and lentils.
- Some cereal grains, such as oats and barley.
- Many fruits and vegetables, such as citrus fruits, pears, apples, grapes, brussels sprouts, and carrots. Fruit juices are not good sources of fiber.

Soluble fiber is found both in and around the cells of plants, where it acts to keep a plant stuck together.

Insoluble fiber includes the structural parts of plants, such as skins and the outer layer of the wheat kernel. You have seen insoluble fiber in the skin of whole-kernel corn and in celery strings. It is found in the following foods:

- Wheat bran
- Whole grains, such as whole wheat and brown rice, and products made with whole grains, such as whole-wheat bread
- Many vegetables and fruits
- Beans and peas
- Seeds

Soluble fiber (oats, legumes, fruits, and vegetables) lowers blood cholesterol and traps carbohydrates to slow digestion and absorption of glucose, thereby helping blood glucose levels stay constant. Insoluble fiber (wheat bran, whole grains, fruits, and vegetables) is important for gastrointestinal health and it helps prevent and treat constipation, diverticulosis, and hemorrhoids.

FIGURE 3-11: Food Sources of Soluble Fiber

Food Source	Soluble Fiber (Grams)	Total Fiber (Grams)
CEREAL GRAINS (½ CUP, COOKED)		
Barley	1	4
Oatmeal	1	2
Oat bran	1	3
SEEDS		
Psyllium seeds, ground (1 tablespoon)	5	6
FRUIT (1 MEDIUM FRUIT)		
Apple	1	4
Banana	1	3
Blackberries (½ cup)	1	4
Citrus fruit (orange, grapefruit)	2	2–3
Nectarine	1	2
Peach	1	2
Pear	2	4
Plum	1	1.5
Prunes (¼ cup)	1.5	3
BEANS (½ CUP, COOKED)		
Black beans	2	5.5
Kidney beans	3	6
Lima beans	3.5	6.5
Navy beans	2	6
Northern beans	1.5	5.5
Pinto beans	2	7
LENTILS (½ CUP, COOKED)		
Lentils, yellow, green, or orange	1	8
PEAS (½ CUP, COOKED)		
Chickpeas (garbanzo beans)	1	6
Black-eyed peas	1	5.5
VEGETABLES (½ CUP, COOKED)		
Broccoli	1	1.5
Brussels sprouts	3	4.5
Carrots	1	2.5

Source: National Cholesterol Education Program.

DIGESTION, ABSORPTION, AND METABOLISM OF CARBOHYDRATES

Cooking carbohydrate-containing foods makes them easier to digest. As mentioned previously, starches gelatinize, making them easier to chew, swallow, and digest. Cooking usually breaks down fiber in fruits and vegetables, also making them easier to chew, swallow, and digest.

Before carbohydrates can be absorbed through the villi of the small intestine, they must be broken down into monosaccharides, or one-sugar units. Starch digestion begins in the mouth, where an enzyme, salivary amylase, starts to break down some starch into small polysaccharides and maltose. In the stomach, salivary amylase is inactivated by stomach acid. An intestinal enzyme, pancreatic amylase, continues the job of breaking down polysaccharides to shorter glucose chains. Maltase completes the breakdown of polysaccharides by splitting maltose into two glucose units. Glucose can now be absorbed.

Through the work of three enzymes made in the intestinal wall (see Figure 3-12), all sugars are broken down into the single sugars: glucose, fructose, and galactose. They are then absorbed and enter the bloodstream, which carries them to the liver. In the liver, fructose and galactose are converted to glucose or further metabolized to make glycogen or fat. Glucose will go where it is most needed: into the bloodstream or to be made into glycogen or fat. The hormone insulin makes it possible for glucose to enter body cells, where it is used for energy or stored as glycogen.

Fiber cannot be digested, or broken down into its components by enzymes, and so it continues down to the large intestine to be excreted. Although human enzymes can't digest most fibers, some bacteria in the large intestine digest soluble fiber and produce small fat particles that are absorbed. Soluble fiber slows the digestive process and also slows the absorption of glucose into the blood. This helps regulate your blood sugar level.

MINI-SUMMARY

1. During digestion, various enzymes break down starch and sugars into monosaccharides, which are then absorbed (Figure 3-12).
2. In the liver, fructose and galactose are converted to glucose or further metabolized to make glycogen or fat. Glucose will go where it is most needed: into the bloodsteam or to be made into glycogen or fat.
3. Fiber can't be digested by human enzymes, but soluble fiber can be digested by intestinal bacteria, producing fat fragments that are absorbed. Soluble fiber slows the digestive process and slows the absorption of glucose into the blood.

FIGURE 3-12: Carbohydrate Digestion

Site	Enzyme	Carbohydrate Acted on	Products Formed
Mouth	Salivary amylase	Starch	Small polysaccharides, maltose
Small intestine	Pancreatic amylase	Starch	Small polysaccharides, maltose
	Sucrase	Sucrose	Glucose, fructose
	Lactase	Lactose	Glucose, galactose
	Maltase	Maltose	Glucose

DIETARY RECOMMENDATIONS FOR CARBOHYDRATES

Americans have eaten about the same amount of carbohydrates daily since the early 1900s. What has changed is the type of carbohydrate eaten. Instead of whole grains and vegetables, Americans are eating more carbohydrates in the form of sugars such as baked goods and refined grains such as white bread. While intake of refined carbohydrates has increased, fiber intake has decreased.

The dietary recommendations for carbohydrates can be summarized as follows.

1. The RDA for carbohydrate (Figure 3-13) is 130 grams per day for children (from 1 year old) and adults. It is based on the minimum amount of carbohydrates needed to supply the brain with enough glucose. We normally eat much more than 130 grams of carbohydrate to meet our total energy needs.

2. The Acceptable Macronutrient Distribution Range (AMDR) for carbohydrate is 45 to 65 percent of total kcalories for anyone over 1 year old. For example, if you eat 2000 kcalories/day, you should get about 900 to 1300 kcalories from carbohydrate, which is 225 to 325 grams. At the low end of this range it is difficult to meet the recommendations for fiber intake.

3. The Dietary Reference Intakes recommend that added sugars not exceed 25 percent of total kcalories. Few people would be able to get enough vitamins and minerals or maintain a healthy body weight if they took in 25 percent of their kcalories as added sugars. The World Health Organization suggests limiting added sugars to 10 percent of total kcalories.

FIGURE 3-13: Dietary Reference Intake Values for Carbohydrate and Total Fiber

Age*	RDA Carbohydrate		AI Total Fiber	
	Male	Female	Male	Female
1–3 years	130 g	130 g	19 g	19 g
4–8 years	130 g	130 g	25 g	25 g
9–13 years	130 g	130 g	31 g	26 g
14–18 years	130 g	130 g	38 g	36 g
19–30 years	130 g	130 g	38 g	25 g
31–50 years	130 g	130 g	38 g	25 g
Over 50 years	130 g	130 g	30 g	21 g
Pregnancy				
14–18 years		175 g		28 g
19–50 years		175 g		28 g
Lactation		210 g		29 g

*Note that infants from 1 to 6 months have an AI of 60 g of carbohydrate per day and infants from 7 to 12 months have an AI of 95 g of carbohydrate per day. There is no AI for fiber for infants from 0 to 12 months old due to insufficient scientific evidence.

Source: Adapted with permission from the Dietary References Intakes for Energy, Carbohydrates, Fiber, Fat, Protein, and Amino Acids (Macronutrients). © 2002 by the National Academy of Sciences. Courtesy of the National Academy Press, Washington, D.C.

4. The Dietary Guidelines for Americans recommend three or more servings daily of whole grains.
5. The Adequate Intake for total fiber is based on 14 grams/1000 kcalories. For women, the AI is 25 grams/day from 19 to 50 years and then goes down to 21 grams/day after age 50. For men, the AI is 38 grams/day from 19 to 50 years and then 30 grams/day after age 50. The actual numbers for the AI were derived from data supporting a decreased risk for developing coronary heart disease.

Dietary Fiber Estimator for Adults

Daily calorie needs	Daily dietary fiber needs
1000	14 grams
1200	17 grams
1400	20 grams
1600	22 grams
1800	25 grams
2000	28 grams
2200	31 grams
2400	34 grams
2600	36 grams
2800	39 grams
3000	42 grams

MINI-SUMMARY

1. The RDA for carbohydrate is 130 grams/day for children and adults. It is based on the minimum amount of carbohydrates needed to supply the brain with enough glucose.
2. The AMDR for carbohydrate is 45 to 65 percent of total kcalories.
3. The Dietary Reference Intakes recommend that added sugars not exceed 25 percent of total kcalories. The World Health Organization suggests limiting added sugars to 10 percent of total kcalories.
4. The AI for total fiber is based on 14 grams/1000 kcalories.
5. The Dietary Guidelines for Americans recommend three or more servings daily of whole grains.

ENDOSPERM
In cereal grains, a large center area high in starch.

GERM
In cereal grains, the area of the kernel rich in vitamins and minerals that sprouts when allowed to germinate.

BRAN
In cereal grains, the part that covers the grain and contains much fiber and other nutrients.

GRAINS

INGREDIENT FOCUS: HIGH-FIBER GRAINS AND LEGUMES

Grains are the edible seeds of annual grasses such as wheat, oats, and barley. When a grain is grown for human consumption, it is called a cereal grain. Examples include wheat, corn, rice, rye, barley, and oats, among others. All cereal grains have a large center area high in starch known as the *endosperm*. The endosperm also contains some protein. At one end of the endosperm is the *germ*, the area of the kernel that sprouts when allowed to germinate. The germ is rich in vitamins and minerals and contains some oil. The *bran*, containing much fiber and other nutrients, covers both the endosperm and the germ. The seed contains everything needed to reproduce the plant: The germ is the embryo, the endosperm contains the nutrients for growth, and the bran protects the entire seed (Figure 3-14).

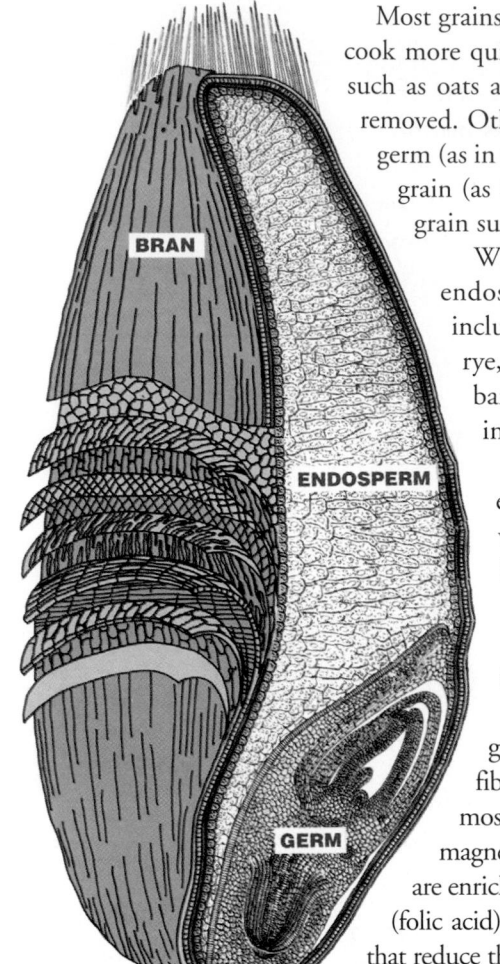

FIGURE 3-14:
A kernel of wheat.
Courtesy of the Wheat
Foods Council.

WHOLE GRAIN
A grain that contains the endosperm, germ, and bran.

REFINED OR MILLED GRAIN
A grain in which the bran and germ are separated (or mostly separated) from the endosperm.

PHYTOCHEMICALS
Minute substances in plants that may reduce the risk of cancer and heart disease when eaten often.

Most grains undergo some type of processing or milling after harvesting to allow them to cook more quickly and easily, make them less chewy, and lengthen their shelf life. Grains such as oats and rice have an outer husk or hull that is tough and inedible, and so it is removed. Other processing steps may include polishing the grain to remove the bran and germ (as in making white flour), cracking the grain (as in cracked wheat), or steaming the grain (as in bulgur) to shorten the cooking time. The process of rolling or grinding a grain such as oats also shortens the cooking time.

Whenever the fiber-rich bran and the vitamin-rich germ are left on the endosperm of a grain, the grain is called a **whole grain**. Examples of whole grains include whole-wheat berries, bulgur (cracked wheat), whole-wheat flour, whole rye, whole oats, rolled oats (including oatmeal), whole cornmeal, whole hulled barley, popcorn, and brown rice. Read the Food Facts in this chapter for more information on a variety of grains.

If the bran and germ are separated (or mostly separated) from the endosperm, the grain is called a **refined or milled grain**. Whereas whole-wheat flour is made from the whole grain, white flour (also called wheat flour) is made only from the endosperm of the wheat kernel. Whole-wheat flour does not stay fresh as long as white flours do. This is due to the presence of the germ, which contains oil. When the oil turns rancid or deteriorates, the flour will turn out a poor-quality product.

When you compare the nutrients in whole grains and refined grains, whole grains are always a far more nutritious choice. They surpass refined grains in their fiber, vitamin, and mineral content. When wheat is refined, over 20 nutrients and most of the fiber are removed. With whole wheat you get more vitamin E, vitamin B_6, magnesium, zinc, potassium, copper, and, of course, fiber. By federal law, refined grains are enriched with five nutrients that are lost in processing: thiamin, riboflavin, niacin, folate (folic acid), and iron. Whole-grain foods also contain **phytochemicals**, substances in plants that reduce the risk of cancer and heart disease when eaten often.

In baking, whole-grain flours produce breads that are denser and chewier. For example, breads made with only whole-wheat flour are heavier and more compact than breads made with only white flour. This is the case because the strands of gluten in the whole-wheat bread are cut by the sharp edges of the bran flakes. Some bakers prefer to use some white flour to strengthen the bread. Many consumers prefer white bread because it lacks the dark color and dense texture of whole-wheat bread, but there are also quite a few health-conscious consumers who appreciate a good-quality whole-grain bread.

It can be difficult to determine from the name of a product whether it is indeed whole grain. Breads simply called "wheat breads" often contain as much as 75 percent enriched white flour with molasses or caramel added to make it look like whole wheat. Here are some tips for finding whole-grain products.

1. Read the ingredient list on the food label. For many whole-grain products, "whole" or "whole grain" will appear before the grain ingredient's name. The whole grain should be the first ingredient listed.
2. Look for the whole-grain health claim on food product labels: "Diets rich in whole-grain foods and other plant foods and low in saturated fat and cholesterol may help reduce the risk of heart disease." Foods that bear the whole-grain health claim must contain 51 percent or more whole grains by weight and be low in fat.
3. Breads labeled as "multi-grain," "seven-grain," or "100% wheat" may not be whole grain. Read the label carefully.

Good whole-grain choices include whole-wheat bread, brown rice, barley, bulgur, whole-wheat pasta, wild rice, oatmeal, barley, whole-kernel corn, and popcorn.

At the store, you will find some whole grains that are stone ground. Grinding by stones is a technique for milling grains that allows them to be processed without the intense heat created by industrialized milling. The high heat causes more loss of flavor as well as nutrients. Stone-ground flours are often more flavorful and nutritious than other types of flours, although more expensive.

It is recommended that you eat at least three servings of whole grains per day. By eating three servings, you can reduce your risk of diabetes, coronary heart disease, and certain cancers. Because the fiber in whole grains make you feel full, they are useful in weight maintenance.

Nutrition

Nutritionally, grains are a powerhouse of nutrients. They are:

- High in vitamins and minerals
- High in starch
- High in fiber (if whole grain)
- Low in fat
- Moderate in protein
- Low or moderate in kcalories

Whole grains are also a source of antioxidants and phytochemicals. Grains are inexpensive and can be quite profitable. Both traditional grains, such as rice, and newer grains, such as quinoa, are being featured more often on menus in side and main dishes.

Culinary Science

Starchy foods in general are not flavorful if eaten raw, and so most are cooked to make them better tasting and more digestible. Starchy foods are used extensively as thickeners in cooking, because starch undergoes a process called *gelatinization* when heated in liquid. When starches gelatinize, the granules absorb water and swell, making the liquid thicken. Around the boiling point, the granules have absorbed a lot of water and burst, letting starch out into the liquid. When this occurs, the liquid quickly becomes still thicker. Gelatinization is a process unique to starches, and so you find starches frequently used as thickeners in soups, sauces, gravies, puddings, and other foods.

The most common starchy thickeners include flour, cornstarch and arrowroot. Others include potato starch and rice flour. Cornstarch and arrowroot will thicken more efficiently than flour since they contain no protein. They have 50 to 100 percent more thickening power than flour and therefore, less of them is needed. They also thicken at a somewhat lower temperature and do not need to be pre-cooked, like roux (a cooked mixture of equal parts fat and flour). However, they do need to be dissolved in liquid first.

> **GELATINIZATION**
> A process in which starches, when heated in liquid, absorb water and swell in size.

CHEF'S TIPS

- See Figure 3-15 for information on flavor, uses, and cooking times for many grains.
- Figures 3-16 to 3-18 showcase many popular grains.
- Figure 3-19 and 3-20 highlight types of rice.
- Grains work very well as main dishes when mixed with each other or with lentils. For example, couscous and wheat berries are attractive, as is barley with quinoa. To either dish you could add lentils, vegetables, and seasonings. Keep in mind that each of the grains must be cooked separately with some bay leaf, onion, and thyme leaf and then strained, cooled, and mixed together with other grains and beans, depending on its application.

(*Text continues on page 108*)

FIGURE 3-15: Grains: Flavors, Uses, and Cooking Times

Grain	Appearance	Flavor	Soaking Required	Cooking Time	Cups Liquid for Cooking	Cups Yield	Uses	Storage
Amaranth	Golden	Sweet, nutlike	No	25 minutes	2½	3½	Hot cereal, pilaf, in baking; can be popped as a snack	Airtight container, in cool place for many months; otherwise 5 months in refrigerator
Barley, pearl	White-tan	Mild, nutty	No (but will reduce cooking time)	35–40 minutes	3	3½	Soups, casseroles, stews, cooked cereals, side dishes, pilafs	Airtight container, 6–9 months at room temperature
Barley, whole hulled	Brownish gray	Nutty, chewy	Yes	60–90 minutes	3	4	Same as above	Airtight container, 1 month at room temperature, 4–5 months in refrigerator
Buckwheat, whole white	Brown-white	Mild	No	20 minutes	2	2½	Side dishes	Airtight container, 1–2 months; better stored in refrigerator
Buckwheat, roasted (kasha)	Brown	Distinct, nutty, chewy	No	10–15 minutes	2	2½	Soups, side dishes, salads, pilaf, stuffing, hot cereal	Same as above
Corn, whole hominy	Yellow or white	Sweet, creamy texture	No	2½–3 hours	2½	3	Soups, stews, casseroles, hot cereal, puddings, baked goods	Airtight container, 1 month at room temperature, 5 months in refrigerator
Corn, hominy grits	Whitish gray	Distinct	No	20–25 minutes	4	3	Hot breakfast cereal	Airtight container, many months at room temperature

Grain	Color	Flavor		Cooking time			Uses	Storage
Millet	Bright gold color, small	Like corn, crunchy	No	30–35 minutes	2	3	Soups, casseroles, meat loaves, porridge, croquettes, pilaf, salads, stuffing, side dishes	Airtight container, 6 months at room temperature
Oats, steel-cut	Off-white	Mild, pleasant	No	45–60 minutes	2	2	Hot cereal	Airtight container, 1 month at room temperature, 6 months in refrigerator
Quinoa	Pale yellow	Nutty	No	12–15 minutes	2	2½	In place of rice	Airtight container, in cool place for 1 month; otherwise 5 months in refrigerator
Rice, regular-milled long grain	White	Mild	No	15–20 minutes	2	3	Side dishes, casseroles, stews, soups, stuffing, salads	Airtight container, many months at room temperature
Rice, regular-milled, medium or short grain	White	Mild	No	20–25 minutes	1½	3	Same as above	Same as above
Rice, parboiled	White	Mild	No	20–25 minutes	2–2½	3–4	Same as above	Same as above
Rice, brown	Tan-brown	Nutty	No	40–50 minutes	2½	4	Same as above	Airtight container, 1 month at room temperature, 6 months in refrigerator

(continued)

FIGURE 3-15: Grains: Flavors, Uses, and Cooking Times (Continued)

Grain	Appearance	Flavor	Soaking Required	Cooking Time	Cups Liquid for Cooking	Cups Yield	Uses	Storage
Rice, wild	Dark brown	Nutty	No (but rinse it)	30–45 minutes	3	3½–4	Side dishes, stuffing, casseroles	Airtight container, many months at room temperature
Rice, basmati	White	Nutty, spicy	No (but rinse it)	25 minutes	1½	3	Side dishes, casseroles	Airtight container, 1 month at room temperature, 6 months in refrigerator
Jasmine rice	White	Aromatic	No	15–20 minutes	2	3	Side dishes, casseroles, stews, soups	Same as above
Texmati rice	White	Nutty	No	15–20 minutes	2	3	Same as above	Same as above
Rye, whole berries	Brown, oval	Distinct rye flavor	No	1½ hours	3	3	Hot cereal, side dishes	Airtight container, 1 month at room temperature, 5 months in refrigerator
Wheat, bulgur	Dark brown	Nutty	No	20–25 minutes	2½	2	Salads, soups, breads, desserts, with rice, meat dishes, in place of rice pilaf, stuffing	Airtight container, 5–6 months in cool place or refrigerator
Wheat, whole berries	Deep brown	Nutty, crunchy	Yes (1 cup to 3½ cups cold water)	1 hour	3	2	Salads, meat loaves, croquettes, breads, side dishes	Airtight container, up to 1 month in cool place, up to 5 months in refrigerator

FIGURE 3-16:
Grains. Top row: wheat berries, black barley, bulgur; bottom row: pearl barley, cracked wheat. Photo by Frank Pronesti.

FIGURE 3-17:
Grains. Top row: steel-cut oats, popcorn, kasha; bottom row: whole oats, rolled oats, rye. Photo by Frank Pronesti.

FIGURE 3-18:
Grains. Top row: millet, red quinoa, kamut; bottom row: spelt, quinoa. Photo by Frank Pronesti.

FIGURE 3-19:
Rice. Top row: converted white rice, short-grain brown rice, wild rice; bottom row: basmati rice, long-grain brown rice, Arborio rice.
Photo by Frank Pronesti.

FIGURE 3-20:
Rice. Top row: purple Thai rice, Chinese black rice, jasmine rice; bottom row: bamboo rice, sushi rice, Himalayan red rice.
Photo by Frank Pronesti.

- Orzo works well with toasted barley or quinoa and a mixture of roasted vegetables and fresh herbs.
- Rice and beans is a very popular and versatile dish using grains and legumes. For appearance, mix purple rice with a variety of beans or wild rice with cranberry beans and fava beans. Each of these mixtures can be seasoned in a number of ways. A balanced dressing can be used for a cold salad as well as a hot accompaniment.
- Don't forget to serve whole-grain cereals hot and cold, from simple steel-cut oatmeal to cold favorites such as corn flakes. Also, dress them up with berries, fruits, and spices. Steel-cut oats are whole-grain groats (the inner portion of the oat kernel) that have been cut into two or three pieces using steel discs. Gold in color, they look like little pieces of rice. Rolled oats are flakes that have been steamed, rolled, and toasted. Due to this process, they have lost some taste and texture.

LEGUMES

Legumes include all sorts of dried beans, peas, and lentils. Dried beans are among the oldest foods and are an important staple for millions of people in other parts of the world. Beans were once considered to be worth their weight in gold—the jeweler's carat owes its origin to a pealike bean on the east coast of Africa.

Nutrition

From a nutritional point of view, legumes are a hit. They are:

- High in complex carbohydrates
- High in fiber
- Low in fat (only a trace, except in a couple of cases)
- Cholesterol-free
- A good source of vitamins and minerals
- Low in sodium

Besides being so nutritious, they are very cost-effective.

CHEF'S TIPS

- See Figure 3-21 for information on flavor, uses, and cooking times for many legumes.
- Figures 3-22 and 3-23 highlight many popular legumes.
- When choosing legumes for a dish, think color and flavor. Make sure the colors you pick will look good when the dish is complete. Also think of other ingredients you will use for flavor. In a salad, for example, black-eyed peas (black and white) go well with flageolets and red adzuki beans. To add a little more color and develop the flavor, you might add chopped tomatoes, fresh cilantro (Chinese parsley), and haricots verts (green beans) or fresh corn.
- Bigger beans, such as gigante white beans, hold their shape well and lend a hearty flavor to stews, ragouts, and salads.
- Chickpeas can be puréed, as in hummus, and used as a dip, a spread, or a sandwich filling; layered with grilled vegetables and mushrooms; or as a filling for pasta, crêpes, or twice-baked potatoes. They can be soaked in changed water for three to four days in the refrigerator, seasoned and puréed, and then pressed into a half-sheet pan and cut into french fry shapes. You can coat them in yogurt and seasoned crumbs and bake at a high temperature as a crisp substitute for french fries.
- To enhance their flavor, cook several types of beans together, such as cranberry, turtle, and white beans, in stock flavored with herbs, shallots, onions, carrots, and celery. Presoaking dried beans helps remove some of the complex sugars they contain that can cause flatulence (gas).
- A number of dried beans are also available fresh: cannellini beans, cranberry beans, fava beans, black-eyed peas, flageolets, lima beans, mung beans, and soybeans. They tend to be expensive but are excellent products. They are plumper and have a fresher flavor than dry beans that you rehydrate.
- Use whole lentils, such as black or French green lentils, in grain dishes or salads because whole lentils hold their shape better. Use split lentils, such as brown, red, or yellow lentils, in soups, where they help thicken the liquid and shape is not as important.
- Pasta and beans work well together, as in pasta e fagioli, a rich Italian vegetable soup with pasta and beans. There are many variations of this classic dish due to its nourishing value and spectacular taste. Many other ethnic groups have similar traditional dishes associated with pasta and beans or rice and beans for their versatility of preparations, health benefits, cost factors, history, and utilization of a region's products.

FIGURE 3-21: Legumes: Flavors, Uses, and Cooking Times

Bean, Pea, or Lentil	Size/Shape/Color	Flavor	Soaking Required	Cooking Time	Cups Liquid for Cooking	Cups Yield*	Uses
Adzuki beans	Small, reddish brown	Nutty, sweet	Yes	1–1½ hours	3	2	Asian cooking
Anasazi beans	Kidney-shaped, white with maroon	Rich, meaty	Yes	2 hours	3	2	Chili and other Mexican dishes
Black beans (turtle beans)	Small, pea-shaped, black	Full, mellow	Yes	1½ hours	4	2	Mediterranean cuisine, soups (black bean soup), chilis, salads, with rice
Black-eyed peas (cowpeas, black-eyed beans)	Small, oval, white with black spot creamy	Earthy, absorb other flavors	No	50–60 minutes	3	2	Casseroles, with rice, with pork, Southern dishes
Chickpeas (garbanzo beans, ceci)	Round, tan, large	Nutty	Yes	2½ hours	4	4	Salads, soups, casseroles, hors d'oeuvres, hummus and other Middle East dishes
Fava beans, whole	Large, round, flat, off-white or tan	Full	Yes	3 hours	2½	4	Soups, casseroles, salads
Great Northern beans	Large, oval, white	Mild	Yes	1½ hours	3½	2	Soups, casseroles, baked beans, and mixing with other varieties
Kidney beans	Large, kidney-shaped red or white (red is much more common)	Rich, meaty, sweet	Yes	1–1½ hours	3	2	Chili, casseroles, salads, soups, a favorite in Mexican and Italian cooking

	Description	Flavor	Soak	Cooking time		Yield*	Uses
Lentils	Small, flat, disk-shaped, green, red, or brown, split or whole	Mild, earthy	No	30–45 minutes	2	2¼	Soups, stews, salads, casseroles, stuffing, sandwiches, spreads, with rice
Lima beans	Flat, oval, cream or greenish, large or baby size	Full (large) mild (baby)	Yes	1½ hours (large); 1 hour (baby)	2	1¼	Soups, casseroles, side dishes
Navy beans (pea beans)	Small to medium, round to oval, white	Mild	Yes	1½ hours	3	2	Baked beans, soups, salads, side dishes, casseroles
Peas, split	Small, flat on one side, green or yellow	Rich, earthy	No	30 minutes	3	2¼	Soups, casseroles
Peas, whole	Small to medium, round, yellow or green	Rich, earthy	Yes	40 minutes	3	2¼	Soups, casseroles, Scandinavian dishes
Pinto beans	Medium, kidney-shaped, pinkish brown	Rich, meaty	Yes	1½ hours	3	2	A favorite for chili, refried beans, other Mexican cooking
Pink beans	Medium, oval, pinkish brown	Rich, meaty	Yes	1 hour	3	2	Popular in barbecue-style dishes
Soybeans	Medium, oval-round, creamy yellow	Distinctive	Yes	3½ hours or more	3	2	Soups, stews, casseroles

*From 1 cup of uncooked beans, peas, or lentils.

FIGURE 3-22:
Beans. Top row: peruano beans, split baby garbanzo beans, black garbanzo beans; bottom row: cranberry beans, runner cannellini beans, anasazi beans.

FIGURE 3-23:
Beans. Top row: chestnut lima beans, French green lentils, scarlet runner beans; bottom row: black calypso beans, petite crimson lentils, butterscotch calypso beans.
Photo by Frank Pronesti.

CHECK-OUT QUIZ

1. Match the food below with the nutrient(s) it is rich in.

Food	Nutrient
White bread	Added sugars
Whole-wheat bread	Natural sugars
Apple juice	Fiber
Baked beans	Starch
Milk	
Bran flakes	
Sugar-frosted whole oats	
Cola drink	
Broccoli	

2. Honey is better for you than sugar.

   a. True
   b. False

3. Carbohydrates prevent protein from being burned for energy so that protein can be used to build and repair the body.

   a. True
   b. False

4. Maltose is made up of galactose and glucose.

   a. True
   b. False

5. Glycemic response refers to how quickly and how high your blood sugar rises after eating.

   a. True
   b. False

6. Eating too much sugar can cause diabetes and hyperactivity in children.

   a. True
   b. False

7. Sugar and starch contribute to dental decay.

   a. True
   b. False

8. Wheat bread is an example of a whole-grain product.

   a. True
   b. False

9. Because insoluble fiber aids digestion and adds bulk to stool, it hastens the passage of fecal material through the gut, thus helping to prevent or alleviate constipation.

   a. True
   b. False

10. Legumes are low in fiber and sodium.

    a. True
    b. False

ACTIVITIES AND APPLICATIONS

1. Carbohydrate Basics

   Use iProfile to complete the chart below. Find out how many grams of sugar, starch, and fiber are found in each food. Which food contains the most sugar? Which food has the most starch? Which food is highest in fiber?

Food	Sugar (grams)	Starch (grams)	Fiber (grams)
1. Hamburger meat, 3 ounces cooked	_____	_____	_____
2. Chicken wing, 1 wing	_____	_____	_____
3. Flounder, 3 ounces cooked	_____	_____	_____
4. Boiled egg, 1	_____	_____	_____
5. American cheese, 1 oz	_____	_____	_____
6. Sour cream, 1 tablespoon	_____	_____	_____
7. White bread, 1 slice	_____	_____	_____
8. Whole-wheat bread, 1 slice	_____	_____	_____
9. Chocolate cake	_____	_____	_____
10. Macaroni, $\frac{1}{2}$ cup	_____	_____	_____
11. Brown rice, $\frac{1}{2}$ cup	_____	_____	_____
12. Split peas, $\frac{1}{4}$ cup	_____	_____	_____
13. Peanuts, $\frac{1}{2}$ ounce	_____	_____	_____
14. Fresh orange, 1 medium	_____	_____	_____
15. Broccoli, $\frac{1}{2}$ cup cooked	_____	_____	_____

2. Self-Assessment

   List how many servings of the following foods you normally eat daily.

Added Sugars	Number of Servings
Sugar in coffee or tea	_____
Sweetened beverages	_____

Sweetened breakfast cereals _____

Candy _____

Commercially made baked goods, including
cakes, pies, cookies, and doughnuts _____

Jam, jelly, pancake syrup _____

Complex Carbohydrates	Number of Servings
Whole-grain breads and rolls	_____
Ready-to-eat and cooked cereals	_____
Pasta, rice, other grains	_____
Dried beans and peas	_____
Potatoes	_____
Fruits	_____
Vegetables	_____

How do you rate? Do you get at least three servings of whole grains a day? Do you get a daily serving of dried beans and peas? Do you get at least five servings per day of fruits and vegetables combined? Compare the number of servings you have daily of foods high in refined sugars and those high in complex carbohydrates. You should not be choosing too many foods from the added sugars part of the chart until you eat enough servings of grains, fruits, and vegetables. The idea is to push complex-carbohydrate intake and minimize added sugars.

3. **Whole Grain or Refined Grain?**

Read the following ingredient labels for breads. Which one is white bread, and which one is whole-grain bread?

A. Made from: Unbromated unbleached enriched wheat flour, corn syrup, partially hydrogenated soybean oil, molasses, salt, yeast, raisin juice concentrate, potato flour, wheat gluten, honey, vinegar, mono- and diglycerides, cultured corn syrup, unbleached wheat flour, xanthan gum, and soy lecithin.

B. Made from: Stone-ground whole-wheat flour, water, high-fructose corn syrup, wheat gluten, yeast, honey, salt, molasses, partially hydrogenated soybean oil, raisin syrup, soy lecithin, mono- and diglycerides.

4. **How Many Teaspoons of Sugar?**

A. One teaspoon of sugar weighs 4 grams. Determine how many teaspoons of sugar are in each of the following foods, as described on their nutrition labels. Which food contains more sugar? Which food contains more fiber?

Nutrition Facts		Nutrition Facts
Amount per serving		Amount per serving
Calories 230		Calories 140
Calories from Fat 140		Calories from Fat 20
Total Fat 16 g		Total Fat 2.5 g
Saturated 6 g		Saturated 0.5 g
Polyunsaturated 1 g		Polyunsaturated 1.0 g
Monounsaturated 7 g		Monounsaturated 0.5 g
Cholesterol 74 g		Cholesterol 0 g

Sodium 180 mg	Sodium 180 mg
Total Carbohydrate 28 g	Total Carbohydrate 21 g
Dietary Fiber 0 g	Dietary Fiber 5 g
Sugars 24 g	Sugars 4 g
Protein 21 g	Protein 8 g

B. In a group of two people, determine the number of teaspoons of sugar in a soft drink or another beverage by using the number of grams of sugar on the Nutrition Facts label. Next, measure out the amount of sugar and place it in a clear glass next to the beverage. Check out how much sugar is in the drinks your other classmates analyzed.

5. **Artificial Sweetener Sleuth**

Check your refrigerator and cupboards to see what kind of foods, and how many, contain artificial sweeteners with names such as Equal, aspartame, saccharin, acesulfame potassium, Sunette, Sweet One, sucralose, Splenda, and neotame. Make a list.

6. **Baking with Alternative Sweeteners**

Use the websites for Sweet 'N Low and Splenda to get some recipes using alternative sweeteners.

www.sweetnlow.com

www.splenda.com

The first website is for the company that manufactures Sweet 'N Low, the artificial sweetener containing saccharin. The second website is for the company that makes Splenda, the artificial sweetener containing sucralose.

On their home pages, click on "Recipes" and choose a recipe. How are these baking recipes different from recipes that use sugar? If possible, make a recipe, such as brownies, using one of these alternative sweeteners and then compare it in taste, texture, and appearance to the same product made with sugar.

7. Using iProfile, click on "Estimating Portion Sizes" at the top of the page. Next, click on "How Many Servings Does My Cereal Bowl Hold? "What Do You Put on Your Bagel?" "What Does a Cup of Pasta or Rice Look Like?" and "How Much Added Sugar Do I Use?"

 # NUTRITION WEB EXPLORER

Joslin Diabetes Center   www.joslin.org/Beginners_guide_2854.asp

Joslin Diabetes Center is an excellent site to learn almost anything about diabetes. Read "Carbohydrate Counting 101" and then complete the exercise "Meal Planning."

National Institutes of Health   www.nlm.nih.gov/medlineplus/hypoglycemia.html
on Hypoglycemia

On this government website, click on "Tutorials: Hypoglycemia" to learn more about this topic.

National Institutes of Health on Lactose Intolerance

http://digestive.niddk.nih.gov/ddiseases/
pubs/lactoseintolerance/

Use this informative site to learn more about the causes and treatment of lactose intolerance. Find out how many grams of lactose are in reduced-fat milk and compare that to yogurt.

Whole Grains Council

www.wholegrainscouncil.org/
whole-grains-101/whole-grains-a-to-z

Read about a huge variety of whole grains. Write up a description of three whole grains that are new to you.

FOOD FACTS: FOODS AND THE GLYCEMIC INDEX

Carbohydrate foods can be grouped according to whether they produce a high, moderate, or low *glycemic response*. Glycemic response refers to how quickly and how high your blood sugar rises after eating. Any number of factors influence how high your blood glucose will rise, such as the amount of carbohydrate eaten, the type of sugar or starch, and the presence of fat and other substances that slow down digestion.

A low glycemic response (meaning that your blood sugar level rises slowly and modestly) is preferable to a high glycemic response (your blood sugar level rises quickly and high). Eating mainly foods with a low glycemic response is definitely important for people with diabetes and seems to decrease the risk of heart disease and type 2 diabetes, as well as may enhance weight management.

The concept of the *glycemic index* was created to identify how selected foods affect your blood glucose level. The higher the glycemic index is, the more the food increases your blood glucose level. Glucose is assigned a value of 100, and white bread is assigned a value of 70 (Figure 3-24). Of course, one shortcoming of the glycemic index is that we usually eat more than one food at a time, and so the numbers are not as meaningful. Also, we may think that a food has a definitive glycemic index, when the glycemic index will vary depending on how the food is processed, stored, cut, and cooked. In any case, foods such as fruits, vegetables, legumes, milk, and whole grains generally have low glycemic index values and are healthy, nutritious choices.

FIGURE 3-24: Glycemic Index of Selected Foods

Glucose
Corn flakes
Waffles, french-fried potatoes, jelly beans
Bagel, white bread
White sugar, cantaloupe
Raisins, tortilla chips, cola soda, ice cream, pizza
Rye bread
Orange juice
Fresh orange, peas, carrots
Fresh peach, old-fashioned oatmeal, apple juice
White rice, spaghetti, apple, pear, tomato soup
Skim milk, low-fat yogurt
Kidney beans
Grapefruit
Soybeans
Peanuts

The introduction of diet soda in the 1950s sparked the widespread use of artificial sweeteners, substitutes for sugar that provide no, or almost no, kcalories. If you drink diet soda, look at the food label and see which artificial sweeteners are present. As of 2005, five artificial sweeteners had been approved for use by the Food and Drug Administration: saccharin, aspartame, acesulfame potassium, sucralose, and neotame. Because artificial sweeteners are considered food additives, the FDA requires that they be tested for safety before going on the market. Besides offering no kcalories, artificial sweeteners are beneficial because they do not cause tooth decay or force insulin levels to rise as do added sugars such as high-fructose corn syrup.

The FDA uses the concept of an Acceptable Daily Intake (ADI) for many food additives, including these artificial sweeteners. The ADI represents an intake level that, if maintained each day throughout a person's lifetime, would be considered safe by a wide margin. For example, the ADI for aspartame is 50 milligrams per kilogram of body weight per day. To take in the ADI for a 150-pound adult, someone would have to drink twenty 12-ounce cans of diet soft drinks daily.

Approved Artificial Sweeteners

SACCHARIN

Saccharin, discovered in 1879, has been consumed by Americans for more than 100 years. Its use in foods increased slowly until the two world wars, when its use increased dramatically due to sugar shortages. Saccharin is about 200 to 700 times sweeter than sucrose and is excreted unchanged directly into the urine. It is approved for use at specific maximum amounts in foods and beverages and as a tabletop sweetener. Known as Sweet 'N Low or Sugar Twin, it is sold in liquid, tablet, packet, and bulk form. Because saccharin leaves some consumers with an aftertaste, it is frequently combined with other artificial sweeteners, such as aspartame.

In 1977, the Food and Drug Administration proposed a ban on its use in foods and allowed its sale as a tabletop sweetener only as an over-the-counter drug. This proposal was based on studies that showed the development of urinary bladder cancer in rats fed the equivalent of hundreds of cans of diet soft drinks a day. The surge of public protest against this proposal (there were no other alternative sweeteners available at that time) led Congress to postpone the ban. In 2000, the National Toxicology Program decided that saccharin should no longer be considered cancer-causing. In 2001, the U.S. Congress repealed the warning labels that had been required on saccharin-containing products.

ASPARTAME

In 1965, aspartame was discovered accidentally. Aspartame is made by joining two protein components, aspartic acid and phenylalanine, and a small amount of methanol. Aspartic acid and phenylalanine are building blocks of protein. Methanol is found naturally in the body and in many foods, such as fruit and vegetable juices. In the intestinal tract, aspartame is broken down into its three components, which are metabolized in the same way as if they had come from food. Aspartame contains 4 kcalories per gram, but so little of it is needed that the kcalorie content is not significant.

Aspartame is approximately 160 to 220 times sweeter than sucrose and has an acceptable flavor with no bitter aftertaste. It is marketed under the brand names NutraSweet and Equal. Aspartame is approved as a general-purpose sweetener and is found in diet sodas, cocoa mixes, pudding and gelatin mixes, fruit spreads and toppings, and other foods. If you drink diet soft drinks, chances are they are sweetened with aspartame. Fountain-made diet soft drinks are more commonly sweetened with a blend of aspartame and saccharin, because saccharin helps provide increased stability. Aspartame may become less sweet after prolonged heating. For stovetop cooking, it is best to add aspartame at the end of cooking or after removing the food from the heat.

The safety of aspartame provoked concerns in the past, and many studies have been done on it. Recent reviews of studies confirm that aspartame consumption is safe over the long term and is not associated with serious health effects.

The only individuals for whom aspartame is a known health hazard are those who have the disease phenylketonuria (PKU), because they are unable to metabolize phenylalanine. For this reason, any product containing aspartame carries a warning label, "Phenylketonurics: Contains phenylalanine." Some other people may also be sensitive to aspartame and need to limit their intake.

ACESULFAME-POTASSIUM

In 1988, the FDA first approved acesulfame-potassium for use in some foods and as a tabletop sweetener. Acesulfame-potassium is now approved as a general-purpose sweetener. Its name is often abbreviated as Acesulfame-K, because K is the chemical symbol for potassium. Marketed under the brand names Sunett and Sweet One, it is about as sweet as aspartame but is more stable and can be used in baking and cooking. Coca-Cola mixes acesulfame-K with aspartame to sweeten one of its diet sodas. Its taste is clean and sweet, with no aftertaste in most products. Acesulfame-K passes through the digestive tract unchanged. It is used in over 50 countries.

SUCRALOSE

Sucralose is the only artificial sweetener made from table sugar. The FDA approved sucralose in 1999 for use as general-purpose sweetener. Sucralose is 600 times sweeter than sugar and actually tastes similar to sugar. Sucralose cannot be digested, and so it adds no kcalories to food. Sucralose has exceptional stability and retains its sweetness over a wide range of conditions, including heat. Sucralose can be used in baking and cooking. It is even used in sweet microwave popcorn.

Sucralose is available as an ingredient for use in a broad range of foods and beverages under the name Splenda Brand Sweetener. Currently, a range of products sweetened with Splenda are on supermarket shelves, including diet soft drinks, low-calorie fruit drinks, maple syrup, and applesauce.

Like aspartame, sucralose is available in a granular form that pours and measures exactly like sugar. Maltodextrin, a starchy powder, is used to give it bulk, and the resulting product has no kcalories.

NEOTAME

In 2002, the Food and Drug Administration approved neotame for use as a general-purpose sweetener. Neotame is a high-intensity sweetener that is manufactured by the same company that first manufactured aspartame. The sweetener is structurally similar to aspartame but is much sweeter and does not cause any problems for people with phenylketonuria. Once ingested, neotame is quickly metabolized and completely eliminated.

Depending on its food application,

neotame is about 7000 to 13,000 times sweeter than sugar. Its strength varies depending on the application and the amount of sweetness needed. It is a white crystalline powder that is heat-stable and can be used as a tabletop sweetener as well as in cooking. Neotame has a clean, sweet taste in foods.

Blending and Cooking with Artificial Sweeteners

Food scientists have discovered that blending certain artificial sweeteners with one another results in products that are much sweeter than expected. This is beneficial because you can reduce the amount of artificial sweeteners used and improve the taste profile. Some artificial sweeteners also work well with traditional caloric sweeteners, such as high-fructose corn syrup and sucralose. When combined with traditional sweeteners, artificial sweeteners not only can improve taste but can also decrease kcalories.

The heating properties of artificial sweeteners are an important consideration if you intend to use them for cooking or baking. Saccharin, sucralose, and acesulfame-K are heat stable. Aspartame will break down during cooking.

Figure 3-25 summarizes information on approved artificial sweeteners.

Other Sweeteners

The artificial sweeteners alitame and cyclamate are awaiting approval from the FDA. Alitame (the brand name is Aclame) is a sweetener made from amino acids (parts of proteins). It is 2000 times sweeter than sucrose and has the potential to be useful in many areas, such as baked goods and as a tabletop sweetener. It is approved for use in Mexico, Australia, and China.

Discovered accidentally in 1937, cyclamate was introduced into beverages and foods in the early 1950s. By the 1960s it dominated the artificial sweetener market. It

is 30 times sweeter than sucrose and is not metabolized by most people. In 1969, cyclamate was banned after studies showed that large doses were associated with an increased risk of bladder cancer. Cyclamate is still banned in the United States but is approved and used in more than 50 other countries worldwide. Cyclamate is again under consideration by the Food and Drug Administration for use in specific products, such as tabletop sweeteners and beverages. It is stable at hot and cold temperatures and has no aftertaste.

Stevioside is a naturally sweet extract from the leaves of the stevia bush found in South America. The extract is 200 times sweeter than sucrose and is used in several countries, such as Japan. In the small amounts in which it is normally used, stevia provides zero kcalories. At the end of 2008, the FDA gave the green light for rebiana A, a sweetener made from the stevia leaf, to be used in food and beverages. *Rebiana* is extracted and purified from stevia leaves using a process similar to that used to extract other natural flavorings, such as vanilla and cinnamon. The finished product contains nothing artificial. The FDA granted rebiana A, at 95 percent purity or higher, GRAS (generally recognized as safe) status as a general purpose sweetener. It is now available as a tabletop sweetener under names such as Truvia™. Beverage companies will be using it in certain drinks such as Sprite Green.

Sugar Replacers

Sugar replacers, also called polyols, are a group of carbohydrates that are sweet and occur naturally in plants. Scientists call them sugar alcohols because part of their structure resembles sugar and part resembles alcohols; however, these sugar-free sweeteners are neither sugars nor alcohols. Many sugar replacers, such as xylitol, have been used for years in products such as sugar-free gums, candy, and fruit spreads.

FIGURE 3-25: Approved Artificial Sweeteners

Artificial Sweetener	Brand Name(s)	Sample Foods Used in	Kcalories/ Gram	Sweetness Compared to Sugar
Saccharin	Sweet 'N Low Sugar Twin* Hermesetas Original (in Europe)	Fountain diet sodas	0	200 to 700 times as sweet

Description

- Not metabolized by body.
- Some consumers report an aftertaste.
- Approved by FDA at specific maximum amounts for use in beverages, tabletop sweeteners, and foods.
- Can be blended with other sweeteners to increase sweetness and decrease kcalories.
- Available in liquid, tablet, packet, and bulk (granular) form. Sweet 'N Low Brown is a sugar substitute for brown sugar.

Aspartame	NutraSweet Equal Canderel (in Europe)	Diet sodas, diet drink mixes, cocoa mixes, pudding and gelatin mixes, fruit spreads and toppings	4[†]	160 to 220 times as sweet

Description

- Made of two protein components and a small amount of methanol (found naturally in many foods).
- Foods containing aspartame must have a warning label because it contains phenylalanine, which a small number of people can't tolerate.
- Loses its sweet flavor with prolonged heating. Add at end of stovetop cooking. Can often be used successfully in baking.
- Can be blended with other sweeteners to increase sweetness and decrease kcalories.
- Available in tablets, packets, and bulk (granular) form. Also available: Equal Spoonful, which measures like sugar and has zero kcalories, and Equal Sugar Lite, which contains sugar and aspartame and provides half the kcalories and carbohydrates of regular sugar.

Acesulfame Potassium (Acesulfame-K)	Sunett Sweet One	Diet sodas	0	200 times as sweet

Description

- Passes through the digestive system unchanged.
- Can be used in baking and cooking.
- Can be blended with other sweeteners to increase sweetness and decrease kcalories.
- Available in liquid, tablets, packets, and bulk (granular) form.

Sucralose	Splenda	Diet sodas, low-calorie fruit drinks, maple syrup, applesauce	0	600 times as sweet

Description

- Cannot be digested.
- Stays sweet during cooking and baking.
- Can be blended with other sweeteners to increase sweetness and decrease kcalories.

(continued)

FIGURE 3-25: (Continued)

Artificial Sweetener	Brand Name(s)	Sample Foods Used in	Kcalories/ Gram	Sweetness Compared to Sugar
• Available in packets and bulk (granular) form. Splenda Granular measures like sugar and has 0 kcalories. Another product, called Splenda Blend for Baking, contains sugar and sucralose. One cup of sugar in a recipe can be replaced by ½ cup of Splenda Blend for Baking.				
Neotame	Not yet available	Products using neotame in the United States are still being developed	0	7000 to 13,000 times as sweet

Description

• Structurally similar to aspartame.

• Can be used in cooking and baking.

• Can be blended with other sweeteners to increase sweetness and decrease kcalories.

*In Canada, Sugar Twin contains cyclamate, not saccharin.

†Because so little aspartame is used, the kcalorie content is very close to zero.

Figure 3-26 lists the eight sugar replacers currently approved for foods in the United States, and tells you how sweet each one is compared to sugar and also how many kcalories each one provides. Each sugar replacer has different characteristics.

Sugar replacers, or polyols, have the following benefits:

• They don't provide as many kcalories as sugar. The average kcalories per gram is 2, compared with 4 kcalories/gram from sugar.

• They don't promote tooth decay.

• They taste sweet, though not as sweet as sugar.

• They cause smaller increases in blood glucose and insulin levels than sugar.

Sugar replacers don't affect your blood glucose level the way sugar does because they are absorbed more slowly and incompletely from the small intestine. The portion that is absorbed is metabolized by processes that require almost no insulin.

The portion that is not absorbed is broken down by bacteria in the large intestine, which can sometimes cause abdominal gas, discomfort, and diarrhea in some individuals. These symptoms are more likely if large amounts of sugar replacers have been consumed. FDA regulations require food labels to use the following warning if reasonable consumption of the food could result in undesirable symptoms: "Excess consumption may have a laxative effect." Most people who experience discomfort can eat a small amount of foods with sugar replacers and then slowly increase these foods in the diet. Not all sugar replacers are equally capable of causing discomfort. For example, sorbitol and mannitol are more likely to cause gas and diarrhea than maltitol is.

You find sugar replacers in a wide assortment of foods: chewing gums, chocolate, candies, frozen desserts such as ice cream, baked goods, salad dressings, beverages, and many foods designed to be lower in kcalories, carbohydrates, and/or fat.

Sugar replacers add bulk and texture and improve the mouth feel of foods. They not only have been used successfully to replace sugar but can replace fat as well. Sugar replacers also enhance the flavor profile, retain moisture in foods, and provide a cooling effect or taste.

Information about sugar replacers is found in two places on the food label. First, the ingredient list must show the name of each sugar replacer the product contains.

Second, the Nutrition Facts panel shows the number of grams of total carbohydrates in a food, which includes the number of grams of any sugar replacers used. A manufacturer is required to give the number of grams of sugar replacers in a serving only if a claim such as "sugar-free" or "no added sugar" appears on the label. FDA regulations specify that the name of the specific sugar replacer must appear only if one sugar replacer is used. If more than one sugar replacer is used, the term "sugar alcohols" must be used.

It is important to have alternatives to sugar in the marketplace, especially for people with diabetes. Although artificial sweeteners are often used as part of a weight control program, research has provided conflicting indications about whether obesity may be a bigger problem among people who use artificial sweeteners. In one study, rats getting artificially sweetened yogurt gained more weight and ate more food than rats eating yogurt sweetened with glucose. The key to losing weight is still to reduce kcalories and increase exercise. For some people, using artificial sweeteners can help them reduce kcalories.

FIGURE 3-26: Sugar Replacers (Polyols)

Sugar Replacer	Kcalories/ Gram	Uses	Description*
Mannitol	1.6	Chewing gum, powdered foods, chocolate coatings	50 to 70 percent as sweet as sugar May cause a laxative effect when 20 grams or more is consumed Does not absorb moisture, and so it works well as a dusting powder for chewing gum so that the gum doesn't stick to the wrapper.
Sorbitol	2.6	Candies, chewing gum, baked goods, frozen desserts	60 percent as sweet as sugar, gum, baked goods May cause a laxative effect when 50 frozen dessert grams or more is consumed Cool, pleasant taste
Xylitol	2.4	Chewing gum, candy	As sweet as sugar
Erythritol	0.2	Beverages, chewing gum, candy, baked goods	Pleasant taste Newest polyol Very heat-stable Much less of a laxative effect than other polyols Works well with other sweeteners to improve flavor and body
Isomalt	2.0	Candies, toffee, fudge, wafers	45 to 60 percent as sweet as sugar Used to add bulk and sweetness to foods Very heat-stable Works well with other sweeteners to improve flavor
Lactitol	2.0	Chocolate, candies, cookies and cakes, frozen dairy desserts	30 to 40 percent as sweet as sugar Mild sweetness with no aftertaste Used to add bulk and sweetness to foods Works well with artificial sweeteners
Maltitol	2.1	No-sugar-added ice cream, low-carb bagels, candy, chewing gum, chocolate, baked goods	90 percent as sweet as sugar Used to add bulk and sweetness to foods
Hydrogenated starch hydrolysates	3.0	Candy, baked goods	25 to 50 percent as sweet as sugar Used as bulk sweetener in low-calorie foods; performs other functions in foods as well Can mask unpleasant off-flavors Blends well with flavors Works well with other sweeteners

*All sugar replacers have the following characteristics:
They occur naturally.
They don't provide as many kcalories as sugar. The average kcalories per gram is 2, compared with 4 kcalories per gram from sugar.
They don't promote tooth decay.
They cause smaller increases in blood glucose and insulin levels than sugar does.

Lipids: Fats and Oils

The word fat is truly all-purpose. We use it to refer to the excess pounds we carry, the blood component that is associated with heart disease, and the greasy foods in our diet that we feel we ought to cut out. To be more precise about the nature of fat, we need to look at fat in more depth.

To begin, *lipid* is the chemical name for a group of compounds that includes fats, oils, cholesterol, and lecithin. Fats and oils are the most abundant lipids in nature and are found in both plants and animals. A lipid is customarily called a *fat* if it is a solid at room temperature, and it is called an *oil* if it is a liquid at that temperature. Lipids obtained from animal sources are usually solids, such as butter and beef fat, whereas oils are generally of plant origin. Therefore, we commonly speak of animal fats and vegetable oils, but we also use the word *fat* to refer to both fats and oils, which is what we will do in this chapter.

Like carbohydrates, lipids are made of carbon, hydrogen, and oxygen. Unlike most carbohydrates, lipids are not long chains of repeating units. Most of the lipids in foods (over 90 percent), and also in the human body, are in the form of *triglycerides*. Therefore, when we talk about fat in food or in the body, we are really talking about triglycerides. Besides triglycerides, there are two other classes of lipids: sterols (such as cholesterol), and phospholipids (such as lecithin in eggs). This chapter will help you to:

- Describe lipids and list their functions in foods and in the body
- Describe the relationship between triglycerides and fatty acids
- Define saturated, monounsaturated, and polyunsaturated fats and list foods in which each one is found
- Describe trans fatty acids and give examples of foods in which they are found
- Identify the two essential fatty acids, list their functions in the body, and give examples of foods in which they are found
- Define cholesterol and lecithin, list their functions in the body, identify where they are found in the body, and give examples of foods in which they are found
- Define rancidity
- Describe how fats are digested, absorbed, and metabolized
- Discuss the relationship between lipids and health conditions such as heart disease and cancer
- State recommendations for dietary intake of fat, saturated fat, trans fat, monounsaturated fat, polyunsaturated fat, and cholesterol
- Distinguish between the percentage of fat by weight and the percentage of kcalories from fat
- Calculate the percentage of kcalories from fat for a food item
- Discuss the nutrition and uses of milk, dairy products, and eggs on the menu

LIPIDS

A group of fatty substances, including triglycerides and cholesterol, that are soluble in fat, not water, and that provide a rich source of energy and structure to cells.

FAT

A lipid that is solid at room temperature.

OIL

A lipid that is usually liquid at room temperature.

TRIGLYCERIDE

The major form of lipid in food and in the body; it is made of three fatty acids attached to a glycerol backbone.

FUNCTIONS OF LIPIDS

Fats have many vital purposes in the body, where they account for about 13 to 30 percent or more of a person's weight. Fat is an essential part of all cells. At least 50 percent of your fat stores are located under the skin, where fat provides insulation (because fat doesn't conduct heat well), optimum body temperature in cold weather, and a cushion around critical organs (like shock absorbers).

Most cells store only small amounts of fat, but specific cells, called fat cells or *adipose cells*, can store loads of fat and actually increase 20 times in weight. The number of fat cells increases most during late childhood and early adolescence. It can also increase during adulthood, when you eat more kcalories than you expend. Fat cells are a compact way to store lots of energy. Remember that 1 gram of fat yields 9 kcalories, compared with 4 kcalories for 1 gram of carbohydrate or protein. Fats provide much of the energy to do the work in your body, especially work involving the muscles. Fat prevents protein from being burned for energy so that protein can do its own important jobs.

Fat is an important part of all cell membranes. Fat also transports the fat-soluble vitamins (A, D, E, and K) throughout the body. Certain fat-containing foods also provide the body with two fatty acids that are considered *essential fatty acids* because the body can't make them. Fatty acids are the major component of triglycerides and are discussed next. The essential fatty acids are needed for normal growth and development in infants and children. They are used to maintain the structural parts of cell membranes, and they play a role in the proper functioning of the immune system. From the essential fatty acids, the body makes hormone-like substances that control a number of body functions, such as blood pressure and blood clotting.

Lipids also include cholesterol and lecithin. Their functions will be discussed later in this chapter.

In foods, fats enhance taste, flavor, aroma, crispness (especially in fried foods), juiciness (especially of meat), and tenderness (especially in baked goods). Fats such as cooking oils do a wonderful job of carrying many flavors, as in an Indian curry. Fats also provide a smooth texture and a creamy feeling in the mouth. The love of fatty foods cuts across all ages (just watch a preschooler devour french fries or an elderly adult eat a piece of chocolate cake) and cultures (where fatty foods are available). Eating a meal with fat makes people feel full, because fat delays the emptying of the stomach. This lasting feeling of fullness is called *satiety*.

ADIPOSE CELL
A cell in the body that readily takes up and stores triglycerides; also called a fat cell.

ESSENTIAL FATTY ACIDS
Fatty acids that the body cannot produce, making them necessary in the diet: linoleic acid and linolenic acid.

SATIETY
A feeling of being full after eating.

Nutrition Science Focus: Lecithin

Lecithin is considered a phospholipid, a class of lipids that are like triglycerides except that one fatty acid is replaced by a phosphate group and choline or another nitrogen-containing group. Phospholipids are unique in that they are soluble in fat and water. As you may know, fats and water (such as oil and vinegar) do not normally stay mixed together. Phospholipids such as lecithin are used by the food industry as emulsifiers, substances that are capable of breaking up the fat globules into small droplets, resulting in a uniform mixture that won't separate. Lecithin is used in foods such as salad dressings and bakery products. In the body, lecithin keeps fats in solution in the blood (and elsewhere), a most important function. Lecithin also functions as a vital component of cell membranes. Egg yolks are especially rich in lecithin and are used as emulsifiers in many baking recipes as well as in mayonnaise and hollandaise sauce.

Although the media have featured lecithin as a wonder nutrient that can burn fat, improve memory, and accomplish other similar feats, none of this is true. Since lecithin is made in the liver, it is not considered an essential nutrient.

LECITHIN
A phospholipid and a vital component of cell membranes that acts as an emulsifier (a substance that keeps fats in solution).

1. Fat accounts for about 13 to 30 percent or more of your body weight.
2. Many fat cells are located just under the skin, where fat provides insulation for the body and a cushion around critical organs.
3. At 9 kcalories per gram, fat stores are a compact way to store lots of energy. Fat spares protein from being burned for energy.
4. Fat is an important part of all cell membranes.
5. Fat also transports the fat-soluble vitamins (A, D, E, and K) throughout the body.
6. The essential fatty acids are needed for normal growth and development in infants and children. They are used to maintain the structural parts of cell membranes, and they play a role in the proper functioning of the immune system. From the essential fatty acids, the body makes hormone-like substances that control a number of body functions, such as blood pressure and blood clotting.
7. Lipids also include cholesterol and lecithin.
8. In foods, fats enhance taste, flavor, aroma, crispness, juiciness, tenderness, and texture. Fats have satiety value.

FATTY ACIDS

Major component of most lipids. Three fatty acids are present in each triglyceride.

GLYCEROL

A derivative of carbohydrate that is part of triglycerides.

POINT OF UNSATURATION

The location of the double bond in unsaturated fatty acids.

SATURATED FAT

A triglyceride made of mostly saturated fatty acids.

MONOUNSATURATED FAT

A triglyceride made of mostly monounsaturated fatty acids.

POLYUNSATURATED FAT

A triglyceride made of mostly polyunsaturated fatty acids.

TRIGLYCERIDES

A triglyceride (Figure 4-1) is made of three *fatty acids* ("tri-" means three) attached to *glycerol*, a derivative of carbohydrate. Glycerol contains three carbon atoms, each attached to one fatty acid. You can think of glycerol as the backbone of a triglyceride.

All fatty acids are molecules composed mostly of carbon and hydrogen atoms. There are three types of fatty acids in the foods you eat:

1. Saturated fatty acids
2. Monounsaturated fatty acids
3. Polyunsaturated fatty acids

A saturated fatty acid has the maximum possible number of hydrogen atoms attached to every carbon atom. It is therefore said to be "saturated" with hydrogen atoms.

Some fatty acids are missing one pair of hydrogen atoms in the middle of the molecule. This gap is called a *point of unsaturation* and the fatty acid is said to be "monounsaturated" because it has one gap.

Fatty acids that are missing more than one pair of hydrogen atoms are called "polyunsaturated." Polyunsaturated fatty acids are of two kinds, omega-3 or omega-6. Scientists tell them apart by where in the molecule the "unsaturations," or missing hydrogen atoms, occur.

Now that you know about the different types of fatty acids, it's time to get back to the concept of triglycerides. From the three types of fatty acids, we get three types of triglycerides, commonly called fats.

1. A saturated triglyceride, also called a *saturated fat*, is a triglyceride in which most of the fatty acids are saturated.
2. A monounsaturated triglyceride, also called a *monounsaturated fat*, is a triglyceride in which most of the fatty acids are monounsaturated.
3. A polyunsaturated triglyceride, also called a *polyunsaturated fat*, is a triglyceride in which most of the fatty acids are polyunsaturated.

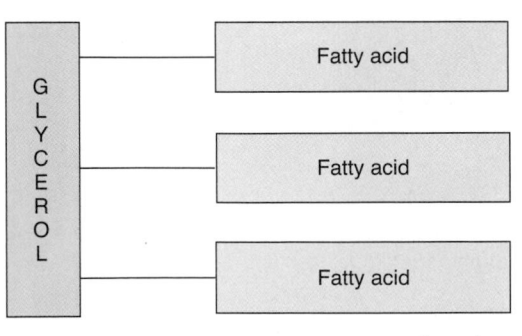

FIGURE 4-1:
A triglyceride.

Nutrition Science Focus: Triglycerides

Fatty acids in triglycerides are made of carbon atoms joined like links in a straight chain. Interestingly, the number of carbons is always an even number. Fatty acids differ from one another in two respects: the length of the carbon chain and the degree of saturation. The length of the chain may be categorized as short (less than 6 carbons), medium (6 to 10 carbons), or long (12 to 24 carbons). Most food lipids contain long-chain fatty acids. The length of the chain influences the ability of the fat to dissolve in water. Generally, triglycerides do not dissolve in water, but the short- and medium-chain fatty acids have some solubility in water; this will have implications later in our discussion on their digestion, absorption, and metabolism.

Fatty acids are referred to as *saturated* or *unsaturated*. To understand this concept, think of each carbon atom in the fatty-acid chain as having hydrogen atoms attached like charms on a bracelet, as you can see in Figure 4-2. Each C represents a carbon atom, each H represents a hydrogen atom, and each O represents an oxygen atom. Each carbon atom can have a maximum of four bonds, and so it can attach to four other atoms. Typically, a carbon atom has one bond each to the two carbon atoms on its sides and one bond each to two hydrogens. If each carbon atom in the chain is filled to capacity with hydrogens, it is considered a saturated fatty acid. That's how saturated fatty acid got its name: It is saturated with hydrogen atoms. When a hydrogen is missing from two neighboring carbons, a double bond forms between the carbon atoms, and this type of fatty acid is considered unsaturated.

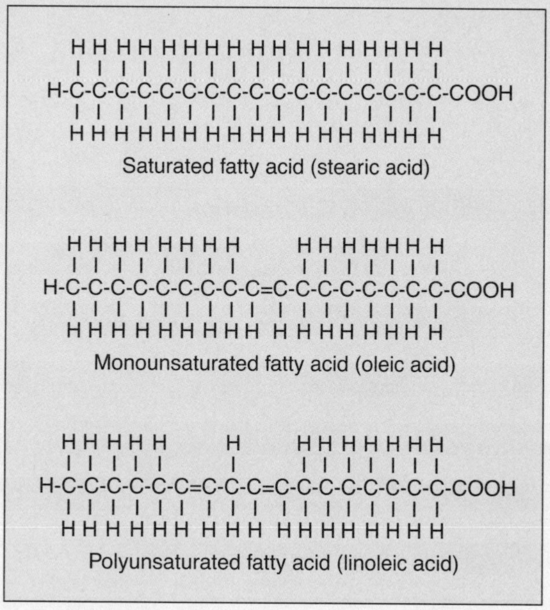

If you look at Figure 4-2, the top fatty acid in the illustration is saturated: It is filled to capacity with hydrogens.

Saturated fatty acid (stearic acid)

Monounsaturated fatty acid (oleic acid)

Polyunsaturated fatty acid (linoleic acid)

FIGURE 4-2:
Types of fatty acids.

By comparison, the middle and lower fatty acids are unsaturated. This is evident because there are empty spaces without hydrogens in the picture. Wherever hydrogens are missing, the carbons are joined by two lines, indicating a double bond. The spot where the double bond is located is called the point of unsaturation.

Unsaturated fatty acids are either *monounsaturated* or *polyunsaturated*. A fatty acid that contains only one ("mono-" means one) point of unsaturation is called monounsaturated; if the chain has two or more points of unsaturation, the fatty acid is called polyunsaturated. Figure 4-2 gives an example of a monounsaturated and a polyunsaturated fatty acid.

SATURATED FATTY ACID
A fatty acid that is filled to capacity with hydrogens.

UNSATURATED FATTY ACID
A fatty acid with at least one double bond.

MONOUNSATURATED FATTY ACID
A fatty acid that contains only one double bond in the chain.

POLYUNSATURATED FATTY ACID
A fatty acid that contains two or more double bonds in the chain.

Now we are ready to see which types of foods the three different fats appear in.

TRIGLYCERIDES IN FOOD

All food fats, animal or vegetable, contain a mixture of saturated and unsaturated fats, as you can see in Figure 4-3. A fat or oil can be classified as saturated, monounsaturated, or polyunsaturated depending on which type of fatty acid predominates. For example, coconut oil contains over 80 percent saturated fat, and so it is classified as mostly saturated fat. In contrast, canola oil contains over 50 percent monounsaturated fat, and so it is considered rich in monounsaturates.

Saturated fats are found mostly in foods of animal origin. Foods of animal origin include meat, poultry, seafood, milk and dairy products such as butter, and eggs. Several plant oils are actually very high in saturated fat; they include coconut oil, palm oil, and palm kernel oil.

Monounsaturated and polyunsaturated fats are found mostly in foods of plant origin and some fish and seafood. Foods of plant origin include fruits, vegetables, dried beans and peas, grains, foods made with grains such as breads and cereals, nuts, seeds, and vegetable oils such as corn oil. The more unsaturated a fat is, the more liquid it is at room temperature.

Before going into more detail on foods that contain the three types of fats, let's see what each food group contributes in terms of overall fat.

FIGURE 4-3:

Fatty acid composition of 1 tablespoon of fats and oils.

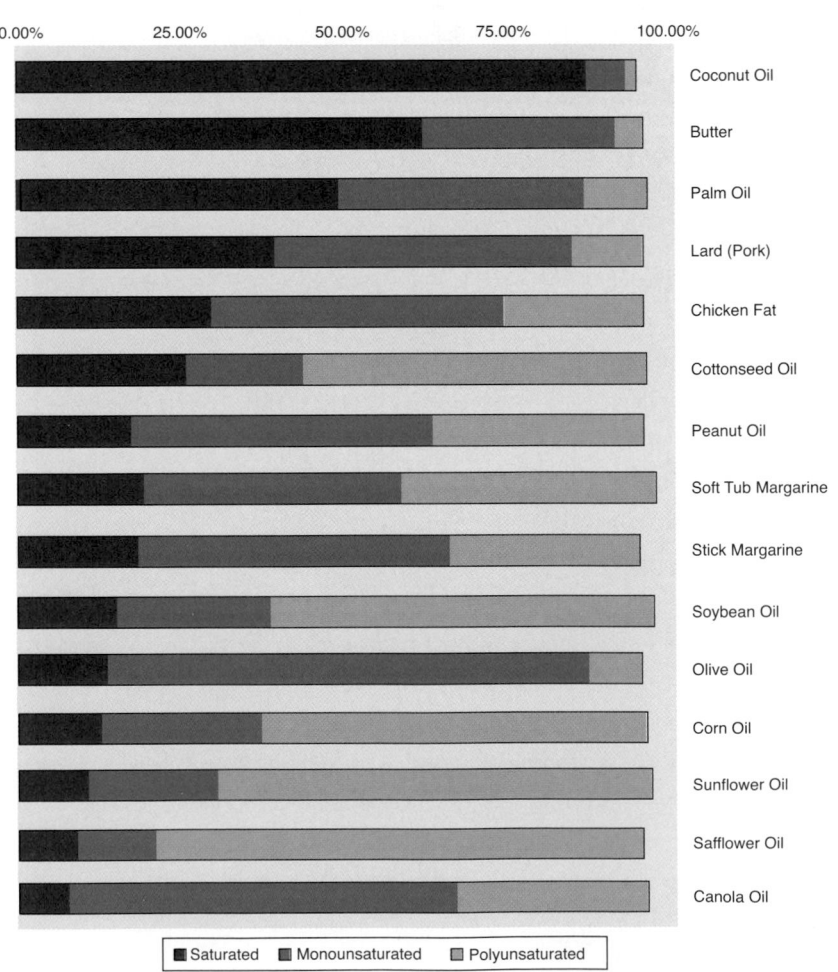

Fatty Acid Composition of 1 Tablespoon of Fats and Oils

■ Saturated ■ Monounsaturated ☐ Polyunsaturated

FIGURE 4-4: Fat, Saturated Fat, and Cholesterol in Different Types of Milk

Type of Milk	Kcalories	Total Fat	Saturated Fat	Cholesterol
Whole milk (3.24 percent)	146	8 grams	4.5 grams	24 milligrams
Reduced-fat (2 percent) milk	122	4.8 grams	3.1 grams	20 milligrams
Low-fat (1 percent) milk	102	2.4 grams	1.5 grams	12 milligrams
Nonfat (skim) milk	83	0.2 gram	0.2 grams	5 milligrams

Source: U.S. Department of Agriculture, Agricultural Research Service, 2004.
USDA Nutrient Database for Standard Reference, Release 17. Nutrient Data Laboratory Home Page,
http//www.nal.usda.gov/fnic/foodcomp.

1. **Fruits and Vegetables.** Whether fresh, canned, or frozen, most fruits and vegetables are practically fat-free. The exceptions are avocados, olives, and coconuts, which contain significant amounts of fat. Also, frozen vegetables that have butter, margarine, or sauces added are often high in fat. Fried vegetables, such as french-fried potatoes, are high in fat.

2. **Breads, Cereals, Rice, Pasta, and Grains.** Most breads and cereals in this group are low in fat. Exceptions include croissants, biscuits, cornbread, and some granolas and crackers. Most baked goods, such as cakes, pies, cookies, and quick breads, are also high in fat, especially when commercially made.

3. **Dry Beans and Peas, Nuts, and Seeds.** Dry beans and peas are very low in fat. Most nuts and seeds, however, such as peanuts and peanut butter, are quite high in fat. However, most of the fat is monounsaturated and polyunsaturated.

4. **Meat, Poultry, and Fish.** Meat and poultry, and to some extent fish, contain a bit of fat. The fat content of meat tends to be higher than that of poultry, and poultry tends to have more fat than seafood does. Of course, within each group there are choices that are quite high in fat and choices that are much more moderate in fat. For example, chicken without the skin is low in fat, because most of the fat is just under the skin.

5. **Dairy Foods.** Most regular dairy foods are high in fat. Luckily, there are plenty of choices with no fat or reduced fat, such as fat-free (skim) milk, fat-free yogurt, and low-fat cheeses (Figure 4-4).

6. **Fats, Oils, and Condiments.** Fats, such as vegetable shortening, and oils are almost all fat. Figure 4-5 lists the total calories and fat in selected fats and oils. Condiments

FIGURE 4-5: Total Kcalories and Fat in Selected Fats and Oils

Fat or Oil	Kcalories/ Tablespoon	Grams Fat/ Tablespoon	Grams Saturated Fat/Tablespoon
Coconut oil	120	13	12
Palm kernel oil	120	13	11
Palm oil	120	13	7
Butter, stick	108	12	7
Lard	115	13	5
Cottonseed oil	120	13	3
Olive oil	119	13	1
Canola oil	120	13	1
Peanut oil	119	13	2

(continued)

FIGURE 4-5: Total Kcalories and Fat in Selected Fats and Oils (*Continued*)

Fat or Oil	Kcalories/ Tablespoon	Grams Fat/ Tablespoon	Grams Saturated Fat/Tablespoon
Safflower oil	120	13	1
Corn oil	120	13	2
Soybean oil	120	13	2
Sunflower oil	120	13	1
Shortening	106	12	3
Margarine, stick	100	11	1–3
Margarine, soft tub	100	11	1–2
Margarine, liquid	90	10	1–2
Margarine, whipped	70	8	1–2
Margarine, spread	60	7	1–2
Margarine, diet	50	6	1

Source: U.S. Department of Agriculture and manufacturers.

such as regular mayonnaise and salad dressings also contain much fat. The main contributors to fat in the American diet are beef, margarine, salad dressings, mayonnaise, cheese, milk, and baked goods. You can't see most of the fat you get in the foods you eat, except of course when you add oils and fats. The fat in meat, milk and cheese, and fried foods is not as obvious as the margarine you spread on bread.

Figure 4-6 shows some foods that are high, moderately high, moderate, and low in fat. In addition to understanding which foods are high in fat, let's look at which foods contain mostly saturated, monounsaturated, and polyunsaturated fat.

FIGURE 4-6: Fat Content of Selected Foods

FOODS HIGH IN FAT (OVER 13 GRAMS/SERVING)

Food	Portion Size	Amount of Fat
Biscuit with egg and sausage	1 biscuit	39
Hamburger, double patty	1 sandwich	32
Pork, spareribs	3 oz.	26
French-fried potatoes	1 medium	25
Danish pastry, cheese	1 pastry	25
Chocolate bar	1 bar (1.75 oz.)	16
Sunflower seeds	$1/4$ cup	16

FOODS MODERATELY HIGH IN FAT (8–13 GRAMS/SERVING)

Food	Portion Size	Amount of Fat
Frankfurter, beef	1 frank	13
Cashew nuts	1 oz.	13
Pie, apple	1 piece	13
Vanilla ice cream	$1/2$ cup	12
Butter	1 Tbsp.	11
Margarine	1 Tbsp.	11
Chicken, drumstick, with skin, fried	1 drumstick	11

(*continued*)

Cake, yellow, with chocolate frosting	1 piece	11
Mayonnaise	1 Tbsp.	11
Doughnuts, cake-type, plain	1 medium	11
Potato chips	1 oz.	10
Pork loin, roasted	3 oz.	9
Cheese, cheddar	1 oz.	9
Cheese, American	1 oz.	9
Tuna salad	½ cup	9
Cookies, chocolate chip	2 cookies	9

FOODS MODERATE IN FAT (3.1–7.9 GRAMS/SERVING)

Food	Portion Size	Amount of Fat
Egg	1 large	7
Yogurt, plain (whole milk)	8-oz. container	7
French dressing	1 Tbsp.	7
Turkey, dark meat, roasted	3 oz.	6
Cake, pound	1 piece	6
Peanut chocolate candies	10 pieces	5
Cream cheese	1 Tbsp.	5
Cottage cheese, creamed	½ cup	5
Swordfish, cooked, dry heat	3 oz.	4
Vanilla pudding	4 oz.	4
Beef, eye of round, roasted	3 oz.	4

FOODS LOW IN FAT (3 GRAMS OR LESS/SERVING)

Food	Portion Size	Amount of Fat
Pancakes	1 pancake	3
Waffles, frozen	1 waffle	3
Chicken, breast, roasted	¼ breast	2
Sour cream	1 Tbsp.	2
Rolls, hamburger or hot dog	1 roll	2
Cheerios	1 cup	2
Brown rice, cooked	½ cup	1
Sherbet, orange	½ cup	1
Carrots, boiled, sliced	½ cup	0
Baked potato	1 medium	0
Apple, raw	1 medium	0

Source: U.S. Department of Agriculture, Agricultural Research Service, 2004. USDA Nutrient Database for Standard Reference, Release 17. Nutrient Data Laboratory Home Page, http://www.nal.usda.gov/fnic/foodcomp.

1. **Saturated Fat.** The biggest sources of saturated fat in the American diet are animal foods: cheese, beef (more than half comes from ground beef), whole milk, fats in baked goods, butter, and margarine. Saturated fat is also found in eggs, poultry skin, and other full-fat dairy products, such as ice cream. Animal fat often contains at least 50 percent saturated fat. Although most vegetable oils are rich in unsaturated fats, don't forget the ones that are high in saturated fat: coconut, palm kernel, and palm oils. They are used in some processed foods, such as baked goods and frozen whipped nondairy toppings. Figures 4-7 and 4-8 show the saturated fat content of selected foods.

FIGURE 4-7: Saturated Fat Content of Selected Foods

FOODS HIGH IN SATURATED FAT (6 GRAMS OR MORE/SERVING)

Food	Portion Size	Amount of Saturated Fat
Cheeseburger, large	1 sandwich	16
Biscuit with egg and sausage	1 biscuit	15
Beef, rib, roasted	3 oz.	10
Chicken pot pie	1 small pie	10
Pork, spareribs, braised	3 oz.	9
Duck, roasted	1/2 duck	9
Cheesecake	1 piece	8
Butter	1 Tbsp.	7
Onion rings, fried	8–9 rings	7
Croissant	1 croissant	7
French-fried potatoes	1 large	7
Beef, ground, 75 percent lean, broiled	3 oz.	6

FOODS MODERATE IN SATURATED FAT (1.1–5.9 MILLIGRAMS/SERVING)

Food	Portion Size	Amount of Saturated Fat
Shrimp, breaded and fried	6–8 shrimp	5
Frankfurter, beef	1 frank	5
Milky Way bar	1 bar (2.15 oz)	5
Milk, whole	1 cup	5
Trail mix, regular with chocolate chips, nuts, and seeds	1/2 cup	4.5
Cheese, American	1 oz.	4

FOODS MODERATE IN SATURATED FAT (1.1–5.9 MILLIGRAMS/SERVING)

Food	Portion Size	Amount of Saturated Fat
Éclair	1 éclair	4
Cookies, chocolate chip	1 cookie	2
Doughnuts, cake-type, plain	1 medium	2
Tuna salad	1/2 cup	1.6

FOODS LOW IN SATURATED FAT (LESS THAN OR EQUAL TO 1 MILLIGRAM/SERVING)

Food	Portion Size	Amount of Saturated Fat
Oil, canola	1 Tbsp.	1.0
Cheese, Parmesan, grated	1 Tbsp.	0.9
Oil, safflower	1 Tbsp.	0.8
Egg substitute	1/4 cup	0.4
Olives, ripe	5 large	0.3
Whole-wheat bread	1 slice	0.3
Chickpeas	1/2 cup	.22
Fat-free milk	1 cup	0.1
Apple	1 medium	0.1
Carrots, boiled	1/2 cup	0

Source: U.S. Department of Agriculture, Agricultural Research Service, 2004. USDA Nutrient Database for Standard Reference, Release 17. Nutrient Data Laboratory Home Page, http://www.nal.usda.gov/fnic/foodcomp.

FIGURE 4-8: A Comparison of Saturated Fat in Some Foods

Food Category	Portion	Saturated Fat Content (Grams)	Kcalories
Cheese			
Regular cheddar cheese	1 oz.	6.0	114
Low-fat cheddar cheese	1 oz.	1.2	49
Ground beef			
Regular ground beef (25 percent fat)	3 oz. (cooked)	6.1	236
Extra-lean ground beef (5 percent fat)	3 oz. (cooked)	2.6	148
Milk			
Whole milk	1 cup	4.6 g. sat. fat	146 kcal.
Low-fat (1 percent) milk	1 cup	1.5	102
Fat-free skim milk	1 cup	0	90
Breads			
Croissant (medium)	1 medium	6.6	231
Bagel, oat bran (4 inches)	4 in.	0.2	227
Frozen desserts			
Regular ice cream	$\frac{1}{2}$ cup	4.9	145
Frozen yogurt, low-fat	$\frac{1}{2}$ cup	2.0	110
Table spreads			
Butter	1 tsp.	2.4	34
Soft margarine with zero trans fat	1 tsp.	0.7	25
Chicken			
Fried chicken (leg with skin)	3 oz. (cooked)	3.3	212
Roasted chicken (breast, no skin)	3 oz. (cooked)	0.9	140
Fish			
Fried fish	3 oz. (cooked)	2.8	195
Baked fish	3 oz. (cooked)	1.5	129

Source: U.S. Department of Agriculture, Agricultural Research Service, 2004. USDA Nutrient Database for Standard Reference, Release 17. Nutrient Data Laboratory Home Page, http://www.nal.usda.gov/fnic/foodcomp.

2. **Monounsaturated Fat.** Good examples of monounsaturated fats include olive oil, canola oil, and peanut oil. Like other vegetable oils, they are used in cooking and in salad dressings. Canola oil is also used to make margarine.

3. **Polyunsaturated Fat.** Polyunsaturated fats are found in the greatest amounts in safflower, corn, soybean, sesame, and sunflower oils. These oils are commonly used in salad dressings and as cooking oils. Nuts and seeds also contain polyunsaturated fats, enough to make them a rather high-calorie snack food depending on serving size.

Too much saturated fat and trans fats are the primary contributors to high blood cholesterol levels, a major risk factor for heart disease.

TRANS FATS

Trans fats, short for *trans fatty acids*, occur naturally at low levels in meat and dairy foods. Most of the trans fats that we eat are a result of a process called hydrogenation. *Hydrogenation*, discovered at the turn of the twentieth century, converts liquid vegetable oils into solid fats by

> **TRANS FATS (TRANS FATTY ACIDS)**
> Unsaturated fatty acids that lose a natural bend or kink so that they become straight (like saturated fatty acids) after being hydrogenated; they act like saturated fats in the body.
>
> **HYDROGENATION**
> A process in which liquid vegetable oils are converted into solid fats (such as margarine) by the use of heat, hydrogen, and certain metal catalysts.

Trans fats

A typical saturated fatty acid

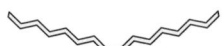

A typical polyunsaturated fatty acid.
At the double bond, both hydrogens are on the same side which makes the fatty acid kinked.

A polyunsaturated fatty acid that has been hydrogenated and is now a trans fatty acid. The hydrogens are now on diagonally opposite sides of the double bond, making the molecule straighter like a saturated fat.

FIGURE 4-9:

Trans fats.

using hydrogen, heat, and certain metal catalysts. The hydrogenation process is used to make vegetable shortening, which is simply vegetable oils that have been partially hydrogenated (Figure 4-9). Vegetable shortening is attractive because it is cheaper to make than butter or lard (pork fat), has a longer shelf life, and works well in many baked goods, such as pie crusts.

The hydrogenation process has also been used to make margarines (hydrogenation makes them easy to spread) and many oils used in deep-fat frying (hydrogenation gives them a high smoking point). Hydrogenation also helps these products stay fresh longer. If you check food labels, you will find partially hydrogenated oils in many baked goods (cookies, cakes, pies, muffins, and the like), fast foods, breads, snack foods, and margarines/shortenings.

During hydrogenation, some of the unsaturated fatty acids become saturated. Other unsaturated fatty acids lose their natural bend or kink and become straight, like saturated fatty acids. These are the trans fatty acids. Because they are straight, they can fit closer together, which makes them more solid. This explains why vegetable shortening is a solid.

Trans fatty acids behave like saturated fats in the body, and have been shown to raise blood cholesterol levels (a major risk factor for heart disease) as much as saturated fat does. This is discussed more in a later section on health and lipids.

Processed foods and oils provide about 80 percent of trans fats in the diet, compared with 20 percent that occur naturally in food from animal sources. Bacteria in cattle produce trans fat that gets into both meat and milk. Following is the contribution of various foods to trans fat intake in the American diet.

Cakes, cookies, pies, bread, crackers, etc.	40 percent of total trans fats consumed
Animal products	21 percent
Margarine	17 percent
Fried potatoes	8 percent
Potato chips, corn chips, popcorn	5 percent
Household shortening	4 percent

The amount of trans fatty acids in selected foods is shown in Figure 4-10.

FIGURE 4-10: Trans Fatty Acids in Selected Foods

Food	Serving Size	Trans Fatty Acids (Grams/Serving)
Vegetable shortening	1 Tbsp.	1.4–4.2
Margarine, stick, 80 percent fat	1 Tbsp.	2.8
Margarine, stick, 60 percent fat	1 Tbsp.	1.9
Margarine, tub, 80 percent fat	1 Tbsp.	1.1
Margarine, tub, 40 percent fat	1 Tbsp.	0.3
Butter	1 Tbsp.	0
Salad dressing	2 Tbsp.	0.06–1.1
Vegetable oils	1 Tbsp.	0
French fries	1 small (fast food)	0–2.9
French fries	1 large (fast food)	0–6.0
Granola bar	1 bar	0–2.5
Sandwich cookies	2 each	0–1.5
Iced cake	1 serving	0–4.5

(continued)

Potato chips	1 oz.	0–2.0
Macaroni and cheese	1 cup	0–1.5
Frozen waffles	2 each	0–2.0

Source: U.S. Department of Agriculture and food companies.

In 2006, food companies were required to list the amount of trans fat per serving on its own line under "Total Fat." As a result, the trans fats content of many processed foods has changed and is likely to continue to change as the food industry continues to reformulate its products. Companies often use canola oil, sunflower oil, and other liquid oils to replace the hydrogenated oils in margarine, many fried foods, and some baked foods. Though trans fat-labeling rules motivated many companies to remove most of the partially hydrogenated oil from most of their processed foods, hundreds of foods still contain trans fats. Read the Hot Topic on page 160 to learn more about trans fats in restaurant foods.

ESSENTIAL FATTY ACIDS: OMEGA-3 AND OMEGA-6 FATTY ACIDS

As mentioned, there are two essential fatty acids that the body can't make. The essential fatty acids are both polyunsaturated fatty acids: *linoleic acid* and *alpha-linolenic acid (ALA)*. Linoleic acid is called an omega-6 fatty acid because its double bonds appear after the sixth carbon in the chain (see Figure 4-2). The major omega-6 fatty acid in the diet is linoleic acid. Alpha-linolenic acid is the leading omega-3 fatty acid found in food, and its double bonds appear after the third carbon in the chain. Adequate Intakes have been set for linoleic and alpha-linolenic acids (Figure 4-11).

LINOLEIC ACID
Omega-6 fatty acid found in vegetable oils such as corn, safflower, soybean, cottonseed, and sunflower oils; this essential fatty acid is vital to growth and development, maintenance of cell membranes, and the immune system.

ALPHA-LINOLENIC ACID
An omega-3 fatty acid found in several oils, notably canola, flaxseed, soybean, walnut, and wheat germ oils (or margarines made with canola or soybean oil); this essential fatty acid is vital to growth and development, maintenance of cell membranes, and the immune system and is inadequate in many Americans' diets.

FIGURE 4-11: Adequate Intake Values for Essential Fatty Acids

Age Group	Linoleic Acid AI (Grams per Day)		Alpha-Linolenic Acid AI (Grams per Day)	
	Male	Female	Male	Female
0–6 months	4.4 grams	4.4 grams	0.5 grams	0.5 grams
7–12 months	4.6	4.6	0.5	0.5
1–3 years	7	7	0.7	0.7
4–8 years	10	10	0.9	0.9
9–13 years	12	10	1.2	1.1
14–18 years	16	11	1.6	1.1
19–30 years	17	12	1.6	1.1
31–50 years	17	12	1.6	1.1
Over 50 years	14	11	1.6	1.1
Pregnancy		13		1.4
Lactation		13		1.3

*Source: Adapted with permission from the Dietary References Intakes for Energy, Carbohydrates, Fiber, Fat, Protein, and Amino Acids (Macronutrients). © 2002 by the National Academy of Sciences. Courtesy of the National Academy Press, Washington, D.C.

Which foods contain linoleic acid or alpha-linolenic acid?

- Linoleic acid (omega-6) is found in vegetable oils such as corn, safflower, soybean, cottonseed, and sunflower oils. Most Americans get plenty of linoleic acid from foods containing vegetable oils, such as margarine, salad dressings, and mayonnaise. Whole grains and vegetables also supply some linoleic (and linolenic) acid.
- Alpha-linolenic acid (omega-3) is found in several oils, notably canola, flaxseed, soybean, and walnut oil or margarines made with these oils. Additional good sources of alpha-linolenic acid include ground flaxseed, walnuts, and soy products.

The body converts a small amount of alpha-linolenic acid (ALA) into two additional omega-3 fatty acids which are heart-healthy:

- Docosahexaenoic acid (DHA)
- Eicosapentaenoic acid (EPA)

Because the conversion of ALA into DHA and EPA is very slow, you need to get DHA and EPA mostly from food. Both DHA and EPA are found in fish such as salmon, mackerel, sardines, halibut, bluefish, trout, and tuna. Note that these fish tend to be fatty fish, not lean fish. White fish like haddock, cod, flounder, sole, and orange roughy contain only small amounts of DHA and EPA. DHA and EPA are not found in any plant foods.

The essential fatty acids serve as part of cell membranes and they play a role in the proper functioning of the immune system. They are also vital to normal growth and development in infants and children. The omega-3 fatty acids DHA and EPA are especially important for proper brain and eye development during pregnancy and infancy. DHA and EPA are excellent for heart health: they help reduce blood pressure, blood clots (which can start a heart attack), heart rate, and blood triglyceride levels.

Whereas Americans generally get plenty of linoleic acid, that is not the case with alpha-linolenic acid, DHA, or EPA. The ratio of omega-3 to omega-6 fatty acids in the diet is important in regulating your blood pressure, blood clotting, and inflammation. Having a healthy ratio of omega-3 to omega-6 fatty acids seems to lower blood pressure, prevent blood clot formation, and reduce inflammation. Inflammation is the body's normal response to injury, but long-term inflammation seems to play a role in chronic diseases like heart disease.

To increase your intake of omega-3 fatty acids, look for good sources of alpha-linolenic acid, DHA and EPA (Figure 4-12). In addition, the American Heart Association recommends that all adults eat fish (especially fatty fish) at least twice a week for enough DHA and EPA to reduce the risk of heart disease.

However, nearly all fish and shellfish contain traces of mercury. For most people, the risk from mercury by eating fish and shellfish is not a health concern. Yet, some fish and shellfish contain higher levels of mercury that may harm an unborn baby or young child's developing nervous system. The risks from mercury in fish and shellfish depend on the amount of fish and shellfish eaten and the levels of mercury in the fish and shellfish. Therefore, the Food and Drug Administration and the Environmental Protection Agency are advising women who may become pregnant, pregnant women, nursing mothers, and young children to avoid some types of fish and eat fish and shellfish that are lower in mercury.

By following these three recommendations for selecting and eating fish or shellfish, women and young children will receive the benefits of eating fish and shellfish and be confident that they have reduced their exposure to the harmful effects of mercury.

- Do not eat shark, swordfish, king mackerel, or tilefish because they contain high levels of mercury.
- Eat up to 12 ounces (2 average meals) a week of a variety of fish and shellfish that are lower in mercury.
- Five of the most commonly eaten fish that are low in mercury are shrimp, canned light tuna, salmon, pollock, and catfish.

FIGURE 4-12: Food Sources of Omega-3 Fatty Acids

ESSENTIAL FATTY ACIDS

Linoleic Acid (Omega-6)

Food sources of linoleic acid:

- Vegetable oils such as corn, safflower, soybean, cottonseed, and sunflower oils.
- Margarines.
- Salad dressings.

Alpha Linolenic Acid (Omega-3)

Food sources of alpha linolenic acid:

- Vegetable oils, notably canola, flaxseed, soybean, and walnut oil or margarines made with these oils.
- Ground flaxseed, walnuts, and soy products.

Food sources of DHA and EPA (other Omega-3 fatty acids):

- Fatty fish such as salmon, mackerel, sardines, halibut, bluefish, trout, and tuna.

*Fatty fish are the richest sources of omega-3s in the diet.

Functions: The essential fatty acids:

- serve as part of cell membranes
- play a role in the proper functioning of the immune system
- vital to normal growth and development in infants and children
- the omega-3 fatty acids DHA and EPA are especially important for proper brain and eye development during pregnancy and infancy.

Dietary Advice:

- The ratio of omega-3 to omega-6 fatty acids in the diet is important in regulating your blood pressure, blood clotting, and inflammation. Having a healthy ratio of omega-3 to omega-6 fatty acids seems to lower blood pressure, prevent blood clot formation, and reduce inflammation.
- Americans get plenty of omega-6 fatty acids and need to concentrate on getting more omega-3 fatty acids especially from fatty fish.

- Another commonly eaten fish, albacore ("white") tuna has more mercury than canned light tuna. So, when choosing your two meals of fish and shellfish, you may eat up to 6 ounces (one average meal) of albacore tuna per week.
- Check local advisories about the safety of fish caught by family and friends in your local lakes, rivers, and coastal areas. If no advice is available, eat up to 6 ounces (one average meal) per week of fish you catch from local waters, but don't consume any other fish during that time.

MINI-SUMMARY

1. A triglyceride is made of three fatty acids attached to glycerol.
2. Fatty acids can be saturated, monounsaturated, or polyunsaturated.
3. All food fats contain a mixture of saturated and unsaturated fats.
4. Most fruits and vegetables are fat-free, and most breads, cereals, rice, and pasta are low in fat.
5. Regular dairy foods are generally high in fat and saturated fat.
6. Saturated fats are found mostly in foods of animal origin, and monounsaturated and polyunsaturated fats are found mostly in foods of plant origin and some fish and seafood.
7. The biggest sources of saturated fat in the American diet are cheese, beef, whole milk, baked goods, butter, and margarine.
8. Olive oil, canola oil, and peanut oil are rich in monounsaturated fats.

9. Polyunsaturated fats are found in the greatest amounts in safflower, corn, soybean, sesame, and sunflower oils.

10. Trans fats, or trans fatty acids, occur naturally in small amounts in some foods and are created by hydrogenation. Hydrogenated oils are found in some baked goods, snack foods, french fries, margarines/shortenings, and chips/popcorn.

11. Both saturated fat and trans fats increase blood cholesterol levels, a risk factor for heart disease.

12. The essential fatty acids are both polyunsaturated fatty acids: linoleic acid (an omega-6 fatty acid) and alpha-linolenic acid (an omega-3 fatty acid). Alpha-linoleic acid is found in vegetable oils such as corn, safflower, soybean, cottonseed, and sunflower oils. Alpha-linolenic acid is found in several oils, notably canola, flaxseed, soybean, and walnut oils (or margarines made with these oils). Additional good sources of alpha-linolenic acid include ground flaxseed, walnuts, and soy products.

13. The body converts linolenic acid into two other omega-3 fatty acids: DHA and EPA. Both DHA and EPA are found in fatty fish such as salmon, mackerel, sardines, halibut, bluefish, trout, and tuna. DHA and EPA are excellent for heart health: they reduce blood pressure, blood clots, heart rate, and blood triglyceride levels.

14. Whereas Americans generally get plenty of linoleic acid, that is not the case with alpha-linolenic acid, DHA, or EPA. Having a healthy ratio of omega-3 to omega-6 fatty acids seems to lower blood pressure, prevent blood clot formation, and reduce inflammation. The American Heart Association recommends that all adults eat fish (especially fatty fish) at least twice a week for enough DHA and EPA to reduce the risk of heart disease.

15. The essential fatty acids serve as part of cell membranes and they play a role in the proper functioning of the immune system. They are also vital to normal growth and development in infants and children. The omega-3 fatty acids DHA and EPA are especially important for proper brain and eye development during pregnancy and infancy.

 # CHOLESTEROL

Triglycerides are one of the three classes of lipids. **Cholesterol** is the most abundant sterol, a second class of lipids. Pure cholesterol is an odorless, white, waxy, powdery substance. You cannot taste it or see it in the foods you eat.

Your body needs cholesterol to function normally. It is part of every cell membrane and is present in every cell in your body, including the brain and nervous system, muscles, skin, liver, and skeleton. The body uses cholesterol to:

- Make *bile acids*, which allow us to digest fat
- Maintain cell membranes
- Make many hormones (hormones are the chemical messengers of the body; they enter the bloodstream and travel to a target organ to influence what the organ does), such as the sex hormones (estrogen and testosterone) and the hormones of the adrenal gland (such as cortisone)
- Make vitamin D

Unfortunately, the cholesterol in your blood builds up in the plaque that clogs arteries and is a risk factor for heart disease. This will be discussed in more detail later in this chapter.

So, which foods contain cholesterol? Cholesterol is found only in foods of animal origin: egg yolks (it's not in the whites), meat, poultry, milk and milk products, and fish (see Figures 4-13 and 4-14). It is not found in foods of plant origin. Egg yolks and organ meats (liver, kidney, sweetbreads, brain) contain the most cholesterol—one egg yolk contains 213 milligrams of cholesterol. About 4 ounces of meat, poultry, or fish (trimmed or untrimmed) contains 100 milligrams of cholesterol, with the exception of shrimp, which

CHOLESTEROL

The most abundant sterol (a category of lipids); a soft, waxy substance present only in foods of animal origin; it is present in every cell in your body.

BILE ACIDS

A component of bile that aids in the digestion of fats in the duodenum of the small intestine.

FIGURE 4-13: Cholesterol in Foods

FOODS HIGH IN CHOLESTEROL (120 MILLIGRAMS OR MORE/SERVING)

Food	Portion Size	Amount of Cholesterol
Biscuit with egg and sausage	1 biscuit	302
Egg, cooked	1 large	212
Shrimp, breaded and fried	6–8 shrimp	200
Duck, roasted	½ duck	197
Éclair	1 éclair	127

FOODS MODERATE IN CHOLESTEROL (21–119 MILLIGRAMS/SERVING)

Food	Portion Size	Amount of Cholesterol
Cake, sponge	1 piece	107
Hamburger, regular, double patty	1 sandwich	103
Salami	3 oz.	90
Beef, ground	3 oz.	77
Chicken, breast, roasted	½ breast	73
Turkey, dark meat, roasted	3 oz.	71
Ice cream, vanilla	½ cup	68
Beef, top sirloin	3 oz.	64
Turkey, light meat, roasted	3 oz.	58
Fish, flounder, cooked, dry heat	3 oz.	58
Cheesecake	1 piece	44
Croissant	1 croissant	38
Cake, yellow with chocolate frosting	1 piece	35
Fish, halibut, cooked	3 oz.	35
Butter, salted	1 Tbsp.	31
Cheese, cheddar	1 oz.	30
Milk, whole	1 cup	24
Frankfurter, beef	1 frank	24

FOODS LOW IN OR NO CHOLESTEROL (20 MILLIGRAMS OR LESS/SERVING)

Food	Portion Size	Amount of Cholesterol
Cheese, provolone	1 oz.	20
Doughnuts, cake-type, plain	1 medium	17
Cheese, cream	1 Tbsp.	16
Tuna, canned in oil	3 oz.	15
Yogurt, plain, low-fat	8-oz. container	14
Cheese, cottage, low-fat	½ cup	9
Cheese, Parmesan, grated	1 Tbsp.	4
Frozen yogurt, chocolate	½ cup	4
Whole-wheat bread	1 slice	0
Skim milk	1 cup	0
Peanut butter	1 Tbsp.	0
Fruits	Any size	0
Vegetables	Any size	0

Source: U.S. Department of Agriculture, Agricultural Research Service, 2004. USDA Nutrient Database for Standard Reference, Release 17. Nutrient Data Laboratory Home Page, http://www.nal.usda.gov/fnic/foodcomp.

is higher in cholesterol. Eggs, meat, and whole milk provide most of the cholesterol we eat, and these sources are also rich in saturated fat.

In milk products, cholesterol is mostly in the fat, and so lower-fat products contain less cholesterol. For example, 1 cup of whole milk contains 24 milligrams of cholesterol, whereas a cup of nonfat milk contains only 5 milligrams (Figure 4-4).

Egg whites and foods that come from plants, such as fruits, nuts, vegetables, grains, cereals, and seeds, have no cholesterol.

We eat about 200 to 400 milligrams of cholesterol daily. The body also manufactures about 700 milligrams of cholesterol daily. The liver makes only 10 to 20 percent of this amount, and the body's cells synthesize the rest. Because the body produces plenty of cholesterol to make bile acids, hormones, and so on, cholesterol is not considered an essential nutrient.

MINI-SUMMARY

1. The body uses cholesterol to make bile acids, cell membranes, many hormones, and vitamin D.
2. Cholesterol is found only in foods of animal origin, such as egg yolks, meat, poultry, fish, milk, and milk products. It is not found in foods of plant origin.
3. The body's cells and the liver produce cholesterol, and so it is not an essential nutrient.

DIGESTION, ABSORPTION, AND METABOLISM

Fats are difficult for the body to digest, absorb, and metabolize. The problem is simple: Fat and water do not mix. Minimal digestion of fats occurs before they reach the upper part of the small intestine. In your mouth, a salivary gland makes an enzyme, called **lingual lipase**, that plays a minor role in fat digestion in adults but an important role in fat digestion in infants. Lingual lipase digests certain fatty acids in milk in infants. In the stomach, the enzyme **gastric lipase** works on breaking down mostly short-chain fatty acids.

Once fats reach the small intestine, the gallbladder is stimulated to release **bile** into the intestine. Bile is made by the liver, stored in the gallbladder, and squirted into the intestinal tract when fat is present. Bile contains bile acids that emulsify fat, meaning that they split fats into small globules or pieces. In this manner, fat-splitting enzymes (including pancreatic lipase and intestinal lipase) can then do their work. The enzymes break down many triglycerides to their component parts—fatty acids and glycerol—so that they can be absorbed across the intestinal wall. **Monoglycerides**, triglycerides with only one fatty acid, are also produced.

Once absorbed into the cells of the small intestine, triglycerides are re-formed. Both shorter-chain fatty acids and glycerol can travel freely in the blood because they are water-soluble. However, triglycerides, monoglycerides, cholesterol, and longer-chain fatty acids would float in clumps and wreak havoc in either the blood or the lymph. Because of this, the body wraps them with protein and phospholipids to make them water-soluble. The resulting substance is called a **lipoprotein**, a combination of fat (lipo-) and protein. Lipoproteins have four components:

- Triglycerides
- Protein
- Cholesterol
- Phospholipids

A lipoprotein called a **chylomicron** carries mostly triglycerides and some cholesterol from the intestine through the blood to the body's cells. The cells can either burn the triglycerides for energy or store them.

LINGUAL LIPASE

An enzyme made in the salivary glands in the mouth that plays a minor role in fat digestion in adults and an important role in fat digestion in infants.

GASTRIC LIPASE

An enzyme in the stomach that breaks down mostly short-chain fatty acids.

BILE

A substance that is stored in the gallbladder and released when fat enters the small intestine because it emulsifies fat.

MONOGLYCERIDES

Triglycerides with only one fatty acid.

LIPOPROTEIN

Protein-coated packages that carry fat and cholesterol through the bloodstream; the body makes four types classified according to their density.

Two main kinds of lipoproteins carry cholesterol in the blood.

- *Low-density lipoprotein,* called simply LDL, carries cholesterol to the body's tissues. Most of the cholesterol in the blood is the LDL form. LDL is called the "bad" cholesterol because the higher the level of LDL in the blood, the greater your risk for heart disease.
- *High-density lipoprotein,* called simply HDL, is called the "good cholesterol" because it takes cholesterol from the tissues to the liver, which removes it from the body. A low level of HDL increases your risk for heart disease.

The next section gives you more information on how the fat you eat can affect your risk for heart disease and cancer.

Nutrition Science Focus: Lipoproteins

The body makes four types of lipoproteins: chylomicrons, very low density lipoproteins, low-density lipoproteins, and high-density lipoproteins. Each is now discussed.

Chylomicron is the name of the lipoprotein responsible for carrying mostly triglycerides, along with some cholesterol, resulting from food digestion in the intestines through the lymph system to the bloodstream. Lymph is similar to blood but has no red blood cells. Lymph vessels are found all around the body, and the lymph transports fat and fat-soluble vitamins to the blood as well as moving fluids found between cells to the bloodstream. The lymph system goes into areas where there are no blood vessels to feed the cells.

In the bloodstream, an enzyme—*lipoprotein lipase*—breaks down the triglycerides in the chylomicrons into fatty acids and glycerol so that they can be absorbed into the body's cells. The cells can either use the fatty acids for energy, which the muscle cells often do, or make triglycerides for storage, which fat cells often do. Once the triglycerides have been broken down and taken up by the cells, what remains of the chylomicron is some protein and cholesterol that is metabolized by the liver.

The primary sites of lipid metabolism are the liver and the fat cells. The liver makes triglycerides and cholesterol that are carried through the body by *very low density lipoprotein (VLDL),* the liver's version of a chylomicron. Half of VLDL is made of triglycerides, which are broken down in the bloodstream with the help of lipoprotein lipase. Again, the body's cells absorb fatty acids and glycerol to be burned for energy or stored as triglycerides. Once the majority of triglycerides are removed, VLDL is converted in the blood into another type of lipoprotein called low-density lipoprotein (LDL).

LDL is the major cholesterol-carrying lipoprotein. LDL also distributes some triglycerides and phospholipids. LDL therefore supplies materials for cells to make new membranes, hormones (chemical messengers), and other substances. Certain cells (especially in the liver) can absorb the entire LDL particle. They play an important role in the control of blood cholesterol concentrations.

The last type of lipoprotein, high-density lipoprotein (HDL), contains much protein and travels throughout the body picking up cholesterol. HDL carries cholesterol back to the liver for breakdown and disposal. Thus, HDL helps remove cholesterol from the blood, preventing the buildup of cholesterol in the arterial walls.

CHYLOMICRON
The lipoprotein responsible for carrying mostly triglycerides, and some cholesterol, from the intestines through the lymph system to the bloodstream.

LOW-DENSITY LIPOPROTEINS (LDL)
Lipoproteins that contain most of the cholesterol in the blood; they carry cholesterol to body tissues.

HIGH-DENSITY LIPOPROTEINS (HDL)
Lipoproteins that contain much protein and carry cholesterol away from body cells and tissues to the liver for excretion from the body.

LIPOPROTEIN LIPASE
An enzyme that breaks down triglycerides from the chylomicron into fatty acids and glycerol so that they can be absorbed in the body's cells.

VERY LOW DENSITY LIPOPROTEINS (VLDL)
Lipoproteins made by the liver to carry triglycerides and some cholesterol through the body.

1. Fats are difficult for the body to digest, absorb, and metabolize because fat and water do not mix.
2. Minimal digestion of fats occurs before they reach the small intestine. Lingual lipase works as an enzyme in the mouth, and gastric lipase acts in the stomach.
3. Once fats reach the small intestine, the gallbladder releases bile. Bile acids emulsify fat so that pancreatic lipase and intestinal lipase can break down triglycerides into their components.
4. Once absorbed into the cells of the small intestine, triglycerides are re-formed. Shorter-chain fatty acids and glycerol can travel in the blood, but the longer-chain fatty acids, monoglycerides, triglycerides, and cholesterol are wrapped into chylomicrons (a lipoprotein) with protein and phospholipids to make them water-soluble. Chylomicrons carry mostly triglycerides and some cholesterol from the intestines to the bloodstream. The cells either use the fatty acids for energy (muscle cells) or make triglycerides for storage (fat cells).
5. Low density lipoprotein (LDL) carries cholesterol to the body's tissues.
6. High density lipoprotein (HDL) takes cholesterol from the tissues to the liver, which removes it from the body.

LIPIDS AND HEALTH

HEART DISEASE

Heart disease is the number-one killer of both men and women in the United States. In 2005, about 800,000 people died due to heart disease. Many Americans have elevated blood cholesterol levels, one of the key risk factors for heart disease. Anyone can develop high blood cholesterol, regardless of age, gender, race, or ethnic background.

Too much circulating cholesterol can build up in the walls of the arteries, especially the heart's arteries (called the coronary arteries), which supply the heart with what it needs to keep pumping. This leads to accumulation of cholesterol-laden *plaque* in blood vessel linings, a condition called *atherosclerosis* or "hardening of the arteries."

This process can happen to blood vessels anywhere in the body, including those of the heart, which are called the coronary arteries. If the coronary arteries become partly blocked by plaque, then the blood may not be able to bring enough oxygen and nutrients to the heart muscle. This can cause chest pain. If the blood supply to a portion of the heart is completely cut off by a blockage, a heart attack (called a *myocardial infarction*) can occur. If the blood supply to part of the brain is cut off, a *stroke* can occur.

When atherosclerosis affects the coronary arteries, the condition is called coronary heart disease. Some cholesterol-rich plaques are unstable—they have a thin covering and can burst, releasing cholesterol and fat into the bloodstream. The release can cause a blood clot to form over the plaque, blocking blood flow through a coronary artery—and cause a heart attack.

LDL cholesterol is the main source of cholesterol buildup and blockage in the arteries. The primary way in which LDL cholesterol levels become too high is through eating too much saturated fat, trans fat, and, to a lesser extent, cholesterol. Dietary factors that lower LDL cholesterol include:

- Replacing saturated and trans fats with polyunsaturated and monounsaturated fats
- Eating soluble fiber and soy protein (to a lesser extent)

PLAQUE
Deposits on arterial walls that contain cholesterol, fat, fibrous scar tissue, calcium, and other biological debris.

ATHEROSCLEROSIS
The most common form of artery disease, characterized by plaque buildup along artery walls.

MYOCARDIAL INFARCTION
Heart attack.

STROKE
Damage to brain cells resulting from an interruption of blood flow to the brain.

Trans fats not only increase your LDL cholesterol but they also decrease your HDL cholesterol.

A variety of other things can affect cholesterol levels. These are things you can do something about:

- Weight. Being overweight is a risk factor for heart disease. It also tends to increase your cholesterol. Losing weight can help lower your LDL and total cholesterol levels as well as raise your HDL and lower your triglyceride levels.
- Physical activity. Not being physically active is a risk factor for heart disease. Regular physical activity can help lower LDL cholesterol and raise HDL cholesterol levels. It also helps you lose weight. You should try to be physically active for 30 minutes on most, if not all, days.

Things you cannot do anything about also can affect cholesterol levels. They include:

- Age and gender. As women and men get older, their cholesterol levels rise. Before the age of menopause, women have lower total cholesterol levels than do men of the same age. After the age of menopause, women's LDL levels tend to rise.
- Heredity. Your genes partly determine how much cholesterol your body makes. High blood cholesterol can run in families.

Besides high LDL cholesterol levels, there are several other risk factors for heart disease. They include cigarette smoking, high blood pressure, diabetes mellitus, obesity, and physical inactivity. If any of these is present in addition to high blood cholesterol, the risk of heart disease is even greater.

The good news is that these factors can be brought under control by changes in lifestyle, such as adopting some dietary changes, losing weight, beginning an exercise program, and quitting a tobacco habit. Drugs also may be necessary for some people. Sometimes one change can help bring several risk factors under control. For example, weight loss can reduce blood cholesterol levels, help control diabetes, and lower high blood pressure.

But some risk factors can't be controlled. They include age (45 or older for men and 55 or older for women) and a family history of early heart disease.

The American Heart Association has developed guidelines to reduce the risk of cardiovascular disease by means of dietary and other practices. Their guidelines and heart disease are discussed in greater depth in Chapter 11.

CANCER

Cancer is the second leading cause of death in the United States after heart disease. The relationship between fat and cancer is not nearly as clear as that between fat and heart disease. Fat may be involved in certain cancers, such as colon or prostate cancer. With breast cancer, being obese and eating excessive kcalories seem to be more related to cancer development than is dietary fat alone. Recent research shows that a diet low in fat may reduce the risk of breast cancer recurrence and also ovarian cancer in healthy postmenopausal women.

What type of fat you eat also can influence your risk of developing cancer. A high intake of saturated fat has been associated with some cancers, such as colon and prostate cancer. Omega-3 fatty acids found in fish, and possibly oleic acid in olive oil, are not associated with increasing cancer risk, and may be protective in terms of cancer.

The American Cancer Society has written guidelines for cancer prevention that are consistent with guidelines from the American Heart Association. You can read more on this in Chapter 11.

1. Heart disease is the number-one killer in the United States. Too much circulating cholesterol can build up in the form of plaque on the walls of the arteries, especially the coronary arteries, a condition called atherosclerosis. Atherosclerosis is a slow, complex disease that starts in childhood. Over time the buildup of plaque in the walls of the arteries causes "hardening of the arteries"—arteries become narrowed, and blood flow to the heart is slowed down or blocked, causing angina, heart attacks, and strokes.

2. The primary way in which LDL cholesterol levels become too high is through eating too much saturated fat, trans fat, and to a lesser extent cholesterol. Dietary factors that lower LDL cholesterol include replacing saturated fats with polyunsaturated and monounsaturated fats and, to a lesser extent, soluble fiber and soy protein.

3. Being overweight and inactive promotes high blood cholesterol levels. Age, gender, and heredity also can affect blood cholesterol levels, but you can't control these factors as you can the other factors mentioned. You can reduce your risk by adopting some dietary changes, losing weight, exercising, and quitting smoking.

4. Fat may be involved in certain cancers, such as prostate cancer.

DIETARY RECOMMENDATIONS

No Dietary Reference Intake has been set for total fat in our diets except for infants. (The AI for infants is 31 grams fat per day for age 1 to 6 months and 30 grams fat per day for age 7 to 12 months.) There is not enough scientific data to determine the level at which inadequacy or disease prevention occurs. However, Acceptable Macronutrient Distribution Ranges (AMDRs) were set for total fat as follows.

Age	AMDR for Fat
1 to 3 years old	30–40 percent of kcalories
4 to 18 years old	25–35 percent of kcalories
Over 18 years old	20–35 percent of kcalories

This range of intake is associated with a reduced risk of chronic disease while providing an adequate intake.

Neither an AI nor an RDA was set for saturated fat or cholesterol because these substances are all made in the body and have no known role in preventing chronic diseases such as heart disease and diabetes. Likewise, no AI or RDA was determined for trans fatty acids because they are not essential and provide no known benefit to health. The Food and Nutrition Board does recommend keeping one's intake of saturated fat, cholesterol, and trans fatty acids as low as possible while eating a nutritionally adequate diet.

The American Heart Association and other health authorities recommend that people consume no more than 1% of total kcalories as trans fat. That's about as much as occurs naturally in milk and meat, leaving virtually no room for trans fat from partially hydrogenated vegetable oil found in french fries, baked goods, and the like.

In 2002, an AI was set for the essential fatty acids: linoleic acid and alpha-linolenic acid (Figure 4-11). There is insufficient evidence to set a Tolerable Upper Intake Level for either essential fatty acid.

Fat intake in the United States as a proportion of total kcalories is lower than it was many years ago (dropping from 45 to 34 percent), but most people still eat too much saturated fat.

Although the percentage of kcalories from fat has decreased, the number of kcalories eaten has increased, and so we are actually eating more grams of fat.

The Dietary Guidelines for Americans (2005) and the American Heart Association's Dietary Guidelines both recommend a diet for healthy Americans that:

- Provides less than 10 percent of total kcalories from saturated fat
- Provides less than 300 milligrams per day of cholesterol
- Replaces most saturated fats with sources of polyunsaturated and monounsaturated fatty acid such as fish, nuts, and vegetable oils

Figure 4-15 shows the maximum number of grams of fat and saturated fat for a variety of calorie levels using these recommendations. Keep in mind that trans fat should be considered part of your saturated fat intake. These recommendations do not apply to children age 2 and under. Children need fat to grow and develop properly. After age 2, children should progressively adopt these recommendations up to age 5 years.

In Chapter 2, we discussed the Mediterranean diet, which includes lots of monounsaturated fat, mostly in the form of olive oil. There seems to be agreement that if fat intake is higher than 30 percent of total calories, the diet should emphasize monounsaturated fats such as olive oil and canola oil.

When looking at fat in food, it is important to distinguish between two different concepts: the percentage of kcalories from fat and the percentage of fat by weight. To explain these two concepts, let's look at an example. In a supermarket, you find sliced turkey breast that is advertised as being "96 percent fat-free." What this means is that if you weighed a 3-ounce serving, 96 percent of the weight would be lean, or without fat. In other words, only 4 percent of its weight is actually fat. The statement "96 percent fat-free" does not tell you anything about how many kcalories come from fat.

Now, if you look at the Nutrition Facts on the label, you read that a 3-ounce serving contains 3 grams of fat, 27 kcalories from fat, and 140 total calories. The label also states that the percentage of kcalories from fat in a serving is 19 percent. To find out the percentage of kcalories from fat in any serving of food, simply divide the number of calories from fat by the number of total kcalories and then multiply the answer by 100, as follows.

$$\frac{\text{Kcalories from fat}}{\text{Total kcalories}} \times 100 = \text{Percentage of kcalories from fat}$$

$$\frac{27 \text{ kcalories from fat}}{140 \text{ kcalories}} \times 100 = 19 \text{ percent}$$

FIGURE 4-15: Recommended Maximum Fat and Saturated Fat Intake		
If Your Total Daily Kcalories Are:	Total Fat (g)	Saturated Fat (g)
1200	40	13
1500	50	17
1800	60	20
2000	67	22
2200	73	24
2400	80	27
2600	86	29
2800	93	31
3000	100	33

This percentage has become more important as many recommendations on fat consumption target 30 to 35 percent or less of total kcalories as a desirable daily total from fat. This does not mean, however, that every food you eat needs to derive only 35 percent of its kcalories from fat. If this were the case, you could not even have a teaspoon of margarine because all of its kcalories come from fat. It is your total fat intake over a few days that is important, not the percentage of fat in just one food or just one meal.

MINI-SUMMARY

1. There is no RDA or AI for fat (except for infants), saturated fat, cholesterol, or trans fatty acids.
2. The AMDR for fat is 20 to 35 percent of kcalories for adults.
3. An AI is set for the essential fatty acids.
4. It is recommended that you keep your intake of saturated fat, cholesterol, and trans fatty acids as low as possible while eating a nutritionally adequate diet.
5. The Dietary Guidelines for Americans (2005) and the American Heart Association's Dietary Guidelines both recommend a diet for healthy Americans that:
   - Provides less than 10 percent of total kcalories from saturated fat
   - Provides less than 300 milligrams per day of cholesterol
   - Replaces most saturated fats with sources of polyunsaturated and monounsaturated fatty acid, such as fish, nuts, and vegetable oils
6. To find out the percentage of calories from fat in any serving of food, simply divide the number of calories from fat by the number of total kcalories and then multiply the answer by 100.

INGREDIENT
FOCUS:
MILK, DAIRY
PRODUCTS,
AND EGGS

NUTRITION

Milk is an excellent source of:

- High-quality protein
- Carbohydrates
- Riboflavin
- Vitamins A and D (if fortified)
- Calcium and other minerals, such as phosphorus, magnesium, and zinc

Cheese is made from various types of milk, with cow's and goat's milk being the most popular. In many parts of the world cheese is produced from different kinds of milk: sheep, reindeer, yak, buffalo, camel, and donkey. Cheese is produced when bacteria or rennet (or both) is added to milk and the milk then curdles. The liquid, known as the whey, is separated from the solid, known as the curd, which is the cheese. It is thought that cheese was discovered by accident thousands of years ago in the Far East.

Cheese is an excellent source of nutrients such as protein and calcium. However, because most cheeses are prepared from whole milk or cream, they are also high in saturated fat and cholesterol. Ounce for ounce, meat, poultry, and most cheeses have about the same amount of cholesterol. But cheeses tend to have much more saturated fat.

Determining which cheeses are high or low in saturated fat and cholesterol can be confusing, because there are so many different kinds on the market: part-skim, low-fat, processed, and so on. Not all reduced-fat or part-skim cheeses are low in fat; they are only lower in fat than similar natural cheeses. For instance, one reduced-fat cheddar gets 56 percent of its kcalories from fat—considerably less than the 71 percent of regular cheddar but not super-lean, either. The trick is to read the label. Figure 4-16 is a guide to fat in cheeses.

FIGURE 4-16: Fat in Cheese[†]

Low Fat 0–3 g fat/oz.	Medium Fat 4–5 g fat/oz.	High Fat 6–8 g fat/oz.	Very High Fat 9–10 g fat/oz.
Natural Cheeses			
*Cottage cheese (¼ cup) dry curd	*Mozzarella, part-skim *Ricotta (¼ c), part-skim	Blue cheese *Brick	Cheddar Colby
Cottage cheese (¼ cup) low-fat 1%	String cheese, part-skim	Brie Camembert	*Cream cheese (1 oz. = 2 Tbsp.)
Cottage cheese (¼ cup) low-fat 2%		Edam Feta	Fontina *Gruyère
Cottage cheese (¼ cup) Creamed 4%		Gjetost Gouda	Longhorn *Monterey Jack
Sapsago		*Light cream cheese (1 oz. = 2 Tbsp.)	Muenster Roquefort
Look for special low-fat brands of mozzarella, ricotta, cheddar, and Monterey jack.	Look for reduced-fat brands of cheddar, colby, Monterey jack, muenster, and Swiss.	Limburger Mozzarella, whole-milk Parmesan (1 oz. = 3 Tbsp.) *Port du Salut Provolone *Ricotta (¼ cup), whole-milk Romano (1 oz. = 3 Tbsp.) *Swiss Tilsit, whole milk	
Modified Cheeses			
Pasteurized process, imitation, and substitute cheeses with 3 g fat/oz or less.	Pasteurized process, imitation, and substitute cheeses with 4–5 g fat/oz.	Pasteurized process Swiss cheese Pasteurized process Swiss cheese food Pasteurized process American cheese Pasteurized process American cheese food American cheese food cold pack Imitation and substitute cheeses with 6–8 g fat/oz.	Some pasteurized process cheeses are found in this category—check the labels.

[†]Check the labels for fat and sodium content. 1 serving = 1 oz. unless otherwise stated.

*These cheeses contain 160 mg or less of sodium per 1 oz.

Source: Reprinted by permission of the American Heart Association, Alameda County Chapter, 11200 Golf Links Road, Oakland, CA 94605.

Eggs are very nutritious and full of high-quality protein, as well as varying amounts of many vitamins and minerals. The concern with overconsumption of eggs stems from the fact that they are very high in cholesterol—215 milligrams per egg (compare that to the suggested maximum of 300 milligrams daily). One egg also contributes 5 grams of fat, of which 2 grams are saturated fat.

Culinary Science

If you leave a bottle of milk, fresh from the cow, to sit for 12 to 24 hours, you will find that the fat has floated up to the top of the bottle. If you were to skim off the contents at the top of the bottle, it would be cream—a liquid milk product with about 20 percent fat. Why does the fat separate? Milk is a mixture of 2 liquids that don't normally mix: fat and water. This type of mixture is called an emulsion. You also see an emulsion any time you try to mix oil and vinegar. The milk you buy at the store does not separate because most milk is homogenized. During homogenization, fat globules are mechanically reduced greatly in size and dispersed permanently in a fine emulsion through the milk.

Cream is made with a number of different fat levels depending on what it will be used for. Light cream must have at least 18 percent fat. Light whipping cream is richer and must contain from 30 to 36 percent fat. Heavy cream contains at least 36 percent fat and whips easily and holds its whipped texture longer than light whipping cream.

To make whipped cream, you will need a cream with at least 30 percent fat. The cream, bowl, and whip should all be well chilled. When you whip cream, you are whipping air into the liquid mixture. The foam that is created when you whip is actually stabilized by the fat. The fat globules form walls around the air bubbles, so you get a stable structure—whipped cream. If the cream, bowl, and whip are not cold enough, the air bubbles will collapse and the whipped cream will not rise very high.

If you whip cream for too long, you will get butter (which is 80 percent fat) and buttermilk, a liquid. This occurs because if you keep beating whipped cream, the fat globules weaken, rupture, and become granular. They then turn into butter.

As a fat used in cooking, butter provides a unique flavor. When used for frying and sautéing, butter works well except that it contains milk solids, which brown and then burn at only 250°F. Clarified butter is butter that has had the milk solids and water removed. A major advantage of clarified butter is that it has a much higher smoke point (about 400°F), so you can cook with it at higher temperatures without it browning and burning.

Cheese is made by coagulating or curdling milk, stirring and heating the curd, draining off the whey (the watery part of milk), collecting and pressing the curd, and in some cases, ripening. The curds are rich in protein and fat. The amount of fat in cheese depends on the fat in the milk it was made from.

When using cheese in sauces and soups, avoid using one that becomes stringy when cooked, such as mozzarella and cheddar. Moist cheeses or grating cheeses such as Parmesan work well. Add the cheese as late as possible in the cooking and keep the temperatures low because high temperatures will cause the protein to become hard and squeeze out the fat. Stir only as needed because stirring can sometimes push the cheese protein into a sticky ball. Starchy ingredients such as flour and cornstarch help to stabilize the cheese during cooking. Aged cheeses melt and blend into foods better than young cheeses.

Like cheese, eggs are full of protein and should not be cooked at high temperatures and long cooking times. The basic principle of cooking eggs it to use a medium to low temperature and time carefully. The egg yolk is high in both fat and protein while the egg white is mostly protein. This is why egg whites coagulate or cook before the yolks do.

When eggs are cooked at too high a temperature or for too long at a low temperature, the egg whites shrink and become tough and rubbery. The yolk becomes tough and rubbery also and the surface turns green. The green color you may find on the yolk of a hard-boiled egg or on the surface of scrambled eggs is due to the iron in the yolk binding to the sulfur found in the egg white. Iron sulfide not only looks bad but it has a strong odor and flavor.

Egg yolks are unique because they contain a number of emulsifiers, which is why egg yolks are so important in making foods such as mayonnaise and hollandaise sauce. Many proteins in egg yolk can act as emulsifiers because they have some amino acids that repel water and some amino acids that attract water.

Rancidity is the deterioration of fat, resulting in undesirable flavors and odors. In the presence of air, fat can lose a hydrogen atom at the point of unsaturation and take on an oxygen atom. This change creates unstable compounds that start a chain reaction, quickly turning a fat rancid. You can tell whether a fat is rancid by its odor and taste.

The greater the number of points of unsaturation, the greater the possibility that rancidity will develop. This explains why saturated fats are more resistant to rancidity than unsaturated fats are. Rancidity is also quickened by heat and ultraviolet light. Luckily, vitamin E is present in plant oils, and it naturally resists deterioration of the oil.

To prevent rancidity, store fats and oils tightly sealed in cool, dark places. For butter, margarine, and milk, check the date on the packaging. For meats, following proper storage time and temperature rules are important. If you refrigerate oils, they sometimes become cloudy and thicker. This usually clears up after they are left at room temperature again or put under warm water.

> **RANCIDITY**
> The deterioration of fat, resulting in undesirable flavors and odors.

CHEF'S TIPS

- When cooking with milk, remember a very important rule: Use moderate heat and heat the milk slowly (but not too long) to avoid curdling—a grainy appearance with a lumpy texture. From a scientific point of view, milk curdles when the casein (protein in milk) becomes solid and separates out of the milk.
- Add other food products to hot milk products slowly, stirring with either a spoon or a wire whisk, if preparing a sauce, to avoid lumps. Be especially careful when adding foods high in acid—milk has a tendency to curdle if not beaten quickly.
- There are several ways to use cheese in balanced cooking:
  - Use a regular cheese with a strong flavor, and use less of it than called for in the recipe.
  - Use less cheese.
  - Substitute low-fat cheeses for regular ones.
  - Use a mixture of half regular cheese and half low-fat cheese.
- Use low heat with cheese. It is best to use as low a heat as possible when cooking with cheese. Cheese has a tendency to toughen when subjected to high heat and long cooking due its moisture evaporating, fat melting, and protein becoming stiff. Avoid boiling at all costs.
- Use short cooking times. Most recipes require the addition of cheese at the end of the recipe to avoid overcooking.
- Grate the cheese. The best way to add cheese to a recipe is to grate it. Grating will break the cheese into small, thin pieces that will melt and blend quickly and evenly into the end product. Grating also creates an image of more cheese when it is melted on top of a product (au gratin).
- Low-fat cheeses are available and have come a long way in flavor and texture. If moderation is your goal and not total elimination, be sure to add an extremely aged and flavorful cheese into your preparation at the end, when your customers will taste it on their first bite.
- To make "whipped cream," drain plain nonfat yogurt in cheesecloth to remove as much liquid as possible. Fold whipped egg whites into the yogurt and add a little honey for flavor. Use frozen pasteurized egg whites to avoid any food safety (salmonella) problems. The other option is to use a dollop of real whipped cream without adding sugar.
- To make an excellent omelet without cholesterol, whip egg whites until they foam. Add a touch of white wine, freshly ground mustard, and chives. Spray a nonstick pan with oil and add the eggs. Cook like a traditional omelet. When the omelet is close to done, put the pan under the broiler to finish. The omelet will puff up. Stuff the omelet, if desired, with vegetables, a shaving of flavorful cheese, or even some fresh fruit filling, and then fold it over and serve. For another interesting appetizer or a dinner or breakfast entrée, make a mixture of egg whites, herbs, and cracked black pepper and pour it over a vegetable-oil-sprayed crème brulée dish filled with mushrooms or other cooked vegetables and a little cheese for a very nice omelet alternative.
- For color and flavor, serve an omelet with spicy salsa poured on a portion of it, or serve it with salsa, black bean relish, and blue corn tortilla chips to create more breakfast options.

CHECK-OUT QUIZ

1. Directions: In the following columns, check off each food that is a significant source of fat and/or cholesterol.

Food	Fat	Cholesterol
1. Butter	_____	_____
2. Margarine	_____	_____
3. Split peas	_____	_____
4. Peanut butter	_____	_____
5. Porterhouse steak	_____	_____
6. Flounder	_____	_____
7. Skim milk	_____	_____
8. Cheddar cheese	_____	_____
9. Chocolate chip cookie made with vegetable shortening	_____	_____
10. Green beans	_____	_____

2. Match each statement on the left with the term on the right that it describes. The terms will be used more than once.

_____ 1. Present in every cell in the body   a. Lecithin

_____ 2. Emulsifies fats   b. Cholesterol

_____ 3. Found only in animal foods

_____ 4. Vital component of cell membranes

_____ 5. Used to make bile

3. Match each numbered statement with the lettered term it describes.

_____ 1. Corn oil is a good source of this fat   a. Rancidity

_____ 2. Lessens possibility of blood clots   b. Chylomicron

_____ 3. Liquid at room temperature   c. Monounsaturated fat

_____ 4. Deterioration of fat in air and heat   d. Oil

_____ 5. Olive oil is a good source of this fat   e. Emulsifier

_____ 6. Breaks up fat globules into small fatty acid droplets   f. Polyunsaturated fat fatty acid

_____ 7. Carries triglycerides and cholesterol through the lymph to the blood   g. Long-chain omega-3

≡€ ACTIVITIES AND APPLICATIONS

1. Self-Assessment

To find out if your diet is high in fat, saturated fat, and cholesterol, check yes or no to the following questions.

DO YOU USUALLY   YES   NO

1. Put butter on popcorn? _____ _____
2. Eat more red meats (beef, pork, lamb) than chicken and fish? _____ _____
3. Leave the skin on chicken? _____ _____
4. Eat whole-milk cheeses, such as cheddar, American, and Swiss, more than three times a week? _____ _____
5. Sauté or fry foods more than once or twice a week? _____ _____
6. Eat regular lunch meats, hot dogs, and bacon more than three times a week? _____ _____
7. Leave visible fat on meat? _____ _____
8. Use regular creamy salad dressings such as Russian, blue cheese, Thousand Island, and creamy French? _____ _____
9. Eat potato chips, nacho chips, and/or cream dips more than twice a week? _____ _____
10. Drink whole milk? _____ _____
11. Eat more than four eggs a week? _____ _____
12. Eat organ meats (liver, kidney, etc.) more than once a week? _____ _____
13. Use mayonnaise, margarine, and/or butter often on your sandwiches? _____ _____
14. Use vegetable shortening in baking or cooking? _____ _____
15. Eat commercially baked goods, including cakes, pies, and cookies, more than twice a week? _____ _____

Ratings: If you answered yes to:

1–3 questions: You are probably eating a diet not too high in fat, saturated fat, and cholesterol.

4–7 questions: You could afford to make some food substitutions, such as skim for regular milk, to reduce your fat and saturated fat intake.

8–15 questions: Your diet is very likely to be high in fat, saturated fat, and cholesterol.

2. **Changing Eating Habits**

If you are eating too much fat, you can make changes a little at a time. Check off one of these things to try (if you are not already doing it) or make up your own:

- The next time, I eat chicken, I will take the skin off.
- I will limit my daily meat and poultry servings to two 3-ounce servings a day. (A 3-ounce serving is about the size of a deck of cards.)
- This week, I will try a new type of fresh or frozen fish.
- I will try a low-fat cheese, such as low-fat Swiss.
- I will switch to low-fat (1 percent) or nonfat (skim) milk.
- I will try sherbet or frozen yogurt for dessert instead of ice cream.
- I will count the number of eggs I eat a week and see whether I meet the recommendations.
- I will try to use a lower-in-fat margarine, salad dressing, or mayonnaise.
- I will keep more fruits and vegetables in the refrigerator so that they will be handy for a snack instead of cookies or chips.
- I will buy pretzels instead of potato chips.

- For breakfast, instead of doughnuts, I will try a hot or cold cereal with skim milk and toast with jelly.
- I will top my spaghetti with stir-fried vegetables instead of a creamy sauce.

3. Reading Food Labels

   Following are food labels from two brands of lasagna. One is heavy on cheese and ground beef. The other is a vegetable lasagna made with moderate amounts of cheese. Using the Nutrition Facts given, can you tell which is which? How did you tell?

Lasagna 1	Lasagna 2
Nutrition Facts	Nutrition Facts
Amount per serving	Amount per serving
Calories 230	Calories 140
Calories from Fat 140	Calories from Fat 20
Total Fat 16 g	Total Fat 2.5 g
Saturated 6 g	Saturated 0.5 g
Polyunsaturated 1 g	Polyunsaturated 1.0 g
Monounsaturated 7 g	Monounsaturated 0.5 g
Cholesterol 74 g	Cholesterol 0 g
Sodium 180 mg	Sodium 180 mg
Total Carbohydrate 0 g	Total Carbohydrate 21 g
Dietary Fiber 0 g	Dietary Fiber 5 g
Sugar 0 g	Sugar 0 g
Protein 21 g	Protein 8 g

4. Meat, Poultry, and Seafood Comparison Using *iProfile*

   Pick out three meat items you eat, three poultry items you eat, and three fish/shellfish items you eat. Using iProfile, make a chart listing the calories, fat, saturated fat, and cholesterol in all these foods. Once the chart is done, ask yourself the following questions:

   - Which food has the least/most fat?
   - Which food has the least/most saturated fat?
   - Which food has the least/most cholesterol?

5. Baking with Fat Substitutes

   In this activity, you will prepare three batches of brownies using mixes. Make the first batch using the instructions on the box. For the second batch, use applesauce or prune puree in place of all the oil called for in the recipe. For the third batch, replace half the oil with applesauce or prune puree. When the brownies are all baked, put some of each type on separate plates and find out if your classmates can tell which product is which. You should also ask your classmates to rank the products in terms of taste, texture, appearance, and overall acceptability.

   If you want to use prune puree in other baked goods, try a recipe from www.sunsweet.com.

6. Using iProfile, click on "Estimating Portion Sizes" at the top of the page. Next click on "How Much Salad Dressing Do I Use?" and also "What Does One Ounce of Cheese Look Like?"

NUTRITION WEB EXPLORER

The American Heart Association   www.deliciousdecisions.org

Visit the nutrition site for the American Heart Association and look at the recipes in their cookbooks. Write down three cooking methods, three seasonings, and three cooking substitutions that are heart-healthy.

Trans Fat Help Center   www.notransfatnyc.org

Visit this website that was designed to help New York City food service establishments. Write two paragraphs on their advice for baking and frying without trans fats.

Golden Valley Flax   www.flaxhealth.com/recipes.htm

Flaxseed is a wonderful source of omega-3 fatty acids. The body cannot break down whole seeds to access the omega-3 containing oil, so most recipes use ground seeds. Read the recipes on this web page and try one out. What is the flavor of flaxseed like? Can you put it on cereal or a salad? Can you bake with it?

Cabot Cheese   www.cabotcheese.com

Cabot makes some excellent cheeses that are lower in fat. Click on "Our Products" and then click on "Reduced-Fat Cheddar." List the products you find. Hopefully you might be able to taste them in class!

FOOD FACTS: OILS AND MARGARINES

There is an ever-widening variety of oils and margarines on the market. They can differ markedly in their color, flavor, uses, and nutrient makeup (Figure 4-17).

When choosing vegetable oils, pick those high in monounsaturated fats, such as olive oil, canola oil, and peanut oil, or high in polyunsaturated fats, such as corn oil, safflower oil, sunflower oil, and soybean oil.

Olive oil contains from 73 to 77 percent monounsaturated fat. The color of olive oil varies from pale yellow to dark green, and its flavor varies from subtle to a full, fruity taste. The color and flavor of olive oil depend on the olive variety, level of ripeness, and processing method. When buying olive oil, make sure you are buying the right product for your intended use.

- Extra virgin olive oil, the most expensive form, has a rich, fruity taste that is ideal for flavoring finished dishes and in salads, vegetable dishes, marinades, and sauces. It is not usually used for cooking because it loses some flavor. It is made by putting mechanical pressure

FIGURE 4-17:
Oils and margarines. Top row: corn oil, safflower oil, liquid margarine in squeeze bottle, olive oil; bottom row: tub margarine, stick margarine. Photo by Frank Pronesti.

on the olives, a more expensive process than using heat and chemicals.

- Olive oil, also called pure olive oil, is golden and has a mild, classic flavor. It is an ideal, all-purpose product that is great for sautéing, stir-frying, salad dressings, pasta sauces, and marinades.

- Light olive oil refers only to color or taste. These olive oils lack the color and much of the flavor found in the other products. Light olive oil is good for sautéing or stir-frying because the oil is used mainly to transfer heat rather than to enhance flavor.

Figure 4-18 gives information on various oils. Be prepared to spend more for exotic oils such as almond, hazelnut, sesame, and walnut. Because these oils tend to be cold-pressed (meaning they are processed without heat), they are not as stable as the all-purpose oils and should be purchased in small quantities. They are strong, so you don't need to use much of them. Don't purchase these oils to cook with—they burn easily.

Vegetable oils are also available in convenient sprays that can be used as nonstick spray coatings for cooking and baking with a minimal amount of fat. Vegetable-oil cooking sprays come in a variety of flavors (butter, olive, Italian, mesquite), and a quick two-second spray adds about 1 gram of fat to the product. To use, first spray the cold pan away from any open flames (the spray is flammable), heat the pan, then add the food.

Margarine was first made in France in the late 1800s to provide an economical fat for Napoleon's army. It didn't become popular in the United States until World

War II, when it was introduced as a low-cost replacement for butter. Margarine must contain vegetable oil and water and/or milk or milk solids. Flavorings, coloring, salt, emulsifiers, preservatives, and vitamins are usually added. The mixture is heated and blended, then firmed by exposure to hydrogen gas at very high temperatures (see information about hydrogenation on pages 135–136). The firmer the margarine, the greater the degree of hydrogenation and the longer the shelf life.

Standards set by the U.S. Department of Agriculture and the Food and Drug Administration require margarine and butter to contain at least 80 percent fat by weight and to be fortified with vitamin A. One tablespoon of either one has approximately 11 grams of fat and 100 kcalories. You can compare the fat profiles of butter and margarine in Figure 4-19. Butter contains primarily saturated fat, whereas margarine is low in saturated fat and rich in monounsaturated and polyunsaturated fat. However, margarine may contain trans fat while butter does not.

Butter must be made from cream and milk. Salt and/or colorings may be added. Margarine must contain vegetable oil and water and/or milk. Salt, food coloring, other vitamins, emulsifying agents such as lecithin, and preservatives may be added to margarine.

If a butter or margarine product does not contain at least 80 percent fat by weight, it can't be called butter or margarine but instead is classified as a spread. The percent of fat (by weight) must appear on the label. Water, gums, gelatins, and various starches are used in spreads to replace some or all of the fat, or air may be whipped into the product.

FIGURE 4-18: Oils

Oil	Characteristics/Uses	Smoke Point
Canola oil (monounsaturated)	Light yellow color Bland flavor Good for frying, sautéing, and in baked goods Good oil for salad dressings	420°F
Corn oil	Golden color Bland flavor Good for frying, sautéing, and in baked goods Too heavy for salad dressings	420°F
Cottonseed oil	Pale yellow color Bland flavor Good for frying, sautéing, and in baked goods Good oil for salad dressings	420°F
Hazelnut oil	Dark amber color Nutty and smoky flavor Not for frying or sautéing as it burns easily Good for flavoring finished dishes and salad dressings Use in small amounts Expensive	*
Olive oil	Varies from pale yellow with sweet flavor to greenish color and fuller flavor to full, fruity taste (color and flavor depend on olive variety, level of ripeness, and how oil was processed) Extra virgin or virgin olive oil—good for flavoring finished dishes and in salad dressings, strong olive taste Pure olive oil—can be used for sautéing and in salad dressings; not as strong an olive taste as extra virgin or virgin Light olive oil—the least flavorful, good for sautéing, stir-frying, or baking	* 280°F
Peanut oil	Pale yellow color Mild nutty flavor Good for frying and sautéing Good oil for salad dressings	420°F
Safflower oil	Golden color Bland flavor Has a higher concentration of polyunsaturated fatty acids than any other oil Good for frying, sautéing, and in baked goods Good oil for salad dressings	470°F
Sesame oil	Light gold flavor Distinctive, strong flavor Good for sautéing Good oil for flavoring dishes and in salad dressings Use in small amounts Expensive	440°F

(continued)

FIGURE 4-18: Oils (*Continued*)

Oil	Characteristics/Uses	Smoke Point
Soybean oil	More soybean oil is produced than any other type; used in most blended vegetable oils and margarine Light color Bland flavor Good for frying, sautéing, and in baked goods Good oil for salad dressings	420°F
Sunflower oil	Pale golden color Bland flavor Good for frying, sautéing, and in baked goods Good oil for salad dressings	340°F
Walnut oil	Medium yellow to brown color Rich, nutty flavor For flavoring finished dishes and in salad dressings Use in small amounts Expensive	*

*Not recommended for cooking.

Margarines basically vary along these dimensions:

- *Physical form.* Margarine comes in either sticks or tubs. Tub margarines contain more polyunsaturated fatty acids than do stick margarines, and so they melt at lower temperatures and are easier to spread. Spreads come in sticks, tubs, liquids, and pumps. Liquid spreads are packaged in squeeze bottles or pump dispensers in which the margarine spread is really liquid, even in the refrigerator. They work well when drizzled on hot vegetables and other cooked dishes. Fat-free sprays also can be used to coat cooking pans.
- *Type of vegetable oil(s) used.* The vegetable oil may be mostly corn oil, safflower oil, canola oil, or another oil. Check the ingredients label to compare how much liquid oil and/or partially hydrogenated oil are used. If the first ingredient is liquid corn oil, more liquid oil is used, meaning that there will be less saturated fat than there is in a product with hydrogenated corn oil as the first ingredient.
- *Percentage of fat by weight and nutrient profile.* Margarine and spreads are available with amounts of fat that vary from 0 percent to 80 percent by weight. Look on the label for the percentage of fat by weight. Also look for terms such as light, diet, and fat-free. Light margarine contains one-third fewer calories or half the fat of the regular product. Diet margarine, also called reduced-calorie margarine, has at least 25 percent fewer calories than the regular product. Fat-free margarine has less than 0.5 gram of fat per serving. If you are wondering how they make a margarine fat-free, part of the answer is gelatin. Water, rice starch, and other fillers are used to make it taste like fat. As mentioned in this chapter, information on the trans fat content of products is on labels. Many margarines without any trans fats are available.

The American Heart Association recommends that consumers choose soft margarines that have liquid vegetable oil as the first ingredient and do not contain any trans fats.

Not all margarines and spreads can be used in the same way. Spreads with lots of water can make bread or toast soggy and may spatter and evaporate quickly in hot pans, causing foods to stick. In baking, low-fat spreads are not recommended because product quality suffers.

In addition to butter and margarine, blends and butter-flavored buds are available. Blends are part margarine and part butter (about 15 to 40 percent). They are made of vegetable oils, milk fat, and other dairy ingredients added to make the product taste like butter. Blends may have as much fat as regular margarine or butter (in other words, at least 80 percent fat), or they may be reduced in fat. Butter-flavored buds are made from carbohydrates and a small amount of dehydrated butter. They are virtually fat-free and cholesterol-free and are designed to melt on hot, moist foods such as a baked potato. When mixed with water, they can make butter-flavored sauces.

1 Tbsp.	Total Fat	Saturated Fat	Trans Fat	Monounsaturated Fat	Polyunsaturated Fat	Cholesterol
Butter	11.5	7.3	0	3.0	0.4	31
Shortening	12.8	5.2	1.4–4.2	5.7	1.4	7
Margarine, Stick, 80% fat	11.0	2.1	2.8	5.2	3.2	0
Margarine, Stick, 60% fat	8.3	1.5	1.9	3.9	2.2	0
Margarine, Tub, 80% fat	11.4	1.8	1.1	5.1	4.0	0
Margarine, Tub, 60% fat	8.6	1.4	0.01	1.9	4.8	0
Soybean oil	13.6	2.0	0	3.2	7.9	0

Source: U.S. Department of Agriculture, Agricultural Research Service, 2004. USDA Nutrient Database for Standard Reference, Release 17. Nutrient Data Laboratory Home Page, http://www.nal.usda.gov/fnic/foodcomp.

Two margarine-like spreads, Take Control (made by Lipton) and Benecol (made by McNeil Consumer Products), use ingredients that lower blood cholesterol levels. Take Control uses a stanol-like ingredient from soybeans. Benecol contains a plant stanol ester that comes from pine trees.

Lipton recommends one to two servings (1 to 2 tablespoons) of Take Control every day as part of a diet low in saturated fat and cholesterol. Take Control is low in saturated fat and free of trans fats, and it contains 6 grams of fat and 1.1 grams of soybean extract per serving.

Benecol is recommended in three daily servings (1½ tablespoons total) of the regular or light spread. Both products cost considerably more than regular margarine. These spreads are examples of functional foods, a topic discussed in detail in Chapter 6.

HOT TOPIC: TRANS FATS IN RESTAURANTS

Over the past two years, restaurants in some jurisdictions have been told to stop serving foods with significant amounts of artificial trans fat. As of 2008, laws have passed in locations such as Baltimore, New York City, Philadelphia, Boston, and King County, Washington (which includes Seattle) phasing out the use of trans fats in restaurant foods. As of 2006, federal rules required trans fat to be listed on food labels, a move that has spurred many large manufacturers of packaged foods, such as Frito Lay, to switch to oils without trans fat. However, restaurants have traditionally not been required to provide nutrition labeling information to customers, although many voluntarily do so using pamphlets, websites, and other tools.

On December 5, 2006, the New York City Board of Health approved an amendment to the Health Code to phase out artificial trans fat in all New York City restaurants and other food service establishments. The phase out of artificial trans fat in restaurant foods was planned in two stages. First, restaurants had until July 1, 2007, to make sure that all oils, shortening, and margarine used for frying or for spreads had less than 0.5 grams of trans fat per serving. Oils and shortening used to deep-fry yeast dough and cake batter were not included in the first deadline. The second deadline was July 1, 2008. By that date, all foods had to have less than 0.5 grams of trans fat per serving. Packaged foods served in the manufacturer's original packaging are exempt.

The New York City Department of Health recommended that food service managers do the following:

1. Change your oils. For cooking and frying, check the ingredients on all oils. If "partially hydrogenated" is listed, switch to a nonhydrogenated oil instead. If there is no ingredients list, ask your supplier or the manufacturer. For baking, use nonhydrogenated oils or shortenings with low or no trans fat.
2. Choose healthy spreads. Instead of stick margarine or butter, use soft tub spreads with low saturated fat and no trans fat.
3. Order prepared foods without trans fat. Check ingredients and ask your supplier for baked products, prefried, and premixed foods that are free of partially hydrogenated vegetable oils.

At the Trans Fat Help Center set up for New York City restaurants, food professionals can learn all about trans fat-free fats that can be used for frying and about where to buy baked goods without trans fats. Many fats are available for frying, such as high oleic canola and safflower oils. They can also learn about shortenings and margarines that are available with no trans fats and can be used to prepare high-quality baked goods such as cakes and pies.

Laws such as those in New York City are controversial, with different organizations taking varying positions. The Center for Science in the Public Interest feels that the Food and Drug Administration should be involved in phasing out the use of partially hydrogenated oils in the United States. As they state, "With no leadership from the Food and Drug Administration on this issue, cities have had to take this important public health issue into their own hands, and we hope other cities will follow suit." Some foodservice chains have voluntarily taken on this issue. Wendy's and McDonald's are each phasing out their use of partially hydrogenated oil. KFC stopped using it for deep-frying in 2007. The National Restaurant Association supports gradually phasing out trans fat in restaurant foods; however, they oppose inflexible bans with unrealistic timetables, which are tough for foodservice operators to implement successfully.

Protein

Have you ever wondered why meat, poultry, and seafood are often considered entrées, or main dishes, whereas vegetables and potatoes are considered side dishes? As recently as 50 years ago, the abundant protein found in meat, poultry, and seafood was considered the mainstay of a nutritious diet. You could say that these foods took center stage or, more accurately, center plate. As a child, I can remember going to visit my grandparents on Sunday and eating a roast beef dinner during our visit. Yes, we had vegetables, too, but the big deal at dinner was a roast that was carefully cooked, sliced, and served (with brown gravy, of course).

Today, protein foods continue to be an important component of a nutritious diet; however, we are much more likely to see foods such as lentils and pasta occupying the center of the plate. For adults who grew up when beef was king (and not nearly as expensive as it is today) and full-fat bologna sandwiches filled many lunchboxes, making spaghetti without meatballs takes a little getting used to, but more and more meatless meals are being served.

So just what is **protein**? Proteins are compounds found in all living cells in animals and plants that play a variety of important roles. The protein found in animals and plants is such an important substance that the term protein is derived from the Greek word meaning "first." Proteins reside in your skin, hair, nails, muscles, and blood, to name just a few places. Whereas carbohydrates and lipids are used primarily for energy, proteins function in a very broad sense to build and maintain your body. This chapter will help you to:

- Identify and describe the building blocks of protein
- List the functions of protein in the body
- Explain how protein is digested, absorbed, and metabolized
- Distinguish between complete protein and incomplete protein and list examples of foods that contain each
- Explain the potential consequences of eating too much or too little protein
- State the dietary recommendations for protein
- Discuss the nutrition and uses of meat, poultry, and fish on the menu
- Describe soy products, their health benefits, and how to use them on the menu
- Discuss the advantages and disadvantages of irradiation

PROTEINS

Major structural parts of the body's cells that are made of nitrogen-containing amino acids assembled in chains; perform other functions as well; particularly rich in animal foods.

STRUCTURE OF PROTEIN

Like carbohydrates and fats, proteins contain carbon, hydrogen, and oxygen. Unlike carbohydrates and fats, proteins contain nitrogen and provide much of the body's nitrogen. Nitrogen is necessary for the body to function; life as we know it wouldn't exist without nitrogen.

Proteins are long chains of **amino acids** strung together the way different railroad cars make up a train. Amino acids are the building blocks of proteins. There are 20 different ones, each consisting of a backbone to which a side group is attached (Figure 5-1). The amino acid backbone is the same for all amino acids, but the side group varies. It is the side group that makes each amino acid unique.

Of the 20 amino acids in proteins (see Figure 5-2), 9 either cannot be made in the body or cannot be made in the quantities needed. They therefore must be obtained in foods for the body to function properly. This is why we call them **essential amino acids**. The remaining 11 can be made in the body, and so they are called **nonessential amino acids**.

Under certain circumstances, one or more nonessential amino acids may not be able to be made in sufficient quantities, and so they become essential amino acids. For example, tyrosine

AMINO ACIDS

The building blocks of protein.

ESSENTIAL AMINO ACIDS

Amino acids that either cannot be made in the body or cannot be made in the quantities needed by the body; must be obtained in foods.

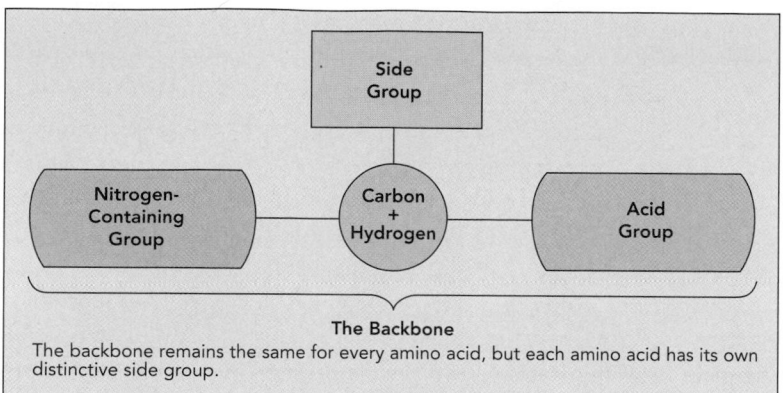

FIGURE 5-1:
An amino acid.

FIGURE 5-2: Amino Acids

Essential Amino Acids

Histidine
Isoleucine
Leucine
Lysine
Methionine
Phenylalanine
Threonine
Tryptophan
Valine

Nonessential Amino Acids

Alanine
Arginine
Asparagine
Aspartic acid
Cysteine
Glutamic acid
Glutamine
Glycine
Proline
Serine
Tyrosine

NONESSENTIAL AMINO ACIDS
Amino acids that can be made in the body.

CONDITIONALLY ESSENTIAL AMINO ACIDS
Nonessential amino acids that may, under certain circumstances, become essential.

PEPTIDE BONDS
The bonds that form between adjoining amino acids.

POLYPEPTIDES
Protein fragments with 10 or more amino acids.

is generally a nonessential amino acid but can become a ***conditionally essential amino acid***. In other words, under certain conditions a nonessential amino acid becomes an essential amino acid.

When the amino-acid backbones join end to end, a protein forms (Figure 5-3). The bonds that form between adjoining amino acids are called ***peptide bonds***. Proteins often contain from 35 to several hundred or more amino acids. Protein fragments with 10 or more amino acids are called ***polypeptides***.

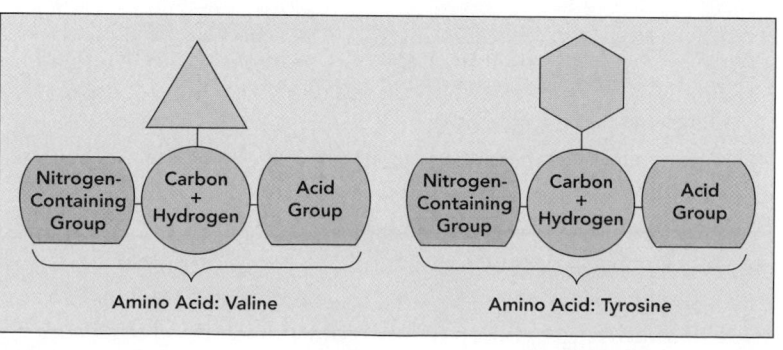

FIGURE 5-3:
A part of a protein.

1. Proteins are the major structural parts of the body's cells and are made of nitrogen-containing amino acids joined end to end by peptide bonds. Proteins often contain from 35 to several hundred or more amino acids.
2. There are 20 different amino acids, each with a different side group.
3. Essential amino acids cannot be made in the body or cannot be made in the quantities needed. Nonessential amino acids can be made in the body. Sometimes a nonessential amino acid becomes essential, and it is then called a conditionally essential amino acid.
4. Protein fragments with 10 or more amino acids are called polypeptides.

FUNCTIONS OF PROTEIN

After reviewing all the jobs proteins perform, you will have a greater appreciation of this nutrient. In brief, protein is part of most body structures; builds and maintains the body; is a part of many enzymes, hormones, and antibodies; transports substances around the body; maintains fluid and acid-base balance; can provide energy for the body; and helps in blood clotting (Figure 5-4). Now let's take a look at each function separately.

Proteins function as part of the body's structure. For example, protein can be found in skin, bones, hair, fingernails, muscles, blood vessels, the digestive tract, and blood (Figure 5-5). Protein appears in every cell.

Proteins are used for building and maintaining body tissues. Worn-out cells are replaced throughout the body at regular intervals. For instance, your skin today will not be the same skin in a few months. A skin cell lives only about one month. Skin is constantly being broken down and rebuilt or remodeled, as are most body cells, including the protein within those cells. The cells that line the gastrointestinal tract are replaced every three to five days.

The greatest amount of protein is needed when the body is building new tissues rapidly, such as during pregnancy or infancy. A newborn boy requires 9.1 grams of protein each day for the first 6 months, which increases to 13.5 for the next 6 months up to his first birthday. By age 9 he needs about 34 grams of protein each day. Additional protein is also needed when body protein is lost or destroyed, as in burns, surgery, or infections.

Proteins are found in many enzymes, some hormones, and all antibodies. Thousands of enzymes have been identified. Almost all the reactions that occur in the body, such as food digestion, involve enzymes. **Enzymes** are catalysts, meaning that they increase the rate of these reactions, sometimes by more than a million times. They do this without being changed in the overall process. Enzymes contain a special pocket called the active site. You can think of the active site as a lock into which only the correct key will fit. Various substances fit into the

ENZYMES
Catalysts in the body that help break down substances, build up substances, and change one substance into another.

FIGURE 5-4: Functions of Protein

- Acts as a structural component of the body
- Builds and maintains the body
- Found in many enzymes and hormones and all antibodies
- Transports iron, fats, minerals, and oxygen
- Maintains fluid and acid-base balance
- Provides energy as last resort
- Helps blood clot

pocket, undergo a chemical reaction, and then exit the enzyme in a new form, leaving the enzyme to perform its function again and again (Figure 5-6). Enzymes help do the following:

1. Break down substances (such as foods during digestion)
2. Build up substances (such as bone)
3. Change one substance into another (such as glucose into glycogen)

Hormones are chemical messengers secreted into the bloodstream by various organs, such as the liver, to travel to a target organ and influence what it does. Hormones regulate certain body activities so that a constant internal environment (called *homeostasis*) is maintained. For example, the hormone insulin is released from the pancreas when your blood sugar level goes up after you eat lunch. Insulin stimulates the transport of sugar from the blood into your cells, resulting in lower, more normal blood sugar levels. Amino acids are components of insulin as well as other hormones.

HORMONES
Chemical messengers in the body.

HOMEOSTASIS
A constant internal environment in the body.

  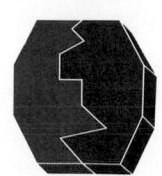 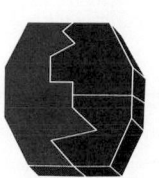  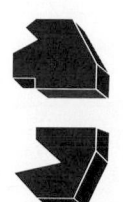

FIGURE 5-6:
How enzymes work.

Antibodies are blood proteins whose job is to bind with foreign bodies or invaders (the scientific name is *antigens*) that do not belong in the body. The invaders could be viruses, bacteria, or toxins. Each antibody fights a specific invader. For example, many different viruses cause the common cold. An antibody that binds with a certain cold virus is of no use to you if you have a different strain of the virus. However, exposure to a cold virus results in increased amounts of the specific type of antibody that can attack it. Next time that particular cold virus comes around, your body remembers and makes the right antibodies. This time the virus is destroyed faster, and your body's response (called the *immune response*) is sufficient to combat the disease.

Proteins also act as taxicabs in the body, transporting iron and other minerals, some vitamins, fats, and oxygen through the blood. For example, hemoglobin carries oxygen from the lungs to the cells of your body.

Protein plays a role in body fluid balance (to be discussed in Chapter 7) and the *acid-base balance* of the blood. Foods that you eat as well as the normal processes that go on in your body produce acids and bases. In reasonable amounts, these acids and bases can be carried away to the kidneys and lungs for excretion. It is crucial that your blood not build up high levels of acids or bases. Your blood has to remain within a neutral range. Otherwise, dangerous conditions known as *acidosis* (above-normal acidity in the blood and body fluids) and *alkalosis* (above-normal alkalinity) can occur and can even result in death. Luckily, some proteins in your blood have the chemical ability to buffer, or neutralize, both acids and bases. Proteins in your cells also can neutralize acids and bases.

In addition, amino acids can be burned to supply energy (4 kcalories per gram) if it is absolutely needed. Of course, burning amino acids for energy takes them away from their vital functions. Some amino acids can also be converted to glucose when necessary to maintain normal blood glucose levels.

Finally, protein is involved in clotting the blood when you cut yourself and blood vessels are injured. Protein fibers known as *fibrin* help form a clot so that bleeding stops.

MINI-SUMMARY

1. Proteins function as part of the body's structure, as in your skin, bones, hair, muscles, blood, blood vessels, and digestive tract.
2. Proteins are used for building and maintaining body tissues. Worn-out cells are replaced at regular intervals.
3. The greatest amount of protein is needed when the body is building new tissues rapidly, such as during pregnancy or infancy.
4. Many enzymes are made of protein. Enzymes are catalysts that help break down substances (such as foods during digestion), build up substances (such as bone), or change one substance into another (such as glucose into glycogen).
5. Some hormones are made of proteins. Hormones are chemical messengers secreted into the bloodstream by various organs to travel to a target organ and influence what it does.
6. Antibodies are blood proteins whose job is to bind with foreign bodies or invaders, such as bacteria, that do not belong in the body. Your body's response to an antigen is called the immune response.
7. Proteins also act as taxicabs in the body, transporting iron and other minerals, some vitamins, fats, and oxygen through the blood.
8. Protein plays a role in body fluid balance and the acid-base balance of the blood.
9. Amino acids can be burned to supply energy if that is absolutely necessary.
10. Protein fibers known as fibrin help form a clot so that bleeding stops.

Nutrition Science Focus: Proteins

Each of the over 50,000 different proteins in the body contains its own unique number and sequence of amino acids. In other words, each protein differs in terms of what amino acids it contains, how many it contains, and the order in which the amino acids appear.

The number and sequence of the amino acids in a protein chain is called the **primary structure**. The number of possible arrangements is as amazing as the fact that all the words in the English language are made of different sequences of 26 letters.

Also, some proteins are made of more than one chain of amino acids. For example, hemoglobin, the protein in red blood cells that carries oxygen, contains four chains of linked amino acids.

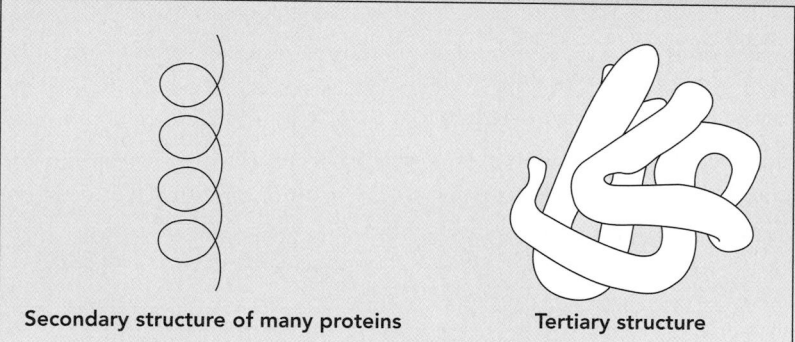

Secondary structure of many proteins   Tertiary structure

FIGURE 5-7:
Secondary and tertiary structures of proteins.

looks like the coil of a telephone cord. Each protein's unique amino acid sequence determines that protein's unique shape.

One more step must take place before the protein can do any work in the body. Due to interactions between the side groups of amino acids in the chain, the protein folds and twists around (Figure 5-7). This process results in the protein's **tertiary structure**. You can think of the tertiary structure as a phone cord that has been folded back on itself and twists around—in other words, a tangled phone cord!

In case you are wondering whether a protein's tertiary structure has any real importance, it does. A protein's tertiary structure—how it folds and twists—makes the protein able to perform its functions in the body. (Proteins with two or more chains of amino acids fold and loop even more, which results in quaternary structure.)

The instructions to make proteins in your body reside in the core, or nucleus, of each of your body's cells. In the nucleus are molecules called deoxyribonucleic acid, or DNA, which contain vital genetic information. DNA exists as two long, paired strands that are spiraled into the famous double helix (Figure 5-8). Each strand is made up of millions of chemical building blocks called bases. There are only four different bases in DNA, but they can be arranged and rearranged in countless ways. The order in

After a protein chain has been made in the body, it does not remain a straight chain. In the instant after a new protein is created, certain parts of the protein either attract or repel each other. This causes the protein to bend in a variety of ways; this is called the protein's **secondary structure**. Figure 5-7 shows a common secondary structure in which the protein

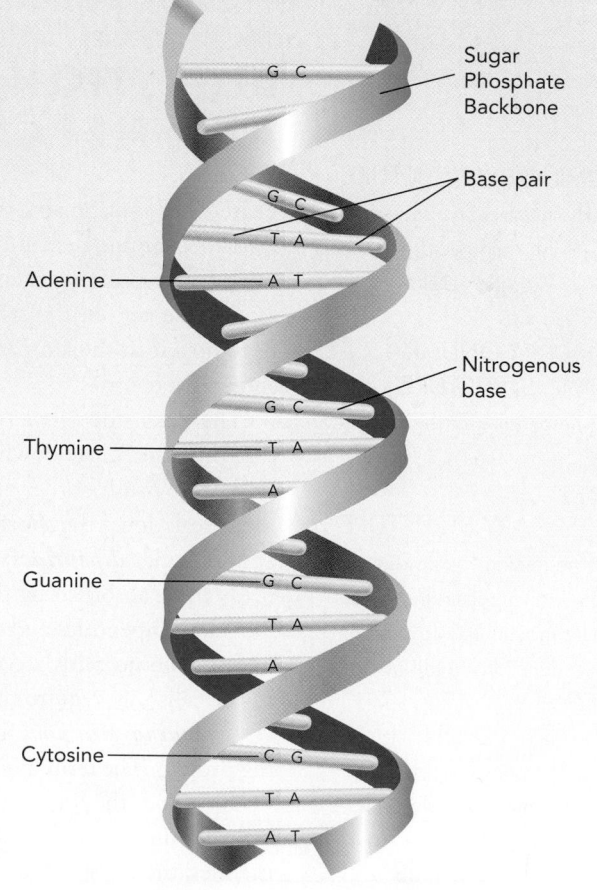

FIGURE 5-8:
Double-helix DNA.

which the bases occur determines the messages to be conveyed, such as how proteins are made. Every human cell (with the exception of mature red blood cells, which have no nuclei) contains the same DNA.

Each DNA molecule in the nucleus is housed in a chromosome. You have 23 pairs of chromosomes in each of your cells (Figure 5-9). Segments of each DNA molecule are called genes. A gene carries a particular set of instructions that allows a cell to produce a specific product—typically a protein such as an enzyme. You have about 25,000 genes in each cell.

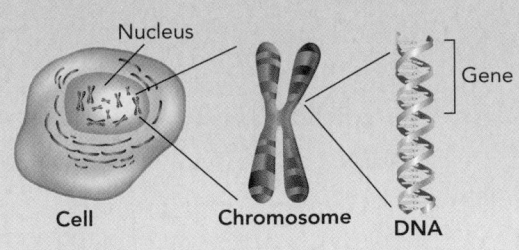

FIGURE 5-9:
Chromosomes in the cell nucleus.

Although each cell contains the instructions to make all possible human proteins, cells use genes selectively. Some genes make proteins needed for basic cell functions, and these genes stay active all the time. More typically, though, a cell activates just the genes it needs at the moment and actively suppresses the rest. The unique selection of genes used by a cell gives that cell its character—making a brain cell, say, different from a bone cell.

A sound body depends on the continuous interplay of thousands of proteins, acting together in just the right amounts and in just the right places. Many diseases have their roots in our genes. Common disorders such as heart disease and most cancers arise from a complex interaction between genes and factors in the environment.

DIGESTION, ABSORPTION, AND METABOLISM

PRIMARY STRUCTURE
The number and sequence of the amino acids in the protein chain.

SECONDARY STRUCTURE
The bending of the protein chain.

TERTIARY STRUCTURE
The folding and twisting of the protein chain that makes the protein able to perform its functions in the body.

PEPSIN
The principal digestive enzyme of the stomach.

PROTEASES
Enzymes that break down protein.

Proteins cannot be absorbed across the intestinal membranes until they are broken down into their amino acid units. Protein digestion starts in the stomach, where stomach acid uncoils the proteins (denaturation) enough to allow enzymes to enter them to do their work. The acid in the stomach, called hydrochloric acid, also converts a substance called pepsinogen to the stomach enzyme **pepsin**. Pepsin splits peptide bonds, making proteins shorter in length.

The next stop is the small intestine, where protein digestion is completed. **Proteases** (enzymes that break down protein) split up the proteins into short peptide chains and amino acids. The brush border of the small intestine produces several **peptidases**, enzymes that further break down the short peptide chains into amino acids or peptides that contain only two amino acids (**dipeptides**) or three amino acids (**tripeptides**). These smaller units are now ready to be absorbed by the microvilli in the walls of the small intestine.

When tripeptides and dipeptides enter the intestinal cells, they are split into amino acids. Because amino acids are water-soluble, they travel easily in the blood to the liver and then to the cells that require them.

An **amino acid pool** in the body provides the cells with a supply of amino acids for making protein. The term amino acid pool refers to the overall amount of amino acids distributed in the blood, the organs (such as the liver), and the body's cells. Amino acids from foods, as well as amino acids from body proteins that have been dismantled, stock these pools. In this manner, the body recycles its own proteins. If the body is making a protein and can't find an essential amino acid for it, the protein can't be completed, and the partially completed protein is disassembled or taken apart. This is important to consider for the next section on protein in food.

1. Protein digestion takes place in the stomach and small intestine.
2. The hydrochloric acid in the stomach converts pepsinogen to the enzyme pepsin, which splits peptide bonds in proteins.
3. Next, in the small intestine, proteases split up proteins into short peptide chains and amino acids.
4. The brush border of the small intestine produces several peptidases, enzymes that break down short peptide chains into amino acids, dipeptides, and tripeptides.
5. When tripeptides and dipeptides enter the intestinal cells, they are split into amino acids. Because amino acids are water-soluble, they travel easily in the blood to the liver and then to the cells that need them.
6. An amino acid pool in the body provides the cells with a supply of amino acids for making protein. If the body is making a protein and can't find an essential amino acid for it, the protein can't be completed, and the partially completed protein is taken apart.

PROTEIN IN FOOD

Protein is found in animal and plant foods (Figure 5-10). Animal foods, such as beef, chicken, fish, and dairy products, have the most protein. Among the plant foods, grains, legumes, and nuts usually contribute more protein than do vegetables and fruits. Protein-rich animal foods are usually higher in fat and saturated fat, and always higher in cholesterol, than plant foods (plant foods have no cholesterol). Protein-rich foods also tend to be the most expensive foods on the menu. One ounce of meat, poultry, or fish is equal to any of the following:

- 1 egg
- ¼ cup cooked dry beans or tofu
- 1 tablespoon peanut butter
- ½ ounce nuts or seeds

To understand the concept of protein quality, you need to recall that 9 of the 20 amino acids either can't be made in the body or can't be made in sufficient quantities.

Food proteins that provide all the essential amino acids in the proportions needed by the body are called high-quality proteins, or **complete proteins**. Examples of complete proteins include the animal proteins, such as meats, poultry, fish, eggs, milk, and other dairy products.

Lower-quality proteins, or **incomplete proteins**, are low in one or more essential amino acids. Plant proteins, including dried beans and peas, grains, vegetables, nuts, and seeds, are incomplete. The essential amino acid in lowest concentration in a protein is referred to as a **limiting amino acid** because it limits the protein's usefulness unless another food in the diet contains it.

Although plant proteins are incomplete, it does not mean they are low-quality. When certain plant foods, such as peanut butter and whole-wheat bread, are eaten over the course of a day, the limiting amino acid in each of these proteins is supplied by the other food. Such combinations are called **complementary proteins**. This is the case when grains, such as whole-wheat bread, are consumed with legumes, such as peanut butter, or when rice is eaten with beans (Figure 5-11). Beans supply plenty of the amino acids lysine and isoleucine, which are both lacking in grains such as rice. Grains have plenty of methionine and trypto-phan, which are lacking in beans and other legumes. Additional examples of legumes and

PEPTIDASES
Enzymes that break down short peptide chains into amino acids or peptides with two or three amino acids.

DIPEPTIDES
A peptide with two amino acids.

TRIPEPTIDES
A peptide with three amino acids.

AMINO ACID POOL
The overall amount of amino acids distributed in the blood, organs, and body cells.

COMPLETE PROTEINS
Food proteins that provide all the essential amino acids in the proportions needed by the body.

INCOMPLETE PROTEINS
Food proteins that contain at least one limiting amino acid.

LIMITING AMINO ACID
An essential amino acid in lowest concentration in a protein that limits the pro-tein's usefulness unless another food in the diet contains it.

COMPLEMENTARY PROTEINS
The ability of two protein foods to make up for the lack of certain amino acids in each other when eaten over the course of a day.

FIGURE 5-10: Protein, Fat, Saturated Fat, Cholesterol, and Fiber in Animal and Plant Foods

Food	Protein	Fat	Saturated Fat	Cholesterol	Fiber
ANIMAL FOODS BEEF					
Ground beef, broiled, 79 percent lean, 3 oz.	21 g	16 g	6 g	74 mg	0 g
Burger King cheeseburger	19	17	8	50	0
Bologna, 2 oz.	7	16	6	31	0
Sirloin steak, broiled, 3 oz.	24	13	5	77	0
Rib roast, 3 oz.	23	13	5	62	0
POULTRY					
Roasted chicken breast, ½ breast, with skin	27	3	1	73	0
Turkey, roasted, light meat, 3 oz., or ground turkey, breast meat only, 3 oz.	25	3	1	59	0
FISH					
Flounder, baked or broiled, 3 oz.	21	1	0.3	58	0
Salmon, baked or broiled, 3 oz.	23	9	2	74	0
Tuna fish, canned, water-pack, 3 oz.	22	1	0.2	26	0
DAIRY					
Milk, reduced-fat, 2 percent, 1 cup	8	5	3	18	0
American cheese, 1 oz.	6	7	4	18	0
Ice cream, vanilla, ½ cup	2	7	5	29	0
PLANT FOODS					
Whole-wheat bread, 1 slice	3	1	0.3	0	2
Cheerios, 1 cup	3	2	0.4	0	3
Whole-wheat spaghetti, cooked, 1 cup	7	1	0.1	0	6
Brown rice, cooked, 1 cup	5	2	0.4	0	4
Red kidney beans, 1 cup	15	1	0.1	0	13
Lentils, 1 cup	18	1	0.1	0	16
Peanut butter, 2 Tbsp.	8	16	3.4	0	2
Banana, 1 raw	1	1	0.2	0	3
Chopped broccoli, cooked, 1 cup	6	Trace	Trace	0	4.5

Source: U.S. Department of Agriculture, Agricultural Research Service, 2004. USDA Nutrient Database for Standard Reference, Release 17. Nutrient Data Laboratory Home Page, www.nal.usda.gov/fnic/foodcomp.

grains include beans and corn or wheat tortillas, lentils and rice, pea soup and bread, and beans and pasta. Nuts/seeds and grains (such as whole-grain bread spread with almond butter) also make a complete protein; so do nuts/seeds and legumes (such as sesame paste with hummus, which contains chickpeas).

Some plant proteins, such as quinoa and protein made from soybeans (isolated soy protein), are complete proteins. In adequate amounts and combinations, plant foods can supply the essential nutrients needed for growth and development and overall health. Many cultures around the world use plant proteins extensively.

Researchers have developed various ways to score the quality of food proteins. They may look at the amino acids found in the protein and how well the proteins support growth or maintenance of body tissue. Animal proteins tend to score higher than vegetable proteins, and animal protein is also more digestible. Protein scores have little use in countries where protein consumption is adequate but are useful to scientists working in countries where protein intakes are low.

MINI-SUMMARY

1. Animal proteins are examples of complete proteins, and most plant proteins are examples of incomplete proteins.
2. By eating complementary plant proteins, such as legumes with grains, you can overcome the problem presented by limiting amino acids and eat a nutritionally adequate diet.
3. In the right amounts and combinations, plant proteins can support growth and maintenance.

HEALTH EFFECTS OF PROTEIN

In the United States, getting enough protein is rarely a problem. Most Americans eat more than the Recommended Dietary Allowance. Getting enough protein in the diet is more of a problem outside the United States. Before we discuss the problems associated with eating too little protein, let's take a look at eating too much.

Eating too much protein has no benefits. It will not result in bigger muscles, stronger bones, or increased immunity. In fact, eating more protein than you need may add kcalories beyond what you require. Extra protein is not stored as protein but is stored as fat if too many kcalories are taken in. High-protein foods also are often rich in fat and therefore high in kcalories.

Diets high in protein can also be a concern if you are eating a lot of high-fat animal proteins such as hamburgers and cheese and few vegetable proteins. Eating too many high-fat animal foods, which contain much saturated fat, raises your blood cholesterol levels. Higher blood cholesterol levels increase your risk of heart disease. Too many high-fat foods also increase the chances of eating too many kcalories and gaining weight. Comparison of the fat and fiber content of animal and vegetable proteins (Figure 5-10) shows that plant sources of protein contain less fat and more fiber (except for nuts, but they contain healthy monounsaturated fats). Plant foods also contain no cholesterol and are rich in vitamins and minerals. As a matter of fact, soy protein (about 25 grams a day) included in a diet low in saturated fat and cholesterol appears to reduce the risk of heart disease by lowering blood cholesterol levels.

High intakes of animal proteins are also associated with certain cancers, such as cancer of the colon. The American Cancer Society advises you to limit consumption of red meats, especially ones that are high in fat and processed.

Studies show that high protein intakes lead to increased calcium losses. This does not necessarily mean that everyone who takes in too much protein is calcium-deficient, since the body will make up for this loss by absorbing more calcium in the intestine. However, if an individual has a high protein intake and a low calcium intake, the increased calcium absorption won't compensate for its loss. Studies also show that high protein increases the workload of the kidneys and can worsen kidney problems in patients with renal (kidney) disease.

High-protein diets have been popular for many years as a way to lose weight. They are discussed in detail in Chapter 12.

In contrast, eating too little protein can cause problems too, such as slowing down the protein rebuilding and repairing process and weakening the immune system. Developing countries have the most problems with *protein-energy malnutrition (PEM)*, also called protein-kcalorie malnutrition. PEM refers to a broad spectrum of malnutrition, from mild to serious cases. PEM can occur in infants, children, adolescents, and adults, although it most often affects children between the ages of 6 months and 5 years. The condition may result from lack of food or from infections that cause loss of appetite while increasing the body's nutrient requirements. PEM can develop slowly or rapidly. Chronic PEM has many effects, including growth retardation, lowered resistance to infection, and increased mortality rates in young children.

In severe cases of PEM, physicians often see the clinical syndromes called kwashiorkor and marasmus. *Kwashiorkor* is usually seen in children who are getting totally inadequate amounts of protein and only marginal amounts of kcalories. This condition is characterized by retarded growth and development, and the child has a protruding abdomen due to edema (swelling), a skin rash, a loss of normal hair color, irritability, and sadness. Kwashiorkor often happens when a child is weaned early from its mother's milk due to the arrival of a new baby or it occurs as a result of infections such as measles or gastroenteritis.

Marasmus is characterized by severe insufficiency of kcalories and protein, which accounts for the child's gross underweight, lack of fat stores, and wasting away of muscles. There is no edema. While marasmus usually is associated with severe food shortage and prolonged semi-starvation, it can also result from chronic or recurring infections with marginal food intake.

PROTEIN-ENERGY MALNUTRITION (PEM)

A broad spectrum of malnutrition from mild to serious cases; also called protein-kcalorie malnutrition.

KWASHIORKOR

A type of PEM associated with children who are getting inadequate amounts of protein and only marginal amounts of kcalories.

MARASMUS

A type of PEM characterized by severe insufficiency of kcalories and protein that accounts for the child's gross underweight and wasting away of muscles.

MINI-SUMMARY

1. Eating too much protein has no benefits. It will not result in bigger muscles, stronger bones, or increased immunity.

2. Diets high in protein can also be a concern if you are eating a lot of high-fat animal proteins such as hamburgers and cheese and few vegetable proteins. Eating too many high-fat animal foods, which contain much saturated fat, raises your blood cholesterol levels. Higher blood cholesterol levels increase your risk of heart disease. Eating too many high-fat foods also increases the chances of eating too many kcalories and gaining weight.

3. High intakes of animal proteins are also associated with certain cancers, such as cancer of the colon.

4. High protein intakes lead to increased calcium losses.

5. Eating too little protein is a problem in many developing countries, which sometimes have high rates of protein-energy malnutrition (PEM), also called protein-kcalorie malnutrition.

6. In severe cases of PEM, physicians often see the clinical syndromes called kwashiorkor and marasmus. Kwashiorkor is usually seen in children who are getting inadequate amounts of protein and

only marginal amounts of kcalories. Kwashiorkor is characterized by retarded growth and development and a protruding abdomen due to edema. Marasmus is characterized by severe insufficiency of kcalories and protein, which accounts for the child's gross underweight, lack of fat stores, and wasting away of muscles.

DIETARY RECOMMENDATIONS FOR PROTEIN

The 2002 RDA for protein for both men and women is 0.8 gram per kilogram of body weight. For healthy adults, the RDA works out to be 0.36 gram of protein per pound of body weight. For example, if you weigh 140 pounds, you need 50 grams of protein each day.

140 pounds × 0.36 gram/pound = 50 grams of protein

This amount allows for adequate protein to make up for daily losses in urine, feces, hair, and so on. In other words, taking in enough protein each day to balance losses results in a state of protein balance called **nitrogen balance**. The RDA for protein is generous and is based on the recommendation that proteins come from both animal and plant foods.

The amount of protein needed daily is proportionally higher during periods of growth such as infancy, childhood, and pregnancy. Accordingly, the RDA for protein is higher than 0.8 gram/kilogram of body weight during these times.

0–6 months old	9.1 grams per day (this is an AI, not an RDA)
7–12 months old	1.5 grams protein/kilogram body weight
1–3 years old	1.1 grams protein/kilogram body weight
4–8 years old	0.95 gram protein/kilogram body weight
9–13 years old	0.95 gram protein/kilogram body weight
14–18 years old	0.85 gram protein/kilogram body weight
Pregnancy and lactation (all age groups)	1.1 grams protein/kilogram body weight OR 25 grams additional protein/day

(Pregnant and lactating women need to get 1.1 grams of protein per kilogram body weight or can eat 25 grams protein above their normal RDA before they were pregnant or lactating.)

During periods of growth when a person needs to eat more protein than is lost, the body is said to be in a state of **positive nitrogen balance**. **Negative nitrogen balance** occurs during starvation and some illnesses, when the body excretes more protein than is taken in.

The 2002 Dietary Reference Intake report established Acceptable Macronutrient Distribution Ranges (AMDR) for protein. Adults should get from 10 to 35 percent of total kcalories from protein. The AMDR for children from 1 to 3 years old is 5 to 20 percent of kcalories, and for children 4 to 18 years old it is 10 to 30 percent. Tolerable upper intake levels for protein and individual amino acids could not be set due to inadequate or conflicting data.

> **NITROGEN BALANCE**
> The difference between total nitrogen intake and total nitrogen loss; a healthy person has the same nitrogen intake as loss, resulting in a zero nitrogen balance.
>
> **POSITIVE NITROGEN BALANCE**
> A condition in which the body excretes less protein than is taken in; this can occur during growth and pregnancy.
>
> **NEGATIVE NITROGEN BALANCE**
> A condition in which the body excretes more protein than is taken in; this can occur during starvation and certain illnesses.

MINI-SUMMARY

1. The RDA for protein for both men and women is 0.8 gram per kilogram of body weight.
2. The amount of protein needed daily is proportionally higher during periods of growth because the body is in a state of positive nitrogen balance.
3. Negative nitrogen balance occurs during starvation and illnesses, when the body excretes more protein than is taken in.
4. The AMDR for protein for adults is 10 to 35 percent of total kcalories.

NUTRITION

Figure 5-10 compares the fat and cholesterol content of a variety of meat, poultry, and fish. To choose nutritious cuts of meat, poultry, or fish, use these guidelines:

- Most fish is lower in fat, saturated fat, and cholesterol than are meat and poultry.
- Chicken breast and turkey breast (meaning white meat) without skin are low in fat—there is only about 3 grams of fat in 3 ounces of chicken or 3 ounces of turkey. By comparison, white meat with skin and dark meat (such as thighs and drumsticks) are much higher in fat. Also, chicken wings may be considered white meat, but they are fattier than the drumstick.
- When buying ground turkey or chicken, make sure it is made from only skinless breast meat to get the least amount of fat. If the product includes skin and dark meat, it will be much higher in fat.
- Trimmed veal is leaner than skinless chicken.
- When choosing beef, you will get the least fat from eye of round, followed by top round and bottom round.

Meat is a good source of many important nutrients, including protein, iron, copper, zinc, and some of the B vitamins, such as B_6 and B_{12}. Meat is also a significant source of fat, saturated fat, and cholesterol.

In comparison to red meats, skinless white-meat chicken and turkey are comparable in cholesterol but lower in total fat and saturated fat. The skin of chicken and turkey contains much of the bird's fat. The skin should be left on during cooking to keep in moisture but can be removed before serving. Chicken and turkey are rich in protein, niacin, and vitamin B_6. They are also good sources of vitamin B_{12}, riboflavin, iron, zinc, and magnesium. Duck and goose are quite fatty in comparison, because they contain all dark meat.

Fish and shellfish are excellent sources of protein and are relatively low in kcalories. Most are also low to moderate in cholesterol and are a good source of certain vitamins, such as vitamins E and K, and minerals, such as iodine and potassium. Certain fish (Figures 5-12 and 5-13) are fattier than others, such as mackerel and herring, but fatty fish are an important source of omega-3 fatty acids.

FIGURE 5-12: Fat Content of Fish

Low-Fat Fish (Fat Content Less Than 2.5 Percent)	Medium-Fat Fish (Fat Content 2.5–5 Percent)	High-Fat Fish (Fat Content Over 5 Percent)
Cod	Bluefish	Albacore tuna
Croaker	Swordfish	Bluefin tuna
Flounder	Yellowfin tuna	Herring
Grouper		Mackerel
Haddock		Sablefish
Pacific halibut		Salmon
Pollock		Sardines
Red snapper		Shad
Rockfish		Trout
Sea bass		Whitefish
Shark		
Sole		
Whiting		

Seafood

Nutrition Facts

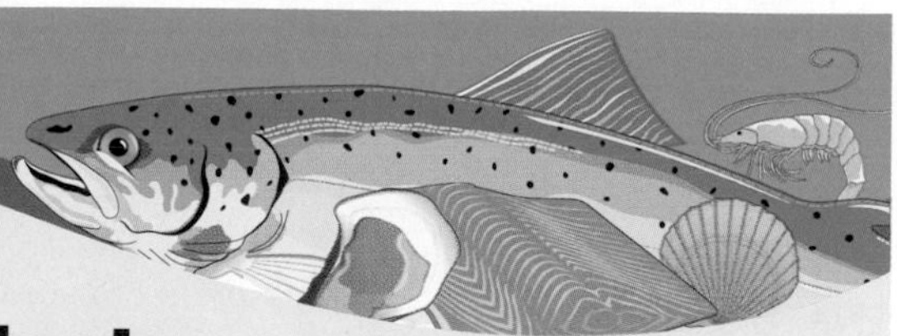

Cooked (by moist or dry heat with no added ingredients), edible weight portion. Percent Daily Values (%DV) are based on a 2,000 calorie diet.

Seafood Serving Size (84 g/3 oz)	Calories	Calories from Fat	Total Fat g	Total Fat %DV	Saturated Fat g	Saturated Fat %DV	Cholesterol mg	Cholesterol %DV	Sodium mg	Sodium %DV	Potassium mg	Potassium %DV	Total Carbohydrate g	Total Carbohydrate %DV	Protein g	Vitamin A %DV	Vitamin C %DV	Calcium %DV	Iron %DV
Blue Crab	100	10	1	2	0	0	95	32	330	14	300	9	0	0	20g	0%	4%	10%	4%
Catfish	130	60	6	9	2	10	50	17	40	2	230	7	0	0	17g	0%	0%	0%	0%
Clams, about 12 small	110	15	1.5	2	0	0	80	27	95	4	470	13	6	2	17g	10%	0%	8%	30%
Cod	90	5	1	2	0	0	50	17	65	3	460	13	0	0	20g	0%	2%	2%	2%
Flounder/Sole	100	15	1.5	2	0	0	55	18	100	4	390	11	0	0	19g	0%	0%	2%	0%
Haddock	100	10	1	2	0	0	70	23	85	4	340	10	0	0	21g	2%	0%	2%	6%
Halibut	120	15	2	3	0	0	40	13	60	3	500	14	0	0	23g	4%	0%	2%	6%
Lobster	80	0	0.5	1	0	0	60	20	320	13	300	9	1	0	17g	2%	0%	6%	2%
Ocean Perch	110	20	2	3	0.5	3	45	15	95	4	290	8	0	0	21g	0%	2%	10%	4%
Orange Roughy	80	5	1	2	0	0	20	7	70	3	340	10	0	0	16g	2%	0%	4%	2%
Oysters, about 12 medium	100	35	4	6	1	5	80	27	300	13	220	6	6	2	10g	0%	6%	6%	45%
Pollock	90	10	1	2	0	0	80	27	110	5	370	11	0	0	20g	2%	0%	0%	2%
Rainbow Trout	140	50	6	9	2	10	55	18	35	1	370	11	0	0	20g	4%	4%	8%	2%
Rockfish	110	15	2	3	0	0	40	13	70	3	440	13	0	0	21g	4%	0%	2%	2%
Salmon, Atlantic/Coho/Sockeye/Chinook	200	90	10	15	2	10	70	23	55	2	430	12	0	0	24g	4%	4%	2%	2%
Salmon, Chum/Pink	130	40	4	6	1	5	70	23	65	3	420	12	0	0	22g	2%	0%	2%	4%
Scallops, about 6 large or 14 small	140	10	1	2	0	0	65	22	310	13	430	12	5	2	27g	2%	0%	4%	14%
Shrimp	100	10	1.5	2	0	0	170	57	240	10	220	6	0	0	21g	4%	4%	6%	10%
Swordfish	120	50	6	9	1.5	8	40	13	100	4	310	9	0	0	16g	2%	2%	0%	6%
Tilapia	110	20	2.5	4	1	5	75	25	30	1	360	10	0	0	22g	0%	2%	0%	2%
Tuna	130	15	1.5	2	0	0	50	17	40	2	480	14	0	0	26g	2%	2%	2%	4%

Seafood provides negligible amounts of *trans* fat, dietary fiber, and sugars.

U.S. Food and Drug Administration
(January 1, 2008)

FIGURE 5-13:
Seafood nutrition chart.
Source: U.S. Department of Agriculture.

Culinary Science

Under certain circumstances, a protein uncoils and loses its shape, causing it to lose its ability to function. This process is called **denaturation**. In most cases, the damage cannot be reversed. Denaturation can occur both to proteins in food and to proteins in our bodies.

Denaturation can be caused by high temperatures (as in cooking), acids, bases, whipping, and a high salt concentration. For example, when you fry an egg, the proteins in the egg white become denatured and turn from clear to white. When you cook eggs or meat, the protein chains become tangled up with each other, which is why they become firmer in texture. Gluten, the protein in flour, denatures during baking to give bread and other baked goods their structure. When fresh vegetables are blanched quickly in boiling water, the heat kills the enzymes that make them spoil. Denaturation of protein can also occur in the body when the blood becomes too acidic or too basic—this is very unhealthy.

Another reaction you see during cooking that involves protein is the Maillard reaction. In the Maillard reaction, sugars and starches combine with proteins to form brown pigments. This is how baked goods get their golden brown color and aromatic odor and how meats brown. When you sear meat, the denatured proteins on the surface of the meat combine with the sugars present, creating a brown color and meaty flavor. The Maillard reaction occurs most readily above 300°F. When meat is cooked, the outside reaches a higher temperature than the inside, triggering the Maillard reaction and creating the strongest flavors on the surface.

CHEF'S TIPS

- The first step is to select a lean cut, such as one of these:
  - Beef: eye of round, inside (top) round, outside (bottom) round, sirloin tip, flank steak, top sirloin butt (Figure 5-14)
  - Veal: any trimmed cut except commercially ground and veal patties
  - Pork: pork tenderloin, pork chop (sirloin), pork chop (top loin), pork chop (loin)
  - Lamb: shank, sirloin, rack of lamb (chop)
  - Poultry: breast (skin removed after cooking)
  - Fish: all fish and shellfish
- Use flavorful rubs, when appropriate, to develop new and creative flavor options. Rubs combine dry ground spices such as chili powders, cumin, and coriander and finely cut herbs such as oregano, basil, and thyme. Rubs may be dry or wet. Wet rubs, also called pastes, use liquid ingredients such as mustards, juices, and vinegars. Pastes produce a crust on the food. Wet or dry seasoning rubs work particularly well with beef and pork because of their density and texture. Rubs can range from a 13 Cajun Spice Rub to a Jamaican Jerk Rub. To make a rub, various seasonings are mixed together and spread or patted evenly on the meat a short time before cooking for delicate items or 24 hours plus in advance for heartier meat cuts (see Figure 5-15 for the ingredients in 13 Cajun Spice Rub). The larger the piece of meat or poultry is, the longer the rub can stay on. The rub flavors the exterior of the meat as it cooks.
- Marinades bring out flavors naturally so that you don't need to drown the food in fat, cream, or sauces. Marinades allow a food to stand on its own with a light dressing, chutney, sauce, or relish (discussed more in Chapter 8). To give marinated foods flavor, try minced fruits and vegetables, low-sodium soy sauce, mustard, fresh herbs, and spices. For example, fruit-juice marinades can be flavored with Asian seasonings such as ginger and lemongrass. As another example, marinate top sirloin butt or

FIGURE 5-14:

Lean cuts of meat. Top row: flank steak, ground hamburger (95% lean meat, 5% fat), eye round roast; bottom row: top round steak, sirloin tip steak, loin pork chop.
Photo by Frank Pronesti.

FIGURE 5-15: 13 Cajun Spice Rub

Ingredients

4 cups paprika	4 Tbsp. thyme
2 cups chili powder	4 Tbsp. oregano
4 Tbsp. cayenne pepper	4 Tbsp. marjoram
4 Tbsp. black and white pepper	4 Tbsp. basil
4 Tbsp. garlic powder	4 Tbsp. gumbo filé
4 Tbsp. onion powder	4 Tbsp. fennel powder
2 cups cumin	

sirloin tip with tomato juice, herbs, and spices. Cut into strips and use in fajitas or stir-fries. Cut into cubes and grill them as kabobs.

- You can marinate fish without any citrus, which ruins the texture of the fish if it is marinated for very long. By eliminating citrus, you can marinate the fish longer, for two hours or even overnight. Try a marinade that includes fish stock, chives, tarragon, thyme, black pepper, and a touch of olive oil. The fish will absorb some liquid, which will keep it moister during cooking.
- Choose a cooking method that will produce a flavorful, moist product and that adds little or no fat to the food. Possibilities include roasting, grilling, broiling, sautéing, poaching, and braising (discussed in detail in Chapter 8).
- Smoking can be used to complement the taste of meat, poultry, or fish. Hardwoods or fruitwoods, such as the following, are best for producing tasty foods:
  - Fruit (apple, cherry, peach). These woods are too strong for light fish but work well with fatty fishes, pork, chicken, and turkey.
  - Hickory and maple. These are great for beef or pork.
  - Mesquite. Mesquite produces an aromatic smoke that works well with beef and pork.
- Organic chicken is very much worth the extra money for its superb sweet taste. When you butcher the whole chicken, there is also less fat under the skin.
- Fish is a very versatile and nutritious food. Anything, such as rice or beans or pasta, goes with fish. Serve fish on top of a vegetable ragout or with a mixture of legumes, beans, diced vegetables, and fresh herbs.
- Cedar-planked fish is another way to add flavor. Soak an untreated cedar plank and then put marinated fish on it and bake in the oven. The cedar plank will impart a unique flavor.
- Fish must be cooked very carefully. Fish is done when it just separates into flakes and turns opaque. Once cooking is completed, serve immediately for the best flavor and texture.

1. Proteins contain nitrogen.

   a. True
   b. False

2. Essential amino acids cannot be made in the body.

   a. True
   b. False

3. Every protein has a unique primary structure.

   a. True
   b. False

4. In denaturation, the protein's shape is distorted, but the protein can still function.

   a. True
   b. False

5. During digestion, proteins are broken down into their tripeptides, dipepetides, and amino acid units, which then are absorbed.

   a. True
   b. False

6. Americans tend to eat just enough protein.

   a. True
   b. False

7. The stomach enzyme pepsinogen aids in the digestion of protein.

   a. True
   b. False

8. You should try to balance your intake of protein from animal and plant sources.

   a. True
   b. False

9. Kwashiorkor is associated with children with insufficient protein intake and marginal amounts of kcalories.

   a. True
   b. False

10. Most plant foods are examples of incomplete proteins.

    a. True
    b. False

ACTIVITIES AND APPLICATIONS

1. Self-Assessment

   Write the number of times per week that you eat the foods listed below in the space provided. Then think about the following questions:

   - Is your protein coming mostly from animal or plant sources, or is it somewhat evenly balanced between the two?
   - Are the meats, poultry, and fish usually the entrées, and are the pasta, rice, vegetables, and dried beans or peas served as side dishes in smaller quantities?
   - Do you get the recommended 3 cups of beans and peas per week if you need about 2000 kcalories each day?
   - What can you do to balance the two sides better, if necessary?

Animal Protein Sources		**Plant Protein Sources**	
Red meats	_____	Dried beans	_____
Poultry	_____	Dried peas	_____
Fish	_____	Bread	_____
Milk	_____	Cereals	_____
Cheese	_____	Pasta	_____
Yogurt	_____	Rice	_____
Eggs	_____	Nuts and seeds	_____
		Vegetables (including potatoes)	_____
Total Number of Servings	_____	Total Number of Servings	_____

2. Reading Food Labels

   Following are food labels from a beef burger and a vegetable burger. Compare and contrast their nutritional content.

BEEF BURGER (3 OZ.)

Nutrition Facts	
Amount per serving	
Calories 230	**Calories from Fat** 140
	% Daily Value
Total Fat 16 g	25%
Saturated Fat 6 g	30%
Polyunsaturated 1 g	
Monounsaturated 7 g	
Cholesterol 74 g	25%
Sodium 180 mg	8%
Total Carbohydrates 0 g	0%
Dietary Fiber 0 g	0%
Sugars 0 g	
Protein 21 g	

VEGETABLE BURGER (2.5 OZ.)

Nutrition Facts	
Amount per serving	
Calories 140	**Calories from Fat** 20
	% Daily Value
Total Fat 2.5 g	4%
Saturated Fat 0.5 g	0%
Polyunsaturated 1.0 g	
Monounsaturated 0.5 g	
Cholesterol 0 g	0%
Sodium 180 mg	8%
Total Carbohydrates 21 g	7%
Dietary Fiber 5 g	20%
Sugars 0 g	
Protein 8 g	

3. **How Much Protein Do You Need?**

   Calculate how many grams of protein you need by multiplying your weight (in pounds) by 0.36.

   Example: 150 pounds × 0.36 grams protein per pound = 54 grams protein

4. **How Much Protein Do You Eat?**

   Write down everything you ate yesterday, including approximate portion sizes. If yesterday was not a typical day, write down what you normally eat during the course of a day. Find the approximate amount of protein in each food by using the following information and/or Appendix A, then total your protein intake for the day.

1 slice bread, ½ English muffin or hamburger roll,	2 grams protein
4 to 6 crackers, ½ cup cooked cereal or pasta,	
¾ cup dry cereal, ½ cup cooked beans,	
½ cup peas or corn, 1 small potato,	
½ cup sweet potato	
Fruit, margarine, butter, salad dressing	0 grams protein
1 cup raw vegetables, ½ cup cooked vegetables,	2 grams protein
½ cup tomato or vegetable juice	
1 ounce cooked meat, poultry, or fish; 1 egg,	7 grams protein
1 ounce cheese, 1 Tbsp. peanut butter	
1 cup milk, 1 cup yogurt	8 grams protein

   Now you can compare how much protein you ate on one day to the RDA. Do you consume too much, too little, or just about the right amount of protein daily? If you are eating too much, what foods would you cut down on and which foods would you eat to replace them?

5. **Meat Diet versus Mostly Plant Diet**

   Using *iProfile*, Appendix A, and/or the information from item 4, find the amount of protein in each food listed below and total each list. Each list represents one day's intake.

   Which diet contained more protein? Do either or both of these diets meet your protein RDA? Is it possible for you to get the protein you need without eating meat, poultry, or seafood?

Meat-Based Diet		**Plant-Based Diet with Dairy**	
2 eggs	_____	1 cup oatmeal	_____
2 slices white toast	_____	½ cup raisins	_____
½ cup orange juice	_____	1 corn muffin	_____
½ cup milk	_____	1 cup milk	_____
1 doughnut	_____	1 apple	_____
3 oz. roast beef	_____	1 vegetarian burger	_____
1 oz. American cheese	_____	1 whole-grain bun	_____
2 slices white bread	_____	Lettuce and tomato slices	_____
1 Tbsp. mayonnaise	_____	1 banana	_____
1-oz. package corn chips	_____	Iced tea	_____
2 cupcakes	_____	1 granola bar	_____
2 slices pizza	_____	1 cup vegetable soup	_____
1 cup vegetable salad	_____	1 cup meatless chili	_____
1 Tbsp. dressing	_____	1 cup vegetable salad	_____
12-oz. soft drink	_____	1 Tbsp. dressing	_____
1 cup vanilla ice cream	_____	½ cup milk	_____
		Peach cobbler	
Total Protein:	_____	Total Protein:	_____

6. Meatless Meals: Food Costs

   Compare recipes for chicken parmesan and eggplant parmesan in terms of their ingredients, food costs, and labor costs.

7. Cooking with Egg Substitutes

   Make the following recipe for a vegetable omelet using fresh eggs and then using Egg Beaters (a brand of egg substitutes). Egg Beaters is 99 percent egg whites and contains no fat or cholesterol. Some coloring and natural flavoring are added to make it look and taste like a whole egg. The product has less than half the calories of whole eggs. Compare how the two dishes look and taste. Next, make the omelet with two whole eggs and substitute four egg whites for the remaining two eggs. Rank the three omelets from most to least desirable.

   Use *iProfile* to do a nutrient analysis for each omelet and compare them.

8. Using *iProfile*, click on "Estimating Portion Sizes" at the top of the page. Next click on "How Much Meat or Fish Am I Eating," and "What's in a Serving of Chicken Nuggets and Fries?"

VEGETABLE OMELET

Yield: 2 servings

Ingredients

2 teaspoons vegetable oil
½ cup sliced fresh mushrooms
¼ cup sliced fresh zucchini
¼ cup fresh broccoli florets
¼ cup sliced fresh red bell pepper
1 teaspoon Italian seasoning
4 eggs, beaten (or 1 cup Egg Beaters)

Steps

1. In a nonstick skillet, heat the oil over medium heat.
2. Add the vegetables and Italian seasoning and sauté until the vegetables are tender.
3. Remove the vegetables from the skillet and keep warm.
4. Pour the eggs (or egg substitute) into the skillet. Cook, lifting the edges to allow the uncooked eggs to flow underneath.
5. When the eggs are almost set, gently spoon the sautéed vegetables into the center of the omelet.
6. Fold the sides of the omelet over the filling and slide the omelet onto the serving plate.

NUTRITION WEB EXPLORER

Nutrition Data   www.nutritiondata.com

On the home page for Nutrition Data, click on "Fast Food Nutrition Facts" to get a nutrient analysis of the foods in many fast-food restaurants. Compare and contrast the nutritional values (kcalories, fat, saturated fat, protein, cholesterol) for five entrées for at least one restaurant chain. Pick entrées that range from being very high in fat to being much lower.

Cattleman's Beef Board   www.beefitswhatsfordinner.com

Click on "Lean Beef" and write down the names of at least 10 lean cuts of beef.

Bass on Hook   www.bassonhook.com/fishforfood/fishcookingtechniques.html

Read "Fish Cooking Techniques." How long does it take to cook fresh fish that is 1 inch thick? Frozen fish that is 1 inch thick? What is the difference between blackening and bronzing?

Georgia Eggs   www.georgiaeggs.org

Click on "Eggs A to Z" and then click on "Egg Products." List 3 different types of egg products. What are the advantages of using these products? Click on "Nutrition" and find out why eggs are a high quality protein.

FOOD FACTS: SOY FOODS AND THEIR HEALTH BENEFITS

The soybean plant was domesticated in China 3000 years ago. The Chinese call it the "yellow jewel" or the "great treasure" for several reasons. Soybeans are easy to farm, and the plants do not deplete the soil. They are inexpensive, contain the most protein among all legumes (with no cholesterol), and are a very versatile food, although when merely boiled, they have a strong taste with a metallic aftertaste. Perhaps due to this problem, the soybean has been used to make a tremendous variety of products (Figure 5-16).

Soybeans are grown in abundance in the United States, but most are sold as animal feed after being processed for their oil. Soybean oil is used extensively in salad dressings, in margarine, and as salad/cooking oil.

In addition to soy oil, another important soybean product, particularly for vegetarians, is tofu, or bean curd. Tofu was invented by a Chinese scholar in 164 B.C.

FIGURE 5-16:
Soyfood products (clockwise from top), soy milk, tofu, edamame, soy flour, and soy nuts.
Courtesy of The Soyfoods Council.
Photo by Karry Hasford.

and is made in a process similar to the one used in making cheese. Soybeans are crushed to produce soymilk, which is then coagulated, causing solid curds (the tofu) and liquid whey to form. Tofu is white in color, soft in consistency, and bland in taste. It readily picks up other flavors, making it a great choice for mixed dishes such as lasagna.

Tofu is available shaped in cakes of varying textures and packed in water, which must be changed daily to keep it fresh. Firm tofu is compressed into blocks and holds its shape during preparation and cooking. Firm tofu can be used for stir-frying, grilling, or marinating. Soft tofu contains much more water and is more delicate. Soft tofu is good to use in blenderized recipes to make dips, sauces, salad dressings, spreads, puddings, cream pies, pasta filling, and cream soups. Silken tofu is even softer and more delicate and works well in creamy desserts. Tofu should be kept refrigerated and used within one week.

Other soybean products include the following:

- *Soy sauce* is made from soybeans that have been fermented. There are several types of soy sauce. Shoyu is made from soybeans and wheat. The wheat is roasted first and contributes both the soy sauce's brown color and its sharp, distinctive flavor. Tamari is made only from soybeans. Teriyaki sauce includes other ingredients, such as sugar, vinegar, and spices.
- *Miso* is similar to soy sauce but is pasty in consistency. It is made by fermenting soybeans with or without rice or other grains. A number of varieties are available, from light-colored and sweet to

dark and robust. It is used in soups and gravies, as a marinade for tofu, as a seasoning, and as a spread on sandwiches and fried tofu.
- *Tempeh* is a white cake made from fermented soybeans. It is a pleasant-tasting, high-protein food that can be cooked quickly to make dishes such as barbecued or fried tempeh or cut into pieces to add to soups. Tempeh is cultured like cheese and yogurt and therefore must be used when fresh or it will spoil.
- *Textured soy protein (TSP)* refers to products made from textured soy flour. It must be rehydrated before being used in recipes. It can replace up to one-quarter of the meat in a recipe without tasting unacceptable. It is a very concentrated source of protein and is almost fat-free. Because of its strong flavor, TSP is most successfully used in highly flavored dishes such as chili, spaghetti sauce, and curries.
- *Meat alternatives* (also called meat analogs) contain soy protein or tofu and other ingredients to simulate various kinds of meat. They are offered in forms such as hamburgers, hot dogs, bacon, ham, and chicken patties and nuggets. They contain little or no fat and no cholesterol but may be high in sodium. Some are fortified with vitamin B_{12} and iron.
- *Soy cheese* is made from soymilk. With its creamy texture, it can be substituted for sour cream or cream cheese.
- *Soy flour* is made from roasted soybeans. Because it has no gluten, it cannot fully replace whole-wheat or white flour in baking. You can replace up to a third of the wheat flour in a recipe with soy flour.
- *Green vegetable soybeans (edamame)* are soybeans that have been harvested when still green and sweet. They are

boiled and then can be served as a vegetable dish, appetizer, or snack. They are high in protein and fiber.

- *Natto* is fermented, cooked soybeans with a sticky coating and a cheesy texture. In traditional preparations, natto is used as a topping for rice, in miso soups, and with vegetables.
- *Soymilk* is made from soaked soybeans that have been ground finely and strained. It is an excellent source of protein and B vitamins. Soymilk is sold in aseptic (nonrefrigerated) containers as well as in the dairy case. Soy yogurt is also available.
- *Soynuts* are roasted and are available in a number of flavors.

Much research is being done on the health effects of soy. Foods containing soy protein may reduce the risk of coronary heart disease. If you consume 25 grams of soy protein per day, as part of a diet low in saturated fat and cholesterol, you may reduce your risk of heart disease by reducing blood cholesterol levels (LDL or "bad"

cholesterol). To get the heart-healthy benefits of soy protein, it is recommended to eat four servings of at least 6.25 grams of soy protein into your daily diet for a total of at least 25 grams each day.

Soybeans contain phytoestrogens, which are chemically similar to estrogen, the female hormone. Other plant foods contain phytoestrogens as well, but soyfoods are by far the richest source. Some possible health effects of soy are due to the fact that phytoestrogens can mildly mimic the actions of estrogen in the body. Some studies suggest that isoflavones, the major group of phytoestrogens found in soyfoods, may reduce hot flashes in women after menopause when natural estrogen is lacking. However, the evidence is not conclusive.

Researchers have also looked at the effect of phytoestrogens on breast cancer. Research in this area started when scientists saw that women from countries with high consumption of soyfoods (such as in Japan and other Asian countries) had lower rates of breast cancer. Soy's possi-

ble role in breast cancer risk is uncertain: studies show that phytoestrogens can protect but also stimulate breast cancer cell growth. Soy appears to help protect women from breast cancer if they were exposed to phytoestrogens in childhood or early adolescence. Eating soy products such as tofu as part of a balanced diet low in saturated fats and high in fruits and vegetables is safe and perhaps even beneficial. However, the long-term use of supplements containing isoflavones is not recommended as they may stimulate breast and uterine cancer. Women with breast cancer should consume only moderate amounts of soy foods as part of a healthy, plant-based diet.

After menopause, women typically lose bone mass because of greatly lowered levels of natural estrogen. Estrogen helps prevent bone loss and works together with calcium and other hormones and minerals to build bones. Although research is not conclusive, it appears that soy may help women promote bone health after menopause.

CHEF'S TIPS

- Marinate tofu with ginger-lime sauce.
- Crumble firm tofu and sauté with chopped onions, bell peppers, and other vegetables, herbs, and seasonings to make tacos and other Mexican dishes.
- Grill tofu and serve as the "meat" in a sandwich with portobello mushrooms.
- Replace part of the cream in creamed soups with blended silken tofu.
- Use blended silken or soft tofu instead of ricotta cheese in Italian dishes and other mixed dishes, such as Indian curry or a hot Thai dish.
- Use soft tofu in place of mayonnaise in salad dressings such as green goddess.

HOT TOPIC: IRRADIATION

Beef is one of the U.S. food industry's hottest sellers, to the tune of billions of pounds a year. However, beef, especially ground beef, has a dark side: It can harbor the bacterium *E. coli* 0157:H7, a pathogen that threatens the safety of the domestic food supply. If not properly prepared, beef tainted with *E. coli* 0157:H7 can make people ill and, in the case of children or the elderly, can even kill.

To help combat this public health problem, the Food and Drug Administration (FDA) approved in 1997 the treatment of red-meat products with a measured dose of radiation. This process, commonly called irradiation, has drawn praise from many food-industry and health organizations because it can control *E. coli* 0157:H7 and several other disease-causing microorganisms. Since 1963, the FDA has allowed the irradiation of a number of foods, such as poultry, fresh fruits and vegetables, dry spices, and seasonings.

The process is similar to sending luggage through a radiation field—gamma rays produced usually from electron-beams or X-rays. The amount of energy is not strong enough to add any radioactivity to the food. Irradiation works by interfering with bacterial genetics so that most bacteria no longer survive or multiply. The same irradiation process is used to sterilize medical products such as bandages, contact lens solutions, and hospital supplies such as gloves and gowns.

Treating raw meat and poultry with irradiation at the slaughter plant could eliminate bacteria commonly found raw meat and raw poultry, such as *E. coli* 0157:H7, *Salmonella*, and *Campylobacter*. These organisms currently cause millions of infections and thousands of hospitalizations in the United States every year. Raw meat irradiation could also eliminate Toxoplasma organisms, which can

be responsible for severe eye and congenital infections. Irradiating prepared ready-to-eat meats like hot dogs and deli meats could eliminate the risk of *Listeria* from such foods. Irradiation could also eliminate bacteria like *Shigella* and *Salmonella* from fresh produce. The potential benefit is also great for those dry foods that might be stored for long times and transported over great distances, such as spices and grains.

Irradiation is a "cold" process that gives off little heat, and so foods can be irradiated within their packaging and remain protected against contamination until opened by users. Because some bacteria survive the process in poultry and meats, it's still important to keep products refrigerated and cook them properly. FDA officials emphasize that although irradiation is a useful tool for reducing the risk of foodborne illness, it complements, but doesn't replace, proper food-handling practices by producers, processors, and consumers.

Additional benefits of irradiation include the fact that it reduces the level of microbes that cause spoilage in meat, poultry, seafood, fruits, and vegetables. For example, irradiated strawberries stay unspoiled for two weeks or more, versus three to five days for untreated berries. When grains, spices, legumes, and dried fruits are irradiated, the process also eliminates any insects that may be present.

Food irradiation has many benefits, but there are some concerns about whether the process is completely safe and how it affects taste and nutrition. The safety of irradiated foods has been studied by feeding them to animals and to people. These extensive studies include animal feeding studies lasting for several generations in several different species, including mice, rats, and dogs. There is no evidence of adverse health effects in these well-controlled trials. In addition, NASA astronauts eat foods

that have been irradiated to the point of sterilization (substantially higher levels of treatment than that approved for general use) when they fly in space. The safety of irradiated foods has been endorsed by the World Health Organization, the Centers for Disease Control and Prevention, and the Assistant Secretary of Health, as well as by the U.S. Department of Agriculture and the Food and Drug Administration.

Some studies have shown that irradiation can affect the taste of beef, poultry, and pork, but the difference in taste is generally subtle. Irradiation may also affect the color of poultry and pork. Certain vitamins are more sensitive to irradiation, such as thiamin, although the losses are similar to those occurring in thermal processes such as canning.

As part of its approval, the FDA requires that irradiated foods include labeling with the statement "treated with radiation" or "treated by irradiation" and the international symbol for irradiation, the radura (Figure 5-17). Irradiation labeling requirements apply only to foods sold in stores. Irradiation labeling does not apply to restaurant foods. Some restaurants serve hamburgers made with irradiated beef, as Dairy Queen has. Irradiated foods are especially useful for people who are most likely to suffer the greatest risk from food-borne illness: the elderly and the immunocompromised, such as individuals with AIDS or on chemotherapy.

FIGURE 5-17:
Radura, the international symbol for irradiation.
Source: Food and Drug Administration.

Vitamins

In the early 1900s, scientists thought they had found the compounds needed to prevent scurvy and pellagra, two diseases caused by vitamin deficiencies. These compounds originally were believed to belong to a class of chemical compounds called amines and were named from the Latin *vita*, or life, plus *amine*—vitamine. Later, the e was dropped when it was found that not all the substances were amines. At first, no one knew what they were chemically, and so vitamins were identified by letters. Later, what was thought to be one vitamin turned out to be many, and numbers were added, such as vitamin B_6. Later on, some vitamins were found to be unnecessary for human needs and were removed from the list; this accounts for some of the numbering gaps. For example, vitamin B_8, adenylic acid, was later found not to be a vitamin.

Vitamins are organic substances that carry out processes in the body that are vital to health. This chapter will help you to:

- State the general characteristics of vitamins
- Identify the functions and food sources of each of the 13 vitamins
- Identify which vitamins are most likely to be deficient in the American diet and which vitamins are most toxic
- List two health benefits of eating a diet rich in fruits and vegetables
- Discuss the use of fruits and vegetables on the menu
- Describe ways to conserve vitamins when handling and cooking fruits and vegetables
- Give examples of functional foods and discuss their role in the diet
- Define phytochemicals and give examples of foods in which they are found

 # CHARACTERISTICS OF VITAMINS

Let's start with some basic facts about vitamins.

1. Very small amounts of vitamins are needed by the human body, and very small amounts are present in foods. Some vitamins are measured in IUs (international units), a measure of biological activity; others are measured by weight, in micrograms or milligrams. To illustrate how small these amounts are, remember that 1 ounce is 28.3 grams. A milligram is $\frac{1}{1000}$ of a gram, and a microgram is $\frac{1}{1000}$ of a milligram.
2. Although vitamins are needed in small quantities, the roles they play in the body are enormously important, as you will see in a moment.
3. Most vitamins are obtained through food. Some are also produced by bacteria in the intestine (and are absorbed into the body), and one (vitamin D) can be produced when the skin is exposed to sunlight, but the body doesn't make enough.
4. There is no perfect food that contains all the vitamins in just the right amounts. The best way to ensure an adequate vitamin intake is to eat a varied and balanced diet of plant and animal foods.
5. Vitamins do not contain kcalories, and so they do not directly provide energy to the body. Vitamins provide energy indirectly because they are involved in extracting energy from carbohydrate, protein, and fat.
6. Some vitamins in foods are not the actual vitamins but are **precursors**. The body chemically changes the precursor to the active form of the vitamin.

Vitamins are classified according to how soluble they are in fat or water. ***Fat-soluble vitamins*** (A, D, E, and K) generally occur in foods containing fat, and they are not readily

PRECURSORS
Forms of vitamins that the body changes chemically to active vitamin forms.

FAT-SOLUBLE VITAMINS
A group of vitamins that generally occur in foods containing fats; include vitamins A, D, E, and K.

excreted (except for vitamin K) from the body. *Water-soluble vitamins* (vitamin C and the B-complex vitamins) are readily excreted from the body (except vitamins B_6 and B_{12}) and therefore don't often reach toxic levels.

MINI-SUMMARY

1. Very small amounts of vitamins are needed by the human body, and very small amounts are present in foods.
2. Although vitamins are needed in small quantities, the roles they play in the body are enormously important.
3. Vitamins must be obtained through foods, because vitamins either are not made in the body or are not made in sufficient quantities. There is no perfect food that contains all the vitamins in just the right amounts.
4. Vitamins have no kcalories.
5. Some vitamins in foods are not the actual vitamins but rather are precursors; the body chemically changes the precursor to the active form of the vitamin.
6. Vitamins are classified according to how soluble they are in fat or water. Fat-soluble vitamins are generally stored in the body, whereas water-soluble vitamins (except vitamins B_6 and B_{12}) are readily excreted.

FAT-SOLUBLE VITAMINS

Fat-soluble vitamins include vitamins A, D, E, and K. They generally occur in foods containing fats and are stored in the body either in the liver or in adipose (fatty) tissue until they are needed. Fat-soluble vitamins are absorbed and transported around the body in the same way as other fats. If anything interferes with normal fat digestion and absorption, these vitamins may not be absorbed. Many require protein carriers to be transported around the body.

Dietary intake of vitamin E is low enough to be of concern for both adults and children. Low intake of vitamin A is a concern for adults. This section will discuss food sources of both vitamins.

Although it is convenient to be able to store these vitamins so that you can survive periods of poor intake, excessive vitamin intake (higher than the Tolerable Upper Intake Level) causes large amounts of vitamins A and D to be stored and may lead to undesirable symptoms.

VITAMIN A

During World War I, many children in Denmark developed eye problems. Their eyes became dry and their eyelids swollen, and eventually blindness resulted. A Danish physician read that an American scientist had given milkfat to laboratory animals to cure similar eye problems in animals. At that time Danish children were drinking skim milk, because all the milkfat was being made into butter and sold to England. When the Danish doctor gave whole milk and butter to the children, they got better. The Danish government later restricted the amount of exported dairy foods. Dr. E. V. McCollum, the American scientist, eventually found vitamin A (the first vitamin to be discovered) to be the curative substance in milkfat.

In the body, vitamin A appears in three forms: retinol, retinal, and retinoic acid. Together they are referred to as *retinoids*. *Retinol* is often called preformed vitamin A and is one of the most active, or usable, forms of vitamin A. Retinol is found in animal foods and can be converted to retinal and retinoic acid in the body.

CAROTENOIDS
A class of pigments that contribute a red, orange, or yellow color to fruits and vegetables; some can be converted to retinol or retinal in the body.

BETA-CAROTENE
A precursor of vitamin A that functions as an antioxidant in the body; the most abundant carotenoid.

XEROSIS
A condition in which the cornea of the eye becomes dry and cloudy; often due to a deficiency of vitamin A.

XEROPHTHALMIA
Hardening and thickening of the cornea that can lead to blindness; usually caused by a deficiency of vitamin A.

NIGHT BLINDNESS
A condition caused by insufficient vitamin A in which it takes longer to adjust to dim lights after seeing a bright light at night; this is an early sign of vitamin A deficiency.

Some plant foods, such as carrots, contain colored pigments called *carotenoids* that can be converted to vitamin A in the body. The carotenoids contribute a red, orange, or yellow color to fruits and vegetables. *Beta-carotene*, the most abundant carotenoid, is the orange pigment you see in carrots and sweet potatoes. It is split in the intestine and liver to make retinol. Other carotenoids, such as alpha-carotene, are also converted to vitamin A, but not as efficiently.

Lycopene and lutein are other carotenoids commonly found in food. They are not sources of vitamin A but may have other health-promoting properties. The Institute of Medicine encourages consumption of carotenoid-rich fruits and vegetables for their health-promoting benefits.

Nutrition Science Focus: Fat-Soluble Vitamins

VITAMIN A

Vitamin A has several roles involving the eyes. First, it is essential for the health of the cornea, the clear membrane covering the eye. Without enough vitamin A, the cornea becomes cloudy. Eventually it dries (called *xerosis*) and thickens, and this can result in permanent blindness (*xerophthalmia*). Vitamin A is also necessary for healthy cells in other parts of the eye, such as the retina.

Vitamin A is crucial to sight for other reasons. The eye converts light energy into nerve impulses that travel to the brain, a process that uses retinal. If you don't take in enough vitamin A, you will experience a problem seeing at night. In *night blindness*, it takes longer to adjust to dim lights after seeing a bright flash of light (such as oncoming car headlights). This is an early sign of vitamin A deficiency. If the deficiency continues, xerosis and xerophthalmia can occur.

Vitamin A is involved in other parts of the body. It is needed to produce and maintain the cells (called epithelial cells) that form the protective linings of the lungs, gastrointestinal tract, urinary tract, and other organs, as well as from the skin. Vitamin A is also essential for specialized epithelial cells called goblet cells to develop and produce mucus. Mucus protects the cells from harmful organisms.

Vitamin A plays a role in reproduction, growth, and development in children, as well as in proper bone growth and tooth development. Vitamin A also helps maintain the integrity of the skin and mucous membranes, which act as a barrier to bacteria and viruses.

Vitamin A helps regulate the immune system. The immune system helps prevent or fight off infections by making white blood cells that destroy harmful bacteria and viruses.

The role of beta-carotene in preventing cancer or cardiovascular disease is being studied. Beta-carotene supplements have not been shown to protect against cancer or heart disease, and are not advisable for the general population.

Functions

Vitamin A has the following functions:

1. Maintains the health of the eye and vision
2. Promotes healthy surface linings of the lungs, gastrointestinal tract, and urinary tract
3. Promotes normal reproduction

4. Promotes growth and development, including bones and teeth
5. Regulates the immune system so it can fight off infection

Some carotenoids, such as beta-carotene found in carrots, may act as antioxidants in the body. **Antioxidants** combine with oxygen so that the oxygen is not available to oxidize, or destroy, important substances in the cell. Antioxidants prevent the oxidation of unsaturated fatty acids in the cell membrane, DNA (the genetic code), and other cell parts that substances called **free radicals** destroy. Free radicals are highly reactive compounds that normally result from cell metabolism and the functioning of the immune system.

In the absence of antioxidants, free radicals destroy cells (possibly accelerating the aging process) and alter DNA (possibly increasing the risk for cancerous cells to develop). Free radicals may also contribute to the development of cardiovascular disease. In the process of functioning as an antioxidant, beta-carotene is itself oxidized or destroyed.

Food sources

Certain plant foods are excellent sources of carotenoids. They include dark green vegetables such as spinach and deep orange fruits and vegetables such as apricots, carrots, and sweet potatoes. Beta-carotene has an orange color seen in many vitamin A-rich fruits and vegetables, but in some cases its orange color is masked by the dark green chlorophyll found in vegetables such as broccoli and spinach. In the United States, 25 to 34 percent of the vitamin A consumed by men and women is provided by carotenoids.

Sources of retinol, also called preformed vitamin A, include animal products such as liver (a very rich source), vitamin A-fortified milk, eggs, and fortified cereals. Most ready-to-eat and instant cereals are fortified with at least 25 percent of the Daily Value for vitamin A. Butter and margarine are also fortified with vitamin A. Retinol is used in fortification. Figure 6-1 lists the vitamin A content of various foods.

The RDA for vitamin A is measured in **retinol activity equivalents (RAEs)**. The concept of the RAE is used because the body obtains vitamin A from retinol and carotenoids. One RAE is equal to:

- 1 microgram of retinol
- 12 micrograms of beta-carotene
- 24 micrograms of other vitamin A precursor carotenoids

Until the RDA for vitamin A was published in 2001, vitamin A was measured in retinol equivalents (REs). Food composition tables and nutrient analysis software sometimes use REs instead of RAEs. RAE and RE values are the same for retinol and preformed vitamin A, but the RAE value for carotenoids is about half the RE value.

Deficiency and toxicity

Vitamin A has been identified as a nutrient that is consumed by American adults in amounts low enough to be of concern. Low intakes of vitamin A tend to reflect low intakes of fruits and vegetables.

Vitamin A deficiency is of most concern in developing countries, where it affects the health of many children and adults, causing night blindness, blindness, poor growth, and other problems. Up to 500,000 children worldwide go blind each year because of vitamin A deficiency. Signs of deficiency include night blindness, dry skin, dry hair, broken fingernails, and decreased resistance to infections. In the United States, vitamin A deficiency is sometimes seen in the elderly, the poor, and preschool children. In children, a mild degree of vitamin A

ANTIOXIDANT
A compound that combines with oxygen to prevent oxygen from oxidizing or destroying important substances; antioxidants prevent the oxidation of unsaturated fatty acids in the cell membrane, DNA, and other cell parts that substances called free radicals try to destroy.

FREE RADICAL
An unstable compound that reacts quickly with other molecules in the body.

RETINOL ACTIVITY EQUIVALENTS (RAE)
The unit for measuring vitamin A. One RAE equals 1 microgram of retinol, 12 micrograms of beta-carotene, or 24 micrograms of other vitamin A precursor carotenoids.

FIGURE 6-1: Food Sources of Vitamin A

FRUITS AND VEGETABLES

Food and Serving Size	Micrograms RAE	RDA Men	RDA Women
Sweet potato, baked in skin, 1 potato	1403	900	700
Carrots, boiled, $1/2$ cup	671	900	700
Spinach, boiled, $1/2$ cup	573	900	700
Collards, boiled, $1/2$ cup	489	900	700
Squash, winter, baked, $1/2$ cup	268	900	700
Lettuce, cos or romaine, raw, 1 cup	162	900	700
Apricots, canned, $1/2$ cup	104	900	700

GRAINS

Food and Serving Size	Micrograms RAE	RDA Men	RDA Women
Cream of Wheat, 1 packet	376	900	700
Special K, 1 cup	230	900	700
Frosted Flakes, 1 cup	213	900	700

DAIRY

Food and Serving Size	Micrograms RAE	RDA Men	RDA Women
Milk, nonfat, 1 cup	150	900	700
Cheese, ricotta, $1/2$ cup	148	900	700
Ice cream, vanilla, $1/2$ cup	135	900	700
Cheese, cheddar, 1 oz.	75	900	700
Cheese, American, 1 oz.	72	900	700
Cheese, cottage, creamed, $1/2$ cup	46	900	700

MEATS, POULTRY, FISH, AND ALTERNATES

Food and Serving Size	Micrograms RAE	RDA Men	RDA Women
Beef, liver, cooked, 3 oz.	6582	900	700
Braunschweiger, (a liver sausage), pork, 2 slices	2392	900	700
Egg, cooked, 1 large	85	900	700

FATS AND SWEETS

Food and Serving Size	Micrograms RAE	RDA Men	RDA Women
Pie, pumpkin, 1 piece	660	900	700
Toaster pastries, fruit, 1 pastry	150	900	700
Margarine, regular, 1 Tbsp.	116	900	700

Source: U.S. Department of Agriculture, Agricultural Research Service, 2004. USDA Nutrient Database for Standard Reference, Release 17. Nutrient Data Laboratory Home Page, www.nal.usda.gov/fnic/foodcomp.

deficiency may increase the risk of developing respiratory and diarrheal infections, decrease the growth rate, slow bone development, and decrease the likelihood of survival from a serious illness.

Prolonged use of high doses of preformed vitamin A (the Tolerable Upper Intake Level is 3000 micrograms/day) may cause symptoms of *hypervitaminosis A* such as hair loss, bone pain and damage, fatigue, skin problems, liver damage, nausea, and vomiting. High doses are particularly dangerous for pregnant women (they may cause birth defects) and the elderly (they can cause joint pain, nausea, muscle soreness, itching, hair loss, and liver and bone damage). Overconsumption of beta-carotene supplements can be quite harmful as well.

VITAMIN D

Vitamin D differs from most other vitamins in that it can be made in the body. When ultraviolet rays shine on your skin, a cholesterol-like compound is converted into a precursor of vitamin D and absorbed into the blood. Over the next one and a half to three days, the precursor is converted to *vitamin D₃*, an inactive form that also is called *cholecalciferol*. Vitamin D₃ is converted into its active form by enzymes in the liver and then the kidneys.

Of course, if you are not in the sun much or if the ultraviolet rays are cut off by heavy clothing, clouds, smog, fog, sunscreen (SPF of 8 or higher), or window glass, less vitamin D will be produced. On the positive side, a light-skinned person needs only about 15 minutes of sun on the face, hands, and arms two to three times per week to make enough vitamin D. A dark-skinned person needs more time in the sun because melanin (dark brown to black pigments in the skin) acts like a sunscreen. Several months' supply of vitamin D can be stored in the body; this is helpful during winter months when the sun is not as strong in northern climates and you need to wear more clothing. That's why you need to get an adequate amount of exposure to sunlight in the spring, summer, and fall to get you through the winter. As you get older, your body makes less active vitamin D.

> **HYPERVITAMINOSIS A**
> A disease caused by prolonged use of high doses of preformed vitamin A that can cause hair loss, bone pain and damage, soreness, and other problems.
>
> **VITAMIN D₃ (CHOLECALCIFEROL)**
> The form of vitamin D found in animal foods.

Nutrition Science Focus: Fat-Soluble Vitamins

VITAMIN D

In its active form, vitamin D functions more as a hormone than as a vitamin. Hormones are substances secreted into the bloodstream that travel to one or more organs. Once the hormone reaches what is called the target organ(s), it affects something that that organ does. The active form of vitamin D travels through the bloodstream to increase calcium (and also phosphorus) absorption in the intestine, decreases the amount of calcium excreted by the kidneys, and pulls calcium out of the bones when necessary. Blood calcium levels must be kept high so that enough calcium is present to build bones and teeth, contract and relax muscles, and transmit nerve impulses. Vitamin D works with other nutrients (such as vitamins A and C) and hormones to make and maintain bone.

Research also suggests that vitamin D may help maintain a healthy immune system and help regulate cell growth and differentiation, the process that determines what a cell is to become. Because of its role in cell growth and differentiation, vitamin D may protect against some cancers.

Functions

The major biologic function of vitamin D is to maintain normal blood levels of calcium and phosphorus. By promoting calcium absorption, vitamin D helps to form and maintain strong bones. Vitamin D also works in concert with a number of other vitamins, minerals, and hormones to promote bone mineralization. Without vitamin D, bones can become thin, brittle, or misshapen.

Food sources

Significant food sources of vitamin D include vitamin D-fortified milk and cereals. If you drink 2 cups of milk each day, you will get about half the RDA of vitamin D (the rest comes from sun exposure and other foods). No vitamin D is added to milk products such as yogurt and cheese. Vitamin D-fortified butter and margarine contain some vitamin D.

Vitamin D was previously measured in international units, and so most nutrient composition tables use IU. The current AI for vitamin D is expressed in micrograms of cholecalciferol. The relationship between IU and micrograms of cholecalciferol is as follows:

1 IU = 0.025 microgram cholecalciferol

The AI for vitamin D assumes that you are not getting any vitamin D from exposure to the sun. Many scientists think the current AI is low, and recommendations will likely increase.

Deficiency and toxicity

Vitamin D deficiency in children causes **rickets**, a disease in which bones do not grow normally, resulting in soft bones and bowed legs. Rickets is rarely seen.

Vitamin D deficiency in adults causes **osteomalacia**, a disease in which bones become soft and hurt. In the elderly, there is an increased risk among people who tend to remain indoors and avoid milk because of lactose intolerance. Vitamin D deficiency is more prevalent in colder climates, where exposure to the sun is more limited.

Vitamin D deficiency can contribute to and worsen the bone disease called **osteoporosis**. Osteoporosis is characterized by fragile bones, which are more likely to fracture. Having normal storage levels of vitamin D in your body is one of several steps needed to keep your bones strong. Osteoporosis is a major public health threat for an estimated 44 million Americans, or 55 percent of people 50 years of age and older. In 2005, 10 million individuals were estimated to have the disease already and almost 34 million more were estimated to have low bone mass, placing them at increased risk.

Vitamin D supplements are often recommended for exclusively breast-fed infants because human milk may not contain adequate supplies. Mothers of infants who are exclusively breast-fed and have limited sun exposure should consult a pediatrician on this issue. Since infant formulas are routinely fortified with vitamin D, formula-fed infants usually have adequate dietary intake of vitamin D.

Vitamin D, when taken in excess of the AI, is the most toxic of all the vitamins. All you need is about four to five times the AI to start feeling symptoms of nausea, vomiting, diarrhea, fatigue, and thirst. It can lead to calcium deposits in the heart, blood vessels, and kidneys that can cause severe health problems and even death. Young children and infants are especially susceptible to the toxic effects of too much vitamin D.

VITAMIN E

Vitamin E exists in eight different forms. Each form has its own biological activity, which is the measure of how powerful it is in the body. **Alpha-tocopherol** is the name of the most active form of vitamin E in humans. It is also a powerful antioxidant.

RICKETS
A childhood disease in which bones do not grow normally, resulting in bowed legs and knock knees; it is generally caused by a vitamin D deficiency.

OSTEOMALACIA
A disease of vitamin D deficiency in adults in which the leg and spinal bones soften and may bend.

OSTEOPOROSIS
The most common bone disease, characterized by loss of bone density and strength; it is associated with debilitating fractures, especially in people 45 years and older, due to a tremendous loss of bone tissue in midlife.

ALPHA-TOCOPHEROL
The most active form of vitamin E in humans; also a powerful antioxidant.

Functions

Vitamin E has an important function in the body as an antioxidant in the cell membrane and other parts of the cell. Antioxidants such as vitamin E act to protect your cells against the effects of free radicals. Free radicals can damage cells and may contribute to the development of cardiovascular disease and cancer. Studies are underway to determine whether vitamin E, through its ability to limit production of free radicals, might help prevent or delay the development of those chronic diseases.

Food sources

Vitamin E is widely distributed in plant foods (Figures 6-2 and 6-3). Rich sources include vegetable oils, margarine and shortening made from vegetable oils, salad dressings made from vegetable oils, nuts, seeds, leafy green vegetables, whole grains, and fortified ready-to-eat cereals. In oils, vitamin E acts like an antioxidant, thereby preventing the oil from going rancid or bad. Unfortunately, vitamin E is easily destroyed by heat and oxygen.

The RDA for vitamin E is expressed in milligrams of alpha-tocopherol (the most active form of vitamin E). Food composition tables usually measure vitamin E in milligrams of tocopherol equivalents, which includes forms of vitamin E in addition to alpha-tocopherol. If you know the number of tocopherol equivalents in a food, multiply that number by 0.8 to come up with the alpha-tocopherol content.

Deficiency and toxicity

Most Americans do not typically consume foods that are especially rich in vitamin E on a daily basis. Although salad dressings, mayonnaise, and oils provide the greatest amount of vitamin E in American diets overall, the oil most commonly used in these products is soybean oil, which is not an especially rich source of vitamin E. Oils containing higher amounts of vitamin E, such as sunflower and safflower oils, are less commonly consumed. The same is true for nuts; almonds and hazelnuts are relatively rich in vitamin E, but peanuts and peanut butter, with lower levels of vitamin E, represent the majority of all nut consumption in the United States.

Toxicity of vitamin E is rare and can cause bleeding problems.

VITAMIN K

Vitamin K is necessary for blood clotting. Every time you cut yourself, a complex series of reactions goes on to stop the bleeding. Vitamin K is essential for these reactions. Vitamin K is also important to build bones.

FIGURE 6-2: Food Sources of Vitamin E

MEATS, POULTRY, FISH, AND ALTERNATES

Food and Serving Size	Milligrams	Adult RDA
Seeds, sunflower, 1/4 cup	8.3	15
Nuts, almonds, 1/2 oz. (12 nuts)	3.7	15
Peanut butter, 2 Tbsp.	2.9	15
Peanuts, 1 oz. (28 nuts)	2.2	15
Chicken breast, cooked, 1/2 breast	1.5	15

FATS AND OILS

Food and Serving Size	Milligrams	Adult RDA
Oil, sunflower, 1 Tbsp.	5.6	15
Oil, safflower, 1 Tbsp.	4.6	15
Oil, olive, 1 Tbsp.	1.9	15
Margarine, 1 Tbsp.	1.3	15
Oil, canola, 1 Tbsp.	1.2	15
French dressing, 1 Tbsp.	0.8	15
Mayonnaise, 1 Tbsp.	0.7	15

GRAINS

Food and Serving Size	Milligrams	Adult RDA
Special K, 1 cup	4.7	15
Whole-wheat bread, 1 slice	0.2	15
White bread, 1 slice	0.0	15

FRUITS AND VEGETABLES

Food and Serving Size	Milligrams	Adult RDA
Spinach, boiled, 1/2 cup	3.4	15
Tomato sauce, 1/2 cup	2.6	15
Sweet potato, canned, 1/2 cup	1.3	15
Broccoli, boiled, 1/2 cup	1.2	15
Mangoes, raw, 1/2 cup	0.9	15
Raspberries, frozen, 1/2 cup	0.9	15

DAIRY

Food and Serving Size	Milligrams	Adult RDA
Soy milk, fluid, 1/2 cup	3.3	15

Source: U.S. Department of Agriculture, Agricultural Research Service, 2004. USDA Nutrient Database for Standard Reference, Release 17. Nutrient Data Laboratory Home Page, www.nal.usda.gov/fnic/foodcomp.

Excellent sources of vitamin K include green leafy vegetables such as kale, collards, spinach, broccoli, brussels sprouts, scallions, cabbage, and iceberg lettuce, as well as oils and margarine.

Vitamin K is also made in the body. Billions of bacteria normally live in the intestines, and some of them make a form of vitamin K. It is thought that the amount of vitamin K produced by bacteria is significant and may meet about half your needs. An infant is normally

FIGURE 6-3:
Good sources of vitamin E (vegetable oils, salad dressings, nuts, seeds, and margarine).
Photo by Frank Pronesti.

given this vitamin after birth to prevent bleeding because the intestine does not yet have the bacteria that produce vitamin K. Food sources of vitamin K provide the balance needed. Vitamin K deficiency is rare but can occur if you have problems absorbing fat or are taking certain drugs, such as antibiotics, that can destroy the bacteria in your intestines that make vitamin K.

Toxicity is normally not a problem. No Tolerable Upper Intake Level has been set for vitamin K.

Nutrition Science Focus: Fat-Soluble Vitamins

VITAMIN K

Vitamin K plays an essential role in the activation of a number of blood-clotting factors, such as prothrombin. Blood clotting prevents excessive blood loss when the skin is broken.

MINI-SUMMARY

1. Fat-soluble vitamins (A, D, E, K) generally occur in foods containing fats and are stored in the body either in the liver or in adipose (fatty) tissue until they are needed. They are absorbed and transported around the body in the same manner as other fats.

2. If anything interferes with normal fat digestion and absorption, these vitamins may not be absorbed.

3. Dietary intake of vitamin E is low enough to be of concern for both adults and children. Low intake of vitamin A is also a concern for adults.

4. Although it is convenient to be able to store these vitamins so that you can survive periods of poor intake, excessive vitamin intake (higher than the Tolerable Upper Intake Level) causes large amounts of vitamins A and D to be stored and may lead to undesirable symptoms.

5. Figure 6-4 summarizes the recommended intake, functions, and sources of the fat-soluble vitamins.

FIGURE 6-4: Summary of Fat-soluble Vitamins

Vitamin	Recommended Intake	Functions	Sources

Vitamin A	RDA:
	Men: 900 micrograms RAE
	Women: 700 micrograms RAE
	Upper Intake Level:
	3000 micrograms/day of
	preformed vitamin A
		Health of eye (especially
		   cornea and retina), vision
		Epithelial cells that form skin and
		   protective linings of lungs,
		   gastrointestinal tract, urinary tract,
		   and other organs
		Reproduction
		Growth and development
		Bone and teeth development
		Immune system function
		Antioxidant
			Preformed:
			Liver, fortified milk, fortified
			   cereals, eggs
			Beta-carotene:
			Dark green vegetables,
			   deep orange fruits and
			   vegetables

Vitamin D	AI:
	5 micrograms cholecalciferol
	(31 to 50 years old)
	10 micrograms cholecalciferol
	(51 to 70+ years old)
	Upper Intake Level:
	50 micrograms cholecalciferol
		Maintenance of blood
		   calcium and phosphorus
		   levels so that calcium can
		   build bones and teeth.
		Bone growth.
			Sunshine, fortified milk, fortified
			   cereals, fatty fish, fortified butter
			   and margarine

Vitamin E	RDA:
	15 mg alpha-tocopherol
	Upper Intake Level:
	1000 mg alpha-tocopherol
	(synthetic forms from supple-
	ments and/or fortified foods)
		Antioxidant; especially helps
		   red blood cells and cells
		   in lungs and brains
			Vegetable oils, margarine, shortening,
			   salad dressing, nuts, seeds, leafy
			   green vegetables, whole grains,
			   fortified cereals

Vitamin K	AI: Men: 120 micrograms Women: 90 micrograms	Blood clotting Healthy bones	Green leafy vegetables, vegetable oils and margarines Made in intestine

WATER-SOLUBLE VITAMINS

Water-soluble vitamins include vitamin C and the B-complex vitamins. The B vitamins include thiamin, riboflavin, niacin, folate, vitamin B_6, Vitamin B_{12}, biotin, and pantothenic acid. The B vitamins work in every body cell, where they function as part of **coenzymes**. A coenzyme combines with an enzyme to make it active. Without the coenzyme, the enzyme is useless. The body stores only limited amounts of water-soluble vitamins (except vitamins B_6 and B_{12}). Due to their limited storage, these vitamins need to be taken in daily. Dietary intakes of vitamin C are low enough to be of concern for many American adults.

Excesses of water-soluble vitamins are excreted in the urine. Even though the excesses are excreted, excessive supplementation of certain water-soluble vitamins can cause toxic side effects.

COENZYME

A molecule that combines with an enzyme and makes the enzyme functional.

VITAMIN C

Scurvy, the name for vitamin C deficiency disease, has been known since biblical times. It was most common on ships, where sailors developed bleeding gums, weakness, loose teeth, and broken capillaries (small blood vessels) under the skin and eventually died. Because sailors' diets included fresh fruits and vegetables only for the first part of a voyage, longer voyages resulted in more cases of scurvy. Once it was discovered that citrus fruits prevent scurvy, British sailors were given daily portions of lemon juice. In those days, lemons were called limes; hence, British sailors got the nickname "limeys."

SCURVY
A vitamin C deficiency disease marked by bleeding gums, weakness, loose teeth, and broken capillaries under the skin.

Nutrition Science Focus: Water-Soluble Vitamins

VITAMIN C

Figure 6-5 shows how vitamin C acts like an antioxidant. In the presence of free radicals, ascorbic acid donates two of its hydrogens (the circled ones) to the free radicals to neutralize them and prevent them from damaging DNA or other substances. Ascorbic acid then becomes dehydroascorbic acid. At a later point, dehydroascorbic acid can readily accept hydrogens to convert back to ascorbic acid.

One additional function of vitamin C is that it is needed to convert certain amino acids into the neurotransmitters serotonin and norepinephrine. These neurotransmitters are required for normal nerve cell communication.

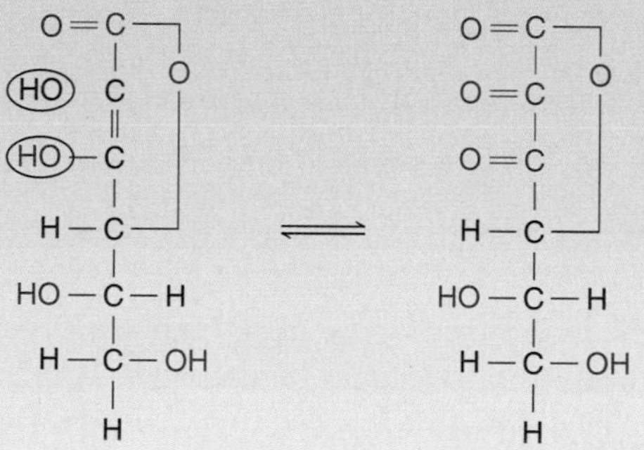

FIGURE 6-5:
Ascorbic acid (on left) as an antioxidant. In the presence of free radicals, ascorbic acid donates two of its hydrogens (the circled ones) to the free radicals to neutralize them. Ascorbic acid then becomes dehydroascorbic acid.

Functions

Vitamin C (its chemical name is ascorbic acid, meaning "no-scurvy acid") has many functions in the body.

1. Vitamin C is important in forming ***collagen,*** a protein substance that provides strength and support to bones, teeth, skin, cartilage, and blood vessels, as well as healing wounds. It has been said that vitamin C acts like cement, holding together cells and tissues.
2. Vitamin C is also important in helping to make some hormones, such as thyroxin, which regulates the metabolic rate.
3. While vitamin C has received much publicity as a cure for the common cold, the most it seems to do is shorten the length of the cold by a day. However, vitamin C

COLLAGEN
The most abundant protein in the body; a fibrous protein that is a component of skin, bone, teeth, ligaments, tendons, and other connective structures.

does have a very important connection to the body's immune system. White blood cells, which defend the body against undesirable invaders, have the highest concentration of vitamin C in the body.

4. Like vitamin E, vitamin C is an important antioxidant in the body. For example, it prevents the oxidation of iron in the intestine so that the iron can be absorbed.

Its antioxidant properties have made variations of vitamin C widely used as a food additive. As a food additive, it may appear on food labels as sodium ascorbate or calcium ascorbate. Neither of these substances has vitamin C activity.

Food sources

Foods rich in vitamin C include citrus fruits (oranges, grapefruits, limes, and lemons), bell peppers, kiwi fruit, strawberries, tomatoes, broccoli, and potatoes. Only foods from the fruit and vegetable groups contribute vitamin C. There is little or no vitamin C in the meat group (except in liver) or the dairy group. Some juices are fortified with vitamin C (if not already rich in vitamin C), as are most ready-to-eat cereals. Many people meet their needs for vitamin C simply by drinking orange juice. This is a good choice because vitamin C is easily destroyed in food preparation and cooking. Figure 6-6 lists the vitamin C content of selected foods.

FIGURE 6-6: Food Sources of Vitamin C

FRUITS AND VEGETABLES

Food and Serving Size	Milligrams	RDA Men	RDA Women
Peppers, sweet, red, boiled, $1/2$ cup	116	90	75
Kiwi fruit, 1 medium	70	90	75
Strawberries, sliced, $1/2$ cup	53	90	75
Apple juice with vitamin C, $1/2$ cup	52	90	75
Broccoli, cooked, $1/2$ cup	51	90	75
Orange juice, $1/2$ cup	48	90	75
Cranberry juice cocktail, $1/2$ cup	45	90	75
Tomato juice, 1 cup	44	90	75
Grapefruit, $1/2$ grapefruit	39	90	75
Melon, cantaloupe, $1/2$ cup	29	90	75
Sweet potato, baked, 1 potato	29	90	75
Tomatoes, red, $1/2$ cup	11	90	75

GRAINS

Food and Serving Size	Milligrams	RDA Men	RDA Women
Product 19, Kellogg's, 1 cup	61	90	75
Total, whole-grain, General Mills, $3/4$ cup	60	90	75
Special K, Kellogg's, 1 cup	21	90	75

Source: U.S. Department of Agriculture, Agricultural Research Service, 2004. USDA Nutrient Database for Standard Reference, Release 17. Nutrient Data Laboratory Home Page, www.nal.usda.gov/fnic/foodcomp.

Deficiency and toxicity

Vitamin C has been identified as a nutrient that is consumed by American adults in amounts low enough to be of concern. Low intakes of vitamin C tend to reflect low intakes of fruits and vegetables. Deficiencies resulting in scurvy are rare. Scurvy may occur in people with poor diets, especially if coupled with alcoholism, drug abuse, or diabetes.

Certain situations require additional vitamin C. They include pregnancy and nursing, growth, fevers and infections, burns, fractures, surgery, and cigarette smoking. Smoking produces oxidants, which deplete vitamin C. The RDA for smokers is 35 milligrams of vitamin C daily in addition to the normal RDA (75 mg for women, 90 mg for men).

The Tolerable Upper Intake Level for vitamin C is 2000 mg or 2 grams per day. Over 2 grams per day can cause gastrointestinal symptoms such as stomach cramps and diarrhea. High doses can interfere with certain clinical laboratory tests.

THIAMIN, RIBOFLAVIN, AND NIACIN

Functions

Thiamin, riboflavin, and niacin all play key roles as part of coenzymes in energy metabolism. They are essential in the release of energy from carbohydrates, fats, and proteins. They are also needed for normal growth.

Nutrition Science Focus: Water-Soluble Vitamins

THIAMIN, RIBOFLAVIN, AND NIACIN

Thiamin also plays a vital role in the normal functioning of the nerves. Riboflavin is part of the coenzymes that help form the vitamin B_6 coenzyme and make niacin in the body from the amino acid tryptophan.

Because thiamin, riboflavin, and niacin all help release food energy, the need for these vitamins increases as kcaloric needs increase.

Food sources

Thiamin is widely distributed in foods, but mostly in moderate amounts (Figure 6-7). Pork is an excellent source of thiamin. Other sources include dry beans, whole-grain and enriched/fortified breads and cereals, peanuts, and acorn squash.

Milk and milk products are the major source of riboflavin in the American diet. Other sources include organ meats such as liver (very high in riboflavin), whole-grain and enriched/fortified breads and cereals, eggs, and some meats (Figure 6-8).

The main sources of niacin are meat, poultry, and fish. Organ meats are quite high in niacin. Whole-grain and enriched/fortified breads and cereals, as well as peanut butter, are also important sources of niacin. All foods containing complete protein, such as those just mentioned and also milk and eggs, are good sources of the precursor of niacin, tryptophan. *Tryptophan*, an amino acid present in some of these foods, can be converted to niacin in the body. This is why the RDA for niacin is stated in **niacin equivalents**. One niacin equivalent is equal to 1 milligram of niacin or 60 milligrams of tryptophan. Less than half the niacin we use is made from tryptophan.

TRYPTOPHAN
An amino acid present in protein foods that can be converted to niacin in the body.

NIACIN EQUIVALENTS
The unit for measuring niacin. One niacin equivalent is equal to 1 milligram of niacin or 60 milligrams of tryptophan.

FIGURE 6-7: Food Sources of Thiamin

GRAINS

Food and Serving Size	Milligrams	RDA Men	RDA Women
Rice Krispies, 1-1/4 cup	0.87	1.2	1.1
Rice, white, long grain, 1/2 cup	0.55	1.2	1.1
Corn Chex, 1 cup	0.38	1.2	1.1
English muffins, 1 muffin	0.25	1.2	1.1
Rice, white, long-grain, cooked, 1/2 cup	0.13	1.2	1.1
White bread, 1 slice	0.12	1.2	1.1
Whole wheat bread, 1 slice	0.10	1.2	1.1

MEATS, POULTRY, FISH, AND ALTERNATES

Food and Serving Size	Milligrams	RDA Men	RDA Women
Pork chops, pan-fried 3 oz.	0.97	1.2	1.1
Pork, cured ham, 3 oz.	0.82	1.2	1.1
Pork, ribs, braised 3 oz.	0.43	1.2	1.1
Beans, navy, cooked, 1/2 cup	0.22	1.2	1.1
Beans, Great Northern, cooked, 1/2 cup	0.14	1.2	1.1

FRUITS AND VEGETABLES

Food and Serving Size	Milligrams	RDA Men	RDA Women
Peas, cooked, 1/2 cup	0.23	1.2	1.1
Orange juice, 1/2 cup	0.14	1.2	1.1
Acorn squash, 1/2 cup	0.12	1.2	1.1

Source: U.S. Department of Agriculture, Agricultural Research Service, 2004. USDA Nutrient Database for Standard Reference, Release 17. Nutrient Data Laboratory Home Page, www.nal.usda.gov/fnic/foodcomp.

Deficiency and toxicity

Deficiencies in thiamin, riboflavin, and niacin are rare in the United States, in large part because breads and cereals are enriched with all three nutrients. General symptoms for B-vitamin deficiencies include fatigue, decreased appetite, and depression. Alcoholism can create deficiencies in these three vitamins due in part to limited food intake.

Toxicity is not a problem except in the case of niacin. Nicotinic acid, a form of niacin, is often prescribed by physicians to lower elevated blood cholesterol levels. Unfortunately, it has some undesirable side effects. Starting at doses of 100 milligrams, typical symptoms include flushing, tingling, itching, rashes, hives, nausea, diarrhea, and abdominal discomfort. Flushing of the face, neck, and chest lasts for about 20 minutes after a person takes a large dose. More serious side effects of large doses include liver damage and high blood sugar levels.

VITAMIN B$_6$

Vitamin B$_6$ is a water-soluble vitamin that exists in three major chemical forms: pyridoxine, pyridoxal, and pyridoxamine.

FIGURE 6-8: Food Sources of Riboflavin

GRAINS

Food and Serving Size	Milligrams	RDA Men	RDA Women
Total, 3/4 cup	2.42	1.3	1.1
Wheaties, 1 cup	0.85	1.3	1.1
White bread, 1 slice	0.09	1.3	1.1
Whole-wheat bread, 1 slice	0.06	1.3	1.1

DAIRY

Food and Serving Size	Milligrams	RDA Men	RDA Women
Yogurt, plain, fat-free milk, 8-oz. container	0.53	1.3	1.1
Milk, all types, 1 cup	0.45	1.3	1.1
Cottage cheese, low-fat, 1/2 cup	0.21	1.3	1.1

FRUITS AND VEGETABLES

Food and Serving Size	Milligrams	RDA Men	RDA Women
Mushrooms, cooked, 1/2 cup	0.23	1.3	1.1
Spinach, cooked, 1/2 cup	0.21	1.3	1.1
Prunes, 1/2 cup	0.12	1.3	1.1

MEATS, POULTRY, FISH, AND ALTERNATES

Food and Serving Size	Milligrams	RDA Men	RDA Women
Beef, liver, cooked, 3 oz.	2.91	1.3	1.1
Shrimp, breaded and fried, 6–8	0.90	1.3	1.1
Egg, cooked, 1	0.26	1.3	1.1
Beef, prime rib, roasted, 3 oz.	0.18	1.3	1.1
Beef, ground, 75% lean, broiled, 3 oz.	0.15	1.3	1.1

Source: U.S. Department of Agriculture, Agricultural Research Service, 2004. USDA Nutrient Database for Standard Reference, Release 17. Nutrient Data Laboratory Home Page, www.nal.usda.gov/fnic/foodcomp.

Functions

Vitamin B_6 has many roles.

1. Vitamin B_6 is part of a coenzyme involved in carbohydrate, fat, and protein metabolism, and is particularly crucial to protein metabolism.
2. Your body needs vitamin B_6 to make hemoglobin. Hemoglobin is found in red blood cells and carries oxygen to the tissues. A vitamin B_6 deficiency can result in a form of anemia that is similar to iron-deficiency anemia.
3. Vitamin B_6 is important to the immune system through its involvement in protein metabolism and cellular growth. It helps maintain the health of the organs that make white blood cells. White blood cells help destroy viruses, bacteria, and fungi that cause infection.

> ## Nutrition Science Focus: Water-Soluble Vitamins
>
> ## VITAMIN B₆
>
> Vitamin B₆ is also used to break down glycogen to glucose, keeping your blood sugar level steady during the night, for instance, when you are not eating. Vitamin B₆ (along with vitamin C) is also needed for the synthesis of neurotransmitters such as serotonin and dopamine. These neurotransmitters are required for normal nerve cell communication.

Food sources

Good sources for vitamin B₆ include meat, poultry, fish, and fortified ready-to-eat cereals. Vitamin B₆ also appears in plant foods; however, it is not as well absorbed from these sources. Good plant sources include potatoes, some fruits (such as bananas and watermelon), and some leafy green vegetables (such as broccoli and spinach). Figure 6-9 gives the amount of vitamin B₆ in many foods.

Deficiency and toxicity

Deficiency of vitamin B₆, which may occur in some women and older adults, can cause symptoms such as fatigue, depression, and skin inflammation. The symptoms can become much more serious if the deficiency continues.

Excessive use of vitamin B₆ can cause irreversible nerve damage to the arms and legs, with symptoms such as numbness in the hands and feet and difficulty walking. The Tolerable Upper Intake Level for B₆ is 100 milligrams. The problem with B₆ is that, unlike other water-soluble vitamins, it is stored in the muscles.

Vitamin B₆ supplementation became popular when it appeared that the vitamin may relieve some of the symptoms of premenstrual syndrome and carpal tunnel syndrome, a condition in which a compressed nerve in the wrist causes much pain. No scientific research has shown that vitamin B₆ helps either condition.

FOLATE

Functions

Folate is a component of coenzymes required to form DNA, the genetic material contained in every body cell. Folate is therefore needed to make all new cells. Much folate is used to produce adequate numbers of red blood cells, white blood cells, and digestive tract cells, since these cells divide frequently.

Folate also is involved in amino acid metabolism.

Food sources

Folate gets its name from the Latin word for leaf, *folium*. Excellent sources of folate include fortified cereals, green leafy vegetables, legumes, orange juice, and fortified breads. Meats and

FIGURE 6-9: Food Sources of Vitamin B₆

MEATS, POULTRY, FISH, AND ALTERNATES

Food and Serving Size	Milligrams	Adult RDA
Tuna, fresh cooked, 3 oz.	0.88	1.3
Beef, liver, cooked, 3 oz.	0.87	1.3
Beef, top sirloin, 3 oz.	0.54	1.3
Chicken breast, roasted, 1/2 breast	0.52	1.3
Fish, halibut, 3 oz.	0.34	1.3
Fish, haddock, 3 oz.	0.32	1.3
Beef, ground 80 percent lean, broiled, 3 oz.	0.31	1.3
Beef, rib, roasted, 3 oz.	0.23	1.3
Fish, flounder, 3 oz.	0.20	1.3

GRAINS

Food and Serving Size	Milligrams	Adult RDA
Total, 3/4 cup	2.82	1.3
Wheaties, 1 cup	1.0	1.3
Corn Chex, 1 cup	0.50	1.3
Instant oatmeal, 1 packet	0.38	1.3

FRUITS AND VEGETABLES

Food and Serving Size	Milligrams	Adult RDA
Potatoes, baked, 1 potato	0.47	1.3
Bananas, 1 banana	0.43	1.3
Prune juice, 1/2 cup	0.28	1.3
Potatoes, mashed, 1/2 cup	0.25	1.3
Brussels sprouts, cooked, 1/2 cup	0.25	1.3
Sweet potato, canned, 1/2 cup	0.24	1.3
Spinach, cooked, 1/2 cup	0.22	1.3
Broccoli, boiled, 1/2 cup	0.15	1.3

DAIRY

Food and Serving Size	Milligrams	Adult RDA
Soy milk, 1 cup	0.24	1.3

Source: U.S. Department of Agriculture, Agricultural Research Service, 2004. USDA Nutrient Database for Standard Reference, Release 17. Nutrient Data Laboratory Home Page, www.nal.usda.gov/fnic/foodcomp.

dairy products contain little folate. Much folate is lost during food preparation and cooking, and so fresh and lightly cooked foods are more likely to contain more folate.

The RDA for folate is stated in micrograms of *dietary folate equivalents (DFEs)*. DFEs take into account the amount of folate that is absorbed from natural and synthetic sources. Synthetic folate, used in supplements and fortified foods such as breads, is absorbed at 1.7 times the rate of folate that naturally occurs in foods such as leafy green vegetables. Whereas 100 micrograms of naturally occurring folate is counted as 100 micrograms DFE, 100 micrograms of synthetic folate is counted as 170 micrograms DFE. The Tolerable Upper Intake Level for folate is 1000 micrograms of the synthetic form found in supplements and fortified foods.

> **DIETARY FOLATE EQUIVALENTS (DFE)**
> The unit for measuring folate; takes into account the amount of folate that is absorbed from natural and synthetic sources.

MEGALOBLASTIC (MACROCYTIC) ANEMIA

A form of anemia caused by a deficiency of vitamin B$_{12}$ or folate and characterized by large, immature red blood cells.

NEURAL TUBE DEFECTS

Diseases in which the brain and/or spinal cord form improperly in early pregnancy.

Deficiency and toxicity

A folate deficiency can cause **megaloblastic (macrocytic) anemia**, a condition in which the red blood cells are larger than normal and function poorly. The red blood cells are large because they have not matured normally due to the fact that DNA synthesis has slowed down. Symptoms include digestive tract problems such as diarrhea, fatigue, mental confusion, and depression. Groups particularly at risk for folate deficiency are pregnant women, low-birth-weight infants, and the elderly.

The need for folate is critical during the earliest weeks of pregnancy, when most women don't know they are pregnant. A folate deficiency may cause **neural tube defects**, which are malformations of the brain, spinal cord, or both during pregnancy that can result in death or lifelong disability.

Because of the importance of folate during pregnancy and the difficulty most women encounter trying to get enough folate in the diet, the Food and Drug Administration requires manufacturers of enriched bread, flour, pasta, cornmeal, rice, and some other foods to fortify their products with a very absorbable form of folate. Women who eat folate-fortified foods should not assume that these foods will meet all their folate needs. They should still seek out folate-rich foods. Folate requirements increase from 400 to 600 micrograms DFE during pregnancy.

The Tolerable Upper Intake Level for folate is 1000 micrograms/day. Intake of supplemental folate should not exceed this level to prevent folate from triggering symptoms of vitamin B$_{12}$ deficiency, which is discussed next.

Nutrition Science Focus: Water-Soluble Vitamins

FOLATE AND VITAMIN B$_{12}$

As mentioned, a deficiency of folate during the early weeks of pregnancy can cause neural tube defects. The neural tube is the tissue in the embryo that develops into the brain and spinal cord. The neural tube closes within the first month of pregnancy. Neural tube defects are diseases in which the brain and spinal cord form improperly in early pregnancy. They affect 1 to 2 of every 1000 babies born each year. Neural tube defects include anencephaly, in which most of the brain is missing, and spina bifida. In one form of spina bifida, a piece of the spinal cord protrudes from the spinal column, causing paralysis of parts of the lower body.

Folate and vitamin B$_{12}$ are both involved with making DNA and new cells. What is most interesting about these vitamins is that they each need each other to be activated in the body. Figure 6-10 shows how folate is trapped in the cells in an inactive form. Vitamin B$_{12}$ activates the folate by removing folate's methyl group. By receiving the methyl group, vitamin B$_{12}$ also becomes activated, and now both coenzymes are available to make DNA and thereby new cells.

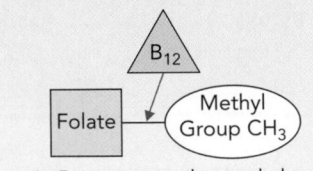

Vitamin B$_{12}$ removes the methyl group.

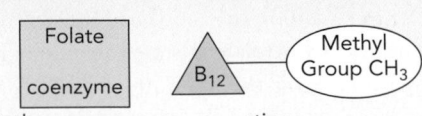

Both coenzymes are now active.

FIGURE 6-10:
Vitamin B$_{12}$ activates folate.

Because a deficiency in either folate or vitamin B_{12} causes macrocytic anemia, a physician may mistakenly administer folate when the problem is really a vitamin B_{12} deficiency. The folate will treat the anemia, but not the deterioration of the nervous system due to a lack of vitamin B_{12}. If untreated, this damage can be significant and sometimes irreversible, although it takes many years to occur. When vitamin B_{12} is deficient due to an absorption problem, injections of the vitamin must be given or a nasal spray with vitamin B_{12} must be used.

VITAMIN B_{12}

Functions

Vitamin B_{12} is called cobalamin because it contains the metal cobalt. It has several very important functions.

1. Vitamin B_{12} functions as part of a coenzyme necessary to make new cells and DNA. It also activates the folate coenzyme so that folate can also make new cells and DNA.
2. Vitamin B_{12} is needed to maintain the protective cover around nerve fibers and ensure the normal functioning of the nervous system.
3. Bone cells also depend on vitamin B_{12}.

Food sources

Vitamin B_{12} differs from other vitamins in that it is found only in animal foods such as meat, poultry, fish, shellfish, eggs, milk, and milk products. Plant foods do not naturally contain any vitamin B_{12}. Many ready-to-eat cereals are fortified with vitamin B_{12}. Unlike the other B vitamins, vitamin B_{12} is easily destroyed when foods containing it are microwaved.

Vitamin B_{12} also differs from other vitamins in that it requires ***intrinsic factor*** (a proteinlike substance produced in the stomach) to be absorbed. Vitamin B_{12} must be separated from protein in food before it can bind with intrinsic factor and be absorbed. Hydrochloric acid in the stomach helps separate the vitamin. As it enters the small intestine, vitamin B_{12} attaches to intrinsic factor and is then carried to the ileum (the last portion of the small intestine), where it is absorbed. Vitamin B_{12} is stored in the liver.

Deficiency and toxicity

A vitamin B_{12} deficiency in the body is usually due not to poor intake but to a problem with absorption. Absorption problems are often due to a lack of intrinsic factor and/or a lack of hydrochloric acid. Both conditions are more likely as you age. In fact, 10 to 30 percent of older people may be unable to absorb vitamin B_{12} in food. Individuals with disorders of the stomach or small intestine may also not absorb enough vitamin B_{12}.

When vitamin B_{12} is not properly absorbed, ***pernicious anemia*** develops. Pernicious means ruinous or harmful, and this type of anemia is marked by a megaloblastic anemia (as with folate deficiency) in which there are too many large, immature red blood cells.

INTRINSIC FACTOR
A proteinlike substance secreted by stomach cells that is necessary for the absorption of vitamin B_{12}.

PERNICIOUS ANEMIA
A type of anemia caused by a deficiency of vitamin B_{12} and characterized by macrocytic anemia and deterioration in the functioning of the nervous system.

Symptoms include extreme weakness and fatigue. Nervous system problems also erupt. The cover surrounding the nerves in the body becomes damaged, making it difficult for impulses to travel along them. This causes a poor sense of balance, numbness and tingling sensations in the arms and legs, and mental confusion. Pernicious anemia can result in paralysis and death if not treated.

Vegetarians who do not eat meats, fish, eggs, milk or milk products, or vitamin B_{12}-fortified foods consume no vitamin B_{12} and are at high risk of developing a deficiency of that vitamin. When adults adopt a vegetarian diet, deficiency symptoms can be slow to appear because it usually takes years to deplete normal body stores. However, severe symptoms of B_{12} deficiency, most often featuring poor neurological development, can show up quickly in children and breast-fed infants of women who follow a strict vegetarian diet. Fortified cereals are one of the few plant food sources of vitamin B_{12} and are an important dietary source for vegetarians who consume no eggs, milk, or milk products. Vegetarian adults who do not eat vitamin B_{12}-fortified foods need to consider taking a supplement. Likewise, vegetarian mothers should consult with a pediatrician regarding appropriate vitamin B_{12} supplementation for their infants and children.

A deficiency of folate, vitamin B_{12}, or vitamin B_6 may increase your level of homocysteine, an amino acid normally found in the blood. Elevated levels of homocysteine have been linked with coronary heart disease and stroke. The evidence suggests that high levels of homocysteine make it harder for blood to flow through blood vessels, damage heart arteries, and make it easier for blood-clotting cells called platelets to clump together and form a clot, which may lead to a heart attack. Folate supplementation has been shown to decrease homocysteine levels and improve the flow of blood through the vessels. The folate fortification program has decreased the prevalence of low levels of folate and high levels of homocysteine in the blood in middle-aged and older adults.

PANTOTHENIC ACID AND BIOTIN

Both pantothenic acid and biotin are parts of coenzymes that are involved in energy metabolism. Pantothenic acid also is involved in steps to make lipids, neurotransmitters, and hemoglobin. As part of a coenzyme, biotin is also involved in making fat and glycogen, as well as metabolizing amino acids.

Both pantothenic acid and biotin are widespread in foods. Good sources of pantothenic acid include fortified ready-to-eat cereals, beef, poultry, mushrooms, potatoes, and tomatoes. Good sources of biotin include egg yolks, whole grains, and soybeans. Intestinal bacteria make considerable amounts of biotin. Deficiency of either vitamin is rare, and toxicity concerns are not known.

CHOLINE AND VITAMINLIKE SUBSTANCES

Choline can be made in the body in small amounts. It is needed to make the neurotransmitter acetylcholine and the phospholipid lecithin, the major component of cell membranes. Lecithin is also a required component of VLDL, the lipoprotein that carries triglycerides and other lipids made in the liver to the body cells. Without enough lecithin, fat and cholesterol accumulate in the liver.

Choline is considered a conditionally essential nutrient, because when the diet contains no choline, the body can't make enough of it and liver damage can result. It is rare for the diet to contain no choline because it is so widespread in foods (rich sources include milk, eggs, and peanuts). The Food and Nutrition Board has set an Adequate Intake and a Tolerable Upper Intake Level for choline based on gender and age.

Vitaminlike substances such as carnitine, lipoic acid, inositol, and taurine are necessary for normal metabolism, but the body makes enough, and so they are not considered vitamins at this time.

Other substances are promoted as being vitamins, or at least important to human nutrition, that are clearly not vitamins or will never be vitamins. Examples include para-aminobenzoic acid (PABA), bioflavonoids (incorrectly called vitamin P), pangamic acid (incorrectly called vitamin B_{15}), and laetrile (incorrectly called vitamin B_{17}—a supposed cancer cure that is, in fact, harmful).

MINI-SUMMARY

1. Water-soluble vitamins include vitamin C and the B-complex vitamins. The B vitamins work in every body cell, where they function as coenzymes. A coenzyme combines with an enzyme to make it active. Without the coenzyme, the enzyme is useless.
2. The body stores only limited amounts of water-soluble vitamins (except vitamins B_6 and B_{12}). Due to their limited storage, these vitamins need to be taken in daily.
3. Dietary intakes of vitamin C are low enough to be of concern for many American adults.
4. Excesses of water-soluble vitamins are excreted in the urine. Even though excesses are excreted, excessive supplementation of certain water-soluble vitamins can cause toxic side effects.
5. Figure 6-11 lists the recommended intakes, functions, and sources of the water-soluble vitamins.

FIGURE 6-11: Summary of Water-soluble Vitamins

Vitamin	Recommended Intake	Functions	Sources
Vitamin C	RDA: Men: 90 mg Women: 75 mg Upper Intake Level: 2000 mg	Collagen formation Wound healing Synthesis of some hormones Healthy immune system Antioxidant Absorption of iron	Citrus fruits, bell peppers, kiwi fruit, strawberries, tomatoes, broccoli, potatoes; fortified juices, drinks, and cereals
Thiamin	RDA: Men: 1.2 mg Women: 1.1 mg Upper Intake Level: None	Part of coenzyme in energy metabolism Nerve function	Pork, dry beans, whole-grain and enriched/fortified breads and cereals, peanuts, acorn squash
Riboflavin	RDA: Men: 1.3 mg Women: 1.1 mg Upper Intake Level: None	Part of coenzymes in energy metabolism	Milk, milk products, organ meats, whole-grain and enriched/fortified breads and cereals, eggs, some meats

(continued)

FIGURE 6-11: Summary of Water-soluble Vitamins (*Continued*)

Vitamin	Recommended Intake	Functions	Sources
Niacin	RDA: Men: 16 mg niacin 　　equivalent Women: 14 mg niacin 　　equivalent Upper Intake Level: 35 mg niacin equivalents 　　(synthetic forms from 　　supplements and/or 　　fortified foods)	Part of coenzymes in 　　energy metabolism	Meat, poultry, fish, 　　organ meats, whole- 　　grain and enriched/ 　　fortified breads and 　　cereals, peanut 　　butter, milk, eggs
Vitamin B$_6$	RDA: 1.3 mg Upper Intake Level: 100 mg	Part of coenzyme involved 　　in carbohydrate, fat, and 　　especially protein metabolism Synthesis of hemoglobin 　　(in red blood cells) and 　　some neuro-transmitters Important for immune system	Meat, poultry, fish, 　　potatoes, fruits such 　　as bananas, some 　　leafy green 　　vegetables, fortified 　　cereals
Folate	RDA: 400 micrograms dietary 　　folate equivalent Upper Intake Level: 1000 micrograms dietary folate 　　equivalents (synthetic forms 　　from supplements and/or 　　fortified foods)	Part of coenzyme required to make 　　DNA and new cells Amino acid metabolism	Fortified cereals, green 　　leafy vegetables, 　　legumes, orange juice, 　　fortified breads
Vitamin B$_{12}$	RDA: 2.4 micrograms Upper Intake Level: None	Part of coenzyme that makes new 　　cells and DNA Conversion of folate into active 　　coenzyme form Normal functioning of nervous system Healthy bones	Animal foods such as 　　meat, poultry, fish, 　　shellfish, eggs, milk, 　　and milk products
Pantothenic Acid	AI: 5 mg Upper Intake Level: None	Part of coenzyme in energy 　　metabolism	Widespread Fortified cereals, beef, 　　poultry, mushrooms, 　　potatoes, tomatoes
Biotin	AI: 30 micrograms Upper Intake Level: None	Part of coenzyme involved in 　　energy metabolism and 　　synthesis of fat and glycogen	Widespread Egg yolks Made in intestine
Choline (Conditionally essential)	AI: Men: 550 mg Women: 425 mg Upper Intake Level: 3500 mg	Synthesis of neurotransmitter Synthesis of lecithin (a 　　phospholipid) found in cell 　　membranes	Widespread Milk, eggs, peanuts

Nutrition: Health Benefits of Fruits and Vegetables

INGREDIENT FOCUS: FRUITS AND VEGETABLES

Fruits and vegetables generally have the following characteristics:

- Low in kcalories
- Low or no fat (except avocados)
- No cholesterol
- Good sources of fiber
- Excellent sources of vitamins and minerals, particularly vitamins A and C (see Figure 6-12)

FIGURE 6-12: Vitamins A and C and Fiber in Fruits and Vegetables

In selecting your daily intake of fruits and vegetables, the National Cancer Institute recommends choosing:

- At least one serving of a vitamin A–rich fruit or vegetable a day.
- At least one serving of a vitamin C–rich fruit or vegetable a day.
- At least one serving of a high-fiber fruit or vegetable a day.
- Several servings of cruciferous vegetables a week. Studies suggest that these vegetables may offer additional protection against certain cancers, although further research is needed.

High in Vitamin A*	High in Vitamin C*	High in Fiber or Good Source of Fiber*	Cruciferous Vegetables
Apricots	Apricots	Apple	Bok choy
Cantaloupe	Broccoli	Banana	Broccoli
Carrots	Brussels sprouts	Blackberries	Brussels sprouts
Kale, collards	Cabbage	Blueberries	Cabbage
Leaf lettuce	Cantaloupe	Brussels sprouts	Cauliflower
Mango	Cauliflower	Carrots	
Mustard greens	Chili peppers	Cherries	
Pumpkin	Collards	Cooked beans and peas (kidney, navy, lima, and pinto beans, lentils, black-eyed peas)	
Romaine lettuce	Grapefruit	Dates	
Spinach	Honeydew melon	Figs	
Sweet potato	Kiwi fruit	Grapefruit	
Winter squash (acorn, Hubbard)	Mango	Kiwi fruit	
	Mustard greens	Orange	
	Orange	Pear	
	Orange juice		
	Pineapple		
	Plum		
	Potato with skin		

(continued)

Spinach	Prunes
Strawberries	Raspberries
Bell peppers	Spinach
Tangerine	Strawberries
Tomatoes	Sweet potato
Watermelon	

*Based on FDA's food labeling regulations.

Source: National Cancer Institute.

Fruits and vegetables contain essential vitamins, minerals, and fiber that protect you from chronic diseases. Regular consumption of fruit and vegetables is associated with reduced risks of cancer, cardiovascular disease, stroke, Alzheimer's disease, cataracts, and some of the functional declines associated with aging. Eating more fruits and vegetables can lower blood pressure as well as lower blood cholesterol.

To get a healthy variety of fruits and vegetables, think color. Eating fruits and vegetables of different colors gives your body a wide range of valuable nutrients, like folate, potassium, vitamin A, and vitamin C. Some examples include green spinach, orange sweet potatoes, black beans, yellow corn, purple plums, red watermelon, and white onions.

• Low in sodium (except for some canned vegetables)

Also, dried fruits such as raisins and apricots provide some iron.

Culinary Science

Fruits and vegetables present different challenges to a chef. Fruits are naturally sweet; they are like the candy of the plant world. Unlike vegetables, fruits soften as they become ripe and change more dramatically in color, taste, and aroma. You have to be careful to prepare them at just the right time to get perfect sweetness and a soft, yet not yet mushy, texture. Vegetables are generally firm, so you will have to cook them to get the right texture. As for flavors, vegetables range from mild to strong—from potatoes to scallions. You will have to use some tricks to get a good flavor and texture in the finished product.

Once fruits and vegetables have been picked, they no longer have their water source. Water makes up most of their weight, on average about 70 percent. The plant cells in picked fruits and vegetables continue to function and take in oxygen for breathing (called respiration in plants). Respiration at this point causes the fruit or vegetable to deteriorate quickly and its quality will go straight downhill. To slow down respiration, you need to keep it cold, keep it moist (this also helps the plant retain its water), and sometimes wrap the produce to prevent it from taking in lots of oxygen.

You may not realize it, but some fruits and vegetables are sprayed with a waxy coating to prevent moisture loss and slow down respiration. The waxy coating can affect your cooked product, so you should wash it or peel it off. Apples, citrus, peppers, and cucumbers are often coated with wax. Certain produce, such as apples and carrots, naturally have lower respiration rates than others such as lettuce and green beans with high respiration rates.

The cells found in plants are different from those found in animals. Plant cells are full of water, which makes them firm. Plant cells also have a cell wall that is flexible and contains fiber, which helps make the structure strong. Once plant cells lose water, the cells shrink in size due to the water loss, the cell walls sag, and the plant looks limp. You have probably seen lettuce in this condition: it is called wilting. Fortunately, you can put lettuce into water for a few hours and its cells will fill up with water and no longer look wilted. Greens will also wilt if you put your salad dressing on too early; the oil in the dressing is the culprit. So put your dressing on just before service, and use cold salad plates to keep the greens crisper a little bit longer.

Fruits and vegetables are mostly carbohydrates and are fairly easy to cook. They cook evenly and become soft and tender, although you have to be careful about getting the right texture. With heat, plant cells die, lose water, and soften (except for some of the fiber in and around the cell wall). The amount of fiber varies in vegetables. For example, carrots have more fiber than spinach, and broccoli stalks have more fiber than the florets.

The fiber in fruits and vegetables can be made firmer by adding sugar, or an acid such as lemon juice, to cooking vegetables. Sugar is sometimes used in cooking fruits, such as poached pears, so the product is not too mushy. Heat and alkalis (such as baking soda) soften fruit and vegetable fiber. Baking soda should not be used when cooking vegetables.

Besides controlling texture changes as you cook, you also need to control flavor changes. The longer a vegetable is cooked, the more flavor is lost into the cooking liquid and into the air through evaporation. To decrease flavor loss, it is best to cook in just enough water to cover and cook for as short a time as possible. The best way to cook vegetables is to steam them; this method cooks fast and helps retain both flavor and nutrients.

Not only does overcooking produce flavor loss, it often creates flavor change. Some vegetables take on new, undesirable flavors when overcooked. Members of the cabbage family—cabbage, turnips, cauliflower, brussels sprouts, broccoli—will develop a strong acrid taste and unpleasant smell if overcooked, owing to chemical changes. These vegetables will taste best when cooked quickly, with the cover off the pot to allow evaporation of the strong-flavored substances.

Finally, you need to control changes in color as the vegetable cooks. Vegetables may be grouped by color into four categories: green, yellow and orange, red, and white. These colors come from substances in vegetables known as pigments. Certain pigments react to heat, acid, or alkali during cooking, undergoing chemical changes that cause a vegetable to change color.

- Green vegetables: Chlorophyll is the pigment present in all green vegetables such as green beans and spinach. Chlorophyll is destroyed by acids, such as lemon juice and vinegar, and by baking soda. Prolonged cooking or overcooking causes green vegetables to turn drab olive green. Steaming is the preferred method for cooking because steam cooks quickly and lessens the loss of nutrients and flavor. If not steaming, cook uncovered to allow plant acids to escape.
- Yellow and orange vegetables: Carotenoids are the yellow and orange pigments found in carrots, corn, sweet potatoes, and winter squash. These pigments are very stable to acids and heat, but some of them can lose color from overcooking.
- Red vegetables: Anthocyanins are the red pigments found only in a few vegetables, such as beets and red cabbage. These red pigments react very strongly to acids and alkalis. Acids make anthocyanins brighter red, and alkalis turn them a blue or blue-green color. So, a small amount of acid gives red beets and red cabbage a bright red color. Because acids toughen vegetables and prolong cooking time, in recipes that call for lemon juice, tomatoes, or other acids, add only a small amount at the beginning of cooking and the remaining toward the end after the vegetables have become tender. Because anthocyanins dissolve easily in water, cook these vegetables quickly in as little water as needed.

- **White vegetables:** Flavones are the white pigments found in potatoes, onions, cauliflower, turnips, and white cabbage. Cook these vegetables for a short time (especially in a steamer) to avoid loss of nutrients, flavor, and color. You can add a little lemon juice to keep vegetables such as cauliflower white. Overcooking and hard water turn white vegetables a dull gray.

Some general rules for cooking vegetables are to cook as close to service time as possible, don't overcook, and steam whenever possible. Microwaving also works well with individual portions.

CHEF'S TIPS

- Fruits work especially well with brunch. For example, there is fruit compote (fresh or dried fruits cooked in their juices and flavored with spices or liqueur), glazed spiced grapefruit, baked apple wrapped with phyllo dough, or steel-oat cut cakes with kiwi salsa.
- Figures 6-13 and 6-14 show the nutritional values for various fruits and vegetables.
- Fruits are a natural in salads with vegetables: combine pineapple, raisins, and carrots or combine white grapes with cucumbers. Feature colorful fruits such as oranges and broccoli in a mixed salad. Cooked fruits and vegetables have many possibilities, too: sweet potatoes with apples and pears or lemon-glazed baby carrots.
- Roasted fruits with shallots are a wonderful base on which to place proteins. For example, place a chicken paillard (a cutlet pounded and grilled or sautéed) on a bed of roasted peaches and mango with jicama or sautéed bok choy, or put monkfish on a bed of roasted pears and fennel. Fruits can also be used to make relishes, chutneys, glazes, and mojo.
- Fruits have always been a natural for dessert: fresh, roasted, or baked into a cobbler with an oatmeal almond crust or into a phyllo strudel.
- Fruits such as pineapple, kiwi, mango, and papaya work well in salsas and relishes.
- Berries, such as blueberries and strawberries, are wonderful when you want vibrant colors in sauces, toppings, and garnishes.
- Vegetables allow you to serve what appears to be a sumptuous portion without the dish being high in kcalories, fat, or cholesterol. What's also wonderful about vegetables is that they are not very expensive when in season.
- When using vegetables, you need to think about what's in season for maximum flavor and about how the dish will look and taste. Think flavor and color. For example, halibut works well with marinated beets, haricots verts, grilled zucchini, summer squash, or eggplant with a tomato relish.
- Also think variety. Serving vegetables doesn't mean switching from broccoli to cauliflower and then back to broccoli. There are many, many varieties of vegetables to choose from. Be adventurous.
- Most of a meal's eye appeal comes from vegetables—use them to your advantage. The length of asparagus and green beans is a great way to bring a single entrée plate or buffet platter together. Be sure to use not only vegetables' colors but their ability to be cut into many different shapes that can accentuate your food.
- Olives add exquisite flavor with even just a small amount. They are included in many of today's eating styles. Here are some great varieties.
  - Picholine: French green, salt brine-cured, delicate flavor, great for salads and relishes.
  - Kalamata: Greek black, harvested fully ripened, deep purple with great flavor, great for vegetable stews, component salads, relishes, and compotes.
  - Nicoise: French black, small, rich nutty flavor; great for roasting with fish, vegetables, or poultry; wonderful in relishes, salads, and ragouts.
  - Manzanilla: Spanish green, available pitted and stuffed, adds zest to a salad or a vegetable dish.
  - Linguiria Black Italian; vibrant flavor; good with roasted vegetables and meats, sautéed vegetables; relishes with fish; and composition salads.
  - Gaeta: Black Italian: wrinkled appearance, usually marinated with rosemary; makes a great stuffing addition as well as a flavoring for stews, ragouts, and sautés.

Nutrition Facts

Raw, edible weight portion. Percent Daily Values (%DV) are based on a 2,000 calorie diet.

Fruits — Serving Size (gram weight/ounce weight)	Calories	Calories from Fat	Total Fat (g)	Total Fat (%DV)	Sodium (mg)	Sodium (%DV)	Potassium (mg)	Potassium (%DV)	Total Carbohydrate (g)	Total Carbohydrate (%DV)	Dietary Fiber (g)	Dietary Fiber (%DV)	Sugars (g)	Protein (g)	Vitamin A (%DV)	Vitamin C (%DV)	Calcium (%DV)	Iron (%DV)
Apple — 1 large (242 g/8 oz)	130	0	0	0	0	0	260	7	34	11	5	20	25g	1g	2%	8%	2%	2%
Avocado — California, 1/5 medium (30 g/1.1 oz)	50	35	4.5	7	0	0	140	4	3	1	1	4	0g	1g	0%	4%	0%	2%
Banana — 1 medium (126 g/4.5 oz)	110	0	0	0	0	0	450	13	30	10	3	12	19g	1g	2%	15%	0%	2%
Cantaloupe — 1/4 medium (134 g/4.8 oz)	50	0	0	0	20	1	240	7	12	4	1	4	11g	1g	120%	80%	2%	2%
Grapefruit — 1/2 medium (154 g/5.5 oz)	60	0	0	0	0	0	160	5	15	5	2	8	11g	1g	35%	100%	4%	0%
Grapes — 3/4 cup (126 g/4.5 oz)	90	0	0	0	15	1	240	7	23	8	1	4	20g	0g	0%	2%	2%	0%
Honeydew Melon — 1/10 medium melon (134 g/4.8 oz)	50	0	0	0	30	1	210	6	12	4	1	4	11g	1g	2%	45%	2%	2%
Kiwifruit — 2 medium (148 g/5.3 oz)	90	10	1	2	0	0	450	13	20	7	4	16	13g	1g	2%	240%	4%	2%
Lemon — 1 medium (58 g/2.1 oz)	15	0	0	0	0	0	75	2	5	2	2	8	2g	0g	0%	40%	2%	0%
Lime — 1 medium (67 g/2.4 oz)	20	0	0	0	0	0	75	2	7	2	2	8	0g	0g	0%	35%	0%	0%
Nectarine — 1 medium (140 g/5.0 oz)	60	5	0.5	1	0	0	250	7	15	5	2	8	11g	1g	8%	15%	0%	2%
Orange — 1 medium (154 g/5.5 oz)	80	0	0	0	0	0	250	7	19	6	3	12	14g	1g	2%	130%	6%	0%
Peach — 1 medium (147 g/5.3 oz)	60	0	0.5	1	0	0	230	7	15	5	2	8	13g	1g	6%	15%	0%	2%
Pear — 1 medium (166 g/5.9 oz)	100	0	0	0	0	0	190	5	26	9	6	24	16g	1g	0%	10%	2%	0%
Pineapple — 2 slices, 3" diameter, 3/4" thick (112 g/4 oz)	50	0	0	0	10	0	120	3	13	4	1	4	10g	1g	2%	50%	2%	2%
Plums — 2 medium (151 g/5.4 oz)	70	0	0	0	0	0	230	7	19	6	2	8	16g	1g	8%	10%	0%	2%
Strawberries — 8 medium (147g/5.3 oz)	50	0	0	0	0	0	170	5	11	4	2	8	8g	1g	0%	160%	2%	2%
Sweet Cherries — 21 cherries; 1 cup (140 g/5.0 oz)	100	0	0	0	0	0	350	10	26	9	1	4	16g	1g	2%	15%	2%	2%
Tangerine — 1 medium (109 g/3.9 oz)	50	0	0	0	0	0	160	5	13	4	2	8	9g	1g	6%	45%	4%	0%
Watermelon — 1/18 medium melon; 2 cups diced pieces (280 g/10.0 oz)	80	0	0	0	0	0	270	8	21	7	1	4	20g	1g	30%	25%	2%	4%

Most fruits provide negligible amounts of saturated fat, *trans* fat, and cholesterol; avocados provide 0.5 g of saturated fat per ounce.

U.S. Food and Drug Administration
(January 1, 2008)

FIGURE 6-13:
Fruits nutrition guide.
Courtesy U.S. Department of Agriculture.

Vegetables
Nutrition Facts

Raw, edible weight portion.
Percent Daily Values (%DV) are
based on a 2,000 calorie diet.

Vegetables Serving Size (gram weight/ounce weight)	Calories	Calories from Fat	Total Fat g / %DV	Sodium mg / %DV	Potassium mg / %DV	Total Carbohydrate g / %DV	Dietary Fiber g / %DV	Sugars g	Protein g	Vitamin A %DV	Vitamin C %DV	Calcium %DV	Iron %DV
Asparagus 5 spears (93 g/3.3 oz)	20	0	0 / 0	0 / 0	230 / 7	4 / 1	2 / 8	2g	2g	10%	15%	2%	2%
Bell Pepper 1 medium (148 g/5.3 oz)	25	0	0 / 0	40 / 2	220 / 6	6 / 2	2 / 8	4g	1g	4%	190%	2%	4%
Broccoli 1 medium stalk (148 g/5.3 oz)	45	0	0.5 / 1	80 / 3	460 / 13	8 / 3	3 / 12	2g	4g	6%	220%	6%	6%
Carrot 1 carrot, 7" long, 1 1/4" diameter (78 g/2.8 oz)	30	0	0 / 0	60 / 3	250 / 7	7 / 2	2 / 8	5g	1g	110%	10%	2%	2%
Cauliflower 1/6 medium head (99 g/3.5 oz)	25	0	0 / 0	30 / 1	270 / 8	5 / 2	2 / 8	2g	2g	0%	100%	2%	2%
Celery 2 medium stalks (110 g/3.9 oz)	15	0	0 / 0	115 / 5	260 / 7	4 / 1	2 / 8	2g	0g	10%	15%	4%	2%
Cucumber 1/3 medium (99 g/3.5 oz)	10	0	0 / 0	0 / 0	140 / 4	2 / 1	1 / 4	1g	1g	4%	10%	2%	2%
Green (Snap) Beans 3/4 cup cut (83 g/3.0 oz)	20	0	0 / 0	0 / 0	200 / 6	5 / 2	3 / 12	2g	1g	4%	10%	4%	2%
Green Cabbage 1/12 medium head (84 g/3.0 oz)	25	0	0 / 0	20 / 1	190 / 5	5 / 2	2 / 8	3g	1g	0%	70%	4%	2%
Green Onion 1/4 cup chopped (25 g/0.9 oz)	10	0	0 / 0	10 / 0	70 / 2	2 / 1	1 / 4	1g	0g	2%	8%	2%	2%
Iceberg Lettuce 1/6 medium head (89 g/3.2 oz)	10	0	0 / 0	10 / 0	125 / 4	2 / 1	1 / 4	2g	1g	6%	6%	2%	2%
Leaf Lettuce 1 1/2 cups shredded (85 g/3.0 oz)	15	0	0 / 0	35 / 1	170 / 5	2 / 1	1 / 4	1g	1g	130%	6%	2%	4%
Mushrooms 5 medium (84 g/3.0 oz)	20	0	0 / 0	15 / 0	300 / 9	3 / 1	1 / 4	0g	3g	0%	2%	0%	2%
Onion 1 medium (148 g/5.3 oz)	45	0	0 / 0	5 / 0	190 / 5	11 / 4	3 / 12	9g	1g	0%	20%	4%	4%
Potato 1 medium (148 g/5.3 oz)	110	0	0 / 0	0 / 0	620 / 18	26 / 9	2 / 8	1g	3g	0%	45%	2%	6%
Radishes 7 radishes (85 g/3.0 oz)	10	0	0 / 0	55 / 2	190 / 5	3 / 1	1 / 4	2g	0g	0%	30%	2%	2%
Summer Squash 1/2 medium (98 g/3.5 oz)	20	0	0 / 0	0 / 0	260 / 7	4 / 1	2 / 8	2g	1g	6%	30%	2%	2%
Sweet Corn kernels from 1 medium ear (90 g/3.2 oz)	90	20	2.5 / 4	0 / 0	250 / 7	18 / 6	2 / 8	5g	4g	2%	10%	0%	2%
Sweet Potato 1 medium, 5" long, 2"diameter (130 g/4.6 oz)	100	0	0 / 0	70 / 3	440 / 13	23 / 8	4 / 16	7g	2g	120%	30%	4%	4%
Tomato 1 medium (148 g/5.3 oz)	25	0	0 / 0	20 / 1	340 / 10	5 / 2	1 / 4	3g	1g	20%	40%	2%	4%

Most vegetables provide negligible amounts of
saturated fat, *trans* fat, and cholesterol.

U.S. Food and Drug Administration
(January 1, 2008)

FIGURE 6-14:
Vegetables nutrition guide.
Courtesy U.S. Department of Agriculture.

CHECK-OUT QUIZ

1. Vitamin E deficiency in adults causes osteomalacia.

   a. True
   b. False

2. Water-soluble vitamins are not toxic when taken in excess of the RDA because they are fully excreted.

   a. True
   b. False

3. Vitamin E is a fat-soluble vitamin.

   a. True
   b. False

4. Vitamins are needed in very small amounts.

   a. True
   b. False

5. Vitamins supply kcalories and energy to the body.

   a. True
   b. False

6. Lycopene is the most abundant carotenoid.

   a. True
   b. False

7. Water-soluble vitamins are more likely than fat-soluble vitamins to be needed in the diet on a daily basis.

   a. True
   b. False

8. Vitamin B_6 requires intrinsic factor to be absorbed.

   a. True
   b. False

9. Thiamin, riboflavin, and niacin all play key roles as coenzymes in energy metabolism.

   a. True
   b. False

10. Good sources of folate include green leafy vegetables, legumes, and orange juice.

    a. True
    b. False

11. Name the vitamins described in the following.

    a. Which vitamin(s) is present only in animal foods?
    b. Which vitamin(s) is found in high amounts in pork and ham?
    c. Which vitamin(s) is found mostly in fruits and vegetables?
    d. Which vitamin(s) needs a compound made in the stomach to be absorbed?
    e. Which vitamin, when deficient, causes osteomalacia?
    f. Which vitamin is made from tryptophan?
    g. Which vitamin(s) is made by intestinal bacteria?
    h. Which vitamin(s) do you need more of if you eat more protein?
    i. Which vitamin(s) is needed for clotting?
    j. Which vitamin(s) is conditionally essential?
    k. Which vitamin(s) is known for forming a cellular cement?
    l. Which vitamin has a precursor called beta-carotene?
    m. Which vitamin is made in the skin?
    n. Which vitamin(s) is an antioxidant?
    o. Which vitamin(s) is needed for bone growth and maintenance?
    p. Which vitamin(s), when deficient, causes night blindness?
    q. Which vitamin(s) is purposely put into milk because there are no other good sources of it available?

ACTIVITIES AND APPLICATIONS

1. **Your Eating Style**

   Using the tables that summarize the fat-soluble and water-soluble vitamins (or a worksheet the instructor hands you), circle any foods that you do not eat at all or that you eat infrequently, such as dairy products and green vegetables. Do you eat most of the foods containing vitamins, or do you hate vegetables and maybe fruits too? In terms of frequency, how often do you eat vitamin-rich foods? The answers to these questions should help you assess whether your diet is adequately balanced and varied, which is necessary to ensure an adequate vitamin intake.

2. **Supermarket Sleuth**

   Check your local pharmacy or supermarket to view the selection of supplements available. How many supply only 100 percent of the RDA? How many supply 500 percent or more of the RDA? Are any "nonvitamins" being sold? Name three.

3. **Vitamin Salad Bar**

   You are to set up a salad bar by using a worksheet the instructor hands out. You may use any foods you like in the salad bar as long as you have a good source of each of the 13 vitamins and fill each of the circles. In each circle on the worksheet, write down the name of the food and which vitamin(s) it is rich in.

4. Rainbow Dinner

Along with your classmates, you are to write a dinner menu of high-vitamin foods that, once prepared, will provide a rainbow of colors and be a good source of all the vitamins discussed in this chapter. Once you have written the menu, check that you have a good source of all the vitamins by doing a nutrient analysis using *iProfile*. If possible, prepare your menu and enjoy the meal with your classmates and others.

NUTRITION WEB EXPLORER

Food and Drug Administration   www.cfsan.fda.gov/~dms/supplmnt.html

This is a special FDA site related to supplements. Click on "Questions and Answers" and find out what a dietary supplement is and where you can get information about a specific dietary supplement.

Centers for Disease Control and Prevention   www.fruitsandveggiesmatter.gov

On this home page for the Fruits and Vegetables campaign, do the exercise "How Many Fruits and Vegetables Do You Need?"

Then click on "What Counts as a Cup?" How many grapes, baby carrots, raisins, or grapefruit count as 1 cup?

Next, click on recipes. Select one recipe and do a nutrient analysis of it using *iProfile*. Which vitamin is highest in your recipe? Compare to others in your class.

FOOD FACTS: FUNCTIONAL FOODS: SUPERFOODS

In contrast to dietary supplements, functional foods are foods found in a usual diet that have biologically active components, such as lycopene found in tomato products, which may provide health benefits beyond basic nutrition. The Food and Nutrition Board has defined a functional food as "any food or food ingredient that may provide a health benefit beyond the traditional nutrients it contains." Many of the health-promoting ingredients in functional foods are phytochemicals. Phytochemicals are bioactive compounds found in plants that are linked to decreased risk of chronic diseases. Phytochemicals are discussed in more detail in this chapter's Hot Topic on page 222. Following are some examples of functional foods, which you might want to think of as superfoods because they are good for your health.

- *Beans.* Beans are an inexpensive way to get lots of nutrients, such as fiber, potassium, and many B vitamins. In addition, beans contain several phytochemicals known as saponins, which are naturally occurring compounds widely distributed in the cells of beans and peas. Studies suggest that saponins may lower blood cholesterol levels and help protect the body from developing cancer.
- *Nuts.* A growing number of clinical studies indicate that the beneficial effect of tree nuts may be due not only to the fact that they contain healthy types of fats—monounsaturated and polyunsaturated fats—but that they contain other

key nutrients, which may provide supplementary health benefits, such as decreasing total cholesterol and LDL. This helps reduce the risk of heart disease.

- *Cocoa.* Cocoa, which is used to make chocolate, is made from cacao beans. Cocoa is a rich source of antioxidants that help protect your blood vessels and heart. Dark chocolate contains more cocoa than milk chocolate and therefore contains more antioxidants.
- *Tea.* In recent years, scientists have investigated the potential benefits of green tea because tea is a rich source of polyphenols, which act as antioxidants in the body. Laboratory studies suggest that green tea may help protect against or slow the growth of certain cancers, but studies in people have shown mixed results. Drinking green tea each day appears to be useful for maintaining cardiovascular health. Green tea goes through a fermentation process in order to make black tea. Black tea also contains polyphenols, but does not contain quite as much as green tea.
- *Spinach.* Spinach contains lutein, a phytochemical that seems to help protect the eyes from cataracts and muscular degeneration, a progressive condition affecting the central part of the retina that leads to the loss of sharpness in vision. Spinach is a powerhouse of antioxidants.
- *Plant stanols and sterols.* Plant stanols and sterols occur naturally in small amounts in many plants. They are added to certain margarines and some other foods, and help block the absorption of

cholesterol from the digestive tract, which helps to lower LDL.

Additional superfoods could include berries, cabbage, broccoli, brussels sprouts, citrus fruits, pumpkin, soy, sweet potatoes, tomatoes, and whole grains.

The food industry has created another category of functional foods by adding vitamins, minerals, phytochemicals, and/or herbs into food and beverages. For example, at the supermarket you will find orange juice with calcium added, margarine with plant stanols to lower cholesterol, enhanced drinks that contain medicinal herbs such as ginseng, and energy drinks with taurine. Beverages have become a popular way for people to consume healthy ingredients.

The use of probiotics and prebiotics is a promising area for functional foods. Probiotics refers to adding health-promoting bacteria to foods such as yogurt. Yogurt often contains live bacterial cultures. The live bacteria improve the balance of intestinal microbes and can result in improved intestinal health and possibly a reduced risk of cancer and heart disease. A prebiotic is a nondigestible food ingredient such as fiber. These ingredients, which can't be digested, stimulate the growth and activity of health-promoting bacteria in the colon. Dietary fibers, sugar replacers (polyols), and other nonabsorbable sugars can all function as prebiotics. For example, inulin, a soluble fiber found in chicory root, is used in a national brand of yogurt.

There are concerns about how effective and safe this category of functional foods is. When manufacturers add herbs to foods, they don't have to disclose how much is added. Is it safe? Do herbs or other added ingredients really work? Will making tea with St. John's wort (a herb) really help depression? Will beta-carotene in a food it normally doesn't appear in still work like the beta-carotene in a carrot? Will any of the added ingredients interact with a medication or dietary supplement you take? Is the fortified food a healthy one, or is it a fortified candy bar?

There are many questions about functional foods. Although it may be possible to isolate specific components of food that may reduce the risk of diseases such as cancer, it is unclear whether phytochemicals added to foods have the same health benefits as whole foods because compounds in foods may act synergistically to impart health benefits. Like dietary supplements, functional foods will not compensate for a poor diet. Whole foods still contain the right amount and balance of nutrients and phytochemicals to promote health.

HOT TOPIC: PHYTOCHEMICALS

Phytochemicals (pronounced fight-o-chemicals) are bioactive compounds that are linked to decreased risk of chronic diseases. Major sources of these bioactive compounds are plants, especially fruits and vegetables. Some foods known for phytochemicals include red wine, which contains reversatrol, which may influence heart health; tomatoes, which contain lycopene, which may have anticancer properties; berries such as blueberries and raspberries, which contain flavonoids, which may act as antioxidants; and green tea, which contains polyphenols, which may have anticancer properties.

The way phytochemicals work and optimum amounts for human consumption are still being investigated. Some act as antioxidants, such as lycopene in tomatoes and phenols in tea, or they may act in other ways.

- Broccoli contains the chemical sulforaphane, which seems to initiate increased production of cancer-fighting enzymes in the cells.
- Isoflavonoids, found mostly in soy foods, are known as plant estrogens or phytoestrogens because they are similar to estrogen and interfere with its actions (estrogen seems to promote breast tumors).
- Members of the cabbage family (cabbage, broccoli, cauliflower, mustard greens, kale), also called cruciferous vegetables, contain phytochemicals such as indoles and dithiolthiones. They activate enzymes that destroy cancer-causing substances. The consumption of cruciferous vegetables has been associated with a reduced risk of cancer of the lung, colon, and rectum.
- Flavonoids, which are found in citrus fruits, onions, apples, grapes, wine, and tea, are thought to be helpful in preventing heart disease and cancer.
- Phytochemicals in grapes may reduce the risk of heart disease risk by preventing blood clotting.
- Garlic is the edible bulb from a plant in the lily family. It has been used as both a medicine and a spice for thousands of years. Garlic contains allicin, a chemical that acts as an antioxidant similar to vitamins A, C, and E, and may help protect the body from free radicals. Some evidence indicates that taking garlic may slightly lower blood cholesterol levels; studies have shown positive effects for short-term (1 to 3 months) use.

Phytochemicals are usually related to the color of fruits and vegetables: green, yellow-orange, red, blue-purple, and white. Figure 6-15 lists phytochemicals according to color. You can benefit from all of them by eating five to nine servings of colorful fruits and vegetables every day.

Phytochemical Names	Good For	Food Sources
Green Vegetables & Fruits		
Lutein and Zeaxanthin	Healthy eyesight	Turnip, collard, and mustard greens; kale; spinach; lettuce; broccoli; green peas; kiwi; honeydew melon

Indoles	Anti-cancer	Broccoli, cabbage, Brussels sprouts, bok choy, arugula, Swiss chard, turnips, rutabaga, watercress, cauliflower, kale
Yellow/Orange Vegetables & Fruits		
Bioflavonoids	Healthy heart, healthy anti-cancer	Oranges, grapefruit, lemons, tangerines, clementines, peaches, papaya, apricots, nectarines, pears, pineapple, yellow raisins, yellow pepper

Red Vegetables & Fruits		
Anthocyanins	Healthy circulation, healthy nerve function, anti-cancer	Raspberries, cherries, strawberries, cranberries, beets, apples, red cabbage, red onion, kidney beans, red beans

Blue/Purple Vegetables and Fruits		
Anthocyanins	Healthy circulation, healthy nerve function, anti-cancer	Blueberries, purple grapes, blackberries, black currants, elderberries.
Phenolics	Healthy cells, anti-cancer	Raisins, prunes, plums, eggplant

White Vegetables and Fruits		
Allium and Allicin	Healthy immune system, healthy cholesterol levels, anti-cancer	Garlic, onions, leeks, scallions, chives

Source: National Cancer Institute

FIGURE 6-15:
Phytochemicals in foods. Source: National Cancer Institute.

CHAPTER

7

Water and Minerals

Water
Functions
How Much Water Do You Need?
Bottled Water

Major Minerals
Calcium and Phosphorus
Nutrition Science Focus: Calcium and Phosphorus
Magnesium
Nutrition Science Focus: Magnesium
Sodium
Potassium
Chloride
Other Major Minerals

Nutrition Science Focus: Water and Electrolytes

Trace Minerals
Iron

Nutrition Science Focus: Iron
Zinc
Nutrition Science Focus: Zinc
Iodine
Selenium
Fluoride
Chromium
Copper
Other Trace Minerals

Ingredient Focus: Nuts and Seeds
Nutrition
Culinary Science

Food Facts: How to Retain Vitamins and Minerals from Purchasing to Serving

Hot Topic: Dietary Supplements

If you weighed all the minerals in your body, they would amount to only 4 or 5 pounds. You need only small amounts of minerals in your diet, but they perform enormously important jobs in your body: building bones and teeth, regulating your heartbeat, and transporting oxygen from the lungs to the tissues, to name a few.

Some minerals are needed in relatively large amounts in the diet—over 100 milligrams daily. These minerals, called *major minerals*, include calcium, chloride, magnesium, phosphorus, potassium, sodium, and sulfur. Other minerals, called *trace minerals* or trace elements, are needed in smaller amounts—less than 100 milligrams daily. Iron, fluoride, and zinc are examples of trace minerals. Figure 7-1 lists the major and trace minerals.

Minerals have some distinctive properties that are not shared by other nutrients. For example, whereas over 90 percent of dietary carbohydrates, fats, and proteins are absorbed into the body, the percentage of minerals that is absorbed varies tremendously. Only 15 percent of the iron in your diet is normally absorbed, about 30 percent of calcium is absorbed, and almost all the sodium you eat is absorbed. Minerals in animal foods tend to be absorbed better than do those in plant foods, because plant foods contain fiber and other substances that bind minerals, preventing them from being absorbed. The degree to which a nutrient is absorbed and is available to be used in the body is called *bioavailability*. Sometimes minerals compete with each other for absorption.

Unlike vitamins, minerals are inorganic elements that are not destroyed in food storage or preparation. They are, however, water-soluble, and so there is some loss in cooking liquids. Like vitamins, minerals can be toxic when consumed in excessive amounts and may interfere with the absorption and metabolism of other minerals. Like vitamins, minerals interact with one another. For example, a high phosphorus intake limits the absorption of magnesium.

Deny someone food and he or she can still live for weeks. But death comes quickly, in a matter of a few days, if you deprive a person of water. Nothing survives without water, and virtually nothing takes place in the body without water playing a vital role.

This chapter will help you to:

- State the general characteristics of minerals
- Identify the percentage of body weight made up of water
- List the functions of water in the body
- Identify the functions and food sources of the major minerals and the trace minerals
- Identify which minerals are most likely to be deficient in the American diet
- Discuss the nutrition of nuts and seeds and how to use them on the menu
- Distinguish between different types of bottled waters
- Explain how dietary supplements are regulated and labeled
- Identify instances when supplements may be necessary

MAJOR MINERALS
Minerals needed in relatively large amounts in the diet—over 100 milligrams daily.

TRACE MINERALS
Minerals needed in smaller amounts in the diet—less than 100 milligrams daily.

BIOAVAILABILITY
The degree to which a nutrient is absorbed and available to be used in the body.

FIGURE 7-1: Major and Trace Minerals

Major Minerals	Trace Minerals
Calcium	Chromium
Chloride	Copper
Magnesium	Fluoride
Phosphorus	Iodine
Potassium	Iron
Sodium	Manganese
Sulfur	Molybdenum
	Selenium
	Zinc

WATTER

The cells in your body are full of water. The ability of water to dissolve so many substances allows your cells to use valuable nutrients, minerals, and chemicals in everyday processes. Water is also sticky and elastic and tends to clump together in drops rather than spread out in a thin film. This allows water (and its dissolved substances) to move through the tiny blood vessels in our bodies.

The average adult's body weight is generally 50 to 60 percent water—enough, if the water were bottled, to fill 40 to 50 quarts. For example, in a 150-pound man, water accounts for about 90 pounds and fat for about 30 pounds, with protein, carbohydrates, vitamins, and minerals making up the balance. Men generally have proportionally more water than women, and a lean person has more than an obese person. Some parts of the body have more water than others. Human blood is about 92 percent water, muscle and brain tissue about 75 percent, and bone 22 percent.

FUNCTIONS

Almost all body cells need and depend on water to perform their functions. Water serves as the medium for many metabolic activities and also participates in some metabolic reactions. Water carries nutrients to the cells and carries away waste materials to the kidneys and out of the body in urine.

Water is needed in each step of the process of converting food into energy and tissue. Water in the digestive secretions softens, dilutes, and liquefies the food to make digestion easier. It also helps move food along the gastrointestinal tract. Differences in the fluid concentration on the two sides of the intestinal wall improve absorption of nutrients.

Over 90 percent of blood is water, so water maintains blood volume in your body. Water in blood also helps maintain normal temperatures. For example, when you exercise, your muscles work extremely hard and create energy and heat. Your body needs to get rid of the heat, and so your blood circulates to the muscles, picks up the heat, and circulates to your skin. Sweating takes place, and you lose some of the water and heat. The end result is that you are cooled down.

Water serves as an important part of body lubricants, helping to cushion the joints and internal organs; keeping tissues in the eyes, lungs, and air passages moist; and surrounding and protecting the fetus during pregnancy.

HOW MUCH WATER DO YOU NEED?

In the United States, women who are adequately hydrated consume 2.7 liters (about 91 ounces) of total water daily from all beverages and foods. Men consume 3.7 liters (about 125 fluid ounces) from all sources.

Typically, about 80 percent of total water intake comes from beverages, including drinking water, coffee, tea, and cola. Contrary to popular belief, there is no convincing evidence that caffeine in coffee, colas, and other beverages is dehydrating. The remaining 20 percent comes from moisture found in foods. Nearly all foods have some water. Milk, for example, is about 87 percent water, eggs about 75 percent, meat between 40 and 75 percent, vegetables from 70 to 95 percent, cereals from 8 to 20 percent, and bread around 35 percent.

The Adequate Intake for total water is based on the average water consumption of people who are adequately hydrated. The Adequate Intakes for total water for adults and the elderly are as follows. Don't forget that these numbers include all fluids and the water in food.

Adequate Intake for Total Water
Men: 3.7 liters/day (about 15$\frac{1}{2}$ cups)
Women: 2.7 liters/day (about 11$\frac{1}{2}$ cups)

The Food and Nutrition Board does not offer any rule of thumb for how many glasses of water people should drink each day. This is because our hydration needs can be met through drinking and eating juice, milk, coffee, tea, soda, fruits, vegetables, and other foods and beverages.

By drinking fluids at meals and between meals when thirst dictates, healthy individuals adequately satisfy their hydration needs. People who engage in strenuous or prolonged physical activity and those who are exposed to hot temperatures may need to consume more total water to replace that lost in sweat.

With aging, thirst declines, as does the ability of the kidneys to conserve water. However, the elderly appear to adequately maintain total body water content from day to day by drinking beverages at meals and drinking fluids when thirsty.

When it is normal and healthy, the body maintains water at a constant level. A number of mechanisms, including the sensation of thirst, operate to keep body water content within narrow limits. There are, of course, conditions in which the various body mechanisms for regulating water balance do not work, such as severe vomiting, diarrhea, excessive bleeding, high fever, burns, and excessive perspiration. In these situations, large amounts of fluids and minerals are lost. These conditions are medical problems that should be managed by a physician.

The body gets rid of the water it doesn't need through the kidneys and skin and, to a lesser degree, from the lungs and gastrointestinal tract. Water is excreted as urine by the kidneys, along with waste materials carried from the cells. About 4 to 6 cups a day is excreted as urine. The amount of urine reflects, to some extent, the amount of an individual's fluid intake. However, the kidneys always excrete a certain amount each day (about 2 cups) to eliminate waste products generated by the body's metabolic actions. In addition to urine, air released from the lungs contains some water, and evaporation that occurs on the skin (when one is sweating or not sweating) contains water as well.

BOTTLED WATER

People buy bottled water for what it does not have: calories, sugar, caffeine, additives, preservatives, and, in most cases, not too much sodium. The Food and Drug Administration (FDA) has established standards of quality for bottled drinking water. Bottled water is defined as water that is intended for human consumption that is sealed in bottles or other containers with no added ingredients except that it may optionally contain safe and suitable antimicrobial agents. Also, fluoride may be added within limits. The FDA has established maximum allowable levels for physical, chemical, microbiological, and radiological contaminants in the standard recommendations for bottled water quality. Similar standards are set for tap water.

Bottled water is different from tap water in that it has more consistent quality and taste. The taste of water has to do with the way it is treated and the quality of its source, including its natural mineral content. One of the key taste differences between tap water and bottled water is due to how the water is disinfected. Tap water may be disinfected with chlorine, chloramines, ozone, or ultraviolet light to kill disease-causing germs. Tap water is often disinfected with chlorine or chloramines while bottled water is often disinfected with ozone. Ozone is preferred by bottlers because it does not leave a taste.

While bottled water originates from protected sources (75 percent from underground springs and aquifers), tap water comes mostly from lakes and rivers. Bottled water is regulated by the FDA as a food product, and tap water is regulated by the U.S. Environmental Protection Agency (EPA).

The FDA has published standard definitions for different types of bottled water to promote honesty and fair dealing in the marketplace.

- Artesian well water is water from a well that taps an aquifer—layers of porous rock, sand, and earth that contain water—which is under pressure from surrounding upper layers of rock or clay. When tapped, the pressure in the aquifer, commonly called artesian pressure, pushes the water above the level of the aquifer, sometimes to the surface. Other means may be used to bring the water to the surface. According to the EPA, water from artesian aquifers often is more pure because the confining layers of rock and clay impede the movement of contamination. However, despite the claims of some bottlers, there is no guarantee that artesian waters are any cleaner than ground water from an unconfined aquifer.
- Mineral water is water from an underground source that contains at least 250 parts per million of total dissolved solids. Minerals and trace elements must come from the source of the underground water. They cannot be added later.
- Spring water is derived from an underground formation from which water flows naturally to the earth's surface. Spring water must be collected only at the spring or through a borehole that taps the underground formation feeding the spring. If some external force is used to collect the water, the water must have the same composition and quality as the water that naturally flows to the surface.
- Well water is water from a hole drilled into the ground which taps into an aquifer. Some bottled water also comes from municipal sources, in other words, the tap.

Municipal water is usually treated before it is bottled. Bottled water that has been treated by one of the following methods and that meets the definition of purified water in the U.S. Pharmacopeia can be labeled as purified water. Examples of water treatments include:

- Reverse osmosis: Water is forced through membranes to remove minerals in the water.
- Absolute 1 micron filtration: Water flows through filters that remove particles larger than 1 micron in size, such as Cryptosporidium, a parasitic protozoan.
- Ozonation: Bottlers of all types of waters typically use ozone gas, an antimicrobial agent, to disinfect the water instead of chlorine, since chlorine can leave a residual taste and odor in the water.

Like all other foods regulated by the FDA, bottled water must be processed, packaged, shipped, and stored in a safe and sanitary manner and be truthfully and accurately labeled.

Bottled water may be used as an ingredient in beverages, such as diluted juices or flavored bottled waters. However, beverages labeled as containing "sparkling water," "seltzer water," "soda water," "tonic water," or "club soda" are not included as bottled water under the FDA's regulations, because these beverages have historically been considered soft drinks.

- Sparkling water is any carbonated water.
- Seltzer is filtered, artificially carbonated tap water that generally has no added mineral salts. It is available with assorted flavor essences, such as black cherry and orange. If seltzer contains sweeteners (and therefore calories), it must be called a flavored soda.
- Club soda, sometimes called soda water or plain soda, is filtered, artificially carbonated tap water to which mineral salts are added to give it a unique taste. Most average 30 to 70 milligrams of sodium per 8 ounces.
- Tonic water is not really water or low in calories. It contains 84 calories per 8 ounces. Diet tonic water uses sugar substitutes.

Different mineral and carbonation levels of waters make them appeal to different customers and appropriate for different eating situations. For instance, a heavily carbonated sparkling water such as Perrier is excellent as an aperitif, yet some customers may prefer the lighter sparkle of San Pellegrino. Still waters such as Evian are generally more popular and appropriate to have on the table during the meal.

Although some people prefer bottled water, it is important to know that the public water supply is regulated by the U.S. Environmental Protection Agency (EPA). All municipal water systems serving 25 or more people are tested regularly for up to 118 chemicals and bacteria specified by the Safe Drinking Water Act (SDWA). Individual states may require additional testing. Everyone who gets their tap water from a public system is therefore assured of regular testing and certain standards.

Besides being cheaper than bottled water, tap water also doesn't need a plastic bottle. Plastic water bottles are the least recycled plastic beverage bottle. Plastic bottles from water (and soda) pile up in our landfills and take hundreds of years to break down. Not only do all these bottles create huge waste problems, they are also produced from petroleum and require energy to manufacture. Be sure to recycle all plastic bottles.

MINI-SUMMARY

1. The average adult's body weight is 50 to 60 percent water. Men generally have proportionally more water than women, and a lean person has more than an obese person.
2. Water serves as the medium for many metabolic activities and also participates in some metabolic reactions.
3. Water transports nutrients to the cells and carries away waste materials to the kidneys and out of the body.
4. Water is needed for digestion and absorption.
5. Water maintains blood volume.
6. Water helps you maintain normal temperatures by sweating.
7. Water serves as an important part of body lubricants.
8. The Adequate Intake for total water is based on the average water consumption of people who are adequately hydrated. About 80 percent comes from beverages, including coffee, tea, and cola. The remaining 20 percent comes from water found in foods.
9. When it is normal and healthy, the body maintains water at a constant level. A number of mechanisms, including the sensation of thirst, operate to keep body water content within certain limits.
10. Bottled water is regulated by the FDA.

MAJOR MINERALS

Among the seven major minerals, three are likely to be consumed in amounts low enough to be of concern. At any age, potassium is likely to be deficient, and increased intakes would be helpful. Increased intakes of calcium and magnesium are also needed for adults and children over 9 years of age.

CALCIUM AND PHOSPHORUS

Calcium and phosphorus are used for building bones and teeth. Approximately 99 percent of total body calcium is found in bones and teeth. Calcium complexes with phosphorus to form part of the crystal called *hydroxyapatite*, which gives bone its strength. Bone is being rebuilt every day, with new bone being formed and old bone being taken apart. The calcium in bones is therefore in a state of constant change.

Calcium plays a similar role in teeth. The turnover of minerals in teeth is not as rapid as it is in bones. Fluoride hardens and stabilizes the crystals of teeth, and this decreases the withdrawal of minerals.

> **HYDROXYAPATITE**
> The main structural component of bone, composed mostly of calcium phosphate crystals.

Nutrition Science Focus

CALCIUM AND PHOSPHORUS

Calcium and phosphorus are very important for bones. Bone undergoes continuous remodeling, with constant resorption (breakdown of bone) and deposition of calcium into newly deposited bone (bone formation). The balance between bone resorption and deposition changes as people age. During childhood there is a higher amount of bone formation and less breakdown. Calcium intake between the ages of 9 and 18 is critical for bone development because most bone mass (bone strength and density) accumulates during this time. Bones stop increasing in density after about age 30. In early and middle adulthood, the breakdown and building up of bone are relatively equal. In aging adults, particularly among postmenopausal women, bone breakdown exceeds its formation, resulting in bone loss, which increases the risk for osteoporosis (a disorder characterized by porous, weak bones discussed in chapter 11).

It is crucial that the body maintain a certain level of calcium in the blood so that muscles contract, nerves transmit impulses, and blood clots. Two hormones, parathyroid hormone and calcitonin, as well as vitamin D, work to keep blood calcium at just the right level. When blood calcium levels are low, parathyroid hormone goes to work by mobilizing calcium from bone, increasing absorption of calcium from the intestines, and preventing calcium from being taken up by the kidney and winding up in the urine. Calcitonin is released into the blood when blood calcium levels go up, and its actions are the opposite of what parathyroid hormone does.

Phosphorus is involved in the metabolic release of energy from fat, protein, and carbohydrates. It also activates many enzymes when a phosphate group is attached. Phosphorus is also present in phospholipids such as lecithin. Phospholipids are a class of lipids that are like triglycerides except that one fatty acid is replaced by a phosphate group and choline or another nitrogen-containing group. Phospholipids are unique in that they are soluble in fat and water. As you may know, fats and water (such as oil and vinegar) do not normally stay mixed together. Phospholipids such as lecithin (found in egg yolks) are used by the food industry as emulsifiers, substances that are capable of breaking up the fat globules into small droplets, resulting in a uniform mixture that won't separate. Lecithin is used in foods such as salad dressings.

Calcium also circulates in the blood, where a constant level is maintained so that it is always available for use. Calcium helps:

- Muscles to contract (including the heart muscle)
- Nerves to transmit impulses
- Blood to clot

Calcium may help in lowering blood pressure. In cases of inadequate dietary intake, calcium is taken out of the bones to maintain adequate blood levels.

Like calcium, phosphorus circulates in the blood and has many functions. Phosphorus:

- Is involved in metabolism
- Is a part of DNA (genetic material) and is therefore needed for growth
- Buffers both acids and bases in all the body's cells

The major sources of calcium are milk and milk products. Not all milk products are as rich in calcium as milk is (see Figure 7-2). As a matter of fact, butter, cream, and cream cheese contain little calcium. One cup of milk or yogurt or 1½ ounces of cheese has a little less than one-third of the AI (Adequate Intake) for most adults.

Without milk or milk products in your diet, it may be difficult to get enough calcium. Other good sources of calcium include fortified soy milk, tofu made with calcium carbonate, calcium-fortified foods such as orange juice, and several greens, such as broccoli, collards, kale, mustard greens, and turnip greens. Other greens, such as spinach, beet greens, Swiss chard, and parsley, are calcium-rich but also contain a binder (*oxalic acid*) that prevents some calcium from being absorbed. Dried beans and peas and whole-wheat bread contain moderate amounts of calcium but are usually not eaten in sufficient quantities to make a significant contribution. **Phytic acid**, a binder found in wheat bran and whole grains, also prevents some calcium from being absorbed.

About 25 to 30 percent of the calcium you eat is absorbed. The body absorbs more calcium (up to 60 percent) during growth and pregnancy, when additional calcium is needed. Once a child's bones stop growing in early adulthood, the absorption rate goes down to about 25 percent, which is normal for adults. Both stomach acid and vitamin D help calcium absorption.

Calcium deficiency is much more common in women than in men and is a major contributing factor in a disease called osteoporosis, which is discussed later in this chapter.

Calcium can be toxic when large doses of supplements are taken. The Tolerable Upper Intake Level for calcium is 2500 milligrams per day. Amounts above that can contribute to the development of calcium deposits in the kidneys and other organs, kidney failure, and other problems.

Phosphorus is widely distributed in foods and is rarely lacking in the diet. Milk and milk products are excellent sources of phosphorus, as they are of calcium. Other good sources of phosphorus are meat, poultry, fish, eggs, and legumes. Fruits and vegetables are generally low in this mineral. Compounds made with phosphorus are used in processed foods, especially soft drinks (phosphoric acid).

MAGNESIUM

Magnesium is found in all body tissues, with about 60 percent in the bones and the remainder in the soft tissues, such as muscles, and the blood. The body works very hard to keep blood levels of magnesium constant. Magnesium is essential to:

- Many enzyme systems responsible for energy metabolism
- Build bones and maintaining teeth
- Muscle relaxation and nerve transmission
- Keep the immune system working properly

OXALIC ACID

An organic acid found in spinach and other leafy green vegetables that can decrease the absorption of certain minerals, such as calcium.

PHYTIC ACID

A binder found in wheat bran and whole grains that can decrease the absorption of certain nutrients, such as calcium and iron.

FIGURE 7-2: Food Sources of Calcium

DAIRY

Food and Serving Size	Milligrams	Adequate Intake Adults 19–50
Milk, nonfat, 1 cup	306	1000
Milk, low-fat, 1 percent, 1 cup	290	1000
Chocolate milk, 1 cup	280	1000
Yogurt, plain, whole milk, 8 oz.	275	1000
Cheese, Swiss, 1 oz.	224	1000
Cheese, cheddar, 1 oz.	204	1000
Cheese, American, 1 oz.	162	1000

GRAINS

Food and Serving Size	Milligrams	Adequate Intake Adults 19–50
General Mills, Total, ¾ cup	1104	1000
General Mills, Golden Grahams, ¾ cup	350	1000
General Mills, Basic 4, 1 cup	196	1000
General Mills, Cheerios, 1 cup	122	1000
Cream of Wheat, ½ cup	57	1000
English muffin, 1 muffin	99	1000
Whole-wheat bread, 1 slice	20	1000

FRUITS AND VEGETABLES

Food and Serving Size	Milligrams	Adequate Intake Adults 19–50
Collards, boiled, ½ cup	179	1000
Spinach, boiled, ½ cup	145	1000
Kale, boiled, ½ cup	90	1000
Tomatoes, stewed, ½ cup	44	1000
Broccoli, boiled, ½ cup	30	1000

MEAT, POULTRY, FISH, AND ALTERNATES

Food and Serving Size	Milligrams	Adequate Intake Adults 19–50
Sardines, canned in oil, 3 oz.	325	1000
Tofu with calcium sulfate, 2½ oz.	163	1000
Beans, baked, ½ cup	71	1000
Beans, navy, boiled, ½ cup	63	1000

Source: U.S. Department of Agriculture, Agricultural Research Service, 2004. USDA Nutrient Database for Standard Reference, Release 17. Nutrient Data Laboratory Home Page, www.nal.usda.gov/fnic/foodcomp.

Evidence suggests that magnesium may play an important role in regulating blood pressure. Diets that provide plenty of fruits and vegetables, which are good sources of magnesium and potassium, are consistently associated with lower blood pressure. The Joint National Committee on Prevention, Detection, Evaluation, and Treatment of High Blood Pressure recommends maintaining an adequate magnesium intake, as well as potassium and calcium intake, as a positive lifestyle modification for preventing and managing high blood pressure.

Nutrition Science Focus

MAGNESIUM

Both calcium and magnesium are involved in muscles. What is interesting is that whereas calcium enhances muscle contraction, magnesium inhibits contraction and promotes relaxation. Research is taking place on magnesium and heart disease. It appears that magnesium is a heart-healthy nutrient.

Magnesium is a part of chlorophyll, the green pigment found in plants, so good sources include green leafy vegetables, potatoes, nuts (especially almonds and cashews), seeds, legumes, and whole-grain breads and cereals. Seafood is also a good source. The magnesium content of refined foods is usually low. Whole-wheat bread, for example, has twice as much magnesium as does white bread because it contains the magnesium-rich germ and bran, which are removed when white flour is processed.

Although magnesium is present in many foods (see Figure 7-3), it usually occurs in small amounts. As with most nutrients, daily needs for magnesium cannot be met from a single food. Eating a wide variety of foods, including at least five servings of fruits and vegetables daily and plenty of whole grains, helps ensure an adequate intake of magnesium.

Even though dietary surveys suggest that many American do not consume magnesium in the recommended amounts, deficiency symptoms are rarely seen in adults in the United States. When magnesium deficiency does occur, it is usually due to disease or medications. Poorly controlled diabetes and a high alcohol intake increase the excretion of magnesium. Signs of chronic magnesium deficiency include muscle twitching, cramps, weakness, depression, blood clots, and other symptoms. If severe, it can cause muscle spasms, hallucinations, and even sudden death.

Very high doses of magnesium supplements can cause diarrhea. Especially in the elderly, they can also cause problems with the kidneys because the kidneys are trying to remove excess magnesium. The elderly are at risk of magnesium toxicity because kidney function declines with age and they are more likely to take magnesium-containing laxatives and antacids.

SODIUM

ELECTROLYTES
Chemical elements or compounds that ionize in solution and can carry an electric current; they include sodium, potassium, and chloride.

ION
An atom or group of atoms carrying a positive or negative electric charge.

Sodium, potassium, and chloride are collectively referred to as *electrolytes* because when dissolved in body fluids, they separate into positively or negatively charged particles called *ions*. Potassium, which is positively charged, is found mainly within the cells. Sodium (positively charged) and chloride (negatively charged) are found mostly in the fluid outside the cells.

FIGURE 7-3: Food Sources of Magnesium

GRAINS

Food and Serving Size	Milligrams	RDA Men	RDA Women
Muffins, oat bran, 3 oz.	89	420	320
Rice brown, cooked, ½ cup	42	420	320
Kellogg's Raisin Bran, ½ cup	42	420	320
Bulgur, cooked, ½ cup	29	420	320
Rice, white, parboiled, ½ cup	28	420	320
Whole-wheat bread, 1 slice	24	420	320

FRUITS AND VEGETABLES

Food and Serving Size	Milligrams	RDA Men	RDA Women
Spinach, cooked, ½ cup	79	420	320
Potato, baked, 1 potato	57	420	320
Lima beans, baby, ½ cup	51	420	320
Raisins, seedless, ½ cup	23	420	320

DAIRY

Food and Serving Size	Milligrams	RDA Men	RDA Women
Yogurt, plain, low-fat, 8-oz. container	39	420	320
Soymilk, ½ cup	31	420	320
Yogurt, plain, whole-milk, 8-oz. container	27	420	320
Whole milk, 1 cup	24	420	320

MEATS, POULTRY, FISH, AND ALTERNATES

Food and Serving Size	Milligrams	RDA Men	RDA Women
Halibut, 3 oz.	91	420	320
Cashew nuts, 1 oz.	74	420	320
Mixed nuts, 1 oz.	64	420	320
Peanuts, 1 oz.	50	420	320
Oysters, cooked, 3 oz.	49	420	320
Baked beans, ½ cup	43	420	320
Haddock, cooked, 3 oz.	43	420	320
Refried beans, canned, ½ cup	42	420	320
Lentils, ½ cup	36	420	320

Source: U.S. Department of Agriculture, Agricultural Research Service, 2004.
USDA Nutrient Database for Standard Reference, Release 17. Nutrient Data Laboratory Home Page, www.nal.usda.gov/fnic/foodcomp.

The electrolytes maintain two critical balancing acts in the body:

- *Water balance*
- *Acid-base balance*

Water balance entails maintaining the proper amount of water in each of the body's three "compartments": inside the cells, outside the cells, and in the blood vessels. Electrolytes maintain water balance by getting water to move into and out of the three compartments as needed. Electrolytes are also able to buffer, or neutralize, various acids and bases in the body.

In addition to its role in water and acid-base balance, sodium is needed for muscle contraction and transmission of nerve impulses.

The major source of sodium in the diet is salt, a compound made of sodium and chloride. Salt by weight is 39 percent sodium, and 1 teaspoon contains 2300 milligrams (a little more than 2 grams) of sodium. Many processed foods have high amounts of sodium added during processing and manufacturing, and it is estimated that these foods provide fully 75 percent of the sodium in most people's diets. The following is a list of processed foods high in sodium:

- Canned, cured, and/or smoked meats and fish such as bacon, salt pork, sausage, scrapple, ham, bologna, corned beef, frankfurters, luncheon meats, canned tuna fish and salmon, and smoked salmon
- Many cheeses, especially processed cheeses such as processed American cheese
- Salted snack foods such as potato chips, pretzels, popcorn, nuts, and crackers
- Food prepared in brine, such as pickles, olives, and sauerkraut
- Canned vegetables, tomato products, soups, and vegetable juices
- Frozen convenience foods such as pizza and entrées
- Prepared mixes for stuffings, rice dishes, and breading
- Dried soup mixes and bouillon cubes
- Certain seasonings, such as salt, soy sauce, garlic salt, onion salt, celery salt, and seasoned salt
- Condiments and sauces such as Worcestershire sauce, horseradish, ketchup, and mustard

Figure 7-4 shows examples of foods that are high in sodium. Salt is also used in food preparation and at the table for seasoning.

In addition to the sodium in salt, sodium appears in monosodium glutamate (MSG), baking powder, and baking soda. Other possible sources of dietary sodium include the sodium in some local water systems. Unprocessed foods also contain natural sodium, but in small amounts (with the exception of milk and some milk products).

The Dietary Guidelines for Americans (2005) recommend that you consume foods with less than 2300 milligrams sodium or about 1 teaspoon salt each day. The AI for sodium is 1500 mg per day for adults. This amount does not apply to highly active individuals, such as endurance athletes and certain workers (such as foundry workers) who lose large amounts of sweat on a daily basis and therefore need more sodium. All adults exceed the AI each day. About 95 percent of adult men and 75 percent of adult women exceed the Tolerable Upper Intake Level of 2300 mg of sodium per day. Figure 7-5 gives strategies for reducing sodium intake.

The most important health issue that is influenced by overconsumption of salt is high blood pressure, also called hypertension. High blood pressure is a major risk factor for heart disease, stroke, and kidney disease. In short, the higher your salt intake, the higher your blood pressure. Individuals with the greatest reductions in blood pressure in response to decreased salt intake are called "salt-sensitive." They include middle-aged and older persons,

FIGURE 7-4: Sodium Content of Foods

FOODS HIGH IN SODIUM (500 OR MORE MILLIGRAMS/SERVING)

Food	Portion Size	Amount of Sodium
Table salt	1 tsp.	2325
Biscuit with egg and sausage	1 biscuit	1141
Ham	3 oz.	1128
Chicken noodle soup	1 cup	1106
Soy sauce	1 Tbsp.	902
Dill pickle	1 pickle	833
Sauerkraut	½ cup	780
Baked beans, canned	½ cup	553
Hamburger, regular	1 sandwich	534
Frankfurter, beef	1 frank	513

FOODS MODERATE IN SODIUM (141–499 MILLIGRAMS/SERVIING)

Food	Portion Size	Amount of Sodium
Cottage cheese, low-fat	½ cup	459
Mushrooms, canned	½ cup	332
Bagel, plain	4-inch bagel	475
Soup, minestrone, canned, ready-to-serve	1 cup	470
Bologna	2 slices (2 oz.)	417
English muffin	1 muffin	264
Raisin bran cereal	½ cup	181
White bread	1 slice	170
Catsup	1 Tbsp.	167

FOODS LOW IN SODIUM (140 MILLIGRAMS OR LESS/SERVING)

Food	Portion Size	Amount of Sodium
Salad dressing, Thousand Island	1 Tbsp.	135
Margarine	1 Tbsp.	133
Yogurt, fruit, low-fat	8-oz. container	132
Fish, halibut	½ fillet (5 oz.)	110
Milk, lowfat	1 cup	107
Peas, green, frozen	½ cup	57
Peanut butter	1 Tbsp.	73
Egg	1 large	70
Beef, ground, cooked	3 oz.	66
Celery, raw	½ cup	48
Carrots, cooked	½ cup	45
Lettuce, iceberg	½ head	27
Banana, raw	1	1
Apple, raw	1	0

Source: U.S. Department of Agriculture, Agricultural Research Service, 2004.
USDA Nutrient Database for Standard Reference, Release 17. Nutrient Data Laboratory Home Page,
www.nal.usda.gov/fnic/foodcomp.

FIGURE 7-5: Strategies for Reducing Sodium Intake

AT THE STORE

- Choose fresh, plain frozen, or canned vegetables without added salt; most often, they are low in salt.
- Choose fresh or frozen fish, shellfish, poultry, and meat most often. They are lower in salt than most canned and processed forms.
- Read the Nutrition Facts Label to compare the amount of sodium in processed foods such as frozen dinners, packaged mixes, cereals, cheese, breads, soups, salad dressings, and sauces. The amount in different types and brands often varies widely.
- Look for labels that say "low sodium." They contain 140 mg or less of sodium per serving.

COOKING AND EATING AT HOME

- If you salt foods in cooking or at the table, add small amounts. Learn to use spices and herbs instead to enhance the flavor of food.
- Go easy on condiments such as soy sauce, ketchup, mustard, pickles, and olives; they can add a lot of salt to your food.
- Leave the salt shaker in a cabinet.

EATING OUT

- Choose plain foods such as grilled or roasted entrées, baked potatoes, and salad with oil and vinegar.
- Batter-fried foods tend to be high in salt, as do combination dishes such as stews or pasta with sauce.
- Ask to have no salt added when the food is prepared.

ANY TIME

- Choose fruits and vegetables often.

African Americans, and individuals with hypertension, diabetes, or chronic kidney disease. There is no benefit to consuming more than 1500 mg per day, especially for members of these groups.

When people sweat, sodium is lost as well as fluid. In most cases, these losses are made up by eating and drinking fluids.

POTASSIUM

Potassium, an electrolyte found mainly in the fluid inside individual body cells, helps:

- Maintain water balance and acid-base balance along with sodium
- Muscles contract, including maintaining a normal heartbeat
- Send nerve impulses

A diet rich in potassium blunts the effects of salt on blood pressure and seems to lower blood pressure.

Potassium is distributed widely in foods, both plant and animal. Unprocessed whole foods such as fruits and vegetables (especially winter squash, potatoes, oranges, grapefruits, and bananas), milk, yogurt, legumes, and meats are excellent sources of potassium (Figure 7-6).

FIGURE 7-6: Sources of Potassium

FRUITS AND VEGETABLES

Food and Serving Size	Milligrams	Adequate Intake Age 14+
Potato, baked, 1 potato	1081	4700
Squash, winter, cooked, ½ cup	448	4700
Potato, boiled, 1 potato	443	4700
Banana, raw, 1 banana	422	4700
Spinach, cooked, ½ cup	420	4700
Prune juice, ½ cup	354	4700
Tomato juice, ½ cup	278	4700
Raisins, seedless, ¼ cup	272	4700
Orange juice, ½ cup	237	4700
Melons, cantaloupe, raw, ½ cup	214	4700
Tomatoes, red, ½ cup	214	4700
Oranges, raw, ½ cup	163	4700

DAIRY

Food and Serving Size	Milligrams	Adequate Intake Age 14+
Yogurt, plain, low-fat, 8-oz. container	531	4700
Milk, fat-free, 1 cup	382	4700
Yogurt, plain, whole milk, 8-oz. container	352	4700
Milk, whole, 1 cup	349	4700

GRAINS

Food and Serving Size	Milligrams	Adequate Intake Age 14+
Kellogg's All-Bran, ½ cup	306	4700
Muffins, oat bran, 1 muffin	289	4700
Granola with raisins, ½ cup	204	4700
Kellogg's Frosted Mini-Wheats, 1 cup	190	4700
Kellogg's Raisin Bran, ½ cup	186	4700

MEATS, POULTRY, FISH, AND ALTERNATES

Food and Serving Size	Milligrams	Adequate Intake Age 14+
Halibut, cooked, 3 oz.	490	4700
Cod, cooked, 3 oz.	439	4700
Beans, pinto, cooked, ½ cup	373	4700
Pork, fresh, loin, 3 oz.	371	4700
Peas, split, cooked, ½ cup	355	4700
Refried beans, canned, ½ cup	337	4700

FATS AND SWEETS

Food and Serving Size	Milligrams	Adequate Intake Age 14+
Potato chips, plain, 1 oz.	361	4700
Pumpkin pie, 1 piece	288	4700

Source U.S. Department of Agriculture, Agricultural Research Service, 2004. USDA Nutrient Database for Standard Reference, Release 17. Nutrient Data Laboratory Home Page, www.nal.usda.gov/fnic/foodcomp.

Potassium deficiency is of concern in the United States, in large part because of overconsumption of processed foods and underconsumption of whole foods. Dehydration, certain diseases, and drugs can also cause a potassium deficiency. Diuretics, a class of blood pressure drugs, cause increased urine output, and some cause an increased excretion of potassium as well. Symptoms of deficiency include muscle cramps, weakness, nausea, and abnormal heart rhythms that can be very dangerous, even fatal.

There is no evidence of chronic excess intakes of potassium in healthy individuals, and therefore no Tolerable Upper Intake Level has been set. Toxic levels of potassium in the body can be caused by diseases or overconsumption of supplements.

CHLORIDE

Chloride, another important electrolyte, helps maintain water balance and acid-base balance. It is also a part of hydrochloric acid, which is highly concentrated in stomach juices. Hydrochloric acid aids in protein digestion, destroys harmful bacteria, and increases the absorption of calcium and iron. The most important source of dietary chloride is sodium chloride, or salt. If sodium intake is adequate, there will be ample chloride as well.

OTHER MAJOR MINERALS

The body doesn't use the mineral sulfur by itself but uses the nutrients it is found in, such as protein, thiamin, and biotin. The protein in hair, skin, and nails is particularly rich in sulfur. There is no DRI for sulfur. Protein foods supply plentiful amounts of sulfur, and deficiencies are not known.

Nutrition Science Focus

WATER AND ELECTROLYTES

The electrolytes are very important to maintain fluid balance. In the body, about two-thirds of your body's fluids are inside your cells. This is called the intracellular fluids (Figure 7-7). Potassium is almost completely found inside the cell.

About one-third of your body's fluids is extracellular fluids—the fluids found outside of your cells. This includes mostly fluids in your blood vessels (called blood plasma—it does not include the red or white cells) and in the interstitial space (the space between the cells that makes up organs). Sodium and chloride are mostly in the extracellular fluids.

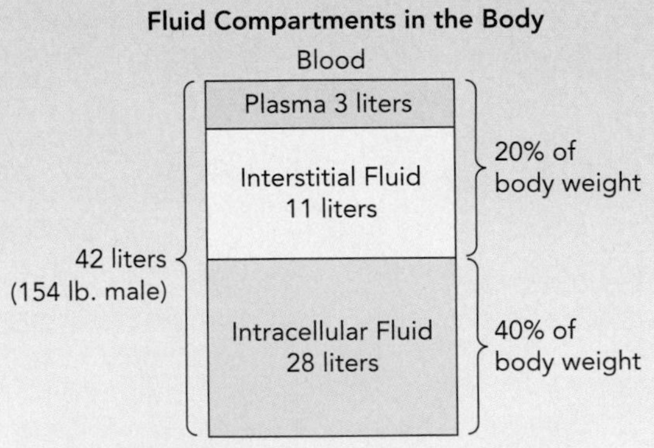

The distribution and balance of the body's fluids and electrolytes in the intracellular and extracellular compartments are essential to the normal functioning of the body. The kidney is one organ that is involved in maintaining this balance. The kidneys adjust the amounts of water and electrolytes by secreting some substances into the urine and holding back others in the bloodstream for use in the body. When the body has too much sodium, thirst is triggered to prompt us to drink fluids and dilute the sodium.

Fluid Compartments in the Body

Blood

Plasma 3 liters

Interstitial Fluid 11 liters — 20% of body weight

42 liters (154 lb. male)

Intracellular Fluid 28 liters — 40% of body weight

FIGURE 7-7:
Fluid compartment in the body.

MINI-SUMMARY

The recommended intakes, functions, and sources of the major minerals are listed in Figure 7-8.

TRACE MINERALS

Trace minerals (Figure 7-1) represent an exciting area for research because our understanding of many trace minerals is still emerging. Like vitamins, minerals can be toxic at high doses. Unlike vitamins, many trace minerals are toxic at levels only several times higher than the recommended levels. Also, trace minerals are highly interactive with each other. For example, taking extra zinc can cause a copper deficiency.

IRON

Iron, one of the most abundant metals in the universe and one of the most important in the body, is a key component of **hemoglobin**, a part of red blood cells that carries oxygen to the cells in the body. Cells require oxygen to break down glucose and produce energy.

HEMOGLOBIN
A protein in red blood cells that carries oxygen to the body's cells.

FIGURE 7-8: Summary of Major Minerals

Mineral	Recommended Intake	Functions	Sources
Calcium	AI: 1000 mg Upper Intake Level: 2500 mg	Building bones and teeth Blood clotting Muscle contraction Nerve transmission May lower blood pressure	Milk and many milk products, calcium- fortified foods, tofu made with calcium carbonate, several greens (broccoli, collards), legumes, whole-wheat bread
Phosphorus	RDA: 700 mg Upper Intake Level: 4000 mg	Building bones and teeth Energy metabolism Part of DNA Buffer of acids and bases	Milk and milk products, meat, poultry, fish, eggs, legumes
Magnesium	RDA: Men: 420 mg Women: 310 mg Upper Intake Level: 350 mg (synthetic forms only)	Energy metabolism Bones and teeth Muscle contraction Nerve transmission Immune system	Nuts, seeds, legumes, green leafy vegetables, potatoes, whole-grain breads and cereals, seafood
Sodium	AI: 1500 mg Upper Intake Level: 2300 mg	Water balance Acid-base balance Muscle contraction Nerve transmission	Salt, processed foods such as luncheon meats, salted snacks, canned foods), soy sauce
Potassium	AI: 4700 mg Upper Intake Level: None	Water balance Acid-base balance Muscle relaxation Nerve transmission	Fruits and vegetables (winter squash, potatoes, oranges, grapefruits, bananas), milk, yogurt, legumes, meat
Chloride	AI: 2300 mg Upper Intake Level: 3600 mg	Water balance Acid-base balance Part of hydrochloric acid in stomach	Salt
Sulfur	None	Part of protein, thiamin, and biotin	Protein foods

MYOGLOBIN

A muscle protein that stores and carries oxygen that the muscles will use to contact.

HEME IRON

The predominant form of iron in animal foods; it is absorbed and used more readily than iron in plant foods.

NONHEME IRON

A form of iron found in all plant sources of iron and also as part of the iron in animal food sources.

Iron is also part of *myoglobin*, a muscle protein that stores and carries oxygen that the muscles use to contract. Iron works with many enzymes in energy metabolism and is therefore necessary for the body to produce energy.

About 15 percent of the body's iron is stored in the bone marrow, spleen, and liver for future needs. It is mobilized when dietary intake is inadequate.

Meat, poultry, and fish are good sources of iron. Whole-grain, enriched, and fortified breads and cereals also supply iron. You also get iron from legumes, green leafy vegetables, and eggs (Figure 7-9).

Only about 15 percent of dietary iron is absorbed in healthy individuals. The greatest influence on iron absorption is the amount stored in the body. Iron absorption significantly increases when body stores are low. When iron stores are high, absorption decreases to help protect against iron overload. The body also absorbs iron more efficiently when there is a high need for red blood cells, such as during growth spurts or pregnancy or when there has been blood loss.

The ability of the body to absorb and use iron from different foods varies. The predominant form of iron in animal foods, called *heme iron*, is absorbed and used twice as readily as is iron in plant foods, called *nonheme iron*. Animal foods also contain some nonheme iron. The presence of vitamin C in a meal increases nonheme iron absorption, as does consuming meat, poultry, and fish. Calcium, substances found in tea and coffee, oxalic acid (in some vegetables, such as spinach), and phytic acid (in grain fiber) can decrease the absorption of nonheme iron. Some proteins found in soybeans also inhibit nonheme iron absorption.

Nutrition Science Focus

IRON

Iron-deficiency anemia is a real problem during pregnancy. Half of all pregnant women develop iron-deficiency anemia because their volume of blood increases and the growing fetus needs iron. At the beginning of pregnancy, iron-deficiency anemia is associated with an increased risk of premature delivery and of giving birth to infants with low birth weight. Iron deficiency during pregnancy may cause lower scores on intelligence, language, and gross motor skill tests in children by five years of age.

Several major health organizations recommend iron supplementation during pregnancy to help pregnant women meet their iron requirements. The Institute of Medicine of the National Academy of Sciences recommends iron supplementation during the second and third trimesters. Obstetricians often monitor the need for iron supplementation during pregnancy and provide individualized recommendations to pregnant women.

IRON DEFICIENCY

A condition in which iron stores are used up.

IRON-DEFICIENCY ANEMIA

A condition in which the size and number of red blood cells are reduced; may result from inadequate iron intake or from blood loss; symptoms include fatigue, pallor, and irritability.

Substantial numbers of adolescent females and women of childbearing age have *iron deficiency*, meaning that their stores of iron are used up. Iron deficiency is the most common form of nutritional deficiency. This condition commonly results from repeated blood loss from menstruation and/or inadequate intake of iron. Iron deficiency results in feelings of being tired, irritable, or depressed.

When iron stores become severely depleted, *iron-deficiency anemia* results. In iron-deficiency anemia, red blood cells are smaller than usual and carry less hemoglobin. Women of childbearing age, pregnant women, infants and young children, and teenage girls are at the greatest risk of developing iron-deficiency anemia because they have the greatest needs. Signs of iron-deficiency anemia include feeling tired and weak, decreased work and school

FIGURE 7-9: Food Sources of Iron

GRAINS

Food and Serving Size	Milligrams Iron	RDA Men	RDA Women
Total cereal, General Mills, ¾ cup	22.4	8	18
Cheerios, General Mills, 1 cup	10.3	8	18
Special K, 1 cup	8.4	8	18
Cream of wheat, ½ cup	5.1	8	18
Rice, white, long-grain, ½ cup	3.3	8	18
White bread, 1 slice	0.8	8	18
Whole-wheat bread, 1 slice	0.9	8	18

FRUITS AND VEGETABLES

Food and Serving Size	Milligrams Iron	RDA Men	RDA Women
Spinach, ½ cup	3.2	8	18
Tomatoes, stewed, ½ cup	1.7	8	18

DAIRY

Food and Serving Size	Milligrams Iron	RDA Men	RDA Women
Soymilk, 1 cup	2.7	8	18

MEATS, POULTRY, FISH, AND ALTERNATES

Food and Serving Size	Milligrams Iron	RDA Men	RDA Women
Beef, chuck roast, cooked, 3 oz.	3.1	8	18
Beans, kidney, ½ cup	2.6	8	18
Beef, rib, roasted, 3 oz.	2.4	8	18
Chickpeas, ½ cup	2.4	8	18
Refried beans, canned, ½ cup	2.1	8	18
Cashew nuts, 1 oz.	1.7	8	18
Halibut, cooked, 3 oz.	0.9	8	18
Egg, boiled, 1 large	0.6	8	18
Chicken breast, roasted, 3 oz.	0.6	8	18
Flounder, 3 oz.	0.3	8	18

FATS AND SWEETS

Food and Serving Size	Milligrams Iron	RDA Men	RDA Women
Pumpkin pie, 1 piece	2.0	8	18
Snack cakes, cream-filled, chocolate with frosting, 1 cupcake	1.7	8	18

Source: U.S. Department of Agriculture, Agricultural Research Service, 2004. USDA Nutrient Database for Standard Reference, Release 17. Nutrient Data Laboratory Home Page, www.nal.usda.gov/fnic/foodcomp.

performance, slow cognitive and social development during childhood, difficulty maintaining body temperature, and decreased immune function.

Although the body generally avoids absorbing huge amounts of iron, some people can absorb large amounts. The problem with iron is that once it is in the body, it is hard to get rid of. In individuals who can absorb much iron, large doses of iron supplements can damage the liver and do other damage, a condition called ***iron overload*** or ***hemochromatosis***. This condition is usually caused by a genetic disorder and is much more prevalent in men than in women. Symptoms include fatigue, joint pain, skin that turns gray or bronze, and symptoms of liver disease such as nausea and stomach pain.

It is especially important to keep iron supplements away from children, because they are so toxic that they can kill. Consuming 1 to 3 grams of iron can be fatal to children under 6, and lower doses can cause severe symptoms such as vomiting and diarrhea. It is important to keep iron supplements tightly capped and away from children's reach.

IRON OVERLOAD (HEMOCHROMA-TOSIS)

A common genetic disease in which individuals absorb about twice as much iron from their food and supplements as other people do.

ZINC

Zinc is in every cell in the body. It is a cofactor (a substance that is necessary for the activity of an enzyme) for nearly 100 enzymes. Some of the functions of zinc are as follows:

- Protein, carbohydrate, and fat metabolism
- DNA synthesis
- Wound healing
- Bone formation
- Development of sexual organs
- Taste perception

The effect of zinc treatments on the severity or duration of cold symptoms is inconclusive. Additional research is needed to determine whether zinc compounds have any positive effect on the common cold.

Protein-containing foods are all good sources of zinc, particularly shellfish (especially oysters), meat, and poultry (see Figure 7-10). Legumes, dairy foods, whole grains, and fortified cereals are

---

Nutrition Science Focus

ZINC

Zinc has a number of additional functions:

- General tissue growth and maintenance
- Vitamin A activity
- Protection of cell membranes from free-radical attacks
- Storage and release of insulin

Zinc is used to produce the active form of vitamin A that is used by the eyes. It also is used to produce the protein that transports vitamin A around the body. Zinc also keeps cell membranes strong so they can resist attack from free-radicals. Although zinc doesn't play a direct role in insulin's actions, zinc does participate in storing and releasing it.

FIGURE 7-10: Food Sources of Zinc

MEATS, POULTRY, FISH, AND ALTERNATES

Food and Serving Size	Milligrams	RDA Men	Women
Oysters, fried, 3 oz.	74.1	11	8
Beef, chuck, blade roast, 3 oz.	7.1	11	8
Beans, canned with pork and tomato sauce, ½ cup	6.9	11	8
Refried beans, ½ cup	6.5	11	8
Crab, Alaska king, 3 oz.	6.5	11	8
Beef, ground, 85 percent lean meat, broiled, 3 oz.	5.4	11	8
Beef round, braised, 3 oz.	5.0	11	8
Pork, spareribs, cooked, 3 oz.	3.9	11	8

GRAINS

Food and Serving Size	Milligrams	RDA Men	Women
Total cereal, ¾ cup	17.5	11	8
Wheaties, ½ cup	7.5	11	8
Cheerios, 1 cup	4.6	11	8
Frosted Mini-Wheats, 1 cup	1.8	11	8
Wild rice, cooked, ½ cup	1.1	11	8
White rice, parboiled, ½ cup	0.9	11	8
Whole wheat bread, 1 slice	0.5	11	8

FRUITS AND VEGETABLES

Food and Serving Size	Milligrams	RDA Men	Women
Spinach, cooked, ½ cup	0.7	11	8
Mushrooms, cooked, ½ cup	0.7	11	8
Corn, cream style, ½ cup	0.7	11	8
Lima beans, forkhook, cooked, ½ cup	0.6	11	8
Peas, green, cooked, ½ cup	0.5	11	8

DAIRY

Food and Serving Size	Milligrams	RDA Men	Women
Yogurt, plain, low-fat, 8-oz. container	2.0	11	8
Yogurt, plain, whole milk, 8-oz. container	1.3	11	8
Soymilk, 1 cup	1.1	11	8
Milk, fat-free, 1 cup	1.0	11	8

Source: U.S. Department of Agriculture, Agricultural Research Service, 2004.
USDA Nutrient Database for Standard Reference, Release 17. Nutrient Data Laboratory Home Page, www.nal.usda.gov/fnic/foodcomp.

good sources as well. Zinc is much more readily available, or is absorbed better, from animal foods. Like iron, zinc is more likely to be absorbed when animal sources are eaten and when the body needs it. Only about 40 percent of the zinc we eat is absorbed into the body. Phytates, which are found in whole-grain breads, cereals, legumes, and other foods, can decrease zinc absorption.

Deficiencies are more likely to show up in pregnant women, the young, the elderly, and vegetarians. Adults deficient in zinc may have symptoms such as poor appetite, diarrhea, skin rash, and hair loss. Signs of severe deficiency in children include growth retardation, delayed sexual maturation, decreased sense of taste, poor appetite, delayed wound healing, and immune deficiencies. Marginal deficiencies occur in the United States.

Long-term ingestion of zinc at or above the Tolerable Upper Intake Level (40 milligrams) results in a copper deficiency. Since zinc supplements can be fatal at lower levels than is the case with many of the other trace minerals, zinc supplements should be avoided unless a physician prescribes them.

IODINE

Iodine is required in extremely small amounts. Once in the body, iodine is chemically changed to iodide (iodine ion). The **thyroid gland**, located in the neck, is responsible for producing two important hormones that maintain a normal level of metabolism and are essential for normal growth and development, body temperature, protein synthesis, and much more. Iodide is a part of both of these hormones.

Most of the iodine in the world is in seawater. Iodine is not found in many foods: mostly saltwater fish and grains and vegetables grown in iodine-rich soil (such as along the coast). Iodized salt was introduced in 1924 to combat iodine deficiencies. Iodine also finds its way accidentally into milk (cows receive iodine-containing drugs, and dairy equipment is sterilized with iodine-containing compounds) and into baked goods through iodine salts used as dough conditioners. Processed foods in the United States do not use iodized salt.

Average intake in the United States is higher than recommended but less than toxic. A deficiency can cause **hypothyroidism**, a condition in which less thyroid hormone is made, leading to a low metabolic rate, a tendency to weight gain, and drowsiness. A deficiency can also cause **simple goiter**, in which the thyroid gland becomes very large and the affected person feels lethargic, gains weight, and has a decreased body temperature. If a woman has a severe iodine deficiency during pregnancy, the development of the fetus will be harmed, and that can cause **cretinism** (**congenital hypothyroidism**), a condition of mental and physical retardation.

SELENIUM

Until 1979, it was not known that selenium is an essential mineral. The first RDA for selenium was announced in 1989. Selenium is an important part of antioxidant enzymes that protect cells against the effects of the free radicals that are produced during normal oxygen metabolism. Antioxidants help control levels of free radicals, which can damage cells and contribute to the development of some chronic diseases. Selenium is also part of an enzyme that activates thyroid hormones.

Researchers know that soils in the high plains of northern Nebraska and the Dakotas have very high levels of selenium. People living in those regions generally have the highest selenium intakes in the United States.

Selenium can also be found in meats and seafood (see Figure 7-11). Vegetables and grains are also important sources of selenium. The amount of selenium in soil, which varies by region, determines the amount of selenium in the plant foods that are grown in that soil. Some nuts, in particular Brazil nuts, are also very good sources of selenium.

THYROID GLAND
A gland found on either side of the trachea that produces and secretes two important hormones that regulate the level of metabolism.

HYPOTHYROIDISM
A condition in which there is less production of thyroid hormones; this leads to symptoms such as low metabolic rate, fatigue, and weight gain.

SIMPLE GOITER
Thyroid enlargement caused by inadequate dietary intake of iodine.

CRETINISM (CONGENITAL HYPOTHYROIDISM)
Mental and physical retardation during fetal and later development caused by iodine deficiency during pregnancy.

FIGURE 7-11: Food Sources of Selenium

Food	Micrograms	% Daily Value
Brazil nuts, unblanched, 1 oz.	840	1200
Tuna, canned in oil, drained, 3½ oz.	78	111
Beef/calf liver, 3 oz.	48	69
Cod, cooked with dry heat, 3 oz.	40	57
Noodles, enriched, boiled, 1 cup	35	50
Macaroni and cheese (box mix), 1 cup	32	46
Turkey breast, oven-roasted, 3-½ oz.	31	44
Macaroni, elbow, enriched, boiled, 1 cup	30	43
Spaghetti with meat sauce, 1 cup	25	36
Chicken, meat only, ½ breast	24	34
Beef chuck roast, lean only, oven-roasted, 3 oz.	23	33
Bread, enriched, whole-wheat, 2 slices	14	20
Rice, enriched, long-grain, cooked, 1 cup	14	20
Cottage cheese, low-fat (2 percent), ½ cup	11	16
Walnuts, black, 1 oz.	5	7
Cheese, cheddar, 1 oz.	4	6

Source: Facts about Dietary Supplements. Office of Dietary Supplements, National Institutes of Health, 2001. Office of Dietary Supplements home page: http://ods.od.nih.gov.

Selenium deficiency is most commonly seen in parts of China where the selenium content in the soil, and therefore selenium intake, is very low. Selenium deficiency is linked to Keshan disease (named after the province in China where it was studied), in which the heart becomes enlarged and does not function properly. Selenium deficiency may also affect thyroid function because selenium is essential for the synthesis of active thyroid hormone.

Selenium intake is good in the United States, and deficiency is rare.

There is a moderate to high health risk associated with too much selenium. High blood levels of selenium can result in a condition called selenosis. Symptoms include gastrointestinal upset, hair loss, white blotchy nails, and mild nerve damage. Selenium toxicity is rare in the United States.

FLUORIDE

Fluoride is the term used for the form of fluorine that appears in drinking water and in the body. In children, fluoride strengthens the mineral composition of the developing teeth so that they resist the formation of dental cavities, and it also strengthens bone. In adults, fluoride helps teeth by decreasing the activity of the bacteria in the mouth that cause dental caries.

The major source of fluoride is drinking water, although fish and most teas contain fluoride as well. Some water supplies are naturally fluoridated, and many supplies have fluoride added, usually at a concentration of one part fluoride to a million parts water. Fluoride levels in water are stated in concentrations of parts per million (ppm). About 1 ppm is ideal; less than 0.7 ppm isn't adequate to protect developing teeth.

More than about 1.5 to 2.0 ppm can lead to mild *fluorosis*, a condition that causes small, white, virtually invisible opaque areas on teeth. In its most severe form, fluorosis causes a distinct brownish mottling or discoloring. Fluorosis can occur only during tooth development. To prevent fluorosis, children should be advised to use only small amounts of fluoride toothpaste

FLUORIDE
The form of fluorine that appears in drinking water and in the body.

FLUOROSIS
A condition in which the teeth become mottled and disordered due to high fluoride ingestion.

and not swallow it. Also, you should monitor the fluoride content of your local water supply and use fluoride supplements as directed by your doctor.

Only fluoride taken internally, whether in drinking water or in dietary supplements, can strengthen babies' and children's developing teeth to resist decay. Once the teeth have erupted, they are beyond help from ingested fluoride. Supplements are often prescribed for the approximately 30 percent of Americans who do not have adequately fluoridated water supplies.

For both children and adults, fluoride applied to the surface of the teeth can add protection, at least to the outer layer of enamel, where it plays a role in reducing decay. The most familiar form, of course, is fluoride-containing toothpaste, which was introduced in the early 1960s. Fluoride rinses are also available, as are applications by dental professionals. They are considered the only useful sources of tooth-strengthening fluoride for teenagers and adults.

CHROMIUM

Chromium works with insulin to transfer glucose from the bloodstream into the body's cells. Chromium deficiency results in a condition much like diabetes, in which the blood glucose level is abnormally high.

Chromium is available in a variety of foods. The best sources are whole, unprocessed foods, such as whole grains, breads and cereals made with whole grains, broccoli, nuts, and egg yolks.

Because it is difficult for scientists to identify who is deficient in chromium, it is not known if chromium deficiency is a concern in otherwise healthy people. Although chromium supplements are advertised as helping you lose weight and put on muscle, research is not conclusive on these effects.

COPPER

Copper works as an important part of many enzymes; for example, it acts with iron to form hemoglobin. It also aids in forming collagen, a protein that gives strength and support to bones, teeth, muscle, cartilage, and blood vessels. As part of many enzymes, it is also involved in energy metabolism.

Copper occurs mostly in unprocessed foods. Good sources include liver, seafood, nuts, seeds, and beans (Figure 7-12). Copper deficiency is rare, but marginal deficiencies do occur. Single doses of copper only four times the recommended level can cause nausea, vomiting, and other symptoms.

OTHER TRACE MINERALS

Manganese is a part of enzymes that help form bones. It also functions as a cofactor for many enzymes involved in the metabolism of carbohydrate, fat, and protein as well as other metabolic processes. Manganese is also part of an enzyme that acts as an antioxidant. It is found in many foods, especially whole grains, dried fruits, nuts, and leafy vegetables. Too much or too little manganese is rare in healthy people.

Molybdenum is a cofactor for several enzymes. It appears in legumes, meat, whole grains, and nuts. Deficiency does not seem to be a problem. Too much molybdenum is rare but may damage the kidneys.

As time goes on, more trace minerals will be recognized as essential to human health. There are several trace minerals essential to animals that are likely to be essential to humans as well. Possible candidates for nutrient status include arsenic, boron, nickel, silicon, and vanadium. Based on adverse effects noted in animal studies, Tolerable Upper Intake Levels have been set for boron, nickel, and vanadium.

FIGURE 7-12: Food Sources of Copper

MEATS, POULTRY, FISH, AND ALTERNATES

Food and Serving Size	Milligrams	Adult RDA
Beef, liver, cooked, 3 oz.	12,400	900
Oyster, breaded, fried, 3 oz.	3,650	900
Lobster, cooked, 3 oz.	1,649	900
Crab, Alaska king, 3 oz.	1,005	900
Cashew nuts, 1 oz.	629	900
Seeds, sunflower seed kernels, 1/4 cup	586	900
Sunflower seed kernels, 1 oz.	519	900
Mixed nuts, 1 oz.	471	900
Beans, with pork and tomato sauce, baked, 1/2 cup	270	900
Lentils, cooked, 1/2 cup	250	900
Beans, baked, 1/2 cup	185	900

GRAINS

Food and Serving Size	Milligrams	Adult RDA
Spaghetti, whole-wheat, cooked, 1 cup	234	900
Corn Chex, 1 cup	232	900
Rice, white, long-grain, 1/2 cup	177	900
Raisin bran, 1/2 cup	100	900
Whole-wheat bread, 1 slice	80	900
White bread, 1 slice	32	900

FRUITS AND VEGETABLES

Food and Serving Size	Milligrams	Adult RDA
Potatoes, baked, 1 potato	335	900
Vegetable juice cocktail, canned, 1/2 cup	240	900
Lima beans, boiled 1/2 cup	220	900
Spinach, 1/2 cup	192	900
Beans, kidney, red, 1/2 cup	192	900
Mushrooms, canned, 1/2 cup	183	900
Raisins, seedless, 1/2 cup	110	900

DAIRY

Food and Serving Size	Milligrams	Adult RDA
Soymilk, fluid, 1 cup	345	900
Shake, fast-food, chocolate, 16 fl. oz.	216	900
Yogurt, fruit, low-fat, 8-oz. container	182	900
Fat-free milk, 1 cup	27	900
Whole milk, 1 cup	24	900

Source: U.S. Department of Agriculture, Agricultural Research Service, 2004.
USDA Nutrient Database for Standard Reference, Release 17. Nutrient Data Laboratory Home Page,
www.nal.usda.gov/fnic/foodcomp.

MINI-SUMMARY

The recommended intakes, functions, and sources of the trace minerals are listed in Figure 7-13.

FIGURE 7-13: Summary of Trace Minerals

Mineral	Recommended Intake	Functions	Sources
Iron	RDA: Men: 8 mg Women: 18 mg Upper Intake Level: 45 mg	Component of hemoglobin and myoglobin Energy metabolism	Beef, poultry, fish, enriched and fortified breads and cereals, legumes, green leafy vegetables, eggs
Zinc	RDA: Men: 11 mg Women: 8 mg Upper Intake Level: 40 mg	Wound healing Bone formation DNA synthesis Protein, carbohydrate, and fat metabolism Development of sexual organs Taste perception	Shellfish, meat, poultry, legumes, dairy foods, whole grains, fortified cereals
Iodine	RDA: 150 micrograms Upper Intake Level: 1100 micrograms	Part of thyroid hormones Normal metabolic rate	Iodized salt, saltwater fish, grains and vegetables grown in iodine-rich soil
Selenium	RDA: 55 micrograms Upper Intake Level: 400 micrograms	Part of antioxidant enzymes Regulates thyroid hormone	Grains and vegetables grown in selenium-rich soil, seafood, meat
Fluoride	AI: Men: 4 mg Women: 3 mg Upper Intake Level: 10 mg	Strong teeth and bones	Fluoridated water, fish, tea
Chromium	AI: Men: 35 micrograms Women: 25 micrograms Upper Intake Level: None	Works with insulin	Unprocessed foods such as whole grains, broccoli, nuts, egg yolks, green beans
Copper	RDA: 900 micrograms Upper Intake Level: 10,000 micrograms	Works with iron to form hemoglobin Synthesis of collagen Energy metabolism	Liver, seafood, nuts, seeds, beans
Manganese	AI: Men: 2.3 mg Women: 1.8 mg Upper Intake Level: 11 mg	Form bone Metabolism of carbohydrate, fat, and protein	Whole grains, dried fruits, nuts, leafy vegetables
Molybdenum	RDA: 45 micrograms Upper Intake Level: 2000 micrograms	Cofactor for several enzymes	Legumes, meat, whole grains, nuts

Source: Facts about Dietary Supplements, Office of Dietary Supplements, National Institutes of Health, 2001. Office of Dietary Supplements Home Page, http://ods.od.nih.gov.

NUTRITION

Nuts and seeds pack quite a few vitamins (such as folate and vitamin E) and minerals, along with fiber and protein, in their small sizes. Nuts in particular also contain quite a bit of fat. Luckily, most of the fat (except in walnuts) is monounsaturated. Walnuts and flaxseed are rich in the omega-3 fatty acid linolenic acid, an essential fatty acid. One ounce of many nuts contains from 13 to 18 grams of fat, making nuts a relatively high-kcalorie food. By comparison, seeds contain less fat and more fiber but still quite a few kcalories, but they can be easily worked into any diet. The fat and fiber they contain will, in any case, make you feel full longer.

Large studies have confirmed the link between eating nuts and a reduction in heart disease. The monounsaturated fat in nuts helps lower low-density lipoprotein cholesterol, the bad kind. Nuts (and seeds) are also excellent sources of phytochemicals.

Culinary Science

Cooking nuts is a quick process that gives them more flavor and softens their texture. As nuts cool after cooking, their texture becomes crispy. You can oven-toast nuts or simply fry them.

Toasting nuts brings out a lot of the flavor. To toast nuts such as almonds, spread them in a single layer in a shallow pan. Bake at 325°F for 5 to 10 minutes or until the nuts are lightly colored. Stir once or twice until lightly toasted. Remove from pan to cool. They will continue to brown slightly after you remove them from the oven. Small seeds, such as sesame seeds, require a cooler oven—about 250–300°F and only about 5 minutes cooking time.

When frying nuts, use low temperatures (250–350°F) and a short cooking time, just a few minutes or longer for large nuts such as Brazil nuts. When the nuts have just turned the right color, pull them off the heat since they will continue to cook a little more. When nuts are warm, they are easier to slice because they are not as brittle.

Nuts and seeds turn rancid easily due to their fat content. To keep them as fresh as possible, store in an air-tight container in the refrigerator for up to six months, or up to one year in the freezer.

- Figures 7-14, 7-15, and 7-16 show a variety of nuts and seeds that can be used in many ways.
- You can dry roast nuts or roast them in a little oil to enhance their flavor as you prefer.
- Nuts are wonderful in muffins, such as honey-almond muffins and walnut-strawberry muffins.
- Nuts and seeds work well in granolas, give crunch and flavor to casseroles, and add interest to salads such as fennel, orange, watercress, and walnut salad.
- Nuts work well in breading for fish or poultry. Mix equal parts seasoned bread crumbs and finely chopped, toasted, mixed nuts. Add herbs/spices of your choice, such as basil, thyme, cayenne pepper, or cumin. Dip poultry or fish into the crumb mixture, pressing to coast. Bake, broil, or grill.
- Pumpkin seeds are common in the cuisines of Austria and parts of Mexico, where people like their zesty flavor. Pumpkin seeds can be coated with olive oil and roasted to bring out their nutty flavor, then tossed on salads. Pumpkin seeds can also be pulverized into a thick powder or paste and used as a thickener or toasted and used as a crust. In Austria, pumpkin seed oil, which has a

(Text continues on page 253)

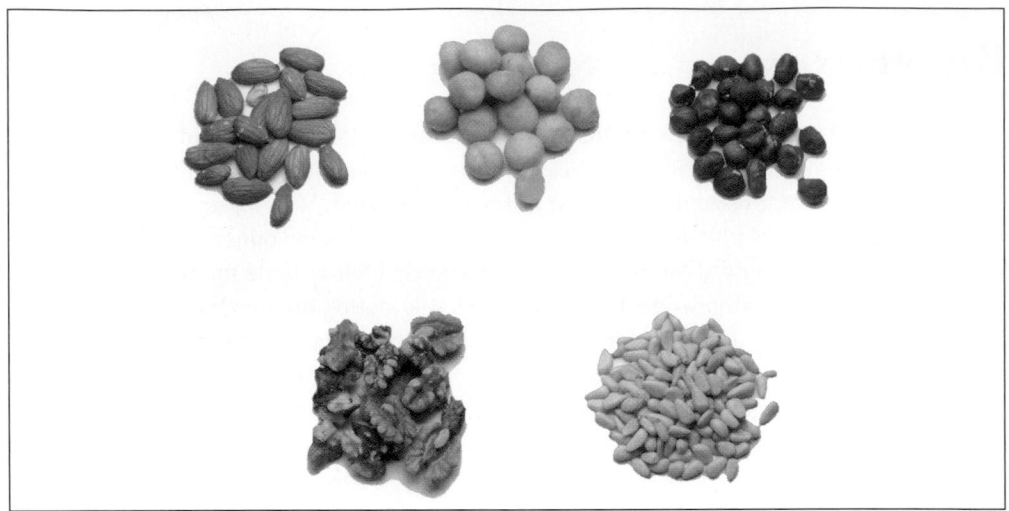

FIGURE 7-14:
Nuts. Top row: almonds, macadamia nuts, filberts; bottom row: walnuts, pine nuts.
Photo by Frank Pronesti.

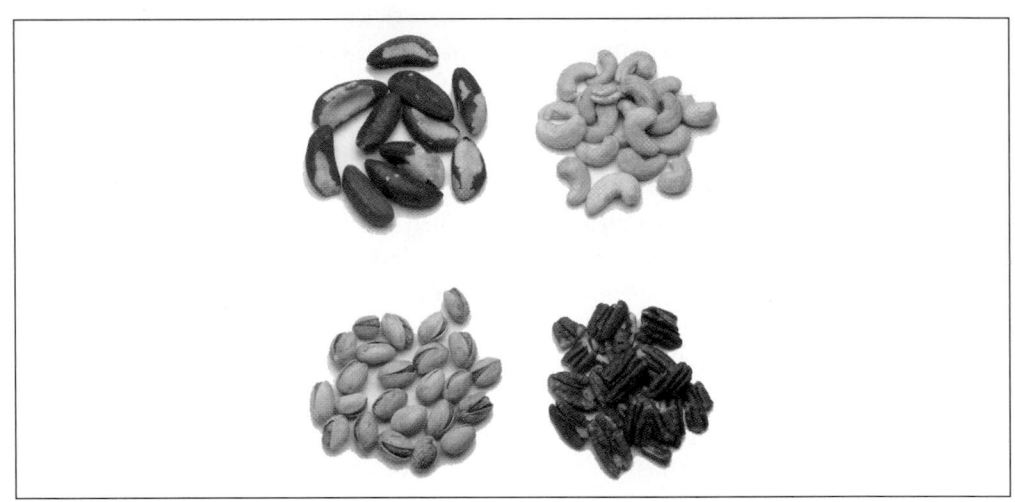

FIGURE 7-15:
Nuts. Top row: Brazil nuts, cashews; bottom row: pistachios, pecans.
Photo by Frank Pronesti.

FIGURE 7-16:
Seeds. Top row: flax seeds, sunflower seeds; bottom row: sesame seeds, pumpkin seeds.
Photo by Frank Pronesti.

very strong flavor, is used in small amounts in salad dressings. It also is used in the United States by some chefs.

- Sunflower seeds are large compared with seeds such as sesame and caraway. They can be used in casseroles, stews, vegetables, stuffings, and salads.
- Sesame seeds and caraway seeds are often used in baking. Toasted sesame seeds can be sprinkled on soups, fish, and cooked vegetables for flavor and texture.
- In cooking, flax seeds add a pleasant, nutty flavor. The attractive, oval reddish-brown seeds of flax (just a little bigger than sesame seeds) add extra texture and good nutrition to breads and other baked goods. That's why flax has been long used in multigrain cereals and snack foods. Flax seed also delivers the benefits of its soluble fiber, lignans, omega-3 fatty acids, and protein. Flax seed is available whole or ground. Whole seeds provide fiber but the body can't break down the whole seed so you don't get any of the benefits of the omega-3 fatty acids and protein inside. You can use flaxseed, whole or ground, in many ways.
- Add whole seeds to bread doughs and pancake, muffin, or cookie mixes. When sprinkled on top of any of these before baking, it adds crunch, taste, and eye appeal.
- For ground flaxseed, grind whole seeds to a smooth consistency. Use it to enhance the flavor of oatmeal, yogurt, or apple sauce.
- Offer flaxseed as toppings for smoothies and yogurt, and at juice bars, fruit bars, and salad bars.
- Use ground flaxseed as a coating ingredient for chicken, fish, and other protein items.
- Ground flaxseed can substitute for fat or eggs in many recipes. Use 3 tablespoons ground flaxseed to replace 1 tablespoon of fat, or use 2 tablespoons ground flaxseed mixed with 3 tablespoons of water to replace 1 egg.

✂ CHECK-OUT QUIZ

1. Our understanding of many trace minerals is still emerging.

   a. True
   b. False

2. Two cups of yogurt contain about as much calcium as 1 cup of milk.

   a. True
   b. False

3. Sodium and potassium are involved in muscle contraction and transmission of nerve impulses.

   a. True
   b. False

4. Canned soft drinks are high in sodium.

   a. True
   b. False

5. Iodine is needed to maintain a normal metabolic rate.

   a. True
   b. False

6. Few minerals are toxic in excess.

   a. True
   b. False

7. Nearly all foods contain water.

   a. True
   b. False

8. The kidneys will always excrete a certain amount each day to eliminate waste products.

   a. True
   b. False

9. Sodium, potassium, and calcium are referred to as electrolytes.

   a. True
   b. False

10. Name the mineral(s) referred to by the following descriptions:

    a. Involved in bone formation.
    b. Found mostly in milk and milk products.
    c. Helps maintain water balance and acid-base balance.
    d. Some diuretics deplete the body of this mineral.
    e. Found in the stomach juices.
    f. High in bananas and oranges.
    g. Found in certain water supplies.
    h. Found in salt.
    i. Found in heme.
    j. Causes a form of anemia.
    k. Occurs in the soil.

ACTIVITIES AND APPLICATIONS

1. Your Eating Style

   Using Figures 7-8 and 7-13 summarizing the major and trace minerals (or a worksheet the instructor hands you), circle any foods that you do not eat at all or that you eat infrequently, such as dairy products or green vegetables. Do you eat most of the foods containing minerals, or do you dislike vegetables and maybe fruits, too? In terms of frequency, how often do you eat mineral-rich foods? The answers to these questions should help you assess whether your diet is adequately balanced and varied, which is necessary to ensure adequate mineral intake.

2. Mineral Salad Bar

   You are to set up a salad bar by using a worksheet the instructor hands out. You may use any foods you like in the salad bar as long as you have a good source of each of the minerals and fill in each of the circles. In each circle on the worksheet, write down the name of the food and which mineral(s) it is rich in.

3. Sodium Countdown

   Using *iProfile*, list the sodium content of 10 of your favorite foods. How much would each contribute to the Adequate Intake for sodium?

4. One-Day Food Record and Nutrient Analysis

   Now that you have learned about all the nutrients, you can see how many nutrients you take in during a typical day. To do so, write down everything you eat and drink (except water) for one day. Include a description of each food and the portion size, such as "1 large apple." Also include on your food record any supplements that you take, including the name of each nutrient and how much is in the supplement. Use iProfile to do the nutrient analysis. First create your Profile, then use the Food Journal to record the foods you have eaten. Look over carefully and print out the following reports:

       My DRI
       Intake Compared to DRI
       MyPyramid

Use these reports to write up an analysis of your diet including which nutrients you get too much or too little of and how well your diet meets the MyPyramid guidelines. Do you need to change some of your food choices to improve the results? If so, describe.

NUTRITION WEB EXPLORER

Oregon Dairy Council   www.oregondairycouncil.org/calcium_checkup/
Click on "Calcium: Are You Getting Enough?" and see how much calcium is in your diet.

American Heart Association   www.americanheart.org/presenter.jhtml?identifier=582
Take the "Sodium Intake Quiz" to determine how much sodium you take in each day.

Medline Plus Health Information on Supplements   www.nlm.nih.gov/medlineplus/
minerals.html

Click on a topic under "Latest News" under "Basics" and write a summary of what you read.

National Center for Complementary and Alternative Medicine   www.nccam.nih.gov
The National Center for Complementary and Alternative Medicine is dedicated to exploring complementary and alternative healing practices in the context of rigorous science. Read "What Is CAM?" to learn about this new area.

Office of Dietary Supplements, National Institutes of Health   http://dietary-supplements.
info.nih.gov/

Click on "Health Information" and then "The Savvy Supplement User." Write down any tips that are useful for you or someone you know, such as a grandmother, who is taking supplements.

Iron Overload Diseases Association   www.ironoverload.org
Click on "Diet for Hemochromatosis, Iron Overload, and Anemia." What foods should be avoided if you have iron overload?

FOOD FACTS: HOW TO RETAIN VITAMINS AND MINERALS FROM PURCHASING TO SERVING

Water-soluble vitamins such as vitamin C and the B vitamins are easily destroyed by excess water, air, heat, and light. They are also affected by the pH balance, meaning too much or too little acid, of the cooking liquid. The fat-soluble vitamins A, D, E, and K are more stable.

GENERAL GUIDELINES

To avoid losing nutrients, be careful with:

- Water. Soaking food in water dissolves water-soluble vitamins and minerals. Avoid it except when absolutely necessary. If foods, such as vegetables, must be soaked or remain in water during cooking, use as small an amount of water as possible and use the leftover cooking liquid in soup or in another product. Or you can steam vegetables in small batches for the best quality.
- Heat. Heating food causes nutrient loss, especially vitamin C. Avoid prolonged overcooking.
- Light. Milk is an excellent source of riboflavin, but if it is allowed to stand open or is exposed to light, considerable destruction of riboflavin can occur. A light-obstructing container, such as a cardboard carton, can help prevent this. If you are using another type of container, be sure to store it away from light.
- pH Balance. Baking soda should *not* be added to green vegetables to retain color during cooking or to dry peas and beans to decrease the cooking time. Baking soda makes the cooking water alkaline, destroying thiamin and vitamin C.
- Air. Vitamins A, C, E, K, and the B vitamins are destroyed by exposure to air.

To reduce nutrient loss:

- Cut and cook vegetables in pieces that are as large as possible.
- Store foods with proper covers.
- Cook vegetables as soon as possible after cutting.
- Cook vegetables until "just tender."
- Prepare food as close to serving time as possible.
- Serve raw vegetables when possible.

PROTECTING FRUITS AND VEGETABLES

Because of advances in food technology, fruits and vegetables are available in many forms—fresh, frozen, or canned. Most frozen fruits and vegetables today are flash-frozen, meaning they were frozen at extremely low temperatures at harvest. Flash freezing stops nutrient loss completely until the fruit or vegetable is thawed out. In any form, fruits and vegetables need to be handled and stored correctly to retain nutrients and ensure food safety. To retain nutrients during purchasing and storage, do the following:

- Buy fresh, high-quality fruits and vegetables.
- Examine fresh fruits and vegetables thoroughly for appropriate color, size, and shape (Figure 7-17).
- Store fresh fruits and vegetables in the refrigerator (except green bananas, avocados, potatoes, and onions) to inhibit enzymes that make fruits and vegetables age and lose nutrients. The enzymes are more active at warm temperatures. Refrigerated goods should be maintained at a temperature of 40°F or

FIGURE 7-17:
Fresh fruits and vegetables.
Courtesy PhotoDisc, Inc.

lower, freezer goods at 0°F or lower. Thermometers should be kept in the refrigerator and the freezer to monitor temperatures.
- Store canned goods in a cool place.
- Foods should not be stored for too long, as nutrient loss occurs during storage.
- When storing fresh fruits and vegetables, close up the wrapping tightly to decrease exposure to the air, which pulls out water and decreases the nutrient content.
- For best results when cooking vegetables:
  - Avoid peeling vegetables when possible. Potatoes and other vegetables that are cooked without being peeled retain many more nutrients than do peeled and cut vegetables.
  - Prepare small amounts. Avoid long exposure to heat. Fresh or frozen vegetables can be cooked by several different methods. You can steam, bake, or sauté them. Regardless of the cooking method you choose, it is better to prepare small amounts than to cook single large batches. Nutritive value is lost and quality is lowered with long exposure to heat. Microwaving is great to prepare several portions.

- To retain nutrients and bright colors, cook "just until tender." Steaming is a good way to cook vegetables.
- Use carefully timed "batch cooking" to avoid having vegetables held too long before serving. A good rule of thumb: The quantity you cook should not exceed the amount you can serve in 15 minutes. This applies both to vegetables served alone and to vegetables used in recipes such as beef and chicken stir-fry.

PROTECTING GRAINS

Health experts encourage Americans to choose grains and breads as a major component of a nutritious diet. This is why grains and breads form the base of MyPyramid. To provide maximum benefit, they must be prepared correctly. To retain the nutrients in grains, remember:

- Rice should not be washed before cooking. Rice is enriched by spraying with vitamins and minerals. When you wash rice, the enrichment is washed off.
- Browning uncooked rice before adding water can destroy a lot of the thiamin content.
- Rinsing cooked grains and pasta causes considerable loss of nutrients and is not recommended.

HOT TOPIC: DIETARY SUPPLEMENTS

Surveys show that more than half the U.S. adult population uses dietary supplements. Annual sales of dietary supplements in the United States are approaching $16 billion. An average of 1000 new products are developed each year (Figure 7-18). Although manufacturers are restricted from claiming that using their products leads to therapeutic benefits, surveys show that many people take supplements for purposes such as treating colds and alleviating depression. According to other survey data, the majority of consumers believe that these products are either reasonably or completely safe.

Traditionally, the term "dietary supplements" referred to products made of one or more of the essential nutrients, such as vitamins, minerals, and protein. But the 1994 Dietary Supplement Health and Education Act (DSHEA) broadened the definition to include, with some exceptions, any product intended for ingestion as a supplement to the diet. In addition to vitamins, minerals, and herbs or other botanicals, dietary supplements may include amino acids and substances such as enzymes, organ tissues, glandulars, and metabolites.

It's easy to spot a supplement, because the DSHEA requires manufacturers to include the words "dietary supplement" on product labels. Manufacturers must also include a list of ingredients. In addition, a Supplement Facts panel is required on the labels of most dietary supplements. This label must identify each dietary ingredient contained in the product.

Ingredients not listed on the Supplement Facts panel must be listed in the "Other Ingredients" statement beneath the panel. The types of ingredients listed there could include the source of dietary ingredients, such as rose hips as the source of vitamin C; other food ingredients, such as water or sugar; and technical additives or processing aids such as starch and colors.

Dietary supplements come in many forms, including tablets, capsules, powders, softgels, gelcaps, and liquids. Though commonly associated with health-food stores, dietary supplements also are sold in grocery, drug, and national discount chain stores, as well as through mail-order catalogs, TV programs, the Internet, and direct sales.

One thing dietary supplements are not is drugs. A drug, which sometimes can be derived from plants used as traditional medicines, is intended to diagnose, cure, relieve, treat, or prevent disease. Before marketing, a drug must undergo clinical studies to determine its effectiveness, safety, possible interactions with other substances, and appropriate dosages, and the FDA must review these data

and authorize a drug's use before it is marketed.

A product sold as a dietary supplement and touted in its labeling as a new treatment or cure for a specific disease or condition would be considered an unauthorized—and thus illegal—drug. Labeling changes consistent with the provisions in the DSHEA would be required to maintain the product's status as a dietary supplement.

By law, the manufacturer is responsible for ensuring that its dietary supplements are safe before they are marketed. Unlike drug products that must be proved to be safe and effective for their intended use before marketing, there are no provisions in the law for the FDA to approve dietary supplements for safety or effectiveness *before* they reach the consumer. In June 2007, the FDA established dietary supplement "current Good Manufacturing Practice" regulations requiring that manufacturers evaluate their products through testing identity, purity, strength, and composition.

Under DSHEA, once a dietary supplement is marketed, the FDA has the responsibility for showing that it is unsafe before it can take action to restrict the product's use or take it off the market. To bolster the FDA's ability to evaluate the safety of dietary supplements, a report in 2004 outlined a new science-based process for assessing supplement ingredients, even when data about a substance's safety in humans are scarce. The process provides a way to identify supplement ingredients that may pose risks, prioritize them on the basis of their level of potential risk, and evaluate them for safety. The report categorizes different kinds of data the FDA can use to assess safety and offers guidelines for determining the

FIGURE 7-18:
Dietary supplements.
Courtesy Andy Washnik for John Wiley & Sons, Inc.

significance of the evidence available on a particular substance.

Claims that tout a supplement's health benefits have always been a controversial feature of dietary supplements. Manufacturers often rely on them to sell their products, but consumers often wonder whether they can trust the claims.

Under the DSHEA and previous food-labeling laws, supplement manufacturers are allowed to use, when appropriate, three types of claims: nutrient claims, health claims, and nutrition support claims, which include "structure-function claims."

Nutrient claims describe the level of a nutrient in a food or dietary supplement and are discussed in Chapter 2. For example, a supplement containing at least 200 milligrams of calcium per serving could carry the claim "high in calcium."

Health claims show a link between a food or substance and a disease or health-related condition. The FDA authorizes these claims based on a review of the scientific evidence. For example, a claim may show a link between folate in the product and a decreased risk of neural tube defects in pregnancy if the supplement contains enough folate. Chapter 2 discusses health claims in more detail.

Nutrition support claims describe a link between a nutrient and the deficiency disease that can result if the nutrient is lacking in the diet. For example, the label of a vitamin C supplement could state that vitamin C prevents scurvy. When these types of claims are used, the label must mention the prevalence of the nutrient-deficiency disease in the United States.

These claims also can refer to the supplement's effect on the body's structure or function, including its overall effect on a person's well-being. These are known

as structure-function claims. Examples include:

- Calcium builds strong bones.
- Antioxidants maintain cell integrity.
- Fiber maintains bowel regularity.

Manufacturers can use structure-function claims without FDA approval. They base their claims on their review and interpretation of the scientific literature. Like all label claims, they must be true and not misleading. They must also be accompanied by the disclaimer "This statement has not been evaluated by the Food and Drug Administration. This product is not intended to diagnose, treat, cure, or prevent any disease."

Manufacturers that plan to use a structure-function claim must inform the FDA of the use of the claim no later than 30 days after the product is first marketed. While the manufacturer must be able to substantiate its claim, it does not have to share the substantiation with the FDA or make it publicly available. If the submitted claim promotes the product as a drug instead of a supplement, the FDA can advise the manufacturer to change or delete the claim.

Dietary supplements also are not replacements for conventional diets. Supplements do not provide all the known—and perhaps unknown—nutritional benefits of conventional food.

To help protect themselves, consumers should:

- Look for ingredients in products with the USP notation, which indicates that the manufacturer followed standards established by the U.S. Pharmacopoeia.
- Avoid substances that are not known nutrients.
- Limit their intake of vitamins and minerals to the doses recommended in the

DRIs. Some vitamins, such as A and D, may be harmful in higher doses.
- Consider the name of the manufacturer or distributor. Supplements made by a nationally known food and drug manufacturer have probably been made under tight controls, because these companies already have in place manufacturing standards for their other products.
- Write to the supplement manufacturer for more information.
- Be sure to consult a doctor before purchasing or taking any supplement.
- Ask the pharmacist about possible interactions between supplements and prescription and over-the-counter (OTC) medicines. Taking a combination of supplements or using these products together with prescription or OTC drugs could, under certain circumstances, produce adverse effects, some of which could be life-threatening. For example, St. John's wort may reduce the effectiveness of prescription drugs for heart disease, depression, seizures, or certain cancers; it may also diminish the effectiveness of oral contraceptives.
- Just because an herb is natural does not mean that it is safe. In addition, herbal products may be contaminated with hazardous substances such as pesticides and disease-causing microorganisms.
- Consumers should also be sure to tell their health-care providers about the supplements they take.

Although most Americans can get needed vitamins and minerals through food, situations occur when supplements may be needed:

- Women in their childbearing years and pregnant or lactating women, who may need iron and/or folate
- People with known nutrient deficiencies, such as iron-deficient women

- Elderly people who are eating poorly, have problems chewing, or have other concerns
- Drug addicts or alcoholics
- People eating less than 1200 kcalories a day—such as dieters—who may need supplements because it is hard to get enough nutrients in such low-calorie diets
- People on certain medications or with certain diseases

If you really feel you need additional nutrients, your best bet is to buy a multivitamin and multimineral supplement that supplies 100 percent of the RDA or AI. It can't hurt, and it may act as a safety net for individuals who eat haphazardly. But keep in mind that more is not always better and that no supplement can take the place of food and serve as a permanent method for improving a poor diet. In other words, use these products to supplement a good diet, not to substitute for a poor diet.

PART TWO

Developing and Marketing Healthy Recipes and Menus

Foundations of Healthy Cooking

Not that long ago, eating out was reserved for special occasions and celebrations. Times have changed. With more dual-income families, fast and convenient meals are a must. Restaurants are now an essential part of the national lifestyle, with Americans spending $0.47 of every food dollar to dine out.

With over 800,000 restaurants in large cities, small towns, rural areas, and every place in between, the restaurant industry presents consumers with more menu choices that can be part of a healthy diet than ever before. The vast majority of operators promote healthy choices, from adding more fruits and vegetables to substituting a sauce that is lower in sodium for a customer on a salt-restricted diet. From healthy salads to decadent desserts, taste and presentation are important for all menu items for all ages (Figure 8-1). This chapter will help you build a strong foundation in healthy cooking by being able to do the following:

- Define seasoning, flavoring, herbs, and spices
- Suggest ingredients and methods to develop flavor
- Identify and suggest healthy cooking methods and techniques

 # FLAVOR

SEASONINGS
Substances used in cooking to bring out a flavor that is already present.

FLAVORINGS
Substances used in cooking to add a new flavor or modify the original flavor.

A solid foundation in foods and cooking is necessary to develop healthy menus and recipes. You are expected to know basic culinary terminology and techniques and have a working knowledge of ingredients, from almonds to zucchini. A basic culinary skill that needs further refinement for cooking healthy is that of flavoring. Because you can't rely on more than moderate amounts of fat, salt, and sugar for taste and flavor, you will need to develop excellent flavor-building skills.

Seasonings and *flavorings* are very important in healthy cooking, because they help replace missing ingredients such as fat and salt. Seasonings are used to bring out flavor that is already present in a dish, whereas flavorings add a new flavor or modify the original one. The difference between them is one of degree.

HERBS AND SPICES

HERBS
The leafy parts of certain plants that grow in temperate climates; they are used to season and flavor foods.

SPICES
The roots, bark, seeds, flowers, buds, and fruits of certain tropical plants; they are used to season and flavor foods.

Herbs and spices are key flavoring ingredients in nutritional menu planning and execution and are the backbone of most menu items, lending themselves to cultural and regional food styles. Good sound nutritional cooking can be virtually equal to classical cooking in terms of technique, creative seasoning, flavor blending, and presentation. It's helpful when one is moderating fat, cholesterol, and sugars to enhance recipes with an abundance of seasonings such as cinnamon, nutmeg, mace, vanilla beans, ginger, fennel, star anise, juniper, and cardamom. These spices provide a sense of sweet satisfaction as well as bold character to recipes.

The use of herbs in recipes changes the flavor direction to whichever herb is prominent. For instance, the use of basil, oregano, and thyme in a tomato vinaigrette points the dish toward an Italian flavor. Take the same dish and add cilantro and lime juice and you move south of the border; add fresh chopped tarragon with shallots and you're in France. There is no end to your creative abilities once you understand the basic format for healthy cooking.

Herbs are the leafy parts of certain plants that grow in temperate climates. *Spices* are the roots, bark, seeds, flowers, buds, and fruits of certain tropical plants. Figures 8-2 and 8-3 show a number of herbs and spices. Herbs are generally available fresh and dried. Spices are mostly available in a dried form.

FIGURE 8-1:
Children should be taught the benefits of healthy eating at an early age.
Courtesy Centers for Disease Control and Prevention, www.fruitsandveggiesmatter.gov

FIGURE 8-2a:
Herbs. Top row, left to right: sage, oregano, dill; bottom row, left to right: Italian parsley, curly parsley, basil.
Photo by Frank Pronesti.

FIGURE 8-2b:
Herbs. Top row: tarragon, thyme, cilantro; bottom row, left to right: chives, mint, rosemary.
Photo by Frank Pronesti.

FIGURE 8-3a:
Spices. Top row, left to right: black peppercorns, green peppercorns, pink peppercorns; bottom row, left to right: white peppercorns, Sichuan peppercorns.
Source: *Professional Cooking, sixth edition*, by Wayne Gisslen. Copyright John Wiley and Sons, Inc. 2007.
Photo by J. Gerard Smith.

FIGURE 8-3b:
Spices. Top row, left to right: cloves, nutmeg, allspice, cinnamon sticks; bottom row, left to right: juniper berries, cardamom, saffron, star anise.
Source: *Professional Cooking, sixth edition,* by Wayne Gisslen. Copyright John Wiley and Sons, Inc. 2007.
Photo by J. Gerard Smith.

Fresh herbs, as opposed to dry, are far superior and more versatile for creating recipes. Herbs commonly available fresh include parsley, cilantro, basil, dill, chives, tarragon, thyme, and oregano. Fresh herbs are great when you need a crisp clean taste and maximum flavor. They can withstand only about 30 minutes of cooking, so they work best for finishing dishes.

When the use of fresh herbs is not always possible, dry herbs can be substituted with better than average results. Dried herbs work well in longer cooking, such as in stocks, stews, and sauces. You can use dried herbs along with fresh herbs toward the end of cooking to get a richer and cleaner flavor.

The real purpose of herbs and spices is not to rescue, remedy, flavor, or season but to build. Spices and herbs are basically flavor builders. This is their proper use in cooking; they should be cooked with the dish as it is being made so that their flavors blend smoothly, giving character and depth to the dish.

Learning to identify the innumerable different herbs and spices requires a keen sense of taste and smell. Simply looking at them is not enough. Taste them, smell them, feel them, use them. The key to most of them is aroma, for in their aroma is about 60 percent of their flavor. Their aromatic quality not only adds flavor to the food as it is eaten but heightens the anticipation of the diner as the food is being cooked and served.

FIGURE 8-3c:
Spices. Top row, left to right: celery seed, dill seed, coriander seed, caraway seed; bottom row, left to right: fennel seed, cumin seed, anise seed.
Source: *Professional Cooking, sixth edition,* by Wayne Gisslen. Copyright John Wiley and Sons, Inc. 2007.
Photo by J. Gerard Smith.

There are many, many herbs and spices. Let's look at those most likely to be found in the kitchen. To help you understand them, we'll sort them into groups, but first let's look at the many forms of pepper.

Pepper comes in three forms: black, white, and green. White and black pepper both come from the oriental pepper plant. Black peppercorns are picked when slightly underripe and then air-dried; this results in their dark color. White peppercorns are fully ripe berries that have been soaked in water and hulled, which produces a slightly fermented taste. Green peppercorns are picked before ripeness and preserved.

Black pepper comes in four forms: whole black peppercorns, crushed, butcher's grind, and table-ground pepper. Whole pepper is used as a flavor builder during cooking, as in making stocks. Crushed black pepper can function as a flavor builder during cooking as well, or it can be added as flavoring to a finished dish. Many Americans enjoy the flavor contrast of fresh crushed peppercorns straight from the pepper mill on a crisp green salad. The flavor of ground black pepper is characteristic of certain cuisines and certain parts of the country. Cooks who cater to these clienteles are likely to add this flavor as they season food.

As a seasoning, black pepper is used only in dark-colored foods; it spoils the appearance of light-colored foods. *White pepper* is used in light-colored foods because its presence is concealed. White pepper comes in two forms: whole peppercorns and ground white pepper. White peppercorns are used in the same ways as black.

Ground white pepper is good for all-around seasoning. It lends itself to white dishes both in appearance and in flavor, and it has the strength necessary to season dark dishes. It also stands up to high heat better than black pepper does. Ground white pepper is chosen by most good cooks as the true seasoning pepper. It is seldom used as a table pepper, as it is expensive.

Green peppercorns are preserved either by packing them in liquid (such as vinegar or brine) or by drying them. They have a fresh taste that's less pungent than that of the other types. They pair well with vegetables such as artichokes and zucchini.

Pink peppercorns are not true peppercorns, but they look like peppercorns and have a sweet, slightly peppery taste. They are native to South America, but they are sometimes mistakenly called Japanese peppers because they are one of the few spices used in Japanese cooking. They are aromatic but not peppery and often are included in mixes mostly for color. Possible adverse reactions to pink peppercorns have been reported when they are added generously to dishes, and so they should be used in small amounts.

Red pepper, also called *cayenne*, is completely unrelated to white or black pepper. It comes from dried pepper pods. It is quite hot and easily overused. When added with restraint in soups and sauces, it can lend a spicy hotness. When used without as much restraint, it creates the hot flavor of many foods from Mexico, South America, and India.

Let's look at nine herbs and spices that are used as often for their distinctive flavors as for general flavor enrichment. As a flavor builder, each goes beyond the subtlety of the stock herbs and spices—even though you can't single it out from the flavor of the dish as a whole. Used in quantities large enough to taste, they become major flavors rather than flavor builders. Basil, oregano, and tarragon are available fresh and also come in the form of crushed dried leaves. They look somewhat alike, but their tastes are very different.

Basil has a warm, sweet flavor that is welcome in many soups, sauces, entrées, relishes, salsas, and dressings as well as with vegetables such as tomatoes, peppers, eggplant, and squash. It blends especially well with tomato, lemons, and oranges. Like many other herbs, it has symbolism: In India it expresses reverence for the dead; in Italy it is a symbol of love.

Oregano belongs to the same herb family as basil, but it makes a very different contribution to a dish—a strong bittersweet taste and aroma you may have met in spaghetti sauce. It is used in many Italian, Mediterranean, Spanish, South American, and Mexican dishes.

Tarragon has a flavor that is somehow light and strong at the same time. It tastes something like licorice. It is used in poultry and fish dishes as well as in salads, sauces, and salad dressings.

Rosemary, like bay leaf, is used in dishes in which a liquid is involved: soups, stocks, sauces, stews, and braised foods. The leaf of an evergreen shrub of the mint family, it has a pungent, hardy flavor and fragrance. Fresh or dried, it looks and feels like pine needles. It is used mostly with meats, game, poultry, mushrooms, and ragouts.

Dill and *mustard* have flavors that will be very familiar to you: dill as in pickle and mustard as in the hot dog condiment. Fresh or dried dill leaves, often called dill weed, are used in soups, fish dishes, stews, salads, and butters. Whole dill seed is used in some soups and sauerkraut. Dry mustard, a powdered spice made from the seed of the mustard plant, comes in three varieties: white, yellow, and brown. The brown variety has the sharpest and most pungent flavor. All are used to flavor sauces, dips, dressings, and entrées. Prepared mustards are made from all the varieties and serve as excellent flavor enhancers.

Paprika is another powdered spice that comes in two flavors: mild and hot. Both kinds are made from dried pods of the same pepper family as red pepper and cayenne and look something like the seasoning peppers, but they do not do the work of seasonings. Hungarian paprika is the hot spicy one; Spanish paprika is mild in flavor, and its red color has lots of eye appeal. Hungarian paprika is used to make goulash and other braised meats and poultry. Spanish paprika is used for coloring, blending rubs, and mild seasoning. Paprikas are sensitive to heat and will turn brown if exposed to direct heat.

Still another branch of this pepper pod family gives us *chili peppers*, the crushed or dried pods of several kinds of Mexican peppers and Asian dried red chilis. Colors range from red to green, and flavors from mild to hot. Chili peppers are used in Mexican, Asian, Thai, Peruvian, Indian, Cuban, and South American cuisines.

Several spice blends are available. For example, *chili powder* is a combination of toasted ground dried chili peppers. Chili powder varies from mild to very hot. It is used, of course, in chili, where it functions as a major flavor, and in many Mexican, South American, Cuban, and Southwestern dishes.

Curry powder is a blend of up to 20 spices. In India, where it originated, cooks blend their own curry powders, which may vary considerably. In the United States, curry powder comes premixed in various blends from mild to hot. Curry powders usually include cloves, black and red peppers, cumin, garlic, ginger, cinnamon, coriander, cardamom, fenugreek, mustard, turmeric (which provides the characteristic yellow color), and sometimes other spices.

A group of powdered sweet aromatic spices from the tropics are used frequently in baking, in dessert cookery, and occasionally in sauces, vegetables, and entrées. Among these spices are cinnamon, nutmeg and its counterpart mace, and ginger. *Cinnamon* comes from the dried bark of the cinnamon or cassia tree, *nutmeg* and *mace* from the seed of the nutmeg tree, and ginger from the dried root of the ginger plant. In hot foods, the nutmeg flavor goes well with potatoes, dumplings, spinach, quiche, and some soups and entrées. Mace, a somewhat paler alternative to nutmeg, has a similar flavor and is used in bratwurst, savory dishes, baked goods, and pâtés. Ground cinnamon and ginger are used in a variety of cuisines, both sweet and savory. Cinnamon is also available in sticks.

Mint is a sweet herb with the familiar flavor you meet in toothpaste and chewing gum. Mint is available in many varieties. The most popular are spearmint and peppermint. Others include chocolate, licorice, orange, and pineapple. The flavor of a mint sauce offers a refreshing complement to lamb. Fresh mint makes a good flavoring and garnish for fruits, vegetables, salsas, relishes, salads, dressings, iced tea, desserts, and sorbets.

To get maximum flavor from spices, you can give them a quick dry toast. The following are some whole spices that can be toasted:

- Mustard seed
- Fennel
- Coriander
- Star anise

- Cardamom
- Caraway
- Cumin
- Juniper
- Allspice

To toast, put a sauté pan on medium to high heat for about 1 minute, add the spices in the pan, and toss quickly until a nutty aroma comes out of the spices. Once cool, fine grind them and use to season marinades, salad dressings, rubs, soups, stews, ragouts, salsas, and relishes.

Many herbs and spices can be combined to produce blends with distinctive flavors. For example, a cattleman's blend with paprika, peppers, chilis, and other dried herbs can be used as a steak seasoning. A seed blend, such as ground cardamom, fennel, anise, star anise, cumin, and coriander, is excellent in soups, marinades, compotes, vegetables, and chutneys. Ethnic blends such as the following also present many flavoring possibilities:

- Italian: garlic, onion, basil, oregano
- Asian: ginger, five-spice powder, garlic, scallion
- French: tarragon, mustard, chive, chervil, shallot
- South American: chili peppers, lime juice, garlic, cilantro
- Indian: ground nutmeg, fennel, coriander, cinnamon, fenugreek, curry
- Mediterranean: oregano, marjoram, thyme, pepper, coriander, onion, garlic

Blends such as these add depth to foundations for starting or finishing a dish. Always check the salt and sugar content of premade blends.

Figure 8-4 is a reference chart for many herbs and spices.

JUICES

Juices can be used as is for added flavor or can be reduced (boiled or simmered down to a smaller volume) to get a more intense flavor, a vibrant color, and a syrupy texture. Reduced juices make excellent sauces and flavorings. Use a good-quality juicer or buy quality premade juices.

For example, orange juice can be reduced (simmered) to orange oil, which is excellent in salad dressings, marinades, and sauces. Also, freshly made beet juice can be used to enhance stocks, glazes, and sauces. Put reduced beet juice and seasonings into a squirt bottle and use it lightly on plates for decoration and flavor, especially with salads, appetizers, and entrées. Other juices that are useful in the kitchen are carrot, fennel, celeriac, pomegranate, ginger, celery, asparagus, bell pepper (yellow, red, orange), herbs (watercress, cilantro, parsley, basil, chive), citrus, and tomato.

Lemon and lime juice are seldom called seasonings, yet their use as seasonings is not unusual. Many recipes call for small amounts of lemon rind or juice, which sparks the flavor of the dish even though the citrus flavor cannot be perceived.

VINEGARS AND OILS

Various types of vinegars can add flavor to a wide variety of dishes, from salads to sauces. They have a light, tangy taste and activate the taste buds without the addition of fat. Popular vinegars include wine vinegars (made from white wine, red wine, rosé wine, rice wine, champagne, or sherry), cider vinegar (made from apples), and balsamic vinegar. True

FIGURE 8-4: Herb and Spice Reference Chart

Product	Market Forms	Description	Uses
Allspice (spice)	Whole, ground	Dried, dark brown berries of an evergreen tree indigenous to the West Indies and Central and South America. Smells of cloves, nutmeg, and cinnamon.	Braised meats, curries, baked goods, puddings, cooked fruit
Anise seed (spice)	Whole, ground	Tiny dried seed from a plant native to eastern Mediterranean. Licorice (sweet) flavor.	Baked goods, fish, shellfish, soups, sauces
Basil (herb)	Fresh, dried: crushed leaves	Bright green tender leaves of an herb in the mint family. Sweet, slightly peppery flavor.	Tomatoes, eggplant, squash, carrots, peas, soups, stews, poultry, red sauces
Bay leaves (herb)	Whole, ground	Long, dark green brittle leaves from the bay tree, a small tree from Asia. Pungent, warm flavor when leaves are broken. Always remove bay leaves before serving (due to toughness).	Stocks, sauces, soups, braised dishes, stews, marinades
Caraway seed (spice)	Whole, ground	Crescent-shaped brown seed of a European plant. Slightly peppery flavor.	German and East European cooking (such as sauerkraut and coleslaw), rye bread, pork
Cardamom (spice)	Whole pod, ground seeds	Tiny seeds inside green or white pods that grow on a bush of the ginger family. Sweet and spicy flavor. Very expensive.	Poultry, curries and other Indian cooking, Scandinavian breads and pastries, puddings, fruits
Cayenne, red pepper (spice)	Ground	Finely ground powder from several hot types of dried red chili peppers. Very hot. Use in small amounts.	In small amounts in meats, poultry, seafood, sauces, and egg and cheese dishes
Celery seed (spice)	Whole, ground, ground mixed with salt or pepper	Small gray-brown seeds produced by celery plant. Distinctive celery flavor.	Dressings, sauces, soups, salads, tomatoes, fish
Chervil (herb)	Fresh, dried: crushed leaves	Fernlike leaves of a plant in the parsley family. Like parsley with slight pepper taste; smells like anise.	Stocks, soups, sauces, salads, egg and cheese dishes, French cooking
Chives (herb)	Fresh, dried	Thin grass like leaves of a plant in the onion family. Mild onion flavor.	Poultry, seafood, potatoes, salads, soups, egg and cheese dishes
Cinnamon (spice)	Stick, ground	Aromatic bark of the cinnamon tree, a small evergreen tree of the laurel family. Sweet, warm flavor.	Baked goods, desserts, fruits, lamb, ham, rice, carrots, sweet potatoes, beverages.
Cloves (spice)	Whole, ground	Dried flower buds of a tropical evergreen tree. Sweetly pungent and very aromatic flavor.	Stocks, marinades, sauces, braised meats, ham, baked goods, fruits

(continued)

FIGURE 8-4: Herb and Spice Reference Chart (*Continued*)

Product	Market Forms	Description	Uses
Coriander seeds (spice)	Whole, ground	Small seeds from the cilantro plant. Mild, slightly sweet and musty flavor.	Pork, pickling, soups, sauces, chutney, casseroles, Indian cooking
Cumin seed (spice)	Whole, ground	Seed of a small plant in the parsley family. Looks like caraway seed but lighter in color. Pungent, strong, earthy flavor.	Used to make curry and chili powders. Cooking of India, Middle East, North Africa, and Mexico; sausage, Muenster cheese, sauerkraut
Curry powder (spice blend)	Ground blend	A blend of up to 20 spices. Often includes black pepper, cloves, coriander, cumin, ginger, mace, and turmeric. Depending on brand, flavor and hotness can vary tremendously.	Indian cooking, eggs, beans, soups, rice
Dill weed (herb)	Fresh, dried; crushed	Delicate, green leaves of dill plant. Dill pickle flavor.	Salads, dressings, sauces, vegetables, fish, dips
Dill seed (spice)	Whole, ground	Small, brown seed of dill plant. Bitter flavor—much stronger than dill weed	Pickling, sour dishes, sauerkraut, fish
Epazote (herb)	Fresh	Coarse leaves of a wild plant that grows in the Americas. Strong, exotic flavor.	Mexican and Southwestern cooking
Fennel seed (spice)	Whole, ground	Oval, green-brown seeds of a plant in the parsley family. Mild licorice flavor	Fish, pork, Italian sausage, tomato sauce, pickles, pastries
Fenugreek seed (spice)	Whole, ground	Small beige seeds of a plant in the pea family. Bittersweet flavor. Smells like curry.	Indian cooking such as curries and chutneys
Ginger (spice)	Fresh whole, dried whole, dry ground (also candied or crystallized)	Dried tan root of the tropical ginger plant. Hot but sweet flavor.	Asian dishes such as curries, baked goods, fruits, beverages
Juniper berries (spice)	Whole	Purple berries of an evergreen bush. Pinelike flavor.	Venison and other game dishes, pork, lamb, marinades
Lemongrass (herb)	Fresh stalks, dried: chopped	Tropical and subtropical scented grass. White leaf stalks and lower part are used. Bright lemon flavor.	Soups, marinades, stir-fries, curries, salads, Southeast Asian cooking
Mace (spice)	Whole blades or ground	Lacy orange covering on nutmeg. Like nutmeg in flavor but less sweet.	Baked products, fruits, pork, poultry, some vegetables
Marjoram (herb)	Fresh, dried; crushed leaves	Leaves from a plant in the mint family. Mild flavor similar to oregano with a hint of mint.	Lamb, poultry, stuffing, sauces, vegetables, soups, stews

(continued)

Mint (herb)	Fresh, dried: crushed leaves	A family of plants that include many species and flavors, such as spearmint, peppermint, and chocolate. Cool, minty flavor.	Lamb, fruits, some vegetables, tea, and other vegetables
Mustard seed (spice)	Whole, ground (prepared mustard)	Tiny seeds of various mustard plants; seed may be white or yellow, brown, or black. The darker the seed, the sharper and more pungent the flavor.	Meats, sauces, dressings, pickling spices, prepared mustard
Nutmeg (spice)	Whole, ground	Large brown seed of the fruit from the nutmeg tree. Sweet, warm flavor.	Baked products, puddings, drinks, soups, sauces, many vegetables
Oregano (herb)	Fresh leaves, dried: crushed	Dark green leaves of oregano plant. Pungent, pepperlike flavor.	Italian foods such as tomato sauce and pizza sauce, meats, sauces, Mexican cooking
Paprika (spice)	Ground	Fine powder from mild varieties of red peppers. Two varieties: Spanish and Hungarian. Hungarian is darker in color and much stronger in flavor.	Spanish used mostly as garnish; Hungarian used in braised meats, sauces, gravies, some vegetables
Parsley (herb)	Fresh, dried flakes	Green leaves and stalks of several varieties of parsley plant. Mild, sweet flavor.	Bouquet garni, fines herbes, almost any food
Peppercorns, black and white	Whole, crushed, ground	Dried, black or white hard berries from same tropical vine that are picked and handled in different ways. Black: pungent earthy flavor. White: similar but milder than black.	Almost any food
Pink peppercorns (spice)	Whole	Dried or pickled red berries of an evergreen. Not related to black pepper. Bitter flavor, not as spicy as black pepper.	Use in small quantities in meat, poultry, and fish dishes and in whole pepper mixtures
Poppy seed (spice)	Whole	Tiny cream-colored or deep blue seeds from the poppy plant. Nutty flavor.	On breads and rolls, in salads and noodles, ground poppy seed in pastries
Rosemary (herb)	Fresh, dried leaves	Stiff green leaves that look like pine needles of a shrub. Strong flavor like pine.	Roasted and grilled meats such as lamb, sauces such as tomato, soup
Saffron (herb)	Whole (threads), ground	Dried flower stigmas of a member of the crocus family. Used in very small amounts, has a bitter yet sweet taste, and colors foods yellow. Most expensive spice. Mix with hot liquid before using.	Paella, risotto, bouillabaisse, seafood, poultry, baked products.

(continued)

FIGURE 8-4: Herb and Spice Reference Chart (*Continued*)

Product	Market Forms	Description	Uses
Sage (herb)	Whole (fresh or dried), rubbed (chopped), ground	Gray-green leaves and blue flowers of a member of the mint family. Strong, musty flavor.	Pork, sausage, stuffing, salads, beans
Savory (herb)	Crushed leaves	Small, narrow leaves of plant in mint family. Summer savory is preferable to winter savory. Bitter flavor	Meat, poultry, sausage, fish, vegetables, beans
Sesame seeds (spice)	Whole	Creamy oval seeds of tall, tropical sesame plant. Nutty flavor.	On breads and rolls; ground seeds used to make tahini
Star anise (spice)	Whole, ground	Dried, star-shaped fruit of an evergreen native to China. Dark red. Licorice-like flavor.	Chinese cooking
Tarragon (herb)	Fresh, dried: crushed	Small plant with long narrow leaves and gray flowers. Delicate sweet flavor with hint of licorice.	Béarnaise sauce, vinegars, dressings, poultry, fish, salads
Thyme (herb)	Fresh, dried: crushed or ground	Tiny leaves and purple flowers of a short, bushy plant. Spicy, slightly pungent flavor.	Bouquet garni, meat poultry, fish, soups, sauces, tomatoes
Turmeric (spice)	Ground	Orange-yellow root of member of ginger family. Musky, peppery flavor. Colors foods yellow.	Curry powder, curries, chutney

balsamic vinegar, a dark brown vinegar with a rich sweet-sour flavor, is made from the juice of a very sweet white grape and is aged in wooden barrels. Vinegars can also be infused, or flavored, with all sorts of ingredients, such as chili peppers, roasted garlic, herbs, vegetables, and fruits. For example, lemon tarragon vinegar works well in salad dressings and cold sauces.

Like vinegar, oils can be infused with ingredients such as ground spices, fresh herbs, juices, and fresh roots. Small amounts of flavored oils can add much flavor to finish sauces, dressings, marinades, relishes, and salads, or they can be used alone to drizzle over foods that are ready to be served. Use a neutral oil such as canola, safflower, or grapeseed oil when you are making your own flavored oil.

To make a ground-spice oil, mix the spice first with water (3 tablespoons of spice to 1 tablespoon of water) to "wake" up the flavor. Then mix in with about 2 cups of oil. Let sit for about four to six days at room temperature, shaking several times a day to fuse the flavors. Filter the oil and reserve for use.

To make a tender-herb oil, quickly blanch the herbs and then shock in ice water. Drain and dry the herbs, then puree them with oil. Let sit for several hours at room temperature, then strain through a fine filter. Keep in the refrigerator for several days, or freeze for a longer shelf life. Herbs with a harder texture (such as rosemary) can be chopped and mixed in a food processor with oil. Let sit several days and then strain. Keep in the refrigerator for several weeks.

To make oil with vegetable and fruit juices, reduce the juice to a syrup, then blend in a food processor with a little oil, Dijon mustard, a touch of honey, and some fresh thyme. The "oil" or syrup is ready to use.

STOCK

Stock, a flavored liquid used in making soups, sauces, stews, sautés, and braised foods, functions as the body of many foods as well as a flavor builder. Body refers to the amount of flavor—its strength or richness. The French call stock fond de cuisine, the base of cooking, which describes its role exactly.

Stocks made with animal products are made by simmering water, bones, regular mirepoix (onion, carrot, celery, and sometimes tomatoes or leeks) or white mirepoix (onion, celery, leek, fennel), herbs, and spices. As the ingredients simmer, the flavor-producing substances are extracted from the bones and flavor builders that dissolve in the water. Gelatin is also drawn from the bones, providing a major source of body. Though it may be imperceptible in a hot stock, gelatin causes the stock to thicken or jell when chilled.

Chicken stock is made using chicken bones. If the mirepoix and bones are caramelized (browned) before being added to the stock, the result will be a brown chicken stock. Without this browning process, the result is a white chicken stock.

Fish stock is made like a white stock, but using cleaned fish bones. Fish stock cooks very quickly, in less than an hour at a rolling simmer. Depending on the types of bones and/or shells used, you can make clam stock, lobster stock, or other types. Crustacean stock is made usually by caramelizing the shells and mirepoix with tomato and paprika or a tomato product for a rich brown-red color. Fish or chicken stock is added. A blend of both stocks creates the most favorable results.

Brown stock is made like white stocks, except that the bones (beef, veal, lamb, or pork) and mirepoix are browned first. Another method is to simmer the bones with onion brulée, bouquet garni, tomato paste, whole carrots, and red wine.

Vegetable stock is made without any animal products. The vegetables are sweated in a touch of oil or stock, and plenty of herbs and spices are added. The vegetables are covered with water and simmered for 1½ hours. Wine may be added. Vegetable stock is normally considered a white stock unless tomatoes and mushrooms are added.

In building a good stock, you need to know the nature of the product you are aiming for. Here are the principal measures of stock quality.

- A good stock is fat-free.
- A good stock is clear—translucent and free of solid matter.
- A good stock is pleasant to the senses of smell and taste.
- A good stock is flavorful, but the flavor is neutral. The flavor of the main ingredient, though predominant, is not overpowering. No single flavor builder is identifiable over the flavor of the main ingredient.

Here are some very important guidelines to observe in making stock.

- Use good raw bones—bones that smell pleasant and fresh. Wash chicken and fish bones.
- Remove excess fat from the bones. Fat will produce grease in the stock, spoiling its flavor and appearance.
- Start with cold liquid. Wash chicken and fish bones. They are naturally filled with blood and other impurities. These impurities will dissolve in cold water, and then as the stock is heated, they will become solid and rise to the surface, where you can skim them off. This is especially true with beef and veal stocks. Therefore, a cold-water start will produce a clear stock, whereas starting with hot water will produce a cloudy one.
- Use a tall, narrow pot to minimize evaporation. A certain amount of flavor is lost in evaporation, and the rate of evaporation depends on the surface area of the liquid.

- After bringing the cold water to a boil, reduce the heat to a simmer (about 185°F, or 85°C). Keep the cooking temperature below the boil. It takes long, slow simmering to extract the flavors you want from the bones and flavor builders. Too high a temperature will increase evaporation and cause the loss of desirable flavors. It will also break down vegetable textures, producing undesirable flavors and a cloudy stock.
- Skim occasionally; that is, remove the impurities that rise to the surface, using a skimmer or ladle.
- Do not salt the stock because it will usually be further cooked—in a sauce or soup, for example. As a stock is cooked further, it reduces, meaning its volume decreases due to evaporation. A stock that tastes lightly salted when prepared will taste much saltier as it is cooked further and reduces in volume.
- If you add kitchen scraps to stock, make sure they are clean and wholesome.
- Degrease the finished stock; that is, remove the fat from the surface. The most effective method is to chill the stock and remove the layer of fat that congeals on top. If you must use the stock immediately, you can skim the hot fat off the top with a ladle.
- Cool the stock quickly in an ice bath and store properly in the refrigerator.
- A stock's shelf life is no more than three to five days in the cooler. Stocks can be frozen without a loss of quality.

To reduce the amount of fat in stocks, use only a small amount of oil to sauté the mirepoix or sweat it in stock, wine, or another liquid.

Stocks are a low-kcalorie way to support flavor in the recipes in which they are used. One cup of stock is about 40 kcalories at most, or only 5 kcalories per fluid ounce. Through the use of herbs, spices, and aromatic vegetables, as well as the flavor from caramelized bones, stocks can be made quite flavorful without adding any kcalories. Also, stocks can be thickened without high-fat roux. Arrowroot or cornstarch does an admirable job of thickening without fat. Pureed vegetables or potatoes can be used to give body, or you can simply reduce the stock to a glaze.

Glazes are basic preparations in classical cookery. They are simply stocks reduced to a thick, gelatinous consistency with flavoring and seasonings. Meat glaze, or glace de viande, is made from brown stock. Glace de volaille is made from chicken stock, and glace de poisson is made from fish stock. To prepare a glaze, you reduce stock over moderate heat, frequently skimming off the scum and impurities that rise to the top. When the stock has reduced by about half, strain it through a fine mesh strainer into a smaller heavy pan. Place it over low heat and continue to reduce it until the glaze forms an even coating on a spoon. Cool, cover, and refrigerate or freeze.

Small amounts of glazes (remember, they are very concentrated) are used to enhance sauces and other items. They make sauces cleaner and fresher-tasting than sauces made with heavy thickeners. They can also be added to soups to improve and intensify flavor. However, they cannot be used to re-create the stock from which they were made; the flavor is not the same after the prolonged cooking at higher temperatures.

Concentrated convenience bases are widely used. The results vary, partly because of the bewildering variety of products on the market and partly because they are often misused. Few of them can function as instant stocks. Compare the taste of a convenience base prepared according to the instructions with the taste of a stock made from scratch and you will see why. Many convenience bases have a high salt content and contain other seasonings and preservatives. This gives them strong and definite tastes that are difficult to work with in building subtle flavors for soups and sauces.

Among the many bases available, take care to choose the highest-quality products, though they are expensive. Look for those that list beef, chicken, or fish extract as the first ingredient (not salt). Avoid those with lots of chemical additives. If you must use a base as a stock, fortify it by following these steps.

1. Dry-sauté a white or regular mirepoix (small dice) until translucent for white stock or carmelized for brown stock.
2. Add base, water, and sachet with fresh herbs, garlic, shallots, bay leaf, and peppercorns.
3. Simmer for 30 minutes.
4. Strain and reserve for use.

Cooking without fresh stock means using your head. If a recipe calls for stock, you will have to analyze the role the stock is to play in that recipe and choose the type of convenience item accordingly. Use with caution, and taste the product as you go. Remember that salt is the major ingredient in nearly every base and adjust the amount of salt in your recipe.

RUBS AND MARINADES

Rubs combine dry ground spices such as coriander, paprika, and chili powder and finely cut herbs such as thyme, cilantro, and rosemary. Rubs may be dry or wet. Wet rubs, also called pastes, use liquid ingredients such as mustard and vinegar. Pastes produce a crust on the food. Wet or dry seasoning rubs work particularly well with beef, chicken, and pork and can range from a smoked paprika barbecue seasoning rub to a Jamaican jerk rub. To make a rub, mix various seasonings together and spread or pat evenly on the meat, poultry, or fish—just before cooking for delicate items and at least 24 hours before for large cuts of meat. The larger the piece of meat or poultry is, the longer the rub can stay on. The rub flavors the exterior of the meat as it cooks.

Marinades, seasoned liquids in which foods are soaked before cooking, are useful for adding flavor as well as for tenderizing meat and poultry. Marinades bring out the biggest flavors naturally so that you don't need to drown the food in fat, cream, or sauces. Marinades allow a food to stand on its own with a light dressing, chutney, sauce, or relish. Fish can also be marinated. Although fish is already tender, a short marinating time (about 30 minutes) can develop a unique flavor. A marinade usually contains an acidic ingredient such as wine, beer, vinegar, citrus juice, or plain yogurt to break down the tough meat or poultry. The other ingredients add flavor. Without the acidic ingredient, you can marinate fish for a few hours to instill flavor. Oil is often used in marinades to carry flavor, but it isn't essential. There are many products available that have balanced ingredients and serve as a good base for marinating. A simple fish marinade consists of fish stock, lemon rind (optional), white wine, tarragon, thyme, dill, black pepper, shallots, Dijon mustard, and a few drops of oil.

To give marinated foods flavor, try citrus zest, diced vegetables, fresh herbs, shallots, garlic, low-sodium soy sauce, mustard, and toasted spices. For example, citrus and pineapple marinades can be flavored with Asian seasonings such as ginger and lemongrass. Tomato juice with allspice, Worcestershire sauce, garlic, cracked black pepper, mustard, fresh herbs, and coriander is great for flank steak. Adding chopped kiwi or pineapple to a marinade helps soften tough meats because these ingredients contain enzymes that break down tough muscle.

AROMATIC VEGETABLES

Figure 8-5 shows examples of aromatic vegetables. Onions and their cousins garlic, scallions, leeks, shallots, and chives are a special category of flavorings that add strong and distinctive flavors and aromas to both cooked and uncooked foods. These bulbous plants of the lily family

---

RUBS
A dry marinade made of herbs and spices (and other seasonings), sometimes moistened with a little oil, that is rubbed or patted on the surface of meat, poultry, or fish (which is then refrigerated and cooked at a later time).

MARINADES
A seasoned liquid used before cooking to flavor and moisten foods; usually based on an acidic ingredient.

FIGURE 8-5:

Aromatic vegetables. Top row: leek, scallions, chives, onion; bottom row: garlic, shallots. Photo by Frank Pronesti.

arrive in the kitchen whole and fresh rather than dried and powdered, though some are available in dried forms. We use them in greater quantity, except for garlic, than the "pinch" or the "few" that is our limit on most herbs and spices.

The onion is the scaly bulb of an herb used since ancient times and grown the world over, the most common and most versatile flavor builder in the kitchen. Raw onion adds a pungent flavor to salads and cold sauces. Cooked, it has a sweet, mellow flavor that blends with almost anything. Cooked onions are also served as a vegetable.

Garlic, the bulb of a plant of the same family, is available as cloves (bulblets), chopped, powdered, granulated, or in juice form, with the fresh clove having by far the best flavor. Garlic is used as a flavor builder in many preparations, including stocks, stews, sauces, salads, and salad dressings. Roasted garlic has a strong flavor that can replace salt in some recipes. When pureed, it gives a binding, creamy texture to sauces, dressings, beans, grains, and soups, as well as a pleasant flavor.

Chives are another bulbous herb of the onion family, the only one whose leaves rather than bulbs are eaten. Chives are usually used raw, since most of their flavor and color is lost if they are cooked. They are clipped from the plant and added, usually finely sliced, to many foods.

The leek, a mild-flavored relative of the onion, has a cylindrical bulb. It is the partner of the onion in the mirepoix, using the white and light green parts. For preparations, the dark green part is cut off because of its bitter flavor. Braised leeks are very popular as a vegetable in France, where they are known as the poor man's asparagus. The leek's culinary triumph is the cold soup known as vichyssoise.

The scallion is a young onion, also known as a green onion or spring onion. It has a mild flavor as onions go. Minced or sliced, it is added to salads, marinades, dressings, salsas, and relishes with the entire white bulb and some of the green top. It can pinch-hit in cooking for the full-grown onion, and its green top, when minced, can substitute for chives in an emergency.

The shallot is a cluster of brown-skinned bulbets similar to garlic. It is somewhere between garlic and onion in both size and flavor but is milder and more delicate than either. One thinks of shallots with wine cookery and with mushrooms in a marvelous stuffing called duxelles, but they are useful to build flavor in a wide variety of dishes.

SAUCE ALTERNATIVES: VEGETABLE PUREES, COULIS, SALSAS, RELISHES, CHUTNEYS, COMPOTES, AND MOJOS

Alternatives to classic sauces, many of which are high in fat, take many forms. Pureeing of vegetables or starchy foods is commonly done as a means to thicken soups, stews, sauces, and other foods without any fat or to create a simple sauce. Sauces made with vegetables are light and low in fat and kcalories. With the addition of herbs and spices, they make a colorful, tasty complement to entrées and side dishes. For example, you can puree roasted yellow peppers with ginger to make a sauce for grilled swordfish or tuna.

Before pureeing, vegetables should be cooked just until tender, using methods such as steaming, roasting, grilling, and sautéing. Roasting caramelizes the vegetables' natural sugars, making them taste rich and sweet. Vegetables that roast well include peppers, eggplant, zucchini, cauliflower, and root vegetables. More delicate vegetables, such as mushrooms, can be sautéed. Other vegetables that make excellent purees are butternut squash, artichokes, parsnips, peas, and carrots.

Coulis, a French term, refers to a sauce made of a puree of vegetables or fruits. A vegetable coulis, such as bell pepper coulis, can be served hot or cold to accompany entrées and side dishes. Fruit coulis is usually served cold as a dessert sauce.

COULIS

A sauce made of a puree of vegetables or fruits.

A vegetable coulis is often made by cooking the main ingredient, such as tomatoes, with typical flavoring ingredients such as onions and herbs in a liquid such as stock. The vegetable and flavorings are then pureed, and the consistency and flavoring of the product are adjusted. A vegetable coulis can also be made by cooking the vegetable with potato or rice (2 ounces per gallon) to give the sauce a silky texture and a smooth consistency when it is pureed and strained.

The texture of a vegetable or fruit coulis is quite variable, depending on the ingredients and how it is to be used. A typical coulis is about the consistency and texture of a thin tomato sauce. The color and flavor of the main ingredient should stand out.

Salsas and relishes are versatile, colorful low-fat sauces. Salsas and relishes are chunky mixtures of vegetables and/or fruits and flavor ingredients. Salsa is a Spanish word, and traditional salsa is made with tomatoes. Most raw salsas consist of a chopped or almost pureed vegetable (most often tomatoes), fruit, or herb to which a strong flavor is added, such as red onion, garlic, and lime juice. Many include several herbs, spices, and chilies. Chilies are frequently cooked to tame their flavor.

Relishes are often spicy and are made by pickling foods. Served cold, they are excellent as sauces for meat, poultry, and seafood. Since salsas and relishes contain little or no fat, they rely on ingredients with an intense flavor, such as cilantro, jalapeño peppers, lime juice, lemon juice, garlic, dill, pickling spice, coriander, and mustard.

Chutney, such as tomato-papaya chutney, is made from fruits, vegetables, and herbs and comes originally from India. Recipe 9-13 explains how to make Papaya and White Raisin Chutney. A *compote* is a dish of fruit, fresh or dried, cooked in syrup flavored with spices or liqueur. It is often served as dessert or as an accompaniment.

A *mojo* is a spicy sauce from the Caribbean and South America. It is traditionally a mixture of sour oranges and their juice, garlic, oil, and fresh herbs.

With their many colors, flavors, and textures, each of these items is a thoroughly contemporary addition to entrées.

WINE AND SPIRITS

Wines, liqueurs, brandy, cognac, and other spirits are often added as flavorings at the end of cooking. Sherry is a popular American flavoring for sauces. Brandy is often poured over a dish and flamed (set afire) at the time of service. This adds some flavor but is done more for show.

In dishes such as sauces, wines and spirits may be added during cooking to become part of the total flavor. They are then flavor builders rather than flavorings. The same product can play one role in one dish and a different role in another dish.

EXTRACTS AND OILS

Extracts and oils from aromatic plants are used in small quantities, primarily in the bakeshop extracts of vanilla, lemon, and almond and in oils such as peppermint and wintergreen.

PUTTING IT ALL TOGETHER: FLAVOR PROFILES

Cooking is cooking is cooking! You must master cooking fundamentals before you can produce consistent flavorful and well-executed products. The many choices for flavor building for balanced dishes may seem extremely overwhelming. These ingredients need to be incorporated

---

SALSAS
Chunky mixtures of vegetables and/or fruits and flavor ingredients.

CHUTNEY
A sauce from India that is made with fruits, vegetables, and herbs.

COMPOTE
A dish of fruit, fresh or dried, cooked in syrup flavored with spices or liqueur; it is often served as an accompaniment or dessert.

MOJO
A spicy Caribbean and South American sauce; it is a mixture of garlic, citrus juice, oil, and fresh herbs.

into your daily style of cooking. You don't need to separate your cooking methods from healthy to fattening. You should basically be using the same methods of preparation as you do for any style of cooking. The exceptions are deep-frying, pan-frying, and sautéing with oil. When you are preparing dishes that limit the amounts of fat, it is smarter to use those fats at the end of your preparation rather then during the cooking process. The best way to go about the dish you are preparing is to start with a basic series of steps. As your skills develop, these steps will become more natural and automatic, just like the basic cooking fundamentals you learn in the beginning of your career.

As you plan your menu item, identify what direction of flavor combinations you want to achieve, such as

- Hints of ginger with soy
- Cilantro with lime and cumin
- Garlic with lemon and basil
- Other combinations that blend well

Now it's time to plan your dish strategically and identify how you are going to achieve the maximum flavor from the list of ingredients you have chosen. In the previous information this was referred to as flavor building. When a dish is created with limited amounts of butter, fat, salt, and sugar, other ingredients are needed to excite the palate, accomplishing an experience of taste and satisfaction. Use the tools provided in this chapter and experiment to get your own style and result, keeping in mind the guidelines of sound cooking and proper techniques. Some key points in this area are toasting spices, reducing liquids to maximize flavor, marinating before cooking, and using fresh herbs and stocks.

Picking the best cooking method is another key way to maximize the potential of each dish. Marinating your meats before cooking them is critical to impart flavor to the meat before being cooked. Caramelizing with a smear of fat in a pan helps to seal in the flavor and add another character to your dish. The addition of a well-prepared stock to braise the item, if that is the cut you have chosen, is perfectly acceptable so long as the meat is well-trimmed and the stock has been defatted. In this case, the addition of a caramelized mirepoix and tomato product with thyme, rosemary, and bay leaf will add an additional level of flavor to your dish. When caramelizing your mirepoix, add spices to give your dish the character you are trying to accomplish. For spices, think of using curry, cumin, chili powder, coriander, garlic, soy, and lemon juice, which are a great combination when braising meats or preparing a sauce.

Your creativity is up to you when you are designing a dish. The same holds true if you are following a classic style recipe. You need to limit the amount of oil, butter, and salt, and add additional spices, herbs, acid, and vegetables to accomplish the same flavor profile. The addition of fat is better at the end before serving so you taste these flavors first. A simple example would be cooking carrots. If you poach them in a flavorful vegetable or chicken stock, they will absorb the flavor. When you are ready to serve, heat them in that same stock, then sauté in reduced stock with the addition of butter to finish, folding in fresh herbs like dill or chives and few twists of fresh black pepper.

When cooking beans, legumes, and grains, use either chicken stock or vegetable stock to impart flavor, as well as shallots, fresh herbs, bay leaf, or thyme. This will help to build flavor in the starches we are preparing as well as help to demonstrate which methods and flavors are needed to create quality dishes.

Your preparation will dictate how successfully your dish comes together. Planning is key to this success. You shouldn't start a project without a detailed plan of execution. This together with your "mise en place" will allow you to create a dish or menu with the finesse of experienced culinarians. You can never be overprepared. You want to have your ingredients at your fingertips, your products prepped and laid out so you can use them in their proper order. If you are preparing something that requires reduced orange juice, make sure that you have already gone through that process and have that set and ready to go. If you need to toast spices, chop herbs, or cut vegetables, all of this should be prepped just as you would for any other dish. Flavor building needs practice, as any cooking method, so read through this chapter time and time again to feel more comfortable with the many methods, styles, and ingredients you can use as fat, salt, and sugar substitutes. The creations you discover can very easily become future menu items. You should include balanced choices on your menus as well as those that naturally contain higher fat, salt, or sugar. Good nutrition is all about balance, portion control, and quality ingredients. Preparing foods that reduce fat, salt, and sugar doesn't mean they are light in flavor. This is the reputation that you as culinary professionals need to eliminate among our clients. We need to be educated and prepare our foods to accommodate special needs, diets, and allergies without creating chaos in our operations. The tools provided in this chapter clearly describe how you can accomplish this mission in your kitchen. Enjoy, and cook with your heart.

MINI-SUMMARY

1. Figure 8-6 lists the powerhouses of flavor described in this section, as well as other possibilities.
2. Flavoring adds a complementary flavor to a dish at the end of its preparation. It creates a blend in which both the original flavor and the added flavor are identifiable, as in the addition of black pepper to a green salad. Most flavorings are products with distinctive tastes that can hold their own in a dish.

FIGURE 8-6: Powerhouses of Flavor

- Fresh herbs
- Toasted spices
- Herbs and spice blends
- Freshly ground pepper
- Citrus juices, citrus juice reductions
- Strong-flavored vinegars and vinaigrettes
- Wines
- Strong-flavored oils such as walnut oil and extra-virgin olive oil
- Infused vinegars and oils
- Reduced stock (glazes)
- Rubs and marinades
- Raw, roasted, or sautéed garlic
- Caramelized onions
- Roasted bell peppers
- Chili peppers
- Grilled or oven-roasted vegetables
- Coulis, salsas, relishes, chutneys, mojos
- Dried foods: tomatoes, cherries, cranberries, raisins
- Fruit and vegetable purees
- Condiments such as sambal, Worcestershire sauce, hot chili sauce, horseradish and Dijon mustard
- Extracts

COOKING METHODS AND TECHNIQUES FOR A HEALTHY EATING STYLE

Healthy cooking methods and techniques can use moderate amounts of primarily monounsaturated or polyunsaturated fats. Before looking at cooking methods, let's review some techniques that are used often to develop flavor in nourishing recipes.

- *Reduction* means boiling or simmering a liquid down to a smaller volume. In reducing, the simmering or boiling action causes some of the liquid to evaporate. The purpose may be to thicken the product, to concentrate the flavor, or both. A soup or sauce is often simmered for one or both reasons. The use of reduction eliminates thickeners and intensifies and increases the flavor so that you can use a smaller portion.
- *Searing* means exposing the surfaces of a piece of meat to high heat before cooking at a lower temperature. Searing, sometimes called browning, can be done in a hot pan in a little oil or in a hot oven. Searing is done to give color and to produce a distinctive flavor. Dry searing can be done over high heat in a nonstick pan, using vegetable oil spray, or brushing with olive or canola oil.
- *Deglazing* means adding liquid to the hot pan used in making sauces and meat dishes. Any browned bits of food sticking to the pan are scraped up, adding flavor and color.
- *Sweating* means cooking slowly in a small amount of fat over low or moderate heat without browning. You can sweat vegetables and other foods without fat. Instead, sweat in stock or wine.
- *Pureeing* of vegetables or starchy foods is commonly done as a way to thicken soups, stews, sauces, and other foods without using any fat. While the food processor is useful for this process because you can slowly pulse the mixture, emulsion blenders and high-speed table blenders can make a very smooth puree with a silky texture and smoothness.

DRY-HEAT COOKING METHODS

Dry-heat cooking methods are acceptable for balanced cooking when heat is transferred with little or no fat and excess fat is allowed to drip away from the food being cooked. Both pan frying and deep frying add varying amounts of fat, kcalories, and perhaps cholesterol, depending on the source of the fat; frying is therefore not an acceptable cooking method. Sautéing can be made acceptable by using nonstick pans and little or no oil.

Roasting

Roasting, cooking with heated dry air, is an excellent method for cooking larger tender cuts of meat, poultry, and fish that will provide multiple servings. When roasting, always place meats and poultry on a rack so that the drippings fall to the bottom of the pan and the meat therefore doesn't cook in its own juices. Also, cooking on a rack allows for air circulation and more even cooking. In addition to meat, poultry, and seafood, vegetables can be oven-roasted to bring out their flavor. The browning that occurs during roasting adds rich flavors to meats, poultry, seafood, and vegetables. For example, potato wedges or slices can be seasoned and roasted. Vegetables and potatoes don't need to be roasted on a rack. Season with herbs, spices, pepper, stock, garlic, onion, and a few sprays of a good flavorful oil.

For an accompaniment to meat or poultry, you may want to simply use the jus. To give the jus additional flavor, add a mirepoix to the roasting pan during the last 30 to 40 minutes. To remove most of the fat from the jus, you can use a fat-separator pitcher or skim off the fat with a ladle. If time permits, you can refrigerate the jus and the fat will congeal at the top.

If you prefer to thicken the natural jus to make jus lié, first remove the fat from the jus. Because you will be using the roasting pan to make this product, also pat out with paper towels the fat from the bottom of the pan. Add some of the jus and a little wine to deglaze the pan. Then add some vegetables and cook at a moderately high heat so that they brown or caramelize. At this point, add more jus so that the vegetables don't burn. Stir the ingredients to release the food from the pan and get its flavor. Continue to add jus, deglaze the pan, and reduce the jus until the color is appropriate. If there is not enough time to reduce the jus to the proper consistency, you can thicken it with a cornstarch slurry (starch and cold water mixed to a syrupy consistency).

Another way to thicken the natural jus is to add starchy vegetables such as carrots, parsnips, or potatoes to the jus during the last 30 to 45 minutes of cooking. These vegetables can then be pureed to naturally bind the jus.

To develop flavor, rubs and marinades for meat, poultry, and fish are two possibilities that already have been discussed. Smoking can also be used to complement the taste of meat, poultry, or fish. Hardwoods or fruitwoods such as the following are best for producing quality results:

- Fruit (apple, cherry, peach) woods work well with light entrées such as poultry and fish.
- Hickory and sugar maple are more flavorful and work better with beef, pork, sausage, and salmon.
- Mesquite produces an aromatic smoke that works well with beef and pork.

To use hardwoods, you need to soak them in water for 30 to 40 minutes and then drain. This way the wood does not burn but smolders.

Smoke-roasting, also called pan-smoking, is done not in the oven but on top of the stove. It works best with smaller tender items such as chicken breast and fish fillet, but it also can be used to add flavor to larger pieces. Place about a half inch of soaked wood chips in the bottom of a roasting pan or hotel pan lined with aluminum foil. Next, place the seasoned food on the rack and reserve. Heat the pan over moderate-high heat until the wood starts to smoke, then lower the heat. Place the rack over the wood and cover the pan. Smoke until the food has the desired smoke flavor, then complete the cooking in the oven if the food is not yet done. Smoking for too long can cause undesirable flavors. Also, be very careful when opening the lid, due to the heat and smoke.

CHEF'S TIPS FOR ROASTING

- Trim excess fat before cooking.
- Roast on a rack and uncovered (otherwise you are steaming the food). Cook at an appropriate temperature to limit drying out. Basting during cooking will also reduce drying. Finally, don't slice the meat until it has rested sufficiently so that you don't lose valuable juices, and be sure to slice across the grain to maximize tenderness.

- To develop flavor in meats, poultry, and fish, use rubs and marinades. You can also stuff the food (with vegetables and grains, for example), sear and/or season the food before cooking, or smoke the food over hardwood chips.
- When smoking, add dried pineapple skins, dried grapevines, or dried rosemary sticks to the wood chips for added flavor.
- Vegetables can be roasted in the oven, which results in wonderful texture and flavor due largely to caramelization of the natural sugars. Root vegetables, such as celery root and sweet potatoes, are hardier and take longer to oven-roast than do vegetables such as peppers, radishes, and pattypan squash. A variety of fresh herbs and spices such as shallots, thyme, rosemary, oregano, cardamom, coriander, fennel seed, and pepper add flavor and uniqueness. Recipe 9-17 features roasted summer vegetables.
- Other accompaniments that add flavor with moderate or no fat are vegetable coulis, chutneys, vinaigrettes, salsas, compotes, and mojos.

Broiling and grilling

Broiling, cooking with radiant heat from above, is wonderful for single servings of steak, chicken breast, and fish with a little more fat such as salmon, tuna, and swordfish that can be served immediately. The more well done you want the product or the thicker it is, the longer the cooking time and the farther from the heat source it should be. Otherwise, the outside of the food will be cooked but the inside will not be done.

Grilling, cooking with radiant heat from below, is also an excellent method for cooking meat, poultry, seafood, and vegetables (Figure 8-7). Like broiling, grilling browns foods, and the resulting caramelization adds flavor.

Once considered a tasty way to prepare hamburgers, steaks, and fish, grilling is now used to prepare a wide variety of dishes from around the world. For example, chicken is grilled to make fajitas (Tex-Mex style of cooking) or to make jerk barbecue (Jamaican style of cooking). Grilling is also an excellent method to bring out the flavors of many vegetables. Grill them with a little oil, vinegar, or lemon juice and selected seasonings.

FIGURE 8-7:
Grilling.
Courtesy of PhotoDisc, Inc.

Grilling foods properly requires much cooking experience to get the grilling temperature and timing just right. Here are some general rules to follow:

1. Cook meat and poultry at higher temperatures than seafood and vegetables.
2. For speedy cooking, use boneless chicken that has been pounded flat and meats that are no more than a half-inch thick. Fat should be trimmed off meats.
3. Don't try to grill thin fish fillets, such as striped bass, because they will fall apart. Firm-fleshed thicker pieces of fish, such as swordfish, salmon, and tuna, do much better on the grill. Don't turn foods too quickly or they'll stick and tear. Mark foods by turning the food around 90 degrees without turning it over.

To flavor foods that will be broiled or grilled, consider marinades, rubs, herbs, and spices. Lean fish can be marinated, sprinkled with Japanese bread crumbs, and glazed in a broiler. If grilling, consider placing soaked hardwood directly on the coals to smoke the food.

During the grilling of beef, poultry, and fish at high temperatures, substances can form on the surface of the food that cause cancerous tumors in animals. The substances, which are called heterocyclic amines (HCAs), are made from the amino acids in protein when the foods are cooked at high temperatures. Also, when fat from the grilled food falls on the hot coals or lava/ceramic bricks, another cancer-causing substance is produced (polycyclic aromatic hydrocarbon, or PAH), and the smoke carries it to the food. Neither HCAs or PAHs have been shown to produce cancer in people. Tips to reduce problems with these substances are as follows.

1. The best way to minimize HCA is to cook at 350°F to 400°F and keep flames from contacting the food.
2. Do not serve blackened or charred foods.
3. Marinate meats, poultry, and fish before grilling. For some reason, marinades reduce the amount of HCAs formed during cooking. Do not take the food out of the marinade over the hot coals. Take it out and let it drain for a minute.
4. Another way to reduce HCA formation is to turn foods often.
5. To reduce PAHs, trim the fat from meat and poultry so that it doesn't drip onto the coals. Also, use foil or a pan to catch the drippings and help eliminate the smoke during grilling.

CHEF'S TIPS FOR BROILING AND GRILLING

- Keep the grill clean and properly seasoned to prevent sticking. The broiler must be kept clean and free of fat buildup; otherwise it will smoke.
- Marinating tender or delicately textured foods for broiling or grilling will firm up their texture so that they are less likely to fall apart during the cooking process.
- Use marinades, rubs, herbs, spices, crumbs, and smoking to add flavor. When grilling, consider using hardwood chips.
- During broiling, butter has traditionally been used to prevent food from drying out from the intense direct heat. In its place, you may spray the food with olive oil or baste it with marinade, stock, wine, or reduced-fat vinaigrette during cooking. To finish, sprinkle a thin layer of Japanese bread crumbs and glaze under the broiler.
- To retain flavor during cooking, turn foods on the broiler or grill with tongs—not a fork, which causes loss of juices.
- If you are making kabobs, soak the wooden skewers in water ahead of time so that they can endure the cooking heat without excessive drying or burning.
- Prepare these foods to order and serve immediately.
- Serve with flavorful sauces such as chutneys, relishes, or salsas.

Sauté and dry sauté

Sautéing, cooking food quickly in a small amount of fat over high heat, can be used to cook tender foods that are in either single portions or small pieces. Sautéing can also be used as a step in a recipe to add flavor to foods such as vegetables by either cooking or reheating. Sautéing adds flavor in large part from the caramelization (browning) that occurs during cooking at relatively high temperatures.

When sautéing, use a shallow pan to let moisture escape and allow space between the food items in the pan. Use a well-seasoned or nonstick pan and add about half a teaspoon or two sprays of oil per serving after preheating the pan. Instead of olive oil, you can use vegetable-oil cooking sprays, which come in a variety of flavors, such as butter, olive, Asian, Italian, and mesquite. A quick two-second spray adds about 1 gram of fat (9 kcalories) to the product. To use these sprays, spray the preheated pan away from any open flame (the spray is flammable) and then add the food. New pump spray bottles are available that allow you to fill them with the oils of your choice.

For an even lower-in-fat cooking method, use the dry sauté technique. To use this technique, heat a nonstick pan, spray with vegetable-oil cooking spray, then wipe out the excess with a paper towel. Heat the pan again, then add the food.

If browning is not important, you can simmer the ingredient in a small amount of fat-free liquid such as wine, vermouth, flavored vinegar, or defatted stock to bring out the flavor. Vegetables naturally high in water content, such as tomatoes and mushrooms, can be cooked with little or no added fluid at a very high heat.

When sautéing or dry sautéing, you can deglaze the pan after cooking with stock, wine, or another low-fat liquid. Then add shallots, garlic, or another seasonings to the sauce.

CHEF'S TIPS FOR SAUTÉING AND DRY SAUTÉING

- To add flavor, use marinades, herbs, or spices.
- Use high heat or moderately high heat and a well-seasoned or nonstick pan used only for dry sautéing.
- Pound pieces of meat or poultry flat to increase the surface area for cooking and thereby reduce the cooking time.
- To sauté vegetables, simmer them with a liquid such as stock, juice, or wine in a nonstick pan and add seasonings such as cardamom, coriander, and fennel. Cook over moderately high heat. As the vegetables start to brown and stick to the pan, add more liquid and deglaze the pan. Once the vegetables are ready, add a little bit of butter or flavorful oil, perhaps 1 teaspoon for four servings, to give the product a rich flavor and a shiny texture without much fat.
- Another sauté method is to blanch vegetables in boiling water to the desired doneness and then shock them in ice water or lay out on towels and place immediately in refrigerator. Drain and dry sauté in a hot nonstick pan with stock, wine, fresh herbs, garlic, and shallots (chopped) and finish with fresh black pepper and extra virgin olive oil, butter, or nut oil (1 teaspoon for four servings).
- Once prepared, serve sautéed foods immediately. They do not hold well.

Stir-frying

Stir-frying, cooking small-size foods over high heat in a small amount of oil, preserves the crisp texture and bright color of vegetables and cooks strips of poultry, meat, or fish quickly. Typically, stir-frying is done in a wok, but a nonstick pan can be used. Steam-jacketed kettles and tilt frying pans can also be used to make large-quantity stir-fry menu items. Cut up the ingredients as appropriate into small pieces, thin strips, or diced portions.

- Coat the cooking surface with a thin layer of oil. Peanut oil works well because it has a strong flavor (so that you can use just a little) and a high smoking point. You can also use vegetable cooking spray and wipe away any excess.
- Have all your ingredients ready and next to you because this process is fast.
- Partially blanch the thick vegetables first (such as carrots or broccoli) so that they will cook completely without excessive browning.
- Preheat the equipment to a high temperature.
- Foods that require the longest cooking times—usually meat and poultry—should be the first ingredients you start to cook.
- Stir the food rapidly during cooking and don't overfill the pan.
- Use garlic, scallion, ginger, rice wine vinegar, low-sodium soy sauce, and chicken/vegetable stock for flavor.
- Add a little sesame oil at the end for taste (about 1 teaspoon per four servings).

MOIST-HEAT COOKING METHODS

Moist-heat cooking methods involve water or a water-based liquid as the vehicle of heat transfer and are often used with secondary cuts of meat and fowl or legs and thighs of poultry. In moist-heat cooking of meat or poultry, the danger is that the fat in the meat or poultry will stay in the cooking liquid. This problem can be resolved to a large extent by chilling the cooking liquid so that the fat separates and is removed before the liquid is used. If the liquid needs to be chilled quickly, place in an ice bath for the quickest results.

Compared to most dry-heat cooking methods, these methods do not add the flavor that dry-heat cooked foods get from browning, deglazing, or reduction. For foods that use moist-heat cooking to be successful, you will need:

- Very fresh ingredients
- Seasoned cooking liquids using fresh stock, wine, fresh herbs, spices, aromatic vegetables, and other ingredients
- Strongly flavored sauces or accompaniments to achieve flavor and balance

Steaming has been the traditional method of cooking vegetables in many high-quantity kitchens because it is quick and retains flavor, moisture, and nutrients. It's healthy too, because it requires no fat. The best candidates for steaming include foods with a delicate texture such as fish, shellfish, chicken breasts, vegetables, and fruits. Be sure to use absolutely fresh ingredients to come out with a quality product.

Fish is great when steamed en papillote (in parchment) or in grape, spinach, or cabbage leaves. The covering also helps retain moisture, flavor, and nutrients. Consider marinating fish beforehand to add moisture and flavor.

It is possible to introduce flavor into steamed foods (as well as poached foods) by adding herbs, spices, citrus juices, and other flavorful ingredients to the water. For example, steam halibut over a broth seasoned with lemon, ginger, and thyme. Steamed foods also continue to cook after they come out of the steamer, so allow for this in your cooking time.

Poaching, cooking a food submerged in liquid at a temperature of 160 to 180°F (71 to 82°C), is used to cook fish, tender pieces of poultry, eggs, and some fruits and vegetables. To add flavor, you can poach the foods in liquids such as chicken stock, fish stock, and wine flavored with fresh herbs, spices, ginger, mirepoix, vegetables such as garlic or shallots, or citrus juices. For sweet poaching liquids, use fruit juices, wine, honey, whole spices, and berries. Serve poached foods with flavorful sauces or accompaniments.

Fish is often poached in a flavored liquid known as court bouillon. The liquid is simmered with vegetables (such as onions, leeks, and celery), seasonings (such as herbs), and an acidic product (such as wine, lemon juice, or vinegar). It may be used to cook fish to be served either hot or cold, but it is not generally used in making sauce. Cook the court bouillon for about 30 minutes, strain out the vegetables and herbs, and then add the fish to infuse these flavors. Place the fish to be poached on a rack or wrap it up in cheesecloth.

Braising, or stewing (stewing usually refers to smaller pieces of meat or poultry), involves two steps: searing or browning the food (usually meat) in a small amount of oil or its own fat and then adding liquid and simmering until done. Foods to be braised are also often marinated before searing to develop flavor and tenderize the meat. When browning meat for braising, sear it in as little fat as possible without scorching and then place it in a covered braising pan to simmer in a small amount of liquid. To add flavor, place roasted vegetables, herbs, spices, and other flavoring ingredients in the bottom of the braising pan before adding the liquid.

Following are the steps for braising meat or poultry.

1. Trim the fat. Season the meat or poultry.
2. In a small amount of oil in a brazier, sear the meat or poultry to brown it and develop flavor.
3. Remove the meat and any excess oil from the pan.
4. Put the pan back on the heat. Add vegetables and caramelize (brown) them.
5. Deglaze the pan with wine and stock, being sure to scrape the fond (the flavorful drippings at the bottom of the pan).
6. Add tomato paste, wine, stock, and aromatic vegetables. Reduce.
7. Return the meat to the brazier, cover, and put into the oven, covered, to simmer. Make sure the meat is three-quarters covered with liquid. This allows the meat to cook more evenly and prevents the bottom from getting scorched.
8. When the meat is done (fork tender), strain the juices. Skim off fat and reduce juices. Puree the vegetables from the juices and use them (or you can use cornstarch) to thicken the jus, making a light gravy.

Vegetables, beans, and fish can also be braised.

Microwaving is another wonderful method for cooking vegetables, because no fat is necessary and the vegetables' color, flavor, texture, and nutrients are retained. Boiling or simmering vegetables is not nearly as desirable as steaming or microwaving, because nutrients are lost in the cooking water and more time is required.

CHEF'S TIPS FOR MOIST-HEAT COOKING

- Marinate the protein several days in advance to add some flavor. A dry rub is best for this, without any salt.
- This is a great method to use with larger cuts of meat that you cook a day ahead of time. Take the cooked protein out of the liquid and cool. Meanwhile, reduce and skim the liquid to the proper consistency or thicken with vegetables. Cool the liquid and pour over the meat to keep moist.
- This is a great method for cooking halibut, swordfish, sea bass, monkfish, and other meaty fish. Place some onions and mushrooms on the bottom of the pan. Season the fish with the rubs or marinade combinations discussed earlier. Top with tomatoes and a few olives or another combination of vegetables. Then finish with a little olive oil, white wine, citrus, and stock and you have a great buffet or à la carte item.

- Remove skin from poultry before using moist-heat cooking methods. The skin will add unnecessary fat to the cooking liquid.
- Use the stock appropriate for the protein you are cooking. Chicken stock has the most neutral flavor for all applications.
- Save the court bouillon stock for several uses within a week's time; it gets more intense with each use. You can reduce this liquid as well and use it to season other fish preparations.

MINI-SUMMARY

1. Healthy cooking methods include roasting, broiling, grilling, stir-frying, sautéing with little or no oil, steaming, boiling, simmering, poaching, microwaving, and braising.
2. Cooking techniques that are especially useful in preparing flavorful and nourishing items include reduction, searing, deglazing, sweating, and pureeing.

1. Sweating vegetables can be done in stock.

   a. True
   b. False

2. Flavorings are used to bring out flavor that is already in a dish.

   a. True
   b. False

3. Ginger is an example of a fresh herb.

   a. True
   b. False

4. Basil, oregano, and garlic are often used in Indian dishes.

   a. True
   b. False

5. Toasted spices can be ground and used in marinades, salad dressings, rubs, and soups.

   a. True
   b. False

6. Gelatin is what gives stock body.

   a. True
   b. False

7. A good stock is brimming with flavor.

   a. True
   b. False

8. The shallot is a green onion.

   a. True
   b. False

9. Chutney refers to a sauce made of a puree of vegetables or fruits.

   a. True
   b. False

10. To reduce PAHs on grilled foods, trim the fat off meat and poultry so that it doesn't drip onto the coals.

    a. True
    b. False

ACTIVITIES AND APPLICATIONS

1. Flavorings

    Pick out five recipes from a traditional cookbook and five recipes from a healthy cookbook. Determine the sources of flavor for each recipe and record them. Compare the ingredients used to flavor each set of recipes. Does either set of recipes tend to use more fat or more herbs and spices?

2. Make Your Own Salsa

    Make a salsa such as Papaya-Plantain Salsa (Recipe 9-14) and look at a number of salsa recipes. Next, design your own salsa recipe, taking into consideration the major ingredients and the herbs and spices you will use. Check the recipe with your instructor before you make it.

3. Herbs and Spices Around the World

    Pick an ethnic cuisine you would like to know more about. For that cuisine, pick five of its commons herbs and spices. For each herb or spice, find out the following (use www.astaspice.org and click on Spice Library).

    • Market forms (whole, ground, fresh, dried)
    • Description
    • Uses
    • Flavor

    Find a recipe that uses your herbs and spices and prepare it. Make the recipe a second time and substitute for two of the herbs or spices with other common herbs or spices. Compare the flavor of both dishes.

NUTRITION WEB EXPLORER

Cooking Light Magazine   www.cookinglight.com

On the home page of this popular magazine on healthy cooking and living, click on "Cooking 101." Then click on "Techniques," read one of the articles, and write a paragraph about it. Finally, click on "Meet the Chef" and read one of the interviews. Write a summary on the Chef you chose.

The Culinary Institute of America's Professional Chef Site   www.ciaprochef.com

On this home page, click on "World of Flavors," then click on "World Culinary Art Series"
    Watch the video online for one of the following cuisines: Indian, Spanish, Mexican, Thai, Singapore, Istanbul, or Southern Spain. What type of flavor profile does the cuisine have?
    Also, click on "Strategies for Chefs" and read one of the articles, such as "Design Menus Seasonally."

Flavor-Online.com   www.flavor-online.com

Enter "Flavor Pyramid" in the "Search Articles" box. Read "Building on the Flavor Pyramid." What are the components of the author's flavor pyramid? Describe each one.

Eating Well Magazine   www.eatingwell.com

Click on "Health" and then "Healthy Cooking." Read the article "Tools for a Healthy Kitchen" and list five tools mentioned.

GigaChef.com   www.GigaChef.com

Register at this website (it's free), then find 5 GigaChef recipes that use a variety of flavor builders discussed in this chapter.

Healthy Menus and Recipes

As the population continues to age, healthy food choices become more important. Although Americans have been accused of asking for healthy foods and then returning to their less healthy alternatives, the tide seems to be turning as consumers embrace healthier restaurant choices. More restaurants are using organic and locally grown produce as well as free range meats to satisfy consumer demand. They're also cooking with healthier oils and eliminating trans fats from their kitchens. In some places, such as New York City, restaurateurs are banned from using trans fats and shortenings, oils, and margarines that contain them.

As a foodservice professional, you have a responsibility to your clients to have an understanding of contemporary eating styles that are balanced, limited in rich ingredients, and nutritious. I prefer to call this a way of life rather then a diet, because a diet indicates a short-term, very bland, deprived way of eating. Your clients are eating out more today than ever. They don't typically cook and eat at home as much as they did years ago, and neither do their children. Very hectic lifestyles force us to eat out at least several times a week. The typical American purchases a meal or snack from a restaurant about five times a week. You have a captive audience of people who need chefs to cook nutritious food for them with the limits and balance they require to maintain their current lifestyles. This menu framework will help you to do the following:

- Provide your clients with healthy selections in each section of the menu
- List the elements to consider when presenting foods at their best
- Select and prepare appropriate garnishes

 ## INTRODUCTION TO HEALTHY MENUS

Balance in the menu is one of the keys to success in a foodservice operation. The concept of cooking—to apply heat or preparation with care and expertise—has been taught for centuries. These concepts and applications that have been taught and practiced for so long have been given a new twist as technology and refinements have changed with the times. The basic methodology is to understand and respect the fine art of cooking and preparation and then to modify the contents to make balanced preparations by using nutritional guidelines. A healthy selection in your menu, although providing a choice for your clientele, should not clash with your other menu selections. Rather, these healthy selections should flow into the mainstream of the menu. The information in these chapters will help you develop well-balanced menu items that are moderate in salt, sugar, fat, saturated fat, and cholesterol and have the most important ingredient in the forefront: taste.

All food tastes good when prepared, cooked, and seasoned correctly. But we sometimes use simple trick ingredients such as butter, sugar, salt, and cream that make a tasty dish but also affect its nutritional value. Excess salt makes us taste more intensely. Sugar helps satisfy a person's sweet tooth. Fat enhances taste and fills our satisfaction, and quantity gratifies our sense of value.

Our views on what we put in our bodies have changed dramatically over the last decade. Diseases of excess, such as heart disease and obesity, touch the lives of most Americans and generate substantial health-care costs. New dietary recommendations and food guides suggest new eating patterns with more emphasis on whole grains, fruits, and vegetables. As foodservice operators we have an opportunity to be instrumental in these new eating patterns. No longer should there be "bad foods" and "healthy foods" on opposite sides of the menu. Healthy items can intermingle with the regular selections on menus. They are no longer reserved for customers looking to lose weight. They should be choices you make depending on what you ate yesterday and what you ate today so that you balance your kcalorie and nutrient intake for the week. People today eat to fit their routines so that they balance their food choices.

So, what is a healthy menu item or meal? Can it be defined? Yes, it can be defined, although in different ways. You may want to define a healthy meal as one that includes whole grains, fruits, vegetables, lean protein, and small amounts of healthy oils. Another way to look at a healthy meal is to look at the nutrients it contains, such as the following:

- 800 kcalories or less
- 35 percent or fewer kcalories from fat, emphasizing oils high in monounsaturated and polyunsaturated fats
  - 10 percent or less of total kcalories from saturated fat
  - no trans fat
  - 100 milligrams or less of cholesterol
- 45 to 65 percent kcalories from carbohydrates
  - 10 grams or more of fiber
  - 10 percent or fewer kcalories from added sugars
- 15 to 25 percent kcalories from protein
- 800 milligrams or less of sodium (about ⅓ teaspoon of salt)

This does not mean that each menu item should follow these guidelines. For example, it is possible to include a cheeseburger, which has more than 35 percent of its kcalories from fat, in a meal. As long as the other components of the meal are lower-fat foods, such as fruits, vegetables, and whole grains, the percentage of total kcalories from fat in the entire meal will meet the goal.

A menu may simply highlight two or more healthy entrées and one or two healthy appetizers and desserts. To develop some healthy menu items, the first step is to look seriously at your existing menu while engaging in some old-fashioned menu planning. You may go in one of three directions.

1. Use existing items on your menu. Certain menu selections, such as fresh vegetable salads and marinated grilled skinless chicken, may meet your needs.
2. Modify existing items to make them more nutritious. For example, pan-sear fish (dry sauté method) instead of pan-frying it with a butter sauce. In general, modification centers on ingredients, preparation, and cooking techniques. Modifying an existing item may simply mean offering a half portion, which would work well for elderly clients who can't eat full portions and those who would prefer this to taking home leftovers. This is a great marketing tool because the customers pay less; however, they do not pay half the price. The food cost is less but the labor and expenses are not.
3. Create new selections. Many resources are available to obtain healthy recipe guidelines and ingredients: cookbooks, magazines, websites. Or draw on your culinary skills and creativity and craft your own recipes.

Whenever you are involved in menu planning, keep in mind the following considerations.

1. Is the menu item tasty? Taste is the key to customer acceptance and the successful marketing of these items. If the food does not taste delicious and have a creative presentation, no matter how nutritious it may be, it is not going to sell.
2. Can each menu item be prepared properly by the cooking staff?
3. Does the menu item blend with and complement the rest of the menu?
4. Does the menu item meet the food habits and preferences of the guests?
5. Is the food cost appropriate for the price that can be charged?
6. Does each menu item require a reasonable amount of preparation time?
7. Is there a balance of color in the foods and in the garnishes?

8. Is there a balance of textures, such as coarse, smooth, solid, and soft?
9. Is there a balance of shape, with different-sized pieces and shapes of food?
10. Are the flavors varied?
11. Are the food combinations acceptable?
12. Are the cooking methods varied?

RECIPE MODIFICATION

You can modify recipes for many reasons, for example, to reduce the amount of kcalories, fat, saturated fat, cholesterol, sodium, or sugar. You may also modify a recipe to get more of a nutrient, such as fiber or vitamin A. Whether modifying a recipe to get more or less, there are four basic ways to go about it.

1. Change/add healthy preparation techniques.
2. Change/add healthy cooking techniques.
3. Change an ingredient by reducing it, eliminating it, or replacing it (see Figure 9-1 for substitution possibilities).
4. Add a new ingredient(s), particularly to build flavor, such as dry rubs, toasted spices, fresh herbs, and condiments.

FIGURE 9-1: Recipe Substitutions

In Place of	Use
Butter	Margarine (Light/lowfat margarines contain more water and may cause a baked product to be tough, so try decreasing regular margarine 1–2 Tbsp first)
2% or whole milk	Nonfat or low fat (1%) milk
Buttermilk	2% buttermilk or 15 tbsp skim milk +2 Tbsp lemon juice
1 cup shortening	¾ cup vegetable oil
1 cup heavy cream	1 cup evaporated skim milk—in soups and casseroles Baking—light cream or Half & Half
1 cup light cream	1 cup evaporated skim milk
1 cup sour cream	1 cup reduced-fat sour cream
Cream cheese	Light cream cheese or Neufchatel (Nonfat cream cheese produces dips and cake frosting that are very runny)
1 ounce baking chocolate	3 tablespoons cocoa and 1 tablespoon vegetable oil
1 egg	¼ cup egg substitute or 2 large egg whites
Whole-milk mozzarella	Part-skim mozzarella, low moisture
Whole-milk ricotta	Part-skim ricotta
Creamed cottage cheese	Low-fat cottage cheese or pot cheese
Cheddar cheese	Low-fat cheddar cheese (nonfat cheese does not melt) or ¾ cup very sharp cheddar cheese for 1 cup cheddar
Swiss cheese	Low-fat Swiss cheese
Ice cream	Ice milk or frozen yogurt
Mayonnaise	Light or nonfat mayonnaise or nonfat plain yogurt (don't use nonfat versions if heating)
1 cup white flour	½ cup whole wheat flour and ½ cup white flour
1 cup oil in quick breads	½ cup pureed fruit +½ cup oil
1 cup chopped nuts	½ cup nuts toasted to bring out flavor
1 cup shredded coconut	½ cup toasted coconut +½ teaspoon coconut extract

If you decide to modify a recipe, follow these steps.

1. Examine the nutritional analysis of the product and decide how and how much you want to change its nutrient profile. For example, in a meatloaf recipe, you may decide to decrease the fat content to less than 40 percent and increase the complex carbohydrate content to 10 grams per serving.

2. Next, you need to consider flavor. What can you do to the recipe to keep maximum flavor? Should you try to mimic the taste of the original version, or will you have to introduce new flavors? Will you be able to produce a tasty dish?

3. Then modify the recipe by using any of the methods discussed in Chapter 8. When modifying ingredients, think about what functions each ingredient performs in the recipe. Is it there for appearance, flavor, texture? What will happen if less of an ingredient is used or a new ingredient is substituted? You also have to consider adding flavoring ingredients in many cases.

4. Evaluate the product to see whether it is acceptable. This step often leads to further modification and testing. Be prepared to test the recipe a number of times and also be prepared for the fact that some modified recipes will never be acceptable.

Of course, if you don't want to go through the trouble of modifying current recipes, you can select and test recipes from healthy cookbooks or other sources or create your own recipes. When developing new recipes, be sure to choose fresh ingredients and cooking methods and techniques that are low in fat. Also, pay attention to developing flavor and cook foods to order as much as possible.

EXAMPLES OF HOW TO MODIFY RECIPES

Meatloaf

TRADITIONAL VERSION

Serves 25

1 pound onions, fine dice
8 ounces celery, fine dice
2 ounces oil
12 ounces soft bread crumbs
12 ounces tomato juice, stock, or milk
5 pounds ground beef
2½ pounds ground pork
7 eggs, beaten slightly
1 tablespoon salt
½ teaspoon black pepper

Steps

1. Sauté the onions and celery in oil until tender. Remove from pan and cool.
2. In a large bowl, soak the bread crumbs in the juice, stock, or milk.
3. Add the sautéed vegetables, the meat, eggs, salt, and pepper. Mix gently until evenly combined.
4. Form the mixture into 2 or 3 loaves in a baking pan, or fill loaf pans.
5. Bake at 350°F about 1½ hours until done (internal temperature of at least 165°F)

HEALTHIER VERSION

Serves 25

1 pound onions, fine dice

8 ounces celery, fine dice

2 ounces oil

12 ounces diced whole-wheat bread

4 ounces skim milk or stock

6 ounces tomato ketchup

5 pounds ground beef, extra lean

2½ pounds ground pork

8 egg whites, beaten to soft peaks

1 tablespoon salt

½ teaspoon black pepper

Steps

1. Sauté the onions and celery in oil until tender. Remove from pan and cool.
2. In a large bowl, soak the bread crumbs in milk or stock .
3. Add the sautéed vegetables, the ketchup, meat, egg whites, salt, and pepper. Mix gently until evenly combined.
4. Form the mixture into 2 or 3 loaves in a baking pan, or fill loaf pans.
5. Bake at 325°F about 1½ hours until done (internal temperature of at least 165°F)

Nutritional Analysis

	Kcalories	Protein (g)	Fat (g)	Carbo (g)	Sodium (mg)	Chol (mg)
Traditional Recipe	463	26	33	13	487	133
Healthier Recipe	398	28	23	11	556	123

CHEF'S NOTES

The major ingredient in meatloaf is obviously the meat. Traditional meatloaf uses ground beef and/or ground pork or veal. Ground pork is quite fatty, and ground veal is very expensive. Beef is available in different ratios of lean to fat, from 75 percent lean to 95 percent lean. I like to use a leaner beef to cut down on fat. Also, you can remove the whole egg (and its cholesterol) and substitute whipped egg whites folded into whole-wheat bread meat mixture. You can also substitute overcooked brown rice instead of the bread for a gluten-free style. Here is another opportunity to indulge in your creativity. Turn your meatloaf slightly oriental with ginger, garlic, and scallions with a hint of hoisin or soy sauce. Or add fresh grated ginger, curry, cinnamon, cardamon, chili powder, and cumin with sautéed onions and fennel, lemon juice, and some golden raisins. Another variation is the popular BBQ meatloaf with Southwest spices, sautéed peppers, onions, and corn, served with a BBQ sauce. You can use turkey, beef, chicken, pork or a combination—the decision is yours.

Beef Stew

TRADITIONAL VERSION:

Serves 25

4 ounces oil

6 pounds beef chuck, boneless and trimmed, cut into 1-inch cubes

1 pound onion, fine dice

2 teaspoons chopped garlic

4 ounces flour

8 ounces tomato puree

2 quarts brown stock

Sachet of bay leaf, thyme, and celery leaves

1 pound celery, large dice

1½ pounds carrots, large dice

1 pound small pearl onions, edible portion

8 ounces tomatoes, canned, drained, and coarsely chopped

8 ounces peas, frozen, thawed

Salt and pepper to taste

Steps

1. Heat the oil in a brazier until very hot. Brown the meat well, stirring occasionally to brown all sides. If necessary, brown the meat in several small batches to avoid overcrowding the pan.
2. Add the diced onion and garlic to the pan and continue to cook until onion is lightly browned.
3. Add the flour to the meat and stir to make a roux. Continue to cook over high heat until the roux is slightly browned.
4. Stir in the tomato puree and stock and bring to a boil. Stir with a spoon as the sauce thickens.
5. Add the sachet. Cover the pot and place in an oven at 325°F. Braise until the meat is tender, about 1½–2 hours.
6. Cook the celery, carrots, and onions separately in boiling salted water until just tender.
8. When meat is tender, remove the sachet and adjust seasoning. Degrease the sauce.
9. Add the celery, carrots, onions, and tomatoes to the stew.
10. Immediately before service, add the peas. Alternatively, garnish the top of each portion with peas. Season with salt and pepper.

HEALTHIER VERSION

Serves 20

MARINADE: 4 cups chicken broth, 1½ cups white wine, 1 ounce olive oil, 1 bay leaf, 1 ounce fresh herbs, 1 lemon, 1 lime

5 pounds filet tip or lean beef

Nonstick cooking spray

SEASONING: 1 ounce granulated garlic, ½ ounce dried thyme, ½ ounce dried oregano, ½ ounce onion powder, 1 tablespoon paprika, 2 teaspoons red pepper flakes (This will make extra that can be reserved and stored in an airtight contain for future use. Or use Mrs. Dash, no-sodium variety.)

10 ounces onions, medium dice

2½ cups red wine

20 ounces carrots, medium dice, steamed

20 ounces zucchini, medium dice, steamed

20 ounces tomatoes, medium dice, seeded, steamed

20 ounces fennel, half-moon sliced, steamed

20 ounces green beans, cut 1½-inch length, steamed

½ cup fresh basil, chiffonade

3 tablespoons fresh thyme, cleaned and chopped

10 cups veal stock

Steps

1. Make marinade by heating the chicken broth and reducing by one-third. Add in wine, olive oil, and herbs. Stir. Squeeze in lemon and lime juices. Simmer 5 minutes. Chill overnight.
2. Trim beef of any fat and cut into 1-inch pieces. Marinate overnight.
3. Spray nonstick skillet with nonstick cooking spray.
4. Season beef with 5 tablespoons vegetable seasoning. Sear beef with onions in the hot skillet. Add red wine. Remove beef. Add steamed vegetables to same skillet and season with basil and thyme. Add veal stock. Reduce halfway. Add beef back to pan and cook quickly. Serve.

Nutritional Analysis

	Kcalories	Protein (g)	Fat (g)	Carbo (g)	Sodium (mg)	Chol (mg)
Traditional Recipe	284	24	14	14	592	71
Healthier Recipe	277	28	9	17	197	67

CHEF'S NOTES

This is where I would argue that classical cooking fundamentals are key to the success of this dish and many others like it. They start with a flavorful marinated cut of meat, a well-crafted stock, and an appropriate vegetable garnish. The procedure for a well-balanced stew and a heavier stew is how the dish is being prepared. Traditional thickeners are eliminated; reductions and light alternatives such as cornstarch and agar-agar are used to emulate traditional roux. Stocks are properly degreased and these stews are rendered and strained before the fat is incorporated into the sauce. There is little or no addition of salt, just the natural flavors achieved by quality of ingredients and cooking methods. There are many varieties and ingredients you can use to create your own signature dish. Try pork, lamb, chicken, or beef with a variety of different vegetables as well as marinades to achieve the flavor or cuisine you are capturing. With the same methodology and different ingredients, you can create many new menu items. Everyone loves stew—it's a comfort food.

Hamburger

TRADITIONAL VERSION

Serves 4

1 pound ground beef
1 egg
2 teaspoons minced garlic
1 tablespoon steak sauce

Steps

1. Mix ingredients together and make into 4 patties.
2. Grill for 10 to 15 minutes, to desired degree of doneness (internal temperature of at least 165°F). Serve on buns.

HEALTHIER VERSION: GRILLED CHICKEN BURGER

Serves 32

10 pounds chicken breast, skinned
1½ pounds cooked potato, small dice
1 cup chopped parsley
2 tablespoons Mrs. Dash garlic herb seasoning
½ teaspoon black pepper
1 cup white onions, diced
½ ounce chicken stock
2 egg whites

Topping

32 whole-grain hamburger buns
32 slices red onion
32 thick slices tomato
1 pound micro greens

Steps

1. In a mixing bowl, add all ingredients. Mix together to incorporate all ingredients.
2. Shape in ring mold, 4-inch diameter to make 5½-ounce chicken patties
3. Grill chicken burger patties slowly until thoroughly cooked
4. Place burger on whole-wheat bun. Arrange onion and tomato slice on plate. Place micro greens on top of tomatoes.

Nutritional Analysis

	Kcalories	Protein (g)	Fat (g)	Carbo (g)	Sodium (mg)	Chol (mg)
Traditional Recipe	441	26	28	19	334	124
Healthier Recipe	303	38	5	27	280	82

CHEF'S NOTES

Classics, tradition, and comfort sell food today, especially when your items are nutritionally balanced. It is imperative as a chef/cook to be versed in all these areas. Most of the recipes in this book are classic methods that you can continue to tweak as you add your own identity and style. Many of the seasonings, preparations, and ingredients can be used in many of your menu items. It is too difficult to think you have to cook two different ways: healthy cooking and not-so-healthy cooking. The cooking is the same, but the ingredients in not-so-healthy cooking contain more sodium, fat, cholesterol, and refined carbohydrates. Even when cooking classics, you should make your brandy cream for a rib eye or filet beginning with a great veal sauce and the addition of reduced cream. We tend to add too much cream, butter, eggs, salt, sugar, and fats, thinking this will make the dish great. On the contrary, people leave your restaurant overstuffed and uncomfortable. Some of the finest restaurants today use vegetable reduction with a hint of butter, or flavored oils that are added just at the end of a preparation, giving a clean distinct flavor to the food you are consuming. There is a lot of passion associated with cooking for people. After all, it is one of the few professions where people ingest what you are selling. For culinarians, that is a tremendous responsibility.

Chicken Quesadillas

TRADITIONAL VERSION

Serves 4

2 tablespoons oil
¼ cup onions, chopped
1 small garlic clove, minced
¼ cup bell peppers, chopped
2 chili peppers, minced
3 plum tomatoes, chopped
1½ cups cooked chicken
Additional oil, for frying
8 flour tortillas
2 cups cheddar cheese, shredded
Salsa
Sour cream

Steps

1. Heat 2 tablespoons oil in a medium-heavy skillet and sauté the onions, garlic, and bell peppers until soft. Add the chili peppers and tomatoes and simmer a few minutes more. Add the chicken and stir well to combine.
2. Heat a thin layer of oil in another pan over medium-low heat. Place a tortilla in the pan and sprinkle with ¼ cup cheese. Add about a quarter of the chicken mixture and top with another ¼ cup cheese. Cover with another tortilla and cook for 2–3 minutes or until golden brown.
3. Flip the quesadilla over and cook an additional 2 minutes. Remove from heat and cut into wedges. Keep warm while frying the remaining quesadillas. Serve with salsa and sour cream on the side.

HEALTHIER VERSION

Serves 3

1 tablespoon olive oil
1 onion, medium sliced
1 large green pepper, cut into ¼-inch strips
1 teaspoon garlic, minced
3 large mushrooms, cleaned and sliced
⅛ teaspoon cumin
⅛ teaspoon chili powder
Pinch cayenne pepper
2 tablespoons sherry vinegar
1 tablespoon white wine
1 teaspoon cilantro, chopped
1 pound skinned chicken breast, cut into ½-inch crosswise strips
½ cup chicken stock
Cooking spray
6 whole-wheat flour tortillas
3 ounces shredded part-skim mozzarella cheese

3 ounces shredded part-skim sharp cheddar cheese

4 cups mixed field greens

½ cup black bean relish

Steps

1. Heat olive oil in saucepan over medium heat. Add onion and green pepper and stir frequently until onion starts to brown, about 5 minutes. Add garlic, mushrooms, cumin, chili powder, and cayenne pepper.

2. Cook, stirring frequently, about 2 minutes. Add sherry vinegar, wine, and cilantro and cook until most of the liquid evaporates, about 2 minutes.

3. Preheat oven to 350°F.

4. In large saucepan, simmer chicken strips in chicken stock for several minutes until just cooked. Remove from heat and drain. Place chicken on plate.

5. Coat large skillet or griddle with cooking spray. Lightly brown and crisp tortillas on each side over medium heat.

6. On each of 3 tortillas, layer ⅙ of the cheeses, ⅓ of vegetable mixture, ⅓ of chicken mixture, and ⅙ of the cheeses. Top each with a second tortilla.

7. Bake quesadillas in oven for 5 minutes or finish on grill.

8. Let quesadillas cool slightly and cut into quarters.

9. Serve with mixed field greens topped with black bean relish.

Nutritional Analysis

	Kcalories	Protein (g)	Fat (g)	Carbo (g)	Sodium (mg)	Chol (mg)
Traditional Recipe	968	43	49	89	1185	108
Healthier Recipe	850	66	34	73	1073	127

CHEF'S NOTES

What's better than quesadillas? With the crisp shell, sautéed vegetables, and melted cheese, this is the ultimate bar snack, light lunch entrée, or late night fun. This is the type of item you should carry on your typical menus to provide an alternative for those who want something great but light. I wouldn't even specifically advertise this as a healthy item—it's good and it's good for you. The addition of black bean salsa, pureed beans, or a little avocado cilantro salad nicely rounds out this dish. My vote is to keep this as a staple on the menu as a light signature item that's better than its original version.

Crab Cakes

TRADITIONAL VERSION

Serves 4

1 pound crabmeat

¼ cup green pepper, chopped

¼ cup finely chopped onion

¼ cup mayonnaise

1 egg

¾ cup bread crumbs

1 tablespoon Old Bay Seasoning

½ teaspoon dry mustard
Dash of Worcestershire sauce
4 tablespoons butter

Steps

1. Combine all ingredients except butter in a large bowl, using just enough bread crumbs to hold together.
2. Make into patties and fry in butter until golden brown.

HEALTHIER VERSION

Serves 5

1¼ pounds jumbo lump crabmeat
1½ teaspoons chives, chopped fine
1½ teaspoons parsley, chopped fine
1 tablespoon Old Bay Seasoning
¾ teaspoon Dijon mustard
¾ teaspoon fresh thyme, chopped
5 ounces potatoes, cooked and riced
1½ egg whites
¾ teaspoon lemon juice
1½ teaspoons white wine
½ egg white
Japanese bread crumbs, as needed
¾ teaspoon chives, chopped fine
Vegetable oil cooking spray

Steps

1. Pick crabmeat to remove bits of shell. Place crab meat in a bowl.
2. In another bowl, mix together chives, parsley, Old Bay Seasoning, mustard, thyme, and riced potatoes. Add crab meat and fold together gently, not breaking up crabmeat.
3. In a stainless-steel bowl, place 1½ egg whites, lemon juice, and white wine. Whip to form stiff peaks.
4. Fold whipped egg whites into crab mixture and mold into 3-ounce crab cakes.
5. In a metal pan, place ½ egg white (whipped slightly to break up). Dip the crab cakes in the egg white, then in the crumbs mixed with chives.
6. In a nonstick pan sprayed with vegetable spray, sauté the crab cakes to a crisp golden brown. Serve with salsa or mojo of your choice with some spicy greens.

Nutritional Analysis

	Kcalories	Protein (g)	Fat (g)	Carbo (g)	Sodium (mg)	Chol (mg)
Traditional Recipe	395	25	25	16	684	176
Healthier Recipe	150	24	1	10	607	48

CHEF'S NOTES

There is so much to say about this dish. Not only is it an American classic, but it is also one of the most widely bought appetizer/entrée selections in restaurants of different styles across the board. This version is delicate and satisfying. You don't miss the mayonnaise or the fried cooking method at all. This dish is all about the crabmeat and the traditional flavorings. The garnishes are endless, from salad greens to shaved fennel to an heirloom tomato salad with basil or cilantro to celery root slaw. This again is where your creativity takes over. We have provided dressings that any of these vegetables can be marinated in to create a new garnish, as well as the many types of dried vegetables and potatoes you can make as a crunchy garnish. This one is a definite on your menu.

Vegetable Lasagna

TRADITIONAL VERSION

Serves 8

20 ounces frozen spinach or broccoli cuts, defrosted and squeezed dry

2 lb ricotta cheese

¾ cup grated Parmesan cheese (save some to sprinkle on top)

1 teaspoon Italian seasoning

9 lasagna noodles, cooked

32 ounces spaghetti sauce

Steps

1. Combine the spinach or broccoli, ricotta, Parmesan cheese, and seasoning. Mix well.
2. In a 9 × 13 inch pan, layer ingredients in the following order: noodles, ½ cheese mixture, then ⅓ sauce. Repeat and top with 3 noodles, sauce, and a sprinkling of Parmesan cheese. Bake 30 minutes at 350°F degrees

HEALTHIER VERSION

Serves 8

MARINADE: 4 cups chicken broth, 1½ cups white wine, 1 ounce olive oil, 1 bay leaf, 1 ounce fresh herbs, 1 lemon, 1 lime

8 slices eggplant

8 slices zucchini

8 slices yellow pepper

8 slices red pepper

2 pounds skimmed ricotta cheese

8 ounces feta cheese

8 tablespoons fresh oregano, chopped

1 tablespoon Mrs. Dash garlic herb seasoning

Cracked black pepper, to taste

8 egg whites

32 ounces tomato sauce

Chopped parsley, for garnish

Steps

1. Combine marinade ingredients and marinate vegetables overnight. Grill.
2. Make cheese mixture by mixing together ricotta, feta, oregano, Mrs. Dash, black pepper, and egg whites until smooth.
3. Make 1 portion at a time to order. Layer vegetables and cheese mixture on plate in this order:
   1 slice eggplant
   2 tablespoons cheese mixture

1 slice zucchini

2 tablespoons cheese mixture

1 slice yellow pepper

2 tablespoons cheese mixture

1 slice red pepper

4. Cover with tomato sauce. Sprinkle with chopped parsley.

Nutritional Analysis

	Kcalories	Protein (g)	Fat (g)	Carbo (g)	Sodium (mg)	Chol (mg)
Traditional Recipe	417	24	19	39	44	64
Healthier Recipe	323	24	16	22	143	61

CHEF'S NOTES

Who doesn't love lasagna, but who can eat this heavy classic more then twice a year? You can easily use this as an entrée accompaniment or as a lunch entrée with an arugula roasted cipollini salad. The vegetables replace the pasta so again this is perfect for a gluten-free appetizer or an addition to their entrée. You get the feeling of cheese and flavoring without the calories of whole-milk mozzarella and ricotta, not to mention rich pasta sheets. Good for the records with a lot of versatility in your menu design.

Chicken Pot Pie

TRADITIONAL VERSION

Serves 6

2 tablespoons butter

1 small onion, chopped

½ cup carrots, chopped

½ cup potatoes, chopped

½ cup celery, chopped

2 tablespoons flour

½ teaspoon salt

½ teaspoon dried thyme

½ teaspoon dried rosemary

Pinch black pepper

1 bay leaf

2 cups chicken stock

1 cup frozen peas

2 cups cooked chicken, chopped

Pastry for 9-inch double pie crust

2 tablespoons milk

Steps

1. Melt butter in a saucepan. Add onions, carrots, potatoes, and celery. Cook until soft, about 5–7 minutes.

2. Whisk in flour over medium heat for 2 minutes. Add seasonings and stock, ½ cup at a time, stirring constantly until it thickens.
3. Turn oven to 375°F.
4. Remove bay leaf. Stir in peas and chicken.
5. Spoon filling into pie crust. Set top layer in place and seal the edges. Make vents in the top and brush with milk.
6. Bake for about 30 minutes or until filling is bubbling and top is golden brown.

HEALTHIER VERSION

Serves 6

2 teaspoons vegetable oil
1 small onion, chopped
½ cup carrots, chopped
½ cup celery, chopped
¼ cup parsnips
½ cup potatoes, chopped blanched in water
2 sprigs of thyme
½ sprig rosemary
1 bay leaf
1 cup chicken stock
1 cup frozen peas
2 cups cooked chicken, medium dice
1 cup velouté base
Salt and pepper, to taste
Pastry: ricotta chive crust

Steps

1. Heat oil in a saucepan. Add onion, carrots, celery, and parsnips. Cook until soft, about 5–7 minutes. Add in potatoes.
2. Make a bouquet garni with thyme, rosemary, and bay leaf. Add stock, ½ cup at a time, to cook vegetables.
3. Turn oven to 375°F.
4. Remove herbs. Stir in peas and chicken and 1 cup of velouté base. Cook an additional 5 minutes to desired consistency. Add salt and pepper to taste, if needed.
5. Spoon filling into a 6-ounce ovenproof dish. Place ricotta dough on top and seal the edges. Make vents in the top and brush lightly with milk.
6. Bake for about 30 minutes or until filling is bubbling and top is golden brown.

Nutritional Analysis

	Kcalories	Protein (g)	Fat (g)	Carbo (g)	Sodium (mg)	Chol (mg)
Traditional Recipe	383	20	20	29	514	55
Healthier Recipe	278	16	15	20	140	43

CHEF'S NOTES

This is another one of those winners. Great flavor, a great look, and a great value. This can easily be a staple on your menu to fit the needs of your clients. Your function as culinarians is to teach America to eat right. Not everything needs to be covered in butter cream, salt, or fat. The variety of vegetables you choose for this dish is up to you. Parsnips, celery root, sweet potatoes, fennel, or butternut squash are all great choices that can work in this application. The choice depends on what your clients prefer, and what you can execute in your kitchen.

Wontons (Filled Dumplings)

TRADITIONAL VERSION: FRIED WONTONS

Serves 4

1 clove garlic
2 sprigs cilantro
1 tablespoon soy sauce
½ cup ground pork
1 pinch ground pepper
24 wonton skins
½ cup vegetable oil for frying

Steps

1. Mince garlic and cilantro. Mix with soy sauce, ground pork, and pepper.
2. Put half a teaspoon of the filling in the middle of the wonton skin. Pick up one corner and fold it to the diagonal corner. Pick up the 2 ends that have 45 degree angle and wet them with a dab of water. Pinch the two corners together. Continue with the remaining wonton skins.
3. Fry the wontons in the oil until golden brown. Keep turning them to cook evenly.

HEALTHIER VERSION: ROAST CHICKEN AND SHREDDED MOZZARELLA TORTELLONIS

Serves 6

4 ounces chicken breast
½ teaspoon canola oil
Pinch fresh oregano, chopped
Pinch fresh basil, chopped
Pinch paprika
½ cup reduced-fat shredded mozzarella
½ cup skim-milk ricotta
¼ teaspoon fresh garlic, chopped
2 pinches freshly ground black pepper
1 teaspoon fresh basil, chopped
½ teaspoon fresh parsley, chopped
18 wonton skins
1 cup water
1 tomato, chopped
18 arugula leaves
1 recipe Red Pepper Coulis

Steps

1. Preheat oven to 350°F.
2. Coat chicken breast with oil, herbs, and paprika. Roast about 5 minutes, or grill over hardwood for 1 minute. Let cool and fine-julienne. Reserve.
3. In bowl, mix cheeses, garlic, pepper, and herbs. Add chicken and mix well.
4. Lay out wonton skins and paint with water.
5. Place 1 full teaspoon of chicken mixture in each skin or more to use up all the filling.
6. Fold skin to make a triangle, then connect ends in opposite direction to make tortelloni shape.
7. Prepare chopped tomato garnish and toss with black pepper. Let juice accumulate. Paint leaves of arugula lightly with juice of tomato for flavor.
8. Poach tortellonis for 3 minutes in lightly boiling water.
9. Arrange each plate with 3 arugula leaves facing out like spokes of a wheel, with a tortelloni between each 2 leaves. Place Red Pepper Coulis and a spoonful of chopped tomato in the middle.

Nutritional Analysis

	Kcalories	Protein (g)	Fat (g)	Carbo (g)	Sodium (mg)	Chol (mg)
Traditional Recipe	339	10	21	29	544	25
Healthier Recipe	151	11	4	17	233	25

CHEF'S NOTES

This is one of the best dishes for ease of preparation, ability to freeze, and delicious flavor. Wontons, tortellinis, egg rolls, dumplings are all produced today using egg roll wrappers and wonton skins. They are light, nutritionally balanced, and easy to work with. The filling of this item can be substituted with hundreds of favorites. Use shredded vegetables with beef, pork, shrimp, and crab for an Asian-style wonton. Use mushroom, chicken, fennel, and caramelized onions with a variety of meats for an Indian, Italian, or Spanish-style cuisine. These ideas are just the tip of the iceberg.

Velouté Sauce

TRADITIONAL VERSION

Makes 2 tablespoons

4 tablespoons butter

4 tablespoons flour

2 cups of chicken, fish, or veal stock (hot)

Steps

1. Heat the butter in a heavy saucepot over low heat. Add the flour and make a blond roux. Cool slightly.
2. Gradually add the hot stock to the roux, beating constantly. Bring to a boil, stirring constantly. Reduce heat to a simmer.
3. Simmer the sauce for 1 hour. Stir occasionally and skim the surface when necessary. Add more stock if needed to adjust consistency.

HEALTHIER VERSION

Makes 2 tablespoons

1 tablespoon arrowroot
2 cups chicken stock
Salt and white pepper

Steps

1. Combine the arrowroot with just enough stock to form a paste.
2. Bring the remaining stock to a boil.
3. Whisk the arrowroot paste into the boiling stock and stir for a few minutes, just until thickened.
4. Salt and pepper to taste.

Nutritional Analysis

	Kcalories	Protein (g)	Fat (g)	Carbo (g)	Sodium (mg)	Chol (mg)
Traditional Recipe	76	2	6	3	136	16
Healthier Recipe	10	1	0.5	1	385	0

CHEF'S NOTES

This type of sauce can be a big part of the cooking foundation. You can use this base sauce in many cooking applications, such as seven onion sauce, corn chowder, and different pasta dishes.

Creamy Dressings

TRADITIONAL VERSION

6 servings

½ cup mayonnaise
2 tablespoons chopped chives
2 tablespoons chopped parsley
1 tablespoon lemon juice
½ teaspoon Worcestershire sauce
¼ teaspoon dry mustard
¼ teaspoon finely chopped garlic

Steps

1. Blend all ingredients thoroughly. Chill.

HEALTHIER VERSION I: GREEN GODDESS DRESSING

6 servings

4 ounces soft tofu, well drained
½ cup cider vinegar
2 stalks celery
½ cup spinach leaves, washed and dried
⅓ cup fresh parsley (no stems)

1 tablespoon lemon juice
2 scallions
Fresh tarragon leaves, to taste
Fresh ground pepper, to taste

Steps

1. Blend all ingredients in food processor until smooth. This dressing will last 2 days in the refrigerator.

HEALTHIER VERSION II: YOGURT DRESSING/SAUCE

6 servings

8 ounces plain nonfat yogurt
1 ounce skim milk
1 ounce white wine
½ ounce lime juice
1 ounce Dijon mustard
½ ounce honey
½ tablespoon curry
1 teaspoon turmeric
½ teaspoon cayenne
½ tablespoon vegetable seasoning
1 tablespoon cilantro, minced
1 teaspoon fresh mint, chopped

Steps

1. Combine all ingredients in blender. Chill.

HEALTHIER VERSION III: CUCUMBER DILL DRESSING

Makes 2 quarts

12 ounces light cream cheese
3 ounces farmer's cheese
1 pint skim milk
1 pound cucumbers, peeled and seeded
2 ounces fresh dill, finely chopped
3 tablespoons Dijon mustard
2 teaspoons garlic
½ teaspoon black pepper
1 teaspoon salt
2 ounces olive oil
3 ounces lemon juice
½ teaspoon Tabasco

Steps

1. Blend cheeses together with skim milk.
2. Add other ingredients and blend until smooth.

HEALTHIER VERSION IV: CARBONARA SAUCE

Makes 1½ quarts

1½ ounces corn-oil margarine
2½ ounces flour
1 quart skim milk
4 ounces white wine
4 ounces skim-milk ricotta cheese
Garlic herb seasoning and cracked black pepper, to taste
2 ounces Parmesan cheese

Steps

1. Make roux with margarine and flour. Add skim milk, wine, ricotta, and seasonings.
2. Bring to a boil, simmering slowly until thickened. Add parmesan and lightly simmer.

Nutritional Analysis

	Kcalories	Protein (g)	Fat (g)	Carbo (g)	Sodium (mg)	Chol (mg)
Traditional Recipe	122	0	13	0	125	7
Healthier Recipe 1	25	2	0	2	147	0
Healthier Recipe 2	45	2	0	6	194	0
Healthier Recipe 3	53	2	4	2	150	7
Healthier Recipe 4	57	3	2	5	77	4

Oil And Vinegar Dressings

TRADITIONAL VERSION: BASIC VINAIGRETTE

Makes 1 quart

1 cup wine vinegar
1 tablespoon salt
1 teaspoon white pepper
3 cups soy oil

Steps

1. Mix together the vinegar, salt, and white pepper until the salt is dissolved.
2. Using a wire whip or blender, mix in the oil.
3. Mix again before using.

HEALTHIER VERSION I: BASIC HERB VINAIGRETTE

Makes 1 gallon

2 quarts plus 1 cup chicken stock or low-sodium vegetable stock
½ ounce cornstarch slurry (cornstarch with cold water)
2 tablespoons fresh chopped garlic
2 teaspoons black pepper, coarsely ground
2 tablespoons fresh thyme, chopped

2 ounces fresh parsley, chopped

2 tablespoons, fresh oregano, chopped

1½ ounces fresh basil, chopped

2 ounces chives, chopped

1 quart balsamic vinegar

4 cups extra-virgin olive oil

Steps

1. Heat stock to a rolling boil.
2. Thicken with cornstarch slurry to a nappe (to coat) consistency.
3. Cool and add garlic, pepper, herbs, and vinegar.
4. Whisk olive oil into the mixture to emulsify.
5. Cool and store.

HEALTHIER VERSION II: ORANGE VINAIGRETTE

Makes 3 cups

1 pint orange juice

1 tablespoon Dijon mustard

1 tablespoon honey

1 tablespoon shallots, finely diced

½ teaspoon coriander, ground

½ teaspoon black pepper, ground

2 ounces white-wine vinegar

2 ounces extra-virgin olive oil

1 teaspoon thyme, fresh, chopped

1 teaspoon chives, fresh, chopped

1 teaspoon basil, fresh, chopped

Steps

1. In a saucepan, reduce orange juice to ½ cup, or use ½ cup concentrate.
2. Place cooled orange syrup in food processor. Add mustard, honey, shallots, coriander, black pepper, and vinegar.
3. Start food processor and slowly add oil to emulsify.
4. Add fresh herbs last, but do not overpuree.
5. Reserve in refrigerator until needed.

Nutritional Analysis

	Kcalories	Protein (g)	Fat (g)	Carbo (g)	Sodium (mg)	Chol (mg)
Traditional Recipe	362	0	40	0	434	0
Healthier Recipe 1	141	1	14	4	15	0
Healtheir Recipe 2	70	0	5	6	31	0

You can substitute different flavored vinegars or virgin nut oil to create alternative dressing options that can easily be adapted to your existing menus. Garnish these dressings with fresh fruits, such as raspberries, mangoes, or strawberries, to accent your menus. Additionally you can add finely chopped vegetables and herbs, such as fennel, basil, and tarragon, to create other dressing options.

CHEF'S NOTES

Carrot Cake

TRADITIONAL VERSION

Serves 12

1½ cups all-purpose flour (about 6¾ ounces)
1⅓ cups granulated sugar
½ cup sweetened flaked coconut
⅓ cup chopped pecans
2 teaspoons baking soda
1 teaspoon salt
2 teaspoons ground cinnamon
3 tablespoons canola oil
2 large eggs
2 cups grated carrot
1½ cups canned crushed pineapple, drained
Cooking spray

Frosting

2 tablespoons butter, softened
1 (8-ounce) block ⅓-less-fat cream cheese, softened
3 cups powdered sugar
2 teaspoons vanilla extract

Steps

1. Preheat oven to 350°F.
2. Combine flour, sugar, coconut, pecans, baking soda, salt, and cinnamon in a large bowl.
3. Combine oil and eggs and stir well.
4. Stir egg mixture, grated carrot, and pineapple into flour mixture.
5. Spoon batter into a 13 × 9-inch baking pan coated with cooking spray. Bake for 35 minutes or until done.
6. To make frosting, combine butter and cream cheese in a large bowl. Beat with a mixer at medium speed until smooth. Beat in powdered sugar and vanilla just until smooth. Spread over cake.

HEALTHIER VERSION

Serves 12

2 cups whole-wheat flour
1½ teaspoon baking powder
⅔ teaspoon cinnamon
¼ teaspoon baking soda
⅛ teaspoon nutmeg
2 eggs
1 cup plus 2 tablespoons canola oil
1¾ cups grated carrots
1½ cups egg whites
1½ cups honey

Naturelle Bavarian Cream

¼ cup apple juice

Steps

1. Stir together dry ingredients. Combine eggs and oil. Add dry ingredients to eggs; whip 3 minutes. Add 1½ cups grated carrots, reserving ¼ cup for garnish.
2. Whip egg whites and honey until medium firm. Fold into carrot mixture. Pour into round cake pan. Bake at 350°F.
3. Slice cake into 8 wedges. Place a wedge on a plate.
4. Mix Naturelle Bavarian Cream with ½ cup grated carrots and apple juice to thin. Drizzle on cake.

Nutritional Analysis

	Kcalories	Protein (g)	Fat (g)	Carbo (g)	Sodium (mg)	Chol (mg)
Traditional Recipe	433	6	14	75	313	51
Healthier Recipe	408	6	21	52	109	35

CHEF'S NOTES

You may have thought that you couldn't make a cake with whole-wheat flour, but it works quite well in this recipe. In most baking recipes, you can successfully replace half of the white flour with whole wheat. The use of egg whites helps to reduce the amount of fat and saturated fat. Canola oil is a great source of monounsaturated fats. The addition of traditional spices, the sweetness of the honey, and the yogurt cream topping prove to be as satisfying as the traditional version.

Oatmeal Raisin Cookies

TRADITIONAL VERSION

Makes 4 dozen

½ pound (2 sticks) margarine or butter, softened

1 cup firmly packed brown sugar

½ cup granulated sugar

2 eggs

1 teaspoon vanilla

1½ cups all-purpose flour

1 teaspoon baking soda

1 teaspoon ground cinnamon

½ teaspoon salt (optional)

3 cups quick-cooking or old-fashioned oats

1 cup raisins

Steps

1. Heat oven to 350°F in large bowl, beat margarine and sugars until creamy. Add eggs and vanilla; beat well. Combine flour, baking soda, cinnamon, and salt; add to mixture and mix well. Add oats and raisins; mix well.
2. Drop dough by rounded tablespoonfuls onto ungreased cookie sheets.
3. Bake 10 to 12 minutes or until light golden brown. Cool 1 minute on cookie sheets; remove to wire rack. Cool completely. Store tightly covered.

HEALTHIER VERSION

Makes 3½ dozen

1 pound brown sugar
14 ounces applesauce
½ ounce canola oil
1 ounce egg whites
12 ounces all-purpose flour
7 ounces oatmeal
1 tablespoon baking soda
2½ cups golden raisins
1 teaspoon cinnamon

Steps

1. Mix together the sugar, applesauce, oil, and egg whites until smooth.
2. Combine the flour, oatmeal, baking soda, raisins, and cinnamon and add to the applesauce mixture. Mix until well blended.
3. Spoon onto a parchment-lined sheet pan and bake at 325°F until golden brown.

Nutritional Analysis

	Kcalories	Protein (g)	Fat (g)	Carbo (g)	Sodium (mg)	Chol (mg)
Traditional Recipe	106	1	4	16	55	19
Healthier Recipe	128	2	0	29	96	0

CHEF'S NOTES

This recipe uses mostly applesauce (with a tiny amount of canola oil) to replace the usual fat. This is an excellent cookie, because it is naturally sweet due to the raisins and spices, as well being a traditional favorite.

MINI-SUMMARY

1. Balance in the menu is one of the keys to success in a foodservice operation. The basic methodology is to understand and respect the fine art of cooking and preparation and then modify the contents to make healthy preparations by using nutritional guidelines.
2. You can define a healthy meal in many ways. One way is to include whole grains, fruit, vegetables, lean protein, and small amounts of healthy oils. Another way is to look at the nutrients a dish contains, such as 800 kcalories or less, 35 percent or fewer kcalories from fat (emphasizing oils high in monounsaturated and polyunsaturated fats), 10 grams or more of fiber, 10 percent or less of total kcalories from saturated and trans fats, 100 milligrams or less of cholesterol, and 800 milligrams or less of sodium.
3. A menu may simply highlight two or more healthy entrées and one or two healthy appetizers and desserts. To develop some healthy menu items, you can use existing items on the menu, modify existing items to make them more nutritious, or create new selections.
4. Whenever you are involved in meal planning, keep in mind the 12 considerations listed in this section.
5. You can modify recipes by changing or adding preparation and cooking techniques or by reducing, replacing, or adding ingredients.
6. To modify a recipe, decide how you want to change its nutritional profile, consider flavor, use your experience to decide how to change ingredients, and evaluate the product.

BREAKFAST

Hot and cold cereals have been the foundation of breakfast for generations in many ethnic cultures, from Swiss birchenmuesli (oats with fresh and dried fruits soaked in milk or cream) to hot Irish oatmeals to contemporary American granolas. There are numerous variations on these classic foundations. For instance, the liquid used to make hot cereals such as oatmeal can be a variety of fruit juices, such as apple, pineapple, or orange. They can also be spiced with cinnamon, nutmeg, ginger, allspice, or cloves. For more adventurous customers, you can use jalapeño jack cheese, star anise, cardamom, lavender, lemon balm, or any fresh herb combination.

The traditional Swiss birchenmuesli is made with raw oats, heavy cream, sugar, nuts, and dried and fresh fruits. To modify this recipe, first use steel-cut oats; more fresh fruits are added to the cereal mixture; skim milk replaces heavy cream; and nonfat yogurt and spices complete the taste needed to make this Old World classic a modern hit.

Pancakes, French toast, and toppings are the apple pie of breakfast. It's hard to imagine a breakfast menu without blueberry pancakes or thick crispy French toast with syrup. A typical pancake batter contains whole eggs, oil, and sugar. To use less of these ingredients, you will have to add other ingredients that the guests can sink their teeth into. For example, you can use wheat germ, steel-cut oats, stone-ground wheat, or millet to create a hearty texture. You also need to include spices and fruit flavorings. By putting leftover berries into batter or using overripe fruit to make syrups, you utilize your inventory while creating a quality product. You can also fine-cut the fruits and toss them with fresh mint or lemon balm to create a sweet salsa or compote.

To create stuffed French toast, layer slices of whole-grain bread with light cream cheese and bananas. Cut into quarters and dip in an egg substitute batter, using honey, Splenda (a no-kcalorie sweetener), or date sugar sparingly instead of white sugar. Add skim milk, cinnamon, and nutmeg to complete the batter. Brown in a nonstick pan and serve with fresh fruit puree or compote.

Quick breads, muffins, and scones are also staple items for breakfast menus as well as buffets. Fruit juices, concentrates, and purees are a wonderful source of flavor in baking quick breads, muffins, and scones as well as in hot cereals and pancake batters. Low-fat spreads might include flavored nonfat ricotta cheese and flavored low-fat cream cheeses.

Some menu possibilities for breakfast include the following:

- Glazed Grapefruit and Orange Slices with Maple Vanilla Sauce (Recipe 9-24) (see Figure 9-2 for a similar recipe)
- Crunchy Cinnamon Granola
- Pineapple Ginger Cream of Wheat
- Jalapeño Jack Cheese Grits
- Spiced Oatmeal with Stewed Dried Fruits
- Stuffed French Toast Layered with Light Cream Cheese and Bananas (Recipe 9-23), topped with strawberry syrup
- Wheat Berry Pancakes with Fresh Fruit Syrup
- Potato Pancakes with Roasted Pear Compote with a Dollop of Nonfat Yogurt
- Maple Yogurt, Fruit, and Granola Parfait (Figure 9-3)

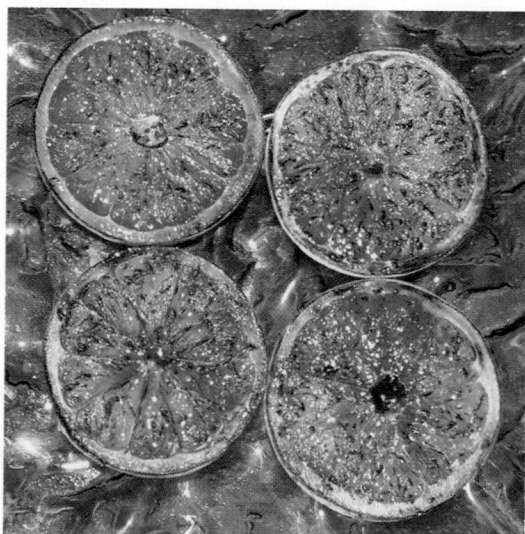

FIGURE 9-2:
Breakfast: Glazed Grapefruits with Vanilla Beans and Honey.
Photo by GigaChef.com

- Whole-Wheat Peach Chimichangas (Recipe 9-25) with Minted Fruit Salad
- Baby Spinach Salad, Egg Whites, Tomatoes, Crisp Mushrooms, and Orange Vinaigrette
- Dried-Cherry Scones, Wild Flower Honey, and Yogurt Sauce
- Shirred Egg Whites, Grilled Eggplant, Zucchini, and Roasted Peppers

Breakfast is a wonderful time for a buffet. It is a time-saver for guests as well as a way of providing variety for the most important meal of the day. Consider a buffet with platters of sliced fresh fruits, Dried-Cherry Scones, Potato Pancakes with Spiced Pear Sauce, Banana-Stuffed French Toast, Shirred Vegetable Frittata, Granola Yogurt Parfaits, Spinach Salad with Egg Whites, Tomatoes and Crisp Mushroom with Orange Vinaigrette or Peach Chimichangas, and Minted Fruit Salad.

CHEF'S TIPS FOR BREAKFASTS

- To make an excellent omelet without cholesterol, whip egg whites until they start to foam. Add a touch of white wine, Dijon mustard, and chives. Whip to a soft peak. Spray a hot nonstick pan with oil and add the eggs. Cook it the same way you do a whole-egg omelet. When the omelet is close to done, put the pan under the broiler to finish. The omelet will puff up. Stuff the omelet, if desired, with grilled, roasted, or sautéed vegetables, wilted spinach, and a little low-fat mozzarella or feta cheese, then fold over and serve.
- For color and flavor, serve an omelet with spicy vegetable relish poured on top of it or place the omelet on a grilled blue corn tortilla and serve with salsa roja (red salsa).
- When writing breakfast menus, make sure to provide balanced, nutritious, and flavorful breakfast choices, such as Glazed Grapefruit and Orange Sections with Maple Vanilla Sauce, Blueberry Wheat Pancakes with Strawberry Syrup, and a side dish of Chicken Hash.
- Breakfast is probably the best time of day to offer freshly squeezed juices. Make sure that they are fresh and that you offer a good variety.

APPETIZERS

Appetizers are a very creative part of the menu. Ingredients might include fresh fruits and vegetables, fresh seafood and poultry, fresh herbs, spices, infused oils, vinegars, and pasta (Figure 9-4).

Recipes for Roast Chicken and Shredded Mozzarella Tortellonis (page 310), Scallop and Shrimp Rolls in Rice Paper (Recipe 9-1), Crab Cakes (page 306), Eggplant Rollatini with Spinach and Ricotta (Recipe 9-2), and Mussels Steamed in Saffron and White Wine (Recipe 9-3) are in this chapter. Additional ideas for appetizers include the following:

- Ricotta Cheese and Basil Dumplings with Salsa Cruda and Arugula
- Napoleon of Grilled Vegetables, Wild Mushrooms, and Goat Cheese with Roasted-Pepper Sauce

FIGURE 9-4:
Appetizer: Crab Cake and Tomato Confit, Shaved Fennel, Red Onions, and Frisée Salad. Photo by GigaChef.com

- Sweet and Yukon Potato and Onion Tart with Forelli Pear Vinaigrette
- Smoked Shrimp with Cucumber Fennel Salad, Micro Greens, and Ginger Mango Sauce
- Spicy Chicken and Jack Cheese Quesadillas with Tomatilla Salsa
- Eggplant or Hummus Dip with Roasted Cauliflower, Raw Vegetables, and Baked Whole-Wheat Pita or Tortilla Chips
- Spicy Chicken Sausage with Roasted Peppers, Grilled Onions, and Tomato Basil Salad
- Maine Crab Cakes with Smoked-Pepper Sauce and Baby Lettuces
- Red Lentil Chili with Baked Spiced Whole-Wheat Tortilla Chips

CHEF'S TIPS FOR APPETIZERS

- Appetizers can often be sized-down entrées. For example, Maine Crab Cakes with Smoked-Pepper Sauce and Baby Lettuces can be made in larger or smaller portions to appear as an entrée or an appetizer.
- Dips and chips are and always will be an American favorite for appetizers. They can be a great selection on your menu with a new twist. Hummus, baba ganoush, white beans and roasted garlic, artichoke, and goat cheese are well-accepted favorite dips that can be accompanied by baked whole-wheat tortilla chips, melba toast, baked multigrain croutons, or a variety of different vegetables.
- There are certain ingredients with which you can make a wide variety of appetizers. Consider using wonton skins and rice paper as wrappers and stuff them with fillings such as white beans and artichokes with roasted garlic or spiced butternut squash, chicken sausage filling, or roasted vegetables. Dried vegetable chips are a wonderful way to add color and crispness to an appetizer. To make red beet chips, for instance, slice red beets thin and dip in simple syrup. Dry on a silk mat, doubled pan in a 275°F oven until crisp, about 1 hour. See pages 331–332 for more information on making garnishes.
- Creative sauces and relishes can help sell appetizers. For example, serve pan-smoked salmon with gingered tomato sauce, asparagus, and malted onion salad or grapefruit-dusted scallops on a quinoa pancake with wilted spinach and cardamom-spiced pear mustard. Toasted spices and fresh herb combinations in conjunction with creative relishes, chutneys, sauces, compotes, and salsas or mojos make tempting and interesting accompaniments to appetizers.
- The most important tip in planning appetizers for your menu is to be in tune with the needs of your guests. There is nothing wrong with an antipasto plate containing grilled and roasted vegetables, a piece of good cheese, and some quality cured meats for certain eating styles (low-carbohydrate, Atkins, or South Beach). An appetizer of fresh roasted artichoke or pickled artichoke with arugula and feta cheese is perfectly acceptable if this fits your clients' taste. What will sell in your establishment, what ingredients are available, what your staff can produce, and what fits with your menu balance are the most important issues in designing a menu change. The items to stay away from when crafting a healthy menu are fried foods, highly processed foods, and foods high in added sugars.

SOUPS

Soups make up some of the most nutritious meals, from the hearty minestrone to a robust butternut squash soup that is creamed with nonfat yogurt and spiced with nutmeg. Soups can be inserted into a meal as an appetizer or given first billing as an entrée.

Soups are a wonderful place to spotlight more than just vegetables. Beans, lentils, split peas, and grains such as rice are also healthy ingredients that work well with soups. Some of these ingredients are starchy enough to be used to thicken soups instead of using a traditional roux (a thickening agent of fat and flour in a one-to-one ratio by weight), which is high in kcalories and fat. Examples of starchy foods that work well as thickeners are beans, lentils, rice, other grains, and pureed vegetables such as potatoes and squash. These foods can be used to make soups such as black bean, lentil, split pea, Pasta e Fagioli (Recipe 9-4), and vegetable chili soup.

To make a cream soup such as cream of broccoli without using cream or roux, start by dry-sautéing broccoli, onion, fresh herbs, garlic, and shallots in chicken stock and white wine. Let reduce by half, then add potatoes and cover with vegetable or chicken stock. Once the potato is done, puree the ingredients to the proper consistency. Garnish the soup with small steamed broccoli florets and seasoned nonfat yogurt.

Rice is an excellent thickener and lends a creamy texture to soups. Rice can be used to thicken corn, carrot, and squash soups. Use about 6 ounces of rice to 1 gallon of stock. See Recipe 9-5, Butternut Squash Bisque, for an example.

CHEF'S TIPS FOR SOUPS

- Strain soups such as broccoli, celery, and asparagus soups through a large-holed china cap to remove fibers.
- Puree bean soups such as black bean and split pea to get a homogeneous product. Next, strain to remove the skins. Add any kind of vinegar or another acid such as lime juice to finish bean soups to bring out a more intense flavor that is pleasing to the palate.
- Rice and potatoes work well as thickeners in many soups.
- Replace ham in bean soups with smoked chiles or your own house-smoked meat, poultry, or veal bacon.
- Garnish soups with an ingredient of the soup whenever possible. For example, put pieces of baked tortillas on top of Mexican soup. The use of fresh vegetables, fruits salsas, or herbs as a garnish adds interest, color, and flavor. Also consider garnishing some soups, as appropriate, with a small amount (such as 1 teaspoon) of cream, roasted nuts, baked wheat croutons, low-fat sour cream, smoked chicken, or avocado cilantro salad.

SALADS AND DRESSINGS

Components of salads go way beyond simple raw vegetables. Consider using grains, beans, lentils, pasta, fresh fruits and juices, oven-dried vegetables, fresh poultry or seafood, game, herbs and spices such as ginger, Kafir lime leaves, star anise, cardamom, curry, lavender, lemon balm, and fresh cinnamon. Salads are a wonderful place to feature high-fiber, good carbs, proteins, and low-fat ingredients.

Recipes for Wild Mushroom Salad (Recipe 9-6) and Baby Mixed Greens with Shaved Fennel and Orange Sections (Recipe 9-7) appear in this chapter. Other possible salad combinations include the following:

- Baby Lentils and Roasted Vegetables
- Yellow and Red Tomato Salad, Fresh Basil, Oregano, and Sweet-Roasted Garlic
- Haricot Verts with Trio of Roasted Peppers

- Carrot, Golden Pineapple, and Dried Pear Salad with Sweet-and-Sour Dressing
- Whole-Wheat Bow-Tie Pasta with Fresh Tuna and Chives, Lemon Basil Dressing
- Orzo Pasta with Tomato, Cilantro, Scallion, and Cucumbers
- Yogurt Chicken, Celery Hearts, Fresh and Dried Fruits
- Wheat Berry Salad with Roasted Vegetables and Golden Raisins
- Organic Baby Lettuce, Marinated Cucumber and Tomato, Classic French Dressing
- Multiple bean and grain salad, roasted or grilled vegetables and fresh herbs, such as Toasted Barley, Gigante or Chestnut Beans, garnished with Haricots Verts (Figure 9-5)
- Fennel, Endive, Frisée, and Orange Salad (Figure 9-6)

FIGURE 9-5:
Salad: Barley, Beans, Tomatoes, and Fresh Herbs with Cut Vegetable and Haricot Garnish.
Photo by GigaChef.com

Dressings are used in much more than salads. They can often be used as an ingredient in entrées, appetizers, relishes, vegetables, and marinades. There are many categories of dressings. The best place to start is the basic vinaigrette, because it is simple and you can use ingredients such as herbs, spices, vegetables, and fruits to create many variations. The best ingredients to use include good-quality vinegars, first-pressed olive and nut oils, and fresh herbs, because you need the strongest flavor with the least amount of fat. Other good ingredients that add flavor without large amounts of kcalories, fats, or carbohydrates include Dijon mustard, shallots or garlic (which may be roasted for a robust flavor), a touch of honey, reduced vinegars, and lemon or lime juice.

For examples of vinaigrette recipes, see the recipes for Basic Herb Vinaigrette and Orange Vinaigrette on pages 314–315. If you look at the Basic Herb Vinaigrette recipe, you will notice that instead of a ratio of 3 parts oils to 1 part vinegar, this recipe uses 1 part oil, 1 part vinegar, and about 2 parts thickened chicken or vegetable stock. This results in a satisfactory product whose flavor profile is boosted with fresh herbs, spices, garlic, vegetables, fruits, high-quality olive oil, and vinegar.

Other salad dressings often fit into one of one of these categories.

1. Creamy dressings: Tofu can be processed to produce a creamy dressing, as in Green Goddess Dressing (page 312). Other creamy ingredients are nonfat or low-fat yogurt, nonfat sour cream, and low-fat ricotta cheese. Pureed fruits and vegetables add a creamy texture and can be used as emulsifiers in salad dressings.

2. Pureed dressings: Examples of pureed dressings include potato vinaigrette, hummus, and smoked-pepper or tomato tarragon coulis. Some of these dressings, such as hummus and green goddess, work well as dips.

FIGURE 9-6:
Salad: Fennel, Endive, Frisée, and Orange Salad.
Photo by GigaChef.com

3. Reduction dressings: Examples include Orange Vinaigrette (page 315), beet, carrot-balsamic, and apple cider dressings. These dressings can be made simply and are powerhouses of flavor. Keep in mind that their sugar content increases with reduction, and so with their intense flavors, they should be used sparingly.

CHEF'S TIPS FOR SALADS AND DRESSINGS

- As elsewhere in the kitchen, use fresh, high-quality ingredients. Choose ingredients for compatibility of flavors, textures, and colors.
- Create a well-balanced dressing that is low in fat and made with extra-virgin olive oil and a good vinegar and finished with fresh herbs and spices. Use this as one of your house dressings so that it is available to prepare a number of different choices.
- Vegetables, both raw and cooked, go well with dressings that have an acid taste, such as vinegar and lemon.
- Legumes make wonderful salads. For example, black-eyed peas go well with flageolets and red adzuki beans. To add a little more color and develop the flavor, you might add chopped tomatoes, fresh cilantro (Chinese parsley), haricots verts (green beans), and roasted peppers. The opportunities for component salad combinations are endless and will add flair to all your outlets' menus.
- Reduction dressings such as reduced beet juice can be put into a squirt bottle and used to decorate the plate for a salad, appetizer, or entrée.
- Plan your presentation carefully in terms of height, color, and composition. Keep it simple. Do not overcolor or overgarnish. When planning your menu selections, add garnishes that are appropriate for the main ingredients. For example, if you have a Southwest-style chicken sausage appetizer, garnish it with a small avocado salad or a black bean tomato relish to complement the theme.

ENTRÉES

Developing balanced entrées will draw on your entire knowledge of products, cooking techniques and methods, and nutritional requirements. There are a variety of books and online resources, plus, of course, hands-on experience to help you create delicious entrées, ranging from traditional to vegetarian to other eating styles of your guests. Examples are as follows:

- Classic Beef Stew with Horseradish Mashed Potatoes
- Grilled Pork Chop Adobo with Spicy Apple Chutney (Recipe 9-10)
- Sautéed Veal Loin with Barley Risotto and Roasted Peppers
- Pan-Seared Louisiana Spiced Breast of Chicken with Creole Tomato and Okra Sauce
- Grilled Chicken Breast and Quinoa Salad with Cucumber, Tomato, Corn, and Peppers
- Glazed Pot Roast, Paysanne Carrots, and Roasted Root Vegetables (shown buffet style in Figure 9-7)
- Grilled Chicken or Turkey Burger with Oven-Baked Chickpea Fries and Tomato Pepper Relish
- Indian Chicken Breast and Sausage, Tomato Chutney, Fava Beans, Fennel, and Cauliflower
- Cedar-Planked Wild Striped Bass with Ratatouille and Cattle Bean Salad
- Corn-Crusted Monkfish with Wheat Berries, Black Beans, Grilled Fennel, Green Beans, and Spicy Tomato Relish
- Everything Crusted Salmon, Glazed Carrots, Green Beans, Purple Pearl Onions, Lentils, Fava Beans, and Roasted Apple Mustard

FIGURE 9-7:
Entrée, buffet style: Glazed Pot Roast, Paysanne Carrots, and Roasted Root Vegetables. Photo by GigaChef.com

Up to this point, you may have noticed the lack of heavy "classical" sauces. Rather than foods covered with rich, heavy sauces, the emphasis in this book is on the taste and appearance of the food itself. In sauce making, this has meant a great change in the techniques of thickening. Followers of the new style often use purees and reductions instead of roux to make sauces, as you have seen in the descriptions of vegetable coulis and meat juices that have been thickened with pureed vegetables. Also, stock can be flavored and reduced to make a quality sauce that can be used in many dishes on the menu. This chapter features the following alternatives to traditional sauces.

- Papaya and White Raisin Chutney (Recipe 9-13)
- Papaya-Plantain Salsa (Recipe 9-14)
- Red Pepper Coulis (Recipe 9-15)
- Hot-and-Sour Sauce (Recipe 9-16)

CHEF'S TIPS FOR ENTRÉES

- For meat, poultry, and fish entrées, about a 5-ounce raw portion is adequate. See Chapter 4 for a list of lean meats, poultry, and fish.
- For marinating meats, many no-salt, no-sugar rubs and seasonings can be used. The addition of salt to your proteins can be done at time of cooking (or not at all) so that when a request for no salt is made, it is easy to accommodate it.
- Use bulgur wheat to extend ground meat. For every pound of meat, add ½ cup of cooked bulgur.
- Fish is a very versatile and nutritious food. Almost any food, such as rice, beans, or grains, goes with fish. Serve fish on top of a vegetable ragout or serve salmon with vegetable curried couscous or asparagus, fennel, and tomato.
- When choosing legumes for a dish, think color and flavor. Make sure the colors you pick will look good when the dish is complete. Also think of other ingredients you will use for flavor.
- Bigger beans, such as gigante white beans, hold their shape well and lend a hearty flavor to stews, ragouts, and salads.
- When using cheese in an entrée, use a small amount of a strong cheese such as Gorgonzola, goat, feta, reggiano, pecorino, or manchego. Also, instead of using cheese throughout the entrée, just use some on the top so that the first taste will include these wonderfully flavorful cheeses. Choose cheese varieties that are low in fat, such as skim-milk mozzarella and ricotta, for fillings and topping to be melted.
- Create new fillings for pasta that don't rely totally on cheese. For example, sweat pureed butternut squash and potato. Add fresh thyme, roasted shallots, and perhaps a little roasted duck for flavor, or puree together cooked artichokes, white beans, roasted garlic or ratatouille, oregano, and imported olives.
- Top casseroles and baked pasta dishes with reduced-fat cheese near the end of the cooking time and heat just until melted. The lower the fat content, the shorter the melting time. Too much heat and/or direct heat may toughen the cheese, so cook reduced-fat cheeses at lower temperatures and for as short a time as possible. Or shave a great hard cheese at the end (sheep or goat cheese) and serve without melting.

SIDE DISHES

There is no end to the variety of substitutions and side dishes for every entrée. The same dish can take on a new face simply by changing the starch or the vegetable. Besides the traditional side dish of vegetables and potatoes, consider using grains such as wheat berries and barley; try legumes such as black beans and lentils with the addition of seared tofu, roasted beets, dried fruits, and fresh herbs. Also consider techniques such as the following:

- Pureeing. For example, puree sweet potatoes, butternut squash, and carrots flavored with cinnamon, fresh grated nutmeg, honey, and thyme.

- Roasting. For example, roast onions with cinnamon, bay leaf, vinegar, and a touch of sugar. Recipe 9-17 shows how to prepare Roasted Summer Vegetables.
- Grilling. For example, grilled portobello mushrooms filled with polenta, garden tomatoes, and roasted elephant garlic. Grilled vegetables are a great side dish and include favorites such as peppers, eggplant, and zucchini and other varieties, such as asparagus, broccolini, and endive.
- Stir-fry. Try Hot and Sour Stir-Fry with Seared Tofu and Fresh Vegetables.

Additional examples of side dishes include these:

- Roasted Garlic and Yogurt Red Mashed Potatoes with Dill
- Couscous with Dried Fruit, Cucumber, and Mint
- Seven-Vegetable Stir-Fried Texmati Rice
- Ratatouille Strudel with Oven-Dried Tomatoes Wrapped in Phyllo
- Oven-Baked French Fries with Cajun Rub
- Swiss-Style Marinated Red Cabbage
- Pan-Seared Wild Mushroom Goulash
- Wheat Berry Risotto with Spring Onions and Pesto
- Portobello Pizza (Tomato, Basil, Wilted Spinach, and Low-Fat Mozzarella)

Grains such as rice are versatile and make excellent side dish ingredients. See Recipe 9-18, which shows how to make Mixed-Grain Pilaf.

CHEF'S TIPS FOR SIDE DISHES

FIGURE 9-8:
Slow-Cooked Pork Roast, Sautéed Cabbage Balls, and Roasted Summer Vegetables.
Photo by GigaChef.com

- When using vegetables, you need to think about what's in season for maximum flavor, how the dish will look (its colors), and how the dish will taste. Consider flavor, color, and whether the combination makes culinary sense. This is where tradition and history come into play no matter what dish you are creating. If you are preparing a roasted pork dish, it would make good sense to serve it with brussels sprouts, cabbage, stewed mushrooms, or hard squashes (see Figure 9-8). That's not to say you couldn't crust it with oregano, basil, and garlic and serve it with roasted peppers and stewed tomatoes. The idea is to think your dishes through from start to finish and know what you want for an end result.
  - Also, think variety. Serving vegetables doesn't mean switching from broccoli to cauliflower and then back to broccoli. There are many, many varieties of vegetables from which to choose. Research, experiment, and practice different vegetable cuts, cooking methods, and flavorings.
    - Blanched vegetables should be reheated in a small amount of seasoned stock and then finished with an oil such as extra virgin olive oil, flavored nut oil, or sweet butter. These delicate coatings will be the first flavor your customers will taste, giving the dish a rich body and taste.
    - Add grains to vegetable dishes, such as brown rice with stir-fried vegetables, wheat berries with roasted golden beets and dried cranberries, or lentils with grilled zucchini, summer squash, peppers, and oven-dried tomatoes.
  - Serve grains and beans. For example, rice and beans is a very popular and versatile dish that uses grains and legumes. Mix Texmati brown rice with white beans or wild rice with cranberry beans for appearance.
- Green salads can often be used as side dishes. There are so many varieties of greens on the market today, as well as a wide variety of garnishes to accent salads, that the choices are limitless. Once you have flavorful dressings to accompany your salads, they become versatile as entrées, appetizers, or side dishes. In any category, they are a great source of income.

DESSERTS

Desserts have always been perceived as the rich, fattening, "bad for you" part of the meal that can never be part of a healthy diet. Newly composed sweet endings can find a place in a diet of limited rich foods in which moderation is the focus. They are not the heavy, sickeningly sweet choices we are accustomed to, but they satisfy a person's sweet tooth without putting on pounds. Creativity and execution are the keys to your client's perception of these desserts. Another great marketing tool is to promote eating a healthy dessert more than just once per week. This can certainly increase your dessert sales and increase the value perception of your customers.

There are more than enough ways to add a limited amount of sugar to appeal to the majority of eating styles today. For example, try a ricotta cheesecake with a crust made of roasted walnuts, spices, and Splenda; a toasted oatmeal chocolate banana pudding with a kiwi or mixed fruit salsa; a buttery phyllo cylinder with maple cream pineapple chutney and berries; a banana polenta soufflé with chocolate sauce and glazed banana slices; or an old-fashioned berry shortcake with fruit sauce berries and rich whipped cream. These desserts may sound too good to be true, but they are very much in line with the guidelines for a healthy eating style.

You can also make a wide variety of desserts with fruits, either fresh cut or as a key ingredient in the baking of quick breads, cobblers, puddings, phyllo strudels, and even some cakes and cookies. The following recipes show how fruit can be used in many forms:

- Oatmeal-Crusted Peach Pie
- Apple Strudel with Caramel Sauce
- Fresh Fruit Sorbets
- Spiced Carrot Cake with Orange Custard Sauce
- Walnut–Dried Cranberry Biscuit Shortcake with Berries and Fresh Cream (Figure 9-9)
- Chocolate Torte with Raspberries, Almond Cookie, and Fresh Cream (Figure 9-10)
- Banana Polenta Soufflé with Chocolate Sauce and Glazed Bananas
- Poached Sickle Pears with Merlot Syrup and Almond Tuile
- Grilled Peppered Pineapple with "Vanilla" Ice Cream and Crunchy Cookies

FIGURE 9-9:
Dessert: Walnut–Dried Cranberry Biscuit Shortcake with Berries and Fresh Cream. Photo by GigaChef.com

FIGURE 9-10:
Desserts: Chocolate Torte with Raspberries, Almond Cookie, and Fresh Cream. Photo by GigaChef.com

- Ricotta Cheesecake with Toasted Walnut Crust and Mango Blackberry Salsa
- Phyllo Crisps with Fresh Berries, Yogurt Bavarian, and Pineapple Compote

Recipe 9-19 highlights a way to use cocoa and bittersweet chocolate to make a chocolate pudding cake that is baked in molds in a bain-marie (hot-water bath). Recipe 9-20 uses soft tofu, egg whites, raspberry puree, and sugar to make ice cream that, with the help of an ice cream machine, can be churned fresh in your kitchen. This raspberry creamed ice goes well with Recipe 9-21, Angel Food Savarin.

CHEF'S TIPS FOR DESSERTS

- To make sorbet without sugar, simply puree and strain the fruits. Make sure the fruits are at the peak of ripeness. Churn in an ice cream machine or place in a freezer and hand stir every 10 minutes until frozen.
- Use angel food cake as a base to build a dessert. Serve it with fresh fruit sorbet, warm sautéed apples and cranberries, pear and ginger compote, mango, mint, papaya salsa, or delicious fruits of the season.
- Phyllo dough is versatile in many ways. For example, stuff it with strawberries and house-made granola like a pouch or bake it molded in a muffin pan, then fill with sautéed spiced apples garnished with dried apple chips to give it some crunch.
- Compote is an additional way to serve fruit. Compote is a dish of fruit, fresh or dried, cooked in syrup flavored with spices or liqueur. When you consider how many fruits you have at your disposal, as well as spices and flavorings, the possibilities are endless.

MORNING AND AFTERNOON BREAKS

MORNING BREAKS

In many food service operations these "morning breaks" are becoming a focus of attention both in their creative design and in their nutritional content. From business and industry to hotel functions, the demand for full-day meeting packages in today's business world has come to represent a large portion of the catering revenue for these food-service operations. Clients who are in all-day workshops and training and brainstorming sessions need to nourish their minds as well as care for their bodies. There are many creative ways to accommodate their needs for balanced foods while matching your quality and standards.

The premise is to provide snacks that are satisfying and well balanced so that the participants can keep their focus on the meeting they are attending. Fresh fruit skewers with wild honey yogurt sauce, citrus suprêmes, or other fruit combinations with rose water marinade; cappuccino chocolate Jell-O with meringue puffs; protein smoothies; low-fat date nut flax seed bread; and banana walnut and blueberry corn meal loaf with pineapple-mango chutney are some of the samplings you can offer as a twist on nutrition and balance to your customers (Figure 9-11). Also, consider High "5" Muffins with tofu fruit spreads, a variety of whole fruits in season, and fresh ground peanut butter with celery sticks and apple slices. Modifying a wide range of snack ideas into a more balanced selection can make the possibilities endless. The ability of your catering department to "up sell" these choices creates a whole new revenue opportunity. You can go even further into your beverage choices with a variety of bottled waters, freshly squeezed vegetable juices, black and green teas, and flavored decaf coffees.

FIGURE 9-11:
Morning breaks: High
"5" Muffins with Tofu
Fruit Spreads, Fresh Fruit
Triangle Skewers with
Wild Honey Yogurt
Sauce, Citrus Suprêmes
with Rose Water
Marinade, Cappuccino
Chocolate Jell-O with
Meringue Puffs, House-
Made Granola Dried Fruit
Bars with Whole Fruits.
Photo by GigaChef.com

AFTERNOON BREAKS

In most foodservice operations that cater to a corporate clientele, the need for 9-to-5, all-day meetings represents a good portion of the business. About 2:30 to 3:00 P.M., there is often a well-needed break time for everyone to charge back up in order to finish the last part of the day's meeting. There is an opportunity for balance and limitations in the contents of what these people should eat. This is typically a time when the eyes start to roll back and the lids get very heavy. Pick-me-ups and light snacks are the way to go to give guests satisfaction and comfort. Choices include a cheese platter, stone-ground wafers, apple and pear chips or toasted seven-grain breads with grapes or green apple slices (skin on for extra fiber and phytochemicals), cru-dités (raw and blanched vegetables) with creative dips such as tofu green goddess, tzatziki (cucumber, yogurt, and mint), hummus (chick pea spread), baba ganoush (roasted eggplant), and a simple spicy tomato salsa (Figure 9-12). For something a little sweet, try oatmeal golden raisin cookies, dried fruits, or some chile-spiced popcorn. Bowls of whole fruits in season add the final color outline and variety to these breaks. Depending on your price point or the client's budget, you can decide on the specific items. The fact that you are creatively offering balanced, limited breaks that have panache and excitement is a great tool to increase catering revenue.

FIGURE 9-12:
Afternoon breaks:
Cheese and Grape
Platter, Assorted Fruits,
Warm Oatmeal Raisin
Cookies, Crudités with
Hummus and Baba
Ganoush, Chile-Spiced
Popcorn.
Photo by GigaChef.com

CHEF'S TIPS FOR MIDMORNING AND AFTERNOON BREAKS

- There is usually about 15 minutes for breaks, so the selections should be varied but not too complicated. The message should get across to the participants that everything is balanced and will help them finish the next two hours before lunch or before they go home without hunger pains or toothpicks to keep the eyes open. The foods are designed to help them stay focused and alert without putting them to sleep as a chocolate cookie or pastry would do.

- These morning and afternoon breaks are avenues for additional revenue. There are many ways to dress up the buffet without adding too many additional expenses. Discount stores are a great place to get interesting serving items such as wood silverware boxes to hold scones or sweet breads with fruit purees. Inexpensive metal dish racks are another way to display homemade granola bars, celery sticks, and apple slices accompanied by fresh-ground peanut butter in a flip-top glass mason jar. Baby martini glasses are a great way to display yogurt parfait or citrus sections with rosehip water. Baskets (both metal and wicker), along with a variety of tier stands for different heights, help make your display even more desirable.

- Break foods should be similar to hors d'oeuvres in size. They are just tastes of different foods to satisfy someone's hunger until lunch is served. Cut or make your items into about two bite-sized pieces. These items should be spread out decoratively on the platters or instruments for display you have chosen to create a bountiful buffet.

- Your beverages could stand some attention as well. Fruit and vegetable juices can be served, for instance, in carafes or in the manufacturer's bottle if you are not making them yourself. Place them in decorative rectangular flower pots with simple garnishes to show what the juice's main ingredient is. Another easy setup would be to fill small glasses with protein smoothies or other beverages and garnish them with whatever they are made from, such as a strawberry hanging off the glass or a pineapple wedge.

- Signage for your displays is important, especially when you are introducing such a beautiful variety and the participants might feel that it is going directly to their waistlines. A low-fat oatmeal cookie looks just as delicious as a traditionally made one. A fresh-ground peanut butter instantly tells you it has no added sugar or fillers. The same is true for whatever choices you create on your menus, and so labeling them is a good way to communicate these balanced breaks to your guest.

MINI-SUMMARY

1. The premise of catering breaks is to provide snacks that are satisfying and well-balanced so that the participants can keep their focus on the meeting they are attending.
2. This section gives chef's tips for morning and afternoon breaks.

PRESENTATION

BASIC PRINCIPLES

Here are the main considerations that underlie the art of presentation:

- Height gives a plate interest and importance. A raised surface or high point calls attention to itself. A flat and level surface is monotonous. When actual height is difficult to attain, implied height, or an illusion of height, can often be achieved by causing the eye to focus on a particular point. This can be done in several ways. One way is by arranging ingredients in a pattern that guides the eye to that point. Another is to use an eye-catching ingredient or garnish to establish a focal point (discussed in the next section).

- Color is very important. Too many colors tend to confuse the eye and dissipate the attention, so don't overdo it.
- Shape is important too. Vary the shapes on one plate. Classic vegetable cuts can play a vital role in your presentation, or simply lay out peeled cooked asparagus in a decorative half circle to change the dynamics of an entrée.
- Match the layout of the menu item with the shape of the plate. For example, salads are usually presented on round plates. This means that the lines, forms, and shapes of the salad ingredients must be arranged in a pattern that fits harmoniously into a circle. The pattern may repeat the curve of the plate's edge, echo its roundness on a smaller scale, or complement it with balance and symmetry. The flow of your food presentation should be tight; this means having food fairly close together to retain heat. The flow from left to right should curve inward to guide the eye back to the middle of the plate. The pattern begins with the rim of the plate, so never place anything on the rim; it is the frame of your design. There are many varieties of plate shapes, ranging from leaf shaped, rectangular, boat-shaped, to square. Pick the plate that highlights the dish you are presenting.
- When planning your dish, keep in mind several key points. What (if any) is the history of the food? Are they combinations that make sense and blend well on the palate? Pork naturally goes with cabbage or root vegetables. Chicken fits great with spinach, artichokes, and tomatoes. But that's not to say to you can't invent another combination; just keep in mind how the items you present taste together.
- The most effective garnish is something bright, eye-catching, contrasting in color, pleasing in shape, and simple in design. It should enhance the plate and not be the focus. At times, sauces, relishes, salsas, or chutneys may act as the garnish.

One of the tricks of presentation with balanced dishes is to make less look like more. When you are serving smaller portions of meat, poultry, or seafood, various techniques can be used to make the portions look larger. By slicing meat or poultry thin, you can fan out the slices on the plate to make an attractive arrangement and one that looks plentiful. You can also arrange a piece of meat, poultry, or seafood on a bed of grains, vegetables, and/or fruits or drape it two-thirds with sautéed vegetables. In addition, serving larger portions of side dishes with the entrée helps make the plate look full. Sauces such as vegetable coulis, salsas, and relishes also help cover the plate and provide eye appeal and color.

A common problem that crops up in plating nourishing foods is that many dishes lose heat quickly and dry out fast. High-fat sauces help keep a dish hot. When meat is sliced for presentation, it loses more juice and heat, and so it dries out quickly. To overcome this problem, chefs often place foods close together on the plate, putting the densest food in the center to keep the other foods warm. When slicing meats for plating, you can slice just part of the meat for appearance and leave the remaining piece whole for the guests to cut.

Keep all garnishes simple. Some dishes, such as angel-food cake with a fruit sauce, have a natural appeal, although you may top them with a slice or two of fresh fruit.

HOW TO MAKE GARNISHES

There are two simple methods you can use for making attractive garnishes for your menu items. The first method is for vegetable and fruit chips. You can use many different vegetables and fruits with this method, but experimenting with your oven conditions and with types of vegetables and fruits will take some work on your part. Other vegetable choices are fennel, celery root, and plantains. Some fruit choices are pears, apples, oranges, and pineapple. Keep in mind that the more water content there is in your garnish, the longer it will take to dry.

To begin, slice your garnish about $\frac{1}{16}$-inch thick on an electric slicer or on a mandoline. Place on silk mats close together without touching. You can also do this on a sheet pan seasoned with vegetable spray, but this takes more time and attention. Paint the slices with a thin coat of simple syrup made with water and honey or date sugar that has been reduced to a light syrup consistency. (Figure 9-13 demonstrates this technique using yellow and gold beets.) This gives the dish some flavor and creates a stronger chip. Place a silk mat over the top and double sheet pan the product. Dry in a 275°F oven for about an hour or until dry. The length of time will depend on your oven and the fruit or vegetable you selected.

The second method is mainly for potatoes because of their starch content. When you thinly slice a potato (a little thicker than paper thin), layer it in different shapes, and bake it using the method just described, the potato will naturally adhere together, creating a very attractive and interesting garnish. These potato garnishes have been around for a long time, only they were baked or fried with richer ingredients. You can cut the potatoes in perfect 1-inch circles and layer them around a 3-inch circle similar to the classic Pommes (potato) Anna.

Lattice style (like a fence) is another very attractive style of garnish that will create not only height but texture for your plates. To make Pommes (potato) Maxime, thinly slice potatoes. Place a fine piece of chive on top of one potato slice and cover with another potato slice. Cut out a leaf shape. Spray a silk mat with vegetable spray and then lay the leaf shapes on the mat. Next spray the top of the potatoes, then cover with another silk mat (Figure 9-14). Double sheet the pan and dry to golden brown in a 275°F oven.

MINI-SUMMARY

1. Pay special attention to height, color, shapes, unity, and garnishes for maximum plate presentation.
2. Two methods are described and illustrated for making garnishes.

CHECK-OUT QUIZ

1. Give two ways to define a healthy meal.

2. Describe two ways to develop healthy menu items.

3. What are four ways to modify a recipe?

4. List five questions you should ask yourself before adding a healthy dish to your menu.

ACTIVITIES AND APPLICATIONS

1. Recipe Modification
   Use an ingredient substitution to prepare a recipe. Compare the flavor, texture, shape, and color of both products. Have a blind taste testing to determine which one tastes better.

2. Nutrient Content of Modified Recipes
   Following is a recipe for Monte Cristo Sandwiches that has been modified to make it more nutritious. Calculate and compare the nutrient content per serving before and after modification.

Original Monte Cristo Sandwiches

Yield:	50 portions
Ham, cooked, boneless	50 1-oz. slices
Turkey, cooked, boneless	50 1-oz. slices
Swiss cheese	50 1-oz. slices
White bread	100 slices
Whole milk	3 cups
Salt	1 teaspoon
Eggs, whole, slightly beaten	1 quart (24 eggs)
Shortening, melted	2 cups

Place one slice each of ham, turkey, and cheese on one slice of bread and top with a second slice. Blend milk, salt, and egg. Dip each side of the sandwich into the egg and milk mixture; drain. Grill each sandwich on a well-greased griddle for about two minutes on each side or until it is golden brown and the cheese is melted.

Modified Monte Cristo Sandwiches

Yield:	50 portions
Plain nonfat yogurt	1¾ cups
Apricot spread	1¾ cups
Low-fat honey ham	50 1-oz. slices
Turkey, oven roasted	50 1-oz. slices
Emmenthaler cheese	50 0.5-oz. slices
8-grain bread	100 slices
Skim milk	3 cups

Egg Beaters	1 quart
Nutmeg, coriander, garlic powder, white pepper	To taste
Butter-flavored cooking spray	As needed

Mix together the yogurt and apricot spread. Spread 15 slices of bread with a tablespoon of the mixture. Place one slice each of ham, turkey, and cheese on top and cover with another bread slice. Blend the milk, Egg Beaters, and seasonings. Dip each side of the sandwich into the egg and milk mixture; drain. Grill each sandwich on a grill sprayed with butter-flavored cooking spray. Cook about two minutes on each side or until it is golden brown and the cheese is melted.

Using nutrient composition information, figure out which ingredient in the recipe for Original Monte Cristo Sandwiches contributes the most fat. Which ingredient contributes the most kcalories? Which ingredient contributes the most carbohydrates?

3. **Menu-Planning Exercise I**

Using a menu from a restaurant or foodservice, recommend two healthy entrées and two healthy desserts that would fit well on this menu. Be ready to explain why you selected these menu items and how they will fit in with the rest of the menu and the clientele.

4. **Menu-Planning Exercise II**

In a small group, plan a meal using the recipes in this chapter. The meal should consist of an appetizer or soup, a salad with dressing, an entrée, a side dish, and a dessert. Use the nutrient analysis information to determine the calories, fat, sodium, and cholesterol. Also, calculate the percentage of calories from fat. Limit fat to 35 percent or less of total kcalories, sodium to 800 milligrams, and cholesterol to 100 milligrams. Each group will write up its menu and nutrient analysis to present and compare with those of the other groups.

NUTRITION WEB EXPLORER

Carrabba's Restaurants   www.carrabbas.com

Cheesecake Factory   www.thecheesecakefactory.com

Cameron Mitchell Restaurants   www.cameronmitchell.com

Go to the website of any of these restaurants and print out its menu. Read the descriptions. Circle the menu items that appear to be directed to nutritionally conscious customers.

For two of the recipes given in this chapter, write menu descriptions that make the foods sound appealing and let guests know that they are balanced.

The Culinary Institute of America's Professional Chef Site   www.ciaprochef.com

Click on "World of Flavors." Next, click on "Worlds of Healthy Flavors Online," then click on "Profiles, Interviews, and Best Practices." Read how one of the Volume Operators, such as Chartwells or Legal Seafoods, are working healthy options into their operations. Also click on "Strategies for Chefs" and learn more on how chefs offer healthy dining.

Eating Well Magazine   www.eatingwell.com

Click on "Recipes," then "Recipe Makeovers" and see how a recipe was made healthier.

RECILES

Appetizers

RECIPE 9-1: SCALLOP AND SHRIMP ROLLS IN RICE PAPER

Category: Appetizer   Yield: 15 portions

Ingredients

1 large carrot

1 large daikon

2 large zucchini

4 tablespoons Ginger Lime Dressing (Recipe 9-8)

1 pound tofu

3 cups bean sprouts

2 whole red peppers, julienne

2 cups spinach leaves

½ cup snow peas, julienne

½ small head of white cabbage, shredded

¼ cup Ginger Lime Dressing (Recipe 9-8)

22 ounces sea scallops (30 scallops), sliced in half

15 sheets rice paper

2 tablespoons cilantro, whole leaves

2 ounces Hot-and-Sour Sauce (Recipe 9-16)

Steps

1. Make carrot and daikon spaghetti with oriental spaghetti machine.
2. Make zucchini noodles with mandoline.
3. Toss vegetable spaghetti and noodles with 2 tablespoons Ginger Lime Dressing.

4. Grill tofu and marinate with 2 tablespoons Ginger Lime Dressing. Dice and reserve for service.
5. Marinate bean sprouts, red pepper, spinach leaves, snow peas, and white cabbage in ¼ cup Ginger Lime Dressing while you do the next step.
6. Grill scallops on one side for color.
7. Soak sheets of rice paper in warm water one at a time for next step.
8. Layer scallops, vegetable noodles, and cilantro leaves on moistened rice paper and fold like an eggroll. Reserve a small amount for presentation. Steam about 3 minutes in a bamboo steamer.
9. In a soup bowl, rest the rice paper roll against the marinated vegetable noodles. Sprinkle tofu around the vegetable noodles. Pour Hot-and-Sour Sauce over mixture.

Nutritional Analysis

Kcalories	Protein (g)	Fat (g)	Carbo (g)	Sodium (mg)	Chol (mg)
107	10	3	11	237	13

RECIPE 9-2: EGGPLANT ROLLATINI WITH SPINACH AND RICOTTA

Category: Appetizer   Yield: 20 servings

Ingredients

3 pounds fresh spinach
1½ pounds eggplant, peeled and cut lengthwise
¼ cup balsamic vinegar
Olive oil spray
Vegetable oil cooking spray
4 ounces onions, finely chopped
2 teaspoons garlic, chopped
1 teaspoon cracked black pepper
1 teaspoon garlic herb seasoning (salt free)
¼ cup Italian parsley, chopped
2 tablespoons fresh basil, chopped
2 ounces feta cheese, crumbled
16 ounces skim milk ricotta
2 tablespoons grated Parmesan cheese
¾ cup whole-wheat bread crumbs
1 egg
1 egg white, slightly beaten
1 quart tomato sauce
1 teaspoon chives, chopped

Steps

1. Steam spinach. Drain well and rough chop. Reserve.
2. Paint eggplant with balsamic vinegar on both sides and spray with olive oil. Place on sheet pan sprayed with vegetable oil cooking spray.
3. Bake in 400°F oven for 10 minutes. Remove and flip over. Finish baking until tender. Reserve.
4. Spray nonstick skillet with vegetable oil cooking spray. Quickly sauté onions and garlic. Add drained spinach. Remove from heat and let cool. Add

black pepper, herb seasoning, fresh herbs, cheeses, and bread crumbs. Mix in egg and beaten egg white. Chill mixture.

5. Place two heaping tablespoons of spinach mixture on each eggplant slice. Roll up and place open end down in casserole dish. Cover with tomato sauce. Bake in 350°F oven for 30 minutes.

6. To serve, place 1 rollatini on plate. Top with tomato sauce and sprinkle with chives.

Nutritional Analysis

Kcalories	Protein (g)	Fat (g)	Carbo (g)	Sodium (mg)	Chol (mg)
121	7	3	18	154	21

RECIPE 9-3: MUSSELS STEAMED IN SAFFRON AND WHITE WINE

Category: Appetizer   Yield: 10 servings

Ingredients

Broth

1 tablespoon garlic, chopped

1 teaspoon olive oil

7 ounces white wine

1 gram saffron

1 ounce fresh lemon juice

2 ounces chicken stock

½ teaspoon black pepper, fresh ground

1 teaspoon fresh thyme, chopped

Mussels

80 fresh New Zealand mussels

Garlic Rouille

2 teaspoons roasted garlic

1 cup Yukon Gold potatoes, cooked and riced

1 tablespoon chicken stock

½ teaspoon fresh lemon juice

1 teaspoon red chili paste

1 teaspoon chives

10 slices of whole-wheat bread, cut into 2½-inch rounds

1 cup carrots, blanched, julienne

1 cup celery, blanched, julienne

2 tablespoons chives

Steps

1. For the broth, sauté garlic in olive oil. Add wine and saffron. Let reduce for 2 minutes.

2. Add lemon juice, stock, black pepper, and fresh thyme. Cook over low heat for 1 minute. Cool and store for service.

3. Scrub mussels and remove beard.

4. Make garlic rouille by pureeing garlic and mixing with potatoes, chicken stock, lemon juice, and red chili paste. Fold in chives, spread on toasted whole-wheat rounds, and toast.

5. At service, steam carrots, celery, and mussels in saffron broth. Place mussels in a large bowl. Finish with chives and broth. Garnish with warm whole-wheat crouton spread with garlic rouille.

Nutritional Analysis

Kcalories	Protein (g)	Fat (g)	Carbo (g)	Sodium (mg)	Chol (mg)
139	5	2	22	448	3

Soups

RECIPE 9-4: PASTA E FAGIOLI

Category: Soup   Yield: 16 servings (approximately 1 cup)

Ingredients

1 large yellow onion, chopped

6 cloves garlic, chopped

2 tablespoons extra-virgin olive oil

2 pounds pinto, kidney, and black beans, dried and soaked overnight

½ cup tomato paste

2 quarts vegetable or chicken stock, defatted

1½ pounds plum tomatoes, peeled, seeded

2 bay leaves

1 tablespoon red pepper flakes

1 tablespoon fresh cracked black pepper

2 tablespoons fresh oregano, chopped

2 tablespoons fresh thyme, chopped

1 tablespoon fresh rosemary, chopped

Steps

1. In a large pot, sauté chopped onion and garlic in olive oil.
2. Drain the beans and add to the pot. Add tomato paste to blend. Sauté.
3. Add 2 quarts of stock and bring to a boil. Reduce to simmer and stir occasionally.
4. Add chopped plum tomatoes, bay leaves, red pepper flakes, and black pepper.
5. Simmer for 1 to 1½ hours until beans are soft. Add herbs during last 30 minutes.
6. Remove bay leaves. Remove 1 cup of beans and puree. Return to pot. Puree more beans to achieve desired thickness.
7. Adjust flavor, if necessary, by adding more herbs or balsamic vinegar. Additions for the soup can be sautéed vegetables (shredded cabbage, diced zucchini, spinach, or escarole) added during the last 10 minutes of cooking or just before service. Option: Serve soup with a side of whole-wheat fettuccini flavored with basil and extra-virgin olive oil.

Nutritional Analysis

Kcalories	Protein (g)	Fat (g)	Carbo (g)	Sodium (mg)	Chol (mg)
242	13	3	42	281	0

RECIPE 9-5: BUTTERNUT SQUASH BISQUE

Category: Soup   Yield: 24 portions

Ingredients

1 ounce sweet butter
2 teaspoons shallots, chopped
1 cup celery, chopped
1 cup onion, chopped
1 tablespoon fresh thyme
2 teaspoons garlic, chopped
4 pounds butternut squash, cleaned and diced
6 ounces rice
2 quarts chicken stock
½ teaspoon cinnamon
½ teaspoon nutmeg
1 bay leaf
Cinnamon, as needed for garnish
Nonfat plain yogurt, as needed for garnish
Chives, as needed for garnish

Steps

1. In soup pot, melt butter. Sauté shallots, celery, onion, garlic, and thyme.
2. Add butternut squash, rice, chicken stock, cinnamon, nutmeg, and bay leaf. Cook until rice is tender.
3. Remove bay leaf. Puree in blender and strain.
4. Garnish with cinnamon, yogurt, and chives.

Nutritional Analysis

Kcalories	Protein (g)	Fat (g)	Carbo (g)	Sodium (mg)	Chol (mg)
101	2	1	15	207	3

Salads and Dressings

RECIPE 9-6: WILD MUSHROOM SALAD

Category: Salad or Appetizer   Yield: 1 portion

Ingredients

Dressing

1 tablespoon sherry vinegar
3 twists black pepper
1 teaspoon Dijon mustard
1 teaspoon white wine
½ teaspoon shallots
¼ teaspoon fresh rosemary, chopped
1 teaspoon apple juice
½ teaspoon fresh thyme, chopped

Salad

1 cup greens: red oak leaf, frisée, mustard greens, or mesclun
½ cup of mushrooms: shiitake mushrooms, oyster mushrooms, domestic mushrooms, or others in season
1 tablespoon diced peppers
1 teaspoon fresh chives, chopped

Steps

1. Incorporate all dressing ingredients together in a bowl.
2. Clean and wash greens, and arrange on plate.
3. Toss mushrooms and peppers in bowl with dressing. Sear on flat-top grill or hot sauté pan.
4. Place warm (or chilled, if preferred) mushrooms in center.
5. Top with fresh chives.

Nutritional Analysis

Kcalories	Protein (g)	Fat (g)	Carbo (g)	Sodium (mg)	Chol (mg)
29	1	1	5	34	0

RECIPE 9-7: BABY MIXED GREENS WITH SHAVED FENNEL AND ORANGE SECTIONS

Category: Salad or Appetizer   Yield: 10 servings

Ingredients

14 cups mixed greens such as baby Bibb, baby romaine, frisée, lolla rosa, tatsoi
3 heads fresh fennel
5 fresh oranges
20 ounces sherry vinaigrette

Steps

1. Prepare salad greens and reserve for service.
2. Cut fennel in half and shave paper-thin on a meat slicer. Place in ice water to crisp. Drain and reserve for service.
3. Dry some fennel slices in a low oven to use as garnish (about 30 minutes).
4. Supreme the oranges by slicing off the top and bottom, then cut off the peel following the curve of the fruit. Slice out each segment by cutting in towards the center of the fruit along the membranes/walls—remove the segment without the membrane.
5. In mixing bowl, toss fennel lightly in sherry vinaigrette.
6. Toss mixed greens with dressing. Use only 2 ounces of dressing per person.
7. Coat orange sections lightly with dressing.
8. Arrange orange sections on outside of plate. Make a tower of lettuce and fennel in the center of the plate. Top with oven-dried fennel slices. Serve immediately.

Nutritional Analysis

Kcalories	Protein (g)	Fat (g)	Carbo (g)	Sodium (mg)	Chol (mg)
63	3	0.5	13	35	0

RECIPE 9-8: GINGER LIME DRESSING

Category: Dressing   Yield: Eight 2-ounce servings

Ingredients

1 cup low-sodium soy sauce
½ cup lime juice
Grated rind from 2 limes
¼ cup scallions, finely sliced
½ cup water
1 teaspoon garlic, chopped
1 teaspoon ginger, chopped

Steps

1. Mix all ingredients together in a bowl. Let sit overnight. Lasts 2 weeks in refrigerator.

Nutritional Analysis

Kcalories	Protein (g)	Fat (g)	Carbo (g)	Sodium (mg)	Chol (mg)
22	2	0	5	800	0

RECIPE 9-9: CAPISTRANO SPICE RUB

Category: Rub   Yield: 5 Cups

Ingredients

1 cup dried oregano
1 cup dried basil
½ cup dried thyme
1 cup fresh rosemary, chopped
½ cup black pepper, butcher's grind
½ cup garlic powder

Steps

1. Blend all ingredients together. Store in an airtight container to preserve freshness.
2. Coat meat, fish, or poultry generously with spice rub before cooking. Can also be used to spice up soups, dressings, and vegetable dishes.

Entrées

RECIPE 9-10: GRILLED PORK CHOP ADOBO WITH SPICY APPLE CHUTNEY

Category: Entrée   Yield: 20 portions

Ingredients

Adobo Spice Rub

3 fresh green chiles (poblano)
2 fresh jalapeño peppers
10 garlic cloves

3 tablespoons fresh oregano

2 tablespoons ground cumin

2 tablespoons freshly ground black pepper

1 teaspoon ground cinnamon

½ pound tomatillos, husks removed

1 cup red wine vinegar

Pork

5 pounds pork loin, center cut, boneless

Chutney

1 chipotle pepper

1 onion, diced

½ tablespoon garlic, chopped

8 apples, cored and diced

½ cup raisins

1 ounce lemon juice

Zest from 1 lemon

1 tablespoon ground cardamom

3 tablespoons sugar

1 cup cider vinegar

½ tablespoon ground fennel seed

¼ teaspoon mace

½ bunch fresh cilantro, chopped

Steps

1. Combine all the adobo spice rub ingredients in a food processor and blend until a paste forms.
2. Generously rub the pork loin with the paste and let marinate overnight in refrigerator.
3. Rehydrate the chipotle pepper in hot water until softened. Cut in half, remove seeds, and chop.
4. In a sauté pot, combine all chutney ingredients except cilantro.
5. Let simmer for 20 minutes until slightly thickened. Finish with cilantro. Cool and serve at room temperature. (The chutney can be made in advance and refrigerated.)
6. Sear pork loin until nicely caramelized. Set up smoking station with wood chips soaked in water. Smoke with medium smoke for about 15 minutes.
7. Remove and finish in a slow-roasting oven at 300°F until pork temperature is 155°F.
8. Let rest 15 minutes before slicing. Serve with apple chutney.

Nutritional Analysis

Kcalories	Protein (g)	Fat (g)	Carbo (g)	Sodium (mg)	Chol (mg)
144	17	3	12	34	48

RECIPE 9-11: SLATES OF SALMON

Category: Entrée   Yield: 1 serving

Ingredients

4 ounces salmon steak, cut very thin on diagonal

1 teaspoon olive oil

Pinch black pepper

1 cup arugula

½ cup endive

1 tablespoon plum tomatoes, seeded and diced

1 tablespoon cucumbers, peeled, seeded, and diced

2 ounces Green Goddess Dressing (page 312)

Steps

1. Paint the salmon with olive oil and black pepper. Grill to desired temperature.
2. Toss arugula and endive with Green Goddess Dressing and place in middle of plate. Place salmon against greens and sprinkle with tomato and cucumbers.

Nutritional Analysis

Kcalories	Protein (g)	Fat (g)	Carbo (g)	Sodium (mg)	Chol (mg)
187	24	9	3	89	59

RECIPE 9-12: BRAISED LAMB

Category: Entrée   Yield: 12 portions

Ingredients

12 lamb hind shanks

1 tablespoon olive oil

4 tablespoons vegetable seasoning

5 cloves garlic, sliced

2 onions, diced

4 carrots, sliced

6 celery stalks, diced

3 bay leaves

2 whole thyme sprigs

2 whole rosemary sprigs

8 ounces red wine

8 ounces fresh tomatoes

½ cup tomato paste

1 gallon lamb stock, defatted

3 tablespoons Worcestershire sauce

1 tablespoon cracked black pepper

36 large shiitake mushroom caps

24 ounces cannellini beans, cooked

3 tablespoons garlic herb seasoning (salt free)

Steps

1. Trim all fat from lamb shanks.
2. Place olive oil into large heated braising pan. Season shanks with vegetable seasoning. Sear in the large pot until brown. Remove shanks.
3. Quickly sauté garlic, onions, carrots, celery, bay leaves, thyme, and rosemary in same pot. Deglaze with red wine, fresh tomatoes, and tomato paste and reduce.
4. Return shanks to pot with defatted lamb stock, Worcestershire sauce, and black pepper. Bring to boil, reduce heat, and simmer gently for 90 minutes. Remove shanks.

Strain and put sauce through a food mill. Reduce to proper consistency and skim. Pour over shanks. Cool.

5. For each order, cook 3 shiitake caps with 2 ounces of cannellini beans in 3 ounces of lamb stock and ½ teaspoon of garlic herb seasoning. Cook lightly to reduce. Heat each shank in sauce slowly.
6. Place 1 piece of shank on plate with a variety of vegetable choices and sauce. Top with shiitakes and beans. This dish can be served with a variety of beans and grains.

Nutritional Analysis

Kcalories	Protein (g)	Fat (g)	Carbo (g)	Sodium (mg)	Chol (mg)
391	40	10	33	471	107

Relishes, Salsas, Coulis, Chutneys, and Sauces

RECIPE 9-13: PAPAYA AND WHITE RAISIN CHUTNEY

Category: Relishes, Salsas, Coulis, and Chutneys   Yield: 16 servings

Ingredients

6 pounds very ripe papaya, peeled and diced
2 cups onion, diced
1 tablespoon garlic, chopped
½ cup brown sugar
½ cup white sugar
1 cup seedless raisins
1 cup white-wine vinegar
1 teaspoon cardamom
1 teaspoon cinnamon
2 bay leaves
2 teaspoons fresh thyme, chopped

Steps

1. In a small saucepot, mix all ingredients except thyme.
2. Reduce to a thick paste. Add thyme while chutney is still hot.
3. Let cool; remove bay leaves. Can be stored up to 2 weeks.

Nutritional Analysis

Kcalories	Protein (g)	Fat (g)	Carbo (g)	Sodium (mg)	Chol (mg)
126	1	0	31	8	0

RECIPE 9-14: PAPAYA-PLANTAIN SALSA

Category: Relishes, Salsas, Coulis, and Chutneys   Yield: 10 servings

Ingredients

1 plantain, ripe, finely diced
1 teaspoon extra-virgin olive oil

1 papaya, peeled and finely diced

½ peppers, red and orange, finely diced

½ red onion, finely diced

2 teaspoons cilantro, chopped

2 teaspoons chives, finely sliced

1 cup white-wine vinegar

2 teaspoons honey

Lime juice from 2 limes

5 teaspoons extra virgin olive oil

Steps

1. In a nonstick sauté pan, toast the diced plantain in 1 teaspoon olive oil until crisp outside and tender inside.
2. In a stainless-steel mixing bowl, place half the papaya, the peppers, and red onion. Add the toasted plantain, cilantro, and chives. Reserve.
3. In a food processor, place the remaining papaya, vinegar, honey, and lime juice. Puree until smooth, adding 5 teaspoons olive oil.
4. Add more vinegar if too thick.
5. Strain through a fine sieve into the reserved plantain-papaya mixture.
6. Reserve in refrigerator for use. Lasts about 5 days.

Nutritional Analysis

Kcalories	Protein (g)	Fat (g)	Carbo (g)	Sodium (mg)	Chol (mg)
75	0.6	3	11	4	0

RECIPE 9-15: RED PEPPER COULIS

Category: Relishes, Salsas, Coulis, and Chutneys   Yield: 36 ounces, or eighteen 2-ounce servings

Ingredients

4 pounds red peppers (or substitute other vegetables)

1 ounce minced shallots

1 tablespoon minced garlic

½ teaspoon minced jalapeño pepper

2 ounces olive oil

2 ounces tomato paste

2 each 90-count potatoes, peeled and diced

18 ounces chicken or vegetable stock

2 tablespoons fresh basil, chopped

2 teaspoons fresh thyme, chopped

2 teaspoons fresh oregano, chopped

1½ ounces balsamic vinegar

Steps

1. Cut peppers in half and remove seeds. Place on oiled sheet pans and roast in hot oven or grill the peppers. Weigh 3¼ pounds of peeled, grilled red peppers. Reserve.
2. Sauté shallots, garlic, and jalapeño pepper in oil.
3. Add tomato paste and sauté. Do not brown.
4. Add potatoes and stock. Simmer until potatoes are almost cooked.

5. Add roasted red peppers and finish cooking potatoes.
6. Add basil, thyme, and oregano. Cook 3 more minutes.
7. Take off stove, then cool for 10 minutes. Puree in a blender.
8. Finish with vinegar and strain through large-hole china cap.

Nutritional Analysis

Kcalories	Protein (g)	Fat (g)	Carbo (g)	Sodium (mg)	Chol (mg)
67	2	3	8	26	0

RECIPE 9-16: HOT-AND-SOUR SAUCE

Category: Sauce   Yield: Sixteen 2-ounce portions

Ingredients

24 ounces chicken stock, defatted
4 ounces white wine
3 ounces rice-wine vinegar
1½ tablespoons cornstarch
1 tablespoon lime juice, fresh squeezed
⅓ teaspoon lime rind
2 ounces scallions, chopped
2 teaspoons ginger, chopped
½ teaspoon jalapeño pepper
¾ teaspoon cumin
1 teaspoon cilantro
2 ounces sesame oil

Steps

1. Heat chicken stock and wine. Prepare cornstarch slurry with vinegar and add to thicken the sauce.
2. Remove from heat and cool. Add remaining ingredients except oil. Whip in oil at the end. Refrigerate for later service.

Nutritional Analysis

Kcalories	Protein (g)	Fat (g)	Carbo (g)	Sodium (mg)	Chol (mg)
54	1	4	3	101	5

Side Dishes

RECIPE 9-17: ROASTED SUMMER VEGETABLES

Category: Side Dish   Yield: 10 servings

Ingredients

2 pounds carrots, peeled and cut oblique
1 pound red bell peppers
1 tablespoon olive oil
2 tablespoons shallots, chopped
4 cloves garlic, chopped

1 pound corn, from husk, cleaned

1½ pounds assorted radishes, sliced ¾-inch thick

½ pound pattypan squash

½ tablespoon thyme

2 bay leaves

Black pepper to taste

4 ounces chicken stock

1 tablespoon basil leaves, chopped

½ tablespoon rub of your choice

Steps

1. Blanch carrots for 2 minutes.
2. Cut peppers in half and remove seeds. Place on oiled sheet pans and roast in hot oven or grill the peppers. Remove skins.
3. Mix together all ingredients except basil and spice rub. Place in hot roasting pan in a 350°F oven. Let roast, stirring occasionally, for 30 minutes. Add more stock if needed to prevent burning. If vegetables are getting too brown, cover with foil.
4. Add basil and spice rub as needed at end for more flavoring.

Nutritional Analysis

Kcalories	Protein (g)	Fat (g)	Carbo (g)	Sodium (mg)	Chol (mg)
123	3	2	26	52	0

RECIPE 9-18: MIXED-GRAIN PILAF

Category: Side Dish   Yield: 10 servings

Ingredients

1 cup wheat berries

2 thyme stems

1 bay leaf

1 cup quinoa

1 cup hot water

1 bay leaf

½ ounce chicken stock, defatted

1 ounce onions, diced

3 ounces brown rice

1 pint chicken stock, defatted

2 cups water

1 bay leaf

1 head roasted garlic

2 teaspoons olive oil

2 ounces onions, chopped

2 tablespoons chopped basil, oregano, and chives

Chicken stock, as needed

Steps

1. In a sauce pot of boiling water, add wheat berries with thyme stems and bay leaf and cook for about 1 hour. Drain and cool.

2. Soak quinoa in cold water to take out any impurities. Steam quinoa in hot water with bay leaf. Cover and let steam for about 20 minutes.

3. Heat $\frac{1}{2}$ ounce of chicken stock in a stockpot. Sweat 1 ounce diced onions until translucent.

4. Add rice, 1 pint stock, thyme, and bay leaf. Simmer and cover pilaf-style in a 350°F oven for about 20 minutes or until done.

5. Puree roasted garlic.

6. Heat oil in stockpot. Sweat 2 ounces chopped onions until translucent. Add pureed garlic, wheat berries, quinoa, brown rice, and herbs.

7. Let simmer for 5 minutes. Remove bay leaf. Adjust consistency with stock if needed.

Nutritional Analysis

Kcalories	Protein (g)	Fat (g)	Carbo (g)	Sodium (mg)	Chol (mg)
211	7	3	39	31	0

Desserts

RECIPE 9-19: WARM CHOCOLATE PUDDING CAKE WITH ALMOND COOKIE AND RASPBERRY SAUCE

Category: Dessert   Yield: 10 portions

Ingredients

1 tablespoon orange zest

3 cups skim milk

$3\frac{1}{2}$ ounces sugar

1 ounce cocoa powder

3 ounces cornmeal

10 egg whites

$3\frac{3}{4}$ ounces sugar

3 ounces bittersweet chocolate

1 pint raspberries

1 ounce Kirschwasser

4 ounces white wine

$2\frac{3}{4}$ ounces honey

$1\frac{1}{2}$ ounces sugar

1 ounce almond paste

$\frac{1}{2}$ ounce bread flour

$\frac{3}{4}$ teaspoon cinnamon

1 egg white

Pinch salt

2 teaspoons cream

2 teaspoons skim milk

Confectioners' sugar, about 1 tablespoon, for dusting cookies

Steps

1. Steep orange zest in milk and bring to a boil. Simmer.
2. Combine $3\frac{3}{4}$ ounces sugar, the cocoa, and cornmeal. Pour in steady stream into simmering milk. Stir until thick and cornmeal is cooked. Allow to cool.
3. Whip egg whites and $3\frac{3}{4}$ ounces sugar. Fold into base, then fold in bittersweet chocolate.
4. Pour mixture into sugared molds and bake in bain-marie about 20 to 24 minutes at 400°F. Keep warm.
5. Puree $\frac{3}{4}$ pint raspberries (reserve others for garnish) in food processor with the Kirschwasser, wine, and honey. Strain and reserve.
6. Cream $1\frac{1}{2}$ ounces sugar, the almond paste, flour, and cinnamon. Add remaining ingredients, except reserved raspberries, reserved raspberry sauce, and confectioners' sugar, and allow to rest. Spread paste in ten 3-inch circles on silicon. Bake in 350°F oven until edges are brown. Curve the circles on a rolling pin while still hot to make a decorative tuile garnish.
7. Cool and dust with confectioners' sugar.
8. Pool sauce on plate. Unmold cake on sauce. Place cookie on top of cake and garnish with raspberries and sauce.

Nutritional Analysis

Kcalories	Protein (g)	Fat (g)	Carbo (g)	Sodium (mg)	Chol (mg)
235	9	2	45	106	3

RECIPE 9-20: RASPBERRY CREAMED ICE

Category: Dessert　　　　　　　　　　Yield: 1 quart or four 1-cup servings

Ingredients

1 pound soft tofu
1 pint raspberry puree (or substitute any other fruit)
6 ounces egg whites
1 ounce turbinado sugar

Steps

1. Cream tofu in processor until smooth.
2. Add raspberry puree slowly to achieve creamy texture.
3. Whip egg whites until they form soft peaks. Add sugar until the meringue forms peaks.
4. Churn in an ice cream machine until the mixture reaches the desired consistency.
5. Freeze in an airtight container.

Nutritional Analysis

Kcalories	Protein (g)	Fat (g)	Carbo (g)	Sodium (mg)	Chol (mg)
148	12	5	17	79	0

RECIPE 9-21: ANGEL FOOD SAVARIN

Category: Dessert　　　　　　　　　　Yield: Eight 4-inch savarin rings

Ingredients

2 ounces cake flour
6 ounces confectioners' sugar

6 ounces egg whites 1 teaspoon cream of tartar
½ teaspoon lemon rind

Steps

1. Sift flour with 3 ounces of sugar. Sift again and set aside.
2. Whip egg whites and cream of tartar to soft peaks, then gradually add remaining 3 ounces sugar. Whip until stiff and glossy.
3. Gently fold in sifted ingredients and lemon rind.
4. Spray nonstick savarin molds with cold water. Pipe mixture into molds.
5. Bake in 350°F oven until light golden brown on top.
6. Cool completely, then remove from mold.

Nutritional Analysis

Kcalories	Protein (g)	Fat (g)	Carbo (g)	Sodium (mg)	Chol (mg)
119	3	0	27	35	0

RECIPE 9-22: FRESH BERRY PHYLLO CONES

Category: Dessert   Yield: 8 portions

Ingredients

Phyllo

Vegetable oil cooking spray
8 phyllo sheets
8 teaspoons melted butter

Bavarian Mix

7 ounces maple syrup
9 ounces skim-milk ricotta cheese
13 ounces nonfat plain yogurt
1 teaspoon vanilla extract
2 teaspoons gelatin
4 teaspoons water

Fruit and Garnish

4 cups fresh berries (strawberries, blueberries, raspberries)
Powdered sugar, for dusting
8 each mint leaves

Steps

1. Make 2½-inch circle cones out of aluminum foil by wrapping foil around a small juice or condiment jar about 1½ inches in diameter.)
2. Spray foil lightly with vegetable-oil spray.
3. Cut phyllo into 2½-inch widths. Paint with melted butter between layers. Use 1 sheet per cone.
4. Wrap phyllo around foil and take the cone off the can.
5. Place cone on baking sheet. Repeat process 8 times. These can be made 1 day in advance.
6. Bake phyllo cones at 375°F for about 10 minutes. Let cool. Take off foil and reserve it for future use.

7. Mix maple syrup, ricotta cheese, yogurt, and vanilla in blender. Whip until smooth.
8. Soften gelatin in water. Warm to dissolve.
9. Add a little of the dessert base to the dissolved gelatin to temper the mixture. Mix the rest of the dessert base into tempered mixture. Fold in 2 cups of fresh berries. Reserve other fruit for garnish.
10. Place a cone on a dessert dish. Scoop 3 ounces of the Bavarian mixture into the cone.
11. Top with fresh berries. Dust with powdered sugar. Garnish with mint leaf.

Nutritional Analysis

Kcalories	Protein (g)	Fat (g)	Carbo (g)	Sodium (mg)	Chol (mg)
262	9	8	39	212	22

RECIPE 9-23: STUFFED FRENCH TOAST LAYERED WITH LIGHT CREAM CHEESE AND BANANAS

Category: Breakfast   Yield: 4 servings

Ingredients

8 slices whole-wheat bread
2 ounces light cream cheese
4 small bananas

Batter

1½ cups egg substitute
¾ cup skim milk
1 teaspoon vanilla
½ teaspoon cinnamon

Syrup

1 cup fresh strawberries, cleaned and cut
¼ cup strawberry all-fruit jam
1 tablespoon lemon juice

Steps

1. Lay out bread and spread all 8 slices evenly with cream cheese.
2. Slice bananas paper-thin and layer on 4 slices of bread, overlapping slightly, and top each slice with other side of bread; press down lightly. Cut in half diagonally.
3. Whip together batter ingredients to a smooth consistency.
4. Heat a nonstick pan and spray with vegetable oil spray.
5. Dip the banana sandwiches carefully in batter on both sides to absorb batter. Place in nonstick pan and brown nicely on both sides, about 1½ to 2 minutes per side. Do not let the pan get too hot or the toast will brown without cooking in the middle. Transfer to an oven-safe dish.
6. Warm slightly in the oven to crisp before serving.
7. While the French toast is in the oven, blenderize strawberries, jam, and lemon juice until smooth. Serve as syrup on the side.

Nutritional Analysis

Kcalories	Protein (g)	Fat (g)	Carbo (g)	Sodium (mg)	Chol (mg)
526	23	11	86	552	10

RECIPE 9-24: GLAZED GRAPEFRUIT AND ORANGE SLICES WITH MAPLE VANILLA SAUCE

Category: Breakfast   Yield: 4 portions

Ingredients

2 grapefruit, peeled and sliced

3 oranges, peeled and sliced

1 ounce maple syrup

2 ounces apple juice

1 ounce water

1 tablespoon granola

4 sprigs mint

Steps

1. Alternate orange slices and grapefruit slices on individual plates like spokes on a wheel until you have 4 portions.

2. Heat maple syrup, apple juice, and water. Reduce glaze by half. Pour lightly over fruits. Top with granola and very thin strips of mint.

Nutritional Analysis

Kcalories	Protein (g)	Fat (g)	Carbo (g)	Sodium (mg)	Chol (mg)
124	2	1	30	2	0

RECIPE 9-25: WHOLE-WHEAT PEACH CHIMICHANGAS

Category: Breakfast   Yield: 6 portions

Ingredients

4 peaches (fresh), diced

2 tablespoons honey

1 pound nonfat cottage cheese

1 teaspoon cinnamon

1/2 teaspoon nutmeg

8 whole-wheat flour tortillas

Steps

1. In a preheated nonstick pan, quickly sauté the peaches. Remove from heat and add honey. Toss to coat peaches and return to heat and toss for about 2 minutes. Remove from heat and let cool.

2. Place cottage cheese and spices in a bowl and mix together. Fold in coated peach mixture and hold until ready to fill tortillas.

3. Warm tortillas on a griddle for a minute, just to make flexible. Cover and reserve for filling process.
4. Spoon about half cup of mixture into center of tortilla. Spread evenly. Fold side a little to keep mixture from oozing out.
5. In a heated pan with vegetable spray, lightly brown and finish in moderate oven (325°F).

Nutritional Analysis

Kcalories	Protein (g)	Fat (g)	Carbo (g)	Sodium (mg)	Chol (mg)
250	18	1.2	43	240	5

Marketing to Health-Conscious Guests

According to the National Restaurant Association, Americans now spend 48 percent of their food dollars away from home, compared to 25 percent in 1955. Nowadays, restaurants offer much more to health-conscious customers than the old-fashioned diet plate consisting of cottage cheese with fruit on top of a lettuce leaf. Menus carry items that range from vegetarian entrees to full-course meals that do it all for under 600 kcalories. When Americans dine out, the nation's over 945,000 restaurants provide an atmosphere in which customers have the opportunity, flexibility, and freedom to choose among a variety of high-quality, healthy, and enjoyable menu items.

Marketing means finding out what your customers need and want and then developing, promoting, and selling the products and services they desire. Keeping in touch with your customers is crucial and can be done without much fuss to make sure you are offering the healthy selections they are looking for. This chapter will help you to:

- Describe two methods a foodservice operator can use to gauge customers' needs and wants
- Give three examples of ways to draw attention to healthy menu options
- Discuss effective ways to communicate and promote healthy menu options
- Explain the importance and extent of staff training needed to implement balanced menu options
- Describe two ways to evaluate healthy menu options
- Respond with menu ideas for special requests from guests
- Discuss how nutrition labeling laws regulate nutrient content or health claims on restaurant menus

> **MARKETING**
>
> The process of finding out what your customers need and want and then developing, promoting, and selling the products and services they desire.

GAUGING CUSTOMERS' NEEDS AND WANTS

Most foodservice operators who have implemented healthy menu options successfully have done so by reviewing eating trends, examining what other operators are doing, and keeping abreast of their customers' requests for healthy foods. To determine customer wants, foodservice operators could interview the waitstaff about customer requests, for example, for light foods such as broiled meat, poultry, or fish; dishes prepared without salt; sauces and gravies removed or served on the side; butter substitutes; reduced-calorie salad dressing; and skim or low-fat milk.

Another way to gauge customer needs is to do a survey, as shown in Figure 10-1. At the same time, answers to the following questions need to be considered.

1. What are the majority of requests made during a particular meal?
2. Which items are most frequently requested?
3. How much time does the cooking staff and waitstaff have available to meet those special requests?
4. Which requests are easy to meet? Which are very time-consuming?

Answers to these questions can help you decide which types of healthy menu items to offer.

If market research demonstrates a sizable need for healthy entrées and there is enough time and staff to commit to this project, this may be the time to do more than meet customers' special requests.

1. How often do you visit this restaurant?
   First visit
   Once or twice a year
   Once every three months
   Once every two months
   Once a month
   Two or three times a month
   Once a week
   More than once a week
2. Today I came for:
   Breakfast
   Lunch
   Dinner
   Snack
3. Are you here during your workday?
   Yes   No
4. Are you here for social reasons?
   Yes   No
5. Have you ever been to a restaurant that offers light and nutritious menu choices?
   Yes   No
6. Would you order light and nutritious foods if they were offered here?
   Yes, frequently
   Yes, sometimes
   No
   Not sure
7. How likely would you be to try the following nutritious menu choices?

Menu Choice	Very Likely	Likely	Unlikely
A. Broiled fish without butter			
B. Reduced-calorie salad dressing			
C. Vegetables with no added salt			

MINI-SUMMARY

Customer interest in balanced menu items can be gauged through waitstaff feedback and customer surveys and/or feedback.

ADDING HEALTHY MENU OPTIONS TO THE MENU

Various personnel are normally involved in the development and implementation phase of new menu items: foodservice operators, directors, and managers; chefs and cooking staff; and nutrition experts such as registered dietitians. Chefs and the cooking staff are valuable resources in modifying recipes or creating new ones and may be given much of this responsibility. Nutrition experts are needed to provide accurate nutrient analysis data as well as suggestions for modifying dishes. In larger companies, personnel responsible for training, advertising and publicity, marketing, menu planning, and recipe development may also be involved.

Chapters 8 and 9 covered the basics of balanced cooking, developing healthy menu items, and modifying recipes. Once you know what you want to offer, you need to think about how to inform your clientele of these options. Here are some suggestions.

1. Simply give a good description of a menu item so that health-conscious customers can see that the item is healthy (Figures 10-2 and 10-3).

FIGURE 10-2: Lunch Menu from Ocean Prime

OCEAN PRIME
FISH · STEAKS · COCKTAILS

Raw Bar
Chef's Selection of East & West Coast Oysters
Shellfish Sampler
Colossal Shrimp Cocktail
Alaskan Red King Crab Legs, served with horseradish cocktail sauce

Soups
French Onion, Brioche Crouton & Aged Swiss
She Crab Bisque & Crab Fritter

Appetizers
Truffled Deviled Eggs
Sweet Chili Point Judith Calamari
Aged Wisconsin Cheddar Fondue
Ahi Tuna, Avocado, Ginger Ponzu
Prime Beef Carpaccio, Creamy Horseradish
Crispy Fried Crab Cake, Yellow Corn Cream

Salads
Crisp Wedge of Iceberg, Red Onion, Smoked Bacon, Grape Tomatoes,
    Bleu Cheese, Cabernet Buttermilk Dressing
Hearts of Romaine Knife & Fork Caesar, Parmesan Garlic
    Dressing & Sourdough Crostini
Ocean Prime House Salad, Romaine, Spinach, Granny Smith
    Apples, Goat Cheese, Walnuts, Sherry Mustard Vinaigrette
Chop Chop Salad, Hard Cooked Egg, Salami,
    Fresh Mozzarella, Smoked Bacon, Club Dressing
Beefsteak Tomato Salad, Arugula, Shaved Onion,
    Crumbled Bleu Cheese Cabernet Buttermilk Dressing

Sandwiches
    All sandwiches served with choice of fries, cup of she crab bisque or house salad
Chicken Club w/toasted brioche, Swiss cheese, smoked bacon
Ocean Prime Steak Burger w/Maytag Blue cheese, caramelized bacon
Maryland Crab Melt w/Tillamook cheddar cheese, jalapeno corn tartar
Soy Glazed Tuna w/pickled cucumber, wasabi mayonnaise

(continued)

Signature Salads

Chicken Chopped Salad w/roasted chicken, asparagus,
  goat cheese, dates, corn, sherry vinaigrette

Crab Wedge w/Alaskan king crab meat, grape tomatoes,
  Red onion, Maytag Blue cheese, crab Louie dressing

Blackened Salmon w/strawberry, cantaloupe, walnuts
  poppy seed dressing

Black & Bleu Caesar w/flat iron steak, Bleu cheese dressing

Prime Fish

Pecan Crusted Mountain Trout w/Skillet Beans,
  Potato Puree, Brown Butter

Colossal Shrimp Sautee w/Angel Hair Pasta, Tabasco Cream Sauce

Ginger Salmon w/Snap Peas, Sticky Rice, Soy Butter Sauce

Blackened Red Snapper, Wilted Spinach, Jalapeno Au gratin, Corn Tartar

Chilean Sea Bass w/Glazed Carrots, Mashed Potato, Champagne
  Truffle Sauce

Chef's Feature

Shell Fish "Cobb" Salad w/Shrimp, Lobster, Crab,
Bacon Gourmet Dressing

Prime Entrees

Park Farms Chicken w/Asparagus, truffle mac & cheese, Lemon Pan Jus

Diver Scallops w/Slow Braised Short Ribs, Wilted Spinach

Flat Iron Steak & Fries w/Forest Mushroom Bordelaise

Bacon Wrapped Bleu Cheese Filet w/Mashed potato, Veal Reduction

NY Strip w/Asparagus, twice baked potato, Cabernet Jus

*Ask your server about menu items that are cooked to order or served raw.

Consuming raw or undercooked meats, poultry, seafood, or shellfish may increase your risk of foodborne illness.

www.cameronmitchell.com

Courtesy: Cameron Mitchell Restaurants.

FIGURE 10-3: Dinner Menu from Ocean Prime

OCEAN PRIME
FISH · STEAKS · COCKTAILS

Raw Bar

Chef's Selection of East & West Coast Oysters

Shellfish Sampler

Colossal Shrimp Cocktail

Alaskan Red King Crab Legs, served with horseradish cocktail sauce

Soups

French Onion, Brioche Crouton & Aged Swiss

She Crab Bisque & Crab Fritter

(continued)

FIGURE 10-3: Dinner Menu from Ocean Prime (*Continued*)

OCEAN PRIME
FISH · STEAKS · COCKTAILS

Appetizers

Truffled Deviled Eggs

Sonoma Goat Cheese Ravioli, Golden Oak Mushrooms

Sweet Chili Point Judith Calamari

Aged Wisconsin Cheddar Fondue

Ahi Tuna, Avocado, Ginger Ponzu

Prime Beef Carpaccio, Creamy Horseradish

Crispy Fried Crab Cake, Yellow Corn Cream

"Surf n Turf," Sea Scallops, Boneless Short Ribs

Classic Oysters Rockefeller

Signature Appetizer

Colossal Shrimp Saute, Tabasco Cream Sauce

Salads

Crisp Wedge of Iceberg, Red Onion, Smoked Bacon, Grape Tomatoes,
    Bleu Cheese, Cabernet Buttermilk Dressing

Hearts of Romaine Knife & Fork Caesar, Parmesan Garlic
    Dressing & Sourdough Crostini

Ocean Prime House Salad, Romaine, Spinach, Granny Smith
    Apples, Goat Cheese, Walnuts, Sherry Mustard Vinaigrette

Chop Chop Salad, Hard Cooked Egg, Salami,
    Fresh Mozzarella, Smoked Bacon, Club Dressing

Beefsteak Tomato Salad, Arugula, Shaved Onion,
    Crumbled Bleu Cheese, Cabernet Buttermilk Dressing

Chef's Specialties

Pecan Crusted Mountain Trout, Skillet Beans, Brown Butter

Ginger Salmon, Stir-Fried Snap Peas, Soy Butter Sauce

Park Farms Amish Chicken, Asparagus, Lemon Pan Jus

Salt n Pepper Tuna, Wild Mushrooms, Green Peppercorn Sauce

Diver Sea Scallops, Green Beans, Whole Grain Mustard Cream

Jumbo Lump Crab Cakes, Asparagus, Sweet Corn Cream

Alaskan Halibut, Crab Succotash, Americana Sauce

Chilean Sea Bass, Glazed Carrots, Champagne Truffle Sauce

Prime Feature

Blackened Snapper, Wilted Spinach, Jalapeno Corn Tartar

Prime Fish	Prime Steaks
All prime cuts are prepared with house made seasoning and broiled at 1200° degrees	
Mahi Mahi	7 oz Petite Filet Mignon
Scallops	10 oz Filet Mignon
Red Snapper	12 oz Bone-In Filet
King Salmon	14 oz New York Strip

(continued)

Chilean Sea Bass
Alaskan Halibut
Australian Lobster Tail
Alaskan King Crab Legs

16 oz Kansas City Strip
18 oz Ribeye
22 oz Porterhouse
12 oz Rack of Lamb

Extras

Béarnaise Sauce

Green Peppercorn & Cognac Sauce

Black Truffle Butter

Maytag Blue Cheese Crust

"Oscar" Style w/Jumbo Lump
   Crabmeat & Béarnaise Sauce

Potatoes

Home Style Potato Latkes w/Apple Sauce & Sour Cream

Creamy Whipped Potatoes

Sea Salt Vinegar Fries

Maytag Blue Cheese Whipped

Jalapeno Au Gratin

Sea Salt Baked Potato, Butter & Sour Cream

Scallion Twice Baked with Cheddar Cheese & Bacon

Candied Yams with Marshmallow Brulee

Supper Club Sides

Creamed Spinach

Jumbo Asparagus with Hollandaise

French Bean Amandine

Sauteed Wild Mushrooms

Chophouse Corn

Steamed Broccoli

Crispy Onion Straws with
   Creamy Horseradish

Sesame Stir-Fried Snap Peas

Glazed Carrots, Brown Sugar Butter

Black Truffle Macaroni & Cheese

*Ask your server about menu items that are cooked to order or served raw.
Consuming raw or undercooked meats, poultry, seafood,
or shellfish may increase your risk of foodborne illness.
Courtesy: Cameron Mitchell Restaurants.

2. Use the waitstaff to offer and describe healthy menu options. In some instances, healthier preparation methods can be suggested for regular menu items. Figures 10-4 and 10-5 show menus samples from popular quality restaurant chains that have a wide variety of selections that can be mixed and matched or served with substitutes to accommodate the needs of customers we have previously outlined. This exercise is to better prepare you for what is the demand out in the industry by clients everywhere.

FIGURE 10-4: How to Modify Existing Menu Items to Respond to Special Requests (The Cheesecake Factory)

APPETIZERS

Roadside Sliders

Bite-sized Burgers on Mini-Buns Served with Grilled Onions, Pickles and Ketchup

This can be an appetizer served with some fresh vegetables, salad greens, and one of the many dips, depending on the needs of the customer. For less fat, you can make it from ground white meat turkey or a combination of ground turkey and lean ground beef. The grilled onions are fine as long as they are lightly coated with a well-balanced vinaigrette.

Fire-Roasted Fresh Artichoke

Fresh Artichoke Fire-Roasted and Topped with a Spicy Vinaigrette. Served with Garlic Dip (Seasonal)

This is a good choice, depending on what is in the garlic dip. A reduced-fat mayo, white beans, or yogurt would be a nice base to complete this dressing.

Vietnamese Shrimp Summer Rolls

Delicate Rice Paper Rolled Around Asparagus, Shiitake Mushrooms, Carrots, Rice Noodles, Green Onion, Cilantro and Shrimp. Served Chilled and not Fried

This is a good appetizer; just be sure there are not too many rice noodles compared to the vegetables. This can also be a choice without the noodles. The rice paper is so thin that it adds a little texture without the high carbohydrate count.

Thai Lettuce Wraps

Create Your Own Thai Lettuce Rolls! Satay Chicken Strips, Carrots, Bean Sprouts, Coconut Curry Noodles and Lettuce Leaves with Three Delicious Spicy Thai Sauces: Peanut, Sweet Red Chili, and Tamarind-Cashew

This is a smart selection for menus, with something for everyone. Because it is rolled to order, you can add or remove ingredients as desired. Keep in mind that people will not mind paying a little extra if they are changing the menu selection, so long as they are adding ingredients like protein or artichoke that they know cost more than a carrot.

Mixed Baby Lettuce Salad

Assorted Field Greens and Baby Lettuces Tossed in a Balsamic Vinaigrette

French Country Salad

Mixed Greens, Grilled Asparagus, Fresh Beets, Goat Cheese, and Candied Pecans

These are the old favorites, a salad as an appetizer. Offer a light dressing that is standardized for low sugar. Choose plain pecans rather than candied.

(continued)

Caesar Salad

The almost traditional recipe, with Croutons, Parmesan Cheese, and our Special Caesar Dressing. Add Chicken

Luau Salad

Fresh Slices of Grilled Chicken Breast Layered with Mixed Greens, Cucumbers, Green Onions, Red and Yellow Peppers, Green Beans, Mango, and Crisp Wontons with Macadamia Nuts and Sesame Seeds, Tossed in Our Vinaigrette

Santa Fe Salad

Lime-Marinated Chicken, Fresh Corn, Black Beans, Cheese, Tortilla Strips, Tomato, and Mixed Lettuces with a Spicy Peanut-Cilantro Vinaigrette

These salads have great ingredients that can satisfy many eating styles for today's customers. Pack in lots of vegetables and put the dressing on the side.

Symphony Salad

A Wonderful Combination of Grilled Chicken, Fresh Tomatoes, Asparagus, Red and Yellow Peppers, Green Beans, and Butter Lettuce Tossed with Parmesan Cheese and Vinaigrette

Herb-Crusted Salmon Salad

Our Wonderful Fresh Herb-Crusted Salmon Served Cold on Top of Baby Lettuces, Vegetables, and Tomato, Tossed in Our Balsamic Vinaigrette

These are great choices as is, as long as the dressings are balanced and the portion size is appropriate.

ENTREES

Veggie Burger

A Blend of Vegetables, Brown Rice, Oats, Black Beans, Garlic, Onion, and Herbs, Topped with Fontina Cheese and Served with Fries

Grilled Portobello on a Bun

A Giant Portobello Mushroom Grilled with Herbs and Served on a Bun with Lettuce, Tomato, Grilled Red Onion, Melted Cheese, and Spicy Mayonnaise, Served with Fries

These are choices that will take little or no work to change. Keep in mind that a kitchen line is like a mad scientist's workshop. As long as all the ingredients are there, changing a plate is fairly simple. Take the fries off and add either stir-fried vegetables or a salad with lettuce, tomato, cucumbers, peppers, and mushrooms. Watch the portion size.

Asian Vegetable Stir-Fry

Topped with Almonds, Served over Rice or Pan-Fried Noodles. Add Chicken

Spicy Cashew Chicken

A Very Spicy Mandarin-Style Dish with Green Onions and Roasted Cashews, Served over Rice

Chicken Brochettes

Skewers of Herb-Basted Chicken, Onions, and Mushrooms on Top of Steamed Rice, Served with a Sauce of Fresh Tomato, Red Pepper, Shallots, Olive Oil, Balsamic, Kalamata Olives, and a Little Goat Cheese

(continued)

FIGURE 10-4: How to Modify Existing Menu Items to Respond to Special Requests (The Cheesecake Factory) (*Continued*)

Good choices for entrée selections. Try some brown rice, perhaps add more stir-fried vegetables. Limit the portion size for low salt. You may need to modify more.

Grilled Pork Chops

Marinated Center-Cut Chops Served with Housemade Apple Sauce, Mashed Potatoes, and Fresh Spinach

Honey-Maple Pork Tenderloin

Slowly Roasted and Glazed, Served with Mashed Potatoes and Fresh Vegetables

Either of these pork cuts are pretty lean if trimmed properly. The glaze on the pork shouldn't be a problem for most unless a strict low-sugar diet is required. (It would be a good idea to have some plain chicken on the menu anyway.) The only other changes would be offering baked potatoes instead of mashed potatoes and extra vegetables and spinach. The applesauce could be a very good item depending on how much sugar was added. In terms of sodium, these items can be cooked without or with minimal salt except for the mashed potatoes.

SIDE DISHES

Fresh Broccoli
Fresh Carrots
Green Beans
Sautéed Spinach
Fresh Asparagus
Sautéed Snow Peas and Vegetables

The best way to heat these would be with a clear chicken or vegetable stock, fresh herbs, and pepper, finished with olive oil and a tiny amount of butter.

FIGURE 10-5: How to Modify Existing Menu Items to Respond to Special Requests (Cameron's American Bistro)

Hot-and-Sour Crab Cake

Watermelon, Radish and Jicama Slaw, Sweet Chili Butter Sauce, and Wasabi Pepper Sauce

Baked Flatbread

Prosciutto, Roasted Potato, Caramelized Onion, Provolone and Truffle Mascarpone

Bloody Mary Shrimp Cocktail

Chilled Jumbo Shrimp and Spicy Cocktail Sauce

The easiest way and the fastest way to satisfy your guest and to keep harmony in your kitchens is to anticipate that on any given night you will need to accommodate a low-fat, low-carbohydrate, vegan, or other special request. A shrimp cocktail appetizer is a protein usually served with some greens or salad and a tomato-based cocktail sauce. It accommodates most diets except low-sodium and vegan. The baked flatbread is fine as an appetizer, but it would be better if the flatbread was made with whole grains. Crab cakes can be a special treat if they are not fried. The jicama radish salad is a perfect accompaniment with some salad greens.

(*continued*)

House Salad

Mixed Greens, Roma Tomatoes, Sweet Onions, Carrots, Croutons, Gorgonzola, and Mustard Seed Balsamic Vinaigrette

Caesar Salad

As it should be, Romaine, Garlic Croutons, Shaved Parmesan, and Lemon Caesar Dressing

Field Greens

Caramelized Apples, Red Onion, Gorgonzola, Spiced Walnuts, and Maple Brandy Vinaigrette

Warm Goat Cheese Salad

Field Greens, Pine Nut-Crusted Goat Cheese, Pancetta, Sun-Dried Cherries, and Port Wine Vinaigrette

Salads are often a great choice. Just be sure your dressings are balanced and the condiments, such as nuts, cheeses, and dried fruits, are moderate in amount. Dressings on the side are usually preferable for guests. If you are pan-frying the goat cheese in little or no oil, this is fine. If not, a plain slice for your guest is just as satisfying.

Low Country Shrimp and Grits

Creole Gulf Shrimp, Stone-Ground Goat Cheese Grits, and Southern-Cooked Broccolini

Soy-Glazed Salmon

Nori-Wrapped Wasabi Mashed Potatoes, Stir-Fried Vegetables, Lemongrass Butter, and Hoisin Sauce

Day Boat Scallops

Pan-Seared, Champagne Beurre Blanc, Tomato Jam, Truffle Potato Chips, and Garlic Mashed Potatoes

Bistro Plate

Crispy Eggplant, Roasted Red Pepper, Oven Dried Tomatoes, Artichoke Hearts, Whipped Ricotta, Goat Cheese, and Cilantro Pesto Drizzle

When planning a dinner menu for your clients, you want to have choices available while keeping your menus creative, exciting, and focused on your style. The basis of a well-crafted dish starts with the main flavoring, the combination and compatibility of ingredients, and the portion size of the protein or main components. The above dishes have a variety of vegetables that can get mixed and matched if necessary for your clients. Sauces can be served on the side. As always, portion size and the use of moderate amounts of appropriate oils are important.

Prosciutto-Wrapped Roasted Chicken

Mushroom and Fontina Stuffed, Yukon Gold Potato Tower, Roasted Red Pepper Relish, and Asparagus

Pine Nut-Crusted Lamb Chops

Balsamic Jus, Goat Cheese Whipped Potatoes, Asparagus, and Plum Tomato Chutney

Grilled Filet Mignon

Garlic Mashed Potatoes, Asparagus, Gorgonzola, Port Wine Sauce, Portabella French Fry, and Horseradish Sauce

Roasted Indiana Duck

Honey Chili Glaze, Horseradish-Cabbage Whipped Potatoes, and Roasted Vegetables

When special eating requests come into the kitchen, it will become increasingly easier to rearrange an appetizer, salad, entrée, or dessert once you grasp the fundamentals of limited versus rich ingredients. Some clients want to splurge on rich foods that are a treat occasionally, some clients don't care at all. But increasingly many are very aware of rich foods high in saturated fats and refined sugar; glazes with sugar; whipped potatoes with butter, cheeses, and cream; proteins with rich stuffing, pan-fried in oils and heavily breaded, will have to be modified for these special requests. Be prepared for what alternatives you can offer without the addition of chaos in an already busy environment.

3. Highlight healthy menu selections with symbols or words such as "fit." For example, put a picture of wheat next to nutrition selections that meet specific nutrition goals, which usually are described at the bottom of the menu.

4. Include a special, separate section on the regular menu. With this format, customers are certain to see the healthy options, and see them as being integrated into the foodservice concept. A heading for this section might be "Fit Fare."

5. Add a clip-on to the regular menu and/or a blackboard or whiteboard. This method requires no alterations to the menu and is particularly useful in that it is flexible and inexpensive. Healthy selections can be changed without spending much time or money. In some operations, treating the nutrition selections like daily specials has increased their selling power.

When designing your menus, use words that are familiar yet cutting-edge and exciting. Terms such as "creamy," "crisp," "spiced," "glazed," "stuffed," and "caramelized" create an alive, exciting menu script.

Another opportunity for increasing sales is to add healthy selections to your à la carte room service and restaurant menus with terms such as "jump start," "great start," and "new continental." These selections can be competitively priced while capturing sales that may not have been there without these choices.

No matter which method you use to include healthy selections on the menu, you must consider how thorough a description is appropriate. In general, customers do not want kcalorie counts and fat, cholesterol, or sodium content on the menu but prefer a good description of the ingredients, portion size, and preparation method. Menu items are more effectively promoted by giving customers this information and by emphasizing quality and variety rather than nutrition.

Marketing healthy menu items can be done in a positive manner. When you tout foods as "heart-healthy," you may be approaching some customers in a negative manner. When you market freshly squeezed fruit juices, you are approaching customers in a positive manner.

> **ADVERTISING**
> Any paid form (such as over radio) of calling public attention to the goods, services, or ideas of a company or sponsor.

 ## PROMOTION

Three methods of promoting healthy options are advertising, sales promotion, and publicity. *Advertising* can be done through magazines and newspapers, radio and television, outdoor displays (posters and signs), indoor table tents and posters, direct mail, and novelties (such as matchboxes). Direct mail works well when it is targeted to current customers.

Advertising messages should say something desirable, beneficial, distinctive, and believable about the balanced dining program. For example, the new menu selections could be advertised as healthy and using only the freshest, most exotic ingredients. Because foodservice operators need to get the best advertising for the money, hiring a reputable advertising company may be the best option.

> **SALES PROMOTION**
> Marketing activities other than advertising and public relations that offer an extra incentive.
>
> **PUBLICITY**
> Obtaining free space or time in various media to get public notice of a program, book, and so on.

Sales promotions can include coupons, point-of-purchase displays (such as a blackboard at the dining room entrance listing the nutrition selections), and contests (such as having customers guess the number of kcalories in a balanced dining entrée to win a free meal).

Publicity involves obtaining free editorial space or time in various media. Many foodservice operators do their own publicity. However, if you wish to obtain the advice of outside publicity consultants, O'Dwyer's Directory includes most public-relations firms. Here are some ideas for publicizing your nutrition program.

1. Send a **_press release_** about your balanced dining options to the appropriate contact person by name, not title, as indicated in the following list:
   a. Television and radio news: Assignment editor or specialty reporters appropriate to your story, such as health and food editors
   b. Television and radio talk shows: Producer
   c. Newspapers: Section editors (food, health and science, lifestyle) or city desk editor for special events
   d. Magazines and trade publications: Managing editor, articles editor, or specialty editors appropriate to your story
   e. Local publications and newsletters: Corporate employee or customer newsletters or supermarket, utility company, bank, school, or church publications
2. Follow up each press release with a phone call. Editors are always looking for article ideas and just may pick up on your story.
3. Offer to write a column on healthy meal preparation for a local newspaper.
4. Offer cooking demonstrations or on-site classes or volunteer to conduct classes for health associations, retail stores, or supermarkets.
5. Invite local media and community leaders for the opening day of your new program and let them taste some healthy menu selections.
6. Contact the foodservice director of a medical center or the public relations director of a health maintenance organization and offer to cosponsor a health or nutrition event such as a bike race or health fair. Check for local health and sporting events in which you can participate.
7. Contact your local American Heart Association and ask if it has a dining-out guide in which you may feature your restaurant.
8. Develop a newsletter for your operation and use it to publicize the new program (include some of your recipes). Newsletters help build loyal customers.

> **PRESS RELEASE**
> A printed announcement by a company about its activities, written in the form of a news article and given to the media to generate publicity.

There are many sources for promotional materials, such as table tents, posters, buttons, menu clip-ons, point-of-sale materials, and artwork (Figure 10-6). Food manufacturers, foodservice distributors, and food marketing boards and associations are excellent sources of promotional materials.

FIGURE 10-6:
The National Peanut Board's nutrition and wellness messaging appeals to and motivates the health conscious consumer. Courtesy The National Peanut Board, www.nationalpeanutboard.org.

The good news is that a handful of peanuts may be good for your heart. There is no bad news.

A friendly reminder from America's Peanut Farmers™
www.nationalpeanutboard.org
Scientific evidence suggests but does not prove that eating 1.5 oz. per day of most nuts, including peanuts, as part of a diet low in saturated fat and cholesterol may reduce the risk of heart disease.

▟ STAFF TRAINING

Staff training centers on the waitstaff and the cooking staff. Before training begins, involve the waitstaff as much as possible in the development of your nutrition program so that they feel a part of it and take some ownership. They can be a valuable resource in designing the program, because they make daily contact with the customers both in selling and serving and in listening to requests, compliments, and complaints. During training, the waitstaff needs to understand:

- The scope and rationale for the nutrition program
- Grand-opening details
- The ingredients, preparation, and service for each menu item
- Some basic food and nutrition concepts so that they can help guests with special dietary concerns, such as low-sodium or gluten-free
- How to handle special customer requests, such as orders for half portions
- Merchandising and promotional details

Figure 10-7 gives specific learning objectives for the service staff.

FIGURE 10-7: Learning Objectives for Service Staff

1. Servers must be able to respond to consumer health concerns by providing menu suggestions that meet their dietary needs. They should be able to make menu suggestions for the following dietary restrictions:
   Low calorie
   Low sodium
   Low cholesterol and low fat
   Low sodium, low cholesterol, and low fat
   High fiber

2. The waitstaff should be able to describe healthful dining options in straightforward, appropriate language to patrons. Servers must be able to provide information on ingredients, methods of preparation, portion sizes, and how the menu items are served.

   Ingredients: The waitstaff should be knowledgeable about details regarding ingredient usage: addition of fat or the type of fat used in cooking, the use of salt or high-sodium seasonings, cuts of meats used, the type of liquids used to prepare menu items, fats and thickening agents used in sauces, and the use of sugar or sugar substitutes.

   Cooking methods: Patrons commonly need to know not only what the composition of a menu item is but how it was prepared. Was the food fried in vegetable oil or animal shortening? Was the food prepared by pan frying, broiling, baking, poaching, or sautéing? Can the item be broiled without added butter? Is fat removed from meat juices or stocks before using them for sauces or soups?

   Presentation and portion: Patrons frequently want to know how the item will be served when making their menu selections. The waitstaff should be prepared to answer the following questions: What is the portion size of the item? What accompaniments are served with the item? Are special food items available to accompany the item? For example, is light syrup available for the light pancakes? Is a fruit spread available instead of jam for the whole-wheat breads? Can toast be served dry instead of buttered? Can salad dressing be served on the side?

   (continued)

---

3. The waitstaff should be able to explain the nutritional basis for menu items designated as light or healthful in terms of caloric, fat, cholesterol, and/or sodium content. They may need to answer questions about the program rationale. They should know the nutritional guidelines (U.S. Dietary Guidelines, American Heart Association guidelines, and/or National Cancer Institute recommendations) that provide the basis for the program.

4. Servers should be able to respond to patron inquiries about the availability of special foods or beverages. Does the restaurant serve brewed decaffeinated coffee? Are diet salad dressings available for the light salad entrées? Is margarine available instead of butter? Are herb seasonings available instead of salt for adjusting seasonings at the table? Does the restaurant serve skim milk?

5. The waitstaff should be able to respond to questions concerning substitutions of meal accompaniments. Can an entrée be served with two vegetables instead of a vegetable and a starch? Can a salad be substituted for the starch or vegetable served with the entrée? Can a fruit appetizer be served for dessert?

6. The waitstaff need to know what special requests the foodservice operation can accommodate. For example, can margarine or vegetable oil be used instead of butter in preparing foods? Can entrées be broiled instead of fried? Are smaller portion sizes available? Can sauces and salad dressing be served on the side?

7. The waitstaff should be able to recommend other foods and beverages that complement menu choices, including appetizers, soups, salads, desserts, and beverages that are light or meet the dietary restrictions of the patron.

8. The waitstaff should be knowledgeable about what is served with light menu items and what the correct portion sizes are for these items. Servers will act as the final quality-control agents prior to the serving of the foods. If light menu items are similar to traditional offerings, the waitstaff should be able to distinguish between the two items.

9. Staff members should respond politely and accurately to guest questions about healthful dining options. It should be emphasized that the proper response to a patron's inquiry is "I can find out for you," not "I don't know." When uncertain of the answer, the staff member should ask the kitchen manager, manager, or chef.

Source: Ganem, Beth Carlson. © 1990. *Nutritional Menu Concepts for the Hospitality Industry.* New York: Van Nostrand Reinhold. Reprinted with permission.

A poorly trained waitstaff will confuse the customer and, quite frankly, doom the program instead of knowledgeably promoting it. Conversely, a properly trained waitstaff can function as excellent sales agents and solicit feedback, including customer recommendations.

The cooking staff also needs training. Their training needs center on:

- The scope and rationale for the nutrition program
- Grand-opening details
- The ingredients, preparation, portion size, and plating of each new menu item
- Some basic food and nutrition concepts so that they can help guests with special dietary concerns
- How to respond to special dietary requests

Figure 10-8 outlines seven ways in which employees learn best. Figure 10-9 is a sample quiz that you can use to assess the basic nutrition knowledge of your staff.

FIGURE 10-8: How Employees Learn Best

1. When employees participate in their own training, they tend to identify with and retain the concepts being taught. To get employees involved, choose appropriate training methods.

2. Employees learn best when training material is practical, relevant, useful, and geared to an appropriate level. Learning is facilitated, too, when the material is well organized and presented in small, easy-to-grasp steps. Adult learners are selective about what they will spend time learning, and learning must be especially pertinent and rewarding for them. Adults also need to be able to master new skills at their own pace.

3. Employees learn best in an informal, quiet, and comfortable setting. Your effort in selecting and maintaining an appropriate training environment shows employees that you think their training is important. When employees are stuffed into a crowded office or a noisy part of the kitchen, or when the trainer is interrupted by phone calls, they may rightly feel that their training isn't really important. Employees like to feel special so find a quiet setting whenever possible. Of course, much training, such as on-the-job training, necessarily takes place in the work environment.

4. Employees learn best when they are being paid for time spent in training.

5. Employees learn best with a good trainer. Although you may not ever find a person with all these qualities, you can use this list to evaluate potential trainers:

   A Successful Trainer
   - Is knowledgeable
   - Displays enthusiasm
   - Has a sense of humor
   - Communicates clearly, concisely, straightforwardly
   - Is sincere, caring, respectful, responsive to employees
   - Encourages employee performance; is patient
   - Sets an appropriate role model
   - Is well organized
   - Maintains control, frequent eye contact with employees
   - Listens well
   - Is friendly and outgoing
   - Keeps calm; is easygoing
   - Tries to involve all employees
   - Facilitates the learning process
   - Positively reinforces employees

6. Employees learn best when they receive awards or incentives. For example, when an employee has completed training for the position of cook, you can send a letter of recognition to the employee, which can also be put into the personnel file. The largest franchisee of Arby's awards employees a progression of bronze, silver, and gold name tags, as well as pay increases, as they learn each area of the restaurant. When the employee has learned all areas, he or she is promoted to the position of crew leader.

7. Employees learn best when they are coached on their performance on the job.

Source: Drummond, Karen. 1992. *Retaining Your Foodservice Employees.*
© 1992 by John Wiley & Sons, Inc. Adapted by permission of John Wiley & Sons, Inc.

FIGURE 10-9: Questions to Quiz the Crew

1. The bread, cereal, rice, and pasta group is a good source of _____.
   a. carbohydrate
   b. fat
   c. protein
   d. calcium

2. Which of the following foods contains the most fiber?
   a. 1 cup cooked white rice
   b. 1 slice white bread
   c. 1 slice whole-wheat bread
   d. 1 cup cooked lentils

3. Which of the following foods contain no added sugar?
   a. regular soda
   b. cupcakes
   c. apple
   d. muffin

4. Which of the following foods is highest in fat and saturated fat?
   a. banana
   b. carrots
   c. T-bone steak
   d. skinless chicken breast

5. Margarine is usually made from
   a. animal fat
   b. vegetable oil
   c. vegetable shortening
   d. lard

6. Cholesterol is in
   a. peanut butter and jelly
   b. milk and cheese
   c. apples and oranges
   d. carrots and broccoli

7. Monounsaturated fats are heart-healthy and include
   a. canola oil
   b. olive oil
   c. peanut oil
   d. all of the above

8. Which is the leanest (contains the least amount of fat)?
   a. chicken breast, no skin
   b. chicken wings
   c. sirloin steak
   d. chicken pot pie

9. Which nutrient builds and maintains the body?
   a. carbohydrate
   b. fat
   c. protein
   d. water

10. A deficiency of which nutrient will cause the most problems?
    a. carbohydrate
    b. fat
    c. protein
    d. water

11. Which nutrient provides the most kcalories per gram?
    a. carbohydrate
    b. fat
    c. protein
    d. water

12. How many kcalories are in 1 tablespoon of oil?
    a. 60
    b. 90
    c. 100
    d. 120

13. Identify four menu items that are low or moderate in sodium.

14. Identify four menu items that are low or moderate in fat, saturated fat, and cholesterol.

15. What desserts can you recommend for someone who is avoiding added sugar?

The cooking staff needs to understand the prime importance of using only the freshest ingredients, using standardized recipes, measuring and weighing accurately, and using attractive presentation.

Training the cooking staff to prepare healthy dishes correctly can be challenging. As managers have found during nutrient-content analysis, cooks do not always prepare recipes exactly as called for. Perhaps a key ingredient was unavailable, time was tight, or the cook forgot a step. Healthy menu items often are more labor-intensive, and more training and coaching are needed. In any case, the cooking staff needs training not only in making new menu items but also in understanding the importance of following recipes and serving the correct portion size.

Studies of healthy menu items have shown that although the items had less fat, fewer kcalories, and more vegetables and fruit than other menu items, sometimes the entrée that was actually served wound up containing a lot more fat than was noted in the nutrition analysis for that recipe. For example, a fajita from the menu of a Mexican restaurant was supposed to have only 17 grams of fat but actually had 30 grams because the appropriate lean meat was not used. Problems such as this point to the importance of training and retraining cooking staff.

MINI-SUMMARY

1. There are a variety of ways of using the menu to inform your clientele of healthy menu items, including simply using good descriptive language about how the item is prepared, what ingredients are used, and the portion size.
2. Healthy menu items should be promoted in a positive manner.
3. Three methods of promoting healthy options are advertising, sales promotion, and publicity (using press releases and other methods).
4. There are many sources for promotional materials from food manufacturers, foodservice distributors, and food marketing boards and associations.
5. Staff training is crucial and centers on the waitstaff and the cooking staff. Training the cooking staff to prepare healthy dishes correctly can be challenging and is very important for success.

PROGRAM EVALUATION

The healthy menu program should be evaluated much like any other program. Key questions for evaluation include:

1. How did the program do operationally? Did the cooks prepare and plate foods correctly? Did the waitstaff promote the program and answer questions well?
2. Did the food look good and taste good?
3. How well did each of the healthy menu options sell? How much did each item contribute to profits? How did the overall program affect profitability?
4. Did the program increase customer satisfaction? What was the overall feedback from customers? Did the program create repeat customers?

Proper program evaluation requires much time observing and talking with staff and customers, as well as going over written records, such as sales records.

Once a program has been evaluated, certain changes to fine-tune the program may be necessary. Here are some suggestions:

- Develop ongoing promotions to maintain customer interest.
- Add, modify, or delete certain menu items.
- Change pricing.
- Improve the appearance of healthy items.
- Listen to customers more to get future menu and merchandising ideas.

MINI-SUMMARY

Evaluation is needed to determine program effectiveness from customer, employee, and management viewpoints.

RESPONDING TO SPECIAL GUEST REQUESTS

Here are some examples of what your customers are asking for:

"What menu items are low in fat?"
"What selections are vegetarian?"
"What do you have that is low in sodium?"

These types of requests are quite common, but there will be other requests such as gluten-free foods. For example, a chef on a cruise ship had a woman who was gluten intolerant so she couldn't have any wheat flour. The menu for lunch once day was pan pizzas, and the server brought one to her and she said to her server, "I can't have that." And they said yes you can because the pastry chef made it with rice flour. The woman started crying. She hasn't had pizza in 30 years and she just started crying. She was just so touched that one person would go out of their way to make something special for her.

So on and on! The days of foodservice establishments serving plain steamed vegetables as a healthy alternative are long gone. Unless you have a customer with a very restricted diet, you will have to be much more creative to accommodate these special requests. Whole grains, carbohydrates with fiber, fresh vegetables, fresh fruits, lean proteins, and the use of healthy oils are not just a fad; they are a way of life for many people seeking a healthy lifestyle that we know can prevent diseases such as heart disease and cancer. With the increasing number of people eating out, the need for balance in foodservice establishments is essential to survival.

To respond to special guest requests, keep in mind these basic preparations:

- When marinating meats there are many no-salt, no-sugar rubs and seasonings that can be used. The addition of salt to your proteins can be done at time of cooking (or not at all), so a request for no salt is easy to accommodate.
- Blanched vegetables should be reheated in a small amount of seasoned stock, then finished with whole butter, an extra-virgin olive oil, or flavored nut oil. These delicate oils will be the first flavor your customers will taste, giving a rich body and the taste fulfillment up front with fewer fats than before.
- Dips and chips are and always will be an American favorite for appetizers. This can still be a great selection to your menu with a new twist. Hummus, baba ghanoush, white bean and roasted garlic, artichoke, and goat cheese are well-accepted favorites dips that can be accompanied by baked whole-wheat tortilla chips, melba toast, baked multigrain croutons, or a variety of different vegetables.
- Create a well-balanced dressing that is low in fat and made with extra-virgin olive oil and good vinegars as we have suggested, and finished with fresh herbs and spices. Use this as one of your house dressings so it is available to prepare a number of different choices.
- Keep a stock or clear broth for reheating vegetables, because many of the new eating habits today recommend starting your meal with a cup of clear broth. This helps to curb

the appetite. This can be a kitchen staple as well as quick-serve dual purpose inexpensive starter on your menu, increasing sales for an appetizer that may not have been purchased otherwise.

- Desserts, desserts, desserts. Creating balanced desserts that your guests can enjoy should not cause stress. There are more than enough ways to add limited sugar to a menu item that can appeal to the majority of eating styles today. A ricotta cheesecake with a roasted walnut, spices, and Splenda crust; a flourless chocolate cake with fresh fruit garnish; a toasted oatmeal, chocolate, banana, pecan pudding with a kiwi or mixed fruit salsa; a buttery phyllo cylinder with maple cream pineapple chutney and berries in season; a banana polenta soufflé with chocolate sauce and glazed banana slices; or an old fashion berry shortcake with fruit sauce berries and rich whip cream. Sounds too good to be true, but very much within the guidelines for a balanced eating style.

As operator/chef, you have incorporated a variety of choices into your menu so the ability to mix and match for special request can be easily remedied. Vegetarian, low fat, and other requests are becoming fairly common. Part of your daily regime as a chef is to be prepared for these special needs without creating kitchen havoc. They must have the same skill, taste, and preparation and presentation of your original menu items so your reputation goes untarnished and your customers are happy. The best approach when designing your menus is to have choices that follow the basic dietary guidelines discussed in this book. With more people eating out regularly because of their demanding schedules and the increase in business travelers, the need for menu selections to be better balanced and limited in rich ingredients is greater than ever. Most diners are not out for that once-in-a-blue-moon special occasion.

The following discussion will coach you into looking at menus in a different light so that you can accommodate any special request from your diners. With the changing habits of your customers, there are some basic menu items that you should consider having so that you can more easily accommodate special requests.

DIET LOW IN FAT, SATURATED FAT, AND CHOLESTEROL

This trio of nutrients appears most often together in the following animal foods: beef, poultry with the skin on, eggs, whole milk, regular cheese, ice cream, fats in baked goods, and butter. The trick here is to serve lean beef, poultry without the skin (it's okay to cook it with the skin on), fish, reduced-fat cheeses, frozen yogurt, and monounsaturated fats. Figure 10-10 lists many appropriate menu choices.

LOW-SODIUM (LOW-SALT) DIET

The major source of sodium in the diet is salt, a compound made of sodium and chloride. Salt by weight is 39 percent sodium, and 1 teaspoon contains 2300 milligrams (a little more than 2 grams) of sodium. Many processed foods have high amounts of sodium added during processing and manufacturing, and it is estimated that these foods provide fully 75 percent of the sodium in most people's diets. The following is a list of processed foods high in sodium:

- Canned, cured, and/or smoked meats and fish, such as bacon, salt pork, sausage, scrapple, ham, bologna, corned beef, frankfurters, luncheon meats, canned tuna fish and salmon, and smoked salmon
- Many cheeses, especially processed cheeses such as processed American cheese

FIGURE 10-10: Menu Choices for Selected Nutrition Goals

	Increase Fiber	Decrease Sugar	Decrease Fat, Saturated Fat, and Cholesterol	Decrease Sodium
Breakfast	Unpeeled fresh fruit	Fresh fruit and juices	Any fruit and juices	Any fruit and juices except canned tomato vegetable juice (unless low-sodium)
	Whole-grain muffins, toast, bagels, rolls, bran muffins	Cereals with less than 4 grams sugar/serving such as corn flakes	Any cold or hot cereals	Cereals with less than 300 mg, sodium/serving such as puffed rice, shredded wheat
		Muffins, toast, bagels, hard rolls	Muffins, toast, bagels, hard rolls	Muffins, toast, bagels, hard rolls
	Whole-grain pancakes, French toast, waffles	Fruit spreads	Egg substitutes	
		Top pancakes with fresh fruit sauces/yogurt	Pancakes, French toast, waffles made with oil, skim milk, and egg sub-; use vegetable oil spray to cook	Pancakes, French toast, waffles made without added salt
			Low-fat or fat-free yogurt/cottage cheese	
	Granola topping on low fat or nonfat yogurt			
Appetizer and soup	Salads using vegetables, fruits, and whole grains	Most appetizers and soups are not high in sugar	Raw vegetables and fruits with yogurt-based dips	Raw vegetables and fruits with homemade dip (no salt or dry soup mix)
			Salads	
			Juice	Juices except canned tomato and vegetable (unless low-sodium)
				Unsalted crackers
			Cheeses low in fat	
			Seafood (except fried)	Seafood
			Broth and vegetable-based soups	Soups made with small amount of salt and no commercial bases
	Soups made with dried beans and fresh vegetables		Cream soup with fat-free or low-fat milk	

(continued)

FIGURE 10-10: Menu Choices for Selected Nutrition Goals (*Continued*)

	Increase Fiber	Decrease Sugar	Decrease Fat, Saturated Fat, and Cholesterol	Decrease Sodium
Salads	Any fresh fruits and vegetables, preferably unpeeled	Most salads are not high in sugar Avoid gelatin salads unless they are artificially sweetened	All green and vegetable salads All gelatin, fruit, and poultry salads Avoid excessive bacon, meat, cheese, eggs Use reduced-calorie or nonfat salad dressings and mayonnaise	All salads Avoid bacon in salads Made from-scratch dressings often have less sodium
Breads	Rye, whole-wheat, and other whole-grain breads and rolls	Most are not high in sugar	All breads except biscuits, croissants, brioche, popovers Use margarine, not butter, for less cholesterol and saturated fat	In moderation, breads do not contribute too much sodium
Entrées	(Meat, poultry, fish contain virtually no fiber) Entrées using beans, peas, brown rice, barley, wild rice, kasha, bulgur Whole-grain pastas Sandwiches and burgers with vegetable toppings Main dish salads using vegetables, beans, peas, rice and fruit	Most entrees are low in sugar	Broiled, baked, stir-fried, steamed, poached lean beef, poultry without skin, or seafood Keep edible portion to 3 to 4 oz. (for meat, poultry or seafood) Entrées as above fixed with wine, lemon juice, some oil Sauces thickened with cornstarch or flour instead of high-fat roux Vegetable puree sauces	Entrees using fresh herbs and spices instead of salt or seasoned salt Fresh meat, poultry, and fish Sauces made without commercial bases or salt

(continued)

Category				
Side dishes	Fresh and frozen vegetables, raw and cooked	Most side dishes are low in sugar	Sandwiches made with fresh roasted turkey, chicken, water-pack tuna, lean roast beef from round, lean ground beef	Sandwiches such as to left (use low-sodium canned tuna fish)
	Potatoes with skin on			Avoid processed and cured cold cuts unless marked low-sodium
	Brown rice, wild rice, kasha, barley, bulgur, buckwheat, couscous, or other whole grains		Steamed, boiled, baked, stir-fried vegetables and starches, with oil or small amount of butter	Any vegetables or starches, except avoid canned and pickled products, and use no salt in cooking
	Fresh and cooked fruits		Rice, noodles, barley, bulgur	
Desserts	Whole-grain cookies and quick breads	Most commercial desserts are high in sugar	Most commercial desserts are high in fat	In moderation, desserts do not contribute too much sodium
		Fresh and cooked fruit	Whole-grain cookies	
		Whole-grain quickbreads	Fresh and cooked fruits	
		Puddings made with sugar substitute	Sorbets, sherbets, fruit ice	
			Pudding made with skim milk	
			Fruit in meringue shell	
			Yogurt-based dessert	
			Angel food cake	
			Fat-free milk	
Beverages	Beverages have little or no fiber (even fruit juices)	Diet sodas and drinks, unsweetened iced tea, mineral waters, freshly squeezed fruit juices, coffee and tea		Most beverages (except for canned tomato and vegetable juice) are low in sodium

- Salted snack foods, such as potato chips, pretzels, popcorn, nuts, and crackers
- Food prepared in brine, such as pickles, olives, and sauerkraut
- Canned vegetables, tomato products, soups, and vegetable juices
- Frozen convenience foods such as pizza and entrées
- Prepared mixes for stuffings, rice dishes, and breading
- Dried soup mixes and bouillon cubes
- Certain seasonings such as salt, soy sauce, garlic salt, onion salt, celery salt, seasoned salt, Worcestershire sauce, horseradish, ketchup, and mustard

Figure 10-10 lists many appropriate menu choices with less sodium.

VEGETARIAN DIET

The number of vegetarians in the United States has been increasing. Instead of eating the meat entrées that have traditionally been the major source of protein in the American diet, these people dine on main dishes emphasizing legumes (dried beans and peas), grains, and vegetables. Vegetarian entrées such as red beans and rice can supply adequate protein with less fat and cholesterol and more fiber than their meat counterparts.

Whereas vegetarians do not eat meat, poultry, or fish, the largest group of vegetarians, referred to as *lacto-ovo vegetarians*, do consume animal products in the form of eggs (ovo-) and milk and milk products (lacto-).

Another group of vegetarians, *lacto vegetarians*, consume milk and milk products but forgo eggs. Most vegetarians are either lacto-ovo vegetarians or lacto vegetarians. *Vegans*, a third group of vegetarians, do not eat eggs or dairy products and therefore rely exclusively on plant foods to meet protein and other nutrient needs. Vegans are a small group; it is estimated that only 4 percent of vegetarians are vegans.

In addition, some vegetarians (*pesco vegetarians*) eat seafood. Also, some vegetarian diets restrict certain foods and beverages, such as highly processed foods containing certain additives and preservatives, foods that contain pesticides and/or have not been grown organically, or caffeinated or alcoholic beverages.

The number-one reason people give for being vegetarian is health benefits. Vegetarians tend to be leaner and to keep their body weight and blood lipid levels closer to desirable levels than nonvegetarians do. Vegetarians tend to have a lower incidence of the following diseases: hypertension, coronary artery disease, cancer (especially colon cancer), and type 2 diabetes. Being vegetarian does not mean that you automatically get these benefits. It is possible to be a lacto-ovo vegetarian and still eat too much fat, saturated fat, and cholesterol. It's probably the exception rather than the rule, but it is still possible. Being vegetarian does not guarantee that your diet will meet current dietary recommendations. Some other reasons for becoming vegetarian include the following:

1. Ecology. For ecological reasons, vegetarians choose plant protein because livestock and poultry require much land, energy, water, and plant food (such as soybeans), which they consider wasteful. According to the North American Vegetarian Society, the grains and soybeans fed to U.S. livestock could feed 1.3 billion people. Livestock also waste loads of water—it takes 2500 gallons of water to produce a pound of meat but only 25 gallons to produce a pound of wheat.
2. Economics. A vegetarian diet is more economical—in other words, less expensive. This can be easily demonstrated by the fact that in a typical foodservice operation, the largest component of food purchases is for meats, poultry, and fish.

LACTO-OVO VEGETARIANS

Vegetarians who do not eat meat, poultry, or fish but do consume animal products in the form of eggs, milk, and milk products.

LACTO VEGETARIANS

Vegetarians who do not eat meat, poultry, or fish but do consume animal products in the form of milk and milk products.

VEGANS

Individuals who eat a type of vegetarian diet in which no eggs or dairy products are eaten; their diet relies exclusively on plant foods.

PESCO VEGETARIANS

Vegetarians who eat fish.

3. **Ethics.** Vegetarians do not eat meat for ethical reasons; they believe that animals should not suffer or be killed unnecessarily. They feel that animals suffer real pain in crowded feed lots and cages and that both their transportation to market and their slaughter are traumatic.

4. **Religious beliefs.** Some vegetarians, such as the Seventh-Day Adventists, practice vegetarianism as a part of their religion, which also encourages exercise and forbids smoking and drinking alcohol.

Vegetarian diets can be nutritionally adequate when varied and adequate in kcalories (except for vegan diets, which need supplementation with vitamin B_{12}). Most vegetarians get enough protein, and their diets are typically lower in fat, saturated fat, and cholesterol. Nutrients that may be of concern include vitamin B_{12}, vitamin D, calcium, iron, and zinc (see Figure 10-11).

FIGURE 10-11: Important Nutrients for Vegetarians

1. **Vitamin B_{12}.** Vitamin B_{12} is found only in animal foods. Lacto-ovo vegetarians usually get enough of this vitamin unless they limit their intake of dairy products and eggs. Vegans definitely need either a supplement or vitamin B_{12}-fortified foods such as most ready-to-eat cereals, most meat analogs, some soy beverages, and some brands of nutritional yeasts.

   Sources: Dairy products, eggs, fortified cereals, meat analogs

2. **Vitamin D.** Milk is fortified with vitamin D, and vitamin D can be made in the skin with sunlight. Generally, only vegans without enough exposure to sunlight need a supplementary source of vitamin D. Some ready-to-eat breakfast cereals and some soy beverages are fortified with vitamin D.

   Sources: Fortified milk, eggs, fortified cereals, soy milk

3. **Calcium.** Lacto vegetarians and lacto-ovo vegetarians generally don't have a problem here, but vegans sometimes do if they don't eat enough calcium-rich foods. Good choices include calcium-fortified soy milk or orange juice and tofu made with calcium sulfate. Some green leafy vegetables (such as spinach, beet greens, Swiss chard, sorrel, and parsley) are rich in calcium, but they also contain a binder (called oxalic acid) that prevents some of the calcium from being absorbed. Dried beans and peas are moderate sources of calcium. Without calcium-fortified drinks or calcium supplements, it can be difficult to consume enough calcium.

   Sources: Milk and milk products, canned salmon and sardines (with bones), oysters, calcium-fortified juice or soy milk, broccoli, collards, kale, greens

4. **Iron.** Interestingly, vegetarians do not experience any more problems with iron-deficiency anemia than do their meat-eating counterparts (don't forget, meat is rich in iron). Iron is widely distributed in plant foods, and its absorption is greatly enhanced by vitamin C-containing fruits and vegetables. Vegetarians get iron from eating dried beans and peas, green leafy vegetables, dried fruits, many nuts and seeds, and enriched and whole-grain products.

   Sources: Liver, meats, breads and cereals, green leafy vegetables, legumes, dried fruits

5. **Zinc.** Zinc is found in many plant foods, such as whole grains, legumes, and nuts and seeds (especially peanut butter). Its absorption into the body is reduced by certain plant substances, such as phytate. Children may need zinc supplements.

   Sources: Whole grains, legumes, nuts and seeds, peanut butter

---

FIGURE 10-12:
Vegetarian food pyramid.

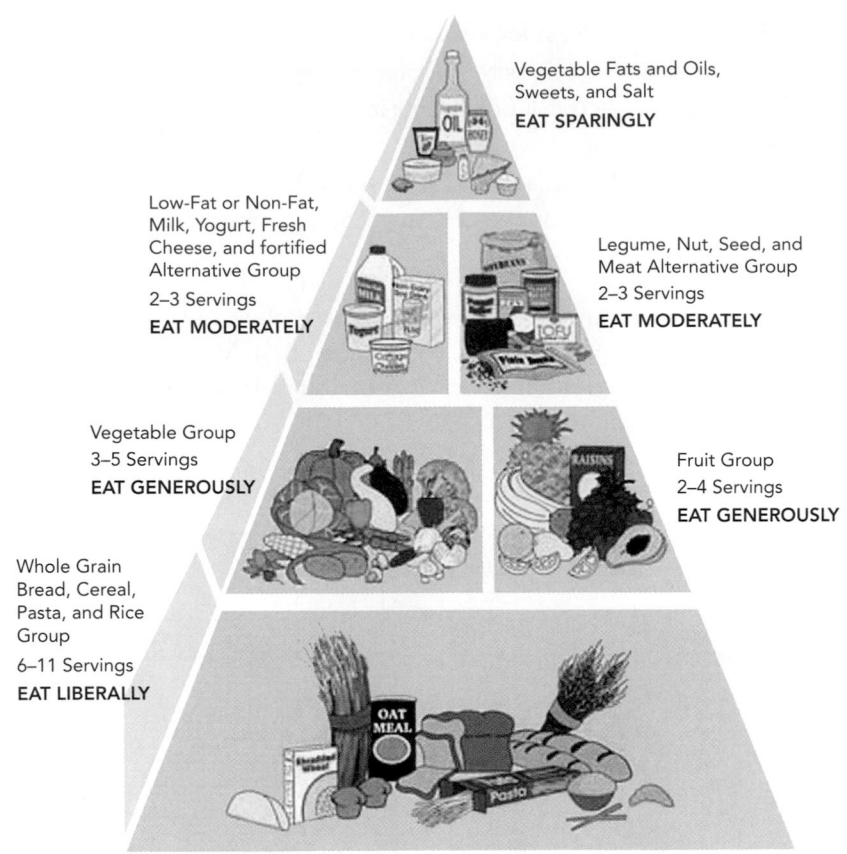

As discussed in Chapter 5, most plant proteins are considered incomplete, but this doesn't mean they are low in quality. When plant proteins are eaten with other foods, the food combinations usually result in complete protein. For example, when peanut butter and whole-wheat bread are eaten over the course of a day, the limiting amino acid in each of these foods is supplied by the other food. Such combinations are called complementary proteins. Eating complementary proteins at different meals during the day generally ensures a balance of dietary amino acids. Some vegetable proteins, such as those found in amaranth, quinoa, and soybeans, are complete proteins.

Figure 10-12 shows a vegetarian food pyramid for lacto vegetarians and vegans.

Variety is a key when planning vegetarian meals.

1. Use a variety of plant protein sources at each meal: legumes, grain products (preferably whole grains), nuts and seeds, and/or vegetables. Vegetarian entrees commonly use cereal grains such as rice and bulgur (precooked and dried whole wheat) in combination with legumes and/or vegetables. Use small amounts of nuts and seeds in dishes.

2. Use a wide variety of vegetables, Steaming, stir-frying, or microwaving vegetables retains flavor, nutrients, and color.

3. Choose low-fat and fat-free varieties of milk and products and limit the use of eggs. This is important to prevent a high intake of saturated fat which is found in whole milk, low-fat milk, regular cheeses, and eggs.

4. Offer dishes made with soybean-based products, such as tofu and tempeh. Soybeans are unique in that they contain the only plant protein that is nutritionally equivalent to animal protein.

5. For menu ideas, don't forget to look at the cuisine of other countries (Figure 10-13).

FIGURE 10-13: Vegetarian Dishes from Around the World

Chile: Quinoa with vegetables
China: Tofu and vegetable stir-fries, noodle bowls
Caribbean: Black beans and rice, spinach and potato croquettes
Ethiopia: Vegetable alecha (spicy vegetable stew)
France: Ratatouille (vegetable casserole based on roasted eggplant)
India: Dal (lentil stew), vegetable curries
Italy: Pasta e fagioli (pasta and bean soup), risotta, vegetable lasagna
Japan: Vegetable tempura
Mexico: Gazpacho, vegetable quesadillas, enchiladas
Middle East: Falafel, tabbouleh, hummus
Morocco: Couscous
Vietnam: Vegetarian spring rolls

HIGH-FIBER DIET

Fiber is abundant in plants, and so legumes (dried beans, peas, and lentils), fruits, vegetables, whole grains, nuts, and seeds provide fiber. Fiber is not found in meat, poultry, fish, dairy products, and eggs. Figure 10-9 lists menu choices for increased fiber.

LOW-LACTOSE DIET

Lactose is present in large amounts in milk, ice cream, ice milk, sherbet, cottage cheese, eggnog, and cream, and so these are the foods customers with this request probably will avoid. Most individuals can drink small amounts of milk without experiencing any symptoms, especially if it is taken with food. Lactose-reduced milk and some other lactose-reduced dairy products are available. Eight fluid ounces of lactose-reduced milk contains only 3 grams of lactose, compared with 12 grams in regular milk.

Yogurt is usually well tolerated because it is cultured with live bacteria that digest lactose. This is not always the case with frozen yogurt, because most brands do not contain nearly the number of bacteria found in fresh yogurt (there are no federal standards for frozen yogurt at this time). Also, some yogurts have milk solids added to them that can cause problems.

Many hard cheeses, such as Swiss or Parmesan, contain very little lactose and usually do not cause symptoms because most of the lactose is removed during processing or is digested by the bacteria used in making cheese.

GLUTEN-FREE DIET

Celiac disease is an inherited autoimmune disease that usually affects several organs in the body before it is diagnosed and treated. When a person with celiac disease consumes any food, beverage, or medication containing wheat, barley, rye, and possibly oats, his or her immune system is "triggered" and responds by damaging the lining of the intestinal tract. As a result, symptoms include recurrent abdominal pain, bloating, diarrhea, weight loss, lactose intolerance, and malnutrition, often accompanied by nonintestinal symptoms such as anemia and fatigue. Some people have no symptoms whatsoever.

It is the gluten in wheat, barley, and rye that causes the disease. Gluten is found in breads and other baked goods, where it provides strength, elasticity, and the structure needed to hold dough together and seal in the gases produced during fermentation, allowing the dough to rise.

Staples of the gluten-free diet include:

- Fruits and vegetables
- Meat
- Milk-based items
- Potatoes, rice, corn, and beans
- Cereals made without wheat or barley malt
- A wide variety of specialty foods (such as pasta, bread, pancakes, and pastries) made with alternative grains (rice, tapioca, potato, and corn flours and starches)

Alternative flours such as rice flour, potato starch, whey powder, cassia, bean flours, and tapioca starch are gluten-free. Since these flours lack the structure and elasticity-providing properties of gluten, baked goods made with them require additional ingredients to stabilize their shape and consistency. Good stabilizers include natural gums such as guar gum and xanthan gum, egg white powder, and fresh egg whites. These substitutions have little effect on taste, but they do change the mixing and baking methods.

Gluten-free bread preparations have an entirely different look and feel than conventional breads made with wheat flour, more like batter than a chewy, elastic dough, and so the mixture must be baked in a walled container such as a loaf pan or muffin tin. If you've ever made a flourless chocolate cake, in which whipped eggs provide the structure, you'll be familiar with the way lightness and shape are brought into many gluten-free baked goods. To prevent particles of gluten from slipping in, use a different rolling pin and work surface than you do for conventional baking. Small amounts of gluten really do make a difference to individuals with celiac disease. To avoid contamination when making gluten-free foods, clean all surfaces thoroughly. Keep gluten-free ingredients and foods separate from gluten-containing foods. For more information on cooking and baking without gluten, go to www.csaceliacs.org.

Pages 385–387 show a variety of low-gluten recipes.

DIET LOW IN ADDED SUGARS

Added sugars include white sugar, high-fructose corn syrup, and other sweeteners added to foods in processing, as well as sugars added to foods at the table. The major sources of added sweeteners in the diet come from soft drinks, candy and sugars, baked goods, fruit drinks, dairy desserts (such as ice cream), and sweetened milk. Figure 10-10 lists menu choices for decreased sugar.

RESTAURANTS AND NUTRITION-LABELING LAWS

Foods prepared and served in restaurants or other foodservice operations are exempt from the mandatory nutrition labels found on packaged foods. However, in the past few years some cities, such as New York City, have started to require limited nutrition information to be provided on-site by restaurant chains.

Restaurants are not exempt from Food and Drug Administration (FDA) rules concerning nutrient claims and health claims (discussed in Chapter 2) that are used on menus, table tents, posters, or signs. In addition, restaurants must have nutrition information available on request for any menu item using nutrient or health claims.

Nutrient content claims such as "good source of calcium" and "fat-free" can appear only if they follow legal definitions (see Figure 2-24 in Chapter 2). For example, a food that is a good source of calcium must provide 10 to 19 percent of the Daily Value for calcium in one serving. Nutrient content claims are based on what the FDA has defined as a standardized

	Total Fat	Saturated Fat	Cholesterol	Sodium
Single serving*	13 g	4 g	60 mg	480 mg
Main dish	19.5	6	90	720
Meal	26	8	120	960

*Per Reference Amount and per 50 g when Reference Amount is 30 grams or less or 2 tablespoons or less.

serving size, called a Reference Amount. Standardized serving sizes or Reference Amounts are frequently measured in grams, milliliters, or cups to be very accurate. For example, the Reference Amount for cookies is 30 grams.

In addition to nutrient content claims based on a single serving, there are nutrient content claims for main dishes, such as lasagna, and meals. A main dish must weigh at least 6 ounces, be represented on the menu as a main dish, and contain no less than 40 grams each of at least three different foods from at least two food groups. Meals are defined as weighing at least 10 ounces and containing no less than 40 grams each of at least three different foods from at least two food groups. In general, for main dishes and meal products the nutrient content claim is based on the nutrient amount per 100 grams of the food. For example, a "low-fat" food must contain 3 grams of fat or less. Main dishes and meals must therefore contain 3 grams of fat or less per 100 grams and not more than 30 percent of kcalories from fat.

Claims that promote a health benefit must meet certain criteria, as described in Chapter 2 and Figure 2-25. In addition, any food being used in a health claim may not contain more than 20 percent of the Daily Value for fat, saturated fat, cholesterol, or sodium. Figure 10-14 shows the maximum amounts of these nutrients that are permitted in a single food serving, main dish, or meal for foods with health claims.

When providing nutrition information for a nutrient or health claim, restaurants do not have to provide the standard nutrition information profile and more exacting nutrient content values required in the Nutrition Facts panel on packaged foods. Instead, restaurants can present the information in any format desired, and they have to provide only information about the nutrient or nutrients to which the claim is referring. Restaurants also are not required to do chemical analyses to determine the nutrient values of their foods. They can use nutrient analysis software, books with nutrient composition information, or cookbooks with reliable nutrient analysis data.

Restaurants may use symbols on the menu to highlight the nutritional content of specific menu items. When doing so, they are required to explain the criteria used for the symbols. Restaurants may also highlight dishes that meet criteria set forth by recognized organizations such as the American Heart Association or a medical center. In these cases, the menu must explain that the items meet the dietary guidelines of that organization. Lastly, menus can use references or symbols to show that a food or meal is based on the Dietary Guidelines for Americans.

MINI-SUMMARY

1. Foods prepared and served in restaurants or other foodservice operations are exempt from the mandatory nutrition labels found on packaged foods.
2. Restaurants are not exempt from FDA rules concerning nutrient claims and health claims (as discussed in Chapter 2) that are used on menus, table tents, posters, or signs.
3. Restaurants must have nutrition information available upon request for any menu items using nutrient or health claims.
4. Restaurants may use symbols on the menu to highlight the nutritional content of specific menu items.

CHECK-OUT QUIZ

1. Marketing includes:

   a. Finding out what consumers want and need
   b. Developing a product consumers want and need
   c. Promoting the product
   d. All of the above

2. When hearing descriptions of healthy menu entrées, most customers want:

   a. Complete nutrient information
   b. Fat, saturated fat, and cholesterol information
   c. Good descriptions of the ingredients, portion size, and method of preparation
   d. None of the above

3. When you tout foods as "heart-healthy," you are approaching customers in a negative manner.

   a. True
   b. False

4. An example of publicity is:

   a. Radio advertising
   b. A press release
   c. Recipes from a food association
   d. A point-of-purchase display

5. When you evaluate the success or failure of healthy menu options, you need to get feedback from:

   a. Customers
   b. Staff
   c. Managers
   d. All of the above

ACTIVITIES AND APPLICATIONS

1. **Restaurant Menu Check**
   Study the menus from five different foodservices, including a quick-service business, and identify any menu items that are healthy. How do menu items appear on the menu? Would customers know they are healthy? What information is included?

2. **Restaurant Promotion**
   Check restaurant advertisements in the newspaper, on radio, or in other media and watch for any advertising of healthy and nutritious foods. What do the advertisements state? To which market segments are these advertisements targeted?

3. **Restaurant Visit**

   Visit a local restaurant/foodservice that offers healthy menu options. Find out how much it sells of its healthy menu options and the profile of the typical customer who buys these items. Also find out how these products are marketed and evaluated.

NUTRITION WEB EXPLORER

Food marketing boards and associations:

American Egg Board   www.aeb.org

Mann Packing   www.broccoli.com

Grains Nutrition Information Center   www.wheatfoods.org

Milk   www.whymilk.com

National Pork Producers Council   www.nppc.org

National Turkey Federation   www.eatturkey.com

Produce Marketing Association   www.pma.com

Pick one of these food marketing board/association websites to visit. Write a brief report about the website, including the name of the board/association, the website address, and a list of items available that a foodservice operator could use (such as recipes), and attach sample material.

Restaurants

www.rockfishseafood.com

www.marcellasristorante.com

www.pfchangs.com

Look at the menus at these restaurant websites and find the menu items that you could use to meet guest special requests. Which restaurant chain provides the easiest-to-use nutritional information?

Gluten-Free Recipes

GLUTEN-FREE BLACK BOTTOM CUPCAKES

Yield: 12 cupcakes

Filling

1 cup cream cheese, softened
½ cup sugar
⅛ teaspoon salt

1 egg
1 cup chocolate chunks

Batter

4 tablespoons rice flour

7 tablespoons tapioca starch

½ cup plus 3 tablespoons soy flour

1 cup plus 1 teaspoon sugar

7 tablespoons cocoa powder

1 teaspoon baking soda

½ teaspoon salt

1 cup water

⅓ cup oil

1 tablespoon vinegar

1 teaspoon vanilla

Steps

1. Preheat oven to 350°F. Grease a full-size 12-muffin tin.
2. Make the cream cheese filling: In a standing mixer with a paddle attachment or by hand with a wooden spoon, cream together the cream cheese, ⅓ cup sugar, and ⅛ teaspoon salt until smooth. Scrape the bowl; add the egg and continue to cream until light in color and as thick as sour cream. Fold in chopped chocolate chunks; set aside.
3. Make the cupcake batter: In a mixing bowl, combine the rice flour, tapioca starch, soy flour, 1 cup plus 1 teaspoon sugar, cocoa powder, baking soda, and ½ teaspoon salt. Create a well in the dry ingredients and pour in the water, oil, vinegar, and vanilla. Mix until smooth.
4. Fill the muffin tins: Put one tablespoon of batter into the bottom of each cup of the greased muffin pan. Add a scoop of cream cheese filing on top of batter, dividing the filling evenly. Distribute the rest of the chocolate batter on top.
5. Bake 24 minutes or until cupcakes hold their shape. Cool in the muffin tin for about 20 minutes before turning out and cooling on racks.

Courtesy: Richard J. Coppedge, Jr., CMB. *Kitchen & Cook*, March 2005.

GLUTEN-FREE PIZZA DOUGH

Ingredients

1½ pounds lukewarm water

4 ounces yeast

1½ ounces honey

1 ounce sugar

3 pounds white rice flour

2 pounds brown rice flour

1 pound 4 ounces tapioca flour

12 ounces potato starch

2½ ounces olive oil

½ ounce sea salt

Steps

1. In lukewarm water, dissolve yeast, honey, and sugar. Let bloom for 10 minutes.
2. Add flours, oil, and salt. Mix with dough hook to a smooth dough. Let rest covered with plastic wrap for 1 hour.
3. Divide into small rounds to desired shell size. Let rest additional ½ hour.
4. Roll into ⅛-inch thick shells, top and bake on pizza stone or in a pizza oven to desired doneness.

Gluten-Free Bread

Yield: 7 loaves

Flour Mix

3 pounds white rice flour
2 pounds brown rice flour
1 pound 4 ounces tapioca flour
12 ounces potato starch
½ ounce salt

Other Ingredients

Kosher salt
2 ounces yeast
1 pound 12 ounces lukewarm water
Vegetable oil cooking spray
Olive oil

Steps

1. First mix together the ingredients under Flour Mix.
2. Mix together 2 pounds of the flour mixture with a pinch of Kosher salt using a dough hook attachment.
3. Dissolve yeast in water. Add water mixture to flour mixture and work until it forms a loose ball.
3. Spray 2 loaf pans with vegetable oil cooking spray and press dough in.
4. Brush with olive oil and sprinkle with kosher salt. Let rise 30 to 45 minutes at room temperature.
5. Bake at 350°F for 30 minutes until golden brown on bottom.

PART THREE

Nutrition's Relationship to Health and Life Span

Nutrition and Health

About two-thirds of all deaths in the United States are due to cardiovascular disease (including stroke), cancer, and diabetes. The financial cost of these diseases is huge.

1. In 2008, the cost of heart disease and stroke was projected to be $448 billion.
2. Cancer costs the nation an estimated $89 billion annually in direct medical costs.
3. The direct and indirect costs of diabetes is $175 billion a year.

Direct costs refer to preventive, diagnostic, and treatment services such as physician visits, medications, and hospital care. Indirect costs are the value of wages lost by people unable to work because of illness or disability as well as the value of future earnings lost because of premature death. Besides huge financial costs, these diseases have emotional costs.

These diseases all have something in common: Their prevention and treatment have a dietary component. This chapter will look at coronary heart disease, stroke, high blood pressure, cancer, and osteoporosis. This chapter will help you to:

- List and describe three common forms of cardiovascular disease
- Explain what atherosclerosis is and how it is related to cardiovascular diseases
- List five risk factors for coronary heart disease
- Distinguish between angina and a heart attack
- Explain how a person's risk for coronary heart disease is assessed
- Explain the two main ways to lower blood cholesterol levels
- Explain how strokes occur
- List five lifestyle modifications for hypertension control
- List five menu-planning guidelines to lower cardiovascular risk
- Define cancer
- Outline the American Cancer Society's four guidelines to reduce cancer risk
- Distinguish between type 1 and type 2 diabetes mellitus and understand the principles of planning meals for people with diabetes
- Define osteoporosis and how to prevent/treat it
- Discuss how to safely use botanicals including herbs
- Analyze the pros and cons of biotechnology used to produce plants for food

NUTRITION AND CARDIOVASCULAR DISEASE

CARDIOVASCULAR DISEASE (CVD)
Diseases of the heart and blood vessels such as coronary artery disease, stroke, and high blood pressure.

Cardiovascular disease (CVD) is a general term for diseases of the heart and blood vessels, as seen in the following list:

- Coronary artery disease
- Stroke
- High blood pressure
- Rheumatic heart disease
- Congenital heart defects
- Congestive heart failure

This section will discuss the first three diseases.

The two medical conditions that lead to most cases of cardiovascular disease are atherosclerosis and high blood pressure. **Atherosclerosis**, a condition characterized by plaque buildup along the artery walls, is the most common form of artery disease. (Arteriosclerosis is a general medical term that includes all diseases of the arteries that involve hardening and blocking of the blood vessels.) Atherosclerosis affects primarily the large and medium-size arteries. In this condition, arterial linings become thickened and irregular with deposits called **plaque**. Plaque contains cholesterol, fat, fibrous scar tissue, calcium, and other biological debris. Why plaque deposits are formed and what role fat and cholesterol play in its formation are questions with only partial answers.

Atherosclerosis develops by a process that is totally silent. At birth, the inside of the blood vessels are clear and smooth. As time goes on, plaque builds up, resulting in narrower passages and less elasticity in the vessel wall, both of which contribute to high blood pressure (Figure 11-1). What's even more dangerous is when the plaque closes off blood flow completely. Also, some cholesterol-rich plaques are unstable—they have a thin covering and can burst, releasing cholesterol and fat into the bloodstream. The release can cause a blood clot to form over the plaque, blocking blood flow through the artery. If the artery takes blood to the brain, a stroke occurs. If the blocked artery is near the heart, a heart attack occurs, the next topic.

CORONARY HEART DISEASE

The heart is like a pump, squeezing and forcing blood throughout the body. With its four chambers, the heart beats about 100,000 times in one day. Like all muscles in the body, the heart must have oxygen and nutrients to do its work. The heart cannot use oxygen and nutrients directly from the blood it pumps within its chambers. Instead, nutrients and oxygen are furnished by the three main blood vessels on the heart, which are referred to as coronary arteries. **Coronary heart disease (CHD)** is a broad term used to describe damage to the heart caused by narrowing or blockage of the coronary arteries. Coronary heart disease is the most common form of cardiovascular disease.

Smoking, high blood cholesterol, and high blood pressure are three major **risk factors** for coronary heart disease. A risk factor is a habit, trait, or condition associated with an increased chance of developing a disease. Preventing or controlling risk factors generally reduces the probability of illness. Research shows that there are definite benefits to controlling these risk factors:

- Cigarette smoking and exposure to tobacco smoke
- High blood cholesterol
- High blood pressure

ATHEROSCLEROSIS
A condition characterized by plaque buildup along the artery walls; it is the most common form of arteriosclerosis.

PLAQUE
Deposits on arterial walls that contain cholesterol, fat, fibrous scar tissue, and other biological debris.

CORONARY HEART DISEASE (CHD)
Damage to the heart caused by narrowing or blockage of the coronary arteries.

RISK FACTOR
A habit, trait, or condition associated with an increased chance of developing a disease.

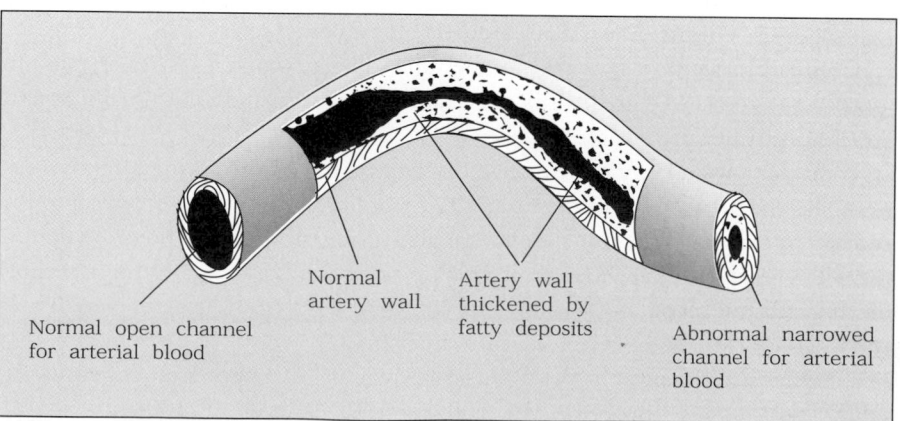

Normal artery wall

Artery wall thickened by fatty deposits

Normal open channel for arterial blood

Abnormal narrowed channel for arterial blood

FIGURE 11-1:
Cross-sectional representation of a coronary artery partially closed with plaque.

- Physical inactivity
- Obesity and overweight
- Diabetes mellitus

Risk factors that you can't control are increasing age (45 and older for men, 55 and older for women), male gender, and a family history of premature heart disease.

In addition, people with at least three of the following conditions in what is known as *metabolic syndrome* are at increased risk of dying from coronary heart disease and cardiovascular diseases:

1. Excessive abdominal obesity: a waist size over 40 inches for men and over 35 inches in women.
2. High blood triglycerides
3. Reduced high-density lipoprotein (HDL) concentrations (less than 40 mg/dL in women and less than 40 mg/dL in men)
4. Elevated fasting glucose (high blood sugar levels after fasting, such as after sleeping)
5. Elevated blood pressure

This syndrome is closely associated with a metabolic disorder called insulin resistance, in which the body can't use insulin efficiently. Insulin is needed for glucose to enter the body's cells.

More than two-thirds of a coronary artery may be filled with fatty deposits without causing symptoms. Symptoms may manifest themselves as chest pain, as in *angina*, or as a heart attack. Angina refers to the symptom of pressing, intense pain in the area of the heart when the heart muscle is not getting enough blood. Sometimes stress or exertion can cause angina.

Most heart attacks are caused by a clot in a coronary artery at the site of narrowing and hardening that stops the flow of blood. In healthy individuals, blood clots normally form and dissolve in response to injuries in the blood vessels. But with atherosclerosis, plaques can rupture, causing blood clots to form. The blood clots may then block blood flow or break off and travel to another part of the body. If either of these things happens and blocks a blood vessel that feeds the heart, it causes a heart attack. (If it blocks a blood vessel that feeds the brain, it causes a stroke.) A heart attack therefore occurs when the blood supply to part of the heart muscle itself—the myocardium—is severely reduced or stopped. The medical term for a heart attack is a *myocardial infarction*.

If an area of the heart is supplied by more than one vessel, that area may live for a period of time even if one vessel becomes blocked. The extent of heart muscle damage after a heart attack depends on which vessel is blocked, whether it is big or small, and the remaining blood supply to that area. As the result of a heart attack, some cells may die and some may be injured. In any case, the heart often loses some of its effectiveness as a pump, because reduced muscle contraction means reduced blood flow.

Coronary heart disease is the number-one killer of women. Whereas most men's heart attacks are experienced at 40 years of age and older, women do not usually experience heart attacks until after menopause.

Your risk for coronary heart disease can be assessed by measuring total blood cholesterol as well as the proportions of the various types of lipoproteins. The only way to find this out is to go to a doctor and have a blood test after fasting for 9 to 12 hours. A lipoprotein profile will reveal your total cholesterol, which is measured in milligrams (mg) of cholesterol per deciliter (dL) of blood. Total cholesterol less than 200 mg/dL is desirable, 200–239 mg/dL is borderline high, and 240 mg/dL or more is high.

Low-density lipoprotein (LDL), also known as "bad cholesterol" because it increases CHD risk, should be less than 100 mg/dL. A level of 100–129 mg/dL is near optimal/above optimal, 130–159 mg/dL is borderline high, 160–189 mg/dL is high, and 190 mg/dL and above is very high.

High-density lipoprotein (HDL), also known as "good cholesterol," protects the arteries by taking cholesterol to the liver for disposal, and so the higher the HDL, the better. An HDL level of 60 mg/dL is desirable, and an HDL level less than 40 mg/dL is considered low for men, or less than 50 mg/dL for women.

The main goal of cholesterol-lowering treatment is to lower the LDL level enough to reduce the risk of developing heart disease or having a heart attack. The higher the risk is, the lower the LDL goal will be. There are two main ways to lower cholesterol.

- Therapeutic Lifestyle Changes (TLC): These include a cholesterol-lowering diet (called the *TLC diet*), physical activity, and weight management. TLC is for anyone whose LDL is above optimum levels.
- Drug treatment: If cholesterol-lowering drugs are needed, they are used together with TLC treatment to help lower LDL.

To reduce the risk for heart disease or keep it low, it is very important to control any other risk factors, such as high blood pressure and smoking.

TLC is a set of things you can do to help lower LDL cholesterol. The main parts of TLC are:

- **The TLC diet.** This is a low-saturated-fat, low-cholesterol eating plan that calls for less than 7 percent of kcalories from saturated fat and less than 200 milligrams of dietary cholesterol per day. The TLC diet recommends only enough kcalories to maintain a desirable weight and avoid weight gain. If LDL is not lowered enough by reducing saturated fat and cholesterol intakes, the amount of soluble fiber in the diet can be increased. Good sources of soluble fiber include oats, certain fruits (such as oranges and pears) and vegetables (such as brussels sprouts and carrots), and dried peas and beans. Certain food products that contain plant stanols or plant sterols (for example, cholesterol-lowering margarines and salad dressings) can be added to the TLC diet to boost its LDL-lowering power. Plant stanols and sterols occur naturally in small amounts in many plants. They are added to certain margarines and some other foods, and help block the absorption of cholesterol from the digestive tract, which helps to lower LDL. Figure 11-2 on page 396 gives tips on using the TLC diet.
- **Soluble fiber** such as barley, oats, psyllium, apples, bananas, berries, citrus fruits, nectarines, peaches, pears, plums, prunes, broccoli, brussels sprouts, carrots, dry beans, peas, soy products
- **Weight management.** Losing weight if you are overweight can help lower LDL and is especially important for those with a cluster of risk factors that includes high triglyceride and/or low HDL levels and being overweight with a large waist measurement (more than 40 inches for men and more than 35 inches for women).
- **Physical activity.** Lack of physical activity is a major risk factor for heart disease. Regular physical activity can help you manage your weight and, in that way, help lower your LDL. It also can help raise HDL and lower triglycerides, improve the fitness of your heart and lungs, and lower blood pressure. Unless your doctor tells you otherwise, try to get at least 30 minutes of a moderate-intensity activity, such as brisk walking on most, and preferably all, days of the week. You can do the activity all at once or break it up into shorter periods of at least 10 minutes each.

> **TLC DIET**
> A low-saturated-fat, low-cholesterol eating plan designed to fight cardiovascular disease and lower LDL; the diet calls for less than 7 percent of kcalories from saturated fat and less than 200 milligrams of cholesterol daily and also recommends only enough kcalories to maintain a desirable weight.

FIGURE 11-2: TLC Diet Tips

BREADS/CEREALS/GRAINS

6 or more servings a day—adjust to kcalorie needs
Foods in this group are high in complex carbohydrates and fiber. They are usually low in saturated fat, cholesterol, and total fat.
Whole-grain breads and cereals, pasta, rice, potatoes, low-fat crackers, and low-fat cookies.

VEGETABLES/DRY BEANS/PEAS

3–5 servings a day
These are important sources of vitamins, fiber, and other nutrients. Dry beans/peas are fiber-rich and good sources of plant protein.
Fresh, frozen, or canned—without added fat, sauce, or salt.

FRUITS

2–4 servings a day
These are important sources of vitamins, fiber, and other nutrients.
Fresh, frozen, canned, dried—without added sugar.

DAIRY PRODUCTS

2–3 servings a day—fat free or low fat (for example, 1% milk)
These foods provide as much or more calcium and protein than whole milk dairy products, but with little or no saturated fat.
Fat-free or low-fat milk, buttermilk, yogurt, sour cream, cream cheese, low-fat cheese (with no more than 3 grams of fat per ounce, such as low-fat cottage cheese)

EGGS

2 or fewer yolks per week—including yolks in baked goods and in cooked or processed foods
Yolks are high in dietary cholesterol. Egg whites or egg substitutes have no cholesterol and fewer kcalories than whole eggs.

MEAT, POULTRY, FISH

5 ounces or less a day
Poultry without skin and fish are lower in saturated fat. Lean cuts of meat have less fat and are rich sources of protein and iron. Be sure to trim any fat from meat and remove skin from poultry.
Lean cuts of beef include sirloin tip, round steak, and rump roast; extra lean hamburger; cold cuts made with lean meat or soy protein; lean cuts of pork are center-cut ham, loin chops, and pork tenderloin.
Strictly limit organ meats, such as liver and kidneys—they are high in cholesterol.
Eat shrimp only occasionally—it is moderately high in cholesterol.

FATS/OILS

Amount depends of daily kcalorie level.
Nuts are high in kcalories and fat, but have mostly unsaturated fat. Nuts can be eaten in moderation on the TLC diet—be sure the amount you eat fits your kcalorie intake.
Unsaturated vegetable oils that are high in unsaturated fat (such as canola, corn, olive, safflower, and soybean); soft or liquid margarines (the first ingredient on the label should be unsaturated liquid vegetable oil, rather than hydrogenated or partially hydrogenated oil) and vegetable oil spreads; salad dressings; seeds; nuts.
Choose products that are labeled "trans fat-free."

DIET OPTIONS

Stanol/sterol-containing food products such as specially labeled margarines and orange juice

Source: National Heart, Lung, and Blood Institute.

Even with drug treatment to lower cholesterol, lifestyle changes are important. This will keep the dose of medicine as low as possible and lower the risk in other ways as well. Several types of drugs are available for cholesterol lowering, such as statins, the most popular drug for lowering LDL and raising HDL.

STROKE

A *stroke* occurs when blood flow to part of the brain is interrupted, which can cause damage to brain cells. The brain must have a continuous supply of blood rich in oxygen and nutrients for energy. Although the brain constitutes only 2 percent of the body's weight, it uses about 25 percent of the oxygen and almost 75 percent of the glucose circulating in the blood. Unlike other organs, the brain cannot store energy. If deprived of blood for more than a few minutes, brain cells die from energy loss and from certain chemical interactions that are set in motion. The functions these cells control—speech, muscle movement, comprehension—die with them. Dead brain cells cannot be revived.

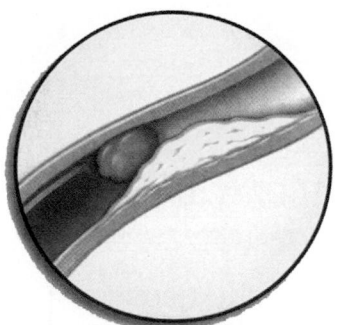

Ischemic Stroke

There are two types of strokes: ischemic and hemorrhagic (Figure 11-3). *Ischemic strokes*, the most common, occur because a blood clot blocks an artery or vessel in the brain. The most common cause is atherosclerosis in which fatty deposits form in the vessel walls of the brain. The process is similar to what happens in the heart for people with heart disease. This is why stroke and heart disease have some of the same controllable risk factors: high blood pressure, cigarette smoking, high cholesterol, diabetes, physical inactivity, and obesity. These factors raise the risk for plaque buildup in the arteries, which in turn raises the risk of the formation of blockages and blood clots. High blood pressure is by far the most important risk factor. A stroke sometimes occurs because plaque develops in the carotid artery, the main blood vessel in the neck that leads to the brain.

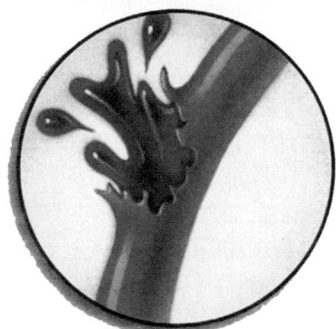

Hemorrhagic Stroke

FIGURE 11-3:
Ischemic and hemorrhagic strokes.
Source: FDA Consumer (March/April 2005).

Hemorrhagic strokes occur because a blood vessel in the brain ruptures, or breaks, and causes bleeding in the surrounding brain tissue. These strokes can be caused by an aneurysm, a thin or weak spot in an artery that bulges and can burst. Other causes include a group of abnormal blood vessels and leakage from a vessel wall that has been weakened by high blood pressure. Hemorrhagic strokes account for less than 20 percent of all types of strokes but are far more lethal, with a death rate over 50 percent. Strokes caused by clots or hemorrhage usually strike suddenly, with little or no warning, and do all their damage in a matter of seconds or minutes.

Sometimes people experience a "mini-stroke," called a transient ischemic attack (TIA). When a TIA occurs, stroke symptoms (Figure 11-4) may last only temporarily and then disappear. Most TIA symptoms disappear within an hour. About 25 percent of people who have a TIA will have a bigger stroke within five years. A TIA is a warning sign that shouldn't be ignored.

Because blood clots play a major role in causing strokes, drugs that inhibit blood coagulation may prevent clot formation. Physicians have several drugs at their disposal, including

---

STROKE
Damage to brain cells resulting from an interruption of blood flow to the brain.

ISCHEMIC STROKE
The most common type of stroke, in which a blood clot blocks an artery or vessel in the brain.

HEMORRHAGIC STROKE
A stroke due to a ruptured brain artery.

FIGURE 11-4: Common Symptoms of Stroke

- Sudden weakness or numbness in the face, arms, or legs, especially on one side of the body
- Sudden confusion or difficulty speaking or understanding speech
- Sudden vision problems, such as blurry vision or a partial or complete loss of vision in one or both eyes
- Sudden dizziness, trouble walking, or loss of balance and coordination
- Sudden severe headache with no known cause

aspirin, to treat those at risk. Aspirin works by preventing blood platelets from sticking together. Controlling blood pressure is also important.

About 10 percent of stroke survivors recover almost completely. According to the National Stroke Association, 25 percent recover with minor impairments and 40 percent experience moderate to severe impairments that require special care. Ten percent require care in a long-term care facility, and 15 percent die shortly after the stroke.

HIGH BLOOD PRESSURE

As many as 50 million Americans have high blood pressure (also called *hypertension*), but about 30 percent of them don't know it. Because high blood pressure usually doesn't give early-warning signs, it is known as the "silent killer." High blood pressure is one of the major risk factors for coronary heart disease and stroke. All stages of hypertension are associated with an increased risk of nonfatal and fatal cerebrovascular disease and renal disease.

Arterial blood pressure is the pressure of blood within arteries as it is pumped through the body by the heart. Whether your blood pressure is high, low, or normal depends mainly on several factors: the output from your heart, the resistance to blood flow by your blood vessels, the volume of your blood, and blood distribution to the various organs.

Everyone experiences hourly and even moment-by-moment changes in blood pressure. For example, your blood pressure will temporarily rise with strong emotions such as anger and frustration, with water retention caused by too much salty food that day, and with heavy exertion, which pushes more blood into your arteries. These transient elevations in blood pressure usually don't indicate disease or abnormality (Figure 11-5).

Blood pressure is represented as a fraction, as in 120/80. The top number, 120, is called the *systolic pressure*—the pressure of blood within arteries when the heart is pumping. The bottom number, 80, is called the *diastolic pressure*—the pressure in the arteries when the heart is resting between beats. Both blood pressure numbers are measured in millimeters of mercury, abbreviated mm Hg.

Normal blood pressure varies from person to person. High blood pressure occurs when the blood pressure stays too high and is defined as a systolic pressure greater than or equal

FIGURE 11-5:
It's important to have your blood pressure checked as part of your regular health maintenance.
Courtesy PhotoDisc, Inc.

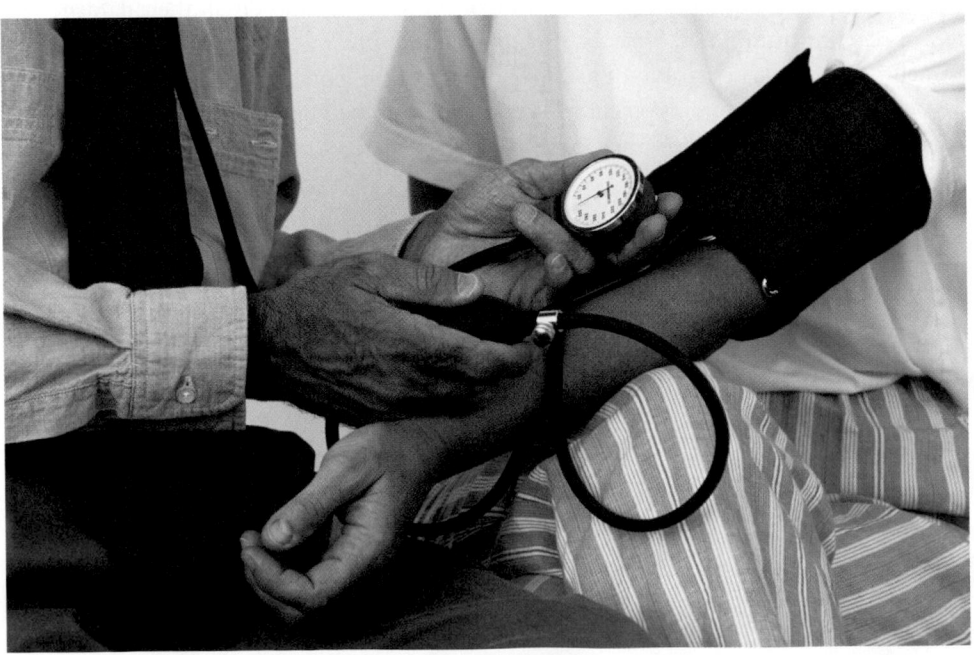

FIGURE 11-6: Classification of Blood Pressure for Adults

Blood Pressure Classification	Systolic Blood Pressure (mm Hg)	Diastolic Blood Pressure (mm Hg)	Result
Normal	<120	<80	Good for you!
Prehypertension	120–139	or 80–89	Your blood pressure could be a problem. Make changes in what you eat and drink, be physically active, and lose extra weight.
Stage 1 Hypertension	140–159	or 90–99	You have high blood pressure; see your doctor
Stage 2 Hypertension	Over 160	or over 100	and find out how to control it.

Source: "The Seventh Report of the Joint National Committee on Prevention, Detection, Evaluation, and Treatment of High Blood Pressure," National Institutes of Health and National Heart, Lung, and Blood Institute, 2003.

to 120 mm Hg or a diastolic pressure greater than or equal to 80 mm Hg. Figure 11-6 classifies blood pressure readings for adults. Individuals with a systolic blood pressure of 120 to 139 mm Hg or a diastolic blood pressure of 80 to 89 mm Hg are considered "prehypertensive" and require health-promoting lifestyle modifications to prevent cardiovascular disease.

Systolic blood pressure is the key determinant for assessing the presence and severity of high blood pressure in middle-aged and older adults. Hypertension occurs in more than two-thirds of individuals after age 65. Controlling systolic hypertension significantly reduces heart attack, heart failure, and stroke.

When persistently elevated blood pressure is due to a medical problem such as a hormonal abnormality or an inherited narrowing of the aorta (the largest artery leading from the heart), it is called **secondary hypertension**. That is, the high blood pressure arises secondary to another condition. Only 5 percent of individuals with hypertension have secondary hypertension. The remaining 95 percent have what is called **primary (essential) hypertension**. The cause of essential hypertension is not well understood. Figure 11-7 lists risk factors for hypertension.

The prevalence of high blood pressure increases with age, is greater for African Americans than for whites, and is especially prevalent and devastating in the lower socioeconomic groups. In young adulthood and early middle age, high blood pressure prevalence is greater for men than for women. Thereafter, more women than men have high blood pressure.

SECONDARY HYPERTENSION
Persistently elevated blood pressure caused by a medical problem.

PRIMARY (ESSENTIAL) HYPERTENSION
A form of hypertension whose cause is unknown.

FIGURE 11-7: Risk Factors for High Blood Pressure

Controllable Risk Factors	Uncontrollable Risk Factors
Obesity	Race (blacks develop it more than whites)
Eating too much sodium/salt	Heredity
Drinking too much alcohol	Increasing age
Lack of physical exercise	

FIGURE 11-8: Lifestyle Modifications to Manage Hypertension

Modification*	Recommendation	Approximate Systolic Blood Pressure Reduction
Weight reduction	Maintain normal body weight.	5–20 mm Hg/22-pound weight loss
Adopt DASH eating plan	Consume a diet rich in fruits, vegetables, and low-fat dairy products with a reduced content of saturated and total fat.	8–14 mm Hg
Dietary sodium reduction	Reduce dietary sodium intake to no more than 2400 mg sodium.	2–8 mm Hg
Physical activity	Engage in regular aerobic physical activity such as brisk walking (at least 30 minutes per day most days of the week)	4–9 mm Hg
Moderation of alcohol consumption	Limit consumption to no more than two drinks (24 oz. beer, 10 oz. wine, or 3 oz. 80-proof whiskey) per day in most men and to no more than one drink per day in women and lighter-weight persons.	2–4 mm Hg

*The effects of implementing these modifications are dose- and time-dependent, and could be greater for some individuals.

Source: "The Seventh Report of the Joint National Committee on Prevention, Detection, Evaluation, and Treatment of High Blood Pressure," National Institutes of Health and National Heart, Lung, and Blood Institute, 2003.

The following lifestyle modifications offer some hope for the prevention of hypertension and are effective in lowering the blood pressure of many people who follow them (Figure 11-8):

- Maintain normal body weight.
- Adopt the DASH (Dietary Approaches to Stopping Hypertension) eating plan (Figure 11-9), which has been shown to reduce blood pressure.
- Reduce dietary sodium intake to no more than 2300 mg.
- Engage in regular aerobic physical activity such as brisk walking (at least 30 minutes per day most days of the week).
- Limit consumption of alcohol to no more than two drinks (24 ounces of beer, 10 ounces of wine, or 3 ounces of 80-proof whiskey) per day for most men and no more than one drink per day for women and lighter-weight persons.

These lifestyle modifications can also reduce other risk factors for premature cardiovascular disease. Because of their ability to improve the cardiovascular risk profile, lifestyle modifications offer many benefits at little cost and with minimal risk.

Excess body weight is correlated closely with increased blood pressure. Weight reduction reduces blood pressure in a large proportion of hypertensive individuals who are more than 10 percent above ideal weight.

Excessive alcohol intake can raise blood pressure and cause resistance to antihypertensive therapy. Hypertensive patients who drink alcohol-containing beverages should be counseled to limit their daily intake to one drink per day for women and two drinks per day for men. One drink equals $1\frac{1}{2}$ ounces of 80-proof whiskey, 5 ounces of wine, or 12 ounces of beer.

Regular aerobic physical activity that is adequate to achieve at least a moderate level of physical fitness may be beneficial for both prevention and treatment of hypertension. Sedentary and unfit individuals with normal blood pressure have a 20 to 50 percent increased risk of developing hypertension during follow-up compared with their more active and fit peers.

FIGURE 11-9: The DASH Eating Plan

The DASH eating plan is based on 2000 calories a day. The number of daily servings in a food group may vary from those listed, depending on your caloric needs. Use this chart to help you plan menus or take it with you when you go to the store.

Food Group	Daily Servings (Except as Noted)	Serving Sizes	Significance of Examples and Notes	Each Food Group to the Dash Eating Plan
Grains and grain products	6–8	1 slice bread 1 oz. dry cereal* ½ cup cooked rice, pasta, or cereal	Whole-wheat bread and rolls, whole wheat pasta, English muffin, pita bread, bagel, cereals, grits, oatmeal, brown rice, unsalted pretzels and popcorn	Major sources of energy and fiber
Vegetables	4–5	1 cup raw leafy vegetable ½ cup cut-up raw or cooked vegetables ½ cup vegetable juice	Broccoli, carrots, collards, green beans, green peas, kale, lima beans, potatoes, spinach, squash, sweet potatoes, tomatoes	Rich sources of potassium, magnesium, and fiber
Fruits	4–5	1 medium fruit ¼ cup dried fruit ½ cup fresh, frozen, or canned fruit ½ cup fruit juice	Apples, apricots, bananas, dates, grapes, oranges, orange juice, grapefruit, grapefruit juice, mangoes, melons, peaches, pineapples, raisins, strawberries, tangerines	Important sources of potassium, magnesium, and fiber
Fat-free or low-fat milk and milk products	2–3	1 cup milk or yogurt 1½ oz. cheese	Fat-free (skim) or low-fat (1%) milk or buttermilk; fat-free, low-fat or reduced-fat cheese; fat-free or low-fat regular or frozen yogurt	Major sources of calcium and protein
Lean meats, poultry, and fish	6 or less	1 oz. cooked meats, poultry or fish 1 egg (Limit egg yolk intake to no more than 4/week) (2 egg whites have the same protein content as 1 oz. of meat)	Select only lean; trim away visible fats; broil, roast, or poach instead of frying; remove skin from poultry	Rich sources of protein and magnesium
Nuts, seeds, and dry beans	4–5 per week	⅓ cup or 1½ oz. nuts 2 Tbsp. peanut butter 2 tablespoon or ½ oz. seeds ½ cup cooked dry beans or peas	Almonds, hazelnuts, mixed nuts, peanuts, walnuts, sunflower seeds, peanut butter, kidney beans, lentils, split peas	Rich sources of energy, magnesium, protein, and fiber

(continued)

FIGURE 11-9: The DASH Eating Plan (*Continued*)

Food Group	Daliy Servings (Except as Noted)	Serving Sizes	Significance of Examples and Notes	Each Food Group to the Dash Eating Plan
Fats and oils[†]	2–3	1 tsp. soft margarine 1 tsp. vegetable oil 1 tablespoon mayonnaise 2 tablespoons salad dressing	Soft margarine, vegetable oil (such as olive, corn, canola, or safflower), low-fat mayonnaise, light salad dressing	DASH has 27 percent of calories as fat, including fat in or added to foods
Sweets	5 or less per week	1 Tbsp. sugar 1 Tbsp. jelly or jam ½ cup sorbet, gelatin 1 cup lemonade	Fruit-flavored gelatin, fruit punch, hard candy, jelly, maple syrup, sorbet and ices, sugar	Sweets should be low in fat

*Serving sizes vary between ½ cup and 1-¼ cups, depending on cereal type. Check the product's Nutrition Facts label.

†Fat content changes serving amount for fats and oils. For example, 1 Tbsp. of regular salad dressing equals 1 serving; 1 Tbsp. of a low-fat dressing equals ½ serving; 1 Tbsp. of a fat-free dressing equals 0 servings.

Source: "Your Guide to Lowering Your Blood Pressure with DASH," National Heart, Lung, and Blood Institute, NIH Publication 06-4082, Revision 2006.

On average, the higher your salt (sodium chloride) intake, the higher your blood pressure. Nearly all Americans consume substantially more salt than they need. Decreasing salt intake is important to reduce the risk of elevated blood pressure. Some individuals tend to be more sensitive to salt than others, including African Americans, middle-aged and older adults, and people with hypertension.

Calcium, potassium, magnesium, and vitamin C also need to be in the diet in adequate supply to normalize blood pressure. In fact, the DASH diet recommends 8 to 10 servings of fruits and vegetables (great sources of potassium, vitamin C, and magnesium) and 2 to 3 servings of dairy products daily (as part of a 2000-kcalorie diet). The DASH diet also emphasizes low intake of fat, saturated fat, cholesterol, and added sugars, and so it is reduced in red meat, sweets, and sugar-containing beverages. Because it is rich in fruits and vegetables, which are naturally lower in sodium than many other foods, the DASH eating plan makes it easier to consume less salt and sodium. Tips to start the DASH diet are listed in Figure 11-10.

When lifestyle modifications do not succeed in lowering blood pressure enough, drugs are the next step. Reducing blood pressure with drugs clearly decreases the incidence of cardiovascular death and disease.

MENU PLANNING FOR CARDIOVASCULAR DISEASES

Menu planning for cardiovascular diseases revolves around offering dishes rich in complex carbohydrates and fiber and using small amounts of fat, saturated fat, cholesterol, and sodium.

General Recommendations
- Decrease or replace salt in recipes by using vegetables, herbs, spices, and flavorings.
- Offer salt-free seasoning blends and lemon wedges.

FIGURE 11-10: Tips on Switching to the DASH Eating Plan

1. Change gradually.
   - If you now eat one or two vegetables a day, add a serving at lunch and another at dinner.
   - If you don't eat fruit now or have only juice at breakfast, try putting fresh fruit or dried fruit on your cereal, eating an apple or another fruit between meals, and making fruit part of desserts, such as strawberry shortcake.
   - Gradually increase your use of fat-free and low-fat dairy products to three servings a day. For example, drink milk with lunch or dinner instead of soda, sugar-sweetened tea, or alcohol. Choose low-fat or fat-free dairy products to reduce your intake of saturated fat, total fat, cholesterol, and kcalories.
   - Read food labels on margarines and salad dressings to choose those lowest in saturated fat and trans fat. Some margarines are now trans fat–free.
2. Treat meat as one part of the whole meal instead of the focus.
   - Limit meat to 6 ounces a day (two servings). Three to four ounces is about the size of a deck of cards.
   - If you now eat large portions of meat, cut them back gradually: a half or a third at each meal.
   - Include two or more vegetarian-style (meatless) meals each week.
   - Increase servings of vegetables, rice, pasta, and dry beans in meals. Try casseroles and pasta and stir-fry dishes, which have less meat and more vegetables, grains, and dry beans.
3. Use fruits or other foods low in saturated fat, cholesterol, and kcalories as desserts and snacks.
   - Fruits and other low-fat foods offer great taste and variety. Use fruits canned in their own juice. Fresh fruits require little or no preparation. Dried fruits are a good choice to carry with you or have ready in the car.
   - Try these snack ideas: unsalted nuts or pretzels mixed with raisins, graham crackers, low-fat and fat-free yogurt and frozen yogurt, popcorn with no salt or butter added, and raw vegetables.
4. Try these other tips.
   - Choose whole-grain foods to get added nutrients, such as minerals and fiber. For example, choose whole-wheat bread or whole-grain cereals.
   - If you have trouble digesting milk, try lactose-free milk.
   - Use fresh, frozen, or no-salt-added canned vegetables.

Source: "Facts About the DASH Eating Plan," National Heart, Lung, and Blood Institute, NIH Publication #03-4082, 2003.

Breakfast

- Offer fresh fruits and juices.
- Almost all cold and hot cereals are great choices. Some granola cereals tend to be high in fat.
- Most breads are low in fat except for croissants, brioche, cheese breads, and many biscuits. Whole-grain bagels, low-fat muffins, and baguettes are good choices.
- Have reduced-fat margarine and light cream cheese available to spread on bagels or toast.
- Serve turkey sausage, low-fat and low-sodium ham slices, and fish as leaner sources of protein than the traditional bacon and pork sausage.
- Offer egg substitutes for scrambled eggs and other egg-based items. Egg substitutes taste better when herbs, flavorings, and/or vegetables are cooked with them. Instead of egg substitutes, you can offer to make scrambled eggs and omelets by mixing one whole egg and two egg whites.

- Serve an omelet with vegetables such as chopped broccoli or spinach and low-fat cheese instead of regular cheese.
- As spreads or toppings on pancakes and waffles, offer sauces combining low-fat or non-fat yogurt with a fruit puree.
- Serve a breakfast buffet with loads of fruits, low-fat dairy products, and cereals.

Appetizers and Soups
- Offer juices and fresh sliced fruits. Fresh sliced fruits can be served with a yogurt dressing flavored with fruit juices.
- Offer raw vegetables with dips that use low-fat yogurt, low-fat cottage cheese, or ricotta cheese as the base rather than dips that use sour cream, cheeses, cream, or cream cheese. Try hummus, a chickpea-based dip, or salsa made from tomatoes, onions, hot peppers, garlic, and herbs.
- Offer grilled chicken, broiled Buffalo-style chicken wings, or steamed seafood such as shrimp.
- Use baked (rather than fried) potato skins and baked corn tortillas for tortilla chips. Sprinkle with grated cheese or garlic, onion, or chili powder.
- Feature soups that use stock as the base and vegetables and grains as the ingredients. Dried beans, peas, and lentils make great soups when cooked and pureed, without using cream or high-fat thickeners such as roux.

Salads
- Offer salads with lots of vegetables and fruits.
- Use only small amounts, if any, of bacon, meat, cheese, eggs, and croutons. Choose cooked beans and peas or low-fat cheeses.
- Offer reduced-calorie or nonfat salad dressings. Place on the side when desired.
- Make tuna fish salad and similar salads with low-fat mayonnaise.
- Use cooked salad dressing that contains little fat for Waldorf and other salads. It has a tarter flavor than mayonnaise.

Breads
- Serve whole-grain breads at each meal.

Entrées
- Serve combination dishes with small amounts of meat, poultry, or seafood with whole grains such as rice, legumes, vegetables, and/or fruits.
- Offer moderate portions of broiled, baked, stir-fried, or poached seafood; white-meat poultry without skin; and lean cuts of meat (see Chapter 5).
- Offer fresh meat, poultry, or seafood instead of canned, cured, smoked, and otherwise salty items (such as ham, corned beef, smoked turkey, dried cod, and most luncheon meats). Cheeses are high in sodium, so use only small amounts in sandwiches.
- Feature freshly made entrées instead of processed or prepared foods.
- Feature one or more meatless entrées, such as vegetarian burgers. Vegetarian burgers either try to imitate beef burgers (usually through the use of soy products) or are real veggie burgers (made of vegetables, especially mushrooms).
- For sauced entrées (or side dishes), feature sauces thickened with flour, cornstarch, or vegetable purees. Salsas, chutneys, relishes, and coulis also work well.
- Offer sandwiches made with roasted turkey, chicken, water-packed tuna fish salad, lean roast beef made from the round, or a spicy bean or lentil spread.
- For sandwich spreads, use reduced-fat mayonnaise, French- or Russian-style salad dressing, or salsa.

- Feature lots of different vegetables in sandwiches.
- Instead of high-sodium accompaniments to sandwiches such as pickles, olives, and potato chips, serve fresh vegetables, coleslaw made with reduced-calorie mayonnaise, or another healthful salad.

Side Dishes
- Side dishes of vegetables, grains, and pasta are good choices.
- Serve grilled potato halves instead of french fries, as well as other grilled vegetables.

Desserts
- Offer fruit-based desserts such as apple cobbler.
- Spotlight sorbets, sherbets, frozen yogurt, and ice milk. All contain less fat than ice cream does.
- Feature desserts made from egg whites (which are fat-free), such as angel food cake and meringues. Serve with a fruit sauce.
- Offer puddings made with skim milk.
- Serve low-fat cookies such as ladyfingers, biscotti, gingerbread, and fruit bars.

Beverages
- Offer fat-free or low-fat milk.

MINI-SUMMARY

1. Cardiovascular disease is the general term used to refer to diseases of the heart and blood vessels, such as coronary artery disease, stroke, and high blood pressure.
2. Atherosclerosis is a condition in which plaque builds up along artery walls.
3. Coronary heart disease is the most common form of cardiovascular disease and describes the damage to the heart caused by narrowing or blockage of the coronary arteries.
4. Smoking, high blood cholesterol, high blood pressure, physical inactivity, obesity and overweight, and diabetes are all risk factors for coronary heart disease.
5. Individuals with metabolic syndrome are at increased risk of dying from cardiovascular diseases.
6. In a heart attack, often a clot in a coronary artery at the site of narrowing and hardening stops the flow of blood.
7. Your risk for coronary heart disease can be assessed by measuring your total blood cholesterol, LDL, and HDL. Total cholesterol less than 200 mg/dL is desirable. Low-density lipoprotein, also known as "bad cholesterol" because it increases CHD risk, should be less than 100 mg/dL. High-density lipoprotein, also known as "good cholesterol," protects the arteries by taking cholesterol to the liver for disposal, and so the higher the HDL, the better. An HDL level of 60 mg/dL is desirable.
8. Therapeutic lifestyle changes to lower cholesterol include the TLC diet, physical activity, and weight management. Drugs may also be needed.
9. The TLC diet is a low-saturated-fat, low-cholesterol eating plan that calls for less than 7 percent of kcalories from saturated fat and less than 200 milligrams of dietary cholesterol per day.
10. There are two types of strokes: ischemic and hemorrhagic (Figure 11-3).
11. Stroke and heart disease have some of the same controllable risk factors: high blood pressure, cigarette smoking, high cholesterol, diabetes, physical inactivity, and obesity. These factors raise the risk for plaque buildup in the arteries, which in turn raises the risk of the formation of blockages and blood clots. High blood pressure is by far the most important risk factor.
12. High blood pressure is one of the major risk factors for coronary heart disease and stroke.

13. Figure 11-6 classifies blood pressure readings for adults.
14. The following lifestyle modifications offer some hope for the prevention of hypertension and are effective in lowering the blood pressure of many people who follow them (Figure 11-8):
    - Maintain normal body weight.
    - Adopt the DASH eating plan (Figure 11-9), which has been shown to reduce blood pressure.
    - Reduce dietary sodium intake to no more than 2300 mg.
    - Engage in regular aerobic physical activity such as brisk walking (at least 30 minutes per day most days of the week)
    - Limit consumption of alcohol to no more than two drinks (24 ounces of beer, 10 ounces of wine, or 3 ounces of 80-proof whiskey) per day for most men and no more than one drink per day for women and lighter-weight persons.
15. Menu-planning guidelines are given for cardiovascular disease in this section.

Caffeine and Health

Check out how much you know or don't know about caffeine. Are the following true or false?

1. Tea has more caffeine than coffee.
2. Brewed coffee has more caffeine than instant coffee.
3. Some nonprescription drugs contain caffeine.
4. Caffeine is a nervous system stimulant.
5. Withdrawing from regular caffeine use causes physical symptoms.

Check your answers as you read on.

Caffeine is a naturally occurring stimulant found in the leaves, seeds, or fruits of over 60 plants around the world. Caffeine appears in the coffee bean in Arabia, the tea leaf in China, the kola nut in West Africa, and the cocoa bean in Mexico. Because of its use throughout all societies, caffeine is the most widely used psychoactive substance in the world. The most common caffeine sources in North America and Europe are coffee and tea. During the last 20 years, extensive research has been conducted on how caffeine affects health. Most experts agree that moderate use of caffeine (300 milligrams or about 3 cups of coffee) is not likely to cause health problems.

HOW CAFFEINE AFFECTS THE BODY

Caffeine is best known for its stimulant, or "wake-up," effect. Once you drink a cup of coffee, the caffeine is readily absorbed by the body and carried around in the bloodstream, where its level peaks in about one hour. Caffeine mildly stimulates the nervous and cardiovascular systems. Caffeine affects the brain and results in an elevated mood, decreased fatigue, and increased attentiveness. You can now think more clearly and work harder. It increases your heart rate, blood flow, respiratory rate, and metabolic rate for several hours. When taken before bedtime, caffeine can interfere with getting to sleep.

Exactly how caffeine will affect you, and for how long, depends on many factors such as the amount of caffeine ingested, whether you are male or female, your height and weight, your age, and whether you are pregnant or smoke. Caffeine is converted by the liver into substances that are excreted in the urine.

Some people are more sensitive than others to the effects of caffeine. With frequent use, tolerance to many of the effects of caffeine develops. At high doses of 600 milligrams (about 6 cups of coffee) or more daily, caffeine can cause nervousness, sweating, tenseness, an upset stomach, anxiety, and insomnia. It can also prevent clear thinking and increase the side effects of certain medications. This level of caffeine intake represents a significant health risk.

Caffeine can be mildly addicting. Even when moderate amounts of caffeine are withdrawn for 18 to 24 hours, you may feel symptoms such as headache, fatigue, irritability, depression, and poor concentration. The symptoms peak on days 1 or 2 and progressively decrease over the course of a week. To minimize withdrawal symptoms, experts recommend reducing caffeine intake gradually.

CAFFEINE IN FOODS AND DRUGS

Figure 11-11 displays the amount of caffeine in foods. An 8-ounce cup of drip-brewed coffee has about 85 milligrams of caffeine, whereas the same amount of brewed tea contains about 47 milligrams. Twelve-ounce cans of cola soft drinks provide about 35 to 45 milligrams of caffeine.

FIGURE 11-11: Caffeine in Foods and Beverages

Food/Beverage	Caffeine (Milligrams)
COFFEE	
Espresso coffee, brewed, 8 fluid ounces	502
Coffee, brewed, 8 fluid ounces	85
Coffee, instant, 8 fluid ounces	62
Coffee, brewed, decaffeinated, 8 fluid ounces	3
Coffee, instant, decaffeinated, 8 fluid ounces	2
TEA	
Tea, brewed, 8 fluid ounces	47
Tea, herbal, brewed, 8 fluid ounces	0
Tea, instant, 8 fluid ounces	29
Tea, brewed, decaffeinated, 8 fluid ounces	3
CHOCOLATE BEVERAGES	
Hot chocolate, 8 fluid ounces	5
Chocolate milk, 8 fluid ounces	5
SOFT DRINKS	
Cola, 12-ounce can	37
Cola, with higher caffeine, 12-ounce can	100
Cola or Dr. Pepper, diet, 12-ounce can	49
Cola or Dr. Pepper, regular or diet, without caffeine, 12-ounce can	0
Lemon-lime soda, regular or diet, 12-ounce can	0
Lemon-lime soda, with caffeine, 12-ounce can	55
Ginger ale, regular or diet, 12-ounce can	0
Root beer, regular or diet, 12-ounce can	0
CHOCOLATE	
Milk chocolate bar, 1.55 ounces	9
M&M milk chocolate candies, 1.69 ounces	5
Dark chocolate, semisweet, 1 ounce	20

Source: U.S. Department of Agriculture National Nutrient Database for Standard Reference, Release 16, July 2003.

The caffeine content of coffee and tea depends on the variety of the coffee bean or tea leaf, the particle size, the brewing method, and the length of brewing or steeping time. Brewed coffee has more caffeine than instant coffee, and espresso has more caffeine than brewed coffee. Espresso is made by forcing hot pressurized water through finely ground dark-roast beans. Because it is brewed with less water, it contains more caffeine than regular coffee per fluid ounce.

In soft drinks, caffeine is both a natural and an added ingredient. About 5 percent of the caffeine in colas and Dr. Pepper is obtained naturally from cola nuts; the remaining 95 percent is added. Caffeine-free drinks contain virtually no caffeine and make up a small part of the soft drink market.

Numerous prescription and nonprescription drugs also contain caffeine. It is often used in headache and pain-relief remedies, cold products, and alertness or stay-awake tablets. When caffeine is an ingredient, it must be listed on the product label. Caffeine increases the ability of aspirin and other painkillers to do their job.

CAFFEINE AND HEALTH

Current research on how caffeine affects a variety of health issues is summarized in this section. Keep in mind that most experts agree that moderate use of caffeine is not likely to cause any health problems. Moderate use of caffeine is considered to be 300 milligrams, which is equal to about 3 cups of coffee daily.

- Studies have looked at the effects of caffeine on heart health. Moderate caffeine consumption does not appear to adversely affect cardiovascular health.
- Caffeine appears to increase the excretion of calcium, a mineral needed for healthy bones. Calcium is particularly important to prevent osteoporosis. Osteoporosis is a bone disease characterized by loss of bone strength that is seen especially in older women, although men get it too. Moderate caffeine intake doesn't seem to cause a problem with calcium as long as you are consuming the recommended amounts (adult men and women should be taking in from 1000 to 1200 milligrams of calcium, depending on age and gender).
- In the past there were concerns that the caffeine in coffee may cause cancer. Research shows that caffeine as present in coffee does not cause breast or intestinal cancer. However, not enough research has been done to determine whether caffeine in coffee is involved in urinary bladder or pancreatic cancer. In moderation, it is unlikely that caffeine causes cancer.
- Evidence suggests that at levels over 500 milligrams a day, caffeine may delay conception. Moderate caffeine consumption does not appear to be of concern to women trying to get pregnant.
- Caffeine had a reputation until recently of being a diuretic, a substance that increases the flow of urine. We now know that caffeine does not flush water out of the body, and so caffeinated beverages do not dehydrate you.
- Because children have developing nervous systems, it is important to moderate their caffeine consumption. For children, major sources of caffeine include soft drinks and chocolate.
- Caffeine may be useful as part of a weight-control program because it increases the rate at which the body burns calories for three or more hours after being consumed.

Caffeine's ability to improve physical performance is well known among well-trained athletes. Through a mechanism that is not completely understood, caffeine seems to increase endurance and speed in some situations. Excessive use of caffeine is restricted in international competitions.

Due to its stimulant properties, caffeine is used around the world in any of its many forms, such as coffee, tea, soft drinks, and chocolate. When used in moderation, caffeine is not likely to cause health problems.

NUTRITION AND CANCER

After heart disease, cancer is the second leading cause of death in the United States. ***Cancer*** is a group of diseases characterized by unrestrained cell division and growth that can disrupt the normal functioning of an organ and spread beyond the tissue in which it started. Cancer is basically a two-step process (Figure 11-12). First, a ***carcinogen***, such as an X-ray, initiates the sequence by altering the genetic material of a cell, deoxyribonucleic acid (DNA), and causing a mutation. Such cells are generally repaired or replaced. When repair or replacement does not occur, however, ***promoters*** such as alcohol and high-kcalorie diets can advance the development of the mutated cell into a tumor. Promoters do not initiate cancer but enhance its development once initiation has occurred. The tumor may disrupt normal body functions and leave the tissue for other sites, a process called ***metastasis***.

Cancer develops as a result of internal and external factors. Internal factors include genetics, hormones, and so forth. Evidence shows that although inherited genes influence the risk of getting cancer, genes alone explain only a fraction of cancer cases. External factors could be tobacco, chemicals, radiation, and other influences. Whereas smoking causes the greatest number of cancer cases, up to one-third of all cancer deaths are very possibly related to nutrition, overweight or obesity, and physical inactivity.

Lung and bronchus, colorectal, prostate, and breast cancers are the four leading causes of cancer death in the United States. (Skin cancer is also prevalent but is not related to diet.) Research suggests that diet and exercise play a role in the cause of certain cancers, as follows:

- Alcohol consumption: associated with increased risk of oral, esophageal, breast, and liver cancers
- Red meat: associated with increased risk of colon cancer
- Physical inactivity: associated with increased risk of colon, breast, and possibly other cancers
- Being overweight: associated with colon, breast (for postmenopausal women), prostate, and possibly other cancers

CANCER
A group of diseases characterized by unrestrained cell division and growth that can disrupt the normal functioning of an organ and also spread beyond the tissue in which it started.

CARCINOGEN
Cancer-causing substance.

PROMOTERS
Substances, such as fat, that advance the development of mutated cells into a tumor.

METASTASIS
The condition when a cancer spreads beyond the tissue in which it started.

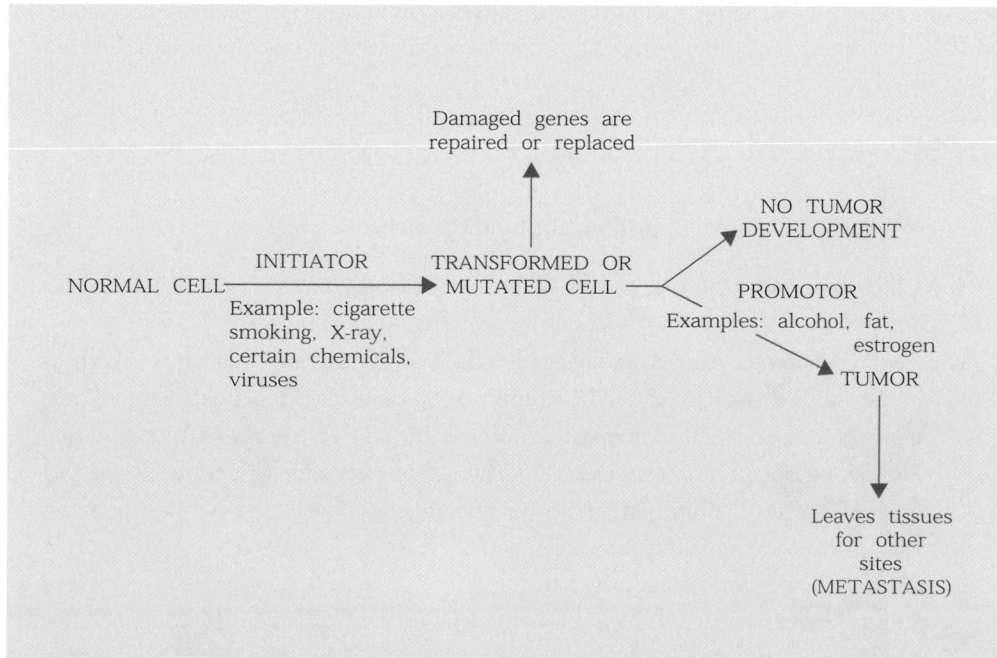

FIGURE 11-12:
Process of Cancer.

The American Cancer Society has four guidelines to reduce cancer risk:

1. Maintain a healthful weight throughout life.
   - Balance kcaloric intake with physical activity.
   - Avoid excessive weight gain as you get older.
   - If you are currently overweight or obese, achieve and maintain a healthy weight.
2. Adopt a physically active lifestyle.
   - Adults: Engage in at least 30 minutes of moderate to vigorous physical activity, above usual activities, on 5 or more days of the week; 45 to 60 minutes of intentional physical activity are preferable.
   - Children and adolescents: Engage in at least 60 minutes per day of moderate to vigorous physical activity at least 5 days per week.
3. Eat a healthy diet with an emphasis on plant sources.
   - Eat 5 or more servings of a variety of vegetables and fruits each day.
   - Choose whole grains in preference to processed (refined) grains.
   - Limit consumption of processed and red meats.
4. If you drink alcoholic beverages, limit consumption to no more than 1 drink per day for women or 2 per day for men.

Fruits and vegetables, as well as other plant foods, contain **phytochemicals**, minute plant compounds that fight cancer formation (see Figure 6-14 on page 216). For instance, broccoli contains the chemical sulforaphane, which seems to initiate increased production of cancer-fighting enzymes in the cells. Isoflavonoids, found mostly in soy foods, are known as plant estrogens or phytoestrogens because they are similar to estrogen and interfere with its actions (estrogen seems to promote breast tumors). Members of the cabbage family (cabbage, broccoli, cauliflower, mustard greens, kale), also called **cruciferous vegetables**, contain phytochemicals such as indoles and dithiolthiones that activate enzymes that destroy carcinogens.

Some consumers are concerned about eating more fruits and vegetables that may contain carcinogenic pesticides. The National Academy of Sciences, along with other organizations, feels that the health benefits of eating fresh fruits and vegetables far outweigh any risk associated with pesticide residues. The federal government strictly regulates the kinds and amounts of pesticides used on field crops grown in the United States. The tiny amounts of pesticide residues found on produce are hundreds of times lower than the amounts that would pose any health threat.

MENU PLANNING TO LOWER CANCER RISK

Use the following guidelines to plan menus to lower cancer risk.

1. Offer lower-fat menu items, such as those based on plant foods. See pages 380 and 381 for menu-planning ideas for vegetarians using plant foods.
2. Avoid salt-cured, smoked, and nitrite-cured foods. These foods, which are also high in fat, include anchovies, bacon, corned beef, dried chipped beef, herring, pastrami, processed lunch meats such as bologna and hot dogs, sausage such as salami and pepperoni, and smoked meats and cheeses. Conventionally smoked meats and fish contain tars that are thought to be carcinogenic due to the smoking process. Nitrites are also thought to be carcinogenic.
3. Offer high-fiber foods. For example:
   - Use beans and peas as the basis for entrées and add them to soups, stews, casseroles, and salads. Nuts and seeds are high in fiber but also contain a significant amount of fat and kcalories, so use them sparingly.

PHYTOCHEMICALS
Minute substances in plants that may reduce the risk of cancer and heart disease when eaten often.

CRUCIFEROUS VEGETABLES
Members of the cabbage family containing phytochemicals that may help prevent cancer.

- Serve whole-grain breads, rolls, crackers, cereals, and muffins. Bran or wheat germ can be added to some baked goods to increase the fiber content.
- High-fiber grains such as brown rice and bulgur (cracked wheat) can be used as side dishes instead of white rice.
- Leave skins on potatoes, fruits, and vegetables as much as possible.
- Offer salads using lots of fresh fruits and vegetables. Omit chopped eggs and bacon bits, both of which contribute fat.

4. Offer lots of dishes that incorporate fruits and vegetables, especially cruciferous vegetables. Cruciferous vegetables contain substances that are thought to be natural anti-carcinogens. These vegetables include broccoli, brussels sprouts, cabbage, cauliflower, bok choy, kale, collards, kohlrabi, mustard, rutabagas, spinach, and watercress. Wash fruits and vegetables well and leave the peel on fruits.

5. Offer foods that are good sources of beta-carotene and vitamins C and E, all of which are antioxidants. Excellent sources of beta-carotene include dark green, yellow, and orange vegetables and fruits such as broccoli, cantaloupe, carrots, spinach, squash, and sweet potatoes. Good sources include apricots, beet greens, brussels sprouts, cabbage, nectarines, peaches, tomatoes, and watermelon. Sources of vitamin C include citrus fruits and juices, any other juices with vitamin C added, berries, tomatoes, broccoli, brussels sprouts, cabbage, melons, cauliflower, and potatoes. Vitamin E is found in vegetable oils and margarines, whole-grain cereals, wheat germ, soybeans, leafy greens, and spinach. To reduce cancer risk, the best advice currently is to consume antioxidants through food sources rather than as supplements.

6. Offer alternatives to alcoholic drinks. Heavy drinkers are more likely to develop cancer in the gastrointestinal tract, such as cancer of the esophagus and stomach.

7. See page 287 for information on how to grill meats to decrease cancer risks.

MINI-SUMMARY

1. Cancer begins as depicted in Figure 11-12.
2. Guidelines to reduce cancer risk include eating a variety of healthful foods with an emphasis on plant sources, choosing whole grains often, limiting consumption of red meats, adopting a physically active lifestyle, maintaining a healthful weight, and drinking alcoholic beverages in moderation.
3. Menu-planning guidelines are given in this section.

NUTRITION AND DIABETES MELLITUS

Diabetes mellitus gets its name from the ancient Greek word for a siphon (tube), because early physicians noted that diabetics tend to be unusually thirsty and to urinate a lot, as if a tube quickly drained out everything they drank. Mellitus is from the Latin version of the ancient Greek word for honey; it was used because doctors in centuries past diagnosed the disease by the sweet taste of the patient's urine.

The number of people diagnosed with diabetes has almost doubled since 1990. In 2005, some 20.8 million people had diabetes, over 7 percent of the population. Among those 20.8 million people, 14.6 million have been diagnosed and over 6 million have diabetes but don't know it yet. Figure 11-13 lists risk factors for diabetes.

Diabetes is a disease in which there is insufficient or ineffective insulin, a hormone that helps regulate the blood sugar level. When the blood sugar rises, such as after eating a meal,

> **DIABETES MELLITUS**
> A disorder of carbohydrate metabolism characterized by high blood sugar levels and inadequate or ineffective insulin.

FIGURE 11-13: Risk Factors for Diabetes

- Age: Risk increases with age.
- Overweight: Body mass index (BMI) of 25 or higher (23 or higher if Asian American, 26 or higher if Pacific Islander)
- Family history of diabetes
- Ethnicity: Black, American Indian, Alaska Native, Asian American, Pacific Islander, or Hispanic heritage
- History of diabetes in pregnancy or giving birth to a baby weighing more than 9 pounds
- Inactive lifestyle: Exercise less than 3 times a week

Source: National Institute of Diabetes and Digestive and Kidney Diseases.

HYPERGLYCEMIA
High levels of blood sugar.

the pancreas releases insulin. The insulin facilitates the entry of glucose into body cells, to be used for energy. If there is no insulin or if the insulin is not working, sugar cannot enter the cells. Thus, high blood sugar levels (called *hyperglycemia*) result, and sugar spills into the urine.

Overall, the risk for death among people with diabetes is about two times that of people without diabetes. People with diabetes are more vulnerable to many kinds of infections and to deterioration of the kidneys, heart, blood vessels, nerves, and vision. Diabetes can lead to serious complications such as blindness, kidney damage, and lower-limb amputations. Working together, people with diabetes and their health-care providers can reduce the occurrence of these and other diabetic complications (Figure 11-14) by controlling the levels of blood glucose, blood pressure, and blood lipids and by receiving other forms of preventive care in a timely manner.

There are two types of diabetes.

TYPE 1 DIABETES
A form of diabetes seen mostly in children and adolescents. These patients make no insulin and therefore require frequent injections of insulin to maintain a normal level of blood glucose.

TYPE 2 DIABETES
A form of diabetes seen most often in overweight adults but increasingly seen in adolescents and children. With type 2 diabetes, either the body does not produce enough insulin or the body's cells do not use insulin properly.

1. *Type 1 diabetes.* This form of diabetes is usually diagnosed in children and adolescents. These patients produce no insulin at all and therefore require frequent doses of insulin delivered by injections or a pump to maintain a normal level of blood glucose. Fewer than 10 percent of Americans who have diabetes have type 1. Type 1 diabetes develops when the body's immune system destroys pancreatic beta cells, the only cells in the body that make insulin. Risk factors for type 1 diabetes may include autoimmune, genetic, and environmental factors.

2. *Type 2 diabetes.* Type 2 diabetes is the most common form of diabetes. People can develop type 2 diabetes at any age, even during childhood. With type 2 diabetes, either the body does not produce enough insulin or the body's cells do not use insulin properly. This form of diabetes usually begins with insulin resistance, a condition in which fat, muscle, and liver cells have problems using insulin to put sugar into the cells. Blood sugar levels start to rise slowly as the pancreas keeps up with the added demand by producing more insulin. In time, however, it loses the ability to secrete enough insulin in response to meals. Being overweight and inactive increases the chances of developing type 2 diabetes. Some of these patients require insulin, but many do not.

Type 2 diabetes accounts for more than 90 percent of cases of diabetes in the United States. Type 2 diabetes is nearing epidemic proportions in the United States due to an

FIGURE 11-14: Complications of Diabetes

HEART DISEASE AND STROKE

- Heart disease is the leading cause of diabetes-related deaths. Adults with diabetes have heart disease death rates about 2 to 4 times higher than those of adults without diabetes.
- The risk for stroke is 2 to 4 times higher among people with diabetes.
- About 65 percent of deaths among people with diabetes are due to heart disease and stroke.

HIGH BLOOD PRESSURE

- About 73 percent of adults with diabetes have blood pressure greater than or equal to 130/80 mm Hg or use prescription medications for hypertension.

BLINDNESS

- Diabetes is the leading cause of new cases of blindness among adults age 20–74 years.
- Diabetic retinopathy causes 12,000 to 24,000 new cases of blindness each year.

KIDNEY DISEASE

- Diabetes is the leading cause of end-stage renal disease, accounting for 44 percent of new cases.

NERVOUS SYSTEM DISEASE

- About 60 to 70 percent of people with diabetes have mild to severe forms of nervous system damage. The results of such damage include impaired sensation or pain in the feet or hands, slowed digestion of food in the stomach, carpal tunnel syndrome, and other nerve problems.
- Severe forms of diabetic nerve disease are a major contributing cause of lower-extremity amputations.

AMPUTATIONS

- More than 60 percent of nontraumatic lower-limb amputations occur among people with diabetes.

DENTAL DISEASE

- Periodontal (gum) disease is more common among people with diabetes. Among young adults, those with diabetes have about twice the risk as those without diabetes.
- Almost one-third of people with diabetes have severe periodontal diseases with loss of attachment of the gums to the teeth measuring 5 millimeters or more.

COMPLICATIONS OF PREGNANCY

- Poorly controlled diabetes before conception and during the first trimester of pregnancy can cause major birth defects in 5 to 10 percent of pregnancies and spontaneous abortions in 15 to 20 percent of pregnancies.
- Poorly controlled diabetes during the second and third trimesters of pregnancy can result in excessively large babies, posing a risk to the mother and the child.

OTHER COMPLICATIONS

- Uncontrolled diabetes often leads to biochemical imbalances that can cause acute life-threatening events such as diabetic ketoacidosis and hyperosmolar (nonketotic) coma.
- People with diabetes are more susceptible to many other illnesses and, once they acquire these illnesses, often have worse prognoses. For example, they are more likely to die from pneumonia or influenza than are people who do not have diabetes.

Source: National Diabetes Fact Sheet, National Center for Chronic Disease Prevention and Health Promotion, 2005.

increased number of older Americans and a greater prevalence of obesity and sedentary lifestyles. Characteristics of type 2 diabetes include the following:

- Age of onset over 40 years old in most cases, although alarming numbers of obese adolescents and children are being diagnosed
- Most frequently occurs in overweight and obese individuals; occasionally occurs in people of normal weight
- Usually slow onset with increased thirst, urination, and appetite, developing over weeks to months or even years
- Can be a "silent" disease
- Most often treated first with diet and exercise, progressing to pills and later to insulin if needed
- Often responds to weight loss and/or changes in diet and exercise

Research has shown that losing 5 to 7 percent of body weight through diet and increased physical activity can prevent or delay insulin resistance from progressing to type 2 diabetes.

Although both types of diabetes are popularly called "sugar diabetes," they are not caused by eating too many sweets. High sugar levels in the blood and urine are a result of these illnesses; their exact causes are unknown.

Symptoms of type 1 diabetes typically appear abruptly and include frequent urination, insatiable hunger, and unquenchable thirst. Unexplained weight loss is also common, as are blurred vision (or other vision changes), nausea and vomiting, weakness, drowsiness, and extreme fatigue.

The most immediate and life-threatening aspect of type 1 diabetes is the formation of poisonous acids called ketone bodies. They occur as an end product of burning fat for energy, because glucose does not get into the cells to be burned for energy. Like glucose, ketone bodies accumulate in the blood and spill into the urine. Ketone bodies are acidic and make the blood acidic, resulting in a dangerous condition called ketoacidosis.

Symptoms of type 2 diabetes may include any or all of those of type 1 except that the problem with ketone bodies is rare. Symptoms are often overlooked because they tend to come on gradually and be less pronounced. Other symptoms are tingling or numbness in the lower legs, feet, or hands; skin or genital itching; and gum, skin, or bladder infections that recur and are slow to clear up. Again, many people fail to connect these symptoms with diabetes.

Measuring glucose levels in samples of the patient's blood is key to diagnosing both types of diabetes. This is first done early in the morning on an empty stomach. For more information, the blood test is repeated before and after the patient has drunk a liquid containing a known amount of glucose.

Treatment for either type seeks to do what the human body normally does naturally: maintain a proper balance between glucose and insulin. The guiding principle is that food makes the blood glucose level rise, whereas insulin and exercise make it fall. The trick is to juggle food, insulin, and exercise to avoid both hyperglycemia (a blood glucose level that is too high) and hypoglycemia (one that is too low). Either condition causes problems for the patient.

The cornerstone of treatment is diet, exercise, and medication. The exact nature of the diet has changed over the years from starvation diets to more liberal diets. The current diet is based on the following principles:

- There is no one diet suitable for every person with diabetes. The diet needs to be based on each person's type of diabetes, food preferences, culture, age, lifestyle, medication, other health concerns, education, nutrition status, medical treatment goals, and other factors.

- The goals of medical nutrition therapy are to maintain the best glucose control possible, keep blood levels of fat and cholesterol in normal ranges, maintain or get body weight within a desirable range, and meet all nutrient needs.
- Sugars can be incorporated into the meal pattern in moderation as part of the total carbohydrate allowance. The total amount of carbohydrate consumed, rather than the source of carbohydrate, should be the priority. Because sugars appear in foods that usually contain a lot of kcalories and fat, sweets can be eaten occasionally.
- Instead of setting rigid percentages of protein, fat, and carbohydrates, guidelines recommend that 60 to 70 percent of kcalories should come from carbohydrates and monounsaturated fats. Saturated fats should be kept at 10 percent or less of total kcalories. The meal pattern should provide sufficient fiber. Kcaloric distribution depends on the individual's nutritional assessment and treatment goals.

The Exchange Lists for Meal Planning have been developed by the American Diabetes Association and the American Dietetic Association for use primarily by people with diabetes, who need to regulate what and how much they eat. There are seven exchange lists of like foods. Each food on a list has approximately the same amount of kcalories, carbohydrate, fat, and protein as another in the portions listed, so that any food on a list can be exchanged, or traded, for any other food on the same list. The seven exchange lists are starch; fruit; milk; sweets, desserts, and other carbohydrates; nonstarchy vegetables; meat and meat substitutes; and fat.

Each exchange list has a typical item with an easy-to-remember portion size:

Starch: 1 slice bread, 80 kcalories
Fruit: 1 small apple, 60 kcalories
Milk: 1 cup fat-free milk, 90 kcalories
Other carbohydrates: 2 small cookies, kcalories vary
Nonstarchy vegetables: ½ cup cooked carrots, 25 kcalories
Meat: 1 ounce lean meat such as chicken, 55 kcalories
Fat: 1 teaspoon margarine, 45 kcalories

The meat exchange is broken down into very lean, lean, medium-fat, and high-fat meat and meat alternatives. Very lean and lean meats are encouraged. The milk exchange contains fat-free or low-fat, reduced-fat, and whole-milk exchanges. Fats are divided into three groups, based on the main type of fat they contain: monounsaturated, polyunsaturated, or saturated. There is also a listing of free foods that contain negligible kcalories.

MINI-SUMMARY

1. There are two classifications of diabetes: type 1, seen mostly in children, and type 2, seen mostly in overweight adults and in growing numbers of overweight adolescents and children.
2. The risk for death among people with diabetes is about two times that of people without diabetes. People with diabetes are more vulnerable to many kinds of infections and to deterioration of the kidneys, heart, blood vessels, nerves, and vision.
3. The cornerstone of treatment is diet, exercise, and medication. Diets are individualized for each person and use an exchange system. Sugar need not be avoided and can be incorporated into the diabetic diet as part of the total carbohydrate allowance.

OSTEOPOROSIS

Osteoporosis, or porous bone, is a disease characterized by low bone mass and structural deterioration of bone tissue, leading to bone fragility and an increased susceptibility to fractures of the hip, spine, and wrist (Figure 11-15). Men as well as women suffer from osteoporosis, a disease that can be prevented and treated.

Osteoporosis is often called the "silent disease" because bone loss occurs without symptoms. People may not know that they have osteoporosis until their bones become so weak that a sudden strain, bump, or fall causes a hip fracture or causes a vertebra to collapse. Collapsed vertebrae may initially be felt or seen in the form of severe back pain, loss of height, or severely stooped posture.

FACTS AND FIGURES

Here are some important facts about osteoporosis.

- Osteoporosis is a major public health threat for 44 million Americans, or 55 percent of people 50 years of age and older.
- In the United States today, 10 million individuals already have osteoporosis and 34 million more have low bone mass, placing them at increased risk for this disease.
- Eighty percent of those affected by osteoporosis are women and twenty percent are men.
- One of every two women and one in four men over age 50 will have an osteoporosis-related fracture in their lifetime.
- Osteoporosis can strike at any age.

FIGURE 11-15:
Osteoporosis: sites of fractures.
Courtesy International Osteoporosis Foundation, www.iofbonehealth.org.

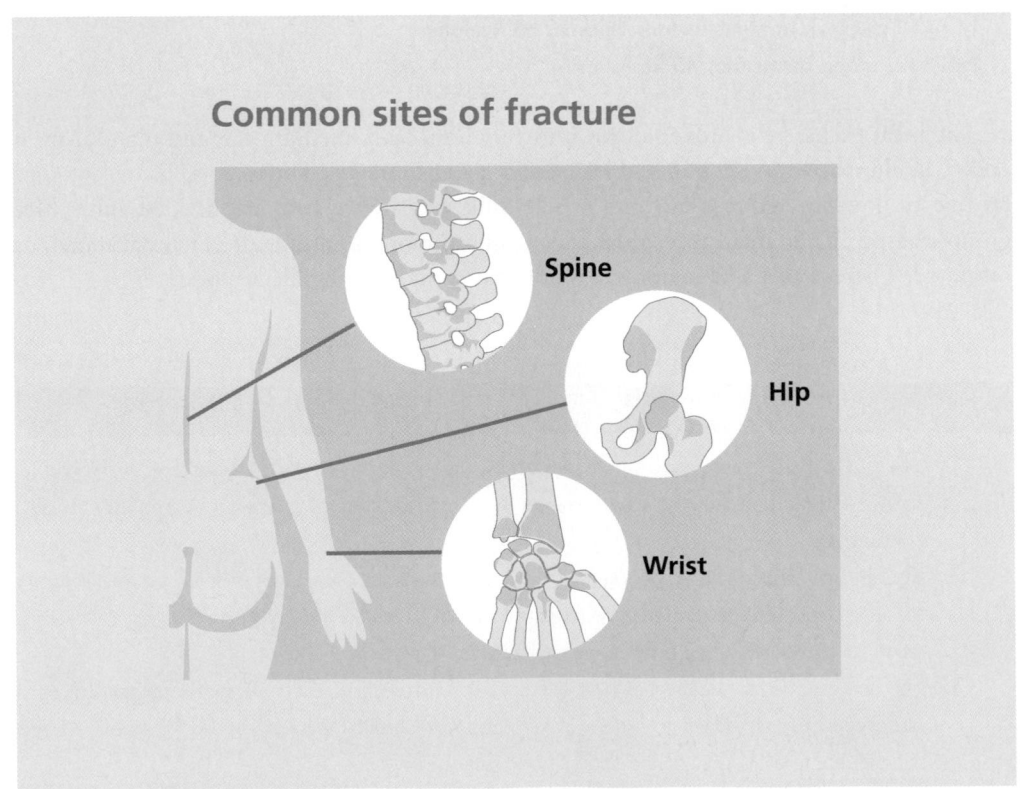

- The total number of fractures due to osteoporosis was about 2 million in 2005, but is expected to rise to more than 3 million by 2025.
- An average of 24 percent of hip fracture patients age 50 and over die in the year following their fracture.

WHAT IS BONE?

Bone is living, growing tissue. It is made mostly of collagen, a protein that provides a soft framework, and calcium phosphate, which adds strength. This combination makes bones strong yet flexible enough to withstand stress.

Throughout your life, old bone is removed (called resorption) and new bone is added to the skeleton. During childhood and the teenage years, new bone is added faster than old bone is removed. As a result, bones become larger, heavier, and denser. Bone formation continues at a pace faster than resorption until peak bone mass (maximum bone density and strength) is reached at around age 30. After age 30, bone resorption slowly begins to exceed bone formation.

Bone loss is most rapid in the first few years after menopause but persists into the post-menopausal years. Osteoporosis is more likely to develop if you did not reach optimal bone mass during your bone-building years.

RISK FACTORS

Certain risk factors contribute to an individual's likelihood of developing osteoporosis. There are some risk factors that you can't change and others that you can. Here are the ones you can't change:

- Gender. Your chances of developing osteoporosis are greater if you are a woman. Women have less bone mass than men and lose bone more rapidly because of the changes involved in menopause.
- Age. The older you are, the greater your risk of osteoporosis. Your bones become less dense and weaker as you age.
- Body size. Small, thin-boned women are at greater risk.
- Ethnicity. White and Asian women have the highest risk. African American and Latino women have a lower but significant risk.
- Family history. Susceptibility to fracture may be in part hereditary. People whose parents have a story of fractures also seem to have reduced bone mass and may be at risk for fractures.

These risk factors you can change:

- A diet low in calcium and vitamin D. For maximum bone health, adequate amounts of calcium and vitamin D need to be taken in. This is especially important before age 30 and during the 5 to 10 years after the beginning of menopause.
- Sedentary lifestyle. Participation in sports and exercise increases bone density in children. To keep bones healthy, exercise must be weight-bearing or involve strength training. In older adults, weight training improves bone density.
- Cigarette smoking or excessive use of alcohol.

PREVENTION

To reach optimal peak bone mass and continue building new bone tissue as you get older, there are several things you can do.

Calcium

An inadequate supply of calcium over the lifetime is thought to play a significant role in the development of osteoporosis. Many published studies show that low calcium intake appears to be associated with low bone mass, rapid bone loss, and high fracture rates. National nutrition surveys have shown that many people consume less than half the amount of calcium recommended to build and maintain healthy bones. Depending on how much calcium you get each day from food, you may need to take a calcium supplement.

Calcium needs change during one's lifetime. The body's demand for calcium is greater during childhood and adolescence, when the skeleton is growing rapidly, and during pregnancy and breast-feeding. Postmenopausal women and older men also need to consume more calcium. This need may be caused by inadequate intake of vitamin D, which is necessary for intestinal absorption of calcium. Also, as you age, your body becomes less efficient at absorbing calcium and other nutrients. Older adults also are more likely to have chronic medical problems and to use medications that may impair calcium absorption.

Vitamin D

Vitamin D plays an important role in calcium absorption and bone health. It is synthesized in the skin through exposure to sunlight. While many people are able to obtain enough vitamin D naturally, studies show that vitamin D production decreases in the elderly, in people who are housebound, and during the winter. These individuals may require vitamin D supplementation to ensure adequate intake.

Exercise

Regular weight-bearing and muscle-strengthening exercise is important. Weight-bearing exercises that force you to work against gravity are good. Examples include walking, hiking, jogging, stair climbing, weight training, tennis, and dancing.

Smoking

Smoking is bad for your bones as well as for your heart and lungs. The mechanism that results in greater bone loss among smokers is unclear. Although smokers are thinner and may tend to exercise less, females who smoke also have lower levels of estrogen and experience an earlier menopause. All of these factors are associated with lower bone density.

Alcohol

Regular consumption of 2 to 3 ounces a day of alcohol may be damaging to the skeleton, even in young women and men. Those who drink heavily are more prone to bone loss and fractures because of both poor nutrition and an increased risk of falling.

Medications

The long-term use of glucocorticoids (medications prescribed for a wide range of diseases, including arthritis, asthma, Crohn's disease, lupus, and diseases of the lungs, kidneys, and liver) can lead to a loss of bone density and fractures. Other forms of drug therapy that can cause bone loss include long-term treatment with certain antiseizure drugs, such as phenytoin

(Dilantin) and barbiturates; gonadotropin-releasing hormone (GnRH) analogs used to treat endometriosis; excessive use of aluminum-containing antacids; certain cancer treatments; and excessive thyroid hormone. It is important to discuss the use of these drugs with your physician and not to stop or alter medication doses on your own.

DETECTION

After a comprehensive medical assessment, your doctor may recommend that you have your bone mass measured. Bone mineral density (BMD) tests measure bone density in the spine, wrist, and/or hip (the most common sites of fractures due to osteoporosis), while other tests measure bone in the heel or hand. These tests use a special kind of X-ray and are safe and painless. Most doctors will recommend a type of BMD test using a central DXA, which stands for dual energy X-ray absorptiometry.

TREATMENT

A comprehensive osteoporosis treatment program includes a focus on proper nutrition (especially for calcium and vitamin D), exercise, and safety practices to prevent falls. In addition, your physician may prescribe a medication to slow bone loss, increase bone density, and reduce fracture risk.

Alendronate (brand name Fosamax) is a medication that is used both to prevent and to treat osteoporosis. In postmenopausal women with osteoporosis, the drug reduces bone loss, increases bone density in both the spine and the hip, and reduces the risk of spine and hip fractures.

MINI-SUMMARY

1. Osteoporosis is a major public health threat for many Americans, mostly the elderly and women.

2. After age 30, bone resorption slowly begins to exceed bone formation. Bone loss is most rapid in the first few years after menopause.

3. Risk factors include gender, age, body size, ethnicity, family history, diet, a sedentary lifestyle, cigarette smoking, and excessive use of alcohol.

4. To prevent osteoporosis, you need to get an adequate supply of calcium and vitamin D over your lifetime, perform weight-bearing exercises, and avoid smoking and excessive consumption of alcohol.

5. Using a special kind of X-ray, a physician can detect osteoporosis.

6. A comprehensive osteoporosis treatment program includes a focus on proper nutrition (especially calcium and vitamin D), exercise, and safety practices to prevent falls. In addition, your physician may prescribe a medication such as Fosamax to slow bone loss, increase bone density, and reduce fracture risk.

CHECK-OUT QUIZ

1. The two medical conditions that lead to most cardiovascular diseases are atherosclerosis and high blood pressure.

    a. True
    b. False

2. Atherosclerosis develops by a process that is totally silent.

    a. True
    b. False

3. The most effective dietary method to lower the level of your blood cholesterol is to eat less cholesterol.

    a. True
    b. False

4. The TLC diet is used to treat diabetes mellitus.

    a. True
    b. False

5. Blood pressure increases with age.

    a. True
    b. False

6. Both heart attacks and strokes are usually due to clots caught in arteries.

    a. True
    b. False

7. Dietary protein probably promotes cancer.

    a. True
    b. False

8. Phytochemicals may protect against cancer.

    a. True
    b. False

9. Symptoms of diabetes include increased hunger and thirst.

    a. True
    b. False

10. Vegetarians enjoy certain health benefits from their diets.

    a. True
    b. False

ACTIVITIES AND APPLICATIONS

1. **A Diet for Disease Prevention**
   You have read about several diseases and the dietary means for preventing and treating each one. What do the dietary guidelines to prevent heart disease, high blood pressure, and cancer have in common?

2. **Vegetarian Meal Planning**
   Using vegetarian cookbooks and magazines or trade publications, suggest two vegetarian entrées and desserts for use at each of the following establishments: a casual-themed restaurant for younger people, a college cafeteria, and a dining room for an investment bank.

3. **Vegetarian Meal**
   With your classmates, plan and carry out a vegetarian meal that uses foods with which you may not be familiar. Plan the meal so that there are items for both lacto-ovo vegetarians and vegans.

4. **Taste Testing: Veggie Burgers**
   A variety of nonmeat burgers are on the market. Ingredients vary and may include vegetables, soy, and other ingredients. Select a number of veggie burgers and prepare according to the package directions. Serve on a bun with lettuce, tomato, and condiments. Find out which one is the class favorite.

5. **Food Costing: Meat and Meatless Wrap Sandwiches**
   Using food costs from your instructor or elsewhere, calculate the food cost/serving of a Spinach Tofu Wrap and a Turkey Wrap Sandwich. How do the food costs compare? If each could be sold for $4.99, compare their food cost percentage.

SPINACH TOFU WRAP SANDWICH

Yield: 2 servings

Ingredients

2 (10-inch) whole wheat tortillas
1 (7.5 ounce) package hickory-flavor baked tofu
¼ cup shredded sharp cheddar cheese
1 cup fresh baby spinach
2 tablespoons light ranch dressing
1 tablespoon grated Parmesan cheese, or to taste

Steps

1. Place the tortillas side by side on a paper plate. Slice tofu, and place slices down the center of each tortilla. Sprinkle cheese over the tofu. Cover with a damp paper towel, and heat in the microwave for about 45 seconds, or until the cheese is melted.

2. Pile some spinach onto each tortilla, and pour on some ranch dressing. Sprinkle with Parmesan cheese, roll tortillas around the filling, and serve.

TURKEY WRAP SANDWICH

Yield: 1 serving

Ingredients

1 Tbsp reduced-fat mayonnaise or other sauce
1 (8-inch) flour tortilla
3 ounces turkey
1 ounce sliced or shredded cheese
½ cup fresh baby spinach
1 ounce diced tomato

Steps

1. Spread sauce or mayonnaise over tortilla.
2. Layer turkey and cheese and sprinkle evenly with spinach and tomato.
3. Roll up tightly and cut in half on a bias. Wrap tightly with sandwich foil and serve.

NUTRITION WEB EXPLORER

National Heart, Lung, and Blood Institute (NHLBI)   http://hp2010.nhlbihin.net/atpiii/calculator.asp

At this website, you can use an interactive tool to assess your 10-year risk for heart disease.

American Cancer Society   www.cancer.org/docroot/PED/PED_0.asp

On the website for the American Cancer Society, click on "Food and Fitness" and then click on "ACS Guideline for Eating Well and Being Active" or "Cooking Smart." What tips do they give to reduce cancer risk?

Center for Science in the Public Interest   www.cspinet.org/quiz

To prevent disease, you need to eat a healthy diet. Click on "Rate Your Diet Quiz." How does your diet rate?

American Diabetes Association   www.diabetes.org.

On the home page of the American Diabetes Association, click on "Nutrition" then click on "Eating Out." What advice does it give if you go to a fast-food restaurant?

National Restaurant Association   www.restaurant.org/foodsafety

Click on "Healthy Dining Finder." What is this program?

FOOD FACTS: BOTANICALS AND HERBS

What is a botanical?

A botanical is a plant or plant part valued for its medicinal or therapeutic properties, flavor, and/or scent. Herbs are a category of botanicals. Products made from botanicals that are used to maintain or improve health may be called herbal or botanical products.

In naming botanicals, botanists use a Latin name made up of the genus and species of the plant. Under this system the herb black cohosh is known as *Actaea racemosa* L. "L" stands for Linneaus, who first described the type of plant specimen. In the Office of Dietary Supplements (ODS) fact sheets, initials such as L are not used because they do not appear on most products used by consumers.

Can botanicals be dietary supplements?

To be classified as a dietary supplement, a botanical must meet the definition given below. Many botanical preparations meet the definition. As defined by Congress in the Dietary Supplement Health and Education Act which became law in 1994, a dietary supplement is a product (other than tobacco) that:

- Is intended to supplement the diet
- Contains one or more dietary ingredients (including vitamins; minerals; herbs or other botanicals; amino acids; and other substances) or their constituents
- Is intended to be taken by mouth as a pill, capsule, tablet, or liquid
- Is labeled on the front panel as being a dietary supplement

How are botanicals commonly sold and prepared?

Botanicals are sold in many forms: as fresh or dried products; liquid or solid extracts; and tablets, capsules, powders, and tea bags. For example, fresh ginger root is often found in the produce section of food stores; dried ginger root is sold packaged in tea bags, capsules, or tablets; and liquid preparations made from ginger root are also sold. A particular group of chemicals or a single chemical may be isolated from a botanical and sold as a dietary supplement, usually in tablet or capsule form. An example is phytoestrogens from soy products.

Common preparations include teas, decoctions, tinctures, and extracts. A *tea*, also known as an infusion, is made by adding boiling water to fresh or dried botanicals and steeping them. The tea may be drunk either hot or cold.

Some roots, bark, and berries require more forceful treatment to extract their desired ingredients. They are simmered in boiling water for longer periods than teas, making a decoction, which also may be drunk hot or cold.

A tincture is made by soaking a botanical in a solution of alcohol and water. Tinctures are sold as liquids and are used for concentrating and preserving a botanical. They are made in different strengths that are expressed as botanical-to-extract ratios. An extract is made by soaking the botanical in a liquid that removes specific types of chemicals. The liquid can be used as is or evaporated to make a dry extract for use in capsules or tablets.

Are botanical dietary supplements standardized?

Standardization is a process that manufacturers may use to ensure batch-to-batch consistency of their products. In some cases, standardization involves identifying specific chemicals (also known as markers) that can be used to manufacture a consistent product. The standardization process can also provide a measure of quality control.

Dietary supplements are not required to be standardized in the United States. In fact, no legal or regulatory definition exists for standardization in the United States as it applies to botanical dietary supplements. Because of this, the term "standardization" may mean many different things. Some manufacturers use the term incorrectly to refer to uniform manufacturing practices; following a recipe is not sufficient for a product to be called standardized. Therefore, the presence of the word "standardized" on a supplement label does not necessarily indicate product quality.

Ideally, the chemical markers chosen for standardization would also be the compounds that are responsible for a botanical's effect in the body. In this way, each lot of the product would have a consistent health effect. However, the components responsible for the effects of most botanicals have not been identified or clearly defined. For example, the sennosides in the botanical senna are known to be responsible for the laxative effect of the plant, but many compounds may be responsible for valerian's (another herb) relaxing effect.

Are botanical dietary supplements safe?

Many people believe that products labeled "natural" are safe and good for them. This is not necessarily true because the safety of a botanical depends on many things, such as its chemical makeup, how it works in the body, how it is prepared, and the dose used.

In the United States, herbal and other botanical supplements are regulated by the U.S. Food and Drug Administration (FDA) as foods. This means that they do not have to meet the same standards as drugs and over-the-counter medications for proof of safety and effectiveness.

The actions of botanicals range from mild to powerful (potent). A botanical with mild action may have subtle effects. Chamomile and peppermint, both mild botanicals, are usually taken as teas to aid digestion and are generally considered safe for self-administration. Some mild botanicals may have to be taken for weeks or months before their full effects are achieved. For example, valerian may be effective as a sleep aid after 14 days of use but it is rarely effective after just one dose. In contrast, a powerful botanical produces a fast result. Kava, as one example, is reported to have an immediate and powerful action affecting anxiety and muscle relaxation.

The dose and form of a botanical preparation also play important roles in its safety. Teas, tinctures, and extracts have different strengths. The same amount of a botanical may be contained in a cup of tea, a few teaspoons of tincture, or an even smaller quantity of an extract. Also, different preparations vary in the relative amounts and concentrations of chemical removed from the whole botanical. For example, peppermint tea is generally considered safe to drink but peppermint oil is much more concentrated and can be toxic if used incorrectly. It is important to follow the manufacturer's suggested directions for using a botanical and not exceed the recommended dose without the advice of a health-care provider.

Herbal supplements can act in the same way as drugs. Therefore, they can cause medical problems if not used correctly or if taken in large amounts. In some cases, people have experienced negative effects even though they followed the instructions on a supplement label. It is important to consult your health-care provider before using an herbal supplement, especially if you are taking any other medications (whether prescription or over-the-counter). Some herbal supplements are known to interact with medications in ways that cause health problems.

The active ingredient(s) in many herbs and other botanical supplements are not known. There may be dozens, even hundreds, of such compounds in an herbal supplement. Some herbal supplements have been found to be contaminated with substances such as metals. Scientists are currently working to identify these active ingredients and analyze products. Identifying the active ingredients in herbs and understanding how herbs affect the body are important research areas for the National Center for Complementary and Alternative Medicine (NCCAM).

Does a label indicate the quality of a botanical dietary supplement product?

It is difficult to determine the quality of a botanical dietary supplement product from its label. The degree of quality control depends on the manufacturer, the supplier, and others in the production process.

What methods are used to evaluate the health benefits and safety of a botanical dietary supplement?

Scientists use several approaches to evaluate botanical dietary supplements for their potential health benefits and safety risks, including their history of use and laboratory studies using cell or animal studies. Studies involving people (individual case reports, observational studies, and clinical trials) can provide information that is relevant to how botanical dietary supplements are used. Researchers may conduct a systematic review to summarize and evaluate a group of clinical trials that meet certain criteria.

What are some additional sources of information on botanical dietary supplements?

For general information about dietary supplements see Dietary Supplements: Background Information (http://ods. od.nih.gov/factsheets/dietarysupplements. asp) from the Office of Dietary Supplements (ODS), available at ods.od.nih.gov. For information on specific herbs, go to Herb Fact Sheets at http://nccam.nih.gov/health/herbsataglance.htm.

If you use herbal supplements it is best to do so under the guidance of a medical professional who has been properly trained in herbal medicine. This is especially important for herbs that are part of a whole medical system, such as traditional Chinese medicine or Ayurvedic medicine.

Sources: Office of Dietary Supplements and National Center for Complementary and Alternative Medicine, National Institutes of Health.

HOT TOPIC: BIOTECHNOLOGY

Background

Potatoes with built-in insecticide. Rice with extra vitamin A. Decaf coffee beans fresh off the tree. What do these foods have in common? They have all been created using biotechnology and genetic engineering. *Biotechnology* is a collection of scientific techniques, including genetic engineering, that are used to create, improve, or modify plants, animals, and microbes (such as bacteria). *Genetic engineering* is a process in which genes are transferred to a plant, animal, or microbe to have a certain effect, such as producing a soybean with built-in insecticide.

Farmers and scientists have been genetically modifying plants for hundreds of years, most often using a process called crossbreeding. Crossbreeding involves cross-fertilizing two related plants, and all 100,000 or so genes of each one, to produce offspring that express the best of both. Since farmers usually want only a few genes transferred, crossbreeding is quite hit-or-miss. In contrast, genetic engineering involves introducing one gene or a group of genes that are known quantities into a plant, animal, or microbe.

To understand genetic engineering, it is necessary to have an understanding of chromosomes, genes, and DNA. Each cell in your body contains a complete copy of your genetic plan or blueprint. The genetic material is packaged into long strands (called chromosomes) made of the chemical substance called DNA (deoxyribonucleic acid). Each chromosome contains genes. Genes contain the information for making proteins, which perform all the critical functions of the cell, and the blueprint for traits such as height, color, and disease resistance.

Some traits are produced from the code contained in one gene; more complex traits depend on several genes. However, not all genes are switched on in every cell. The genes active in a liver cell are different from the genes active in a brain cell because the cells have different functions.

The language of DNA is common to all organisms. Humans share 7000 genes with a worm called *C. elegans*! The main difference between organisms lies in their total number of genes, how the genes are arranged, and which ones are turned on or off in different cells. A gene from the Arctic flounder that keeps the fish from freezing was introduced into strawberries to extend their growing season in northern climates. This did not make the strawberries fishy because the gene that was introduced is a cold-tolerance gene, not a flounder gene.

Did you know that several million people with diabetes use insulin produced by genetically modified (GM) bacteria? When given a copy of the gene for human insulin, a certain type of bacteria can make insulin that is purified and used to treat diabetes. In 1982, human insulin became the first GM product made commercially and approved for use. Since then, genetic engineering has developed additional medicines and vaccines.

Plant Applications

Genetic engineering has been used with plants to make:

1. Fruits or vegetables that ripen differently
2. Plants that are resistant to disease, pests, selected herbicides, or environmental conditions (such as drought)
3. Plant foods with desirable nutritional characteristics

One of the first GM foods to appear in the supermarket was the fresh tomato, called Flav Savr. If picked when ripe, tomatoes rot quickly, and so they are usually picked when green. The Flav Savr tomato was engineered to remain on the vine longer to ripen to full flavor before harvest. Once harvested, it did not soften as quickly as other tomatoes and was still firm and flavorful when it arrived at the supermarket. Eventually, the Flav Savr tomato was withdrawn from the market due to shipping problems and flavor problems because of the tomato varieties used.

Plants can also be made resistant to disease, pests, selected herbicides, or environmental conditions. For example, GM squash and potatoes are protected from viruses that normally affect those plants.

For years, organic farmers used a spray containing the microbe *Bacillus thuringiensis* as an effective pesticide. The microbe, which naturally occurs in soil, makes a protein (known as Bt) that poisons certain insect pests but is harmless to humans, animals, and most other insects. Scientists transferred the Bt gene that makes the insect-killing protein to corn, cotton, and soybeans. In this manner, farmers use less pesticide because the plant that once was a food source for the insect now kills it. Most corn and soybean crops in the United States are fed to animals.

Much of the American soybean crop carries a gene that made it resistant to a popular herbicide used to control weeds. When the farmer sprays the fields to kill weeds, the plant is not harmed. This gene has been introduced into corn, canola, and cotton seeds as well. Farmers can till (or plow) the soil less often (a common technique for reducing weeds), which results in less soil erosion.

Another use of genetic engineering has been to enhance the nutritional profile of foods. For example, by inserting two genes from a daffodil and one from a bacterium into a rice plant, scientists have created a rice with beta-carotene, which the body makes into vitamin A. Called golden rice, it is still being tested and could eventually be used in developing countries where vitamin A deficiency is common.

Another example involves a soybean with a different fat profile that works well in fryers, can be repeatedly reheated without changing the flavor of foods, and contains no trans fats. This product (called low-linolenic acid soybeans) is needed by fast-food chains and manufacturers who want to stop using partially hydrogenated vegetable oil, which contains trans fats. Soybean oil made from low-linolenic soybeans does not require hydrogenation.

Animal Applications

Using the same principle of gene transfer that gives plants more desirable traits, scientists have started working with GM animals, also called transgenic animals. Animal biotechnology research has largely focused on producing transgenic animals for the study of human disease or to produce drugs. There are many more barriers to using genetic engineering in the breeding and production of animal foods than there are with plant foods, such as difficulties working with living tissues and life cycles.

Currently seeking Food and Drug Administration approval is a variety of Atlantic salmon that grows to market weight in about 18 months, compared with the 24 to 30 months it normally takes for a fish to reach that size. New genes cause a continuous supply of

salmon growth hormones that accelerates the fish's development. For fish farmers, raising these salmon is cheaper and faster because it takes less feed and about half the time to produce a crop they can send to market.

One future goal of research is to produce transgenic cows with differing milk nutritional quality. For example, cow's milk could be made to have reduced lactose or a high whey-to-casein ratio so that the milk is more similar to breast milk and could be used to make infant formulas.

Regulations

Bioengineered foods are regulated by three federal agencies: the Food and Drug Administration (FDA), the U.S. Department of Agriculture (USDA), and the Environmental Protection Agency (EPA). The FDA is the lead agency for assessing the safety of GM plants and animals. Although the FDA only has to formally approve GM animals, companies that produce GM crops are asked to voluntarily submit test results, which is normally done. The USDA ensures that new plant varieties pose no threat to agriculture or to the environment during cultivation. The USDA requires extensive field trials that are controlled and contained. The EPA regulates any pesticides that may be present in food and sets tolerance levels for them. For new plants that protect themselves against insects or disease, the EPA is responsible for assessing the safety of the protein or trait for human and animal consumption as well as for the environment and nontarget organisms.

The FDA requires labeling of biotech foods significantly different from their conventional counterparts. The FDA considers whether a GM orange, for example, is "substantially equivalent" to a traditional orange. If the orange has a higher or lower

level of vitamin C, for example, the FDA requires that the label state that the food is genetically engineered.

Issues: Pros And Cons

The whole idea of using biotechnology brings up many issues and pros and cons. Here is a partial listing for you to consider:

- Environmental. Outcrossing occurs when a domestic crop breeds with a related "wild" species. Wild weeds can incorporate bioengineered genes, potentially making the weeds stronger and more resistant to pests and/or herbicides. Luckily, corn, soybeans, and cotton have no wild relatives with which to interbreed, but other crops, such as squash and canola, do have wild relatives in the United States. Likewise, if transgenic salmon escaped and cross-bred with their wild cousins, this could adversely affect the wild populations as the transgenic fish might outproduce (they are bigger and might get more food and mates) and push aside wild salmon. There is also the possibility of contaminating non-GM crops, including organic crops, with GM crops (due to pollen carried by the wind, birds, or the like) to adversely affect wildlife such as butterflies and worms.

- Health. GM foods pose the same risk to human health as do other foods: allergens, toxins (produced by the gene or the protein made by the gene), and compounds known as antinutrients that inhibit the absorption of nutrients. Allergic reactions are an important consideration for bioengineered foods because there is some possibility that a new protein in a food could be an allergen. To date, all new proteins in bioengineered foods have been shown to lack the characteristics of food allergens. If fewer pesticides are used, there will be less pesticide residue on plants. The accidental mixing of StarLink corn, a GM product approved as animal feed only, with corn intended for human consumption demonstrates another concern.

- Agricultural. Using biotechnology, it is possible to have increased crop yield, greater flexibility in growing environments, less erosion due to less tillage, and decreased use of pesticides and herbicides.
- Nutrition. Genetic engineering can certainly enhance the nutrient profile of foods, but there is also potentially the possibility of decreased nutrients in a product, such as a banana that takes longer to ripen but loses nutrients.
- Ethical and moral. Some people have religious and philosophical concerns regarding the transfer of genes into plants or animals and accuse the food industry of playing God.

Weight Management and Exercise

Overweight and obesity are among the most pressing new health challenges in the United States. Just take a look at these statistics:

- As of 2004, 66 percent of American adults were either overweight or obese. That means that two out of every three Americans are overweight or obese.
- From 1963–1970, about 5 percent of U.S. children were obese. This number shot up to 17 percent by 2004.
- In 1963, 10-year-old boys weighed 74.2 pounds on average; by 2002, the group's average weight was nearly 85 pounds. Similarly, the average weight for 10-year-old girls in 1963 was 77.4 pounds; by 2002, their average weight was nearly 88 pounds.
- American children and adolescents today are less physically active as a group than were previous generations.
- The obesity epidemic affects some racial and ethnic groups much more than others. For example, obesity is much more prevalent among black girls and Mexican American boys. Rates have also increased in Native American and Asian American youths.
- Overall, poverty has been associated with more obesity among adolescents.

As the prevalence of overweight and obesity has increased in the United States, so have related health-care costs, both direct and indirect. Direct health-care costs refer to preventive, diagnostic, and treatment services such as physician visits, medications, and hospital care. Indirect costs are the value of wages lost by people who are unable to work because of illness or disability, as well as the value of future earnings lost because of premature death. The cost of overweight and obesity is over $100 billion in direct medical costs and indirect costs such as lost wages due to illness.

Obesity is a disease, not a moral failing. It is a very complex disease that has social, behavioral, cultural, physiological, metabolic, and genetic factors. It occurs when energy intake exceeds the amount of energy expended over time. In a small number of cases, obesity is caused by illnesses such as hypothyroidism or results from taking medications that can cause weight gain.

Because so many factors affect how much or how little food a person eats and how food is metabolized by the body, losing weight is not simple. This chapter discusses obesity's impact on health, theories about what causes obesity, and treatments to lose weight and maintain weight loss. Exercise, which is important for both obese and nonobese individuals, is also examined, along with nutrition for athletes. This chapter will help you to:

- Define obesity and overweight
- List one advantage and one disadvantage of each of the three methods of measuring obesity
- List the health implications of obesity
- Explain possible causes of obesity
- List the six components of a comprehensive treatment program for obesity
- Describe basic concepts of nutrition education to consider in planning weight-reducing eating plans
- List five benefits of exercise
- Explain how behavior and attitude modification can be used to help a person lose weight
- Explain when drugs and surgery may be used to treat obesity
- Give an advantage and a disadvantage of using drugs to lose weight
- Identify strategies that support weight maintenance
- Evaluate diet books
- Identify nutrient needs for athletes and plan menus for athletes

"HOW MUCH SHOULD I WEIGH?"

The best way to determine if you have a healthy weight is to use the **body mass index** (BMI) chart in Figure 12-1. BMI is a direct calculation based on height and weight, and it applies to both men and women. For children and adolescents, weight status is determined by the comparison of the individual's BMI with age- and gender-specific percentile values (see the BMI growth curves in Appendix D). BMI does not directly measure the percentage of body fat, but it is more accurate at approximating body fat than is measuring body weight alone. BMI is found by dividing a person's weight in kilograms by that person's height in meters squared.

The National Institutes of Health defines **overweight** for adults as a BMI of 25 to 29.9. A BMI of 30 or greater is considered **obese**. These cutoff points were chosen because studies show that as BMI rises above 25, blood pressure and blood cholesterol rise, HDL levels fall, and maintaining normal blood sugar levels becomes more difficult. Thus, individuals with a BMI of 25 or higher run a greater risk of coronary artery disease, high blood pressure, heart attacks, stroke, type 2 diabetes, and certain cancers.

> **BODY MASS INDEX**
> A method of measuring the degree of obesity that is a more sensitive indicator than height-weight tables.
>
> **OVERWEIGHT**
> Having a body mass index of 25 or greater.
>
> **OBESE**
> Having a body mass index of 30 or greater.

FIGURE 12-1: Body Mass Index Chart

Locate the height of interest in the left-most column and read across the row for that height to the weight of interest. Follow the column of the weight up to the top row that lists the BMI. BMI of 19–24 is the healthy weight range, BMI of 25–29 is the overweight range, and BMI of 30 and above is in the obese range.

BMI	19	20	21	22	23	24	25	26	27	28	29	30	31	32	33	34	35
Height							**Weight in Pounds**										
4'10"	91	96	100	105	110	115	119	124	129	134	138	143	148	153	158	162	167
4'11"	94	99	104	109	114	119	124	128	133	138	143	148	153	158	163	168	173
5'	97	102	107	112	118	123	128	133	138	143	148	153	158	163	168	174	179
5'1"	100	106	111	116	122	127	132	137	143	148	153	158	164	169	174	180	185
5'2"	104	109	115	120	126	131	136	142	147	153	158	164	169	175	180	186	191
5'3"	107	113	118	124	130	135	141	146	152	158	163	169	175	180	186	191	197
5'4"	110	116	122	128	134	140	145	151	157	163	169	174	180	186	192	197	204
5'5"	114	120	126	132	138	144	150	156	162	168	174	180	186	192	198	204	210
5'6"	118	124	130	136	142	148	155	161	167	173	179	186	192	198	204	210	216
5'7"	121	127	134	140	146	153	159	166	172	178	185	191	198	204	211	217	223
5'8"	125	131	138	144	151	158	164	171	177	184	190	197	203	210	216	223	230
5'9"	128	135	142	149	155	162	169	176	182	189	196	203	209	216	223	230	236
5'10"	132	139	146	153	160	167	174	181	188	195	202	209	216	222	229	236	243
5'11"	136	143	150	157	165	172	179	186	193	200	208	215	222	229	236	243	250
6'	140	147	154	162	169	177	184	191	199	206	213	221	228	235	242	250	258
6'1"	144	151	159	166	174	182	189	197	204	212	219	227	235	242	250	257	265
6'2"	148	155	163	171	179	186	194	202	210	218	225	253	241	249	256	264	272
6'3"	152	160	168	176	184	192	200	208	216	224	232	240	248	256	264	272	279
			Healthy Weight					**Overweight**						**Obese**			

Source: Evidence Report of Clinical Guidelines on the Identification, Evaluation and Treatment of Overweight and Obesity in Adults, 1998. NIH/National Heart, Lung and Blood Institute (NHLBI).

Using BMI is simple, quick, and inexpensive, but it does have limitations. One problem with using BMI as a measurement tool is that very muscular people may fall into the "overweight" category when they are actually healthy and fit. Another problem is that people who have lost muscle mass, such as the elderly, may be in the healthy weight category when they actually have reduced nutritional reserves. Further evaluation of a person's weight and health status is necessary.

Because BMI doesn't tell you how much of your excess weight is fat and where that fat is located, the National Institutes of Health has asked physicians to measure patients' waistlines. Studies show that excessive abdominal fat is more health-threatening than is fat in the hips or thighs. A woman whose waist measures more than 35 inches (88 cm) and a man whose waist measures more than 40 inches (100 cm) may be at particular risk for developing health problems. Studies indicate that increased abdominal fat is related to the risk of developing heart disease, high blood pressure, and stroke. Body fat concentrated in the lower body (around the hips, for example) seems to be less harmful.

Another indicator of obesity is the waist circumference measurement. Waist circumferences greater than 40 inches (100 cm) in men, and greater than 35 inches (88 cm) in women are good indicators of the presence of visceral fat. Visceral fat is fat that wraps around the vital organs of the liver, heart, kidneys, and intestine.

Total body fat measurements also give an indication of obesity. For men, a desirable percentage of body fat is under 25 percent; for women, under 35 percent. Body fat is most often measured by using special calipers to measure the skinfold thickness of the triceps and other parts of the body. Because half of all your fat is under the skin, this method is quite accurate when performed by an experienced professional. Other methods of estimating body fatness include air displacement by BODPOD, dual-energy X-ray absorptiometry (DEXA), and bioelectrical impedance.

MINI-SUMMARY

1. Body mass index (BMI) is a direct calculation based on height and weight. BMI does not directly measure the percentage of body fat, but it is more accurate at approximating body fat than is measuring body weight alone.

2. The National Institutes of Health defines overweight as a BMI of 25 to 29.9. A BMI of 30 or greater is considered obese.

3. BMI has some limitations: It may place very muscular people into the overweight category and place people who have lost muscle mass into the healthy weight category.

4. Studies show that excessive abdominal fat is more health-threatening than is fat in the hips or thighs. A woman whose waist measures more than 35 inches and a man whose waist measures more than 40 inches may be at particular risk for developing heart disease, high blood pressure, and stroke.

HEALTH IMPLICATIONS OF OBESITY

An estimated 300,000 deaths per year may be attributable to obesity. Obese persons have more than 10 times the risk of developing type 2 diabetes and 3 times the risk of developing coronary heart disease compared to those who are lean.

An obese individual is at increased risk for:

- Type 2 diabetes
- High blood cholesterol levels

- Hypertension (high blood pressure)
- Cardiovascular disease
- Stroke
- Sleep apnea (interrupted breathing during sleeping) and respiratory problems
- Certain types of cancer
- Gallbladder disease
- Osteoarthritis
- Complications in pregnancy and childbirth

Conditions aggravated by obesity include arthritis, varicose veins, and gallbladder disease. In addition, surgery is riskier for obese individuals.

Obesity creates a psychological burden that in terms of suffering may be its greatest adverse effect. In American and other Westernized societies there are powerful messages that people, especially women, should be thin and that to be fat is a sign of poor self-control. Negative attitudes about the obese have been reported in children and adults, health-care professionals, and the overweight themselves. People's negative attitudes toward the obese often translate into discrimination in employment opportunities, job earnings, and other areas. Losing weight often decreases blood pressure and blood cholesterol levels and brings diabetes under better control. Although obesity does not cause these medical conditions, losing weight can help reduce some of their negative effects.

MINI-SUMMARY

1. An obese individual is at increased risk for type 2 diabetes, high blood cholesterol levels, high blood pressure, cardiovascular disease, stroke, sleep apnea, and certain types of cancer.
2. Obesity (BMI over 30) creates a psychological burden and discrimination.
3. Losing weight—and keeping it off—improves health.

THEORIES OF OBESITY

As researchers try to figure out why some people get fat and others don't, it is becoming increasingly apparent that obesity is caused by an interaction of genetic (inherited), environmental (social and cultural), metabolic (physical and chemical), and behavioral (psychological and emotional) factors; therefore, no single cure is available.

The body has an almost limitless capacity to store fat. When calorie intake exceeds expenditures, each fat cell can balloon to more than six times its original size. If the available cells get filled to the brim, new ones will be created. As the body stores more fat, weight and girth increase. Losing weight causes fat cells to shrink in size, but not in number.

For many individuals, genetics influences the development of overweight and obesity. Studies suggest that about 50 percent of the variance in body weight in any person depends on genetic factors.

The environment is also a major determinant of overweight and obesity. Environmental influences on overweight and obesity are primarily related to food intake and physical activity behaviors. In countries such as the United States, there is an overall abundance of tasty, kcalorie-dense food. In addition, aggressive and sophisticated food marketing promotes high kcalorie consumption. Many of our sociocultural traditions promote overeating and the preferential consumption of high-kcalorie foods. For many people, even when kcaloric intake is not above the recommended level, the number of calories expended in physical activity is insufficient to offset consumption. Many people are stuck in daily routines that are completely sedentary.

Have you ever noticed that some people can eat lots of food and never gain a pound, while others gain weight easily? Everyone has a different basal metabolic rate, which amounts to roughly 70 percent of his or her total energy expenditure and is influenced by genetics, body composition, and age. The remaining 30 percent of daily energy expenditure is due to contributions from the thermic effect of food eaten and physical activity (exercise and nonexercise activity referred to as fidgeting), with body composition and age moderating these effects. One way to increase your total daily energy expenditure is to get regular physical exercise, particularly aerobic exercise. Also, as you replace fat tissue with muscle tissue, your body will burn more calories because muscle is more metabolically active than is fat.

MINI-SUMMARY

Obesity is caused by an interaction of genetic (inherited), environmental (social and cultural), metabolic (physical and chemical), and behavioral (psychological and emotional) factors; therefore, no single cure is available.

TREATMENT OF OBESITY

Before discussing the treatment of obesity, it is a good idea to discuss treatment goals. The goals of most weight-loss programs focus on short-term weight loss. Critics of this type of goal feel that short-term weight loss is not a valid measure of success, because long-term weight loss is associated with many more health benefits. Also, weight-loss goals tend to reinforce the American preoccupation with being slender, especially for women. Finally, critics point out that weight-loss goals are often set too high, since even a small weight loss can reduce the risk of developing chronic diseases. Weight loss of only 5 to 10 percent of body weight may improve many of the problems associated with overweight, such as the following:

- Lower blood pressure and thereby the risks of high blood pressure
- Reduce abnormally high levels of blood glucose associated with diabetes
- Bring down blood levels of cholesterol and triglycerides associated with cardiovascular disease
- Reduce sleep apnea or irregular breathing during sleep
- Decrease the risk of osteoarthritis in the weight-bearing joints
- Decrease depression
- Increase self-esteem

To gain long-term results, treatment of obesity generally consists of one or more of the following: eating plan, exercise, behavior modification, and drug therapy. Treatment programs often concentrate on two aspects of losing weight: eating plan and exercise. Through reducing kcalories moderately and increasing physical activity, you can lose about 5 to 10 percent of your body weight over six months. However, maintaining this weight loss is difficult, and many people regain some, if not all, of the weight.

The next section will discuss each of the following components of treatment: eating plan and nutrition education, exercise, behavior and attitude modification, social support, maintenance support, drugs, and surgery. But first, let's take a look at a newer approach to treating obesity that doesn't use diets.

More health professionals are adopting a nondieting approach (that is, a nonrestriction healthy eating approach) to treating obesity that includes acceptance of body size and lifestyle changes such as exercising more. This new health-centered approach steers clear of dieting and emphasizes helping obese people adopt a healthier lifestyle. In many cases, diets that restrict a wide variety of food and food groups simply don't work. By creating a condition of "forbidden" food, some diets can cause some dieters to become obsessed with food, which may lead to binge eating, a condition that occurs in roughly 2 to 5 percent of the adult population with obesity

EATING PLAN AND NUTRITION EDUCATION

Basic nutrition education is crucial for anyone who wants to lose weight. These individuals need education about fat, carbohydrate, and protein in foods and about balancing them in a lower-kcalorie diet. They need to understand variety, moderation, and nutrient density, particularly because they have to pack the same amount of nutrients into fewer kcalories.

These eight basic concepts of nutrition education are important.

1. Kcalories should not be overly restricted because this practice decreases the likelihood of success. If you normally eat 2500 kcalories daily and decide to eat only 1200 kcalories a day, that is less than half of what you normally eat. Reducing kilocalories by 500–1000 each day amounts to about 1 pound lost in a week (1 pound of fat is equivalent to 3500–4200 kcalories). In any case, kcalories should not be restricted below 1200 without medical supervision, because getting adequate nutrients is impossible below that level. A progressive weight loss of 1 to 2 pounds a week is considered safe.

2. Fat should be restricted to about 30 percent or less of total kcalories, protein to 15 percent, and carbohydrate to 55 percent or more. During weight loss, attention should be given to maintaining an adequate intake of vitamins and minerals.

3. Don't forget that a healthy eating plan is one that:
   - Emphasizes fruits, vegetables, whole grains, and fat-free or low-fat milk and milk products
   - Includes lean meats, poultry, fish, beans, eggs, and nuts
   - Is low in saturated fats, trans fats, cholesterol, salt, and added sugars

4. No foods should be forbidden, because that only makes them more attractive.

5. Eating three meals and one or two snacks each day is crucial to minimizing the possibility of getting hungry. People tend to overeat when they are hungry.

6. Portion control is vital. Measuring and weighing foods is important, because "eyeballing" is not always accurate See Figure 2-27 on page 68 for hints on how to estimate portion sizes.

7. Variety, balance, and moderation are crucial to satisfying all nutrient needs.

While *fad diets*, popular eating plans that promise quick weight loss, may help you take off some pounds initially, a diet as described above is more likely to help you take weight off and keep it off.

While losing weight, regular monitoring of your weight will be essential to help you maintain your lower weight. When keeping a record of your weight, a graph may be more informative than a list of your weights. When weighing yourself, remember that one day's diet and exercise patterns won't have a measurable effect on your weight the next day. Today's weight is not a true measure of how well you followed your program yesterday, because your body's water weight will change much more from day to day than will your fat weight, and water changes are often the result of things that have nothing to do with your weight-management efforts.

See Figure 12-2 for information on weight-loss myths.

FIGURE 12-2: Weight-loss Myths

MYTH: Fad diets work for permanent weight loss.

FACT: Fad diets are not the best way to lose weight and keep it off. Fad diets often promise quick weight loss or tell you to cut certain foods out of your diet. You may lose weight at first on one of these diets. But diets that strictly limit kcalories or food choices are hard to follow. Most people quickly get tired of them and regain any lose weight.

Fad diets may be unhealthy because they often do not provide all of the nutrients your body needs. Diets that provide less than 800 kcalories per day also could result in heart rhythm abnormalities, which can be fatal.

TIP: Research suggest that losing $\frac{1}{2}$ to 2 pounds a week by making healthy food choices, eating moderate portions, and building physical activity into your daily life is the best way to lose weight and keep it off. By adopting healthy eating and physical activity habits, you may also lower your risk for developing type 2 diabetes, heart disease, and high blood pressure.

MYTH: High-protein, low-carbohydrate diets are a healthy way to lose weight.

FACT: The long-term health effects of a high-protein, low-carbohydrate diet are unknown. But getting most of your daily kcalories from high-protein foods like meat, eggs, and cheese is not a balanced eating plan. You may be eating too much fat and cholesterol and too few fruits, vegetables, and whole grains. Following the diet may make you feel nauseous, tired, weak, and constipated.

TIP: High-protein, low-carbohydrate diets are often low in kcalories because food choices are strictly limited, so they may cause short-term weight loss. By following a lower-in-kcalories eating plan with recommended amounts of carbohydrate, protein, and fat, you will not have to stop eating whole classes of foods, such as whole grains, fruits, and vegetables. You may also find it easier to stick with an eating plan that includes a wide variety of foods.

MYTH: Starches are fattening and should be limited when trying to lose weight.

FACT: Many foods high in starch, like bread, rice, pasta, cereals, beans, and fruits, are low in fat and kcalories. They become high in fat and kcalories when eaten in large portion sizes or when covered with high-fat toppings like butter, sour cream, or mayonnaise. Foods high in starch are an important source of energy for your body.

TIP: When eating bread, rice, pasta, and cereals, choose whole grains for even more nutrients and fiber. For example, substitute a whole-grain product for a refined product—such as eating whole-wheat bread instead of white bread or brown rice instead of white rice.

MYTH: Certain foods, like grapefruit, celery, or cabbage soup, can burn fat and make you lose weight.

FACT: No foods can burn fat. Some foods with caffeine may speed up your metabolism for a short time, but the number of kcalories burned is usually not significant, so you do not lose weight.

TIP: The best way to lose weight is to cut back on the number of kcalories you eat and be more physically active.

MYTH: Natural or herbal weight-loss products are safe and effective.

FACT: A weight-loss product that claims to be "natural" or "herbal" is not necessarily safe. These products are not usually scientifically tested to provide that they work. For example, herbal products containing ephedra (now banned in the United States) have caused serious health problems and even death.

TIP: Talk with your health care provider before using a weight-loss product.

(continued)

MYTH: Low-fat or fat-free means no kcalories.

FACT: A low-fat or fat-free food is often lower in kcalories than the same size portion of the full-fat product. But many processed low-fat or fat-free foods have just as many kcalories as the full-fat version—or even more. They may contain added sugar, flour, or starch thickeners to improve flavor and texture after fat is removed. These ingredients add kcalories.

TIP: Read the Nutrition Facts on a food package to find out how many kcalories are in a serving. Check the serving size too—it may be less than you are used to eating.

MYTH: Fast foods are always an unhealthy choice and you should not eat them when dieting.

FACT: Fast foods can be part of a healthy weight-loss program with a little bit of know-how.

TIP: Avoid supersize combo meals, or split one with a friend. Choose grilled foods, like a grilled chicken sandwich or a small hamburger. Try a "fresco" taco (with salsa instead of cheese). Order fried foods, like french fries or fried chicken, only occasionally and order a small portion. Use small amounts of high-fat toppings such as regular mayonnaise and salad dressings, as well as bacon and cheese.

MYTH: Skipping meals is a good way to lose weight.

FACT: Studies show that people who skip breakfast tend to be heavier than people who eat breakfast most days. This may be because people who skip meals tend to feel hungrier later on, and eat more than they normally would. It may also be that eating many small meals throughout the day helps people control their appetites.

TIP: Eat small meals throughout the day that include a variety of foods.

MYTH: Eating after 8 p.m. causes weight gain.

FACT: It does not matter what time of day you eat. It is how many kcalories you eat and how much physical activity you get during the whole day that determines whether you gain, lose, or maintain your weight.

TIP: If you want to have a snack before bedtime, think first about how many kcalories you have eaten that day. And try to avoid snacking in front of the television at night—it may be easier to overeat when you are distracted by the television.

MYTH: Nuts are fattening and you should not eat them if you want to lose weight.

FACT: In small amounts, nuts can be part of a healthy weight-loss program. Nuts are high in kcalories and fat. However, most nuts contain healthy fats that do not clog arteries. Nuts are also good sources of protein, dietary fiber, and minerals including magnesium and copper.

TIP: Enjoy small portions of nuts.

MYTH: Lifting weights is not good to do if you want to lose weight, because it will make you "bulk up."

FACT: Lifting weight or doing strengthening activities like push-ups and crunches on a regular basis can actually help you maintain or lose weight. These activities can help you build muscle, and muscle burns more kcalories than body fat. So, if you have more muscle, you burn more kcalories—even when sitting still. Doing strengthening activities 2 or 3 days a week will not bulk you up. Only intense strength training, combined with a certain genetic background, can build very large muscles.

TIP: In addition to doing at least 30 minutes of moderate-intensity physical activity (like walking 2 miles in 30 minutes) on most days of the week, try to do strengthening activities 2 to 3 days a week. You can lift weights, use resistance bands, do push-ups or sit-ups, or do household or garden tasks that make you lift or dig.

EXERCISE

Exercise is a vital component of any weight-control program (Figure 12-3). Research consistently shows that time spent exercising is a major predictor of long-term weight loss. Exercise not only facilitates weight loss through direct energy expenditure but burns fat both during and after exercise. Regular exercise also helps control or suppress appetite and builds and tones muscles, which in turn raises your basal metabolic rate. But regular exercise has many benefits beyond simply losing pounds and keeping them off. The additional advantages of regular physical activity include:

1. Improves functioning of the cardiovascular system and reduced levels of blood lipids associated with cardiovascular disease
2. Helps control blood pressure
3. Lowers risk factors for diabetes and colon cancer
4. Increases ability to cope with stress, anxiety, and depression
5. Increases stamina
6. Increases resistance to fatigue
7. Improves psychological well-being and self-image

A consistent pattern of exercise is vital to achieving these beneficial results.

The Dietary Guidelines for Americans recommend daily exercise as follows.

- To reduce the risk of chronic disease in adulthood, engage in at least 30 minutes of moderate-intensity physical activity, above the usual activity, on most days of the week. The recommendation is 60 minutes for children.

FIGURE 12-3: Kcalories per Hour Expended in Common Physical Activities

Moderate Physical Activity	Approximate Kcalories/Hour for a 154-Pound Person
Hiking	370
Light gardening/yard work	330
Dancing	330
Golf (walking and carrying clubs)	330
Bicycling (<10 mph)	290
Walking (3.5 mph)	280
Weight lifting (general light workout)	220
Stretching	180

Vigorous Physical Activity	Approximate Kcalories/Hour for a 154-Pound Person
Running/jogging (5 mph)	590
Bicycling (>10 mph)	590
Swimming (slow freestyle laps)	510
Aerobics	480
Walking (4.5 mph)	460
Heavy yard work (chopping wood)	440
Weight lifting (vigorous effort)	440
Basketball (vigorous)	440

Source: Adapted from the 2005 Dietary Guidelines Advisory Committee Report.

- For most people, greater health benefits can be obtained by engaging in physical activity of more vigorous intensity or longer duration.
- To help manage body weight and prevent gradual, unhealthy body weight gain in adulthood, engage in approximately 60 minutes of moderate- to vigorous-intensity activity on most days of the week while not exceeding caloric intake requirements.
- To sustain weight loss in adulthood, participate in at least 60 to 90 minutes of daily moderate-intensity physical activity while not exceeding kcaloric intake requirements. Some people may need to consult with a health-care provider before participating in this level of activity.

The key to a successful exercise program is choosing an enjoyable activity. Some questions people need to answer to develop a good exercise program include:

1. Do you like to exercise alone or with others?
2. Do you prefer to exercise outdoors or indoors?
3. Do you prefer to exercise at home?
4. What activities do you find particularly enjoyable?
5. How much money are you willing to spend for sports equipment or facilities if needed?
6. When can you best fit the activity into your schedule?

An obese person may resist starting an exercise program for several reasons. Many people have had bad experiences in school physical education classes and want to avoid such activity. Some tend to be self-conscious and may not want to be seen exercising. Activity is also harder for obese people and requires more effort. Recent studies suggest that this may be due to the fact that obese individuals have a 50 percent reduction in the mitochondrial content of their cells, the energy units of the cell. Over time, exercise can help to increase the energy units in cells and exercise becomes easier to sustain.

Aerobic activities such as walking, jogging, cycling, and swimming are ideal as the major component of an exercise program. Aerobic activities must be brisk enough to raise heart and breathing rates, and they must be sustained, meaning that they must be done for at least 15 to 30 minutes without interruption. Activities such as baseball, bowling, and golf are not vigorous or sustained. They have certain benefits—they can be enjoyable, help improve coordination and muscle tone, and help relieve tension—but they are not aerobic.

Any exercise program for sedentary and overweight people must be started slowly, with enjoyment and commitment as the major goals. A buildup in intensity and duration should be gradual and progressive, depending to a large extent on how overweight the individual is. Aerobic capacity improves when exercise increases the heart rate to a target zone of 65 to 80 percent of the maximum heart rate, which is the fastest the heart can beat. The maximum heart rate can be calculated by subtracting your age from 220. One goal of the exercise program should be to build up to this intensity of exercise. Exercise that increases the heart rate to between 65 and 80 percent of its maximum conditions the heart and lungs besides burning kcalories (Figure 12-4). To determine whether your heart rate is in the target zone, take your pulse as follows:

1. When you stop exercising, quickly place the tip of your third finger lightly over one of the blood vessels on your neck to the left or right of your Adam's apple. (Another convenient pulse spot is the inside of your wrist just below the base of the thumb.)
2. Count your pulse for 30 seconds and multiply by 2.

FIGURE 12-4:
Exercise has many
benefits.
Courtesy of PhotoDisc,
Inc.

3. If your pulse is below your target zone, exercise a little harder the next time. If you are above your target zone, take it a little easier. If it falls within the target zone, you are doing fine.
4. Once you are exercising within your target zone, you should check your pulse at least once a week.

Exercise should take place most days and include a warm-up period, the exercise itself, and then a cooling-down period. To warm up before exercising, do stretching exercises slowly and in a steady, rhythmic way. Start at a medium pace and gradually increase it. Next, begin jumping rope or jogging in place slowly before starting any vigorous activities to ease the cardiovascular system into the aerobic exercise. The exercise part of the session should burn at least 300 kcalories, which can be achieved with 15 to 30 minutes of aerobic activity in the target zone or 40 to 60 minutes of lower-intensity activity, such as leisurely walking. After exercising, slowing down the exercise or changing to a less vigorous activity for 5 to 10 minutes is important to allow the body to relax gradually.

Beyond the exercise program, obese people should be encouraged to schedule more activity into their daily routines. For instance, they should use stairs, both up and down, instead of elevators or escalators. They can also park or get off public transportation farther away from their destination to allow more walking.

BEHAVIOR AND ATTITUDE MODIFICATION

Behavior modification deals with identifying and changing behaviors that affect weight gain, such as raiding the refrigerator at midnight. It also involves setting specific, achievable goals.

Setting the right goals is an important first step. Most people trying to lose weight focus on just that one goal: weight loss. However, the most productive areas to focus on are the eating and exercise changes that will lead to that long-term weight change. Effective goals are:

1. Specific and measurable
2. Attainable and realistic
3. Forgiving—less than perfect

"Exercise more" is a commendable ideal, but it's not specific. "Walk five miles every day" is specific and measurable, but is it realistic if you are just starting out? "Walk 30 minutes every day" is more attainable, but what happens if you're held up at work one day and there's a thunderstorm during your walking time another day? "Walk 30 minutes, five days each week" is specific, attainable, and forgiving. In short, a great goal!

Shaping is a behavioral technique in which you select a series of short-term goals that get closer and closer to the ultimate goal. For example, you might initially reduce your fat intake from 40 percent of kcalories to 35 percent of kcalories, then later reduce your fat intake to 30 percent of kcalories. Shaping is based on the concept that "nothing succeeds like success." Shaping uses two important behavioral principles: consecutive goals that move you ahead in small steps are the best way to reach a distant point, and frequent rewards keep the overall effort invigorated.

Additional elements of behavior and attitude modification can be grouped into several categories: self-monitoring, stimulus or cue control, eating behaviors, reinforcement or self-reward, self-control, and attitude modification. Let's take a look at each category.

Self-monitoring involves keeping a food diary or daily record of types and amounts of foods and beverages consumed, as well as time and place of eating, mood at the time, and degree of hunger felt. Its purpose is, of course, to increase awareness of what is actually being eaten and whether the eating is in response to hunger or other stimuli. Once harmful patterns that encourage overeating are identified, negative behaviors can be changed to more positive ones. Figure 12-5 shows a sample food diary page.

Through self-monitoring of your eating behaviors, cues or stimuli to overeating can be identified. For example, passing a bakery may be a cue for someone to stop and buy a dozen cookies. Examples of behavioral modification techniques for cue or stimulus control follow.

Food Purchasing, Storage, and Cooking

1. Plan meals a week or more ahead.
2. Make a shopping list.
3. Do food shopping after eating, on a full stomach.
4. Do not shop for food with someone who will pressure you to buy foods you do not need.
5. If you feel you must buy high-kcalorie foods for someone else in the family who can afford the kcalories, buy something you do not like or let that person buy, store, and serve the particular food.
6. Store food out of sight and limit storage to the kitchen.
7. Keep low-kcalorie snacks on hand and ready to eat.
8. When cooking, keep a small spoon such as a half-teaspoon measure on hand to use if you must taste while cooking.

Mealtime

1. Do not serve food at the table or leave serving dishes on the table.
2. Leave the table immediately after eating.

Holidays and Parties

1. Eat and drink something before you go to a party.
2. Drink fewer alcoholic beverages.

FIGURE 12-5: Food Diary						
Time?	What Was Eaten?	Where?	How Much?	Hungry?	With Whom?	Mood?

3. Bring a low-kcalorie food to the party.
4. Decide what you will eat before the meal.
5. Stay away from the food as much as possible.
6. Concentrate on socializing.
7. Be polite but firm and persistent when refusing another portion or drink.

Eating behaviors need to be modified to discourage overeating. First, one or two eating areas, such as the kitchen and dining-room tables, need to be set up so that eating occurs only in these designated locations. Eat only while sitting at the table and make the environment as attractive as possible. Do not read, watch television, or do anything else while you eat, because you can easily form associations between certain activities and food, such as television and snacking. Do not eat while standing at cabinets or refrigerators.

Second, plan three meals and at least two snacks daily, preferably at specified times of the day. Third, eat slowly by putting your fork down between bites, eating your favorite foods first, talking to others at the table, eating a high-fiber food that requires time to chew, and drinking a no-kcalorie beverage to help you fill up. In addition, take smaller bites, savor each bite, use a smaller plate to make the food look bigger, and leave a bite or two on the plate. If you clean your plate, you are responding to the sight of food, not to real hunger. When you want a snack, postpone it for 10 minutes.

Reward yourself for positive steps taken to lose weight but do not use food as a reward. An effective reward is something that is desirable and timely. The reward might be tangible, such as a movie or music CD or a payment toward buying a more costly item, or intangible, such as an afternoon off from work. Numerous small rewards, delivered for meeting smaller goals, are more effective than bigger rewards, requiring a long, difficult effort.

Overeating sometimes occurs in reaction to stressful situations, emotions, or cravings. The food diary is very useful in identifying these situations. Then you can handle these situations in new ways. You can express your feelings verbally if you are overeating in response to frustration or similar stressful emotions. You can exercise or use relaxation techniques to relieve stress, or you can switch to a new activity, such as taking a walk, knitting, reading, or engaging in a hobby to help take your mind off food. If you allow yourself five minutes before getting something to eat, you often will go on to something else and forget about the food. Positive self-talk is important for good self-control. Instead of repeating a negative statement such as "I cannot resist that cookie," say, "I will resist that cookie."

The most common attitude problem obese people have is thinking of themselves as either on or off a diet. Being on a diet implies that at some point the diet will be over, resulting in weight gain if old habits are resumed. Dieting should not be so restrictive or have such unrealistic goals that the person cannot wait to get off the diet. When combined with exercise, behavior and attitude modification, social support, and a maintenance plan, dieting is really a plan of sensible eating that allows for periodic indulgences.

Another attitude that needs modification revolves around using words such as always, never, and every. The following are examples of unrealistic statements using these terms.

- I will always control my desire for chocolate.
- I will never eat more than 1500 kcalories each day.
- I will exercise every day.

Goals stated in this manner decrease the likelihood that you will ever accomplish them and thus result in discouragement and feelings of failure.

Even with reasonable goals, occasional lapses in behavior occur. This is when having a constructive attitude is critical. After a person eats and drinks too much at a party one night, for example, feelings of guilt and failure are common. However, they do nothing to help people get back on their feet. Instead, the dieter must stay calm, realize that what is done is done, and understand that no one is perfect.

Two other attitudes that often need correcting concern hunger and foods that are "bad for you." Hunger is a physiological need for food, whereas appetite is a psychological need. Eating should be in response to hunger, not to appetite. Although people frequently regard certain foods as "good" or "bad," they must realize that no food is inherently good or bad. Some foods do contain more nutrients per kcalorie, and some are mostly empty kcalories with few nutrients. However, no food is so bad that it can never be eaten.

SOCIAL SUPPORT

Obese people are more likely to lose weight when their families and friends are supportive and involved in their weight-loss plans. As social support increases, so do a person's chances of maintaining weight loss. When possible, obese people need to enlist the help of someone who is easy to talk to, understands and empathizes with the problems in losing weight, and is genuinely interested in helping. Supporters can model good eating habits and give praise and encouragement. An obese person needs to tell others exactly how to be supportive by, for example, not offering high-kcalorie snacks. Requests need to be specific and positive.

MAINTENANCE SUPPORT

Not enough is known about factors associated with weight-maintenance success or what support is needed during the first few months of weight maintenance, when a majority of dieters begin to relapse. Being at a normal, or more normal, weight can bring about stress as adjustments are made. Food is no longer a focal point, and old friends and activities may not fit into the new lifestyle. Support and encouragement from significant others will probably diminish. A formal maintenance program can help deal with those issues as well as others.

Strategies that appear to support weight maintenance include:

1. Determining how many kcalories are needed for weight maintenance and working out a livable diet
2. Learning skills for dealing with high-risk situations when a lapse in eating behavior may occur and knowing what to do when a relapse occurs
3. Continued self-monitoring, including weighing
4. Continued exercise that's enjoyable
5. Continued social support
6. Continued use of other strategies that were useful during weight loss
7. Dealing with unrealistic expectations about being thin

These strategies can be used during formal maintenance programs and after treatment is terminated.

The National Weight Control Registry (NWCR) is helping researchers find out more about how people maintain their weight loss. Anyone who has lost at least 30 pounds and kept it off at least a year can join the NWCR. Those who enroll fill out questionnaires about how they lost weight, how they are keeping it off, and other aspects of their health. There are now over 6,000 people in the registry. People who successfully control their weight tend to:

- Eat a low-fat diet
- Eat a healthy breakfast
- Watch their total kcalories

- Eat fast food less than once a week and eat out no more than three times a week
- Exercise regularly and watch less than 10 hours of TV per week (much less than average)
- Weigh themselves daily
- Eat consistently whether on vacation or on the weekends

Registry members have lost an average of about 70 pounds and have maintained their weight loss for an average of 5.7 years.

DRUGS

Weight-loss drugs approved by the Food and Drug Administration for long-term use may be used by people with a BMI of 30 or more or people with a BMI of 27 or more who also have diseases such as diabetes or other risk factors. Drugs are used only as part of a program that includes an eating plan, physical activity, and behavior therapy. Because of the tendency to regain weight after weight loss, the use of long-term medication to aid in the treatment of obesity may be indicated for some people. On average, individuals who use prescribed weight-loss drugs lose about 5 to 10 percent of their original weight. Unfortunately, some individuals stop taking these medications in part due to their side effects.

All the prescription weight-loss drugs work by suppressing the appetite. Meridia (the brand name for sibutramine) was approved in 1997, and it increases the levels of certain brain chemicals that help reduce appetite. Because it may increase blood pressure and heart rate, Meridia should not be used by people with a history of heart disease, stroke, or uncontrolled high blood pressure. Other common side effects of Meridia include headache, dry mouth, constipation, and insomnia.

Meridia is approved for long-term, but not indefinite, use. Other antiobesity prescription drugs include products such as Bontril (phendimetrazine tartrate); these products are to be taken for only a few weeks.

A former prescription drug, Xenical (the brand name for orlistat) was approved for over-the-counter use as a weight loss aid in February 2007 under the name Alli. It is the first over-the-counter drug in a new class of antiobesity drugs known as lipase inhibitors. Lipase is the enzyme that breaks down dietary fat. Alli interferes with lipase function, decreasing dietary fat absorption. Since undigested fats are not absorbed, there is less caloric intake, which may help in controlling weight. The main side effects of Alli are cramping, diarrhea, flatulence, intestinal discomfort, and leakage of oily stool. Alli is only approved for use in adults 18 and over in conjunction with a calorie-reduced, low fat diet. Since lipase inhibitors can reduce absorbance of fat-soluble vitamins, a multivitamin should be taken daily while using Alli.

Americans spend over $1.3 billion each year on dietary supplements for weight loss (such as bitter orange), most of which have scant research showing that they work or are even safe to use. Unlike true "drugs" that must be proved to be safe and effective for their intended use before marketing, there is no provision in the law for the Food and Drug Administration to approve dietary supplements for safety or effectiveness before they reach the consumer.

Many herbal weight-loss products now use bitter orange peel in place of ephedra. However, bitter orange contains the chemical synephrine, which is similar to the main chemical in ephedra. The U.S. Food and Drug Administration banned ephedra because it raises blood pressure and is linked to heart attacks and strokes; it is unclear whether bitter orange has similar effects. There is currently little evidence that bitter orange is safer to use than ephedra.

Another popular weight-loss supplement is green tea extract. Some evidence suggests that the use of green tea preparations improves mental alertness, most likely because of its caffeine content. There are not enough reliable data to determine if it can aid in weight control. Green tea extract can cause jitteriness from the caffeine in some individuals, as well as indigestion, vomiting, and diarrhea.

Herbal preparations are not recommended as part of a weight-loss program. These preparations have unpredictable amounts of active ingredients and unpredictable—and potentially harmful—effects.

SURGERY

Weight-loss surgery is an option for people with clinically severe obesity, which is defined as a BMI of 40 or higher or a BMI of 35 or higher with other risk factors. This treatment is used only for people for whom other methods of treatment have failed. Weight-loss surgery provides medically significant sustained weight loss for more than 5 years in most people.

Obesity surgery alters the digestive process. The operations promote weight loss by closing off parts of the stomach to make it smaller. Operations that only reduce stomach size are known as restrictive operations because they restrict the amount of food the stomach can hold. Restrictive operations for obesity include adjustable gastric banding and vertical banded gastroplasty. Although restrictive operations lead to weight loss in almost all patients, they are less successful than malabsorptive operations, which are more common.

Gastric bypass is a very popular malabsorptive operation in which a small stomach pouch is created to restrict food intake. Next, a section of the small intestine is attached to the pouch to allow food to bypass the lower stomach and part of the upper small intestine. This bypass reduces the amount of kcalories and nutrients the body absorbs. Malabsorptive operations produce more weight loss than do restrictive operations and are more effective in reversing the health problems associated with severe obesity.

An integrated program that provides guidance on diet, physical activity, and psychosocial concerns before and after surgery is necessary. Most patients fare well, with control of hypertension, marked improvement in mobility, and other improvements.

MINI-SUMMARY

1. A comprehensive treatment plan for obesity needs to include an eating plan and nutrition education, exercise, behavior and attitude modification, social support, and maintenance support.

2. Kcalories should not be overly restricted. A reduction of 500–1000 kcalories a day will result in about 1 pound lost in a week.

3. A healthy eating plan emphasizes fruits, vegetables, whole grains, and fat-free or low-fat milk and milk products and includes meat, poultry, fish, beans, eggs, nuts, and foods generally lower in saturated fat, trans fat, cholesterol, salt, and added sugars.

4. No foods should be forbidden.

5. Portion control is vital, as are variety, balance, and moderation.

6. Frequent weighing is usually helpful.

7. Exercise can help control or suppress appetite; build muscles, which burn more kcalories; improve functioning of the cardiovascular system; reduce levels of blood lipids associated with heart disease; and increase a person's stamina, self-image, and ability to cope with stress, anxiety, and depression.

8. To sustain weight loss in adulthood, participate in at least 60 to 90 minutes of daily moderate-intensity physical activity while not exceeding kcaloric intake requirements. Pick an activity you enjoy.

9. Behavior modification deals with identifying and changing behaviors that affect weight gain, such as raiding the refrigerator at midnight. It also involves setting specific, achievable goals. Setting realistic goals,

program. Through goal setting, complex behavior changes can be broken down into a series of small, successive steps. Goals need to be reasonable and be stated in a positive, behavioral manner.

10. Elements of behavior and attitude modification can be grouped into several categories: self-monitoring, stimulus or cue control, eating behaviors, reinforcement or self-reward, self-control, and attitude modification.

11. Obese people are more likely to lose weight when their families and friends are supportive and involved in their weight-loss plans.

12. Strategies that appear to support weight maintenance include working out a realistic eating plan, learning skills for dealing with high-risk situations, continued self-monitoring, exercise, social support, use of other strategies that worked during weight loss, and dealing with unrealistic expectations about being thin.

13. Weight-loss drugs may be used for people with a BMI of 30 or more or people with a BMI of 27 or more who also have diseases such as diabetes or other risk factors. Drugs are used only along with an eating plan, physical activity, and behavior therapy.

14. Most drugs work by decreasing appetite or decreasing fat absorption. They do have side effects. There is scant research that any dietary supplements for weight loss are safe or effective.

15. Weight-loss surgery is normally reserved for patients who are very obese.

MENU PLANNING FOR WEIGHT LOSS AND MAINTENANCE

Balanced menus and recipes appropriate for weight loss and maintenance are discussed in depth in Chapter 9. Figure 12-6 shows some lower-kcalorie alternatives.

THE PROBLEM OF UNDERWEIGHT

A person is considered underweight if he or she has a BMI below 18.5. Although anyone who has seriously dieted may think an underweight person is problem-free, this is hardly the case. Underweight persons who have trouble gaining weight have very real concerns. Just as some people cannot seem to lose weight, other people have trouble putting on a few extra pounds. The cause could be genetics, metabolism, or environment. However, some thin people, if they were to gain weight, would feel uncomfortable. Anyone who is underweight due to wasting diseases such as cancer or eating disorders also has a problem: malnutrition.

The following list contains tips on gaining weight.

1. If you find you cannot eat large meals, do not get discouraged. You can increase your intake by eating smaller meals frequently.

2. Avoid drinking low-kcalorie beverages such as coffee, tea, and water, especially with meals. Try fruit juices, milk, and milkshakes for more calories.

3. Add calories to your meals by using margarine, mayonnaise, oil, salad dressing, or other fats that do not contain much saturated fat. For example, spread margarine on bread or use oil-packed tuna fish.

FIGURE 12-6: Lower-kcalorie, Lower-fat Alternatives

	Instead of . . .	Replace with . . .

Dairy Products

Evaporated whole milk	Evaporated fat-free (skim) or reduced fat (2%) milk
Whole milk	Low-fat (1%), reduced fat (2%), or fat-free (skim) milk
Ice cream	Sorbet, sherbet, low-fat or fat-free frozen yogurt, or ice milk (check label for calorie content)
Whipping cream	Imitation whipped cream (made with fat-free [skim] milk) or low-fat vanilla yogurt
Sour cream	Plain low-fat yogurt
Cream cheese	Neufchatel or "light" cream cheese or fat-free cream cheese
Cheese (cheddar, American, Swiss, jack)	Reduced calorie cheese, low calorie processed cheeses, etc.; fat-free cheese
Regular (4%) cottage cheese	Low-fat (1%) or reduced fat (2%) cottage cheese
Whole milk mozzarella cheese	Part skim milk, low moisture, mozzarella cheese
Whole milk ricotta cheese	Part skim milk ricotta cheese
Coffee cream (half and half) or nondairy creamer (liquid, powder)	Low-fat (1%) or reduced fat (2%) milk or fat-free dry milk powder

Cereals, Grains, and Pasta

Ramen noodles	Rice or noodles (spaghetti, marcaroni, etc.)
Pasta with cream sauce (alfredo)	Pasta with red sauce (marinara)
Pasta with cheese sauce	Pasta with vegetables (primavera)
Granola	Bran flakes, crispy rice, etc.
	Cooked grits or oatimeal
	Whole grains (e.g., couscous, barley, bulgur, etc.)
	Reduced fat granola

Meat, Fish, and Poultry

Cold cuts or lunch meats (bologna, salami, liverwurst, etc.)	Low-fat cold cuts (95% to 97% fat-free lunch meats low-fat pressed meats)
Hot dogs (regular)	Lower-fat hot dogs
Bacon or sausage	Canadian bacon or lean ham
Regular ground beef	Extra lean ground beef such as ground round or ground turkey (read labels)
Chicken or turkey with skin, duck, or goose	Chicken or turkey without skin (white meat)
Oil-packed tuna	Water-packed tuna (rinse to reduce sodium content)
Beef (chuck, rib, brisket)	Beef (round loin) trimmed of external fat (choose select grades)
Pork (spareribs, untrimmed loin)	Pork tenderloin or trimmed, lean smoked ham
Frozen breaded fish or fried fish (homemade or commercial)	Fish or shellfish, unbreaded (fresh, frozen, canned in water)
Whole eggs	Egg whites or egg substitutes
Frozen dinners (containing more than 13 grams of fat per serving)	Frozen dinners (containing less than 13 grams of fat per serving and lower in sodium)
Chorizo sausage	Turkey sausage, drained well (read label)
	Vegetarian sausage (made with tofu)

(continued)

FIGURE 12-6: Lower-kcalorie, Lower-fat Alternatives (*Continued*)

	Instead of . . .	Replace with . . .

Baked Goods

Croissants, brioches, etc.

Donuts, sweet rolls, muffins, scones, or pastries

Crackers

Cake (pound, chocolate, yellow)

Cookies

Hard french rolls or soft "brown 'n serve" rolls

English muffins, bagels, reduced fat or fat-free muffins or scones

Low-fat crackers (choose lower in sodium)

Saltine or soda crackers (choose lower in sodium)

Cake (angel food, white, gingerbread)

Reduced fat or fat-free cookies (graham crackers, ginger snaps, fig bars) (compare calorie level)

Snacks and Sweets

Nuts

Ice cream, e.g., cones or bars

Custards or puddings (made with whole milk)

Popcorn (air-popped or light microwave), fruits, vegetables

Frozen yogurt, frozen fruit, or chocolate pudding bars

Puddings (made with skim milk)

Fats, Oils, and Salad Dressings

Regular margarine or butter

Regular mayonnaise

Regular salad dressings

Butter or margarine on toast or bread

Oils, shortening, or lard

Light-spread margarines, diet margarine, or whipped butter, tub or squeeze bottle

Light or diet mayonnaise or mustard

Reduced calorie or fat-free salad dressings, lemon juice, or plain, herb-flavored, or wine vinegar

Jelly, jam, or honey on bread or toast

Nonstick cooking spray for stir-frying or sautéing

As a substitute for oil or butter, use applesauce or prune puree in baked goods.

Miscellaneous

Canned cream soups

Canned beans and franks

Gravy (homemade with fat and/or milk)

Fudge sauce

Avocado on sandwiches

Guacamole dip or refried beans with lard

Canned broth-based soups

Canned baked beans in tomato sauce

Gravy mixes made with water or homemade with the fat skimmed off and fat-free milk included

Chocolate syrup

Cucumber slices or lettuce leaves

Salsa

Source: Aim for a Healthy Weight. NIH Publ. No. 05-5213, National Institutes of Health.

4. Add skim milk powder to soups, sauces, gravies, casseroles, scrambled eggs, and hot cereals. It adds both kcalories and protein. It can also be blended with milk at the rate of 2 to 4 tablespoons of powder to 1 cup of milk.

5. Add cheese to your favorite sandwiches. Use grated cheese on top of casseroles, salads, soups, sauces, and baked potatoes.

6. Try breaded foods.

7. Eat regular yogurt, peanut butter or cheese with crackers, nuts, milkshakes, and whole-grain cookies and muffins as snacks.
8. Add regular cottage cheese to casseroles or egg dishes such as quiche, scrambled eggs, and soufflés. Add it to spaghetti or noodles.
9. Make every mouthful count!

NUTRITION FOR THE ATHLETE

Many athletes (Figure 12-7) require 3000 to 6000 kcalories daily. The amount of energy required by an athlete depends on the type of activity and its duration, frequency, and intensity. In addition, the athlete's basal metabolic rate, body composition, age, and environment must be taken into account.

Carbohydrate (from glycogen and blood glucose) and fat are the primary fuel sources for exercise. Protein plays a minor role. The availability of carbohydrate—more specifically, the amount of glycogen stores—heavily influences athletic performance. Glucose is the main source of energy for intense exercise, while fat is the main source of energy during low to moderate exercise. An appropriate diet for many athletes consists of 60 to 65 percent of kcalories as carbohydrates, 30 percent or less as fat, and enough protein to provide 1.2 to 1.6 grams per kilogram of body weight for endurance athletes and 1.6 to 1.7 grams for power (strength or speed) athletes.

Although many athletes take vitamin and mineral supplements, these supplements will not enhance performance unless there is a deficiency. Most athletes get plenty of vitamins and minerals in their regular diets, although young athletes and women need to pay special attention to iron and calcium.

Water is the most crucial nutrient for athletes. They need about 1 liter of water for every 1000 kcalories consumed. Athletes need 16 ounces of fluid about 2 to 3 hours before beginning a workout, an additional 1 to 2 cups 15 minutes before the workout, and then about 4 to 8 ounces every 15 minutes during the workout. One gulp is about 1 ounce. For persons performing prolonged physical activity (especially in hot weather), sports drinks are recommended (see Food Facts on page 455).

Each person's fluid needs are different. Your fluid loss will depend on how big you are (larger people sweat more), your fitness level (the more fit you are, the more you sweat), and genetics. A good way to determine how much fluid to replace after exercising is to weigh in before and after exercise. For every pound that is lost, you need to drink 2 cups, or 16 ounces, of water.

For endurance events, some carbohydrates, such as glucose polymers, in fluids taken before and during competition may be helpful in maintaining normal blood-sugar levels. See Food Facts: Sports Drinks in this chapter for more information.

Carbohydrate or glycogen loading is a regimen involving three or more days of decreasing amounts of exercise and increased consumption of carbohydrates before an event to increase glycogen stores. The theory is that increasing glycogen stores by 50 to 80 percent will enhance performance by providing more energy during lengthy competition. It is most appropriate for endurance athletes.

Here are some menu-planning guidelines for athletes.

1. Include a variety of foods from MyPyramid.
2. Good sources of complex carbohydrates to emphasize on menus include pasta, rice, other grain products such as whole-grain breads and cereals, legumes, and fruits

FIGURE 12-7:
Many athletes require more daily kcalorie intake. Courtesy Corbis Digital Stock.

and vegetables. On the eve of the New York City Marathon each year, marathon officials typically host a pasta dinner for runners that features spaghetti with marinara sauce and cold pasta primavera. Complex carbohydrates such as pasta also provide needed B vitamins, minerals, and fiber. Whole-grain products such as whole-wheat bread contain more nutrients than do refined products such as white bread. If using refined products, be sure they are enriched (the thiamin, riboflavin, niacin, and iron have been replaced). Here are some ways to include complex carbohydrates in your menu:

- At breakfast, offer a variety of pancakes, waffles, cold and hot cereals, breads, and rolls.
- At lunch, make sandwiches with different types of bread, such as pita pockets, raisin bread, onion rolls, and brown bread. Also, have a variety of breads and rolls available for nonsandwich items.
- Serve pasta and rice as a side or main dish with, for example, chicken and vegetables. Cold pasta and rice salads are great, too.
- Potatoes, whether baked, mashed, or boiled, are an excellent source of carbohydrates.
- Always have available as many types of fresh fruits and salads as possible.
- Don't forget to use beans and peas in soups, salads, entrées, and side dishes.
- Nutritious desserts that emphasize carbohydrates include frozen yogurt with fruit toppings, oatmeal cookies, and fresh fruit.

3. Don't offer too much protein and fat in the belief that athletes need the extra kcalories. They do, but many of those extra calories should come from complex carbohydrates such as whole grains, fruits, and vegetables. The days of steak-and-egg dinners are over for athletes. The protein and fat present in these meals do nothing to improve performance. Here are some ways to moderate the amount of fat and protein:
- Use lean, well-trimmed cuts of beef.
- Offer chicken, turkey, and fish—all lower in fat than beef. Broiling, roasting, and grilling are the preferred cooking methods, with frying being acceptable occasionally.
- Offer larger serving sizes of meat, poultry, and fish, perhaps 1 to 2 ounces more, but don't overdo it.
- Offer fried food in moderation.
- Offer low-fat and fat-free milk.
- Offer high-fat desserts, such as ice cream and many types of sweets, in moderation. Frozen yogurt and ice milk generally contain less fat than ice cream and can be topped with fruit or crushed oatmeal cookies. Fruit ice and sorbet contain no fat.

4. Offer a variety of fluids, not just soft drinks and other sugared drinks. Good beverage choices include fruit juices, iced tea and iced coffee (preferably freshly brewed decaffeinated), plain and flavored mineral and seltzer water, spritzers (fruit juice and mineral water), and smoothies made with yogurt and fruit. Soft drinks and juice drinks—both are loaded with sugar—should be offered in moderation.

5. Make sure iodized salt is on the table.

6. Be sure to include sources of iron, calcium, and zinc at each meal. Good iron sources include liver, red meats, legumes, and iron-fortified breakfast cereal. Moderate iron sources include raisins, dried fruit, bananas, nuts, and whole-grain and fortified grain products. Be sure to include good vitamin C sources at each meal, as vitamin C assists in iron absorption. Vitamin C sources include citrus fruits and juices, cantaloupe, strawberries, broccoli, potatoes, and brussels sprouts. Calcium (found in milk and dairy products) and zinc (found in shellfish, meat and poultry, legumes, dairy foods, whole grains, and fortified cereals) are also important.

7. The most important meal is the one closest to the competition, commonly called the *precompetition meal*. The functions of this meal include getting the athlete fueled up both physically and psychologically, helping to settle the stomach, and preventing hunger. The meal should consist of mostly complex carbohydrates (they are digested faster and more easily and help maintain blood sugar levels) and should be moderate in protein and low in fat. High-fat foods take longer to digest and can cause sluggishness. Substantial precompetition meals are usually served three to four hours before the competition to allow enough time for stomach emptying (to avoid cramping and discomfort during the competition). Menus might include cereals with low-fat or skim milk topped with fresh fruit, low-fat yogurt with muffins and juice, or one or two eggs with toast and jelly, and juice. The meal should include 2 to 3 cups of fluid for hydration and typically provide 300 to 1000 kcalories. Smaller precompetition meals may be served two to three hours before competition. Many athletes have specific comfort foods they enjoy before competition. It is recommended that the stomach be fairly empty during competition so that blood can be sent to the working muscles instead of to the stomach to help in digestion.

8. After competition and workouts, again emphasize complex carbohydrates to ensure glycogen restoration. It takes about 20 hours to fill glycogen stores without interruption from another practice or competition. Protein also needs to be replenished, as well as minerals lost in sweating.

> **PRECOMPETITION MEAL**
> The meal closest to the time of a competition or event.

MINI-SUMMARY

1. Athletes have increased needs for kcalories. Carbohydrate (from glycogen and blood glucose) is the primary fuel source for exercise. Glucose is the main source of energy for intense exercise, while fat is the main source of energy during low to moderate exercise.

2. An appropriate diet for many athletes consists of 60 to 65 percent of kcalories as carbohydrates, 30 percent or less as fat, and enough protein to provide 1.2 to 1.7 grams per kilogram of body weight depending on the type of athletic activity.

3. Although many athletes take vitamin and mineral supplements, these supplements will not enhance performance unless there is a deficiency.

4. Water is the most crucial nutrient for athletes. They need about 1 liter of water for every 1000 kcalories consumed. Athletes need 16 ounces of fluid about 2 to 3 hours before beginning a workout, and additional 1 to 2 cups 15 minutes before the workout, and then 4 to 8 ounces every 15 minutes during the workout.

5. Carbohydrate or glycogen loading is a regimen involving three or more days of decreasing amounts of exercise and increased consumption of carbohydrates before an event to increase glycogen stores.

6. Menu-planning guidelines are given in this section, including tips on the precompetition meal.

CHECK-OUT QUIZ

1. Obesity is due simply to overeating.

   a. True
   b. False

2. An obese person is at increased risk for coronary heart disease.

   a. True
   b. False

3. Obesity is defined as having a BMI greater than 25.

   a. True
   b. False

4. When trying to lose weight, omitting junk foods that you like from your diet is a good idea.

   a. True
   b. False

5. Carbohydrate loading is a regimen involving three days of increased exercise and increased intake of carbohydrates.

   a. True
   b. False

6. Serious athletes need about 8 grams of protein per kilogram of body weight.

   a. True
   b. False

7. BMI is a direct calculation based on height and weight.

   a. True
   b. False

8. An example of using stimulus control to lose weight is to leave the table immediately after eating.

   a. True
   b. False

9. A person is considered underweight if he or she has a BMI less than 18.5.

   a. True
   b. False

10. Most weight-loss drugs work by suppressing the appetite or decreasing fat absorption.

    a. True
    b. False

ACTIVITIES AND APPLICATIONS

1. Your Desirable Weight

   Using Figure 12-1, determine your BMI. Are you overweight (a BMI of 25 or more)? If so, find out if you have a family history of any of the medical conditions discussed in the section on health and obesity.

2. Balanced Menu Planning

   A local steak and seafood restaurant has asked you to design lower-kcalorie menu items as follows: two appetizers, three entrées, and one dessert. Their emphasis is on freshly made traditional American cooking. Provide recipes and do a nutrient analysis for each recipe using the "Recipe Builder" in iProfile.

3. Using a Food Diary

   Using the food diary form shown in Figure 12-5, complete a three-day food diary. Then examine it to increase your awareness of how much and how often you are eating and whether you are eating in response to moods, people, and/or activities. Write down two insights you gained about your eating habits.

4. Box Lunches

   Design three complete box lunches for kcalorie- and health-conscious customers of a gourmet take-out deli with complete kitchen facilities in San Francisco. The meal must be well balanced and provide a main dish, a side dish, a dessert, and a beverage. Use iProfile to get a nutrient analysis for each box lunch.

5. Precompetition Meals

   Devise a menu for a noontime precompetition meal for long-distance runners on a university track team who will compete at 4 p.m. Have two selections for each category you choose.

NUTRITION WEB EXPLORER

National Institute of Diabetes and Digestive and Kidney Diseases   www.niddk.nih.gov

On the home page for the NIDDK, under "Health and Disease Topics," click on "Weight Control." Next, click on "Publication" and choose one to read and write a paragraph about.

U.S. Department of Agriculture

www.ars.usda.gov/is/AR/archive/mar06/diet0306.htm

Read this article and find out the strongest predictor for weight loss.

Sports Science News

www.sportsci.org

On the home page, click on the index on the left for "Sports Nutrition." Read any one of the articles listed. Summarize in one paragraph what you read.

Shape Up America!

www.shapeup.org/shape/steps.php

Read about a program that encourages participants to use a pedometer and walk 10,000 steps a day for physical fitness.

FOOD FACTS: SPORTS DRINKS

A topic of great interest to athletes is whether sports drinks such as Gatorade or Powerade are needed during an event or workout (Figure 12-8). Sports drinks contain a dilute mixture of carbohydrate and electrolytes. Most contain 50 to 100 kcalories per cup, about 3–4 teaspoons of carbohydrate per cup, and small amounts of sodium and potassium. Sodium, which is lost during sweating, helps maintain fluid balance in your body and also helps you absorb fluid in your intestines.

Sports drinks are purposely made to be weak solutions so that they can empty faster from the stomach, and the nutrients they contain are therefore available to the body more quickly. They are primarily designed to be used during exercise.

During exercise lasting 60 minutes or more, sports drinks can help replace water and electrolytes and provide some carbohydrates for energy. During an endurance event or workout, you increasingly rely on blood sugar for energy as your muscle glycogen stores diminish. Carbohydrates taken during exercise can help you maintain a normal blood-sugar level and enhance (as well as lengthen) performance. Athletes often consume $^1/_2$ to 1 cup of sports fluids every 15 to 20 minutes during exercise to get both the build and the carbohydrates for endurance.

Most sports drinks contain several carbohydrate sources: such as glucose, fructose, and sucrose. Sports drinks that contain several different carbohydrates may improve the amount of carbohydrate that gets to the muscles as fuel. Carbohydrate concentrations over 8 percent can cause stomach cramps and diarrhea. Be aware that fitness waters don't contain as much carbohydrate as sports drinks. They can keep you hydrated but will not boost endurance.

Although sports drinks clearly can help an athlete in lengthy events or workouts, are there other products that can do much the same thing? Long before sports drinks were available, athletes had their own homemade sports drinks: diluted juices with a pinch of salt, tea with honey, and diluted lemonade. Which works best? Whichever satisfies the athlete both physically and psychologically.

You have probably seen energy drinks, such as Red Bull. They are not at all like sports drinks. Energy drinks contain large doses of caffeine and other legal stimulants like ginseng. Energy drinks may contain as much as 80 mg of caffeine, the equivalent of a cup of coffee. Compared to the 37 mg. of caffeine in a Mountain Dew, or the 23 mg. in a Coca-Cola Classic, that's a big punch. These drinks are marketed to people under 30, especially to college students, who are looking for a stimulus to get through studying or other activities.

FIGURE 12-8:
Sports drinks.
Courtesy Corbis Digital Stock.

HOT TOPIC: DIET BOOKS

It is estimated that over 60 million Americans are trying to lose weight. Some succeed in taking weight off, but far fewer—under 5 percent—manage to keep the weight off over the long term. With half the adult population in the United States considered overweight, it's little wonder that consumers are constantly searching for a "magic bullet" to help them lose weight quickly and effortlessly (Figure 12-9).

There is little scientific research to corroborate the theories expounded in the majority of diet books currently on the market. Many promise weight-loss programs that are easy, allow favorite foods or foods traditionally limited in weight-loss diets without limitations, and do not require a major shift in exercise habits. Authors may simplify or expand upon biochemistry and physiology to help support their theories and provide a plethora of scientific jargon that people do not understand but that seems to make sense. Few, if any, offer solid scientific support for their claims in the form of published research studies. Instead, most evidence is based on anecdotal findings, theories, and testimonials about short-term results.

Low-carbohydrate, high-protein diets have been popular for many years and promise significant weight loss. These diets are nothing more than low-kcalorie diets in disguise, but with some potentially serious consequences. Following a low-carbohydrate, high-protein diet will encourage the body to burn its own fat. Without carbohydrates, however, fat is not burned completely and substances called ketones are formed and released into the bloodstream. Abnormally high ketone levels in the body, or ketosis, may indeed make dieting easier, since they typically decrease appetite and may cause nausea. Although these diets may not be harmful when used by healthy people for a short period, they restrict healthful foods that provide essential nutrients. Individuals who follow these diets for more than just a short period run the risk of compromised vitamin and mineral intake, as well as potential health risks. There are no long-term studies that have found these diets to be effective and safe.

Here's a rundown on some of the more popular diet books:

1. *Dr. Atkins' New Diet Revolution* by Robert Atkins, MD
   Premise/Theory: Excess carbohydrate intake prevents the body from burning fat efficiently. Eating too many carbohydrates causes production of excessive amounts of insulin, leading to obesity and a variety of other health problems. Drastically decreasing dietary intake of carbohydrates forces the body to burn reserves of stored fat for energy, causing a buildup of ketones that lead to decreased hunger.

   Dietary Recommendations:
   - The Atkins diet limits carbohydrates to 20 grams per day at the start of the diet and 0–60 grams per day in the ongoing weight-loss phase. Carbohydrate intake ranges from 25 to 90 grams per day in the maintenance diet.
   - Unlimited quantities of protein foods and fat—steak, bacon, eggs, chicken, fish, butter, and vegetable oil—are allowed. Avoid or limit carbohydrates, specifically breads, pasta, most fruits and vegetables, milk, and yogurt.

   Concerns:
   - Offers extremely limited food choices. Diet is nutritionally unbalanced and excessively high in protein, fat, saturated fat, and cholesterol. Dieters lose weight

FIGURE 12-9:
Bookstores have entire sections devoted to diet books.
Courtesy PhotoDisc, Inc.

because they are bored and eat fewer kcalories.
- Promotes ketosis as a means of weight loss.
- There is no definitive long-term data on its safety.
- Dehydration is possible if large amounts of water are not consumed.
- Diet is low in calcium, magnesium, potassium, vitamin C, and folate (dietary supplements are recommended).

What's Good
- Atkins Diet does help some people lose weight.

2. *South Beach Diet* by Arthur Agatston, MD
Premise: Dr. Agatston believes that excess consumption of so-called "bad carbohydrates," such as the rapidly absorbed carbohydrates found in foods with a high glycemic index, creates an insulin resistance syndrome—an impairment of the hormone insulin's ability to properly process fat or sugar. In addition, he believes along with many physicians that excess consumption of "bad fats," such as saturated fat and trans fat, contributes to an increase in cardiovascular disease. To prevent these two conditions, Agatston's diet minimizes consumption of bad fats and bad carbohydrates and encourages increased consumption of good fats and good carbohydrates.

Dietary Recommendations:
- Three phases:
Phase 1 (14 days): Most vegetables, some nuts, some cheeses, lean meat, fish, poultry. No fruit, grains, fast food, or alcohol. No specific quantities.

Phase 2: Whole grains, fruits, and dairy are slowly reintroduced to the diet but in small portion sizes. Phase 3 (weight maintenance/3 meals daily plus snacks): Only 3 servings of whole grains and 3 fruit servings each day in addition to lots of vegetables; lean meats, poultry, and fish; and low fat dairy; return to phase 1 if you gain weight or overindulge.

Concerns:
- Restricts foods such as carrots, bananas, pineapple and watermelon
- Menus only have about 1200 kcalories/day—too low for most people.

What's Good:
- Encourages high fiber carbohydrates and foods low in saturated and trans fats.
- Restricts fatty meats and cheeses, as well as desserts and sweets.

3. *Volumetrics* by Barbara Rolls, PhD
Premise/Theory: People feel full because of the amount of food they eat—not because of the number of kcalories, or the amount of carbohydrate, fat, or protein. The trick is to fill up on foods that don't have a lot of kcalories.

Dietary Recommendations:
- No foods are banned. Foods are evaluated based on their energy density.
- Relies heavily on foods with a high water content such as many fruits and vegetables, nonfat milk, and soup broth.
- Encourages high-fiber foods, adequate lean proteins, and some healthy fats such as in fish.
- Also encourages small amounts of energy-dense foods such as fats and sweets.

- Use low-fat cooking techniques.
- First course of broth-based soup of low-kcalorie salad will take the edge off the appetite.

Concerns:
- Recipes are somewhat time-consuming to prepare.
- Hard to use if you eat out a lot or eat mostly packaged foods.

What's Good:
- Commonsense, healthy eating plan.
- Gives choices and variety.

4. *Enter the Zone* by Barry Sears, PhD
Premise/Theory: The "zone" is a metabolic state in which the mind is relaxed and focused and the body is strong and works at peak efficiency. A person in the "zone" will allegedly experience permanent body-fat loss, optimal health, greater athletic performance, and improved mental productivity. Insulin is released as a result of eating carbohydrates and leads to weight gain. Because food has a potent druglike effect on the hormonal systems that regulate the body's physiological processes, eating the right combination of foods leads to a metabolic state in which the body works at peak performance and experiences decreased hunger, weight loss, and increased energy. The diet is low in carbohydrates.

Dietary Recommendations:
- To get into the "zone," rigid quantities of food, apportioned in blocks and at prescribed times, are recommended in a distribution of 40 percent carbohydrate, 30 percent protein, and 30 percent fat. Meals should provide no more than 500 kcalories, and snacks less than 100 kcalories.

- Food should be treated like a medical prescription or drug.
- Menus suggest lots of egg whites, nuts, olives, peanut butter, and monounsaturated fats and large amounts of allowable fruits and vegetables. Alcohol is okay in moderation, but "zone" followers are advised to avoid or limit carbohydrates, especially pasta, bread, and fruits and vegetables such as carrots and bananas, which cause blood sugar to rise more than others do. Saturated fat should also be avoided.
- Diet averages 1300 kcalories per day, although some menus may run as low as 850 kcalories.

Concerns:
- Oversimplifies complicated body physiology.
- Exaggerates evidence that the diet may cure a number of diseases.
- Relies on unproven claims based on case histories, testimonials, and uncontrolled studies that are not published in peer-review journals.

What's Good
- Lean meat, fruits, vegetables, nuts, and low-fat dairy are good choices.

Source: Sections were adapted from "Fad Diets: Look Before You Leap," Food Insight: Current Topics in Food Safety and Nutrition (March–April 2000), published by the International Food Information Council.

Nutrition Over the Life Cycle

Although much of the nutrition advice we hear on television or read in magazines is for adults—indeed, the first part of this book is mostly for adults—other groups have their own special nutrition needs and concerns. What do pregnant women, babies, children, and even teenagers have in common? They are all growing, and growth requires more nutrients. Did you know that compared with an adult, a one-month-old baby needs twice the amount, proportionally, of kcalories, protein, and many vitamins and minerals? At the other end of the age spectrum, the fastest-growing age group in the United States is the over-85 group. As people age, many new factors affect their nutrition status: the aging process, the onset of chronic diseases such as heart disease, living alone, dentures, and the inability to get out to shop for food.

This chapter takes you from pregnancy through infancy, childhood, and adolescence and on to the golden years. Along the way, we will explore the nutritional needs and factors that affect nutrition status for each group, along with menu-planning guides. This chapter will help you to:

- Explain the benefits of good nutrition for mother and baby during pregnancy
- Identify nutrients of special concern during pregnancy and their food sources
- Explain the possible effects of alcohol, fish, caffeine, and artificial sweeteners during pregnancy
- Plan menus for women during pregnancy and lactation
- Describe what an infant should be fed during the first year, including the progression of solid foods
- Give five reasons why breastfeeding is preferable to bottle feeding
- Describe how to ensure enjoyable mealtimes with young children and teach them good eating habits
- Plan menus for preschool and school-age children
- Identify the nutrients that children and adolescents are most likely to be lacking and their food sources
- Describe influences on children's and adolescents' eating habits
- Plan menus for adolescents
- Distinguish among anorexia nervosa, bulimia nervosa, binge eating disorder, and female athlete triad
- Describe factors that influence the nutrition status of older adults
- Identify nutrients of concern for older adults and their food sources
- Plan menus for healthy older adults
- Describe ways to prevent the development of obesity during childhood and obesity

---

FETUS

The infant in the mother's uterus from eight weeks after conception until birth.

AMNIOTIC SAC

The protective bag, or sac, that cushions and protects the fetus during pregnancy.

EMBRYO

The name of the fertilized egg from conception to the eighth week.

PLACENTA

The organ that develops during the first month of pregnancy, which provides for exchange of nutrients and wastes between fetus and mother and secretes the hormones necessary to maintain pregnancy.

PREGNANCY

From a modest one-cell beginning, an actual living and breathing baby is born after 40 weeks. From the second to the eighth week after conception, the infant is called an *embryo*. From eight weeks after conception until birth, the infant in the mother's uterus is called a *fetus*. At eight weeks the fetus is about 1½ inches long and has a beating heart and gastrointestinal and nervous systems. To cushion and protect the fetus, it floats in a protected bag, or sac, called the *amniotic sac*. During the first month of pregnancy, an organ called the *placenta* develops. The placenta provides oxygen and nutrients and then picks up wastes from the fetus to be removed. The placenta also secretes the hormones necessary to maintain pregnancy.

After the second week of pregnancy, the cells of the embryo multiply and begin to take on specific functions. This process is called differentiation, and it produces the varied cell types that make up a human being (such as blood cells, kidney cells, and nerve cells).

There is rapid growth, and the baby's main external features begin to take form. It is during this critical period of differentiation (most of the first trimester) that the growing baby is most susceptible to damage from nutritional deficiencies, alcohol, and other things. The following list describes some of the changes by week.

- Week 3
  - Beginning development of the brain, spinal cord, and heart.
  - Beginning development of the gastrointestinal tract.
- Weeks 4 to 5
  - Formation of tissue that develops into the vertebrae and some other bones.
  - Further development of the heart, which now beats at a regular rhythm.
  - The brain develops into five areas; arm and leg buds are visible.
- Week 6
  - Beginning of formation of the lungs.
  - Hands and feet have digits but may still be webbed.
- Week 7
  - All essential organs have at least begun to form.

Pregnancy lasts for nine months. The first three-month period is called the first trimester. Likewise, the next three months is the second trimester and the final three months is the third trimester.

NUTRITION DURING PREGNANCY

The nutritional status of women before and during pregnancy influences both the mother's and the baby's health. Factors that place a woman at nutritional risk during pregnancy include an inadequate diet, smoking, and other influences described in Figure 13-1.

Figure 13-2 shows optimum weight gain during pregnancy. Underweight women (BMI less than 18.5) must either gain weight before or gain more weight during pregnancy. Underweight women have a higher chance of having babies that are born early, weigh below average, and/or die within the first year. Overweight women need to lose weight before pregnancy or gain less weight during pregnancy; otherwise they are at greater risk of having problems such as diabetes during pregnancy (called gestational diabetes) and high blood pressure, both of which can cause

FIGURE 13-1: Nutrition Risk Factors for Pregnant Women

Prepregnancy weight below BMI of 18.5 or BMI of 25 or higher

Inadequate kcalories intake

Inadequate intake of nutrient(s)

Alcohol use

Teenager

Woman over age 35

Chronic disease such as diabetes or high blood pressure

Poverty and/or food insecurity

Multiple births (twins, triplets, etc.)

Prepregnancy Weight	Recommended Weight Gain
Underweight (BMI less than 18.5)	28–40 pounds
Healthy weight (BMI 18.5 to 24.9)	25–35 pounds
Overweight (BMI 25 to 29.9)	15–25 pounds
Obese (BMI greater than 30)	15 pounds at least

Source: Adapted from *Nutrition During Pregnancy* (1990), National Academy of Sciences, Washington, D.C.: National Academy Press.

LOW-BIRTH-WEIGHT BABY

A newborn who weighs less than 5½ pounds; these infants are at higher risk for disease.

---

complications for the mother and the baby. Figure 13-3 shows the components of the weight gained; about 8 pounds is actually the baby, with the rest serving to support the baby's growth.

Both prepregnancy weight and weight gain during pregnancy directly influence infant birth weight. Newborn weight and health status tend to increase as weight gain increases during pregnancy. The newborn's weight is the number one indicator of his or her future health status. A newborn who weighs less than 5½ pounds is referred to as a **_low-birth-weight baby_**. Low-birth-weight babies, as well as babies born prematurely before 37 weeks of gestation, are at higher risk for complications and experience more difficulties surviving the first year. Often the mother of a low-birth-weight baby has a history of poor nutrition status before and/or during pregnancy. Other factors associated with low birth weight are smoking, alcohol use, drug use, and certain disease conditions.

Kcalories

To have healthy babies, pregnant women need to eat more kcalories, but not a whole lot more. If a pregnant woman "eats for two" during pregnancy, she is likely to put on too much weight. Pregnancy does increase daily kcalorie needs, but only by an additional 340 kcalories during the second trimester and an additional 450 kcalories during the third trimester. Within that measly 340 to 450 kcalories, however, a pregnant woman must pack more carbohydrate, fiber, essential fatty acids, and protein, besides more of most vitamins and minerals! See Appendix B for a comparison of the Dietary Reference Intakes for nonpregnant and pregnant women and Figure 13-4 for the MyPyramid Plan for Moms.

FIGURE 13-3:

Components of weight gain during pregnancy.

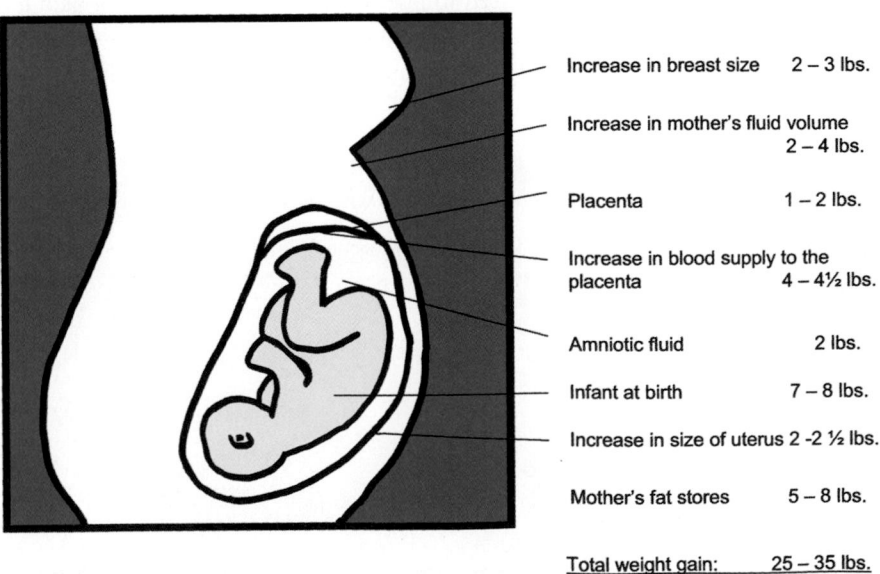

Increase in breast size	2 – 3 lbs.
Increase in mother's fluid volume	2 – 4 lbs.
Placenta	1 – 2 lbs.
Increase in blood supply to the placenta	4 – 4½ lbs.
Amniotic fluid	2 lbs.
Infant at birth	7 – 8 lbs.
Increase in size of uterus	2 - 2 ½ lbs.
Mother's fat stores	5 – 8 lbs.
Total weight gain:	**25 – 35 lbs.**

What Should I Eat?
MyPyramid Plan for Moms

When you are pregnant, you have special nutritional needs. Follow the MyPyramid Plan for Moms below to help you and your baby stay healthy.

- Eat these amounts from each food group daily.
- The calories and amounts of food you need change with the stage of pregnancy.
- The Plan shows different amounts of food for different trimesters, to meet your changing nutritional needs.

Food Group	1st Trimester	2nd and 3rd Trimesters	What counts as 1 cup or 1 ounce?	Remember to...
	Eat this amount from each group daily.*			
Fruits	2 cups	2 cups	1 cup fruit or juice, ½ cup dried fruit	*Focus on fruits*— Eat a variety of fruit.
Vegetables	2½ cups	3 cups	1 cup raw or cooked vegetables or juice, 2 cups raw leafy vegetables	*Vary your veggies*— Eat more dark green and orange vegetables and cooked dry beans.
Grains	6 ounces	8 ounces	1 slice bread; ½ cup cooked pasta, rice, cereal; 1 ounce ready-to-eat cereal	*Make half your grains whole*— Choose whole instead of refined grains.
Meat and Beans	5½ ounces	6½ ounces	1 ounce lean meat, poultry, fish; 1 egg; ¼ cup cooked dry beans; ½ ounce nuts; 1 tablespoon peanut butter	*Go lean with protein*—Choose low-fat or lean meats and poultry.
Milk	3 cups	3 cups	1 cup milk, 8 ounces yogurt, 1½ ounces cheese, 2 ounces processed cheese	*Get your calcium-rich foods*—Go low-fat or fat-free when you choose milk, yogurt, and cheese.

*These amounts are for an average pregnant woman. You may need more or less than the average. Check with your doctor to make sure you are gaining weight as you should.

In each food group, choose foods that are low in "extras." Pregnant women and women who may become pregnant should not drink alcohol. Any amount of alcohol during pregnancy could cause problems for your baby.

Most doctors recommend that pregnant women take a prenatal vitamin and mineral supplement every day **in addition to** eating a healthy diet. This is so you and your baby get enough folic acid, iron, and other nutrients. But don't overdo it. Taking extra can be harmful.

Get a MyPyramid Plan for Moms designed just for you on the MyPyramid website. Go to www.mypyramid.gov. Choose "For Pregnancy and Breastfeeding" on the left side menu.

FIGURE 13-4:

What should I eat? MyPyramid Plan for moms.
Courtesy U.S. Department of Agriculture.

During the first 13 weeks of pregnancy, the total weight gain is between 2 and 4 pounds. Thereafter, about 1 pound per week is normal. Corresponding with the timing of weight gain, it makes sense that the greatest need for kcalories begins at the end of the first trimester and continues until birth. The two major factors that influence kcalorie requirements during pregnancy are the woman's activity level and basal metabolic rate (BMR). The BMR increases to support the growth of the fetus. Pregnancy is no time to diet or, especially, to follow a fad diet, which could have dangerous implications for the fetus.

Protein

Protein needs increase 25 grams above prepregnancy needs for pregnant women. The requirements for protein are generous during pregnancy and probably high for some women. Meeting protein needs is rarely a problem.

Essential fatty acids

The requirements for the essential fatty acids, linoleic and alpha-linolenic acids, increase during pregnancy. These fatty acids are especially important to growth and development of the brain. Linoleic acid is found in vegetable oils such as corn, safflower, soybean, cottonseed, and sunflower oils. Most Americans get plenty of linoleic acid from foods containing vegetable oils, such as margarine, salad dressings, and mayonnaise. Alpha-linolenic acid is found in several oils, notably canola, flax seed, soybean, walnut, and wheat germ oils (or margarines made with these oils). Other good sources of alpha-linolenic acid include ground flax seed, walnuts, and soy products.

The body converts alpha-linolenic acid into two other omega-3 fatty acids: docosahexaenoic acid (DHA) and eicosapentaenoic acid (EPA). DHA is incorporated into the retina of the developing baby's eye at a high rate during the final months of pregnancy. Both DHA and EPA are found in fish such as salmon, mackerel, sardines, halibut, bluefish, trout, and tuna. Notice that these fish tend to be fatty fish, not lean fish. Whereas Americans generally get plenty of linoleic acid, that is not the case with alpha-linolenic acid, DHA, or EPA.

Nutrients for bones and teeth

During pregnancy, calcium, vitamin D, phosphorus, and magnesium are necessary for the proper development of the skeleton and teeth. In adequate amounts, calcium may help reduce the incidence of **preeclampsia**, a sometimes deadly disorder marked by high blood pressure (called hypertension), and **edema** (swelling due to an abnormal accumulation of fluid in the intercellular spaces). On the positive side, much more calcium (about double) is absorbed through the intestine. On the negative side, many women do not eat enough calcium-rich foods during pregnancy or lactation. The need for calcium (1000 milligrams) can be met by having at least three servings from the dairy group or eating calcium-fortified foods each day. Vitamin D can also be obtained through milk (it must be fortified with vitamin D), as well as regular exposure to sunlight. Magnesium is found in green leafy vegetables, nuts, seeds, legumes, and whole grains.

Folate, vitamin B₁₂, and iron

The need for folate and iron increases 50 percent during pregnancy. This makes perfect sense when you realize that folate is needed to sustain the growth of new cells and the increased blood volume that occur during pregnancy, while iron is necessary to make new red blood cells.

PREECLAMPSIA

Hypertension during pregnancy that can cause serious complications.

EDEMA

Swelling due to an abnormal accumulation of fluid in the intercellular spaces.

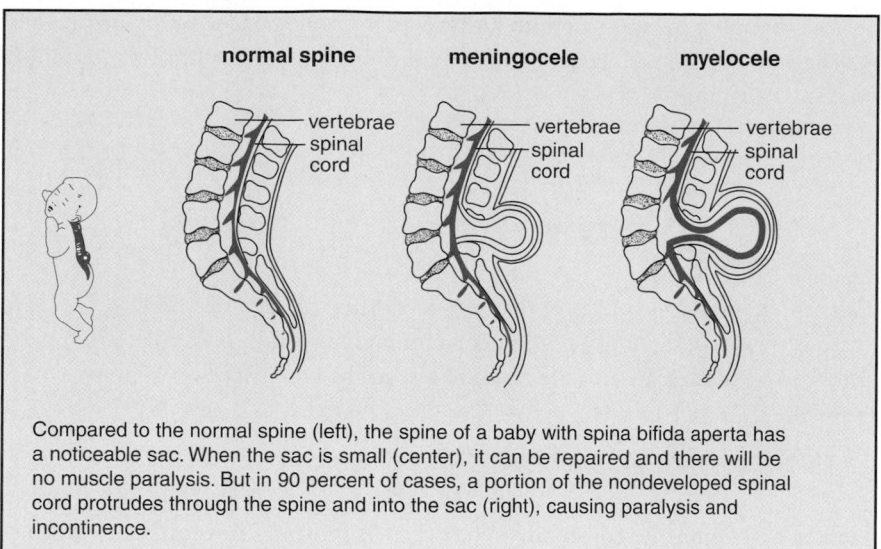

FIGURE 13-5:
Spina bifida.

Compared to the normal spine (left), the spine of a baby with spina bifida aperta has a noticeable sac. When the sac is small (center), it can be repaired and there will be no muscle paralysis. But in 90 percent of cases, a portion of the nondeveloped spinal cord protrudes through the spine and into the sac (right), causing paralysis and incontinence.

Folate is critical between the twenty-first and twenty-seventh days after conception (when most women don't even know they are pregnant) because this is when the **neural tube**, the tissue that develops into the brain and spinal cord, forms. Without enough folate, birth defects of the brain and spinal cord, such as **spina bifida**, can occur. In spina bifida, parts of the spinal cord are not properly fused, and so gaps are present (Figure 13-5).

Folate is also vital during the entire pregnancy. The RDA for folate during pregnancy is 600 micrograms of Dietary Folate Equivalents. The dietary folate equivalent (DFE) was developed to help account for the differences in absorption of naturally occurring dietary folate (found in green leafy vegetables, legumes, fruits such as orange juice, and seeds) and that of the more bioavailable synthetic folic acid that is added to most breads and cereals. It is recommended that women take in 400 micrograms of synthetic folic acid per day from fortified foods and/or dietary supplements and 200 micrograms daily from folate-rich foods.

Since vitamin B_{12} works with folate to make new cells, increased amounts of this vitamin are also needed. As long as animal products such as meat and milk are being consumed, vitamin B_{12} deficiency is not a concern.

Iron also helps in the formation of blood—it is necessary for hemoglobin in both maternal and fetal red blood cells. After 34 weeks of pregnancy, a woman's blood volume has increased 50 percent from the time of conception. Although iron absorption increases during pregnancy, whether the diet can supply enough iron is questionable. The National Academy of Sciences recommends iron supplements during the second and third trimesters to prevent anemia (a decreased number of red blood cells in the blood).

Sodium

In the past it was thought that sodium restriction was necessary for women with edema, or tissue swelling. It is now known that moderate swelling is normal during pregnancy and that sodium restriction is unnecessary and could actually be harmful to healthy pregnant women.

Prenatal supplements

Women who are trying to or may become pregnant should consider taking a supplement of folate. Once she becomes pregnant, the only nutrient suggested for all pregnant women is

NEURAL TUBE
The embryonic tissue that develops into the brain and spinal cord.

SPINA BIFIDA
A birth defect in which parts of the spinal cord are not fused together properly, and so gaps are present where the spinal cord has little or no protection.

iron, although a prenatal vitamin and mineral supplement is prescribed routinely by most physicians. The supplement usually contains iron, folate, vitamin B6, vitamin C, zinc, copper, and calcium.

DIET-RELATED CONCERNS DURING PREGNANCY

Certain diet-related concerns that occur during pregnancy need to be discussed. They include nausea and vomiting, food cravings and aversions, constipation, heartburn, and the intake of alcohol, seafood, caffeine, and artificial sweeteners.

Nausea and vomiting

Nausea and vomiting, commonly referred to as morning sickness, can occur at any time of the day. Nausea and vomiting are common in early pregnancy, affecting 70 to 85 percent of pregnant women. Morning sickness typically begins within the first nine weeks of pregnancy, with symptoms ranging from mild to severe. Severe morning sickness occurs in about 0.5 to 2 percent of pregnancies and can result in hospitalization.

No one is sure why morning sickness occurs, but the hormonal changes of early pregnancy seem to be related in some way. Mild cases may be resolved with lifestyle and dietary change, and safe and effective treatments are available for more severe cases.

Dietary advice in the past concentrated on small, carbohydrate-rich meals and tea and crackers. For many women, this dietary advice doesn't work. More recent advice centers on eating whatever foods you can keep down, even foods that aren't terribly nutritious, such as potato chips. The logic behind this recommendation is that tastes change when you are sick and you often crave something when you feel ready to eat. It's better to eat that food and keep it down than to eat something that is not appealing and throw it up. During pregnancy, women often develop a fine-tuned sense of smell, which can add to their nausea. The smell of foods and cooking can make them sick.

The American College of Obstetricians and Gynecologists recommends the following to prevent and treat nausea and vomiting.

- Taking a multivitamin at the time of conception may decrease the severity of symptoms.
- Taking vitamin B6 or vitamin B6 plus doxylamine (an antihistamine) is safe and effective (under a physician's direction).

Food cravings and aversions

Pregnant women commonly report changes in taste and smell. They may prefer saltier foods and crave sweets and dairy products such as ice cream. Foods they may have aversions to include alcohol, caffeinated drinks, and meats. Their cravings and aversions do not necessarily reflect actual physiological needs.

Constipation

Constipation is not uncommon, due to the relaxation of gastrointestinal muscles. It can be counteracted by eating more high-fiber foods, drinking more fluids, and getting additional exercise. Fiber is abundant in legumes (dried beans, peas, and lentils), whole grains, fruits, vegetables, nuts, and seeds.

Heartburn

Heartburn is a common complaint toward the end of pregnancy, when the growing uterus crowds the stomach. This condition has nothing to do with the heart but is actually a painful burning sensation in the esophagus. It occurs when stomach contents, which are acidic, flow back into the lower esophagus. Possible solutions include eating small and frequent meals, eating slowly and in a relaxed atmosphere, avoiding caffeine, wearing comfortable clothes, and not lying down after eating.

Alcohol

Alcohol and pregnancy don't mix. Pregnant women and women who may become pregnant are warned to abstain from alcohol consumption to eliminate the chance of giving birth to a baby with any of the harmful birth defects seen in *Fetal Alcohol Spectrum Disorders* (FASD). FASD range from mild and subtle changes, such as a slight learning disability and/or physical abnormality through full-blown *Fetal Alcohol Syndrome* (FAS). FAS can include severe learning disabilities, mental retardation, growth deficiencies, abnormal facial features, and central nervous system disorders.

Based on current science, we now know the following.

- Alcohol consumed during pregnancy increases the risk of alcohol-related birth defects, including growth deficiencies, facial abnormalities, central nervous system impairment, behavioral disorders, and mental impairments.
- No amount of alcohol consumption can be considered safe during pregnancy, not even one drink a day. Pregnant women who have seven drinks per week risk having children with problems in growth, behavior, arithmetic, language, memory, visual-spatial abilities, attention, and speed of information processing.
- Drinking habits known to place a fetus at greatest risk include binge drinking, defined as having five or more drinks at one time, and drinking seven or more drinks per week.
- Alcohol can damage a fetus at any stage of a pregnancy. Damage can occur in the earliest weeks of pregnancy, even before a woman knows she is pregnant.
- The cognitive deficits and behavioral problems resulting from prenatal alcohol exposure are lifelong.
- Alcohol-related birth defects are completely preventable.

For these reasons a pregnant woman or a woman who is considering becoming pregnant should abstain from alcohol. A pregnant woman who has already consumed alcohol during her pregnancy should stop to minimize further risk.

Seafood

Seafood is also a concern due to possible high levels of methylmercury. Nearly all fish and shellfish contain traces of methylmercury, a type of mercury found in water than can be harmful, especially to unborn babies and young children, whose nervous systems are still developing. Some types of fish and shellfish contain higher levels of mercury. The risks depend on the amount of fish and shellfish eaten and the levels of mercury in the seafood.

By following these three recommendations for selecting and eating fish or shellfish, pregnant women, women who may become pregnant, and nursing mothers will receive the benefits of eating fish and shellfish and can be confident that they have reduced their exposure to the harmful effects of mercury.

HEARTBURN
A painful burning sensation in the esophagus caused by acidic stomach contents flowing back into the lower esophagus.

FETAL ALCOHOL SPECTRUM DISORDERS
A variety of physical changes and/or brain damage associated with fetal exposure to alcohol during pregnancy.

FETAL ALCOHOL SYNDROME (FAS)
A set of symptoms occurring in newborn babies that are due to alcohol use by the mother during pregnancy; symptoms may include mental retardation and brain damage.

1. Pregnant women (and also nursing mothers) should not eat shark, swordfish, king mackerel, or tilefish. These long-lived larger fish contain the highest levels of methylmercury.
2. Eat up to 12 ounces, two average meals, a week of a variety of fish and shellfish that are lower in mercury. Five of the most commonly eaten fish that are low in mercury are shrimp, canned light tuna, salmon, pollock, and catfish. Albacore ("white") tuna has more mercury than does canned light tuna. When choosing your two meals of fish and shellfish, you may eat up to 6 ounces, one average meal, of albacore tuna per week.
3. Check to see if advisories exist concerning the safety of fish caught in local lakes, rivers, and coastal areas. If no advice is available, eat up to 6 ounces per week of fish you catch from local waters, but don't eat any other fish during that week.

Caffeine

Caffeine should be used moderately during pregnancy. The fetus can't detoxify caffeine, and large amounts of caffeine (5 cups of coffee per day or more) may increase the chance of miscarriage in early pregnancy and the possibility of having a low-birth-weight baby. During pregnancy, caffeine-containing beverages should be limited to 3 to 4 cups a day.

Artificial Sweeteners

Studies on the effects of artificial sweeteners such as aspartame show that they are safe during pregnancy within acceptable daily intakes.

MENU PLANNING DURING PREGNANCY

Figure 13-6 is a daily food guide for pregnancy. An individualized plan can be calculated using the MyPyramid for Moms tool at www.MyPyramid.com (click on "Pregnancy and Breastfeeding"). Problems can arise when an individual omits entire groups or substantial parts of certain groups. For instance, vegetarians who do not eat any food of animal origin need varied, adequate diets and supplements to obtain adequate vitamin B_{12}, calcium, zinc, and, unless they are getting adequate sunshine, vitamin D. Individuals who avoid the dairy group may need calcium and vitamin D supplements unless they eat foods fortified with these nutrients, such as calcium-fortified orange juice and calcium- and vitamin D-fortified soymilk.

The following are some menu-planning guidelines for pregnant (and lactating) women.

1. Offer a varied and balanced selection of nutrient-dense foods. Because energy needs increase less than nutrient needs, empty kcalories are rarely an acceptable choice.
2. In addition to traditional meat entrées, choose entrées based on legumes and/or grains and dairy products. Beans, peas, rice, pasta, and cheese can be used in many entrées. Beans and peas, as well as whole-grain rice and pasta, are excellent sources of fiber. Chapter 10 covers vegetarianism and has much information on meatless entrées.
3. Be sure to offer dairy products made with nonfat or low-fat milk.
4. Use a variety of whole-grain and enriched breads, rolls, cereals, rice, pasta, and other grains.
5. Use assorted fruits and vegetables in all areas of the menu, including appetizers, salads, entrées, side dishes, and desserts.
6. Be sure to have good sources of problem nutrients: the essential fatty acids, calcium, vitamin D, magnesium, folate, vitamin B_{12}, and iron (see Figure 13-7).
7. Be sure to use iodized salt.

FIGURE 13-6: Daily Food Guide for Pregnancy and Lactation Based on 2200 Kcalories

Food Group	Servings	Serving Size
Meat/meat alternative	6 ounces (pregnancy, to include 1 serving legumes) 6 ounces (lactation)	2 ounces cooked lean meat, poultry, or fish 2 eggs 2 ounces cheese ½ cup cottage cheese 1 cup dried beans or peas 4 tablespoons peanut butter
Milk and dairy	3 cups (pregnancy and lactation) 4 cups (pregnant/lactating teenagers)	1 cup milk, yogurt, pudding, or custard 1½ ounces cheese 1½ to 2 cups cottage cheese
Vegetables*	3 cups	½ cup cooked or juice 1 cup raw
Fruits*	2 cups	Portion commonly served, such as a medium apple or banana
Grain	7 ounces (at least half should be whole grains)	1 slice whole-grain or enriched bread 1 cup ready-to-eat cereal ½ cup cooked cereal or pasta ½ bagel or hamburger roll 6 crackers 1 small roll ½ cup rice or grits
Fats and sweets†		Includes butter, margarine, salad dressings, mayonnaise, oils, candy, sugar, jams, jellies, syrups, soft drinks, and any other fats and sweets

*A source rich in vitamin C (citrus, strawberries, melons, tomatoes) is needed daily; a source rich in vitamin A (dark green and deep yellow vegetables) is needed every other day.

†In general, the amount of these foods to use depends on the number of calories you require. Get your essential nutrients in the other food groups before choosing foods from this group.

FIGURE 13-7: Food Sources of Nutrients of Concern During Pregnancy

Essential fatty acids	Vegetable oils such as canola, soybean, corn, and walnut oils Margarines made with vegetable oils Salad dressings and mayonnaise made with vegetable oils Ground flax seed Soy products Walnuts Fish
Calcium	Milk, yogurt Cheese Fortified cereals and orange juice Leafy greens (collards, spinach, kale, broccoli) Tofu with calcium sulfate

(continued)

Vitamin D	Vitamin D-fortified milk
	Vitamin D-fortified cereals
Magnesium	Beans and peas
	Nuts (especially almonds and cashews)
	Green leafy vegetables
	Potatoes
	Seeds
	Whole-grain breads and cereals
Folate	Fortified breads and cereals
	Green leafy vegetables
	Legumes
	Orange juice
Vitamin B$_{12}$	Animal foods such as meat, poultry, fish, eggs, milk
Iron	Beef, poultry, fish
	Enriched and fortified breads and cereals
	Beans and peas
	Green leafy vegetables
	Eggs

MINI-SUMMARY

1. During most of the first trimester, when critical periods of differentiation are occurring, the growing baby is most susceptible to damage from nutritional deficiencies, alcohol, and other influences.

2. Women's nutritional status before and during pregnancy influences both the mother's and the baby's health.

3. Both prepregnancy weight and weight gain during pregnancy directly influence infant birth weight, the most important indicator of a baby's future health status.

4. Underweight women must either gain weight before or gain more weight during pregnancy. Overweight women need to lose weight before pregnancy or gain less weight during pregnancy.

5. A woman at a healthy weight should gain 25 to 35 pounds during pregnancy.

6. Although a pregnant woman should consume only 340 or 450 additional kcalories per day (during the second and third trimesters, respectively), she must take in more nutrients.

7. Nutrients of special concern during pregnancy include the essential fatty acids; calcium, vitamin D, phosphorus, and magnesium for bones and teeth; and folate, vitamin B$_{12}$, and iron for new cells such as red blood cells.

8. Prenatal vitamin and mineral supplements with iron are routinely prescribed.

9. Nausea and vomiting are common in early pregnancy. Taking a multivitamin at the time of conception may decrease the severity of symptoms. Taking vitamin B$_6$ may also help.

10. Pregnant women commonly report changes in taste and smell. Their cravings and aversions do not necessarily reflect actual physiological needs.

11. During pregnancy, the gastrointestinal tract slows down, and so fiber, fluids, and exercise are important to prevent constipation.

12. Solutions to heartburn include eating small and frequent meals, eating slowly and in a relaxed atmosphere, avoiding caffeine, wearing comfortable clothes, and not lying down after eating.

13. Pregnant women and women who may become pregnant are warned to abstain from alcohol consumption to eliminate the chance of giving birth to a baby with any of the harmful birth defects seen in fetal alcohol spectrum disorders.

14. Pregnant women should not eat shark, swordfish, king mackerel, or tilefish because of the mercury levels. They can eat up to 12 ounces a week of a variety of fish and shellfish lower in mercury, such as shrimp, canned light tuna, salmon, pollock, and catfish.

15. During pregnancy, caffeine-containing beverages should be limited to 3 to 4 cups a day.

16. Artificial sweeteners are safe during pregnancy.

17. Figure 13-6 is a daily food guide for pregnancy.

18. Menu-planning guidelines for pregnancy are given in this section.

NUTRITION AND MENU PLANNING DURING LACTATION

For the first four to six months of life, the baby's only source of nutrients is either breast milk or formula. Breastfeeding, also referred to as lactation or nursing, is recommended for all infants in the United States from birth to 12 months.

Figure 13-6 shows the daily food guide for breastfeeding mothers. During lactation, the period of milk production, an intake of 330 additional kcalories is necessary for the first six months; this increases to 400 additional kcalories from the seventh to the twelfth month. Actually, more than 330 extra kcalories is needed during the first six months, but some are supplied by extra fat stored during pregnancy. Protein needs continue to be 25 grams above the normal RDA.

The need for certain vitamins and minerals increases during breastfeeding.

Vitamins	Minerals
Vitamins A, C, and E	Magnesium
Thiamin, riboflavin, and niacin	Zinc
Folate and vitamin B_{12}	Selenium
Biotin and pantothenic acid	Iodine

The need for extra nutrients demonstrates that lactating mothers need to make healthy food choices.

Most lactating women can obtain all the nutrients they need from a well-balanced diet without taking supplements. However, some women have depleted stores of iron after pregnancy and may need iron supplements. Likewise, lactating vegetarian mothers who eat no food of animal origin need to pay special attention to getting enough kcalories, calcium, vitamin D, vitamin B_{12}, iron, and zinc.

Lactating mothers, who normally produce about 25 ounces of milk a day, also need at least 3 to 4 quarts of fluids each day to prevent dehydration. Caffeine does cross to the baby through breast milk and can cause irritability. Moderate use, the equivalent of about 3 cups of coffee or less per day, is okay.

A balanced, varied, and adequate diet (at least 1800 kcalories per day) is critical to successful breast-feeding and infant health. If the mother is not eating properly, any nutritional deficiencies are more likely to affect the quantity rather than the quality of the milk she makes.

Menu-planning guidelines for lactating women are the same as those for pregnant women, with an emphasis on fluids, dairy products, fruits, and vegetables. Regular consumption of alcohol is not advised because alcohol passes directly into the breast milk. Figure 13-8 shows the MyPyramid Plan for Moms when breastfeeding.

MINI-SUMMARY

1. Figure 13-6 shows the daily food guide for breastfeeding mothers.
2. During the first six months of lactation, mothers need 330 extra kcalories a day. During the seventh to twelfth months, 400 additional kcalories a day are needed. Lactating mothers also need plenty of fluids, at least 3 to 4 quarts daily.
3. Moderate use of caffeine is okay. Regular consumption of alcohol is not advised.
4. If the mother is not eating a nutritionally adequate diet, the quantity of milk will be adversely affected.
5. Menu-planning guidelines are the same as those for pregnant women, with an emphasis on fluids, dairy products, fruits, and vegetables.

INFANCY: THE FIRST YEAR OF LIFE

The growth rate during the first year of life will never be duplicated again. It is most rapid during the first six months then slows a little during the second six months. The nutrient needs of infants are about double those of an adult when viewed in proportion to their weights. Little wonder, considering that infants generally double their birth weight in the first four to five months and then triple it by the first birthday. An infant will also grow 50 percent in length by the first birthday. (In other words, a baby who was 20 inches at birth grows to 30 inches by one year of age.)

NUTRITION DURING INFANCY

Newborns need a plentiful supply of all nutrients, especially those necessary for growth, such as protein, vitamins C and D, folate, vitamin B_{12}, calcium, and iron. A DRI has been established for two age categories for infants: 0–6 months and 7–12 months. These nutrient recommendations reflect the decreased nutrient needs during the second half of the first year.

For the first four to six months of life, the source of all nutrients is breast milk or formula. Breast milk is recommended for all infants in the United States from birth to 12 months. A baby needs breast milk for the first year of life and can have it as long as desired after that.

The number of women who are choosing to breast-feed is increasing. Current estimates are that almost 75 percent of American mothers breastfeed their babies in the hospital, but only 39 percent are still breastfeeding six months later. The following list shows the advantages of breastfeeding compared with formula feeding:

1. Breast milk is nutritionally superior to any formula or other type of feeding. The balance of nutrients in breast milk closely matches the infant's requirements for growth and development. Breast milk provides more of the essential fatty acids compared with formula. The composition of breast milk changes to meet the needs of the growing infant.
2. Breast milk helps the infant build up immunities to infectious disease, because it contains the mother's antibodies to disease. Breast-fed infants are much less likely to

What Should I Eat?
MyPyramid Plan for Moms

When you are breastfeeding, you have special nutritional needs. Follow the MyPyramid Plan for Moms below to help you and your baby stay healthy.

- Eat these amounts from each food group daily.
- The calories and amounts of food you need differ if you are only breastfeeding or breastfeeding and giving formula.
- Choose the Plan that is right for you.
- In each food group, choose foods and beverages that are low in "extras."

Food Group	Breastfeeding only	Breastfeeding plus formula	What counts as 1 cup or 1 ounce?	Remember to...
	Eat this amount from each group daily.*			
Fruits	2 cups	2 cups	1 cup fruit or juice, ½ cup dried fruit	*Focus on fruits*—Eat a variety of fruit.
Vegetables	3 cups	3 cups	1 cup raw or cooked vegetables or juice, 2 cups raw leafy vegetables	*Vary your* veggies—Eat more dark green and orange vegetables and cooked dry beans.
Grains	8 ounces	7 ounces	1 slice bread; ½ cup cooked pasta, rice, cereal; 1 ounce ready-to-eat cereal	*Make half your grains whole*—Choose whole instead of refined grains.
Meat and Beans	6½ ounces	6 ounces	1 ounce lean meat, poultry, fish; 1 egg; ¼ cup cooked dry beans; ½ ounce nuts; 1 tablespoon peanut butter	*Go lean with protein*—Choose low-fat or lean meats and poultry.
Milk	3 cups	3 cups	1 cup milk, 8 ounces yogurt, 1½ ounces cheese, 2 ounces processed cheese	*Get your calcium-rich foods*—Go low-fat or fat-free when you choose milk, yogurt, and cheese.

*These amounts are for an average breastfeeding woman. You may need more or less than the average. Check with your doctor to make sure you are losing the weight you gained during pregnancy.

Get a MyPyramid Plan for Moms designed just for you on the MyPyramid website. Go to www.mypyramid.gov. Choose "For Pregnancy and Breastfeeding" on the left side menu.

FIGURE 13-8:

MyPyramid meal guidelines for breastfeeding mothers.
Courtesy U.S. Department of Agriculture.

develop both routine illnesses, such as ear infections, and serious respiratory and gastrointestinal illnesses.

3. Newborns are less apt to be allergic to breast milk than to any other food. Cow's milk contains a type of protein different from that in breast milk, and some infants have difficulty digesting it.

4. Suckling promotes the development of the infant's jaw and teeth. It's harder work to get milk from a breast than from a bottle; the exercise strengthens the jaw and encourages the growth of straight, healthy teeth. The baby at the breast can control the flow of milk by sucking and stopping. With a bottle, the baby must constantly suck or react to the pressure of the nipple in the mouth.

5. Breastfeeding promotes a close relationship—a bonding between mother and child. At birth, infants see only 12 to 15 inches, the distance between a nursing baby and its mother's face.

6. Breastfeeding is less expensive, creates less impact on the environment, and is more convenient than formula feeding.

7. Breast milk is less likely to be mishandled. Some formulas require accurate dilutions, and all are much more apt to be mishandled, which can result in food-borne illness.

8. Breastfed infants have lower rates of hospital admissions, ear infections, diarrhea, and other medical problems than do bottle-fed babies.

9. Breastfeeding is associated with decreased risk for developing chronic diseases later in life, including diabetes and obesity.

10. Breastfeeding has benefits for the mother, including an earlier return to prepregnancy weight and a decreased risk of breast cancer and ovarian cancer.

Breastfeeding is not recommended if the mother uses addictive drugs, drinks more than a minimal amount of alcohol, is on certain medications, or is HIV-positive (has the virus that causes AIDS).

Breast milk is not a good source of vitamin D; therefore, breast-fed infants should receive vitamin D supplements beginning at two months of age and until they begin taking at least 17 fluid ounces of vitamin D-fortified milk daily.

Formula feeding is an acceptable substitute for breastfeeding. Some women find formula feeding more convenient (others find breastfeeding more convenient). Other family members can take part in formula feeding. For some women who are uncomfortable with breast-feeding, even after education, formula feeding is the method of choice.

All infant formulas must meet nutrient standards set by the American Academy of Pediatrics. The three forms of formulas on the market are ready-to-feed formula, liquid concentrate that needs to be mixed with equal amounts of water, and powdered formulas, which also has to be mixed with water. All formulas must be handled in a sanitary manner to prevent contamination and possible food poisoning.

Cow-milk-based formulas are normally used unless the baby is allergic to the protein or sugar in milk. In that case, a soy-based formula is used. For a baby who is allergic to both milk-based and soy-based formulas, predigested formulas are available. Symptoms of allergies usually include diarrhea and/or vomiting.

The Food and Drug Administration approved the addition of two fatty acids to infant formula: docosahexaenoic acid (DHA) and arachidonic acid (AA), which are made in the body from the essential fatty acids. DHA and AA are present in breast milk and are thought to enhance the mental and visual development of infants. Major formula manufacturers now have brands that include DHA and AA. These formulas are more expensive than their traditional counterparts.

Regular cow's milk has too much protein and minerals and too little essential fatty acids, vitamin C, and iron. Therefore, it is not recommended until 12 months of age, when the baby is less likely to be allergic to it. Babies are normally switched slowly from breast milk or formula to cow's milk.

Whether infants are breastfed or formula fed, their iron stores are relatively depleted by age four to six months, at which time they typically start to eat iron-fortified cereals. Fluoride supplements may also be prescribed for infants after six months, unless the infant gets enough fluoride from fluoridated water.

FEEDING THE INFANT

Successful infant feeding requires a cooperative relationship between the mother and her baby. Feeding time should be a pleasurable period for both parent and child, and so a calm, relaxed environment is ideal. The feeding schedule should ideally be based on reasonable self-regulation by the baby. By the end of the first week of life, most infants want eight to 12 feedings a day. Breastfed babies feed about every two or three hours, and formula-fed babies about every four hours.

Breastfeeding mothers should put the baby to the breast as soon as possible after delivery, ideally within the first hour after delivery, to enhance success. *Colostrum*, a yellowish fluid, is the first secretion to come from the breast a day or so after delivery. It is rich in proteins, antibodies, and other factors that protect against infectious disease. Colostrum changes to *transitional milk* between the third and sixth days, and mature milk is produced about 10 days after the infant's birth.

Babies are ready to eat semisolid foods such as cereal when they can sit up and open their mouths. This usually occurs between four and six months of age with minimal support. Other signs that babies are ready for spoon-feeding are when they:

- Have doubled their birth weight
- Drink more than a quart of formula per day
- Seem hungry often
- Open their mouths in response to seeing food coming
- Can move the tongue from side to side without moving the head

Although some parents think that feeding of solids will help a baby sleep through the night, this is not often the case. Feeding of solid food before a baby is ready can create problems, because the baby's digestive system is not ready for it. Feeding solids early also increases the risk of allergies and the chance of choking and may encourage overfeeding.

Most babies can digest starchy foods, like infant cereal, fruits, or vegetables, at around four months of age. Once a baby starts on solid foods, it is important to make sure that he or she gets sufficient fluids. Up to this point, breast milk or formula will meet the baby's need for fluids. Now, however, drinking water or other fluids is needed to prevent dehydration. Proportionally, babies have more water in their bodies than adults do. They can become dehydrated very quickly due to hot weather, diarrhea, or vomiting, and so extra fluids need to be offered at these times.

Although eating solid food is certainly simple for an adult, it involves a number of difficult steps for a baby. First, the infant must have enough muscle control to close his or her mouth over the spoon, scrape the food from the spoon with the lips, and then move the food from the front to the back of the tongue. By about 16 weeks, a baby generally has these skills, but probably has no teeth! The baby's first teeth will cut through the gums between 6 and 10 months of age.

COLOSTRUM
A yellowish fluid that is the first secretion to come from the mother's breast a day or so after the delivery of a baby; it is rich in proteins, antibodies, and other factors that protect against infectious disease.

TRANSITIONAL MILK
The type of breast milk produced from about the third to the tenth day after childbirth.

FIGURE 13-9: Feeding the Baby for the First Year

Baby's Age:	When Babies Can:	Serve:

Birth through 3 months

FIGURE 13-9A

- Only suck and swallow

LIQUIDS ONLY
- Breast milk
- Infant formula with iron

4 months through 7 months

FIGURE 13-9B

- Draw in upper or lower lip as spoon is removed from mouth
- Move tongue up and down
- Sit up with support
- Swallow semisolid foods without choking
- Open the mouth when they see food
- Drink from a cup with help, with spilling

ADD SEMISOLID FOODS
- Infant cereal with iron
- Pureed then textured vegetables
- Pureed then textured fruit
- Fruit juice (start at 6 months, dilute at first)

8 months through 11 months

FIGURE 13-9C

- Move tongue from side to side
- Begin spoon feeding themselves with help
- Begin to chew and have some teeth
- Begin to hold food and use their fingers to feed themselves
- Drink from a cup with help, with less spilling

ADD MODIFIED TABLE FOODS
- Mashed or diced soft fruit
- Mashed or soft cooked vegetables
- Mashed egg yolk
- Finely cut meat/poultry
- Mashed cooked beans or peas
- Cottage cheese, yogurt, or cheese strips
- Pieces of soft bread
- Crackers
- Breast milk, iron-fortified formula, or fruit juice in a cup

Courtesy of U.S. Department of Agriculture.

Foods are generally introduced as follows. Keep in mind that the order of introducing different types and textures of foods is tied to the baby's developmental stages (Figure 13-9).

The first solid food is iron-fortified baby cereal mixed with breast milk or formula. Usually, rice cereal is offered first because it is the least likely to cause an allergic reaction. Barley and oatmeal cereals follow. Once single ingredient cereals have been given with no reaction, mixed cereals can be offered. The iron in these cereals is very important to meet an infant's high iron needs. Avoid putting cereals or any other solids into the infant's bottle since it may cause choking or overfeeding.

Once the baby is used to various cereals, pureed or mashed vegetables and fruits can be tried. It is a good idea to start with vegetables so that the baby does not become accustomed to the sweet taste of fruits (babies like sweets) and then reject the vegetables. When adding new foods to an infant's diet, always do so one food at a time (and in small quantities) so that if there is an allergic reaction (such as hives or diarrhea), you will know which food caused it. Introduce new vegetables and fruits about one week apart. Babies adjust differently to new tastes and new textures. If the baby does not like a certain food, offer it a few days later. If you offer new foods when the baby is hungry, such as at the beginning of a meal, he or she is more likely to eat them.

Fruit juice that is fortified with vitamin C can be started about the sixth month. Although some babies get two or more bottles a day of apple juice (or another type of juice), it is a good idea to limit juice to a half to ¾ cup, or 4–6 fluid ounces, daily. Make sure the juice has no sugar added and dilute it with water initially.

More than a half cup of juice daily can result in growth failure if it is substituted for breast milk or formula. Another problem with fruit juice can occur when you let your baby go to sleep with a bottle in his or her mouth. The natural sugars in the juice can cause serious tooth decay, called baby bottle tooth decay. Letting a baby go to bed with a bottle of formula, cow's milk, or breast milk will also cause **baby bottle tooth decay**. Too much juice can also give babies diarrhea.

Before a baby can move on to finger foods, he or she has to be able to grab them. At about eight months, a baby discovers and starts to use the thumb and forefinger together to pick things up (called the **pincer grasp**). From about six months, the baby has been using the palm (called the **palmar grasp**) to do this. Suitable finger foods include chopped ripe bananas, dry cereal, and pieces of cheese. About this time infants can also start eating protein foods. Meat and poultry will have to be chopped or cut very fine and possibly moistened.

Between 10 and 12 months of age, babies may have four to six sharp teeth, and many are eating soft, chopped foods with the family. At this time it is appropriate to let a child begin drinking from a cup. It takes time, but sooner or later the child will get the idea. By one year of age, infants can enjoy cut-up table foods as well as whole milk. By 12 months, some babies may be almost entirely self-feeding. Children should not be switched to low-fat milk until they are at least two years old, because they need the fat in whole milk for proper growth and development.

Because honey and liquid corn syrup can be contaminated with spores that cause botulism, these foods may cause food poisoning or food-borne illness in children younger than one year.

Certain foods are also more apt than others to cause choking and block a child's airway. These foods are unsafe for infants and toddlers and include nuts, seeds, raisins, hot dogs, popcorn, whole grapes, hard candies (including jelly beans), peanut butter, cherry tomatoes, raw carrots and celery, cherries with pits, and large chunks of any food, including meat, potatoes, and raw vegetables and fruit. Grapes and cherry tomatoes can be cut in quarters. Hot dogs can be cut into tiny pieces.

Foods that are more apt to cause allergies include milk, eggs, wheat, peanuts and other nuts, chocolate, and shellfish. Whole milk and eggs are usually introduced at about 12 months.

BABY BOTTLE TOOTH DECAY
Serious tooth decay in babies caused by letting a baby go to bed with a bottle of juice, formula, cow's milk, or breast milk.

PINCER GRASP
The ability of a baby at about eight months of age to use the thumb and forefinger together to pick things up.

PALMAR GRASP
The ability of a baby from about six months of age to grab objects with the palm of the hand.

MINI-SUMMARY

1. The growth rate during the first year will never be duplicated again.

2. An infant will triple the birth weight by the first birthday and grow 50 percent in length as well.

3. For the first four to six months, the only food an infant should get is either breast milk or iron-fortified formula. Breast milk is recommended for many reasons. It is nutritionally superior to formula, and newborns are less likely to be allergic to it. Breast milk contains antibodies to help babies build up immunities, and breastfeeding promotes bonding between mother and infant. Breast-feeding also helps the mother return to her prepregnancy weight and decreases her risk of breast and ovarian cancers.

4. Breast milk is not a good source of vitamin D. Breastfed babies should be given vitamin D supplements beginning at two months of age until they begin drinking at least 17 fluid ounces of vitamin D-fortified milk daily.

5. For formula-fed babies, the formula is generally cow-milk based. If the baby is allergic to it, a soy-based formula is used.

6. Successful infant feeding requires a cooperative relationship between the mother and her baby.

7. The breast-feeding process starts with the infant using a suckling action that stimulates the mother's hormones to move milk into the ducts of the breast (milk letdown).

8. Between four and seven months, a baby usually shows several signs that he or she is ready to eat semisolid foods such as cereal.

9. Figure 13-9 shows the progression of introducing solid foods to infants.

10. Whole milk, whole eggs, and peanut butter are not recommended until 12 months because of possible allergic reactions.

11. Between 10 and 12 month of age, babies may have four to six sharp teeth, and many are eating soft, chopped foods with the family. At this time, it is appropriate to let a child begin drinking from a cup.

12. Avoid honey and liquid corn syrup, as well as foods that cause choking, such as hot dogs and whole grapes.

CHILDHOOD

At around age one, a baby's growth rate decreases markedly. Yearly weight gain now approximates 4 to 6 pounds per year. Children can be expected to grow about 3 inches per year between ages one and five and then 2 inches per year until the pubertal growth spurt. The pubertal growth spurt begins around age 10 or 11 for girls and 12 or 13 for boys. Until adolescence, growth will come in spurts, during which the child will grow more and eat more.

Appendix D shows growth charts for boys and girls that compare a child's height and weight to national percentiles. If a girl is in the 90th percentile height for age, it means that, out of 100 girls her age, she is taller than 89 and shorter than 10. These measures are commonly used by physicians as an indicator of a child's overall health and the adequacy of calorie intake for children and adolescents from 2 to 20 years of age. Growth charts are available for ages birth to 36 months and then 2 to 20 years of age.

After age one, children start to lose baby fat and become leaner, with muscle accounting for a larger percentage of body weight. The legs become longer, and the baby starts to look like a child and to walk, run, and jump like a child. A child's head size in relation to body size starts to decrease and look more normal. The brain continues to grow and mature during childhood and well into adolescence.

By about age one a child has six to eight teeth, and by age two the baby teeth are almost all in. Between ages 6 and 12, these teeth are gradually replaced with permanent teeth. After the first birthday, as children's physical capabilities and desire for independence increase, they are more capable of feeding themselves. By age 18 months, many children can use a spoon without too much spilling and can eat many food textures, including raw fruits and vegetables. By 24 months many children can drink properly from a cup.

NUTRITION DURING CHILDHOOD

Figure 13-10 shows the estimated energy requirement (EER) for kcalories and the RDA for protein for children. A two-year-old needs about 1000 kcalories a day. By age six, a child needs closer to 1700 kcalories daily. Energy needs of children of similar age, sex, and size can vary due to differing BMRs (basal metabolic rates), growth rates, and activity levels. Energy and protein needs decline gradually per pound of body weight.

FIGURE 13-10: Energy and Macronutrients for Children

ESTIMATED ENERGY REQUIREMENT (EER) AND RECOMMENDED DIETARY ALLOWANCE (RDA) FOR PROTEIN

Gender and age	Height	Weight	EER*	Protein
Male 1–3 years	34 inches	27 pounds	1046 kcal	1.1 grams/kilogram body weight
Female 1–3 years	34	27	992	1.1
Male 4–8 years	45	44	1742	0.95
Female 4–8 years	45	44	1642	0.95

*The EER is for healthy moderately active Americans and Canadians.

ACCEPTABLE MACRONUTRIENT DISTRIBUTION RANGES

Age	Carbohydrate	Fat	Protein
1–3 years	45–65%	30–40%	5–20%
4–18 years	45–65%	25–35%	10–30%
Over 18 years	45–65%	20–35%	10–35%

Source: Adapted with permission from the Dietary References Intakes for Energy, Carbohydrates, Fiber, Fat, Protein, and Amino Acids (Macronutrients). © 2002 by the National Academy of Sciences. Courtesy of the National Academy Press, Washington, D.C.

For children two to three years of age, fat intake should be 30 to 35 percent of total kcalories. Children and adolescents from 4 to 18 years of age should keep fat intake between 25 and 35 percent of kcalories, with most fats coming from sources of polyunsaturated and monounsaturated fatty acids, such as vegetable oils, nuts, and fish. Children over two years old should limit consumption of saturated and trans fats to 10 percent or less of total kcalories.

Problem nutrients for children include vitamin E, potassium, iron, and fiber. For a variety of reasons, children are not eating enough foods rich in these nutrients. Figure 13-14 shows foods rich in each of these nutrients.

Lack of iron can cause fatigue and affect behavior, mood, and attention span. A balanced diet with adequate consumption of iron-rich foods such as lean meat (ground meat is easier for younger children to chew), enriched breads and cereals, and legumes is important to get enough iron. A source of vitamin C, such as citrus fruits, increases the amount of iron absorbed.

The Adequate Intake for calcium increases from 500 mg for one- to three-year-olds to 800 mg for four- to eight-year-olds. Children from two to eight years old should drink 2 cups per day of skim or low-fat milk or equivalent. It is recommended that children from one to five years old should drink no more than 3 cups of milk a day because of its low iron content. Drinking more than 3 cups a day often causes children to eat too few high-iron foods, which can result in iron-deficiency anemia.

During **growth spurts**, the requirements for kcalories and nutrients are greatly increased. Appetite fluctuates tremendously, with a good appetite during growth spurts (periods of rapid growth) and a seemingly terrible appetite during periods of slow growth. Parents may worry and force a child to eat more than needed at such times, when the child appears to be "living on air." A decreased appetite in childhood is perfectly normal. As long as the child is choosing nutrient-dense kcalories, nutritional problems are unlikely. In preparation for the adolescent growth spurt, children accumulate stores of nutrients, such as calcium, that will be drawn on later, as intake cannot meet all the demands of this intensive growth spurt.

Preschoolers exhibit some food-related behaviors that drive their parents crazy, such as **food jags** (eating mostly one food for a period of time). Food jags usually don't last long

GROWTH SPURTS
Periods of rapid growth.

FOOD JAGS
A habit of young children in which they have favorite foods they want to eat frequently.

enough to cause any harm. Preschoolers often pick at foods or refuse to eat vegetables or drink milk. Lack of variety, erratic appetites, and food jags are typical of this age group. Toddlers (ages one to three) tend to be pickier eaters than older preschoolers (ages four to five). Toddlers are just starting to assert their independence and love to say no to parental requests. They may wage a control war, and parents need to set limits without being too controlling or rigid. Here are several ways to make mealtimes less stressful and teach good eating habits.

1. Make mealtimes as relaxing and enjoyable as possible.
2. Don't nag, bribe, force, or cajole a child to eat. Stay calm. Pushing or prodding children almost always backfires. Children learn to hate the foods they are encouraged to eat and to desire the foods used as rewards, such as cake, ice cream, and candy. Once children know that you won't allow eating to be made into an issue of control, they will eat when they're hungry and stop when they're full. Your child is the best judge of when he or she is full.
3. Allow children to choose what they will eat from two or more healthy choices. You are responsible for choosing which foods are offered, and the child is responsible for deciding how much he or she wants of those foods.
4. Let children participate in food selection and preparation. Figure 13-11 lists cooking activities for children of various ages.
5. Respect your child's preferences when planning meals but don't make your child a quick peanut butter sandwich, for instance, if he or she rejects your dinner.
6. Make sure your child has appropriate-size utensils and can reach the table comfortably.
7. Preschoolers love rituals, so start them early with the habit of eating three meals plus 1–2 snacks each day at fairly regular times. Also, eat with your preschooler and model good eating habits.
8. Expect preschoolers to reject new foods at least once, if not many times. Simply continue presenting the new food, perhaps prepared differently, and one day they will try it (usually after 12 to 15 exposures).
9. Serve small portions.
10. Do not use desserts as a reward for eating meals. Make dessert a normal part of the meal and make it nutritious.
11. Ask children to try new foods (just a little bite) and praise them when they try something different. Encourage them by telling them about someone who really likes the food or relating the food to something they think is fun. Realize, though, that some children are less likely to try new things, including new foods.
12. Be a good role model.
13. Be consistent at mealtimes.
14. If all else fails, keep in mind that children under six have more taste buds (which may explain why youngsters are such picky eaters) and that this, too, will pass.

Luckily, school-age children are much better eaters. Although they generally have better appetites and will eat a wider range of foods, they often dislike vegetables and casserole dishes. Both preschoolers and school-age children learn about eating by watching others: their parents, siblings, friends, and teachers. Parents, siblings, and friends provide role models for children and influence children's developing food patterns. Parents' interactions with their children also influence what foods they will or will not accept.

FIGURE 13-11: Age-appropriate Cooking Activities

2½–3-YEAR-OLDS

- Wash fruits and vegetables
- Peel bananas
- Stir batters
- Slice soft foods with table knife (cooked potatoes, bananas)
- Pour
- Fetch cans from low cabinets
- Spread with a knife (soft onto firm)
- Use rotary egg beater (for a short time)
- Measure (e.g., chocolate chips into 1-cup measure)
- Make cracker crumbs

4–5-YEAR-OLDS

- Grease pans
- Open packages
- Peel carrots
- Set table (with instruction)
- Shape dough for cookies/hamburger patties*
- Snip fresh herbs for salads or cooking
- Wash and tear lettuce for salad, separate broccoli, cauliflower
- Place toppings on pizza or snacks

6–8-YEAR-OLDS

- Take part in planning part of or entire meal
- Set table (with less supervision)
- Make a salad
- Find ingredients in cabinet or spice rack
- Shred cheese or vegetables
- Garnish food
- Use microwave, blender, or toaster oven (with previous instruction)
- Measure ingredients
- Present prepared food to family at table
- Roll and shape cookies

9–12-YEAR-OLDS

- Depending on previous experience, plan and prepare an entire meal

*Children should not put their hands in their mouths while handling raw hamburger meat or dough with eggs. It can carry harmful bacteria. They should wash hands after shaping patties.

When children go to school, their peers influence their eating behaviors as well as what they eat for lunch. Lunch for school-age children often consists of the school lunch (Figure 13-12) or a packed lunch from home.

Having breakfast makes a difference in how children perform at school. Breakfast also makes a significant contribution to the child's intake of kcalories and nutrients for the day. Children who skip breakfast usually don't make up for calories at other meals.

Preschoolers and school-age children also learn about food by watching television. Research shows that children who watch a lot of television are more apt to be overweight. It

not only takes them away from more robust activities but also exposes them to commercials for sugared cereals, candy, and other empty-calorie foods. Both obesity and inactivity are currently on the rise among school-age children, especially adolescents.

So what can parents do to make sure their children eat nutritious diets and get exercise? Be a good role model by eating a well-balanced and varied diet. Have nutritious food choices readily available at home and serve a regular, nutritious breakfast. Maintain regular family meals as much as possible. Family meals are an appropriate time to model healthy eating habits and try out new foods. Also, limit television watching and encourage physical activity. Eating behaviors are learned during childhood (and adolescence) and are maintained into adulthood.

MENU PLANNING FOR CHILDREN

By the time children are four years old, they can eat amounts that count as regular servings eaten by older family members—that is, $\frac{1}{2}$ cup of fruit or vegetable, $\frac{3}{4}$ cup of juice, 1 slice of bread, and 2 to 3 ounces of cooked lean meat, poultry, or fish. Children two to three years of age need the same variety of foods as four- to six-year-olds but may need slightly smaller portions (except for milk). Two- to six-year-old children need a total of two full servings from the milk group each day. MyPyramid for Kids is shown in Figure 13-13.

Following are additional menu-planning guidelines.

Preschoolers
1. Offer simply prepared foods and avoid casseroles or any foods that are mixed together, as children need to be able to identify what they are eating. Some toddlers and preschoolers may not want their foods to touch on the plate and may prefer a plate with divided sections.
2. Children learn to like new foods by being presented with them repeatedly, as many as 12 or more times. Put a small amount of a new food along with a meal and don't require the child to eat it if he or she doesn't want to. The more times you do this, the more likely it is that the child will eventually eat the food.
3. Offer at least one colorful food, such as carrot sticks.

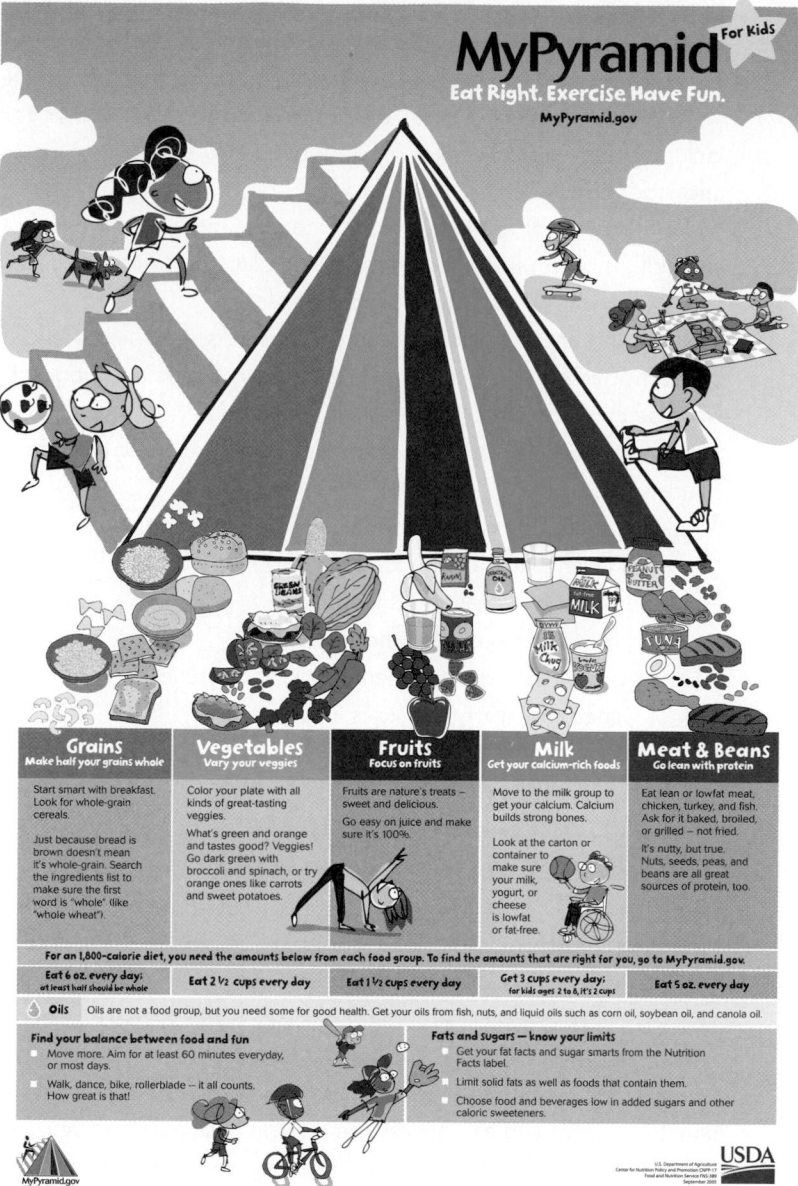

FIGURE 13-13:
MyPyramid for kids.
Courtesy of the U.S.
Department of Agriculture.

4. Preschoolers like nutritious foods in all food groups but may be reluctant to eat vegetables. Part of this concern may be due to the difficulty involved in getting them onto a spoon or fork. Vegetables are more likely to be accepted if served raw and cut up as finger foods. However, when serving celery, be sure to take off the strings. Serve cooked vegetables somewhat undercooked so that they are a little crunchy. Brightly colored, mild-flavored vegetables such as peas and corn are more popular with children.

5. Provide at least one soft or moist food that is easy to chew at each meal. A crisp or chewy food is important, too, to develop chewing skills.

6. Avoid strong-flavored and highly salted foods. Children have more taste buds than adults, and so these foods taste too strong to them.

7. Preschoolers love carbohydrate foods, including cereals, breads, and crackers, as they are easy to hold and chew. Try whole grains.

8. Smooth-textured foods such as pea soup and mashed potatoes should not have any lumps—children find this unusual.

9. Before age four, when food-cutting skills start to develop, a child needs to have food served in bite-size pieces that are eaten as finger foods or with utensils. For example, cut meat into strips or use ground meat, cut fruit into wedges or slices, and serve pieces of raw vegetables instead of a mixed salad. Other good finger foods include cheese sticks, wedges of hard-boiled egg, dry ready-to-eat cereal, fish sticks, arrowroot biscuits, and graham crackers.

10. Serve foods warm, not hot; a child's mouth is more sensitive to hot and cold than an adult's is. Also, little children need little plates, utensils, and cups, as well as seats that allow them to reach the table comfortably.

11. Cut-up fruits and vegetables make good snacks. Let preschoolers spread peanut butter on crackers or use a spoon to eat yogurt. Snacks are important to preschoolers because they need to eat more often than adults.

12. Serve good sources of the nutrients children are most likely to be lacking (see Figure 13-14).

13. To minimize choking hazards for children under age four:
    - Avoid large chunks of any food.
    - Slice hot dogs in quarters lengthwise.
    - Shred hard raw vegetables and fruits.
    - Remove pits from apples, cherries, plums, peaches, and other fruits.
    - Cut grapes and cherry tomatoes in quarters.
    - Spread peanut butter thin.
    - Chop nuts and seeds fine.
    - Check to make sure fish is boneless.
    - Avoid popcorn and hard or gummy candies.

School-Age Children

1. Serve a wide variety of foods, including children's favorites: tuna fish, pizza (use vegetable toppings), macaroni and cheese, hamburgers (use lean beef combined with ground turkey breast), and peanut butter.

2. Good snack choices are important, as children do not always have the desire or the time to sit down and eat. Snacks can include fresh fruits and vegetables, dried fruits, unsweetened fruit juices, bread, cold cereals, popcorn (without excessive fat), pretzels, tortillas, muffins, milk, yogurt, cheese, pudding, sliced lean meats and poultry, and peanut butter.

3. Balance menu items that are higher in fat with those containing less fat.

4. Pay attention to serving sizes.

5. Offer iron-rich foods such as meat in hamburgers, peanut butter, baked beans, chili, dried fruits, and fortified dry cereals.

6. As children grow, they need to eat more high-fiber foods such as fruits, vegetables, beans and peas, and whole-grain foods.

7. Serve good sources of the additional nutrients children are most likely to be lacking (see Figure 13-14).

All children up to age 10 need to eat every four to six hours to keep their blood glucose at a desirable level; therefore, snacking is necessary between meals. Nutritious snack choices for both preschoolers and school-age children are noted above.

Snacks as well as meals should provide good sources of calcium, such as dairy foods. If a child drinks little or no milk, try adding flavorings such as chocolate or strawberry or make

FIGURE 13-14: Problem Nutrients for Children, Adolescents, and/or Adults

Nutrient	Problem Nutrient for	Food Sources
Iron	Children	Beef, poultry, fish
		Enriched and fortified breads and cereals
		Beans and peas
		Green leafy vegetables
		Eggs
Vitamin E		Vegetable oils
		Margarine and shortening made from vegetable oils
		Salad dressings made from vegetable oils
		Nuts and seeds
		Leafy green vegetables
		Whole grains
		Fortified cereals
Potassium	Children	Fruits and vegetables, especially winter squash, potatoes,
	Adolescents	oranges, grapefruits, and bananas
	Adults	Milk, yogurt
		Beans and peas
		Meats
Fiber	Children	Beans and peas
	Adolescents	Vegetables
	Adults	Fruits
		Whole grains, such as whole-grain breads, cereals, and pasta
Calcium	Adolescents	Milk, yogurt
	Adults	Cheese
		Fortified cereals and orange juice
		Leafy greens (collards, spinach, kale, broccoli)
		Tofu with calcium sulfate
		Beans
Magnesium	Adolescents	Beans and peas
	Adults	Nuts (especially almonds and cashews)
		Green leafy vegetables
		Potatoes
		Seeds
		Whole-grain breads and cereals
		Fish
Vitamin A	Adults	Fortified milk
		Fortified cereals
		Dark green vegetables
		Deep orange fruits and vegetables
		Eggs
Vitamin C	Adults	Citrus fruits, kiwi fruit, strawberries
		Bell peppers, tomatoes, broccoli
		Potatoes
		Fortified juices, drinks, and cereals

cocoa, milkshakes, puddings, and custards. Milk can be fortified with powdered milk by blending 2 cups of fluid milk with $\frac{1}{3}$ cup of powdered milk. One cup of fortified milk is equal to $1\frac{1}{2}$ cups of regular milk. Powdered milk can also be added in baking and to casseroles, soups, sauces, gravies, ground meats, mashed potatoes, and scrambled eggs. Cheese and yogurt, of course, are also good sources of calcium.

MINI-SUMMARY

1. By a child's first birthday, the growth rate decreases markedly, and yearly weight gain until puberty is about 4 to 6 pounds.
2. Growth charts for boys and girls are used to compare a child's height and weight to national percentiles.
3. By age two, the baby teeth are almost all in and many children can use a cup and spoon well.
4. For children two to three years of age, fat intake should be 30 to 35 percent of total kcalories. For children and adolescents from 4 to 18 years of age, keep fat intake between 25 and 35 percent of kcalories, with most fats coming from sources of polyunsaturated and monounsaturated fatty acids, such as vegetable oils, nuts, and fish. Children over two years old should limit consumption of saturated and trans fats to 10 percent or less of total kcalories.
5. Problem nutrients for children include vitamin E, potassium, iron, and fiber.
6. During growth spurts, children's appetites are good; otherwise, their appetites may seem poor.
7. Preschoolers can be fussy eaters, often have food jags, and can take the pleasure out of mealtimes.
8. Guidelines for eating with preschoolers and menu planning for them are detailed in this section.
9. The Food Guide Pyramid for Young Children (two to six years old) appears in Figure 13-13.
10. School-age children are much better eaters than preschoolers.
11. Children's eating habits are influenced by family, friends, teachers, availability of school breakfast and lunch programs, and television.
12. By the time children are four years old, they can eat amounts that count as servings eaten by older family members.
13. Menu-planning guidelines for school-age children are detailed in this section.

ADOLESCENCE

Adolescence is the period of life between about 11 and 21 years of age. The beginning of adolescence is marked by puberty, the process of physically developing from a child to an adult. Puberty starts at about age 10 or 11 for girls and 12 or 13 for boys. The growth spurt is intense for 2 to $2\frac{1}{2}$ years, and then there are a few more years of growth at a slower place. The timing of puberty and rates of growth show much individual variation. During the five to seven years of pubertal development, adolescents gain about 20 percent of adult height and 50 percent of ideal adult body weight. Most of the body organs double in size, and almost half of total bone growth occurs.

Whereas before puberty the proportion of fat and muscle was similar in males and females, males now put on twice as much muscle as females, and females gain proportionately more fat. In adolescent girls, an increasing amount of fat is being stored under the skin, particularly in the abdominal area. The male also experiences a greater increase in bone mass than does the female.

NUTRITION DURING ADOLESCENCE

Figure 13-15 compares the EER for kcalories and the RDA for protein for adolescents with adult needs. Energy requirements for adolescents vary greatly depending on gender, body composition, level of physical activity, and current rate of growth. The highest levels of nutrients are for individuals growing at the fastest rate.

Adolescent males now need more kcalories, protein, magnesium, and zinc for muscle and bone development than do females; however, females need increased iron due to the onset of menstruation. Owing to their big appetites and high kcalorie needs, teenage boys are more likely than girls to get sufficient nutrients. Females have to pack more nutrients into fewer kcalories, and this can become difficult if they decrease their food intake to lose weight.

Three problem nutrients for children continue to be problems for adolescents as well: vitamin E, potassium, and fiber. In addition, children age nine years and older need increased intakes of calcium and magnesium. Figure 13-14 lists sources of these problem nutrients.

Calcium is very important for adolescents because half of their peak bone mass is built during this time. The greatest capacity to absorb calcium occurs in early adolescence, when the body can absorb almost four times more calcium than it can during young adulthood. The Adequate Intake for calcium increases from 800 mg at age 8 to 1300 mg from ages 9 to 18 for both males and females. Adolescents need 4 glasses of milk or milk equivalents to meet their calcium needs. Unfortunately, calcium intake drops as age increases during adolescence.

Magnesium is very important so that the body can make protein and fat, make bones, and maintain a healthy immune system. Magnesium also works with calcium to contract muscles and transmit nerve impulses.

With their increased independence, adolescents assume responsibility for their own eating habits. Teenagers are not fed; they make most of their own food choices. They eat more meals away from home, such as at fast-food restaurants, and skip more meals than they did previously. Irregular meals and snacking are common due to busy social lives and after-school

---

FIGURE 13-15: Energy and Macronutrients for Adolescents

ESTIMATED ENERGY REQUIREMENT (EER) AND RECOMMENDED DIETARY ALLOWANCE (RDA) FOR PROTEIN

Gender and Age	Height	Weight	EER*	Protein
Male 9–13 years	57 inches	79 pounds	2279 kcal	0.95 grams/kilogram body weight
Female 9–13 years	57	81	2071	0.95
Male 14–18 years	68	134	3152	0.85
Female 14–18 years	64	119	2368	0.85
Male 19–30 years	70	154	3067	0.80
Female 19–30 years	64	126	2403	0.80

*The EER is for healthy moderately active Americans and Canadians.

ACCEPTABLE MACRONUTRIENT DISTRIBUTION RANGES

Age	Carbohydrate	Fat	Protein
4–18 years	45–65%	25–35%	10–30%
Over 18 years	45–65%	20–35%	10–35%

Source: Adapted with permission from the Dietary References Intakes for Energy, Carbohydrates, Fiber, Fat, Protein, and Amino Acids (Macronutrients). © 2002 by the National Academy of Sciences. Courtesy of the National Academy Press, Washington, D.C.

activities and jobs. Adolescents are more likely to skip breakfast than either younger children or adults. Teenagers will tell you that they lack the time or discipline to eat right, although many are pretty well informed about good nutrition practices. Ready-to-eat foods such as cookies, chips, and soft drinks are readily available, and teenagers pick them up as snack foods.

Adolescents often have a variety of lunch options when they are at school. These choices may include leaving the school to buy lunch, eating a lunch from the National School Lunch Program, buying à la carte foods in the school cafeteria that do not qualify as a lunch in the National School Lunch Program, and buying food from a school store or vending machines. State, local, or school rules may close vending machines during mealtimes and other times during the school day. Two professional associations, the American Dietetic Association and the School Nutrition Association, are concerned that the foods sold in vending machines, school stores, and à la carte cafeterias discourage students from eating meals provided by the National School Lunch and Breakfast programs.

The media have a powerful influence on adolescents' eating patterns and behaviors. Advertising for not-so-nutritious foods and fast foods permeates television, radio, and billboards. In addition, questionable eating habits are portrayed on television shows. Studies show that the prevalence of obesity increases as the number of hours of watching television increases.

A typical meal at a fast-food restaurant—a 4-ounce hamburger, french fries, and a regular soft drink—is high in calories, fat, and sodium. However, more nutritious choices are available at fast-food restaurants. Smaller hamburgers, milk, salads, and grilled chicken sandwiches are examples of more nutritious options.

Parents can positively influence adolescents' eating habits by being good role models and by having dinner and nutritious breakfast and snack foods available at home. Adolescents can become involved in food purchasing and preparation. Parents can also influence their children's fitness level by limiting their sedentary activities and encouraging exercise.

Both adolescent boys and girls are influenced by their body images. Adolescent boys may take nutrition supplements and fill up on protein in hopes of becoming more muscular. Adolescent girls who feel they are overweight may skip meals and modify their food choices in hopes of losing weight. Whether overweight or not, teens need regular exercise.

MENU PLANNING FOR ADOLESCENTS

Here are some tips about planning menus for adolescents.

1. Emphasize complex carbohydrates containing fiber, such as whole-grain breads and cereals, fruits, vegetables, potatoes, brown rice, and dried beans and peas. These foods supply kcalories along with needed nutrients. Whole-grain products contain more fiber and other nutrients than do their refined counterparts.
2. Offer well-trimmed lean beef, poultry, and fish. Don't think that just because adolescents need more calories, fatty meats are in order. Their fat calories should come mostly from foods that are moderate or low in saturated fat.
3. Low-fat milk and nonfat milk need to be offered at all meals. Girls are more likely to need to select nonfat milk than are boys. Other forms of calcium also need to be available, such as pizza, macaroni and cheese, and other entrées using cheese, along with yogurt, frozen yogurt, ice milk, and puddings made with nonfat or low-fat milk
4. Have nutritious choices available for hungry on-the-run adolescents looking for a snack. Nutritious snack choices include fresh fruit, muffins and other quick breads,

crackers or rolls with low-fat cheese or peanut butter, vegetable-stuffed pita pockets, yogurt or cottage cheese with fruit, and fig bars.

5. Emphasize quick and nutritious breakfasts, such as whole-grain pancakes or waffles with fruit, juices, whole-grain toast or muffins with low-fat cheese, cereal topped with fresh fruit, and bagels with peanut butter.

6. Serve good sources of the nutrients adolescents are most likely to be lacking (see Figure 13-14).

MINI-SUMMARY

1. Adolescence is the period from 11 to 21 years of age.

2. Puberty starts at about age 10 or 11 for girls and 12 or 13 for boys. The growth spurt is intense for 2 to 2½ years, and then there are a few more years of growth at a slower pace.

3. Males now put on twice as much muscle as females do, and females gain proportionately more fat.

4. Adolescent males now need more kcalories, protein, magnesium, and zinc for muscle and bone development than females do; however, females need increased iron due to the onset of menstruation.

5. Owing to their big appetites and high kcalorie needs, teenage boys are more likely than girls to get sufficient nutrients. Females have to pack more nutrients into fewer kcalories.

6. Problem nutrients and their sources are listed in Figure 13-14.

7. Adolescents need 4 glasses of milk or milk equivalents to meet their calcium needs.

8. Teenagers make most of their own food choices. They eat more meals away from home and skip more meals than previously.

9. Parents can positively influence adolescents' eating habits by being good role models, serving dinner, and having nutritious foods available at home.

10. Menu-planning guidelines are given in this section.

EATING DISORDERS

Each year millions of Americans develop serious and sometimes life-threatening eating disorders. The vast majority—more than 90 percent—of those afflicted with eating disorders are adolescent and young adult women. One reason women in this age group are particularly vulnerable to eating disorders is their tendency to go on strict diets to achieve an "ideal" figure. Researchers have found that such stringent dieting can play a key role in triggering eating disorders. The actual cause of eating disorders is not entirely understood, but many risk factors have been identified. Those risk factors may include a high degree of perfectionism, low self-esteem, genetics, and a family preoccupation with dieting and weight.

Eating-disorder patients deal with two sets of issues: those surrounding their eating behaviors and those surrounding their interactions with others and with themselves. Eating disorders are considered a psychological disorder, and both psychotherapy and medical nutrition therapy are cornerstones of treatment.

Approximately 1 percent of adolescent girls and young women develop ***anorexia nervosa***, a dangerous condition in which they can literally starve themselves to death. Another 2 to 3 percent of adolescent girls and young women develop ***bulimia nervosa***, a destructive pattern of excessive overeating followed by vomiting or other "purging" behaviors to control

ANOREXIA NERVOSA
An eating disorder most prevalent in adolescent females who starve themselves.

BULIMIA NERVOSA
An eating disorder characterized by a destructive pattern of excessive overeating followed by vomiting or other "purging" behaviors to control weight.

their weight. A more recently recognized eating disorder, ***binge eating disorder***, could turn out to be the most common (about 3 percent). With this disorder, binges are not followed by purges, and so these individuals often become overweight. Eating disorders also occur in men and older women, but much less frequently.

The consequences of eating disorders can be severe, with 1 in 10 cases of anorexia nervosa leading to death from starvation, cardiac arrest, or suicide over the course of 10 years. The outlook is better for bulimia; anorexia patients tend to relapse more. Many patients with anorexia or bulimia also suffer from other psychiatric illnesses, such as clinical depression, anxiety, obsessive-compulsive disorder, and substance abuse. Fortunately, increasing awareness of the dangers of eating disorders—sparked by medical studies and extensive media coverage of these illnesses—has led many people to seek help. Nevertheless, some people with eating disorders refuse to admit that they have a problem and do not get treatment. Family members and friends can help recognize the problem and encourage these persons to seek treatment. The earlier treatment is started, the better the chance of a full recovery is.

ANOREXIA NERVOSA

People who intentionally starve themselves suffer from anorexia nervosa. This disorder, which usually begins in young people around the time of puberty, involves extreme weight loss—at least 15 percent below an individual's normal body weight. Many people with the disorder look emaciated but are convinced they are overweight. Sometimes they must be hospitalized to prevent starvation.

The typical symptoms include:

- Refusal to maintain body weight at or above a minimally normal weight for height, body type, age, and activity level
- Intense fear of weight gain or being "fat"
- Feeling "fat" or overweight despite dramatic weight loss
- Loss of menstrual periods
- Extreme concern with body weight and shape
- Excessive exercising

Let's look at a typical case.

Deborah developed anorexia nervosa when she was 16. A rather shy, studious teenager, she tried hard to please everyone. She had an attractive appearance but was slightly overweight. Like many teenage girls, she was interested in boys but concerned that she wasn't pretty enough to get their attention. When her father jokingly remarked that she would never get a date if she didn't take off some weight, she took him seriously and began to diet relentlessly—never believing she was thin enough, even when she became extremely underweight.

Soon after the pounds started dropping off, Deborah's menstrual periods stopped. As anorexia tightened its grip, she became obsessed with dieting and food and developed strange eating rituals. Every day she weighed all the food she would eat on a kitchen scale, cutting solids into minuscule pieces and precisely measuring liquids. She would put her daily ration in small containers, lining them up in neat rows. She also exercised compulsively even after she weakened and became faint.

No one could convince Deborah that she was in danger. Finally, her doctor insisted that she be hospitalized and carefully monitored for treatment of her illness. While in the hospital, she secretly continued her exercise regimen in the bathroom. It took several hospitalizations and a good deal of individual and family outpatient therapy for Deborah to face and solve her problems.

One of the most frightening aspects of the disorder is that people with anorexia continue to think they are overweight even when they are bone-thin. Food and weight become obsessions. For some, the compulsiveness shows up in strange eating rituals or refusal to eat in front of others. It is not uncommon for anorexics to collect recipes and prepare gourmet feasts for family and friends but not partake in the meals themselves.

In patients with anorexia, starvation can damage vital organs such as the heart and brain. To protect itself, the body shifts into "slow gear": Menstrual periods stop, and breathing, pulse, and blood pressure rates drop. Nails and hair become brittle. The skin dries, yellows, and becomes covered with soft hair called *lanugo*. Reduced body fat leads to lowered body temperature and the inability to withstand cold.

Mild anemia, swollen joints, reduced muscle mass, and light-headedness are also common. If the disorder becomes severe, patients may lose calcium from the bones, making them brittle and prone to breakage. They may also experience irregular heart rhythms and heart failure.

LANUGO
Downy hair on the skin.

BULIMIA NERVOSA

People with bulimia nervosa consume large amounts of food and then engage in a variety of compensatory behaviors, such as vomiting, abusing laxatives or diuretics, taking enemas, and exercising obsessively. Because many individuals with bulimia engage in these behaviors alone and maintain normal or above-normal body weight, they can often hide their problem for years.

Typical symptoms include:

- Repeated episodes of bingeing and purging
- Feeling out of control during a binge and eating beyond the point of comfortable fullness
- Compensatory behaviors after a binge (typically by self-induced vomiting; abuse of laxatives, diet pills, and/or diuretics; excessive exercise; or fasting)
- Frequent dieting
- Extreme concern with body weight and shape

To be diagnosed as bulimic, the person must engage in binge eating and compensatory behaviors at least twice a week on average for three months.

Let's take a look at Lisa.

Lisa developed bulimia at age 18. As with Deborah, her strange eating behavior began when she started to diet. She dieted and exercised to lose weight, but unlike Deborah, she regularly ate huge amounts of food and maintained her normal weight by forcing herself to vomit. Lisa often felt like an emotional powder keg—angry, frightened, and depressed.

Unable to understand her own behavior, she thought no one else would either, and so she felt isolated and lonely. Typically, when things were not going well, she would be overcome with an uncontrollable desire for sweets. She would eat pounds of candy and cake at a time and often not stop until she was exhausted or in severe pain. Then, overwhelmed with guilt and disgust, she would make herself vomit.

While recuperating in a hospital from a suicide attempt, Lisa was referred to an eating disorders clinic, where she got into group therapy. She also received medications to treat the illness and the understanding and help she desperately needed from others who had the same problem.

Individuals with this disorder may binge and purge once or twice a week or as much as several times a day. Dieting stringently between episodes of bingeing and purging is also common.

As with anorexia, bulimia typically begins during adolescence. The condition occurs most often in women but is also found in men. Many individuals with bulimia, ashamed of their strange habits, do not seek help until they reach their thirties or forties. By that time, their eating behavior is deeply ingrained and more difficult to change.

Bulimic patients—even those of normal weight—can severely damage their bodies by frequent binge eating and purging. Vomiting causes serious problems: The acid in vomit wears down the outer layer of the teeth and can cause scarring on the backs of the hands when fingers are pushed down the throat to induce vomiting. Further, the esophagus becomes inflamed and the glands near the cheeks become swollen.

BINGE EATING DISORDER

Binge eating disorder resembles bulimia in that it is characterized by episodes of uncontrolled eating, or bingeing. However, binge eating disorder differs from bulimia in that its sufferers do not engage in vomiting or other methods of purging. Binge eating is not a recent development—it's been around for a long time, sometimes called compulsive overeating—but only recently has it been categorized as a mental disorder.

Binge eaters feel that they lose control of themselves when eating. They eat large quantities of food, although not as much as bulimics do, and don't stop eating until they are uncomfortably full. To be diagnosed as having binge eating disorder, an individual must binge at least twice a week on average for six months.

Binge eaters usually have more difficulty losing weight and keeping it off than do people with other serious weight problems. Most people with this disorder are obese and have a history of weight fluctuations. Binge eating disorder is found in about 3 to 5 percent the general population, more often in women than in men.

FEMALE ATHLETE TRIAD

FEMALE ATHLETE TRIAD
An eating disorder found among female college athletes in which they have disordered eating, osteoporosis, and no menstruation.

Some physically active girls and women, especially competitive athletes, are at risk for *female athlete triad*, which is characterized by the following.

- *Disordered eating*—The disordered eating may take different forms, such as restrictive eating, bingeing, or purging.
- *Amenorrhea*—Poor eating, low body weight, low body fat, and continuous training all contribute to amenorrhea: lack of menstruation.
- *Osteoporosis (weak bones)*—The decreased estrogen levels associated with the absence of a period may be the cause of premature osteoporosis in female athletes. They start to lose strength in their bones after only a few months with no periods and over time have an increased risk of stress fractures.

Athletes in sports such as gymnastics, figure skating, ballet, diving, swimming, and distance running, which emphasize a lean physique, appear to be most at risk of developing the female athlete triad.

Treatment consists of increasing meals and snacks to establish regular menstrual periods and increase bone density. Principles of treatment include reducing the intensity of training until menstruation resumes, increasing kcalorie intake, ensuring adequate calcium and vitamin D intakes, and possibly starting hormone replacement therapy.

TREATMENT

The sooner a disorder is diagnosed, the better the chances are that treatment can work. The longer abnormal eating behaviors persist, the more difficult it is to overcome the disorder and its effects on the body. In some cases, long-term treatment is required.

Once an eating disorder is diagnosed, the clinician must determine whether the patient is in immediate medical danger and requires hospitalization. Although most patients can be treated as outpatients, some need hospital care, as in the case of severe purging or risk of suicide. Eating-disorder patients commonly work with a treatment team that includes an internist, a nutritionist, a psychotherapist, and someone who is knowledgeable about the psychoactive medications used in treating these disorders. Treatment usually includes individual psychotherapy, family therapy, cognitive-behavioral therapy, medical nutrition therapy, and possibly medications such as antidepressant drugs.

Eating disorders, unfortunately, have a very high death rate; 1 of every 10 patients will die prematurely from the disease. With that in mind, prevention of these diseases needs to be seriously examined. Research has identified the community groups most important to reach: middle school and high school students, coaches, and parents.

Figure 13-16 lists questions to help individuals determine whether they have an eating disorder.

MINI-SUMMARY

1. Most people afflicted with these problems are adolescent girls and young women.
2. Approximately 1 percent of adolescent girls and young women develop anorexia nervosa, a dangerous condition in which they can literally starve themselves to death. Another 2 to 3 percent of adolescent girls and young women develop bulimia nervosa, a destructive pattern of excessive overeating followed by vomiting or other compensatory behaviors to control their weight.
3. A more recently recognized eating disorder, binge eating disorder, could turn out to be the most common (about 3 percent). With this disorder, binges are not followed by purges, and so these individuals often become overweight.
4. Eating disorders also occur in adolescent boys and older women, but much less frequently.
5. The sooner the disorder is diagnosed, the better the chances are for successful treatment. Treatment usually includes individual psychotherapy, family therapy, cognitive-behavior therapy, medical nutrition therapy, and possibly medications.

FIGURE 13-16: Do You Have an Eating Disorder?

A positive answer to one or more of these questions may indicate an eating disorder.

1. Do you eat large amounts of food in a very short period while feeling out of control and are by yourself?
2. Do you frequently eat a lot of food when you are not hungry and usually when you are alone?
3. Do you feel guilty after overeating?
4. Do you make yourself vomit or use laxatives or diuretics to purge yourself?
5. Do you carefully make sure you eat only a small number of calories each day, such as 500 kcalories or less, and exercise a lot?
6. Do you avoid going out to maintain your eating and exercise schedule?
7. Do you feel that food controls your life?

OLDER ADULTS

The older population—persons age 65 years or older—numbered 36.3 million in 2004. These persons represented 12 percent of the U.S. population, about one in every eight Americans. In 2000, persons reaching age 65 had an average life expectancy of an additional 17.9 years (19.2 years for women and 16.3 years for men).

The older population will grow dramatically between the years 2010 and 2030, when the baby boom generation reaches age 65. Minority populations are projected to represent 25.4 percent of the elderly population in 2030, up from 16.4 percent in 2000.

Before looking at nutrition during aging, let's take a look at what happens when we age. Studies suggest that maximum efficiency of many organ systems occurs between ages 20 and 35. After age 35, the functional capability of almost every organ system declines. Similar changes occur in all adults as they age, but the rate of decline shows great individual variation. Both genetics and environmental factors such as nutrition affect the rate of aging. Conversely, changes brought about by the aging process affect nutrition status. Of particular importance are changes that affect digestion, absorption, and metabolism of nutrients.

The basal metabolic rate declines between 8 and 12 percent from age 30 to age 70. This change is largely the result of loss of muscle mass as we age. From the mid-twenties to the mid-seventies, people slowly lose about 24 pounds of muscle, which is replaced by 22 pounds of fat. Combined with a general decrease in activity level, these factors clearly indicate a need for decreased kcalorie intake, which generally does take place during aging. But the elderly need not lose all that muscle mass. Studies have shown that when the elderly do regular weight-training exercises, they increase muscular strength, increase basal metabolism, improve appetite, and improve blood flow to the brain.

Overall, the functioning of the cardiovascular system declines with age. The workload of the heart increases due to atherosclerotic deposits and less elasticity in the arteries. The heart does not pump as hard as before, and cardiac output is reduced in elderly people who do not remain physically active. Blood pressure increases normally with age. Pulmonary capacity decreases by about 40 percent throughout life. This decrease does not restrict the normal activity of healthy older persons but may limit vigorous exercise. Kidney function deteriorates over time, and the aging kidney is less able to excrete waste. Adequate fluid intake is important, as is avoiding megadoses of water-soluble vitamins because they put a strain on the kidneys to excrete them. Loss of bone occurs normally during aging, and osteoporosis is common (see Chapter 7).

FACTORS AFFECTING NUTRITION STATUS

The nutrition status of an elderly person is greatly influenced by many variables, including physiological, psychosocial, and socioeconomic factors.

Physiological Factors
- Disease. The presence of disease, both acute and chronic, and the use of modified diets can affect nutrition status. Most older persons have at least one chronic condition, and many have multiple conditions, such as arthritis, hypertension, heart disease, and diabetes. Their chronic conditions often require modified diets. Certain chronic diseases, such as gastrointestinal disease, congestive heart failure, renal disease, and cancer, are associated with *anorexia* (lack of appetite). Other diseases, such as stroke, are not associated with anorexia but can cause the individual to take in little food.

ANOREXIA

Lack of appetite.

- Less muscle mass. With aging, there is less muscle mass, and so the basal metabolic rate decreases. As the BMR slows down, the number of kcalories needed decreases.
- Activity level. Because active individuals tend to eat more calories than their sedentary counterparts do, they are more likely to ingest more nutrients.
- Dentition. Approximately 30 percent of Americans have lost their teeth by age 65. Despite widespread use of dentures, chewing still presents problems for many of the elderly.
- Functional disabilities. Functional disabilities interfere with the ability of the elderly to perform daily tasks such as the purchasing and preparation of food and eating. These disabilities may be due to diseases and conditions such as arthritis or rheumatism, stroke, visual impairment, heart trouble, and dementia.
- Taste and smell. Sensitivity to taste and smell declines slowly with age. The taste buds become less sensitive, and the nasal nerves that register aromas need extra stimulation to detect smells. That's why seniors may find ordinarily seasoned foods too bland. Medications also may alter an individual's ability to taste.
- Changes in the gastrointestinal tract. The movement of food through the gastrointestinal tract slows down over time, causing problems such as constipation, a common complaint among older people. Constipation may also be related to low fiber and fluid intake, medications, and lack of exercise. Other common complaints include nausea, indigestion, and heartburn. (Heartburn, a burning sensation in the area of the throat, has nothing to do with the heart. It occurs when acidic stomach contents are pushed into the lower part of the esophagus or throat.)
- Medications. More than half of all seniors take at least one medication daily, and many take six or more a day. Medications may alter appetite or the digestion, absorption, and metabolism of nutrients (Figure 13-17).
- Thirst. Many of the elderly suffer a diminished perception of thirst—especially problematic when they are not feeling well. Because the aging kidney is less able to concentrate urine, more fluid is lost, setting the stage for dehydration.

Psychosocial Factors
- Cognitive functioning. Poor cognitive functioning may affect nutrition, or perhaps poor nutritional status contributes to poor cognitive functioning.
- Social support. An individual's nutritional health results in part from a series of social acts. The purchasing, preparing, and eating of foods are social events for most people. For example, elderly people may rely on one another for rides to the supermarket, cooking, and sharing meals. The benefits of social networks or support are largely due to the

FIGURE 13-17: Nutrients Depleted by Selected Drugs

Drug Group	Drug	Nutrients Depleted
Analgesics	Uncoated aspirin	Iron
Antacids	Aluminum or magnesium hydroxide	Phosphate, calcium, and folate
	Sodium bicarbonate	Calcium, folate
Antiulcer drugs	Cimetidine	B_{12}
Chemotherapeutic agents	Methotrexate	Folate
Cholesterol-lowering agents	Cholestyramine	Fat, vitamins A and K
Diuretics	Lasix	Potassium

companionship and emotional support they provide. It is anticipated that this has a positive effect on appetite and dietary intake.

Socioeconomic Factors
- **Education.** Higher levels of education are positively associated with increased nutrient intakes.
- **Income.** Almost one in five elderly has an income at or below the federal poverty level. Lower-income elderly persons tend to eat fewer calories and fewer servings from different food groups than do higher-income elderly persons.
- **Living arrangements.** The elderly, particularly women, are more likely to be widowed. The trend has been for widows and widowers in the United States to live alone after the spouse dies. Research focusing on the impact of living arrangements on dietary quality showed that living alone is a risk factor for dietary inadequacy for older men, especially those over age 75 years of age, and for women in the youngest age group (55 to 64).
- **Availability of federally funded meals.** The availability of nutritious meals through federal programs such as Meals on Wheels, in which meals are delivered to the home, is crucial to the nutritional health of many elderly people. Another popular elderly feeding program is the Congregate Meals Program, in which the elderly go to a senior center to eat.

NUTRITION FOR OLDER ADULTS

Older adults do eat fewer kcalories as they age. Between decreased physical activity and a decreased basal metabolic rate, they need 20 percent fewer kcalories to maintain their weight. This means that there is less room in the diet for empty-kcalorie foods such as sweets, alcohol, and fats. At a time when good nutrition is so important to good health, there are many obstacles to fitting more nutrients into fewer kcalories, such as medical conditions, dentures, and medications.

Nutrients of concern to the elderly include the following:

- **Water.** Due to decreased thirst sensation and other factors, fluid intake is more important for older adults than it is for younger people. Water is also important to prevent constipation along with eating foods high in fiber.
- **Vitamin B_{12} and folate.** The elderly often have a problem with vitamin B_{12}, even if they take in enough. The stomach of an elderly person secretes less gastric acid and pepsin, both of which are necessary to break vitamin B_{12} from its polypeptide linkages in food. Older adults also often lack enough of the intrinsic factor that is needed for absorption. The result is that less vitamin B_{12} is absorbed. Vitamin B_{12} is necessary to convert folate into its active form so that folate can do its job of making new cells, such as new red blood cells. When vitamin B_{12} is not properly absorbed, pernicious anemia develops. Pernicious means ruinous or harmful, and this type of anemia is marked by a megaloblastic anemia (as with folate deficiency), also called macrocytic anemia, in which there are too many large, immature red blood cells. Symptoms include extreme weakness and fatigue. Nervous system problems also erupt. The cover surrounding the nerves becomes damaged, making it difficult for impulses to travel along them. This causes a poor sense of balance, numbness and tingling sensations in the arms and legs, and mental confusion. Pernicious anemia can result in paralysis and death if it is not treated. Because deficiencies in folate or vitamin B_{12} cause macrocytic anemia, a physician may mistakenly administer folate when the problem is really a vitamin B_{12} deficiency. The

folate would treat the anemia, but not the deterioration of the nervous system due to a lack of vitamin B_{12}. If untreated, this damage can be significant and sometimes irreversible, although it takes many years for this to occur. When vitamin B_{12} is deficient due to an absorption problem, injections of the vitamin are usually given.

- Vitamin D. Several factors adversely affect the vitamin D status of the elderly. First, the elderly tend to be outside less, and so they make less vitamin D from exposure to the sun. Also, they have less of the vitamin D precursor in the skin that is necessary to make vitamin D. Vitamin D–fortified milk and cereals are important to getting enough vitamin D. If milk intake is low, supplements may be recommended.

- Calcium. Current intakes for calcium are below the recommendations for individuals over 65 years of age (1200 milligrams). To meet this recommendation, an elderly person would need to eat four servings of dairy products or other calcium-rich foods daily. Because this can be difficult, supplements may be recommended.

- Zinc. Because the elderly are at risk for taking in less than the RDA for zinc and due to the importance of zinc in cell production, wound healing, the immune system, and taste, attention needs to be paid to getting enough of this mineral.

Here are some ways in which the elderly can improve the nutrient content of what they eat.

1. Eat with other people. This usually makes mealtime more enjoyable and stimulates the appetite. Taking a walk before eating also stimulates the appetite.
2. Prepare larger amounts of food and freeze some for heating up at a later time. This saves cooking time and is helpful for someone who is reluctant to cook.
3. If big meals are too much, eat small amounts more frequently during the day. Eat regular meals.
4. If getting to the supermarket is a bother, go at a time when it is not busy or engage a delivery service (this is more expensive, though).
5. Use unit pricing, sales, and coupons to cut back on the amount of money spent on food.
6. Take advantage of community meal programs for the elderly such as Meals on Wheels and meals at senior centers.
7. To perk up a sluggish appetite, increase use of herbs, spices, lemon juice, vinegar, and garlic.

MENU PLANNING FOR OLDER ADULTS

Figure 13-18 shows a modified Food Guide Pyramid for people over 70 developed by researchers at the USDA Human Nutrition Research Center on Aging at Tufts University. This pyramid is designed for healthy individuals over 70 years of age. You will notice the following differences.

1. This Pyramid is narrower than the traditional Pyramid to illustrate that older adults have decreased kcalorie needs and therefore fewer food selections. Their selections must be more nutrient-dense than those for other groups of people.
2. A small supplement flag at the top of the Pyramid suggests supplements for nutrients (calcium, vitamin D, vitamin B_{12}) that may be insufficient in the diet due to smaller and fewer servings of food and medical conditions such as lactose intolerance.
3. Below the food groups in the Pyramid you will find a horizontal band with the suggestion to drink at least eight servings of water (fluid) daily.

FIGURE 13-18:
Modified food pyramid
for adults age 70 and
over.
© 2007 Tufts University.
Reprinted with permission.

Modified MyPyramid for Older Adults

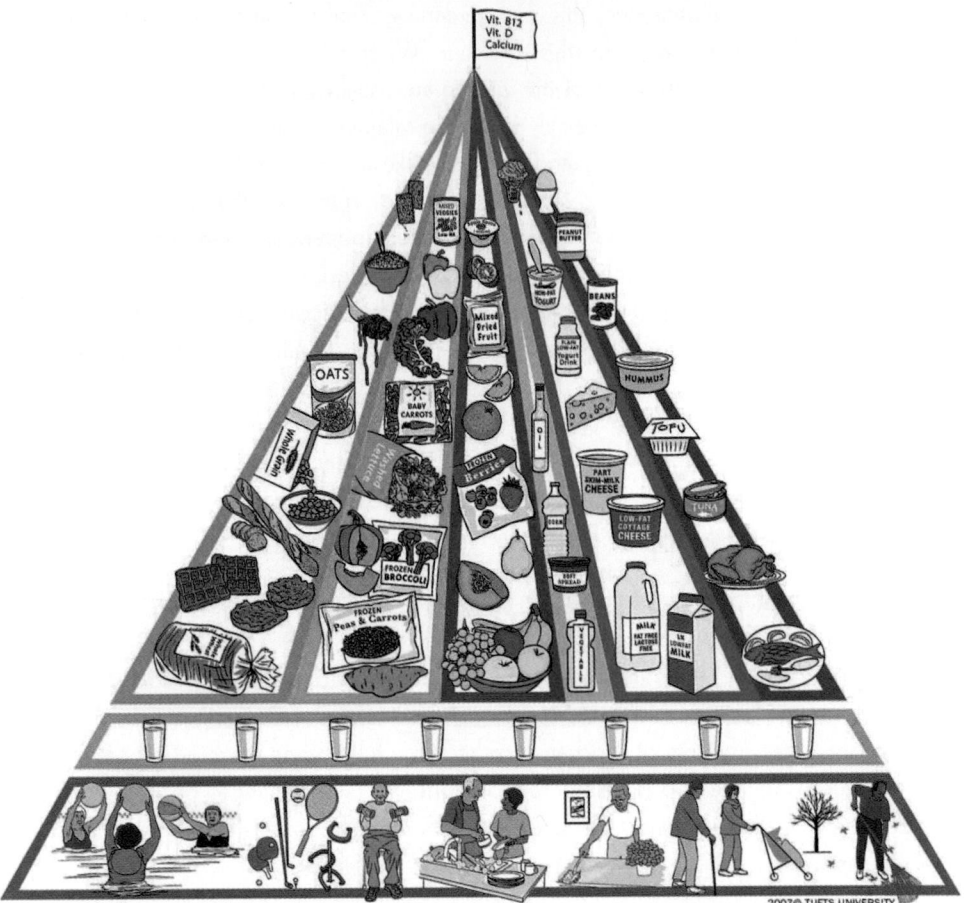

4. At the base of the Pyramid, you see a band emphasizing the importance of physical activity for older adults.

When planning meals for older adults, use these guidelines.

1. Offer moderately sized meals and half portions. Older adults frequently complain when given too much food because they hate to see waste. Restaurants might reduce the size of their entrées by 15 to 25 percent.
2. Emphasize high-fiber foods such as fruits, vegetables, whole grains, and beans. Older people requiring softer diets may have problems chewing some high-fiber foods. High-fiber foods that are soft in texture include cooked beans and peas, bran cereals soaked in milk, canned prunes and pears, and cooked vegetables such as potatoes, corn, green peas, and winter squash.
3. Moderate the use of fat. Many seniors don't like to see an entrée swimming in a pool of butter. Use lean meats, poultry, or fish and sauces prepared with vegetable or fruit purées. Have low-fat dairy products available, such as nonfat or low-fat milk.
4. Dairy products such as milk and yogurt are important sources of calcium, vitamin D, protein, potassium, and vitamin B$_{12}$.
5. Offer adequate protein but not too much. Use a variety of both animal and vegetable sources. Providing protein on a budget, as in a nursing home, need not be a problem. Lower-cost protein sources include beans and peas, cottage cheese, macaroni and cheese, eggs, chicken, and ground beef.

6. Moderate the use of salt. Many seniors are on low-sodium diets and recognize a salty soup when they taste it. Avoid highly salted soups, sauces, and other dishes. It is better to let seniors season food to taste.

7. Use herbs and spices to make foods flavorful. Seniors are looking for tasty foods just like anyone else, and they may need them more than ever!

8. Offer a variety of foods, including traditional menu items and cooking from other countries and regions of the United States.

9. Fluid intake is critical, so offer a variety of beverages. Diminished sensitivity to dehydration may cause older adults to drink less fluid than needed. Special attention must be paid to fluids, particularly for those who need assistance with eating and drinking. Beverages such as water, milk, juice, coffee, and tea and foods such as soup contribute to fluid intake.

10. If chewing or swallowing is a problem, softer foods can be chosen to provide a well-balanced diet. Following are some guidelines for soft diets.
    - Use tender meats and, if necessary, chop or grind them. Ground meats can be used in soups, stews, and casseroles. Cooked beans and peas, soft cheeses, and eggs are additional softer protein sources.
    - Cook vegetables thoroughly and dice or chop them by hand if necessary after cooking.
    - Serve mashed potatoes or rice, with gravy if desired.
    - Serve chopped salads.
    - Soft fruits such as fresh or canned bananas, berries, peaches, pears, and melon, as well as applesauce, are some good choices.
    - Soft breads and rolls can be made even softer by dipping them briefly in milk.
    - Puddings and custard are good dessert choices.
    - Many foods that are not soft can be easily chopped by hand or puréed in a blender or food processor to provide additional variety.

MINI-SUMMARY

1. During aging, the functional capability of almost every organ declines.
2. From the mid-twenties to the mid-seventies, you slowly lose about 24 pounds of muscle, which is replaced by 22 pounds of fat.
3. Kcalorie needs decrease about 20% but nutrient needs remain the same.
4. The nutrition status of an elderly person is greatly influenced by many variables: presence of disease, decreased muscle mass, activity level, quality of dentition, functional disabilities, decline in taste and smell acuity, changes in the gastrointestinal tract, use of medications, diminished sense of thirst, level of cognitive functioning, available social support, level of education and income, living arrangements, and availability of federally funded meals.
5. Nutrients of concern to the elderly include water, vitamin B_{12}, folate, vitamin D, calcium, and zinc.
6. The modified Food Guide Pyramid for people over 70, as well as the guidelines given in this section, will help you plan menus for older adults.

✦ CHECK-OUT QUIZ

1. Low-birth-weight babies are at increased health risk.

   a. True
   b. False

2. Morning sickness occurs only in the morning.

   a. True
   b. False

3. Nutritional deficiencies during lactation are more likely to affect the quantity than the quality of milk the mother makes.

   a. True
   b. False

4. For the first year, the newborn's only source of nutrients is either breast milk or formula.

   a. True
   b. False

5. A deficiency of iron during the first weeks of pregnancy can cause birth defects.

   a. True
   b. False

6. Breast-feeding is considered to be more nutritious than formula feeding.

   a. True
   b. False

7. Moderate drinking during pregnancy may cause fetal alcohol syndrome.

   a. True
   b. False

8. A baby's first food is strained vegetables.

   a. True
   b. False

9. School-age children tend to be better eaters than preschoolers.

   a. True
   b. False

10. After one year of age, children start to lose baby fat and become leaner.

    a. True
    b. False

11. A good way to get a child to finish dinner is to promise dessert once all foods are eaten.

    a. True
    b. False

12. A child will ask for and need more food during growth spurts than during slower periods of growth.

    a. True
    b. False

13. Breast-fed babies need supplementation with vitamin A.

    a. True
    b. False

14. Children from two to eight years old should drink 2 cups per day of milk.

    a. True
    b. False

15. During adolescence, females put on proportionately more muscle than fat.

    a. True
    b. False

16. The elderly are major users of modified diets.

    a. True
    b. False

17. Certain drugs can impair the absorption and metabolism of certain nutrients.

    a. True
    b. False

18. After age 35, the functional capability of almost every organ system declines.

    a. True
    b. False

19. As you get older, energy needs decrease.

    a. True
    b. False

20. Two vitamins of concern to the elderly are thiamin and riboflavin.

a. True
b. False

ACTIVITIES AND APPLICATIONS

1. **Media Watch**
   On a Saturday morning, watch children's television for one hour and record the name of each featured product. How many of the total number of advertisers were selling food? Were the majority of the advertised foods healthy foods or junk foods?

2. **Childhood Eating Habits**
   Think back to when you were a child and teenager. What influenced what you put in your mouth? Consider influences such as home, school, friends, and relatives. Which positively influenced your eating style? Which negatively? Discuss this with someone else in your class who is close in age.

3. **Preschoolers' Eating Habits**
   Visit a preschool or a day-care center when a meal is being served. Observe the children while they eat. Ask the caregivers about the children's food preferences and eating habits. Ask how well the children accept new foods and how the caregivers introduce new foods.

4. **School Lunch Menu**
   Write a five-day lunch menu for elementary, middle school, or high school students. The menu must provide one-third of the DRI for protein, vitamins C and D, calcium, and iron. Only 30 percent of total calories from fat is allowed. Saturated fat can provide no more than 10 percent of total calories. Use the resource "A Menu Planner for Healthy School Meals" available at schoolmeals.nal.usda.gov/Recipes/menuplan/menuplan.html. Use iProfile to do a nutrient analysis.

5. **Eldercare Menu**
   Visit or phone the foodservice director of a local continuing-care retirement community, congregate meals feeding center, or nursing home. Ask about the type of menu being used (restaurant-style or cycle menu) and the meals and foods being offered. What are the major meal-planning considerations used in planning meals for the elderly? What special circumstances come up that are unique to them?

NUTRITION WEB EXPLORER

Medline Plus   www.nlm.nih.gov/medlineplus/pregnancy.html

Click on one of the articles listed under "Nutrition." Read the article and write a one-paragraph summary of what you read.

Food and Nutrition Services, USDA   www.fns.usda.gov/fns

On this home page, click on "Nutrition Education, then click on "Team Nutrition." Find out what Team Nutrition is and what it does.

National Eating Disorders Association

www.nationaleatingdisorders.org

At the top of this home page, click on "Information and Resources" and read one of the articles. Write a one-paragraph summary of what you read.

Federal Interagency Forum on Aging-Related Statistics

www.agingstats.gov

On the home page, click on "Older Americans Update 2008: Key Indicators of Well-Being." Then click on "Introduction Section." This Table of Contents lists 31 indicators of well-being for older adults. Write down any indicators that you think are nutrition-related.

FirstGov for Seniors

www.seniors.gov

This portal site of FirstGov includes many topics of interest to seniors. Under "Health for Seniors," click on "Staying Healthy As A Senior" and then click on "Nutrition and Aging." Find out what it tells seniors about water, fiber, and fat.

FOOD FACTS: FOOD ALLERGIES

Do you start itching whenever you eat peanuts? Does seafood cause your stomach to churn? Symptoms like those cause millions of Americans to suspect they have a *food allergy*, when indeed most probably have a *food intolerance*. True food allergies affect a relatively small percentage of people. Experts estimate that only 2 percent of adults and from 4 to 8 percent of children are truly allergic to certain foods. So what's the difference between a food intolerance and a food allergy? A food allergy involves an abnormal immune-system response. If the response doesn't involve the immune system, it is called a food intolerance. Symptoms of food intolerance may include gas, bloating, constipation, dizziness, and difficulty sleeping.

Food allergy symptoms are quite specific. Food allergens, the food components that cause allergic reactions, are usually proteins. These food protein fragments are not broken down by cooking or by stomach acids or enzymes that digest food. When an allergen passes from the mouth into the stomach, the body recognizes it as a foreign substance and produces antibodies to halt the invasion. As the body fights off the invasion, symptoms begin to appear throughout the body. The most common sites (Figure 13-19) are the mouth (swelling of the lips or tongue, itching lips), the digestive tract (stomach cramps, vomiting, diarrhea), the skin (hives, rashes, or eczema), and the airways (wheezing or breathing problems). Allergic reactions to foods usually begin within minutes to a few hours after eating.

Food intolerance may produce symptoms similar to those of food allergies, such as abdominal cramping. But whereas people with true food allergies must avoid the offending foods

altogether, people with food intolerance can often eat small amounts of the offending foods without experiencing symptoms.

Food allergies are much more common in infants and young children, who may outgrow them later. Cow's milk, peanuts, eggs, wheat, and soy are the most common food allergies in children. Children typically outgrow their allergies to milk, egg, soy, and wheat, while allergies to peanuts, tree nuts, fish, and shrimp usually are not outgrown. In general, the more severe the first allergic reaction is, the longer it takes to outgrow it. The most common foods to cause allergies in adults are shrimp, lobster, crab, and other shellfish; peanuts (one of the chief foods responsible for severe anaphylaxis, described in a moment); walnuts and other tree nuts; fish; and eggs. Adults usually do not lose their allergies.

Most cases of allergic reactions to foods are mild, but some are violent and life-threatening. The greatest danger in food allergy comes from *anaphylaxis*, also known as anaphylactic shock, a rare allergic reaction involving a number of body parts simultaneously. Like less serious allergic reactions, anaphylaxis usually occurs after a person is exposed to an allergen to which he or she was sensitized by previous exposure. That is, it does not usually occur the first time a person eats a particular food. Anaphylaxis can produce severe symptoms in as little as 5 to 15 minutes. Signs of such a reaction include difficulty breathing, swelling of the mouth and throat, a drop in blood pressure, and loss of consciousness. The sooner anaphylaxis

is treated, the greater the person's chance of surviving is.

Although any food can trigger anaphylaxis, peanuts, nuts, shellfish, milk, eggs, and fish are the most common culprits. Peanuts are the leading cause of death from food allergies. As little as $\frac{1}{5}$ to $\frac{1}{5000}$ of a teaspoon of the offending food has caused death.

There is no specific test to predict the likelihood of anaphylaxis, although allergy testing may help determine which foods a person is allergic to and provide some guidance as to the severity of the allergy. Experts advise people who are prone to anaphylaxis to carry medication—usually injectable epinephrine—with them at all times and to check the medicine's expiration date regularly.

Diagnosing a food allergy begins with a thorough medical history to identify the suspected food, the amount that must be eaten to cause a reaction, the amount of time between food consumption and the development of symptoms, how

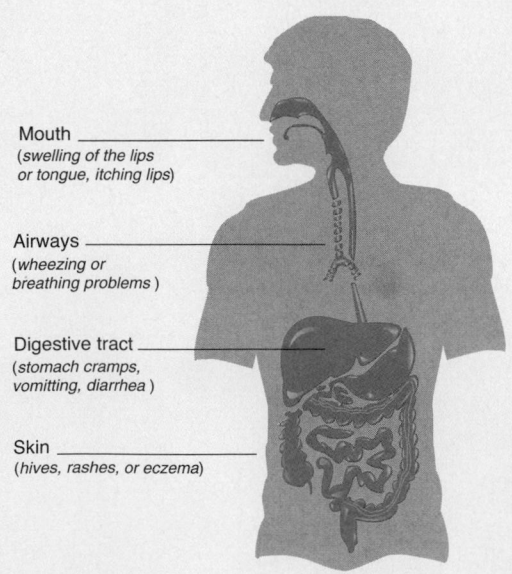

Mouth _____
(*swelling of the lips or tongue, itching lips*)

Airways _____
(*wheezing or breathing problems*)

Digestive tract _____
(*stomach cramps, vomitting, diarrhea*)

Skin _____
(*hives, rashes, or eczema*)

FIGURE 13-19:
Common sites for allergic reactions.

often the reaction occurs, and other detailed information. A complete physical examination and selected laboratory tests are conducted to rule out underlying medical conditions not related to food allergy. Several tests, such as skin testing and blood tests, are available to determine whether a person's immune system is sensitized to a specific food. In skin-prick testing, a diluted extract of the suspected food is placed on the skin, which is then scratched or punctured. If no reaction at the site occurs, the skin test is negative and an allergy to the food is unlikely. If a bump surrounded by redness (similar to a mosquito bite) forms within 15 minutes, the skin test is positive and the person may be allergic to the tested food.

Once diagnosed, most food allergies are treated by avoidance of the food allergen. Avoiding allergens is relatively easy to do at home, where you can read food labels. Manufacturers must list each allergen by its recognizable name, such as "milk" instead of "casein." However, this task becomes harder when you go out to eat. Even in the home, the most diligent label readers and ingredient checkers probably will be inadvertently exposed to proteins that elicit an allergic response at some point. For example, commercially produced foods that do not contain peanuts or eggs may contain traces of these foods through improper cleaning of utensils.

As a foodservice professional, you can do the following to help customers with food allergies.

1. Have recipe/ingredient information available for customers with food allergies. Some restaurants designate a person on each shift who knows this information well to be in charge of discussing allergy concerns with customers. To be useful, this information must be accurate and updated regularly. A manufacturer may change the ingredients in salsa, for example, and these changes are important to note.

2. If you are not sure what is in a menu item, don't give the customer false reassurances. It's better to say "I don't know—why not pick something else" than to give false information that could result in the customer's death and hundreds of thousands of dollars in fines and lawsuits.

3. Staff need training in the nature of food allergies, the foods commonly involved in anaphylactic shock, the restaurant's procedure for identifying and handling customers with food allergies, and emergency procedures.

4. Kitchen staff need training in avoiding ingredient substitutions. They also need to be trained to prepare and serve foods without contacting the foods most likely to cause anaphylactic shock: nuts, peanuts, fish, and shellfish. This means that all preparation and cooking equipment should be thoroughly cleaned after working with these foods. Remember that even minute amounts of the offending food, sometimes even a strong smell of the food, can cause anaphylactic reactions.

The prevalence of food allergy is growing and probably will continue to grow. Take it seriously when a customer asks whether there are walnuts in your Waldorf salad. Some customers who ask such questions are probably just trying to avoid an upset stomach, but for some it's a much more serious, and possibly life-threatening, matter. Since you don't know which customer really has food allergies, take every customer seriously.

Figure 13-20 lists foods to omit for specific allergies.

FIGURE 13-20: Foods to Omit for Specific Allergies

Food Allergen	Foods to Omit	Check Food Labels for
Milk	All fluid milk including buttermilk, evaporated or condensed milk, nonfat dry milk, all cheeses, all yogurts, ice cream and ice milk, butter, many margarines, most nondairy creamers and whipped toppings, hot cocoa mixes, creamed soups, many breads, crackers and cereals, pancakes, waffles, many baked goods such as cakes and cookies (check the label), fudge, instant potatoes, custards, puddings, some hot dogs and luncheon meats	Instant nonfat dry milk, nonfat milk, milk solids, whey, curds, casein, caseinate, milk, lactose-free milk, lactalbumin, lactoglobulin, sour cream, butter, cheese, cheese food, butter, milk chocolate, buttermilk

(continued)

FIGURE 13-20: Foods to Omit for Specific Allergies (*Continued*)

Food Allergen	Foods to Omit	Check Food Labels for
Eggs	All forms of eggs, most egg substitutes, eggnog, any baked goods made with eggs such as muffins and cookies or glazed with eggs such as sweet rolls, ice cream, sherbet, custards, meringues, cream pies, puddings, French toast, pancakes, waffles, some candies, some salad dressings and sandwich spreads such as mayonnaise, any sauce made with egg such as hollandaise, soufflés, any meat or potato made with egg, all pastas unless egg-free, soups made with egg noodles, soups made with stocks that were cleared with eggs, marshmallows	Eggs, albumin, globulin, livetin, ovalbumin, ovomucin, ovomucoid, ovoglobulin egg albumin, ovovitellin, vitellin
Gluten	All foods containing wheat, oats, barley, or rye as flour or in any other form, salad dressings, gravies, malted beverages, postum, soy sauce, instant puddings, distilled vinegar, beer, ale, some wines, gin, whiskey, vodka	Flour* (unless from sources noted below), modified food starch, monosodium glutamate, hydrolyzed vegetable protein, cereals, malt or cereal extracts, food starch, vegetable gum, wheat germ, wheat bran, bran, semolina, malt flavoring, distilled vinegar, emulsifiers, stabilizers

*Corn, rice, soy, arrowroot, tapioca, and potato do not contain gluten, and so they are safe to use.

HOT TOPIC: CHILDHOOD OBESITY

Obesity is a serious health problem for children and adolescents. The prevalence of obesity in U.S. children has increased from about 5 percent in 1963–1970 to 17 percent in 2003 to 2004. Data from a national nutrition survey comparing data from 1971 to 1974 to data from 2003 to 2004 show increases in overweight among all age groups:

Among preschool-aged children, aged 2–5 years, the prevalence of overweight increased from 5.0 percent to 13.9 percent.

Among school-aged children, aged 6–11 years, the prevalence of overweight increased from 4.0 percent to 18.8 percent.

Among school-aged adolescents, aged 12–19 years, the prevalence of overweight increased from 6.1 percent to 17.4 percent.

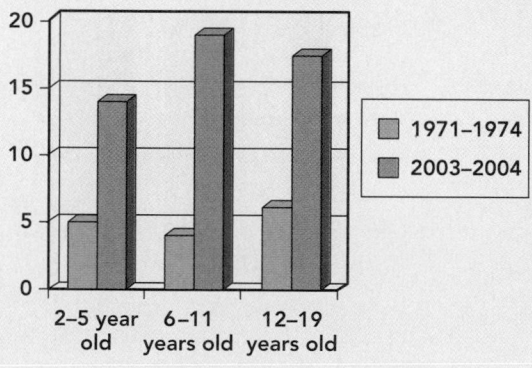

Percent of Children Who Are Overweight, 1971–1974 versus 2003–2004.

Until 2008, overweight was defined as a BMI at or above the 95th percentile using the CDC growth charts entitled "Body Mass Index-For-Age Percentiles: Boys (Girls), 2 to 20 years. (Appendix D). Overweight is now defined as a BMI between the 85th to 94th percentile for children of the same age and sex, and obesity is defined as a BMI at or above the 95th percentile.

Overweight and obese children and adolescents are at risk for health problems during their youth and as adults. For example, during their youth, they are more likely to have risk factors associated with cardiovascular disease (such as high blood pressure, high cholesterol, and type 2 diabetes) than are other children and adolescents. They are also more likely to become obese as adults. For example, one study found that approximately 80 percent of children who were overweight at aged 10–15 years were obese adults at age 25 years.

Contributing Factors

At the individual level, childhood overweight is the result of an imbalance between the kcalories a child consumes as food and beverages and the kcalories a child uses to support normal growth and development, metabolism, and physical activity. The imbalance between calories consumed and calories used can result from the influences and interactions of a number of factors, including genetic, behavioral, and environmental factors. It is the interactions among these factors—rather than any single factor— that is thought to cause overweight.

Studies indicate that certain genetic characteristics may increase an individual's susceptibility to overweight. However, this genetic susceptibility may need to exist in conjunction with contributing environmental factors (such as the ready availability of high-kcalorie foods) and behavioral factors (such as minimal physical activity) to have a significant effect on weight. Genetic factors alone can play a role in specific cases of overweight. Yet the rapid rise in the rates of overweight and obesity in the general population in recent years cannot be attributed solely to genetic factors. The genetic characteristics of the human population have not changed in the last three decades, but the prevalence of overweight has tripled among school-aged children during that time.

Because the factors that contribute to childhood overweight interact with each other, it is not possible to specify one behavior as the "cause" of overweight. However, certain behaviors can be identified as potentially contributing to an energy imbalance and, consequently, to overweight.

Energy intake: Evidence is limited on specific foods or dietary patterns that contribute to excessive energy intake in children and teens. However, large portion sizes for food and beverages, eating meals away from home, frequent snacking on energy-dense foods and consuming beverages with added sugar are often hypothesized as contributing to excess energy intake of children and teens. In the area of consuming sugar-sweetened drinks, evidence is growing to suggest an association with increased overweight in children and adolescents. Children may not compensate at meals for the calories they have consumed in sugar-sweetened drinks, although this may vary by age. Also, liquid forms of energy may be less satiating than solid forms and lead to higher caloric intake.

Physical activity: Participating in physical activity is important for children and teens as it may have beneficial effects not only on body weight, but also on blood pressure and bone strength. Physically active children are also more likely to remain physically active throughout adolescence and possibly into adulthood. Children may be spending less time engaged in physical activity during school. Daily participation in school physical education among adolescents dropped 14 percentage points over the last 13 years, from 42 percent in 1991 to 28 percent in 2003. In addition, less than one-third (28 percent) of high school students meet currently recommended levels of physical activity.

Sedentary behavior: Children spend a considerable amount of time with media. One study found that time spent watching TV, videos, DVDs, and movies averaged slightly over 3 hours per day among children aged 8–18 years. Several studies have found a positive association between the time spent viewing television and increased prevalence of overweight in children. Media use, and specifically television viewing, may displace time children spend in physical activities, contribute to increased energy consumption through excessive snacking and eating meals in front of the TV, influence children to make unhealthy food choices through exposure to food advertisements, and lower children's metabolic rate.

Home, child care, school, and community environments can influence children's behaviors related to food intake and physical activity.

Within the home: Parent-child interactions and the home environment can affect the behaviors of children and youth related to calorie intake and physical activity. Parents are role models for their children who are likely to develop habits similar to those of their parents.

Within schools: Because the majority of young people aged 5–17 years are enrolled in schools and because of the amount of time that children spend at school each day, schools provide an ideal setting for teaching children and teens to adopt healthy eating and physical activity behaviors. According to the Institute of Medicine (IOM), schools and school districts are, increasingly, implementing innovative programs that focus on improving the nutrition and increasing physical activity of students.

Within the community: The environment within communities influences access to physical activity opportunities and access to affordable and healthy foods. For example, a lack of sidewalks, safe bike paths, and parks in neighborhoods can discourage children from walking or biking to school as well as from participating in physical activity. Additionally, lack of access to affordable, healthy food choices in neighborhood food markets can be a barrier to purchasing healthy foods.

Consequences

Childhood overweight is associated with various health-related consequences. Overweight children and adolescents may experience immediate health consequences and may be at risk for weight-related health problems in adulthood.

Some consequences of childhood and adolescent overweight are psychosocial. Overweight children and adolescents are targets of jokes from other children. The psychological stress of being made fun of can cause low self-esteem which, in turn, can hinder academic and social functioning, and persist into adulthood.

Overweight children and teens have been found to have risk factors for cardiovascular disease (CVD), including high cholesterol levels, high blood pressure, and abnormal glucose tolerance. Less common health conditions associated with increased weight include asthma, type 2 diabetes, and sleep apnea.

Type 2 diabetes is increasingly being reported among children and adolescents who are overweight. While diabetes and glucose intolerance, a precursor of diabetes, are common health effects of adult obesity, only in recent years has type 2 diabetes begun to emerge as a health-related problem among children and adolescents. Onset of diabetes in children and adolescents can result in advanced complications such as CVD and kidney failure.

Tips for Parents

To help your child maintain a healthy weight, balance the kcalories your child consumes from foods and beverages with the calories your child uses through physical activity and normal growth. Remember that the goal for overweight children and teens is to reduce the rate of weight gain while allowing normal growth and development. Children and teens should *not* be placed on a weight reduction diet without the consultation of a health-care provider.

One part of balancing calories is to eat foods that provide adequate nutrition and an appropriate number of calories. You can help children learn to be aware of what they eat by developing healthy eating habits, looking for ways to make favorite dishes healthier, and reducing kcalorie-rich temptations. There's no great secret to healthy eating. To help your children and family develop healthy eating habits:

- Provide plenty of vegetables, fruits, and whole-grain products.
- Include low-fat or non-fat milk or dairy products.
- Choose lean meats, poultry, fish, lentils, and beans for protein.
- Serve reasonably sized portions.
- Encourage your family to drink lots of water.
- Limit sugar sweetened beverages.
- Limit consumption of sugar and saturated fat.
- Remember that small changes every day can lead to a recipe for success!
- Look for ways to make favorite dishes healthier. The recipes that you may prepare regularly, and that your family enjoys, with just a few changes can be healthier and just as satisfying. For new ideas about how to add more fruits and vegetables to your daily diet check out the recipe database from the FruitsandVeggiesMatter.gov.

Although everything can be enjoyed in moderation, reducing the calorie-rich temptations of high-fat, high-sugar, or salty snacks can also help your children develop healthy eating habits. Instead only allow your children to eat them sometimes, so that they truly will be treats! Here are examples of easy-to-prepare, low-fat and low-sugar treats that are 100 calories or less:

A medium-size apple
A medium-size banana
1 cup blueberries
1 cup grapes
1 cup carrots, broccoli, or bell peppers with 2 tbsp. hummus

Another part of balancing calories is to engage in an appropriate amount of physical activity and avoid too much sedentary time. In addition to being fun for children and teens, regular physical activity has many health benefits, including helping with weight management and increasing self-esteem. Children and teens should participate in at least 60 minutes of moderate intensity physical activity most days of the week, preferably daily. Remember that children imitate adults. Start adding physi-

FIGURE 13-21:
Children should eat healthily and enjoy fun physical activities.
Courtesy Jim Gathany and Centers for Disease Control and Prevention.

cal activity to your own daily routine and encourage your child to join you. Some examples of moderate intensity physical activity include:

• Brisk walking
• Playing tag
• Jumping rope
• Playing soccer
• Swimming
• Dancing

In addition to encouraging physical activity (Figure 13-21), help children avoid too much sedentary time. Although quiet time for reading and homework is fine, limit the time your children watch television, play video games, or surf the web to no more than 2 hours per day. Instead, encourage your children to find fun activities to do with family members or on their own that simply involve more activity.

APPENDIX A
Nutritive Value of Foods

Appendix A lists the nutritive values of foods commonly consumed in the United States. The data source is USDA Nutrient Database for Standard Reference, Release 13 (U.S. Department of Agriculture, Agricultural Research Service 2000). Most of the differences in values between this table and the Standard Reference are due to rounding.

Foods are grouped under the following headings:

Beverages
Dairy products
Eggs
Fats and oils
Fish and shellfish
Fruits and fruit juices
Grain products
Legumes, nuts, and seeds
Meat and meat products
Mixed dishes and fast foods
Poultry and poultry products
Soups, sauces, and gravies
Sugars and sweets
Vegetables and vegetable products
Miscellaneous items

Most of the foods listed are in ready-to-eat form. Some are basic products widely used in food preparation, such as flour, oil, and cornmeal. Most snack foods are found under Grain Products.

Measures and Weights

The approximate measure given for each food is in cups, ounces, pounds, some other well-known unit, or a piece of a specified size. The measures do not necessarily represent a serving, but the unit given may be used to calculate a variety of serving sizes. For example, nutrient values are given for 1 cup of applesauce. If the serving you consume is ½ cup, divide the values by 2 or multiply by 0.5.

For fluids, the cup measure refers to the standard measuring cup of 8 fluid ounces. The ounce is one-sixteenth of a pound unless "fluid ounce" is indicated. The weight of a fluid ounce varies according to the food. If the household measure of a food is listed as 1 ounce, the nutrients are based on a weight of 28.35 grams, rounded to 28 grams in the table. All measure weights are actual weights or are rounded to the nearest whole number.

The table gives the weight in grams for an approximate measure of each food. The weight applies to only the edible portion (the part of food normally eaten), such as the banana pulp without the peel. Some poultry descriptions provide weights for the whole part, such as a drumstick, including the skin and/or bone. Keep in mind that the nutritive values are only for the edible portions indicated in the description. For example, item 877, roasted chicken drumstick, indicates a weight of 2.9 oz. (82 grams) with the bone and skin. But note that the weight of one drumstick, meat only, is listed as 44 grams (about 1½ oz.), and so the skin and bone equal 38 grams (82 minus 44). Nutrient values are always given for the gram weight listed in the column Weight—in this case, 44 grams.

Food Values

Values are listed for water; calories; protein; total fat; saturated, monosaturated, and polyunsaturated fatty acids; cholesterol; carbohydrate; total dietary fiber; four minerals (calcium, iron, potassium, and sodium); and five vitamins (vitamin A, thiamin, riboflavin, niacin, and ascorbic acid, or vitamin C). Water content is included because the percentage of moisture is helpful for identification and comparison of many food items. For example, to identify whether the cocoa listed is powder or prepared, you could check the water value, which is much less for cocoa powder. Values are in grams or milligrams except for water, calories, and vitamin A.

Food energy is reported as calories. A calorie is the unit of measure for the amount of energy that protein, fat, and carbohydrate furnish the body. Alcohol also contributes to the calorie content of alcoholic beverages. The official unit of measurement for food energy is actually kilocalories (kcal), but the term *calories* is commonly used in its place. In fact, "calories" is printed on the food label.

Vitamin A is reported in two different units: International Units (IU) are used on food labels and in the past were used for expressing vitamin A activity; Retinol Equivalents (REs) are the units released in 1989 by the Food and Nutrition Board for expressing the RDAs for vitamin A.

Values for calories and nutrients shown in the table are the amounts in the part of the item that is customarily eaten—corn without cob, meat without bones, and peaches without pits. Nutrient values are averages for the products presented here. Values for some nutrients may vary more widely for specific food items. For example, the vitamin A content of beef liver varies widely, but the values listed in the table represent an average for that food.

In some cases, as with many vegetables, values for fat may be trace (Tr), yet there will be numerical values listed for some of the fatty acids. The values for fat have been rounded to whole numbers unless they are between 0 and 0.5; then they are listed as trace. This definition of trace also applies to the other nutrients in the table that are rounded to whole numbers. Other uses of "trace" in the table are:

- For nutrients rounded to one decimal place, values falling between 0 and 0.05 are trace.
- For nutrients rounded to two decimal places, values falling between 0 and 0.005 are trace.

Thiamin, riboflavin, niacin, and iron values in enriched white flours, white bread and rolls, cornmeals, pastas, farina, and rice are based on the current enrichment levels established by the Food and Drug Administration. Enrichment levels for riboflavin in rice were not in effect at press time and are not used in the table. Enriched flour is used in most home-prepared and commercially prepared baked goods.

Niacin values given are for preformed niacin that occurs naturally in foods. The values do not include additional niacin that may be formed in the body from tryptophan, an essential amino acid in the protein of most foods.

Nutrient values for many prepared items were calculated from the ingredients in typical recipes. Examples are biscuits, cornbread, mashed potatoes, white sauce, and many dessert foods. Adjustments were made for nutrient losses during cooking.

Nutrient values for toast and cooked vegetables do not include any added fat, either during preparation or at the table. Cutting or shredding vegetables may destroy part of some vitamins, especially ascorbic acid. Since such losses are variable, no deduction has been made.

Values for cooked dry beans, vegetables, pasta, noodles, rice, cereal, meat, poultry, and fish are without salt added. If hot cereals are prepared with salt, the sodium content ranges from about 324 to 374 mg for Malt-O-Meal, Cream of Wheat, and rolled oats. The sodium value for corn grits is about 540 mg; sodium for Wheatena is about 238 mg. Sodium values for canned vegetables labeled as "no salt added" are similar to those listed for the cooked vegetables.

The mineral contribution of water was not considered for coffee, tea, soups, sauces, or concentrated fruit juices prepared with water. Sweetened items contain sugar unless identified as artificially sweetened.

Several manufactured items—including some milk products, ready-to-eat breakfasts, imitation cream products, fruit drinks, and various mixes—are included in the table. Such foods may be fortified with one or more nutrients; the label will describe any fortification. Values for these foods may be based on products from several manufacturers, and so they may differ from the values provided by any one source. Nutrient values listed on food labels may also differ from those in the table because of rounding on labels.

Nutrient values represent meats after they have been cooked and drained of the drippings. For many cuts, two sets of values are shown: meat including lean and fat parts, and lean meat from which the outer fat layer and large fat pads have been removed either before or after cooking.

In the entries for cheeseburger and hamburger in Mixed Dishes and Fast Foods, "condiments" refers to ketchup, mustard, salt, and pepper; "vegetables" refers to lettuce, tomato, onion, and pickle; and "regular" is a 2-oz. patty and large is a 4-oz. patty (precooked weight).

Source: The Nutritive Value of Foods. USDA Home and Garden Bulletin no. 72, 2002.

Food No.	Food Description	Measure of edible portion	Weight (g)	Water (%)	Calories (kcal)	Protein (g)	Total fat (g)	Fatty acids Saturated (g)	Mono-unsaturated (g)	Poly-unsaturated (g)

Beverages

Alcoholic

Beer

| 1 | Regular | 12 fl oz | 355 | 92 | 146 | 1 | 0 | 0.0 | 0.0 | 0.0 |
| 2 | Light | 12 fl oz | 354 | 95 | 99 | 1 | 0 | 0.0 | 0.0 | 0.0 |

Gin, rum, vodka, whiskey

3	80 proof	1.5 fl oz	42	67	97	0	0	0.0	0.0	0.0
4	86 proof	1.5 fl oz	42	64	105	0	0	0.0	0.0	0.0
5	90 proof	1.5 fl oz	42	62	110	0	0	0.0	0.0	0.0
6	Liqueur, coffee, 53 proof	1.5 fl oz	52	31	175	Tr	Tr	0.1	Tr	0.1

Mixed drinks, prepared from recipe

| 7 | Daiquiri | 2 fl oz | 60 | 70 | 112 | Tr | Tr | Tr | Tr | Tr |
| 8 | Pina colada | 4.5 fl oz | 141 | 65 | 262 | 1 | 3 | 1.2 | 0.2 | 0.5 |

Wine

Dessert

| 9 | Dry | 3.5 fl oz | 103 | 80 | 130 | Tr | 0 | 0.0 | 0.0 | 0.0 |
| 10 | Sweet | 3.5 fl oz | 103 | 73 | 158 | Tr | 0 | 0.0 | 0.0 | 0.0 |

Table

| 11 | Red | 3.5 fl oz | 103 | 89 | 74 | Tr | 0 | 0.0 | 0.0 | 0.0 |
| 12 | White | 3.5 fl oz | 103 | 90 | 70 | Tr | 0 | 0.0 | 0.0 | 0.0 |

Carbonated*

| 13 | Club soda | 12 fl oz | 355 | 100 | 0 | 0 | 0 | 0.0 | 0.0 | 0.0 |
| 14 | Cola type | 12 fl oz | 370 | 89 | 152 | 0 | 0 | 0.0 | 0.0 | 0.0 |

Diet, sweetened with aspartame

15	Cola	12 fl oz	355	100	4	Tr	0	0.0	0.0	0.0
16	Other than cola or pepper type	12 fl oz	355	100	0	Tr	0	0.0	0.0	0.0
17	Ginger ale	12 fl oz	366	91	124	0	0	0.0	0.0	0.0
18	Grape	12 fl oz	372	89	160	0	0	0.0	0.0	0.0
19	Lemon lime	12 fl oz	368	90	147	0	0	0.0	0.0	0.0
20	Orange	12 fl oz	372	88	179	0	0	0.0	0.0	0.0
21	Pepper type	12 fl oz	368	89	151	0	Tr	0.3	0.0	0.0
22	Root beer	12 fl oz	370	89	152	0	0	0.0	0.0	0.0

Chocolate flavored beverage mix

| 23 | Powder | 2-3 heaping tsp | 22 | 1 | 75 | 1 | 1 | 0.4 | 0.2 | Tr |
| 24 | Prepared with milk | 1 cup | 266 | 81 | 226 | 9 | 9 | 5.5 | 2.6 | 0.3 |

Cocoa

Powder containing nonfat dry milk

| 25 | Powder | 3 heaping tsp | 28 | 2 | 102 | 3 | 1 | 0.7 | 0.4 | Tr |
| 26 | Prepared (6 oz water plus 1 oz powder) | 1 serving | 206 | 86 | 103 | 3 | 1 | 0.7 | 0.4 | Tr |

Powder containing nonfat dry milk and aspartame

| 27 | Powder | ½-oz envelope | 15 | 3 | 48 | 4 | Tr | 0.3 | 0.1 | Tr |
| 28 | Prepared (6 oz water plus 1 envelope mix) | 1 serving | 192 | 92 | 48 | 4 | Tr | 0.3 | 0.1 | Tr |

Coffee

29	Brewed	6 fl oz	178	99	4	Tr	0	Tr	0.0	Tr
30	Espresso	2 fl oz	60	98	5	Tr	Tr	0.1	0.0	0.1
31	Instant, prepared (1 rounded tsp powder plus 6 fl oz water)	6 fl oz	179	99	4	Tr	0	Tr	0.0	Tr

*Mineral content varies depending on water source.

Choles-terol (mg)	Carbo-hydrate (g)	Total dietary fiber (g)	Calcium (mg)	Iron (mg)	Potas-sium (mg)	Sodium (mg)	Vitamin A (IU)	Vitamin A (RE)	Thiamin (mg)	Ribo-flavin (mg)	Niacin (mg)	Ascor-bic acid (mg)	Food No.
0	13	0.7	18	0.1	89	18	0	0	0.02	0.09	1.6	0	1
0	5	0.0	18	0.1	64	11	0	0	0.03	0.11	1.4	0	2
0	0	0.0	0	Tr	1	Tr	0	0	Tr	Tr	Tr	0	3
0	Tr	0.0	0	Tr	1	Tr	0	0	Tr	Tr	Tr	0	4
0	0	0.0	0	Tr	1	Tr	0	0	Tr	Tr	Tr	0	5
0	24	0.0	1	Tr	16	4	0	0	Tr	0.01	0.1	0	6
0	4	0.0	2	0.1	13	3	2	0	0.01	Tr	Tr	1	7
0	40	0.8	11	0.3	100	8	3	0	0.04	0.02	0.2	7	8
0	4	0.0	8	0.2	95	9	0	0	0.02	0.02	0.2	0	9
0	12	0.0	8	0.2	95	9	0	0	0.02	0.02	0.2	0	10
0	2	0.0	8	0.4	115	5	0	0	0.01	0.03	0.1	0	11
0	1	0.0	9	0.3	82	5	0	0	Tr	0.01	0.1	0	12
0	0	0.0	18	Tr	7	75	0	0	0.00	0.00	0.0	0	13
0	38	0.0	11	0.1	4	15	0	0	0.00	0.00	0.0	0	14
0	Tr	0.0	14	0.1	0	21	0	0	0.02	0.08	0.0	0	15
0	0	0.0	14	0.1	7	21	0	0	0.00	0.00	0.0	0	16
0	32	0.0	11	0.7	4	26	0	0	0.00	0.00	0.0	0	17
0	42	0.0	11	0.3	4	56	0	0	0.00	0.00	0.0	0	18
0	38	0.0	7	0.3	4	40	0	0	0.00	0.00	0.1	0	19
0	46	0.0	19	0.2	7	45	0	0	0.00	0.00	0.0	0	20
0	38	0.0	11	0.1	4	37	0	0	0.00	0.00	0.0	0	21
0	39	0.0	19	0.2	4	48	0	0	0.00	0.00	0.0	0	22
0	20	1.3	8	0.7	128	45	4	Tr	0.01	0.03	0.1	Tr	23
32	31	1.3	301	0.8	497	165	311	77	0.10	0.43	0.3	2	24
1	22	0.3	92	0.3	202	143	4	1	0.03	0.16	0.2	1	25
2	22	2.5	97	0.4	202	148	4	0	0.03	0.16	0.2	Tr	26
1	9	0.4	86	0.7	405	168	5	1	0.04	0.21	0.2	0	27
2	8	0.4	90	0.7	405	173	4	0	0.04	0.21	0.2	0	28
0	1	0.0	4	0.1	96	4	0	0	0.00	0.00	0.4	0	29
0	1	0.0	1	0.1	69	8	0	0	Tr	0.11	3.1	Tr	30
0	1	0.0	5	0.1	64	5	0	0	0.00	Tr	0.5	0	31

Food No.	Food Description	Measure of edible portion	Weight (g)	Water (%)	Calories (kcal)	Protein (g)	Total fat (g)	Fatty acids Saturated (g)	Mono-unsaturated (g)	Poly-unsaturated (g)

Beverages (continued)

Fruit drinks, noncarbonated, canned or bottled, with added ascorbic acid

Food No.	Food Description	Measure	Weight	Water	Calories	Protein	Total fat	Sat	Mono	Poly
32	Cranberry juice cocktail.........	8 fl oz 253		86	144	0	Tr	Tr	Tr	0.1
33	Fruit punch drink	8 fl oz 248		88	117	0	0	Tr	Tr	Tr
34	Grape drink	8 fl oz 250		88	113	0	0	Tr	0.0	Tr
35	Pineapple grapefruit juice drink...................	8 fl oz 250		88	118	1	Tr	Tr	Tr	0.1
36	Pineapple orange juice drink...................	8 fl oz 250		87	125	3	0	0.0	0.0	0.0
	Lemonade									
37	Frozen concentrate, prepared	8 fl oz 248		89	99	Tr	0	Tr	Tr	Tr
	Powder, prepared with water									
38	Regular	8 fl oz 266		89	112	0	0	Tr	Tr	Tr
39	Low calorie, sweetened with aspartame...............	8 fl oz 237		99	5	0	0	0.0	0.0	0.0
	Malted milk, with added nutrients									
	Chocolate									
40	Powder..................................	3 heaping tsp 21		3	75	1	1	0.4	0.2	0.1
41	Prepared	1 cup 265		81	225	9	9	5.5	2.6	0.4
	Natural									
42	Powder..................................	4-5 heaping tsp....21		3	80	2	1	0.3	0.2	0.1
43	Prepared	1 cup 265		81	231	10	9	5.4	2.5	0.4
	Milk and milk beverages. See Dairy Products.									
44	Rice beverage, canned (RICE DREAM)	1 cup 245		89	120	Tr	2	0.2	1.3	0.3
	Soy milk. See Legumes, Nuts, and Seeds.									
	Tea									
	Brewed									
45	Black	6 fl oz 178		100	2	0	0	Tr	Tr	Tr
	Herb									
46	Chamomile	6 fl oz 178		100	2	0	0	Tr	Tr	Tr
47	Other than chamomile.......	6 fl oz 178		100	2	0	0	Tr	Tr	Tr
	Instant, powder, prepared									
48	Unsweetened	8 fl oz 237		100	2	0	0	0.0	0.0	0.0
49	Sweetened, lemon flavor	8 fl oz 259		91	88	Tr	0	Tr	Tr	Tr
50	Sweetened with saccharin, lemon flavor....................	8 fl oz 237		99	5	0	0	0.0	0.0	Tr
51	Water, tap..................................	8 fl oz 237		100	0	0	0	0.0	0.0	0.0

Dairy Products

Butter. See Fats and Oils.
Cheese
Natural

Food No.	Food Description	Measure	Weight	Water	Calories	Protein	Total fat	Sat	Mono	Poly
52	Blue	1 oz 28		42	100	6	8	5.3	2.2	0.2
53	Camembert (3 wedges per 4-oz container)...............	1 wedge 38		52	114	8	9	5.8	2.7	0.3
	Cheddar									
54	Cut pieces	1 oz 28		37	114	7	9	6.0	2.7	0.3
55		1 cubic inch 17		37	68	4	6	3.6	1.6	0.2
56	Shredded	1 cup 113		37	455	28	37	23.8	10.6	1.1

Choles-terol (mg)	Carbo-hydrate (g)	Total dietary fiber (g)	Calcium (mg)	Iron (mg)	Potas-sium (mg)	Sodium (mg)	Vitamin A		Thiamin (mg)	Ribo-flavin (mg)	Niacin (mg)	Ascor-bic acid (mg)	Food No.
							(IU)	(RE)					
0	36	0.3	8	0.4	46	5	10	0	0.02	0.02	0.1	90	32
0	30	0.2	20	0.5	62	55	35	2	0.05	0.06	0.1	73	33
0	29	0.0	8	0.4	13	15	3	0	0.01	0.01	0.1	85	34
0	29	0.3	18	0.8	153	35	88	10	0.08	0.04	0.7	115	35
0	30	0.3	13	0.7	115	8	1,328	133	0.08	0.05	0.5	56	36
0	26	0.2	7	0.4	37	7	52	5	0.01	0.05	Tr	10	37
0	29	0.0	29	0.1	3	19	0	0	0.00	Tr	0.0	34	38
0	1	0.0	50	0.1	0	7	0	0	0.00	0.00	0.0	6	39
1	18	0.2	93	3.6	251	125	2,751	824	0.64	0.86	10.7	32	40
34	29	0.3	384	3.8	620	244	3,058	901	0.73	1.26	10.9	34	41
4	17	0.1	79	3.5	203	85	2,222	668	0.62	0.75	10.2	27	42
34	28	0.0	371	3.6	572	204	2,531	742	0.71	1.14	10.4	29	43
0	25	0.0	20	0.2	69	86	5	0	0.08	0.01	1.9	1	44
0	1	0.0	0	Tr	66	5	0	0	0.00	0.02	0.0	0	45
0	Tr	0.0	4	0.1	16	2	36	4	0.02	0.01	0.0	0	46
0	Tr	0.0	4	0.1	16	2	0	0	0.02	0.01	0.0	0	47
0	Tr	0.0	5	Tr	47	7	0	0	0.00	Tr	0.1	0	48
0	22	0.0	5	0.1	49	8	0	0	0.00	0.05	0.1	0	49
0	1	0.0	5	0.1	40	24	0	0	0.00	0.01	0.1	0	50
0	0	0.0	5	Tr	0	7	0	0	0.00	0.00	0.0	0	51
21	1	0.0	150	0.1	73	396	204	65	0.01	0.11	0.3	0	52
27	Tr	0.0	147	0.1	71	320	351	96	0.01	0.19	0.2	0	53
30	Tr	0.0	204	0.2	28	176	300	79	0.01	0.11	Tr	0	54
18	Tr	0.0	123	0.1	17	105	180	47	Tr	0.06	Tr	0	55
119	1	0.0	815	0.8	111	701	1,197	314	0.03	0.42	0.1	0	56

Food No.	Food Description	Measure of edible portion	Weight (g)	Water (%)	Calories (kcal)	Pro-tein (g)	Total fat (g)	Fatty acids Satu-rated (g)	Mono-unsatu-rated (g)	Poly-unsatu-rated (g)

Dairy Products (continued)

Cheese (continued)
Natural (continued)
Cottage
Creamed (4% fat)

Food No.	Food Description	Measure	Weight (g)	Water (%)	Calories (kcal)	Pro-tein (g)	Total fat (g)	Satu-rated (g)	Mono-unsatu-rated (g)	Poly-unsatu-rated (g)
57	Large curd	1 cup	225	79	233	28	10	6.4	2.9	0.3
58	Small curd	1 cup	210	79	217	26	9	6.0	2.7	0.3
59	With fruit	1 cup	226	72	279	22	8	4.9	2.2	0.2
60	Low fat (2%)	1 cup	226	79	203	31	4	2.8	1.2	0.1
61	Low fat (1%)	1 cup	226	82	164	28	2	1.5	0.7	0.1
62	Uncreamed (dry curd, less than ½% fat)	1 cup	145	80	123	25	1	0.4	0.2	Tr
	Cream									
63	Regular	1 oz	28	54	99	2	10	6.2	2.8	0.4
64		1 tbsp	15	54	51	1	5	3.2	1.4	0.2
65	Low fat	1 tbsp	15	64	35	2	3	1.7	0.7	0.1
66	Fat free	1 tbsp	16	76	15	2	Tr	0.1	0.1	Tr
67	Feta	1 oz	28	55	75	4	6	4.2	1.3	0.2
68	Low fat, cheddar or colby	1 oz	28	63	49	7	2	1.2	0.6	0.1
	Mozzarella, made with									
69	Whole milk	1 oz	28	54	80	6	6	3.7	1.9	0.2
70	Part skim milk (low moisture)	1 oz	28	49	79	8	5	3.1	1.4	0.1
71	Muenster	1 oz	28	42	104	7	9	5.4	2.5	0.2
72	Neufchatel	1 oz	28	62	74	3	7	4.2	1.9	0.2
73	Parmesan, grated	1 cup	100	18	456	42	30	19.1	8.7	0.7
74		1 tbsp	5	18	23	2	2	1.0	0.4	Tr
75		1 oz	28	18	129	12	9	5.4	2.5	0.2
76	Provolone	1 oz	28	41	100	7	8	4.8	2.1	0.2
	Ricotta, made with									
77	Whole milk	1 cup	246	72	428	28	32	20.4	8.9	0.9
78	Part skim milk	1 cup	246	74	340	28	19	12.1	5.7	0.6
79	Swiss	1 oz	28	37	107	8	8	5.0	2.1	0.3
	Pasteurized process cheese American									
80	Regular	1 oz	28	39	106	6	9	5.6	2.5	0.3
81	Fat free	1 slice	21	57	31	5	Tr	0.1	Tr	Tr
82	Swiss	1 oz	28	42	95	7	7	4.5	2.0	0.2
83	Pasteurized process cheese food, American	1 oz	28	43	93	6	7	4.4	2.0	0.2
84	Pasteurized process cheese spread, American	1 oz	28	48	82	5	6	3.8	1.8	0.2
	Cream, sweet									
85	Half and half (cream and milk)	1 cup	242	81	315	7	28	17.3	8.0	1.0
86		1 tbsp	15	81	20	Tr	2	1.1	0.5	0.1
87	Light, coffee, or table	1 cup	240	74	469	6	46	28.8	13.4	1.7
88		1 tbsp	15	74	29	Tr	3	1.8	0.8	0.1
	Whipping, unwhipped (volume about double when whipped)									
89	Light	1 cup	239	64	699	5	74	46.2	21.7	2.1
90		1 tbsp	15	64	44	Tr	5	2.9	1.4	0.1
91	Heavy	1 cup	238	58	821	5	88	54.8	25.4	3.3
92		1 tbsp	15	58	52	Tr	6	3.5	1.6	0.2
93	Whipped topping (pressurized)	1 cup	60	61	154	2	13	8.3	3.9	0.5
94		1 tbsp	3	61	8	Tr	1	0.4	0.2	Tr

Choles-terol (mg)	Carbo-hydrate (g)	Total dietary fiber (g)	Calcium (mg)	Iron (mg)	Potas-sium (mg)	Sodium (mg)	Vitamin A (IU)	Vitamin A (RE)	Thiamin (mg)	Ribo-flavin (mg)	Niacin (mg)	Ascor-bic acid (mg)	Food No.
34	6	0.0	135	0.3	190	911	367	108	0.05	0.37	0.3	0	57
31	6	0.0	126	0.3	177	850	342	101	0.04	0.34	0.3	0	58
25	30	0.0	108	0.2	151	915	278	81	0.04	0.29	0.2	0	59
19	8	0.0	155	0.4	217	918	158	45	0.05	0.42	0.3	0	60
10	6	0.0	138	0.3	193	918	84	25	0.05	0.37	0.3	0	61
10	3	0.0	46	0.3	47	19	44	12	0.04	0.21	0.2	0	62
31	1	0.0	23	0.3	34	84	405	108	Tr	0.06	Tr	0	63
16	Tr	0.0	12	0.2	17	43	207	55	Tr	0.03	Tr	0	64
8	1	0.0	17	0.3	25	44	108	33	Tr	0.04	Tr	0	65
1	1	0.0	29	Tr	25	85	145	44	0.01	0.03	Tr	0	66
25	1	0.0	140	0.2	18	316	127	36	0.04	0.24	0.3	0	67
6	1	0.0	118	0.1	19	174	66	18	Tr	0.06	Tr	0	68
22	1	0.0	147	0.1	19	106	225	68	Tr	0.07	Tr	0	69
15	1	0.0	207	0.1	27	150	199	54	0.01	0.10	Tr	0	70
27	Tr	0.0	203	0.1	38	178	318	90	Tr	0.09	Tr	0	71
22	1	0.0	21	0.1	32	113	321	85	Tr	0.06	Tr	0	72
79	4	0.0	1,376	1.0	107	1,862	701	173	0.05	0.39	0.3	0	73
4	Tr	0.0	69	Tr	5	93	35	9	Tr	0.02	Tr	0	74
22	1	0.0	390	0.3	30	528	199	49	0.01	0.11	0.1	0	75
20	1	0.0	214	0.1	39	248	231	75	0.01	0.09	Tr	0	76
124	7	0.0	509	0.9	257	207	1,205	330	0.03	0.48	0.3	0	77
76	13	0.0	669	1.1	308	307	1,063	278	0.05	0.46	0.2	0	78
26	1	0.0	272	Tr	31	74	240	72	0.01	0.10	Tr	0	79
27	Tr	0.0	174	0.1	46	406	343	82	0.01	0.10	Tr	0	80
2	3	0.0	145	0.1	60	321	308	92	0.01	0.10	Tr	0	81
24	1	0.0	219	0.2	61	388	229	65	Tr	0.08	Tr	0	82
18	2	0.0	163	0.2	79	337	259	62	0.01	0.13	Tr	0	83
16	2	0.0	159	0.1	69	381	223	54	0.01	0.12	Tr	0	84
89	10	0.0	254	0.2	314	98	1,050	259	0.08	0.36	0.2	2	85
6	1	0.0	16	Tr	19	6	65	16	0.01	0.02	Tr	Tr	86
159	9	0.0	231	0.1	292	95	1,519	437	0.08	0.36	0.1	2	87
10	1	0.0	14	Tr	18	6	95	27	Tr	0.02	Tr	Tr	88
265	7	0.0	166	0.1	231	82	2,694	705	0.06	0.30	0.1	1	89
17	Tr	0.0	10	Tr	15	5	169	44	Tr	0.02	Tr	Tr	90
326	7	0.0	154	0.1	179	89	3,499	1,002	0.05	0.26	0.1	1	91
21	Tr	0.0	10	Tr	11	6	221	63	Tr	0.02	Tr	Tr	92
46	7	0.0	61	Tr	88	78	506	124	0.02	0.04	Tr	0	93
2	Tr	0.0	3	Tr	4	4	25	6	Tr	Tr	Tr	0	94

Dairy Products (continued)

Food No.	Food Description	Measure of edible portion	Weight (g)	Water (%)	Calories (kcal)	Protein (g)	Total fat (g)	Saturated (g)	Mono-unsaturated (g)	Poly-unsaturated (g)
	Cream, sour									
95	Regular	1 cup	230	71	493	7	48	30.0	13.9	1.8
96		1 tbsp	12	71	26	Tr	3	1.6	0.7	0.1
97	Reduced fat	1 tbsp	15	80	20	Tr	2	1.1	0.5	0.1
98	Fat free	1 tbsp	16	81	12	Tr	0	0.0	0.0	0.0
	Cream product, imitation (made with vegetable fat)									
	Sweet									
	Creamer									
99	Liquid (frozen)	1 tbsp	15	77	20	Tr	1	0.3	1.1	Tr
100	Powdered	1 tsp	2	2	11	Tr	1	0.7	Tr	Tr
	Whipped topping									
101	Frozen	1 cup	75	50	239	1	19	16.3	1.2	0.4
102		1 tbsp	4	50	13	Tr	1	0.9	0.1	Tr
103	Powdered, prepared with whole milk	1 cup	80	67	151	3	10	8.5	0.7	0.2
104		1 tbsp	4	67	8	Tr	Tr	0.4	Tr	Tr
105	Pressurized	1 cup	70	60	184	1	16	13.2	1.3	0.2
106		1 tbsp	4	60	11	Tr	1	0.8	0.1	Tr
107	Sour dressing (filled cream type, nonbutterfat)	1 cup	235	75	417	8	39	31.2	4.6	1.1
108		1 tbsp	12	75	21	Tr	2	1.6	0.2	0.1
	Frozen dessert									
	Frozen yogurt, soft serve									
109	Chocolate	½ cup	72	64	115	3	4	2.6	1.3	0.2
110	Vanilla	½ cup	72	65	114	3	4	2.5	1.1	0.2
	Ice cream									
	Regular									
111	Chocolate	½ cup	66	56	143	3	7	4.5	2.1	0.3
112	Vanilla	½ cup	66	61	133	2	7	4.5	2.1	0.3
113	Light (50% reduced fat), vanilla	½ cup	66	68	92	3	3	1.7	0.8	0.1
114	Premium low fat, chocolate	½ cup	72	61	113	3	2	1.0	0.6	0.1
115	Rich, vanilla	½ cup	74	57	178	3	12	7.4	3.4	0.4
116	Soft serve, french vanilla	½ cup	86	60	185	4	11	6.4	3.0	0.4
117	Sherbet, orange	½ cup	74	66	102	1	1	0.9	0.4	0.1
	Milk									
	Fluid, no milk solids added									
118	Whole (3.3% fat)	1 cup	244	88	150	8	8	5.1	2.4	0.3
119	Reduced fat (2%)	1 cup	244	89	121	8	5	2.9	1.4	0.2
120	Lowfat (1%)	1 cup	244	90	102	8	3	1.6	0.7	0.1
121	Nonfat (skim)	1 cup	245	91	86	8	Tr	0.3	0.1	Tr
122	Buttermilk	1 cup	245	90	99	8	2	1.3	0.6	0.1
	Canned									
123	Condensed, sweetened	1 cup	306	27	982	24	27	16.8	7.4	1.0
	Evaporated									
124	Whole milk	1 cup	252	74	339	17	19	11.6	5.9	0.6
125	Skim milk	1 cup	256	79	199	19	1	0.3	0.2	Tr
	Dried									
126	Buttermilk	1 cup	120	3	464	41	7	4.3	2.0	0.3
127	Nonfat, instant, with added vitamin A	1 cup	68	4	244	24	Tr	0.3	0.1	Tr
	Milk beverage									
	Chocolate milk (commercial)									
128	Whole	1 cup	250	82	208	8	8	5.3	2.5	0.3
129	Reduced fat (2%)	1 cup	250	84	179	8	5	3.1	1.5	0.2
130	Lowfat (1%)	1 cup	250	85	158	8	3	1.5	0.8	0.1

*The vitamin A values listed for imitation sweet cream products are mostly from beta-carotene added for coloring.

Choles-terol (mg)	Carbo-hydrate (g)	Total dietary fiber (g)	Calcium (mg)	Iron (mg)	Potas-sium (mg)	Sodium (mg)	Vitamin A (IU)	Vitamin A (RE)	Thiamin (mg)	Ribo-flavin (mg)	Niacin (mg)	Ascor-bic acid (mg)	Food No.
102	10	0.0	268	0.1	331	123	1,817	449	0.08	0.34	0.2	2	95
5	1	0.0	14	Tr	17	6	95	23	Tr	0.02	Tr	Tr	96
6	1	0.0	16	Tr	19	6	68	17	0.01	0.02	Tr	Tr	97
1	2	0.0	20	0.0	21	23	100	13	0.01	0.02	Tr	0	98
0	2	0.0	1	Tr	29	12	13*	1*	0.00	0.00	0.0	0	99
0	1	0.0	Tr	Tr	16	4	4	Tr	0.00	Tr	0.0	0	100
0	17	0.0	5	0.1	14	19	646*	65*	0.00	0.00	0.0	0	101
0	1	0.0	Tr	Tr	1	1	34*	3*	0.00	0.00	0.0	0	102
8	13	0.0	72	Tr	121	53	289*	39*	0.02	0.09	Tr	1	103
Tr	1	0.0	4	Tr	6	3	14*	2*	Tr	Tr	Tr	Tr	104
0	11	0.0	4	Tr	13	43	331*	33*	0.00	0.00	0.0	0	105
0	1	0.0	Tr	Tr	1	2	19*	2*	0.00	0.00	0.0	0	106
13	11	0.0	266	0.1	380	113	24	5	0.09	0.38	0.2	2	107
1	1	0.0	14	Tr	19	6	1	Tr	Tr	0.02	Tr	Tr	108
4	18	1.6	106	0.9	188	71	115	31	0.03	0.15	0.2	Tr	109
1	17	0.0	103	0.2	152	63	153	41	0.03	0.16	0.2	1	110
22	19	0.8	72	0.6	164	50	275	79	0.03	0.13	0.1	Tr	111
29	16	0.0	84	0.1	131	53	270	77	0.03	0.16	0.1	Tr	112
9	15	0.0	92	0.1	139	56	109	31	0.04	0.17	0.1	1	113
7	22	0.7	107	0.4	179	50	163	47	0.02	0.13	0.1	1	114
45	17	0.0	87	Tr	118	41	476	136	0.03	0.12	0.1	1	115
78	19	0.0	113	0.2	152	52	464	132	0.04	0.16	0.1	1	116
4	22	0.0	40	0.1	71	34	56	10	0.02	0.06	Tr	2	117
33	11	0.0	291	0.1	370	120	307	76	0.09	0.40	0.2	2	118
18	12	0.0	297	0.1	377	122	500	139	0.10	0.40	0.2	2	119
10	12	0.0	300	0.1	381	123	500	144	0.10	0.41	0.2	2	120
4	12	0.0	302	0.1	406	126	500	149	0.09	0.34	0.2	2	121
9	12	0.0	285	0.1	371	257	81	20	0.08	0.38	0.1	2	122
104	166	0.0	868	0.6	1,136	389	1,004	248	0.28	1.27	0.6	8	123
74	25	0.0	657	0.5	764	267	612	136	0.12	0.80	0.5	5	124
9	29	0.0	741	0.7	849	294	1,004	300	0.12	0.79	0.4	3	125
83	59	0.0	1,421	0.4	1,910	621	262	65	0.47	1.89	1.1	7	126
12	35	0.0	837	0.2	1,160	373	1,612	483	0.28	1.19	0.6	4	127
31	26	2.0	280	0.6	417	149	303	73	0.09	0.41	0.3	2	128
17	26	1.3	284	0.6	422	151	500	143	0.09	0.41	0.3	2	129
7	26	1.3	287	0.6	426	152	500	148	0.10	0.42	0.3	2	130

Food No.	Food Description	Measure of edible portion	Weight (g)	Water (%)	Calories (kcal)	Protein (g)	Total fat (g)	Fatty acids Saturated (g)	Fatty acids Monounsaturated (g)	Fatty acids Polyunsaturated (g)
Dairy Products (continued)										
	Milk beverage (continued)									
131	Eggnog (commercial)	1 cup	254	74	342	10	19	11.3	5.7	0.9
	Milk shake, thick									
132	Chocolate	10.6 fl oz	300	72	356	9	8	5.0	2.3	0.3
133	Vanilla	11 fl oz	313	74	350	12	9	5.9	2.7	0.4
	Sherbet. See Dairy Products, frozen dessert.									
	Yogurt									
	With added milk solids									
	Made with lowfat milk									
134	Fruit flavored	8-oz container	227	74	231	10	2	1.6	0.7	0.1
135	Plain	8-oz container	227	85	144	12	4	2.3	1.0	0.1
	Made with nonfat milk									
136	Fruit flavored	8-oz container	227	75	213	10	Tr	0.3	0.1	Tr
137	Plain	8-oz container	227	85	127	13	Tr	0.3	0.1	Tr
	Without added milk solids									
138	Made with whole milk, plain	8-oz container	227	88	139	8	7	4.8	2.0	0.2
139	Made with nonfat milk, low calorie sweetener, vanilla or lemon flavor	8-oz container	227	87	98	9	Tr	0.3	0.1	Tr
Eggs										
	Egg									
	Raw									
140	Whole	1 medium	44	75	66	5	4	1.4	1.7	0.6
141		1 large	50	75	75	6	5	1.6	1.9	0.7
142		1 extra large	58	75	86	7	6	1.8	2.2	0.8
143	White	1 large	33	88	17	4	0	0.0	0.0	0.0
144	Yolk	1 large	17	49	59	3	5	1.6	1.9	0.7
	Cooked, whole									
145	Fried, in margarine, with salt	1 large	46	69	92	6	7	1.9	2.7	1.3
146	Hard cooked, shell removed	1 large	50	75	78	6	5	1.6	2.0	0.7
147		1 cup, chopped	136	75	211	17	14	4.4	5.5	1.9
148	Poached, with salt	1 large	50	75	75	6	5	1.5	1.9	0.7
149	Scrambled, in margarine, with whole milk, salt	1 large	61	73	101	7	7	2.2	2.9	1.3
150	Egg substitute, liquid	¼ cup	63	83	53	8	2	0.4	0.6	1.0
Fats and Oils										
	Butter (4 sticks per lb)									
151	Salted	1 stick	113	16	813	1	92	57.3	26.6	3.4
152		1 tbsp	14	16	102	Tr	12	7.2	3.3	0.4
153		1 tsp	5	16	36	Tr	4	2.5	1.2	0.2
154	Unsalted	1 stick	113	18	813	1	92	57.3	26.6	3.4
155	Lard	1 cup	205	0	1,849	0	205	80.4	92.5	23.0
156		1 tbsp	13	0	115	0	13	5.0	5.8	1.4
	Margarine, vitamin A-fortified, salt added									
	Regular (about 80% fat)									
157	Hard (4 sticks per lb)	1 stick	113	16	815	1	91	17.9	40.6	28.8
158		1 tbsp	14	16	101	Tr	11	2.2	5.0	3.6
159		1 tsp	5	16	34	Tr	4	0.7	1.7	1.2
160	Soft	1 cup	227	16	1,626	2	183	31.3	64.7	78.5
161		1 tsp	5	16	34	Tr	4	0.6	1.3	1.6

Choles-terol (mg)	Carbo-hydrate (g)	Total dietary fiber (g)	Calcium (mg)	Iron (mg)	Potas-sium (mg)	Sodium (mg)	Vitamin A (IU)	Vitamin A (RE)	Thiamin (mg)	Ribo-flavin (mg)	Niacin (mg)	Ascor-bic acid (mg)	Food No.
149	34	0.0	330	0.5	420	138	894	203	0.09	0.48	0.3	4	131
32	63	0.9	396	0.9	672	333	258	63	0.14	0.67	0.4	0	132
37	56	0.0	457	0.3	572	299	357	88	0.09	0.61	0.5	0	133
10	43	0.0	345	0.2	442	133	104	25	0.08	0.40	0.2	1	134
14	16	0.0	415	0.2	531	159	150	36	0.10	0.49	0.3	2	135
5	43	0.0	345	0.2	440	132	16	5	0.09	0.41	0.2	2	136
4	17	0.0	452	0.2	579	174	16	5	0.11	0.53	0.3	2	137
29	11	0.0	274	0.1	351	105	279	68	0.07	0.32	0.2	1	138
5	17	0.0	325	0.3	402	134	0	0	0.08	0.37	0.2	2	139
187	1	0.0	22	0.6	53	55	279	84	0.03	0.22	Tr	0	140
213	1	0.0	25	0.7	61	63	318	96	0.03	0.25	Tr	0	141
247	1	0.0	28	0.8	70	73	368	111	0.04	0.29	Tr	0	142
0	Tr	0.0	2	Tr	48	55	0	0	Tr	0.15	Tr	0	143
213	Tr	0.0	23	0.6	16	7	323	97	0.03	0.11	Tr	0	144
211	1	0.0	25	0.7	61	162	394	114	0.03	0.24	Tr	0	145
212	1	0.0	25	0.6	63	62	280	84	0.03	0.26	Tr	0	146
577	2	0.0	68	1.6	171	169	762	228	0.09	0.70	0.1	0	147
212	1	0.0	25	0.7	60	140	316	95	0.02	0.22	Tr	0	148
215	1	0.0	43	0.7	84	171	416	119	0.03	0.27	Tr	Tr	149
1	Tr	0.0	33	1.3	208	112	1,361	136	0.07	0.19	0.1	0	150
248	Tr	0.0	27	0.2	29	937	3,468	855	0.01	0.04	Tr	0	151
31	Tr	0.0	3	Tr	4	117	434	107	Tr	Tr	Tr	0	152
11	Tr	0.0	1	Tr	1	41	153	38	Tr	Tr	Tr	0	153
248	Tr	0.0	27	0.2	29	12	3,468	855	0.01	0.04	Tr	0	154
195	0	0.0	Tr	0.0	Tr	Tr	0	0	0.00	0.00	0.0	0	155
12	0	0.0	Tr	0.0	Tr	Tr	0	0	0.00	0.00	0.0	0	156
0	1	0.0	34	0.1	48	1,070	4,050	906	0.01	0.04	Tr	Tr	157
0	Tr	0.0	4	Tr	6	132	500	112	Tr	0.01	Tr	Tr	158
0	Tr	0.0	1	Tr	2	44	168	38	Tr	Tr	Tr	Tr	159
0	1	0.0	60	0.0	86	2,449	8,106	1,814	0.02	0.07	Tr	Tr	160
0	Tr	0.0	1	0.0	2	51	168	38	Tr	Tr	Tr	Tr	161

Food No.	Food Description	Measure of edible portion	Weight (g)	Water (%)	Calories (kcal)	Pro-tein (g)	Total fat (g)	Satu-rated (g)	Mono-unsatu-rated (g)	Poly-unsatu-rated (g)
									Fatty acids	

Fats and Oils (continued)

Margarine, vitamin A-fortified,
  salt added (continued)
  Spread (about 60% fat)

Food No.	Food Description	Measure of edible portion	Weight (g)	Water (%)	Calories (kcal)	Pro-tein (g)	Total fat (g)	Saturated (g)	Mono-unsaturated (g)	Poly-unsaturated (g)
162	Hard (4 sticks per lb)	1 stick ...115		37	621	1	70	16.2	29.9	20.8
163		1 tbsp ...14		37	76	Tr	9	2.0	3.6	2.5
164		1 tsp ...5		37	26	Tr	3	0.7	1.2	0.9
165	Soft	1 cup ...229		37	1,236	1	139	29.3	72.1	31.6
166		1 tsp ...5		37	26	Tr	3	0.6	1.5	0.7
167	Spread (about 40% fat)	1 cup ...232		58	801	1	90	17.9	36.4	32.0
168		1 tsp ...5		58	17	Tr	2	0.4	0.8	0.7
169	Margarine butter blend	1 stick ...113		16	811	1	91	32.1	37.0	18.0
170		1 tbsp ...14		16	102	Tr	11	4.0	4.7	2.3
	Oils, salad or cooking									
171	Canola	1 cup ...218		0	1,927	0	218	15.5	128.4	64.5
172		1 tbsp ...14		0	124	0	14	1.0	8.2	4.1
173	Corn	1 cup ...218		0	1,927	0	218	27.7	52.8	128.0
174		1 tbsp ...14		0	120	0	14	1.7	3.3	8.0
175	Olive	1 cup ...216		0	1,909	0	216	29.2	159.2	18.1
176		1 tbsp ...14		0	119	0	14	1.8	9.9	1.1
177	Peanut	1 cup ...216		0	1,909	0	216	36.5	99.8	69.1
178		1 tbsp ...14		0	119	0	14	2.3	6.2	4.3
179	Safflower, high oleic	1 cup ...218		0	1,927	0	218	13.5	162.7	31.3
180		1 tbsp ...14		0	120	0	14	0.8	10.2	2.0
181	Sesame	1 cup ...218		0	1,927	0	218	31.0	86.5	90.9
182		1 tbsp ...14		0	120	0	14	1.9	5.4	5.7
183	Soybean, hydrogenated	1 cup ...218		0	1,927	0	218	32.5	93.7	82.0
184		1 tbsp ...14		0	120	0	14	2.0	5.8	5.1
185	Soybean, hydrogenated and cottonseed oil blend	1 cup ...218		0	1,927	0	218	39.2	64.3	104.9
186		1 tbsp ...14		0	120	0	14	2.4	4.0	6.5
187	Sunflower	1 cup ...218		0	1,927	0	218	22.5	42.5	143.2
188		1 tbsp ...14		0	120	0	14	1.4	2.7	8.9
	Salad dressings									
	Commercial									
	Blue cheese									
189	Regular	1 tbsp ...15		32	77	1	8	1.5	1.9	4.3
190	Low calorie	1 tbsp ...15		80	15	1	1	0.4	0.3	0.4
	Caesar									
191	Regular	1 tbsp ...15		34	78	Tr	8	1.3	2.0	4.8
192	Low calorie	1 tbsp ...15		73	17	Tr	1	0.1	0.2	0.4
	French									
193	Regular	1 tbsp ...16		38	67	Tr	6	1.5	1.2	3.4
194	Low calorie	1 tbsp ...16		69	22	Tr	1	0.1	0.2	0.6
	Italian									
195	Regular	1 tbsp ...15		38	69	Tr	7	1.0	1.6	4.1
196	Low calorie	1 tbsp ...15		82	16	Tr	1	0.2	0.3	0.9
	Mayonnaise									
197	Regular	1 tbsp ...14		15	99	Tr	11	1.6	3.1	5.7
198	Light, cholesterol free	1 tbsp ...15		56	49	Tr	5	0.7	1.1	2.8
199	Fat free	1 tbsp ...16		84	12	0	Tr	0.1	0.1	0.2
	Russian									
200	Regular	1 tbsp ...15		35	76	Tr	8	1.1	1.8	4.5
201	Low calorie	1 tbsp ...16		65	23	Tr	1	0.1	0.1	0.4
	Thousand island									
202	Regular	1 tbsp ...16		46	59	Tr	6	0.9	1.3	3.1
203	Low calorie	1 tbsp ...15		69	24	Tr	2	0.2	0.4	0.9

Cholesterol (mg)	Carbohydrate (g)	Total dietary fiber (g)	Calcium (mg)	Iron (mg)	Potassium (mg)	Sodium (mg)	Vitamin A (IU)	Vitamin A (RE)	Thiamin (mg)	Riboflavin (mg)	Niacin (mg)	Ascorbic acid (mg)	Food No.
0	0	0.0	24	0.0	34	1,143	4,107	919	0.01	0.03	Tr	Tr	162
0	0	0.0	3	0.0	4	139	500	112	Tr	Tr	Tr	Tr	163
0	0	0.0	1	0.0	1	48	171	38	Tr	Tr	Tr	Tr	164
0	0	0.0	48	0.0	68	2,276	8,178	1,830	0.02	0.06	Tr	Tr	165
0	0	0.0	1	0.0	1	48	171	38	Tr	Tr	Tr	Tr	166
0	1	0.0	41	0.0	59	2,226	8,285	1,854	0.01	0.05	Tr	Tr	167
0	Tr	0.0	1	0.0	1	46	171	38	Tr	Tr	Tr	Tr	168
99	1	0.0	32	0.1	41	1,014	4,035	903	0.01	0.04	Tr	Tr	169
12	Tr	0.0	4	Tr	5	127	507	113	Tr	Tr	Tr	Tr	170
0	0	0.0	0	0.0	0	0	0	0	0.00	0.00	0.0	0	171
0	0	0.0	0	0.0	0	0	0	0	0.00	0.00	0.0	0	172
0	0	0.0	0	0.0	0	0	0	0	0.00	0.00	0.0	0	173
0	0	0.0	0	0.0	0	0	0	0	0.00	0.00	0.0	0	174
0	0	0.0	Tr	0.8	0	Tr	0	0	0.00	0.00	0.0	0	175
0	0	0.0	Tr	0.1	0	Tr	0	0	0.00	0.00	0.0	0	176
0	0	0.0	Tr	0.1	Tr	Tr	0	0	0.00	0.00	0.0	0	177
0	0	0.0	Tr	Tr	Tr	Tr	0	0	0.00	0.00	0.0	0	178
0	0	0.0	0	0.0	0	0	0	0	0.00	0.00	0.0	0	179
0	0	0.0	0	0.0	0	0	0	0	0.00	0.00	0.0	0	180
0	0	0.0	0	0.0	0	0	0	0	0.00	0.00	0.0	0	181
0	0	0.0	0	0.0	0	0	0	0	0.00	0.00	0.0	0	182
0	0	0.0	0	0.0	0	0	0	0	0.00	0.00	0.0	0	183
0	0	0.0	0	0.0	0	0	0	0	0.00	0.00	0.0	0	184
0	0	0.0	0	0.0	0	0	0	0	0.00	0.00	0.0	0	185
0	0	0.0	0	0.0	0	0	0	0	0.00	0.00	0.0	0	186
0	0	0.0	0	0.0	0	0	0	0	0.00	0.00	0.0	0	187
0	0	0.0	0	0.0	0	0	0	0	0.00	0.00	0.0	0	188
3	1	0.0	12	Tr	6	167	32	10	Tr	0.02	Tr	Tr	189
Tr	Tr	0.0	14	0.1	1	184	2	Tr	Tr	0.02	Tr	Tr	190
Tr	Tr	Tr	4	Tr	4	158	3	Tr	Tr	Tr	Tr	0	191
Tr	3	Tr	4	Tr	4	162	3	Tr	Tr	Tr	Tr	0	192
0	3	0.0	2	0.1	12	214	203	20	Tr	Tr	Tr	0	193
0	4	0.0	2	0.1	13	128	212	21	0.00	0.00	0.0	0	194
0	1	0.0	1	Tr	2	116	11	4	Tr	Tr	Tr	0	195
1	1	Tr	Tr	Tr	2	118	0	0	0.00	0.00	0.0	0	196
8	Tr	0.0	2	0.1	5	78	39	12	0.00	0.00	Tr	0	197
0	1	0.0	0	0.0	10	107	18	2	0.00	0.00	0.0	0	198
0	2	0.6	0	0.0	15	190	0	0	0.00	0.00	0.0	0	199
3	2	0.0	3	0.1	24	133	106	32	0.01	0.01	0.1	1	200
1	4	Tr	3	0.1	26	141	9	3	Tr	Tr	Tr	1	201
4	2	0.0	2	0.1	18	109	50	15	Tr	Tr	Tr	0	202
2	2	0.2	2	0.1	17	153	49	15	Tr	Tr	Tr	0	203

Food No.	Food Description	Measure of edible portion	Weight (g)	Water (%)	Calories (kcal)	Pro-tein (g)	Total fat (g)	Fatty acids Satu-rated (g)	Mono-unsatu-rated (g)	Poly-unsatu-rated (g)

Fats and Oils (continued)

Salad dressings (continued)
Prepared from home recipe

204	Cooked, made with margarine	1 tbsp	16	69	25	1	2	0.5	0.6	0.3
205	French	1 tbsp	14	24	88	Tr	10	1.8	2.9	4.7
206	Vinegar and oil	1 tbsp	16	47	70	0	8	1.4	2.3	3.8
207	Shortening (hydrogenated soybean and cottonseed oils)	1 cup	205	0	1,812	0	205	51.3	91.2	53.5
208		1 tbsp	13	0	113	0	13	3.2	5.7	3.3

Fish and Shellfish

209	Catfish, breaded, fried	3 oz	85	59	195	15	11	2.8	4.8	2.8
	Clam									
210	Raw, meat only	3 oz	85	82	63	11	1	0.1	0.1	0.2
211		1 medium	15	82	11	2	Tr	Tr	Tr	Tr
212	Breaded, fried	¾ cup	115	29	451	13	26	6.6	11.4	6.8
213	Canned, drained solids	3 oz	85	64	126	22	2	0.2	0.1	0.5
214		1 cup	160	64	237	41	3	0.3	0.3	0.9
	Cod									
215	Baked or broiled	3 oz	85	76	89	20	1	0.1	0.1	0.3
216		1 fillet	90	76	95	21	1	0.1	0.1	0.3
217	Canned, solids and liquid	3 oz	85	76	89	19	1	0.1	0.1	0.2
	Crab									
	Alaska king									
218	Steamed	1 leg	134	78	130	26	2	0.2	0.2	0.7
219		3 oz	85	78	82	16	1	0.1	0.2	0.5
220	Imitation, from surimi	3 oz	85	74	87	10	1	0.2	0.2	0.6
	Blue									
221	Steamed	3 oz	85	77	87	17	2	0.2	0.2	0.6
222	Canned crabmeat	1 cup	135	76	134	28	2	0.3	0.3	0.6
223	Crab cake, with egg, onion, fried in margarine	1 cake	60	71	93	12	5	0.9	1.7	1.4
224	Fish fillet, battered or breaded, fried	1 fillet	91	54	211	13	11	2.6	2.3	5.7
225	Fish stick and portion, breaded, frozen, reheated	1 stick (4" x 1" x ½")	28	46	76	4	3	0.9	1.4	0.9
226		1 portion (4" x 2" x ½")	57	46	155	9	7	1.8	2.9	1.8
227	Flounder or sole, baked or broiled	3 oz	85	73	99	21	1	0.3	0.2	0.5
228		1 fillet	127	73	149	31	2	0.5	0.3	0.8
229	Haddock, baked or broiled	3 oz	85	74	95	21	1	0.1	0.1	0.3
230		1 fillet	150	74	168	36	1	0.3	0.2	0.5
231	Halibut, baked or broiled	3 oz	85	72	119	23	2	0.4	0.8	0.8
232		½ fillet	159	72	223	42	5	0.7	1.5	1.5
233	Herring, pickled	3 oz	85	55	223	12	15	2.0	10.2	1.4
234	Lobster, steamed	3 oz	85	76	83	17	1	0.1	0.1	0.1
235	Ocean perch, baked or broiled	3 oz	85	73	103	20	2	0.3	0.7	0.5
236		1 fillet	50	73	61	12	1	0.2	0.4	0.3
	Oyster									
237	Raw, meat only	1 cup	248	85	169	17	6	1.9	0.8	2.4
238		6 medium	84	85	57	6	2	0.6	0.3	0.8
239	Breaded, fried	3 oz	85	65	167	7	11	2.7	4.0	2.8
240	Pollock, baked or broiled	3 oz	85	74	96	20	1	0.2	0.1	0.4
241		1 fillet	60	74	68	14	1	0.1	0.1	0.3
242	Rockfish, baked or broiled	3 oz	85	73	103	20	2	0.4	0.4	0.5
243		1 fillet	149	73	180	36	3	0.7	0.7	0.9

Choles-terol (mg)	Carbo-hydrate (g)	Total dietary fiber (g)	Calcium (mg)	Iron (mg)	Potas-sium (mg)	Sodium (mg)	Vitamin A (IU)	Vitamin A (RE)	Thiamin (mg)	Ribo-flavin (mg)	Niacin (mg)	Ascor-bic acid (mg)	Food No.
9	2	0.0	13	0.1	19	117	66	20	0.01	0.02	Tr	Tr	204
0	Tr	0.0	1	Tr	3	92	72	22	Tr	Tr	Tr	Tr	205
0	Tr	0.0	0	0.0	1	Tr	0	0	0.00	0.00	0.0	0	206
0	0	0.0	0	0.0	0	0	0	0	0.00	0.00	0.0	0	207
0	0	0.0	0	0.0	0	0	0	0	0.00	0.00	0.0	0	208
69	7	0.6	37	1.2	289	238	24	7	0.06	0.11	1.9	0	209
29	2	0.0	39	11.9	267	48	255	77	0.07	0.18	1.5	11	210
5	Tr	0.0	7	2.0	46	8	44	13	0.01	0.03	0.3	2	211
87	39	0.3	21	3.0	266	834	122	37	0.21	0.26	2.9	0	212
57	4	0.0	78	23.8	534	95	485	145	0.13	0.36	2.9	19	213
107	8	0.0	147	44.7	1,005	179	912	274	0.24	0.68	5.4	35	214
40	0	0.0	8	0.3	439	77	27	9	0.02	0.04	2.1	3	215
42	0	0.0	8	0.3	465	82	29	9	0.02	0.05	2.2	3	216
47	0	0.0	18	0.4	449	185	39	12	0.07	0.07	2.1	1	217
71	0	0.0	79	1.0	351	1,436	39	12	0.07	0.07	1.8	10	218
45	0	0.0	50	0.6	223	911	25	8	0.05	0.05	1.1	6	219
17	9	0.0	11	0.3	77	715	56	17	0.03	0.02	0.2	0	220
85	0	0.0	88	0.8	275	237	5	2	0.09	0.04	2.8	3	221
120	0	0.0	136	1.1	505	450	7	3	0.11	0.11	1.8	4	222
90	Tr	0.0	63	0.6	194	198	151	49	0.05	0.05	1.7	2	223
31	15	0.5	16	1.9	291	484	35	11	0.10	0.10	1.9	0	224
31	7	0.0	6	0.2	73	163	30	9	0.04	0.05	0.6	0	225
64	14	0.0	11	0.4	149	332	60	18	0.07	0.10	1.2	0	226
58	0	0.0	15	0.3	292	89	32	9	0.07	0.10	1.9	0	227
86	0	0.0	23	0.4	437	133	48	14	0.10	0.14	2.8	0	228
63	0	0.0	36	1.1	339	74	54	16	0.03	0.04	3.9	0	229
111	0	0.0	63	2.0	599	131	95	29	0.06	0.07	6.9	0	230
35	0	0.0	51	0.9	490	59	152	46	0.06	0.08	6.1	0	231
65	0	0.0	95	1.7	916	110	285	86	0.11	0.14	11.3	0	232
11	8	0.0	65	1.0	59	740	732	219	0.03	0.12	2.8	0	233
61	1	0.0	52	0.3	299	323	74	22	0.01	0.06	0.9	0	234
46	0	0.0	116	1.0	298	82	39	12	0.11	0.11	2.1	1	235
27	0	0.0	69	0.6	175	48	23	7	0.07	0.07	1.2	Tr	236
131	10	0.0	112	16.5	387	523	248	74	0.25	0.24	3.4	9	237
45	3	0.0	38	5.6	131	177	84	25	0.08	0.08	1.2	3	238
69	10	0.2	53	5.9	207	354	257	77	0.13	0.17	1.4	3	239
82	0	0.0	5	0.2	329	99	65	20	0.06	0.06	1.4	0	240
58	0	0.0	4	0.2	232	70	46	14	0.04	0.05	1.0	0	241
37	0	0.0	10	0.5	442	65	186	56	0.04	0.07	3.3	0	242
66	0	0.0	18	0.8	775	115	326	98	0.07	0.13	5.8	0	243

Food No.	Food Description	Measure of edible portion	Weight (g)	Water (%)	Calories (kcal)	Pro-tein (g)	Total fat (g)	Satu-rated (g)	Mono-unsatu-rated (g)	Poly-unsatu-rated (g)

Fatty acids span the last three columns (Saturated, Mono-unsaturated, Poly-unsaturated).

Fish and Shellfish (continued)

Food No.	Food Description	Measure of edible portion	Weight (g)	Water (%)	Calories (kcal)	Pro-tein (g)	Total fat (g)	Saturated (g)	Mono-unsaturated (g)	Poly-unsaturated (g)
244	Roughy, orange, baked or broiled	3 oz	85	69	76	16	1	Tr	0.5	Tr
	Salmon									
245	Baked or broiled (red)	3 oz	85	62	184	23	9	1.6	4.5	2.0
246		½ fillet	155	62	335	42	17	3.0	8.2	3.7
247	Canned (pink), solids and liquid (includes bones)	3 oz	85	69	118	17	5	1.3	1.5	1.7
248	Smoked (chinook)	3 oz	85	72	99	16	4	0.8	1.7	0.8
249	Sardine, Atlantic, canned in oil, drained solids (includes bones)	3 oz	85	60	177	21	10	1.3	3.3	4.4
	Scallop, cooked									
250	Breaded, fried	6 large	93	58	200	17	10	2.5	4.2	2.7
251	Steamed	3 oz	85	73	95	20	1	0.1	0.1	0.4
	Shrimp									
252	Breaded, fried	3 oz	85	53	206	18	10	1.8	3.2	4.3
253		6 large	45	53	109	10	6	0.9	1.7	2.3
254	Canned, drained solids	3 oz	85	73	102	20	2	0.3	0.2	0.6
255	Swordfish, baked or broiled	3 oz	85	69	132	22	4	1.2	1.7	1.0
256		1 piece	106	69	164	27	5	1.5	2.1	1.3
257	Trout, baked or broiled	3 oz	85	68	144	21	6	1.8	1.8	2.0
258		1 fillet	71	68	120	17	5	1.5	1.5	1.7
	Tuna									
259	Baked or broiled	3 oz	85	63	118	25	1	0.3	0.2	0.3
	Canned, drained solids									
260	Oil pack, chunk light	3 oz	85	60	168	25	7	1.3	2.5	2.5
261	Water pack, chunk light	3 oz	85	75	99	22	1	0.2	0.1	0.3
262	Water pack, solid white	3 oz	85	73	109	20	3	0.7	0.7	0.9
263	Tuna salad: light tuna in oil, pickle relish, mayo type salad dressing	1 cup	205	63	383	33	19	3.2	5.9	8.5

Fruits and Fruit Juices

Food No.	Food Description	Measure of edible portion	Weight (g)	Water (%)	Calories (kcal)	Pro-tein (g)	Total fat (g)	Saturated (g)	Mono-unsaturated (g)	Poly-unsaturated (g)
	Apples									
	Raw									
264	Unpeeled, 2¾" dia (about 3 per lb)	1 apple	138	84	81	Tr	Tr	0.1	Tr	0.1
265	Peeled, sliced	1 cup	110	84	63	Tr	Tr	0.1	Tr	0.1
266	Dried (sodium bisulfite used to preserve color)	5 rings	32	32	78	Tr	Tr	Tr	Tr	Tr
267	Apple juice, bottled or canned	1 cup	248	88	117	Tr	Tr	Tr	Tr	0.1
268	Apple pie filling, canned	⅛ of 21-oz can	74	73	75	Tr	Tr	Tr	0.0	Tr
	Applesauce, canned									
269	Sweetened	1 cup	255	80	194	Tr	Tr	0.1	Tr	0.1
270	Unsweetened	1 cup	244	88	105	Tr	Tr	Tr	Tr	Tr
	Apricots									
271	Raw, without pits (about 12 per lb with pits)	1 apricot	35	86	17	Tr	Tr	Tr	0.1	Tr
	Canned, halves, fruit and liquid									
272	Heavy syrup pack	1 cup	258	78	214	1	Tr	Tr	0.1	Tr
273	Juice pack	1 cup	244	87	117	2	Tr	Tr	Tr	Tr
274	Dried, sulfured	10 halves	35	31	83	1	Tr	Tr	0.1	Tr
275	Apricot nectar, canned, with added ascorbic acid	1 cup	251	85	141	1	Tr	Tr	0.1	Tr
	Asian pear, raw									
276	2¼" high x 2½" dia	1 pear	122	88	51	1	Tr	Tr	0.1	0.1
277	3⅜" high x 3" dia	1 pear	275	88	116	1	1	Tr	0.1	0.2

Choles-terol (mg)	Carbo-hydrate (g)	Total dietary fiber (g)	Calcium (mg)	Iron (mg)	Potas-sium (mg)	Sodium (mg)	Vitamin A (IU)	Vitamin A (RE)	Thiamin (mg)	Ribo-flavin (mg)	Niacin (mg)	Ascor-bic acid (mg)	Food No.
22	0	0.0	32	0.2	327	69	69	20	0.10	0.16	3.1	0	244
74	0	0.0	6	0.5	319	56	178	54	0.18	0.15	5.7	0	245
135	0	0.0	11	0.9	581	102	324	98	0.33	0.27	10.3	0	246
47	0	0.0	181	0.7	277	471	47	14	0.02	0.16	5.6	0	247
20	0	0.0	9	0.7	149	666	75	22	0.02	0.09	4.0	0	248
121	0	0.0	325	2.5	337	429	190	57	0.07	0.19	4.5	0	249
57	9	0.2	39	0.8	310	432	70	20	0.04	0.10	1.4	2	250
45	3	0.0	98	2.6	405	225	85	26	0.09	0.05	1.1	0	251
150	10	0.3	57	1.1	191	292	161	48	0.11	0.12	2.6	1	252
80	5	0.2	30	0.6	101	155	85	25	0.06	0.06	1.4	1	253
147	1	0.0	50	2.3	179	144	51	15	0.02	0.03	2.3	2	254
43	0	0.0	5	0.9	314	98	116	35	0.04	0.10	10.0	1	255
53	0	0.0	6	1.1	391	122	145	43	0.05	0.12	12.5	1	256
58	0	0.0	73	0.3	375	36	244	73	0.20	0.07	7.5	3	257
48	0	0.0	61	0.2	313	30	204	61	0.17	0.06	6.2	2	258
49	0	0.0	18	0.8	484	40	58	17	0.43	0.05	10.1	1	259
15	0	0.0	11	1.2	176	301	66	20	0.03	0.10	10.5	0	260
26	0	0.0	9	1.3	201	287	48	14	0.03	0.06	11.3	0	261
36	0	0.0	12	0.8	201	320	16	5	0.01	0.04	4.9	0	262
27	19	0.0	35	2.1	365	824	199	55	0.06	0.14	13.7	5	263
0	21	3.7	10	0.2	159	0	73	7	0.02	0.02	0.1	8	264
0	16	2.1	4	0.1	124	0	48	4	0.02	0.01	0.1	4	265
0	21	2.8	4	0.4	144	28	0	0	0.00	0.05	0.3	1	266
0	29	0.2	17	0.9	295	7	2	0	0.05	0.04	0.2	2	267
0	19	0.7	3	0.2	33	33	10	1	0.01	0.01	Tr	1	268
0	51	3.1	10	0.9	156	8	28	3	0.03	0.07	0.5	4	269
0	28	2.9	7	0.3	183	5	71	7	0.03	0.06	0.5	3	270
0	4	0.8	5	0.2	104	Tr	914	91	0.01	0.01	0.2	4	271
0	55	4.1	23	0.8	361	10	3,173	317	0.05	0.06	1.0	8	272
0	30	3.9	29	0.7	403	10	4,126	412	0.04	0.05	0.8	12	273
0	22	3.2	16	1.6	482	4	2,534	253	Tr	0.05	1.0	1	274
0	36	1.5	18	1.0	286	8	3,303	331	0.02	0.04	0.7	137	275
0	13	4.4	5	0.0	148	0	0	0	0.01	0.01	0.3	5	276
0	29	9.9	11	0.0	333	0	0	0	0.02	0.03	0.6	10	277

Fruits and Fruit Juices (continued)

Food No.	Food Description	Measure of edible portion	Weight (g)	Water (%)	Calories (kcal)	Protein (g)	Total fat (g)	Fatty acids Saturated (g)	Monounsaturated (g)	Polyunsaturated (g)
	Avocados, raw, without skin and seed									
278	California (about ⅕ whole)....	1 oz 28		73	50	1	5	0.7	3.2	0.6
279	Florida (about ⅒ whole)	1 oz 28		80	32	Tr	3	0.5	1.4	0.4
	Bananas, raw									
280	Whole, medium (7" to 7⅞" long)...................................	1 banana 118		74	109	1	1	0.2	Tr	0.1
281	Sliced	1 cup 150		74	138	2	1	0.3	0.1	0.1
282	Blackberries, raw	1 cup 144		86	75	1	1	Tr	0.1	0.3
	Blueberries									
283	Raw	1 cup 145		85	81	1	1	Tr	0.1	0.2
284	Frozen, sweetened, thawed	1 cup 230		77	186	1	Tr	Tr	Tr	0.1
	Cantaloupe. See Melons.									
	Carambola (starfruit), raw									
285	Whole (3⅝" long)..................	1 fruit...................91		91	30	Tr	Tr	Tr	Tr	0.2
286	Sliced	1 cup 108		91	36	1	Tr	Tr	Tr	0.2
	Cherries									
287	Sour, red, pitted, canned, water pack..........................	1 cup 244		90	88	2	Tr	0.1	0.1	0.1
288	Sweet, raw, without pits and stems..................................	10 cherries...........68		81	49	1	1	0.1	0.2	0.2
289	Cherry pie filling, canned	⅛ of 21-oz can 74		71	85	Tr	Tr	Tr	Tr	Tr
290	Cranberries, dried, sweetened...	¼ cup.................28		12	92	Tr	Tr	Tr	Tr	0.1
291	Cranberry sauce, sweetened, canned (about 8 slices per can).....................	1 slice 57		61	86	Tr	Tr	Tr	Tr	Tr
	Dates, without pits									
292	Whole...............................	5 dates 42		23	116	1	Tr	0.1	0.1	Tr
293	Chopped	1 cup 178		23	490	4	1	0.3	0.3	0.1
294	Figs, dried............................	2 figs..................38		28	97	1	Tr	0.1	0.1	0.2
	Fruit cocktail, canned, fruit and liquid									
295	Heavy syrup pack	1 cup 248		80	181	1	Tr	Tr	Tr	0.1
296	Juice pack	1 cup 237		87	109	1	Tr	Tr	Tr	Tr
	Grapefruit									
	Raw, without peel, membrane and seeds (3¾" dia)									
297	Pink or red.............................	½ grapefruit 123		91	37	1	Tr	Tr	Tr	Tr
298	White	½ grapefruit 118		90	39	1	Tr	Tr	Tr	Tr
299	Canned, sections with light syrup	1 cup 254		84	152	1	Tr	Tr	Tr	0.1
	Grapefruit juice									
	Raw									
300	Pink	1 cup 247		90	96	1	Tr	Tr	Tr	0.1
301	White	1 cup 247		90	96	1	Tr	Tr	Tr	0.1
	Canned									
302	Unsweetened	1 cup 247		90	94	1	Tr	Tr	Tr	0.1
303	Sweetened	1 cup 250		87	115	1	Tr	Tr	Tr	0.1
	Frozen concentrate, unsweetened									
304	Undiluted...............................	6-fl-oz can 207		62	302	4	1	0.1	0.1	0.2
305	Diluted with 3 parts water by volume...........................	1 cup 247		89	101	1	Tr	Tr	Tr	0.1
306	Grapes, seedless, raw	10 grapes 50		81	36	Tr	Tr	0.1	Tr	0.1
307		1 cup 160		81	114	1	1	0.3	Tr	0.3

Choles-terol (mg)	Carbo-hydrate (g)	Total dietary fiber (g)	Calcium (mg)	Iron (mg)	Potas-sium (mg)	Sodium (mg)	Vitamin A (IU)	(RE)	Thiamin (mg)	Ribo-flavin (mg)	Niacin (mg)	Ascor-bic acid (mg)	Food No.
0	2	1.4	3	0.3	180	3	174	17	0.03	0.03	0.5	2	278
0	3	1.5	3	0.2	138	1	174	17	0.03	0.03	0.5	2	279
0	28	2.8	7	0.4	467	1	96	9	0.05	0.12	0.6	11	280
0	35	3.6	9	0.5	594	2	122	12	0.07	0.15	0.8	14	281
0	18	7.6	46	0.8	282	0	238	23	0.04	0.06	0.6	30	282
0	20	3.9	9	0.2	129	9	145	15	0.07	0.07	0.5	19	283
0	50	4.8	14	0.9	138	2	101	9	0.05	0.12	0.6	2	284
0	7	2.5	4	0.2	148	2	449	45	0.03	0.02	0.4	19	285
0	8	2.9	4	0.3	176	2	532	53	0.03	0.03	0.4	23	286
0	22	2.7	27	3.3	239	17	1,840	183	0.04	0.10	0.4	5	287
0	11	1.6	10	0.3	152	0	146	14	0.03	0.04	0.3	5	288
0	21	0.4	8	0.2	78	13	152	16	0.02	0.01	0.1	3	289
0	24	2.5	5	0.1	24	1	0	0	0.01	0.03	Tr	Tr	290
0	22	0.6	2	0.1	15	17	11	1	0.01	0.01	0.1	1	291
0	31	3.2	13	0.5	274	1	21	2	0.04	0.04	0.9	0	292
0	131	13.4	57	2.0	1,161	5	89	9	0.16	0.18	3.9	0	293
0	25	4.6	55	0.8	271	4	51	5	0.03	0.03	0.3	Tr	294
0	47	2.5	15	0.7	218	15	508	50	0.04	0.05	0.9	5	295
0	28	2.4	19	0.5	225	9	723	73	0.03	0.04	1.0	6	296
0	9	1.4	14	0.1	159	0	319	32	0.04	0.02	0.2	47	297
0	10	1.3	14	0.1	175	0	12	1	0.04	0.02	0.3	39	298
0	39	1.0	36	1.0	328	5	0	0	0.10	0.05	0.6	54	299
0	23	0.2	22	0.5	400	2	1,087	109	0.10	0.05	0.5	94	300
0	23	0.2	22	0.5	400	2	25	2	0.10	0.05	0.5	94	301
0	22	0.2	17	0.5	378	2	17	2	0.10	0.05	0.6	72	302
0	28	0.3	20	0.9	405	5	0	0	0.10	0.06	0.8	67	303
0	72	0.8	56	1.0	1,002	6	64	6	0.30	0.16	1.6	248	304
0	24	0.2	20	0.3	336	2	22	2	0.10	0.05	0.5	83	305
0	9	0.5	6	0.1	93	1	37	4	0.05	0.03	0.2	5	306
0	28	1.6	18	0.4	296	3	117	11	0.15	0.09	0.5	17	307

Food No.	Food Description	Measure of edible portion	Weight (g)	Water (%)	Calories (kcal)	Protein (g)	Total fat (g)	Fatty acids Saturated (g)	Fatty acids Monounsaturated (g)	Fatty acids Polyunsaturated (g)

Fruits and Fruit Juices (continued)

	Grape juice									
308	Canned or bottled	1 cup	253	84	154	1	Tr	0.1	Tr	0.1
	Frozen concentrate, sweetened, with added vitamin C									
309	Undiluted	6-fl-oz can	216	54	387	1	1	0.2	Tr	0.2
310	Diluted with 3 parts water by volume	1 cup	250	87	128	Tr	Tr	0.1	Tr	0.1
311	Kiwi fruit, raw, without skin (about 5 per lb with skin)	1 medium	76	83	46	1	Tr	Tr	Tr	0.2
312	Lemons, raw, without peel (2⅛" dia with peel)	1 lemon	58	89	17	1	Tr	Tr	Tr	0.1
	Lemon juice									
313	Raw (from 2⅛"-dia lemon)	juice of 1 lemon	47	91	12	Tr	0	0.0	0.0	0.0
314	Canned or bottled, unsweetened	1 cup	244	92	51	1	1	0.1	Tr	0.2
315		1 tbsp	15	92	3	Tr	Tr	Tr	Tr	Tr
	Lime juice									
316	Raw (from 2"-dia lime)	juice of 1 lime	38	90	10	Tr	Tr	Tr	Tr	Tr
317	Canned, unsweetened	1 cup	246	93	52	1	1	0.1	0.1	0.2
318		1 tbsp	15	93	3	Tr	Tr	Tr	Tr	Tr
	Mangos, raw, without skin and seed (about 1½ per lb with skin and seed)									
319	Whole	1 mango	207	82	135	1	1	0.1	0.2	0.1
320	Sliced	1 cup	165	82	107	1	Tr	0.1	0.2	0.1
	Melons, raw, without rind and cavity contents									
	Cantaloupe (5" dia)									
321	Wedge	⅛ melon	69	90	24	1	Tr	Tr	Tr	0.1
322	Cubes	1 cup	160	90	56	1	Tr	0.1	Tr	0.2
	Honeydew (6"-7" dia)									
323	Wedge	⅛ melon	160	90	56	1	Tr	Tr	Tr	0.1
324	Diced (about 20 pieces per cup)	1 cup	170	90	60	1	Tr	Tr	Tr	0.1
325	Mixed fruit, frozen, sweetened, thawed (peach, cherry, raspberry, grape and boysenberry)	1 cup	250	74	245	4	Tr	0.1	0.1	0.2
326	Nectarines, raw (2½" dia)	1 nectarine	136	86	67	1	1	0.1	0.2	0.3
	Oranges, raw									
327	Whole, without peel and seeds (2⅝" dia)	1 orange	131	87	62	1	Tr	Tr	Tr	Tr
328	Sections without membranes	1 cup	180	87	85	2	Tr	Tr	Tr	Tr
	Orange juice									
329	Raw, all varieties	1 cup	248	88	112	2	Tr	0.1	0.1	0.1
330		juice from 1 orange	86	88	39	1	Tr	Tr	Tr	Tr
331	Canned, unsweetened	1 cup	249	89	105	1	Tr	Tr	0.1	0.1
332	Chilled (refrigerator case)	1 cup	249	88	110	2	1	0.1	0.1	0.2
	Frozen concentrate									
333	Undiluted	6-fl-oz can	213	58	339	5	Tr	0.1	0.1	0.1
334	Diluted with 3 parts water by volume	1 cup	249	88	112	2	Tr	Tr	Tr	Tr
	Papayas, raw									
335	½" cubes	1 cup	140	89	55	1	Tr	0.1	0.1	Tr
336	Whole (5⅛" long x 3" dia)	1 papaya	304	89	119	2	Tr	0.1	0.1	0.1

*Sodium benzoate and sodium bisulfite added as preservatives.

Choles-terol (mg)	Carbo-hydrate (g)	Total dietary fiber (g)	Calcium (mg)	Iron (mg)	Potas-sium (mg)	Sodium (mg)	Vitamin A (IU)	(RE)	Thiamin (mg)	Ribo-flavin (mg)	Niacin (mg)	Ascor-bic acid (mg)	Food No.
0	38	0.3	23	0.6	334	8	20	3	0.07	0.09	0.7	Tr	308
0	96	0.6	28	0.8	160	15	58	6	0.11	0.20	0.9	179	309
0	32	0.3	10	0.3	53	5	20	3	0.04	0.07	0.3	60	310
0	11	2.6	20	0.3	252	4	133	14	0.02	0.04	0.4	74	311
0	5	1.6	15	0.3	80	1	17	2	0.02	0.01	0.1	31	312
0	4	0.2	3	Tr	58	Tr	9	1	0.01	Tr	Tr	22	313
0	16	1.0	27	0.3	249	51*	37	5	0.10	0.02	0.5	61	314
0	1	0.1	2	Tr	16	3*	2	Tr	0.01	Tr	Tr	4	315
0	3	0.2	3	Tr	41	Tr	4	Tr	0.01	Tr	Tr	11	316
0	16	1.0	30	0.6	185	39*	39	5	0.08	0.01	0.4	16	317
0	1	0.1	2	Tr	11	2*	2	Tr	Tr	Tr	Tr	1	318
0	35	3.7	21	0.3	323	4	8,061	805	0.12	0.12	1.2	57	319
0	28	3.0	17	0.2	257	3	6,425	642	0.10	0.09	1.0	46	320
0	6	0.6	8	0.1	213	6	2,225	222	0.02	0.01	0.4	29	321
0	13	1.3	18	0.3	494	14	5,158	515	0.06	0.03	0.9	68	322
0	15	1.0	10	0.1	434	16	64	6	0.12	0.03	1.0	40	323
0	16	1.0	10	0.1	461	17	68	7	0.13	0.03	1.0	42	324
0	61	4.8	18	0.7	328	8	805	80	0.04	0.09	1.0	188	325
0	16	2.2	7	0.2	288	0	1,001	101	0.02	0.06	1.3	7	326
0	15	3.1	52	0.1	237	0	269	28	0.11	0.05	0.4	70	327
0	21	4.3	72	0.2	326	0	369	38	0.16	0.07	0.5	96	328
0	26	0.5	27	0.5	496	2	496	50	0.22	0.07	1.0	124	329
0	9	0.2	9	0.2	172	1	172	17	0.08	0.03	0.3	43	330
0	25	0.5	20	1.1	436	5	436	45	0.15	0.07	0.8	86	331
0	25	0.5	25	0.4	473	2	194	20	0.28	0.05	0.7	82	332
0	81	1.7	68	0.7	1,436	6	588	60	0.60	0.14	1.5	294	333
0	27	0.5	22	0.2	473	2	194	20	0.20	0.04	0.5	97	334
0	14	2.5	34	0.1	360	4	398	39	0.04	0.04	0.5	87	335
0	30	5.5	73	0.3	781	9	863	85	0.08	0.10	1.0	188	336

						Pro-	Total		Fatty acids Mono-	Poly-
Food No.	Food Description	Measure of edible portion	Weight (g)	Water (%)	Calories (kcal)	tein (g)	fat (g)	Satu- rated (g)	unsatu- rated (g)	unsatu- rated (g)

Fruits and Fruit Juices (continued)

Peaches
Raw

337	Whole, 2½" dia, pitted (about 4 per lb)	1 peach	98	88	42	1	Tr	Tr	Tr	Tr
338	Sliced	1 cup	170	88	73	1	Tr	Tr	0.1	0.1

Canned, fruit and liquid

339	Heavy syrup pack	1 cup	262	79	194	1	Tr	Tr	0.1	0.1
340		1 half	98	79	73	Tr	Tr	Tr	Tr	Tr
341	Juice pack	1 cup	248	87	109	2	Tr	Tr	Tr	Tr
342		1 half	98	87	43	1	Tr	Tr	Tr	Tr
343	Dried, sulfured	3 halves	39	32	93	1	Tr	Tr	0.1	0.1
344	Frozen, sliced, sweetened, with added ascorbic acid, thawed	1 cup	250	75	235	2	Tr	Tr	0.1	0.2

Pears

345	Raw, with skin, cored, 2½" dia	1 pear	166	84	98	1	1	Tr	0.1	0.2

Canned, fruit and liquid

346	Heavy syrup pack	1 cup	266	80	197	1	Tr	Tr	0.1	0.1
347		1 half	76	80	56	Tr	Tr	Tr	Tr	Tr
348	Juice pack	1 cup	248	86	124	1	Tr	Tr	Tr	Tr
349		1 half	76	86	38	Tr	Tr	Tr	Tr	Tr

Pineapple

350	Raw, diced	1 cup	155	87	76	1	1	Tr	0.1	0.2

Canned, fruit and liquid
Heavy syrup pack

351	Crushed, sliced, or chunks	1 cup	254	79	198	1	Tr	Tr	Tr	0.1
352	Slices (3" dia)	1 slice	49	79	38	Tr	Tr	Tr	Tr	Tr

Juice pack

353	Crushed, sliced, or chunks	1 cup	249	84	149	1	Tr	Tr	Tr	0.1
354	Slice (3" dia)	1 slice	47	84	28	Tr	Tr	Tr	Tr	Tr
355	Pineapple juice, unsweetened, canned	1 cup	250	86	140	1	Tr	Tr	Tr	0.1

Plantain, without peel

356	Raw	1 medium	179	65	218	2	1	0.3	0.1	0.1
357	Cooked, slices	1 cup	154	67	179	1	Tr	0.1	Tr	0.1

Plums

358	Raw (2⅛" dia)	1 plum	66	85	36	1	Tr	Tr	0.3	0.1

Canned, purple, fruit and liquid

359	Heavy syrup pack	1 cup	258	76	230	1	Tr	Tr	0.2	0.1
360		1 plum	46	76	41	Tr	Tr	Tr	Tr	Tr
361	Juice pack	1 cup	252	84	146	1	Tr	Tr	Tr	Tr
362		1 plum	46	84	27	Tr	Tr	Tr	Tr	Tr

Prunes, dried, pitted

363	Uncooked	5 prunes	42	32	100	1	Tr	Tr	0.1	Tr
364	Stewed, unsweetened, fruit and liquid	1 cup	248	70	265	3	1	Tr	0.4	0.1
365	Prune juice, canned or bottled	1 cup	256	81	182	2	Tr	Tr	0.1	Tr

Raisins, seedless

366	Cup, not packed	1 cup	145	15	435	5	1	0.2	Tr	0.2
367	Packet, ½ oz (1½ tbsp)	1 packet	14	15	42	Tr	Tr	Tr	Tr	Tr

Raspberries

368	Raw	1 cup	123	87	60	1	1	Tr	0.1	0.4
369	Frozen, sweetened, thawed	1 cup	250	73	258	2	Tr	Tr	Tr	0.2
370	Rhubarb, frozen, cooked, with sugar	1 cup	240	68	278	1	Tr	Tr	Tr	0.1

Choles-terol (mg)	Carbo-hydrate (g)	Total dietary fiber (g)	Calcium (mg)	Iron (mg)	Potas-sium (mg)	Sodium (mg)	Vitamin A (IU)	Vitamin A (RE)	Thiamin (mg)	Ribo-flavin (mg)	Niacin (mg)	Ascor-bic acid (mg)	Food No.
0	11	2.0	5	0.1	193	0	524	53	0.02	0.04	1.0	6	337
0	19	3.4	9	0.2	335	0	910	92	0.03	0.07	1.7	11	338
0	52	3.4	8	0.7	241	16	870	86	0.03	0.06	1.6	7	339
0	20	1.3	3	0.3	90	6	325	32	0.01	0.02	0.6	3	340
0	29	3.2	15	0.7	317	10	945	94	0.02	0.04	1.4	9	341
0	11	1.3	6	0.3	125	4	373	37	0.01	0.02	0.6	4	342
0	24	3.2	11	1.6	388	3	844	84	Tr	0.08	1.7	2	343
0	60	4.5	8	0.9	325	15	710	70	0.03	0.09	1.6	236	344
0	25	4.0	18	0.4	208	0	33	3	0.03	0.07	0.2	7	345
0	51	4.3	13	0.6	173	13	0	0	0.03	0.06	0.6	3	346
0	15	1.2	4	0.2	49	4	0	0	0.01	0.02	0.2	1	347
0	32	4.0	22	0.7	238	10	15	2	0.03	0.03	0.5	4	348
0	10	1.2	7	0.2	73	3	5	1	0.01	0.01	0.2	1	349
0	19	1.9	11	0.6	175	2	36	3	0.14	0.06	0.7	24	350
0	51	2.0	36	1.0	264	3	36	3	0.23	0.06	0.7	19	351
0	10	0.4	7	0.2	51	Tr	7	Tr	0.04	0.01	0.1	4	352
0	39	2.0	35	0.7	304	2	95	10	0.24	0.05	0.7	24	353
0	7	0.4	7	0.1	57	Tr	18	2	0.04	0.01	0.1	4	354
0	34	0.5	43	0.7	335	3	13	0	0.14	0.06	0.6	27	355
0	57	4.1	5	1.1	893	7	2,017	202	0.09	0.10	1.2	33	356
0	48	3.5	3	0.9	716	8	1,400	140	0.07	0.08	1.2	17	357
0	9	1.0	3	0.1	114	0	213	21	0.03	0.06	0.3	6	358
0	60	2.6	23	2.2	235	49	668	67	0.04	0.10	0.8	1	359
0	11	0.5	4	0.4	42	9	119	12	0.01	0.02	0.1	Tr	360
0	38	2.5	25	0.9	388	3	2,543	255	0.06	0.15	1.2	7	361
0	7	0.5	5	0.2	71	Tr	464	46	0.01	0.03	0.2	1	362
0	26	3.0	21	1.0	313	2	835	84	0.03	0.07	0.8	1	363
0	70	16.4	57	2.8	828	5	759	77	0.06	0.25	1.8	7	364
0	45	2.6	31	3.0	707	10	8	0	0.04	0.18	2.0	10	365
0	115	5.8	71	3.0	1,089	17	12	1	0.23	0.13	1.2	5	366
0	11	0.6	7	0.3	105	2	1	Tr	0.02	0.01	0.1	Tr	367
0	14	8.4	27	0.7	187	0	160	16	0.04	0.11	1.1	31	368
0	65	11.0	38	1.6	285	3	150	15	0.05	0.11	0.6	41	369
0	75	4.8	348	0.5	230	2	166	17	0.04	0.06	0.5	8	370

Food No.	Food Description	Measure of edible portion	Weight (g)	Water (%)	Calories (kcal)	Pro-tein (g)	Total fat (g)	Fatty acids Satu-rated (g)	Mono-unsatu-rated (g)	Poly-unsatu-rated (g)

Fruits and Fruit Juices (continued)

Strawberries
Raw, capped

371	Large (1⅛" dia).................... 1 strawberry	18		92	5	Tr	Tr	Tr	Tr	Tr
372	Medium (1¼" dia) .... 1 strawberry	12		92	4	Tr	Tr	Tr	Tr	Tr
373	Sliced......................... 1 cup	166		92	50	1	1	Tr	0.1	0.3
374	Frozen, sweetened, sliced, thawed.............................. 1 cup	255		73	245	1	Tr	Tr	Tr	0.2

Tangerines

375	Raw, without peel and seeds (2⅜" dia) .... 1 tangerine...........84			88	37	1	Tr	Tr	Tr	Tr
376	Canned (mandarin oranges), light syrup, fruit and liquid.................................. 1 cup	252		83	154	1	Tr	Tr	Tr	0.1
377	Tangerine juice, canned, sweetened .... 1 cup	249		87	125	1	Tr	Tr	Tr	0.1

Watermelon, raw (15" long x 7½" dia)

378	Wedge (about ⅟₁₆ of melon)................................. 1 wedge 286			92	92	2	1	0.1	0.3	0.4
379	Diced .... 1 cup	152		92	49	1	1	0.1	0.2	0.2

Grain Products

Bagels, enriched

380	Plain .... 3½" bagel...........71			33	195	7	1	0.2	0.1	0.5
381		4" bagel 89		33	245	9	1	0.2	0.1	0.6
382	Cinnamon raisin .... 3½" bagel...........71			32	195	7	1	0.2	0.1	0.5
383		4" bagel 89		32	244	9	2	0.2	0.2	0.6
384	Egg .... 3½" bagel...........71			33	197	8	1	0.3	0.3	0.5
385		4" bagel 89		33	247	9	2	0.4	0.4	0.6
386	Banana bread, prepared from recipe, with margarine .... 1 slice 60			29	196	3	6	1.3	2.7	1.9

Barley, pearled

387	Uncooked .... 1 cup	200		10	704	20	2	0.5	0.3	1.1
388	Cooked .... 1 cup	157		69	193	4	1	0.1	0.1	0.3

Biscuits, plain or buttermilk, enriched

389	Prepared from recipe, with 2% milk .... 2½" biscuit...........60			29	212	4	10	2.6	4.2	2.5
390		4" biscuit 101		29	358	7	16	4.4	7.0	4.2

Refrigerated dough, baked

391	Regular .... 2½" biscuit...........27			28	93	2	4	1.0	2.2	0.5
392	Lower fat .... 2¼" biscuit...........21			28	63	2	1	0.3	0.6	0.2

Breads, enriched

393	Cracked wheat .... 1 slice 25			36	65	2	1	0.2	0.5	0.2
394	Egg bread (challah) .... ½" slice 40			35	115	4	2	0.6	0.9	0.4
395	French or vienna (includes sourdough) .... ½" slice 25			34	69	2	1	0.2	0.3	0.2
396	Indian fry (navajo) bread........ 5" bread 90			27	296	6	9	2.1	3.6	2.3
397		10½" bread........160		27	526	11	15	3.7	6.4	4.1
398	Italian .... 1 slice 20			36	54	2	1	0.2	0.2	0.3

Mixed grain

399	Untoasted .... 1 slice 26			38	65	3	1	0.2	0.4	0.2
400	Toasted .... 1 slice 24			32	65	3	1	0.2	0.4	0.2

Oatmeal

401	Untoasted .... 1 slice 27			37	73	2	1	0.2	0.4	0.5
402	Toasted .... 1 slice 25			31	73	2	1	0.2	0.4	0.5
403	Pita .... 4" pita 28			32	77	3	Tr	Tr	Tr	0.1
404		6½" pita...............60		32	165	5	1	0.1	0.1	0.3

Choles-terol (mg)	Carbo-hydrate (g)	Total dietary fiber (g)	Calcium (mg)	Iron (mg)	Potas-sium (mg)	Sodium (mg)	Vitamin A (IU)	Vitamin A (RE)	Thiamin (mg)	Ribo-flavin (mg)	Niacin (mg)	Ascor-bic acid (mg)	Food No.
0	1	0.4	3	0.1	30	Tr	5	1	Tr	0.01	Tr	10	371
0	1	0.3	2	Tr	20	Tr	3	Tr	Tr	0.01	Tr	7	372
0	12	3.8	23	0.6	276	2	45	5	0.03	0.11	0.4	94	373
0	66	4.8	28	1.5	250	8	61	5	0.04	0.13	1.0	106	374
0	9	1.9	12	0.1	132	1	773	77	0.09	0.02	0.1	26	375
0	41	1.8	18	0.9	197	15	2,117	212	0.13	0.11	1.1	50	376
0	30	0.5	45	0.5	443	2	1,046	105	0.15	0.05	0.2	55	377
0	21	1.4	23	0.5	332	6	1,047	106	0.23	0.06	0.6	27	378
0	11	0.8	12	0.3	176	3	556	56	0.12	0.03	0.3	15	379
0	38	1.6	53	2.5	72	379	0	0	0.38	0.22	3.2	0	380
0	48	2.0	66	3.2	90	475	0	0	0.48	0.28	4.1	0	381
0	39	1.6	13	2.7	105	229	52	0	0.27	0.20	2.2	Tr	382
0	49	2.0	17	3.4	132	287	65	0	0.34	0.25	2.7	1	383
17	38	1.6	9	2.8	48	359	77	23	0.38	0.17	2.4	Tr	384
21	47	2.0	12	3.5	61	449	97	29	0.48	0.21	3.1	1	385
26	33	0.7	13	0.8	80	181	278	72	0.10	0.12	0.9	1	386
0	155	31.2	58	5.0	560	18	44	4	0.38	0.23	9.2	0	387
0	44	6.0	17	2.1	146	5	11	2	0.13	0.10	3.2	0	388
2	27	0.9	141	1.7	73	348	49	14	0.21	0.19	1.8	Tr	389
3	45	1.5	237	2.9	122	586	83	23	0.36	0.31	3.0	Tr	390
0	13	0.4	5	0.7	42	325	0	0	0.09	0.06	0.8	0	391
0	12	0.4	4	0.6	39	305	0	0	0.09	0.05	0.7	0	392
0	12	1.4	11	0.7	44	135	0	0	0.09	0.06	0.9	0	393
20	19	0.9	37	1.2	46	197	30	9	0.18	0.17	1.9	0	394
0	13	0.8	19	0.6	28	152	0	0	0.13	0.08	1.2	0	395
0	48	1.6	210	3.2	67	626	0	0	0.39	0.27	3.3	0	396
0	85	2.9	373	5.8	118	1,112	0	0	0.69	0.49	5.8	0	397
0	10	0.5	16	0.6	22	117	0	0	0.09	0.06	0.9	0	398
0	12	1.7	24	0.9	53	127	0	0	0.11	0.09	1.1	Tr	399
0	12	1.6	24	0.9	53	127	0	0	0.08	0.08	1.0	Tr	400
0	13	1.1	18	0.7	38	162	4	1	0.11	0.06	0.8	0	401
0	13	1.1	18	0.7	39	163	4	1	0.09	0.06	0.8	Tr	402
0	16	0.6	24	0.7	34	150	0	0	0.17	0.09	1.3	0	403
0	33	1.3	52	1.6	72	322	0	0	0.36	0.20	2.8	0	404

Grain Products (continued)

Food No.	Food Description	Measure of edible portion	Weight (g)	Water (%)	Calories (kcal)	Protein (g)	Total fat (g)	Saturated (g)	Monounsaturated (g)	Polyunsaturated (g)
	Breads, enriched (continued)									
	Pumpernickel									
405	Untoasted	1 slice	32	38	80	3	1	0.1	0.3	0.4
406	Toasted	1 slice	29	32	80	3	1	0.1	0.3	0.4
	Raisin									
407	Untoasted	1 slice	26	34	71	2	1	0.3	0.6	0.2
408	Toasted	1 slice	24	28	71	2	1	0.3	0.6	0.2
	Rye									
409	Untoasted	1 slice	32	37	83	3	1	0.2	0.4	0.3
410	Toasted	1 slice	24	31	68	2	1	0.2	0.3	0.2
411	Rye, reduced calorie	1 slice	23	46	47	2	1	0.1	0.2	0.2
	Wheat									
412	Untoasted	1 slice	25	37	65	2	1	0.2	0.4	0.2
413	Toasted	1 slice	23	32	65	2	1	0.2	0.4	0.2
414	Wheat, reduced calorie	1 slice	23	43	46	2	1	0.1	0.1	0.2
	White									
415	Untoasted	1 slice	25	37	67	2	1	0.1	0.2	0.5
416	Toasted	1 slice	22	30	64	2	1	0.1	0.2	0.5
417	Soft crumbs	1 cup	45	37	120	4	2	0.2	0.3	0.9
418	White, reduced calorie	1 slice	23	43	48	2	1	0.1	0.2	0.1
	Bread, whole wheat									
419	Untoasted	1 slice	28	38	69	3	1	0.3	0.5	0.3
420	Toasted	1 slice	25	30	69	3	1	0.3	0.5	0.3
	Bread crumbs, dry, grated									
421	Plain, enriched	1 cup	108	6	427	14	6	1.3	2.6	1.2
422		1 oz	28	6	112	4	2	0.3	0.7	0.3
423	Seasoned, unenriched	1 cup	120	6	440	17	3	0.9	1.2	0.8
	Bread crumbs, soft. See White bread.									
424	Bread stuffing, prepared from dry mix	½ cup	100	65	178	3	9	1.7	3.8	2.6
425	Breakfast bar, cereal crust withfruit filling, fat free	1 bar	37	14	121	2	Tr	Tr	Tr	0.1
	Breakfast Cereals									
	Hot type, cooked									
	Corn (hominy) grits									
	Regular or quick, enriched									
426	White	1 cup	242	85	145	3	Tr	0.1	0.1	0.2
427	Yellow	1 cup	242	85	145	3	Tr	0.1	0.1	0.2
428	Instant, plain	1 packet	137	82	89	2	Tr	Tr	Tr	0.1
	CREAM OF WHEAT									
429	Regular	1 cup	251	87	133	4	1	0.1	0.1	0.3
430	Quick	1 cup	239	87	129	4	Tr	0.1	0.1	0.3
431	Mix'n Eat, plain	1 packet	142	82	102	3	Tr	Tr	Tr	0.2
432	MALT O MEAL	1 cup	240	88	122	4	Tr	0.1	0.1	Tr
	Oatmeal									
433	Regular, quick or instant, plain, nonfortified	1 cup	234	85	145	6	2	0.4	0.7	0.9
434	Instant, fortified, plain	1 packet	177	86	104	4	2	0.3	0.6	0.7
	QUAKER instant									
435	Apples and cinnamon	1 packet	149	79	125	3	1	0.3	0.5	0.6
436	Maple and brown sugar	1 packet	155	75	153	4	2	0.4	0.6	0.7
437	WHEATENA	1 cup	243	85	136	5	1	0.2	0.2	0.6
	Ready to eat									
438	ALL BRAN	½ cup	30	3	79	4	1	0.2	0.2	0.5
439	APPLE CINNAMON CHEERIOS	¾ cup	30	3	118	2	2	0.3	0.6	0.2
440	APPLE JACKS	1 cup	30	3	116	1	Tr	0.1	0.1	0.2

Choles-terol (mg)	Carbo-hydrate (g)	Total dietary fiber (g)	Calcium (mg)	Iron (mg)	Potas-sium (mg)	Sodium (mg)	Vitamin A (IU)	Vitamin A (RE)	Thiamin (mg)	Ribo-flavin (mg)	Niacin (mg)	Ascor-bic acid (mg)	Food No.
0	15	2.1	22	0.9	67	215	0	0	0.10	0.10	1.0	0	405
0	15	2.1	21	0.9	66	214	0	0	0.08	0.09	0.9	0	406
0	14	1.1	17	0.8	59	101	0	0	0.09	0.10	0.9	Tr	407
0	14	1.1	17	0.8	59	102	Tr	0	0.07	0.09	0.8	Tr	408
0	15	1.9	23	0.9	53	211	2	Tr	0.14	0.11	1.2	Tr	409
0	13	1.5	19	0.7	44	174	1	0	0.09	0.08	0.9	Tr	410
0	9	2.8	17	0.7	23	93	1	0	0.08	0.06	0.6	Tr	411
0	12	1.1	26	0.8	50	133	0	0	0.10	0.07	1.0	0	412
0	12	1.2	26	0.8	50	132	0	0	0.08	0.06	0.9	0	413
0	10	2.8	18	0.7	28	118	0	0	0.10	0.07	0.9	Tr	414
Tr	12	0.6	27	0.8	30	135	0	0	0.12	0.09	1.0	0	415
Tr	12	0.6	26	0.7	29	130	0	0	0.09	0.07	0.9	0	416
Tr	22	1.0	49	1.4	54	242	0	0	0.21	0.15	1.8	0	417
0	10	2.2	22	0.7	17	104	1	Tr	0.09	0.07	0.8	Tr	418
0	13	1.9	20	0.9	71	148	0	0	0.10	0.06	1.1	0	419
0	13	1.9	20	0.9	71	148	0	0	0.08	0.05	1.0	0	420
0	78	2.6	245	6.6	239	931	1	0	0.83	0.47	7.4	0	421
0	21	0.7	64	1.7	63	244	Tr	0	0.22	0.12	1.9	0	422
1	84	5.0	119	3.8	324	3,180	16	4	0.19	0.20	3.3	Tr	423
0	22	2.9	32	1.1	74	543	313	81	0.14	0.11	1.5	0	424
Tr	28	0.8	49	4.5	92	203	1,249	125	1.01	0.42	5.0	1	425
0	31	0.5	0	1.5	53	0	0	0	0.24	0.15	2.0	0	426
0	31	0.5	0	1.5	53	0	145	15	0.24	0.15	2.0	0	427
0	21	1.2	8	8.2	38	289	0	0	0.15	0.08	1.4	0	428
0	28	1.8	50	10.3	43	3	0	0	0.25	0.00	1.5	0	429
0	27	1.2	50	10.3	45	139	0	0	0.24	0.00	1.4	0	430
0	21	0.4	20	8.1	38	241	1,252	376	0.43	0.28	5.0	0	431
0	26	1.0	5	9.6	31	2	0	0	0.48	0.24	5.8	0	432
0	25	4.0	19	1.6	131	2	37	5	0.26	0.05	0.3	0	433
0	18	3.0	163	6.3	99	285	1,510	453	0.53	0.28	5.5	0	434
0	26	2.5	104	3.9	106	121	1,019	305	0.30	0.35	4.1	Tr	435
0	31	2.6	105	3.9	112	234	1,008	302	0.30	0.34	4.0	0	436
0	29	6.6	10	1.4	187	5	0	0	0.02	0.05	1.3	0	437
0	23	9.7	106	4.5	342	61	750	225	0.39	0.42	5.0	15	438
0	25	1.6	35	4.5	60	150	750	225	0.38	0.43	5.0	15	439
0	27	0.6	3	4.5	32	134	750	225	0.39	0.42	5.0	15	440

Food No.	Food Description	Measure of edible portion	Weight (g)	Water (%)	Calories (kcal)	Pro-tein (g)	Total fat (g)	Satu-rated (g)	Mono-unsatu-rated (g)	Poly-unsatu-rated (g)

Grain Products (continued)

Breakfast Cereals (continued)
Ready to eat (continued)

441	BASIC 4	1 cup ...55		7	201	4	3	0.4	1.0	1.1
442	BERRY BERRY KIX	¾ cup...30		2	120	1	1	0.2	0.5	0.1
443	CAP'N CRUNCH	¾ cup...27		2	107	1	1	0.4	0.3	0.2
444	CAP'N CRUNCH'S CRUNCHBERRIES	¾ cup...26		2	104	1	1	0.3	0.3	0.2
445	CAP'N CRUNCH'S PEANUT BUTTER CRUNCH	¾ cup...27		2	112	2	2	0.5	0.8	0.5
446	CHEERIOS	1 cup...30		3	110	3	2	0.4	0.6	0.2
	CHEX									
447	Corn	1 cup...30		3	113	2	Tr	0.1	0.1	0.2
448	Honey nut	¾ cup...30		2	117	2	1	0.1	0.4	0.2
449	Multi bran	1 cup...49		3	165	4	1	0.2	0.3	0.5
450	Rice	1¼ cup...31		3	117	2	Tr	Tr	Tr	Tr
451	Wheat	1 cup...30		3	104	3	1	0.1	0.1	0.3
452	CINNAMON LIFE	1 cup...50		4	190	4	2	0.3	0.6	0.8
453	CINNAMON TOAST CRUNCH	¾ cup...30		2	124	2	3	0.5	0.9	0.5
454	COCOA KRISPIES	¾ cup...31		2	120	2	1	0.6	0.1	0.1
455	COCOA PUFFS	1 cup...30		2	119	1	1	0.2	0.3	Tr
	Corn Flakes									
456	GENERAL MILLS, TOTAL	1⅓ cup...30		3	112	2	Tr	0.2	0.1	Tr
457	KELLOGG'S	1 cup...28		3	102	2	Tr	0.1	Tr	0.1
458	CORN POPS	1 cup...31		3	118	1	Tr	0.1	0.1	Tr
459	CRISPIX	1 cup...29		3	108	2	Tr	0.1	0.1	0.1
460	Complete Wheat Bran Flakes	¾ cup...29		4	95	3	1	0.1	0.1	0.4
461	FROOT LOOPS	1 cup...30		2	117	1	1	0.4	0.2	0.3
462	FROSTED FLAKES	¾ cup...31		3	119	1	Tr	0.1	Tr	0.1
	FROSTED MINI WHEATS									
463	Regular	1 cup...51		5	173	5	1	0.2	0.1	0.6
464	Bite size	1 cup...55		5	187	5	1	0.2	0.2	0.6
465	GOLDEN GRAHAMS	¾ cup...30		3	116	2	1	0.2	0.3	0.2
466	HONEY FROSTED WHEATIES	¾ cup...30		3	110	2	Tr	0.1	Tr	Tr
467	HONEY NUT CHEERIOS	1 cup...30		2	115	3	1	0.2	0.5	0.2
468	HONEY NUT CLUSTERS	1 cup...55		3	213	5	3	0.4	1.8	0.4
469	KIX	1⅓ cup...30		2	114	2	1	0.2	0.1	Tr
470	LIFE	¾ cup...32		4	121	3	1	0.2	0.4	0.6
471	LUCKY CHARMS	1 cup...30		2	116	2	1	0.2	0.4	0.2
472	NATURE VALLEY Granola	¾ cup...55		4	248	6	10	1.3	6.5	1.9
	100% Natural Cereal									
473	With oats, honey, and raisins	½ cup...51		4	218	5	7	3.2	3.2	0.8
474	With raisins, low fat	½ cup...50		4	195	4	3	0.8	1.3	0.5
475	PRODUCT 19	1 cup...30		3	110	3	Tr	Tr	0.2	0.2
476	Puffed Rice	1 cup...14		3	56	1	Tr	Tr	Tr	Tr
477	Puffed Wheat	1 cup...12		3	44	2	Tr	Tr	Tr	Tr
	Raisin Bran									
478	GENERAL MILLS, TOTAL	1 cup...55		9	178	4	1	0.2	0.2	0.2
479	KELLOGG'S	1 cup...61		8	186	6	1	0.0	0.2	0.8
480	RAISIN NUT BRAN	1 cup...55		5	209	5	4	0.7	1.9	0.5
481	REESE'S PEANUT BUTTER PUFFS	¾ cup...30		2	129	3	3	0.6	1.4	0.6
482	RICE KRISPIES	1¼ cup...33		3	124	2	Tr	0.1	0.1	0.2

Choles-terol (mg)	Carbo-hydrate (g)	Total dietary fiber (g)	Calcium (mg)	Iron (mg)	Potas-sium (mg)	Sodium (mg)	Vitamin A (IU)	(RE)	Thiamin (mg)	Ribo-flavin (mg)	Niacin (mg)	Ascor-bic acid (mg)	Food No.
0	42	3.4	310	4.5	162	323	1,250	375	0.37	0.42	5.0	15	441
0	26	0.2	66	4.5	24	185	750	225	0.38	0.43	5.0	15	442
0	23	0.9	5	4.5	35	208	36	4	0.38	0.42	5.0	0	443
0	22	0.6	7	4.5	37	190	33	5	0.37	0.42	5.0	Tr	444
0	22	0.8	3	4.5	62	204	37	4	0.38	0.42	5.0	0	445
0	23	2.6	55	8.1	89	284	1,250	375	0.38	0.43	5.0	15	446
0	26	0.5	100	9.0	32	289	0	0	0.38	0.00	5.0	6	447
0	26	0.4	102	9.0	27	224	0	0	0.38	0.44	5.0	6	448
0	41	6.4	95	13.7	191	325	0	0	0.32	0.00	4.4	5	449
0	27	0.3	104	9.0	36	291	0	0	0.38	0.02	5.0	6	450
0	24	3.3	60	9.0	116	269	0	0	0.23	0.04	3.0	4	451
0	40	3.0	135	7.5	113	220	16	2	0.63	0.71	8.4	Tr	452
0	24	1.5	42	4.5	44	210	750	225	0.38	0.43	5.0	15	453
0	27	0.4	4	1.8	60	210	750	225	0.37	0.43	5.0	15	454
0	27	0.2	33	4.5	52	181	0	0	0.38	0.43	5.0	15	455
0	26	0.8	237	18.0	34	203	1,250	375	1.50	1.70	20.1	60	456
0	24	0.8	1	8.7	25	298	700	210	0.36	0.39	4.7	14	457
0	28	0.4	2	1.9	23	123	775	233	0.40	0.43	5.2	16	458
0	25	0.6	3	1.8	35	240	750	225	0.38	0.44	5.0	15	459
0	23	4.6	14	8.1	175	226	1,208	363	0.38	0.44	5.0	15	460
0	26	0.6	3	4.2	32	141	703	211	0.39	0.42	5.0	14	461
0	28	0.6	1	4.5	20	200	750	225	0.37	0.43	5.0	15	462
0	42	5.5	18	14.3	170	2	0	0	0.36	0.41	5.0	0	463
0	45	5.9	0	15.4	186	2	0	0	0.33	0.39	4.7	0	464
0	26	0.9	14	4.5	53	275	750	225	0.38	0.43	5.0	15	465
0	26	1.5	8	4.5	56	211	750	225	0.38	0.43	5.0	15	466
0	24	1.6	20	4.5	85	259	750	225	0.38	0.43	5.0	15	467
0	43	4.2	72	4.5	171	239	0	0	0.37	0.42	5.0	9	468
0	26	0.8	44	8.1	41	263	1,250	375	0.38	0.43	5.0	15	469
0	25	2.0	98	9.0	79	174	12	1	0.40	0.45	5.3	0	470
0	25	1.2	32	4.5	54	203	750	225	0.38	0.43	5.0	15	471
0	36	3.5	41	1.7	183	89	0	0	0.17	0.06	0.6	0	472
1	36	3.7	39	1.7	214	11	4	1	0.14	0.09	0.8	Tr	473
1	40	3.0	30	1.3	169	129	9	1	0.15	0.06	0.9	Tr	474
0	25	1.0	3	18.0	41	216	750	225	1.50	1.71	20.0	60	475
0	13	0.2	1	4.4	16	Tr	0	0	0.36	0.25	4.9	0	476
0	10	0.5	3	3.8	42	Tr	0	0	0.31	0.22	4.2	0	477
0	43	5.0	238	18.0	287	240	1,250	375	1.50	1.70	20.0	0	478
0	47	8.2	35	5.0	437	354	832	250	0.43	0.49	5.6	0	479
0	41	5.1	74	4.5	218	246	0	0	0.37	0.42	5.0	0	480
0	23	0.4	21	4.5	62	177	750	225	0.38	0.43	5.0	15	481
0	29	0.4	3	2.0	42	354	825	248	0.43	0.46	5.5	17	482

Food No.	Food Description	Measure of edible portion	Weight (g)	Water (%)	Calories (kcal)	Protein (g)	Total fat (g)	Fatty acids: Saturated (g)	Fatty acids: Monounsaturated (g)	Fatty acids: Polyunsaturated (g)
	Grain Products (continued)									
	Breakfast Cereals (continued)									
	Ready to eat (continued)									
483	RICE KRISPIES TREATS cereal	¾ cup	30	4	120	1	2	0.4	1.0	0.2
484	SHREDDED WHEAT	2 biscuits	46	4	156	5	1	0.1	NA	NA
485	SMACKS	¾ cup	27	3	103	2	1	0.3	0.1	0.2
486	SPECIAL K	1 cup	31	3	115	6	Tr	0.0	0.0	0.2
487	QUAKER Toasted Oatmeal, Honey Nut	1 cup	49	3	191	5	3	0.5	1.2	0.7
488	TOTAL, Whole Grain	¾ cup	30	3	105	3	1	0.2	0.1	0.1
489	TRIX	1 cup	30	2	122	1	2	0.4	0.9	0.3
490	WHEATIES	1 cup	30	3	110	3	1	0.2	0.2	0.2
	Brownies, without icing									
	Commercially prepared									
491	Regular, large (2¾" sq x ⅞")	1 brownie	56	14	227	3	9	2.4	5.0	1.3
492	Fat free, 2" sq	1 brownie	28	12	89	1	Tr	0.2	0.1	Tr
493	Prepared from dry mix, reduced calorie, 2" sq	1 brownie	22	13	84	1	2	1.1	1.0	0.2
494	Buckwheat flour, whole groat	1 cup	120	11	402	15	4	0.8	1.1	1.1
495	Buckwheat groats, roasted (kasha), cooked	1 cup	168	76	155	6	1	0.2	0.3	0.3
	Bulgur									
496	Uncooked	1 cup	140	9	479	17	2	0.3	0.2	0.8
497	Cooked	1 cup	182	78	151	6	Tr	0.1	0.1	0.2
	Cakes, prepared from dry mix									
498	Angelfood (1/12 of 10" dia)	1 piece	50	33	129	3	Tr	Tr	Tr	0.1
499	Yellow, light, with water, egg whites, no frosting (1/12 of 9" dia)	1 piece	69	37	181	3	2	1.1	0.9	0.2
	Cakes, prepared from recipe									
500	Chocolate, without frosting (1/12 of 9" dia)	1 piece	95	24	340	5	14	5.2	5.7	2.6
501	Gingerbread (1/9 of 8" square)	1 piece	74	28	263	3	12	3.1	5.3	3.1
502	Pineapple upside down (1/9 of 8" square)	1 piece	115	32	367	4	14	3.4	6.0	3.8
503	Shortcake, biscuit type (about 3" dia)	1 shortcake	65	28	225	4	9	2.5	3.9	2.4
504	Sponge (1/12 of 16-oz cake)	1 piece	63	29	187	5	3	0.8	1.0	0.4
	White									
505	With coconut frosting (1/12 of 9" dia)	1 piece	112	21	399	5	12	4.4	4.1	2.4
506	Without frosting (1/12 of 9" dia)	1 piece	74	23	264	4	9	2.4	3.9	2.3
	Cakes, commercially prepared									
507	Angelfood (1/12 of 12-oz cake)	1 piece	28	33	72	2	Tr	Tr	Tr	0.1
508	Boston cream (1/6 of pie)	1 piece	92	45	232	2	8	2.2	4.2	0.9
509	Chocolate with chocolate frosting (1/8 of 18-oz cake)	1 piece	64	23	235	3	10	3.1	5.6	1.2
510	Coffeecake, crumb (1/9 of 20-oz cake)	1 piece	63	22	263	4	15	3.7	8.2	2.0
511	Fruitcake	1 piece	43	25	139	1	4	0.5	1.8	1.4
	Pound									
512	Butter (1/12 of 12-oz cake)	1 piece	28	25	109	2	6	3.2	1.7	0.3
513	Fat free (3¼" x 2¾" x ⅝" slice)	1 slice	28	31	79	2	Tr	0.1	Tr	0.1

Choles-terol (mg)	Carbo-hydrate (g)	Total dietary fiber (g)	Calcium (mg)	Iron (mg)	Potas-sium (mg)	Sodium (mg)	Vitamin A (IU)	Vitamin A (RE)	Thiamin (mg)	Ribo-flavin (mg)	Niacin (mg)	Ascor-bic acid (mg)	Food No.
0	26	0.3	2	1.8	19	190	750	225	0.39	0.42	5.0	15	483
0	38	5.3	20	1.4	196	3	0	NA	0.12	0.05	2.6	0	484
0	24	0.9	3	1.8	42	51	750	225	0.38	0.43	5.0	15	485
0	22	1.0	5	8.7	55	250	750	225	0.53	0.59	7.0	15	486
Tr	39	3.3	27	4.5	185	166	500	150	0.37	0.42	5.0	6	487
0	24	2.6	258	18.0	97	199	1,250	375	1.50	1.70	20.1	60	488
0	26	0.7	32	4.5	18	197	750	225	0.38	0.43	5.0	15	489
0	24	2.1	55	8.1	104	222	750	225	0.38	0.43	5.0	15	490
10	36	1.2	16	1.3	83	175	39	3	0.14	0.12	1.0	0	491
0	22	1.0	17	0.7	89	90	1	Tr	0.03	0.04	0.3	Tr	492
0	16	0.8	3	0.3	69	21	0	0	0.02	0.03	0.2	0	493
0	85	12.0	49	4.9	692	13	0	0	0.50	0.23	7.4	0	494
0	33	4.5	12	1.3	148	7	0	0	0.07	0.07	1.6	0	495
0	106	25.6	49	3.4	574	24	0	0	0.32	0.16	7.2	0	496
0	34	8.2	18	1.7	124	9	0	0	0.10	0.05	1.8	0	497
0	29	0.1	42	0.1	68	255	0	0	0.05	0.10	0.1	0	498
0	37	0.6	69	0.6	41	279	6	1	0.06	0.12	0.6	0	499
55	51	1.5	57	1.5	133	299	133	38	0.13	0.20	1.1	Tr	500
24	36	0.7	53	2.1	325	242	36	10	0.14	0.12	1.3	Tr	501
25	58	0.9	138	1.7	129	367	291	75	0.18	0.18	1.4	1	502
2	32	0.8	133	1.7	69	329	47	12	0.20	0.18	1.7	Tr	503
107	36	0.4	26	1.0	89	144	163	49	0.10	0.19	0.8	0	504
1	71	1.1	101	1.3	111	318	43	12	0.14	0.21	1.2	Tr	505
1	42	0.6	96	1.1	70	242	41	12	0.14	0.18	1.1	Tr	506
0	16	0.4	39	0.1	26	210	0	0	0.03	0.14	0.2	0	507
34	39	1.3	21	0.3	36	132	74	21	0.38	0.25	0.2	Tr	508
27	35	1.8	28	1.4	128	214	54	16	0.02	0.09	0.4	Tr	509
20	29	1.3	34	1.2	77	221	70	21	0.13	0.14	1.1	Tr	510
2	26	1.6	14	0.9	66	116	9	2	0.02	0.04	0.3	Tr	511
62	14	0.1	10	0.4	33	111	170	44	0.04	0.06	0.4	0	512
0	17	0.3	12	0.6	31	95	27	8	0.04	0.08	0.2	0	513

Food No.	Food Description	Measure of edible portion	Weight (g)	Water (%)	Calories (kcal)	Protein (g)	Total fat (g)	Fatty acids		
								Saturated (g)	Monounsaturated (g)	Polyunsaturated (g)

Grain Products (continued)

Cakes, commercially prepared (continued)
Snack cakes

514	Chocolate, creme filled, with frosting	1 cupcake	50	20	188	2	7	1.4	2.8	2.6
515	Chocolate, with frosting, low fat	1 cupcake	43	23	131	2	2	0.5	0.8	0.2
516	Sponge, creme filled	1 cake	43	20	155	1	5	1.1	1.7	1.4
517	Sponge, individual shortcake	1 shortcake	30	30	87	2	1	0.2	0.3	0.1
	Yellow									
518	With chocolate frosting	1 piece	64	22	243	2	11	3.0	6.1	1.4
519	With vanilla frosting	1 piece	64	22	239	2	9	1.5	3.9	3.3
520	Cheesecake (⅙ of 17-oz cake)	1 piece	80	46	257	4	18	7.9	6.9	1.3
521	Cheese flavor puffs or twists	1 oz	28	2	157	2	10	1.9	5.7	1.3
522	CHEX mix	1 oz (about ⅔ cup)	28	4	120	3	5	1.6	NA	NA
	Cookies									
523	Butter, commercially prepared	1 cookie	5	5	23	Tr	1	0.6	0.3	Tr
	Chocolate chip, medium (2¼"-2½" dia) Commercially prepared									
524	Regular	1 cookie	10	4	48	1	2	0.7	1.2	0.2
525	Reduced fat	1 cookie	10	4	45	1	2	0.4	0.6	0.5
526	From refrigerated dough (spooned from roll)	1 cookie	26	3	128	1	6	2.0	2.9	0.6
527	Prepared from recipe, with margarine	1 cookie	16	6	78	1	5	1.3	1.7	1.3
528	Devil's food, commercially prepared, fat free	1 cookie	16	18	49	1	Tr	0.1	Tr	Tr
529	Fig bar	1 cookie	16	17	56	1	1	0.2	0.5	0.4
	Molasses									
530	Medium	1 cookie	15	6	65	1	2	0.5	1.1	0.3
531	Large (3½"-4" dia)	1 cookie	32	6	138	2	4	1.0	2.3	0.6
	Oatmeal Commercially prepared, with or without raisins									
532	Regular, large	1 cookie	25	6	113	2	5	1.1	2.5	0.6
533	Soft type	1 cookie	15	11	61	1	2	0.5	1.2	0.3
534	Fat free	1 cookie	11	13	36	1	Tr	Tr	Tr	0.1
535	Prepared from recipe, with raisins (2⅝" dia)	1 cookie	15	6	65	1	2	0.5	1.0	0.8
	Peanut butter									
536	Commercially prepared	1 cookie	15	6	72	1	4	0.7	1.9	0.8
537	Prepared from recipe, with margarine (3" dia)	1 cookie	20	6	95	2	5	0.9	2.2	1.4
	Sandwich type, with creme filling									
538	Chocolate cookie	1 cookie	10	2	47	Tr	2	0.4	0.9	0.7
	Vanilla cookie									
539	Oval	1 cookie	15	2	72	1	3	0.4	1.3	1.1
540	Round	1 cookie	10	2	48	Tr	2	0.3	0.8	0.8
	Shortbread, commercially prepared									
541	Plain (1⅝" sq)	1 cookie	8	4	40	Tr	2	0.5	1.1	0.3
	Pecan									
542	Regular (2" dia)	1 cookie	14	3	76	1	5	1.1	2.6	0.6
543	Reduced fat	1 cookie	16	5	73	1	3	0.6	1.6	0.4

Choles- terol (mg)	Carbo- hydrate (g)	Total dietary fiber (g)	Calcium (mg)	Iron (mg)	Potas- sium (mg)	Sodium (mg)	Vitamin A (IU)	Vitamin A (RE)	Thiamin (mg)	Ribo- flavin (mg)	Niacin (mg)	Ascor- bic acid (mg)	Food No.
9	30	0.4	37	1.7	61	213	9	3	0.11	0.15	1.2	0	514
0	29	1.8	15	0.7	96	178	0	0	0.02	0.06	0.3	0	515
7	27	0.2	19	0.5	37	155	7	2	0.07	0.06	0.5	Tr	516
31	18	0.2	21	0.8	30	73	46	14	0.07	0.08	0.6	0	517
35	35	1.2	24	1.3	114	216	70	21	0.08	0.10	0.8	0	518
35	38	0.2	40	0.7	34	220	40	12	0.06	0.04	0.3	0	519
44	20	0.3	41	0.5	72	166	438	117	0.02	0.15	0.2	Tr	520
1	15	0.3	16	0.7	47	298	75	10	0.07	0.10	0.9	Tr	521
0	18	1.6	10	7.0	76	288	41	4	0.44	0.14	4.8	13	522
6	3	Tr	1	0.1	6	18	34	8	0.02	0.02	0.2	0	523
0	7	0.3	3	0.3	14	32	Tr	0	0.02	0.03	0.3	0	524
0	7	0.4	2	0.3	12	38	Tr	0	0.03	0.03	0.3	0	525
7	18	0.4	7	0.7	52	60	15	4	0.04	0.05	0.5	0	526
5	9	0.4	6	0.4	36	58	102	26	0.03	0.03	0.2	Tr	527
0	12	0.3	5	0.4	18	28	Tr	NA	0.01	0.03	0.2	Tr	528
0	11	0.7	10	0.5	33	56	5	1	0.03	0.03	0.3	Tr	529
0	11	0.1	11	1.0	52	69	0	0	0.05	0.04	0.5	0	530
0	24	0.3	24	2.1	111	147	0	0	0.11	0.08	1.0	0	531
0	17	0.7	9	0.6	36	96	5	1	0.07	0.06	0.6	Tr	532
1	10	0.4	14	0.4	20	52	5	1	0.03	0.03	0.3	Tr	533
0	9	0.8	4	0.2	23	33	0	0	0.02	0.03	0.1	0	534
5	10	0.5	15	0.4	36	81	96	25	0.04	0.02	0.2	Tr	535
Tr	9	0.3	5	0.4	25	62	1	Tr	0.03	0.03	0.6	0	536
6	12	0.4	8	0.4	46	104	120	31	0.04	0.04	0.7	Tr	537
0	7	0.3	3	0.4	18	60	Tr	0	0.01	0.02	0.2	0	538
0	11	0.2	4	0.3	14	52	0	0	0.04	0.04	0.4	0	539
0	7	0.2	3	0.2	9	35	0	0	0.03	0.02	0.3	0	540
2	5	0.1	3	0.2	8	36	7	1	0.03	0.03	0.3	0	541
5	8	0.3	4	0.3	10	39	Tr	Tr	0.04	0.03	0.3	0	542
0	11	0.2	8	0.5	15	55	1	Tr	0.05	0.03	0.4	Tr	543

Food No.	Food Description	Measure of edible portion	Weight (g)	Water (%)	Calories (kcal)	Protein (g)	Total fat (g)	Fatty acids Saturated (g)	Monounsaturated (g)	Polyunsaturated (g)

Grain Products (continued)

Cookies (continued)
Sugar

Food No.	Food Description	Measure of edible portion	Weight (g)	Water (%)	Calories (kcal)	Protein (g)	Total fat (g)	Saturated (g)	Monounsaturated (g)	Polyunsaturated (g)
544	Commercially prepared	1 cookie	15	5	72	1	3	0.8	1.8	0.4
545	From refrigerated dough	1 cookie	15	5	73	1	3	0.9	2.0	0.4
546	Prepared from recipe, with margarine (3" dia)	1 cookie	14	9	66	1	3	0.7	1.4	1.0
547	Vanilla wafer, lower fat, medium size	1 cookie	4	5	18	Tr	1	0.2	0.3	0.2
	Corn chips									
548	Plain	1 oz	28	1	153	2	9	1.3	2.7	4.7
549	Barbecue flavor	1 oz	28	1	148	2	9	1.3	2.7	4.6
	Cornbread									
550	Prepared from mix, piece 3¾" x 2½" x ¾"	1 piece	60	32	188	4	6	1.6	3.1	0.7
551	Prepared from recipe, with 2% milk, piece 2½" sq x 1½"	1 piece	65	39	173	4	5	1.0	1.2	2.1
	Cornmeal, yellow, dry form									
552	Whole grain	1 cup	122	10	442	10	4	0.6	1.2	2.0
553	Degermed, enriched	1 cup	138	12	505	12	2	0.3	0.6	1.0
554	Self rising, degermed, enriched	1 cup	138	10	490	12	2	0.3	0.6	1.0
555	Cornstarch	1 tbsp	8	8	30	Tr	Tr	Tr	Tr	Tr
	Couscous									
556	Uncooked	1 cup	173	9	650	22	1	0.2	0.2	0.4
557	Cooked	1 cup	157	73	176	6	Tr	Tr	Tr	0.1
	Crackers									
558	Cheese, 1" sq	10 crackers	10	3	50	1	3	0.9	1.2	0.2
	Graham, plain									
559	2½" sq	2 squares	14	4	59	1	1	0.2	0.6	0.5
560	Crushed	1 cup	84	4	355	6	8	1.3	3.4	3.2
561	Melba toast, plain	4 pieces	20	5	78	2	1	0.1	0.2	0.3
562	Rye wafer, whole grain, plain	1 wafer	11	5	37	1	Tr	Tr	Tr	Tr
	Saltine									
563	Square	4 crackers	12	4	52	1	1	0.4	0.8	0.2
564	Oyster type	1 cup	45	4	195	4	5	1.3	2.9	0.8
	Sandwich type									
565	Wheat with cheese	1 sandwich	7	4	33	1	1	0.4	0.8	0.2
566	Cheese with peanut butter	1 sandwich	7	4	34	1	2	0.4	0.8	0.3
	Standard snack type									
567	Bite size	1 cup	62	4	311	5	16	2.3	6.6	5.9
568	Round	4 crackers	12	4	60	1	3	0.5	1.3	1.1
569	Wheat, thin square	4 crackers	8	3	38	1	2	0.4	0.9	0.2
570	Whole wheat	4 crackers	16	3	71	1	3	0.5	0.9	1.1
571	Croissant, butter	1 croissant	57	23	231	5	12	6.6	3.1	0.6
572	Croutons, seasoned	1 cup	40	4	186	4	7	2.1	3.8	0.9
	Danish pastry, enriched									
573	Cheese filled	1 danish	71	31	266	6	16	4.8	8.0	1.8
574	Fruit filled	1 danish	71	27	263	4	13	3.5	7.1	1.7
	Doughnuts									
575	Cake type	1 hole	14	21	59	1	3	0.5	1.3	1.1
576		1 medium	47	21	198	2	11	1.7	4.4	3.7
577	Yeast leavened, glazed	1 hole	13	25	52	1	3	0.8	1.7	0.4
578		1 medium	60	25	242	4	14	3.5	7.7	1.7
579	Eclair, prepared from recipe, 5" x 2" x 1¾"	1 eclair	100	52	262	6	16	4.1	6.5	3.9
	English muffin, plain, enriched									
580	Untoasted	1 muffin	57	42	134	4	1	0.1	0.2	0.5
581	Toasted	1 muffin	52	37	133	4	1	0.1	0.2	0.5

Cholesterol (mg)	Carbohydrate (g)	Total dietary fiber (g)	Calcium (mg)	Iron (mg)	Potassium (mg)	Sodium (mg)	Vitamin A (IU)	Vitamin A (RE)	Thiamin (mg)	Riboflavin (mg)	Niacin (mg)	Ascorbic acid (mg)	Food No.
8	10	0.1	3	0.3	9	54	14	4	0.03	0.03	0.4	Tr	544
5	10	0.1	14	0.3	24	70	6	2	0.03	0.02	0.4	0	545
4	8	0.2	10	0.3	11	69	135	35	0.04	0.04	0.3	Tr	546
2	3	0.1	2	0.1	4	12	1	Tr	0.01	0.01	0.1	0	547
0	16	1.4	36	0.4	40	179	27	3	0.01	0.04	0.3	0	548
0	16	1.5	37	0.4	67	216	173	17	0.02	0.06	0.5	Tr	549
37	29	1.4	44	1.1	77	467	123	26	0.15	0.16	1.2	Tr	550
26	28	1.9	162	1.6	96	428	180	35	0.19	0.19	1.5	Tr	551
0	94	8.9	7	4.2	350	43	572	57	0.47	0.25	4.4	0	552
0	107	10.2	7	5.7	224	4	570	57	0.99	0.56	6.9	0	553
0	103	9.8	483	6.5	235	1,860	570	57	0.94	0.53	6.3	0	554
0	7	0.1	Tr	Tr	Tr	1	0	0	0.00	0.00	0.0	0	555
0	134	8.7	42	1.9	287	17	0	0	0.28	0.13	6.0	0	556
0	36	2.2	13	0.6	91	8	0	0	0.10	0.04	1.5	0	557
1	6	0.2	15	0.5	15	100	16	3	0.06	0.04	0.5	0	558
0	11	0.4	3	0.5	19	85	0	0	0.03	0.04	0.6	0	559
0	65	2.4	20	3.1	113	508	0	0	0.19	0.26	3.5	0	560
0	15	1.3	19	0.7	40	166	0	0	0.08	0.05	0.8	0	561
0	9	2.5	4	0.7	54	87	1	0	0.05	0.03	0.2	Tr	562
0	9	0.4	14	0.6	15	156	0	0	0.07	0.06	0.6	0	563
0	32	1.4	54	2.4	58	586	0	0	0.25	0.21	2.4	0	564
Tr	4	0.1	18	0.2	30	98	5	1	0.03	0.05	0.3	Tr	565
Tr	4	0.2	6	0.2	17	69	22	2	0.03	0.02	0.5	Tr	566
0	38	1.0	74	2.2	82	525	0	0	0.25	0.21	2.5	0	567
0	7	0.2	14	0.4	16	102	0	0	0.05	0.04	0.5	0	568
0	5	0.4	4	0.4	15	64	0	0	0.04	0.03	0.4	0	569
0	11	1.7	8	0.5	48	105	0	0	0.03	0.02	0.7	0	570
38	26	1.5	21	1.2	67	424	424	106	0.22	0.14	1.2	Tr	571
3	25	2.0	38	1.1	72	495	16	4	0.20	0.17	1.9	0	572
11	26	0.7	25	1.1	70	320	104	32	0.13	0.18	1.4	Tr	573
81	34	1.3	33	1.3	59	251	53	16	0.19	0.16	1.4	3	574
5	7	0.2	6	0.3	18	76	8	2	0.03	0.03	0.3	Tr	575
17	23	0.7	21	0.9	60	257	27	8	0.10	0.11	0.9	Tr	576
1	6	0.2	6	0.3	14	44	2	1	0.05	0.03	0.4	Tr	577
4	27	0.7	26	1.2	65	205	8	2	0.22	0.13	1.7	Tr	578
127	24	0.6	63	1.2	117	337	718	191	0.12	0.27	0.8	Tr	579
0	26	1.5	99	1.4	75	264	0	0	0.25	0.16	2.2	0	580
0	26	1.5	98	1.4	74	262	0	0	0.20	0.14	2.0	Tr	581

Food No.	Food Description	Measure of edible portion	Weight (g)	Water (%)	Calories (kcal)	Protein (g)	Total fat (g)	Fatty acids Saturated (g)	Fatty acids Monounsaturated (g)	Fatty acids Polyunsaturated (g)

Grain Products (continued)

French toast
Prepared from recipe, with 2% milk, fried in

Food No.	Food Description	Measure	Weight	Water	Calories	Protein	Total fat	Saturated	Monounsaturated	Polyunsaturated
582	margarine	1 slice 65		55	149	5	7	1.8	2.9	1.7
583	Frozen, ready to heat .... 1 slice 59			53	126	4	4	0.9	1.2	0.7
	Granola bar									
584	Hard, plain .... 1 bar.................28			4	134	3	6	0.7	1.2	3.4
	Soft, uncoated									
585	Chocolate chip .... 1 bar.................28			5	119	2	5	2.9	1.0	0.6
586	Raisin................................... 1 bar.................28			6	127	2	5	2.7	0.8	0.9
587	Soft, chocolate-coated, peanut butter .... 1 bar.................28			3	144	3	9	4.8	1.9	0.5
588	Macaroni (elbows), enriched, cooked .... 1 cup 140			66	197	7	1	0.1	0.1	0.4
589	Matzo, plain.............................. 1 matzo.............28			4	112	3	Tr	0.1	Tr	0.2
	Muffins									
	Blueberry									
590	Commercially prepared (2¾" dia x 2")................. 1 muffin 57			38	158	3	4	0.8	1.1	1.4
591	Prepared from mix (2¼" dia x 1¾") .... 1 muffin 50			36	150	3	4	0.7	1.8	1.5
592	Prepared from recipe, with 2% milk..................... 1 muffin 57			40	162	4	6	1.2	1.5	3.1
593	Bran with raisins, toaster type, toasted .... 1 muffin 34			27	106	2	3	0.5	0.8	1.7
	Corn									
594	Commercially prepared (2½" dia x 2¼") .... 1 muffin 57			33	174	3	5	0.8	1.2	1.8
595	Prepared from mix (2¼" dia x 1½") .... 1 muffin 50			31	161	4	5	1.4	2.6	0.6
596	Oat bran, commercially prepared (2½" dia x 2¼")................................. 1 muffin 57			35	154	4	4	0.6	1.0	2.4
597	Noodles, chow mein, canned .... 1 cup 45			1	237	4	14	2.0	3.5	7.8
	Noodles (egg noodles), enriched, cooked									
598	Regular .... 1 cup 160			69	213	8	2	0.5	0.7	0.7
599	Spinach.................................... 1 cup 160			69	211	8	3	0.6	0.8	0.6
600	NUTRI GRAIN Cereal Bar, fruit filled .... 1 bar.....................37			15	136	2	3	0.6	1.9	0.3
	Oat bran									
601	Uncooked .... 1 cup 94			7	231	16	7	1.2	2.2	2.6
602	Cooked .... 1 cup 219			84	88	7	2	0.4	0.6	0.7
603	Oriental snack mix .... 1 oz (about ¼ cup) 28			3	156	5	7	1.1	2.8	3.0
	Pancakes, plain (4" dia)									
604	Frozen, ready to heat............... 1 pancake.............36			45	82	2	1	0.3	0.4	0.3
605	Prepared from complete mix .... 1 pancake.............38			53	74	2	1	0.2	0.3	0.3
606	Prepared from incomplete mix, with 2% milk, egg and oil .... 1 pancake.............38			53	83	3	3	0.8	0.8	1.1
	Pie crust, baked									
	Standard type									
607	From recipe .... 1 pie shell 180			10	949	12	62	15.5	27.3	16.4
608	From frozen.......................... 1 pie shell 126			11	648	6	41	13.3	19.8	5.1
609	Graham cracker .... 1 pie shell 239			4	1,181	10	60	12.4	27.2	16.5

Choles-terol (mg)	Carbo-hydrate (g)	Total dietary fiber (g)	Calcium (mg)	Iron (mg)	Potas-sium (mg)	Sodium (mg)	Vitamin A (IU)	(RE)	Thiamin (mg)	Ribo-flavin (mg)	Niacin (mg)	Ascor-bic acid (mg)	Food No.
75	16	0.7	65	1.1	87	311	315	86	0.13	0.21	1.1	Tr	582
48	19	0.7	63	1.3	79	292	110	32	0.16	0.22	1.6	Tr	583
0	18	1.5	17	0.8	95	83	43	4	0.07	0.03	0.4	Tr	584
Tr	20	1.4	26	0.7	96	77	12	1	0.06	0.04	0.3	0	585
Tr	19	1.2	29	0.7	103	80	0	0	0.07	0.05	0.3	0	586
3	15	0.8	31	0.4	96	55	37	10	0.03	0.06	0.9	Tr	587
0	40	1.8	10	2.0	43	1	0	0	0.29	0.14	2.3	0	588
0	24	0.9	4	0.9	32	1	0	0	0.11	0.08	1.1	0	589
17	27	1.5	32	0.9	70	255	19	5	0.08	0.07	0.6	1	590
23	24	0.6	13	0.6	39	219	39	11	0.07	0.16	1.1	1	591
21	23	1.1	108	1.3	70	251	80	22	0.16	0.16	1.3	1	592
3	19	2.8	13	1.0	60	179	58	16	0.07	0.10	0.8	0	593
15	29	1.9	42	1.6	39	297	119	21	0.16	0.19	1.2	0	594
31	25	1.2	38	1.0	66	398	105	23	0.12	0.14	1.1	Tr	595
0	28	2.6	36	2.4	289	224	0	0	0.15	0.05	0.2	0	596
0	26	1.8	9	2.1	54	198	38	4	0.26	0.19	2.7	0	597
53	40	1.8	19	2.5	45	11	32	10	0.30	0.13	2.4	0	598
53	39	3.7	30	1.7	59	19	165	22	0.39	0.20	2.4	0	599
0	27	0.8	15	1.8	73	110	750	227	0.37	0.41	5.0	0	600
0	62	14.5	55	5.1	532	4	0	0	1.10	0.21	0.9	0	601
0	25	5.7	22	1.9	201	2	0	0	0.35	0.07	0.3	0	602
0	15	3.7	15	0.7	93	117	1	0	0.09	0.04	0.9	Tr	603
3	16	0.6	22	1.3	26	183	36	10	0.14	0.17	1.4	Tr	604
5	14	0.5	48	0.6	67	239	12	3	0.08	0.08	0.7	Tr	605
27	11	0.7	82	0.5	76	192	95	27	0.08	0.12	0.5	Tr	606
0	86	3.0	18	5.2	121	976	0	0	0.70	0.50	6.0	0	607
0	62	1.3	26	2.8	139	815	0	0	0.35	0.48	3.1	0	608
0	156	3.6	50	5.2	210	1,365	1,876	483	0.25	0.42	5.1	0	609

Grain Products (continued)

Food No.	Food Description	Measure of edible portion	Weight (g)	Water (%)	Calories (kcal)	Protein (g)	Total fat (g)	Saturated (g)	Monounsaturated (g)	Polyunsaturated (g)
	Pies									
	Commercially prepared (⅙ of 8" dia)									
610	Apple	1 piece	117	52	277	2	13	4.4	5.1	2.6
611	Blueberry	1 piece	117	53	271	2	12	2.0	5.0	4.1
612	Cherry	1 piece	117	46	304	2	13	3.0	6.8	2.4
613	Chocolate creme	1 piece	113	44	344	3	22	5.6	12.6	2.7
614	Coconut custard	1 piece	104	49	270	6	14	6.1	5.7	1.2
615	Lemon meringue	1 piece	113	42	303	2	10	2.0	3.0	4.1
616	Pecan	1 piece	113	19	452	5	21	4.0	12.1	3.6
617	Pumpkin	1 piece	109	58	229	4	10	1.9	4.4	3.4
	Prepared from recipe (⅛ of 9" dia)									
618	Apple	1 piece	155	47	411	4	19	4.7	8.4	5.2
619	Blueberry	1 piece	147	51	360	4	17	4.3	7.5	4.5
620	Cherry	1 piece	180	46	486	5	22	5.4	9.6	5.8
621	Lemon meringue	1 piece	127	43	362	5	16	4.0	7.1	4.2
622	Pecan	1 piece	122	20	503	6	27	4.9	13.6	7.0
623	Pumpkin	1 piece	155	59	316	7	14	4.9	5.7	2.8
624	Fried, cherry	1 pie	128	38	404	4	21	3.1	9.5	6.9
	Popcorn									
625	Air popped, unsalted	1 cup	8	4	31	1	Tr	Tr	0.1	0.2
626	Oil popped, salted	1 cup	11	3	55	1	3	0.5	0.9	1.5
	Caramel coated									
627	With peanuts	1 cup	42	3	168	3	3	0.4	1.1	1.4
628	Without peanuts	1 cup	35	3	152	1	5	1.3	1.0	1.6
629	Cheese flavor	1 cup	11	3	58	1	4	0.7	1.1	1.7
630	Popcorn cake	1 cake	10	5	38	1	Tr	Tr	0.1	0.1
	Pretzels, made with enriched flour									
631	Stick, 2¼" long	10 pretzels	3	3	11	Tr	Tr	Tr	Tr	Tr
632	Twisted, regular	10 pretzels	60	3	229	5	2	0.5	0.8	0.7
633	Twisted, dutch, 2¾" x 2⅝"	1 pretzel	16	3	61	1	1	0.1	0.2	0.2
	Rice									
634	Brown, long grain, cooked	1 cup	195	73	216	5	2	0.4	0.6	0.6
	White, long grain, enriched									
	Regular									
635	Raw	1 cup	185	12	675	13	1	0.3	0.4	0.3
636	Cooked	1 cup	158	68	205	4	Tr	0.1	0.1	0.1
637	Instant, prepared	1 cup	165	76	162	3	Tr	0.1	0.1	0.1
	Parboiled									
638	Raw	1 cup	185	10	686	13	1	0.3	0.3	0.3
639	Cooked	1 cup	175	72	200	4	Tr	0.1	0.1	0.1
640	Wild, cooked	1 cup	164	74	166	7	1	0.1	0.1	0.3
641	Rice cake, brown rice, plain	1 cake	9	6	35	1	Tr	0.1	0.1	0.1
642	RICE KRISPIES Treat Squares	1 bar	22	6	91	1	2	0.3	0.6	1.1
	Rolls									
643	Dinner	1 roll	28	32	84	2	2	0.5	1.0	0.3
644	Hamburger or hotdog	1 roll	43	34	123	4	2	0.5	0.4	1.1
645	Hard, kaiser	1 roll	57	31	167	6	2	0.3	0.6	1.0
	Spaghetti, cooked									
646	Enriched	1 cup	140	66	197	7	1	0.1	0.1	0.4
647	Whole wheat	1 cup	140	67	174	7	1	0.1	0.1	0.3
	Sweet rolls, cinnamon									
648	Commercial, with raisins	1 roll	60	25	223	4	10	1.8	2.9	4.5
649	Refrigerated dough, baked, with frosting	1 roll	30	23	109	2	4	1.0	2.2	0.5

Choles-terol (mg)	Carbo-hydrate (g)	Total dietary fiber (g)	Calcium (mg)	Iron (mg)	Potas-sium (mg)	Sodium (mg)	Vitamin A (IU)	Vitamin A (RE)	Thiamin (mg)	Ribo-flavin (mg)	Niacin (mg)	Ascor-bic acid (mg)	Food No.
0	40	1.9	13	0.5	76	311	145	35	0.03	0.03	0.3	4	610
0	41	1.2	9	0.4	59	380	164	40	0.01	0.04	0.4	3	611
0	47	0.9	14	0.6	95	288	329	63	0.03	0.03	0.2	1	612
6	38	2.3	41	1.2	144	154	0	0	0.04	0.12	0.8	0	613
36	31	1.9	84	0.8	182	348	114	28	0.09	0.15	0.4	1	614
51	53	1.4	63	0.7	101	165	198	59	0.07	0.24	0.7	4	615
36	65	4.0	19	1.2	84	479	198	53	0.10	0.14	0.3	1	616
22	30	2.9	65	0.9	168	307	3,743	405	0.06	0.17	0.2	1	617
0	58	3.6	11	1.7	122	327	90	19	0.23	0.17	1.9	3	618
0	49	3.6	10	1.8	74	272	62	6	0.22	0.19	1.8	1	619
0	69	3.5	18	3.3	139	344	736	86	0.27	0.23	2.3	2	620
67	50	0.7	15	1.3	83	307	203	56	0.15	0.20	1.2	4	621
106	64	2.2	39	1.8	162	320	410	109	0.23	0.22	1.0	Tr	622
65	41	2.9	146	2.0	288	349	11,833	1,212	0.14	0.31	1.2	3	623
0	55	3.3	28	1.6	83	479	220	22	0.18	0.14	1.8	2	624
0	6	1.2	1	0.2	24	Tr	16	2	0.02	0.02	0.2	0	625
0	6	1.1	1	0.3	25	97	17	2	0.01	0.01	0.2	Tr	626
0	34	1.6	28	1.6	149	124	27	3	0.02	0.05	0.8	0	627
2	28	1.8	15	0.6	38	73	18	4	0.02	0.02	0.8	0	628
1	6	1.1	12	0.2	29	98	27	5	0.01	0.03	0.2	Tr	629
0	8	0.3	1	0.2	33	29	7	1	0.01	0.02	0.6	0	630
0	2	0.1	1	0.1	4	51	0	0	0.01	0.02	0.2	0	631
0	48	1.9	22	2.6	88	1,029	0	0	0.28	0.37	3.2	0	632
0	13	0.5	6	0.7	23	274	0	0	0.07	0.10	0.8	0	633
0	45	3.5	20	0.8	84	10	0	0	0.19	0.05	3.0	0	634
0	148	2.4	52	8.0	213	9	0	0	1.07	0.09	7.8	0	635
0	45	0.6	16	1.9	55	2	0	0	0.26	0.02	2.3	0	636
0	35	1.0	13	1.0	7	5	0	0	0.12	0.08	1.5	0	637
0	151	3.1	111	6.6	222	9	0	0	1.10	0.13	6.7	0	638
0	43	0.7	33	2.0	65	5	0	0	0.44	0.03	2.5	0	639
0	35	3.0	5	1.0	166	5	0	0	0.09	0.14	2.1	0	640
0	7	0.4	1	0.1	26	29	4	Tr	0.01	0.01	0.7	0	641
0	18	0.1	1	0.5	9	77	200	60	0.15	0.18	2.0	0	642
Tr	14	0.8	33	0.9	37	146	0	0	0.14	0.09	1.1	Tr	643
0	22	1.2	60	1.4	61	241	0	0	0.21	0.13	1.7	Tr	644
0	30	1.3	54	1.9	62	310	0	0	0.27	0.19	2.4	0	645
0	40	2.4	10	2.0	43	1	0	0	0.29	0.14	2.3	0	646
0	37	6.3	21	1.5	62	4	0	0	0.15	0.06	1.0	0	647
40	31	1.4	43	1.0	67	230	129	38	0.19	0.16	1.4	1	648
0	17	0.6	10	0.8	19	250	1	0	0.12	0.07	1.1	Tr	649

Food No.	Food Description	Measure of edible portion	Weight (g)	Water (%)	Calories (kcal)	Pro-tein (g)	Total fat (g)	Fatty acids Satu-rated (g)	Mono-unsatu-rated (g)	Poly-unsatu-rated (g)
Grain Products (continued)										
650	Taco shell, baked	1 medium	13	6	62	1	3	0.4	1.2	1.1
651	Tapioca, pearl, dry	1 cup	152	11	544	Tr	Tr	Tr	Tr	Tr
	Toaster pastries									
652	Brown sugar cinnamon	1 pastry	50	11	206	3	7	1.8	4.0	0.9
653	Chocolate with frosting	1 pastry	52	13	201	3	5	1.0	2.7	1.1
654	Fruit filled	1 pastry	52	12	204	2	5	0.8	2.2	2.0
655	Low fat	1 pastry	52	12	193	2	3	0.7	1.7	0.5
	Tortilla chips									
	Plain									
656	Regular	1 oz	28	2	142	2	7	1.4	4.4	1.0
657	Low fat, baked	10 chips	14	2	54	2	1	0.1	0.2	0.4
	Nacho flavor									
658	Regular	1 oz	28	2	141	2	7	1.4	4.3	1.0
659	Light, reduced fat	1 oz	28	1	126	2	4	0.8	2.5	0.6
	Tortillas, ready to cook (about 6" dia)									
660	Corn	1 tortilla	26	44	58	1	1	0.1	0.2	0.3
661	Flour	1 tortilla	32	27	104	3	2	0.6	1.2	0.3
	Waffles, plain									
662	Prepared from recipe, 7" dia	1 waffle	75	42	218	6	11	2.1	2.6	5.1
663	Frozen, toasted, 4" dia	1 waffle	33	42	87	2	3	0.5	1.1	0.9
664	Low fat, 4" dia	1 waffle	35	43	83	2	1	0.3	0.4	0.4
	Wheat flours									
	All purpose, enriched									
665	Sifted, spooned	1 cup	115	12	419	12	1	0.2	0.1	0.5
666	Unsifted, spooned	1 cup	125	12	455	13	1	0.2	0.1	0.5
667	Bread, enriched	1 cup	137	13	495	16	2	0.3	0.2	1.0
668	Cake or pastry flour, enriched, unsifted, spooned	1 cup	137	13	496	11	1	0.2	0.1	0.5
669	Self rising, enriched, unsifted, spooned	1 cup	125	11	443	12	1	0.2	0.1	0.5
670	Whole wheat, from hard wheats, stirred, spooned	1 cup	120	10	407	16	2	0.4	0.3	0.9
671	Wheat germ, toasted, plain	1 tbsp	7	6	27	2	1	0.1	0.1	0.5
Legumes, Nuts, and Seeds										
	Almonds, shelled									
672	Sliced	1 cup	95	5	549	20	48	3.7	30.5	11.6
673	Whole	1 oz (24 nuts)	28	5	164	6	14	1.1	9.1	3.5
	Beans, dry									
	Cooked									
674	Black	1 cup	172	66	227	15	1	0.2	0.1	0.4
675	Great Northern	1 cup	177	69	209	15	1	0.2	Tr	0.3
676	Kidney, red	1 cup	177	67	225	15	1	0.1	0.1	0.5
677	Lima, large	1 cup	188	70	216	15	1	0.2	0.1	0.3
678	Pea (navy)	1 cup	182	63	258	16	1	0.3	0.1	0.4
679	Pinto	1 cup	171	64	234	14	1	0.2	0.2	0.3
	Canned, solids and liquid									
	Baked beans									
680	Plain or vegetarian	1 cup	254	73	236	12	1	0.3	0.1	0.5
681	With frankfurters	1 cup	259	69	368	17	17	6.1	7.3	2.2
682	With pork in tomato sauce	1 cup	253	73	248	13	3	1.0	1.1	0.3
683	With pork in sweet sauce	1 cup	253	71	281	13	4	1.4	1.6	0.5
684	Kidney, red	1 cup	256	77	218	13	1	0.1	0.1	0.5
685	Lima, large	1 cup	241	77	190	12	Tr	0.1	Tr	0.2
686	White	1 cup	262	70	307	19	1	0.2	0.1	0.3

Choles-terol (mg)	Carbo-hydrate (g)	Total dietary fiber (g)	Calcium (mg)	Iron (mg)	Potas-sium (mg)	Sodium (mg)	Vitamin A (IU)	Vitamin A (RE)	Thiamin (mg)	Ribo-flavin (mg)	Niacin (mg)	Ascor-bic acid (mg)	Food No.
0	8	1.0	21	0.3	24	49	0	0	0.03	0.01	0.2	0	650
0	135	1.4	30	2.4	17	2	0	0	0.01	0.00	0.0	0	651
0	34	0.5	17	2.0	57	212	493	112	0.19	0.29	2.3	Tr	652
0	37	0.6	20	1.8	82	203	500	NA	0.16	0.16	2.0	0	653
0	37	1.1	14	1.8	58	218	501	2	0.15	0.19	2.0	Tr	654
0	40	0.8	23	1.8	34	131	494	49	0.15	0.29	2.0	2	655
0	18	1.8	44	0.4	56	150	56	6	0.02	0.05	0.4	0	656
0	11	0.7	22	0.2	37	57	52	6	0.03	0.04	0.1	Tr	657
1	18	1.5	42	0.4	61	201	105	12	0.04	0.05	0.4	1	658
1	20	1.4	45	0.5	77	284	108	12	0.06	0.08	0.1	Tr	659
0	12	1.4	46	0.4	40	42	0	0	0.03	0.02	0.4	0	660
0	18	1.1	40	1.1	42	153	0	0	0.17	0.09	1.1	0	661
52	25	0.7	191	1.7	119	383	171	49	0.20	0.26	1.6	Tr	662
8	13	0.8	77	1.5	42	260	400	120	0.13	0.16	1.5	0	663
9	15	0.4	20	1.9	50	155	506	NA	0.31	0.26	2.6	0	664
0	88	3.1	17	5.3	123	2	0	0	0.90	0.57	6.8	0	665
0	95	3.4	19	5.8	134	3	0	0	0.98	0.62	7.4	0	666
0	99	3.3	21	6.0	137	3	0	0	1.11	0.70	10.3	0	667
0	107	2.3	19	10.0	144	3	0	0	1.22	0.59	9.3	0	668
0	93	3.4	423	5.8	155	1,588	0	0	0.84	0.52	7.3	0	669
0	87	14.6	41	4.7	486	6	0	0	0.54	0.26	7.6	0	670
0	3	0.9	3	0.6	66	Tr	0	0	0.12	0.06	0.4	Tr	671
0	19	11.2	236	4.1	692	1	10	1	0.23	0.77	3.7	0	672
0	6	3.3	70	1.2	206	Tr	3	Tr	0.07	0.23	1.1	0	673
0	41	15.0	46	3.6	611	2	10	2	0.42	0.10	0.9	0	674
0	37	12.4	120	3.8	692	4	2	0	0.28	0.10	1.2	2	675
0	40	13.1	50	5.2	713	4	0	0	0.28	0.10	1.0	2	676
0	39	13.2	32	4.5	955	4	0	0	0.30	0.10	0.8	0	677
0	48	11.6	127	4.5	670	2	4	0	0.37	0.11	1.0	2	678
0	44	14.7	82	4.5	800	3	3	0	0.32	0.16	0.7	4	679
0	52	12.7	127	0.7	752	1,008	434	43	0.39	0.15	1.1	8	680
16	40	17.9	124	4.5	609	1,114	399	39	0.15	0.15	2.3	6	681
18	49	12.1	142	8.3	759	1,113	314	30	0.13	0.12	1.3	8	682
18	53	13.2	154	4.2	673	850	288	28	0.12	0.15	0.9	8	683
0	40	16.4	61	3.2	658	873	0	0	0.27	0.23	1.2	3	684
0	36	11.6	51	4.4	530	810	0	0	0.13	0.08	0.6	0	685
0	57	12.6	191	7.8	1,189	13	0	0	0.25	0.10	0.3	0	686

Food No.	Food Description	Measure of edible portion	Weight (g)	Water (%)	Calories (kcal)	Protein (g)	Total fat (g)	Saturated (g)	Mono-unsaturated (g)	Poly-unsaturated (g)

Legumes, Nuts, and Seeds (continued)

Food No.	Food Description	Measure of edible portion	Weight (g)	Water (%)	Calories (kcal)	Protein (g)	Total fat (g)	Saturated (g)	Mono-unsaturated (g)	Poly-unsaturated (g)
	Black eyed peas, dry									
687	Cooked	1 cup	172	70	200	13	1	0.2	0.1	0.4
688	Canned, solids and liquid	1 cup	240	80	185	11	1	0.3	0.1	0.6
689	Brazil nuts, shelled	1 oz (6-8 nuts)	28	3	186	4	19	4.6	6.5	6.8
690	Carob flour	1 cup	103	4	229	5	1	0.1	0.2	0.2
	Cashews, salted									
691	Dry roasted	1 oz	28	2	163	4	13	2.6	7.7	2.2
692	Oil roasted	1 cup	130	4	749	21	63	12.4	36.9	10.6
693		1 oz (18 nuts)	28	4	163	5	14	2.7	8.1	2.3
694	Chestnuts, European, roasted, shelled	1 cup	143	40	350	5	3	0.6	1.1	1.2
	Chickpeas, dry									
695	Cooked	1 cup	164	60	269	15	4	0.4	1.0	1.9
696	Canned, solids and liquid	1 cup	240	70	286	12	3	0.3	0.6	1.2
	Coconut									
	Raw									
697	Piece, about 2" x 2" x ½"	1 piece	45	47	159	1	15	13.4	0.6	0.2
698	Shredded, not packed	1 cup	80	47	283	3	27	23.8	1.1	0.3
699	Dried, sweetened, shredded	1 cup	93	13	466	3	33	29.3	1.4	0.4
700	Hazelnuts (filberts), chopped	1 cup	115	5	722	17	70	5.1	52.5	9.1
701		1 oz	28	5	178	4	17	1.3	12.9	2.2
702	Hummus, commercial	1 tbsp	14	67	23	1	1	0.2	0.6	0.5
703	Lentils, dry, cooked	1 cup	198	70	230	18	1	0.1	0.1	0.3
704	Macadamia nuts, dry roasted, salted	1 cup	134	2	959	10	102	16.0	79.4	2.0
705		1 oz (10-12 nuts)	28	2	203	2	22	3.4	16.8	0.4
	Mixed nuts, with peanuts, salted									
706	Dry roasted	1 oz	28	2	168	5	15	2.0	8.9	3.1
707	Oil roasted	1 oz	28	2	175	5	16	2.5	9.0	3.8
	Peanuts									
	Dry roasted									
708	Salted	1 oz (about 28)	28	2	166	7	14	2.0	7.0	4.4
709	Unsalted	1 cup	146	2	854	35	73	10.1	36.0	22.9
710		1 oz (about 28)	28	2	166	7	14	2.0	7.0	4.4
711	Oil roasted, salted	1 cup	144	2	837	38	71	9.9	35.2	22.4
712		1 oz	28	2	165	7	14	1.9	6.9	4.4
	Peanut butter									
	Regular									
713	Smooth style	1 tbsp	16	1	95	4	8	1.7	3.9	2.2
714	Chunk style	1 tbsp	16	1	94	4	8	1.5	3.8	2.3
715	Reduced fat, smooth	1 tbsp	18	1	94	5	6	1.3	2.9	1.8
716	Peas, split, dry, cooked	1 cup	196	69	231	16	1	0.1	0.2	0.3
717	Pecans, halves	1 cup	108	4	746	10	78	6.7	44.0	23.3
718		1 oz (20 halves)	28	4	196	3	20	1.8	11.6	6.1
719	Pine nuts (pignolia), shelled	1 oz	28	7	160	7	14	2.2	5.4	6.1
720		1 tbsp	9	7	49	2	4	0.7	1.6	1.8
721	Pistachio nuts, dry roasted, with salt, shelled	1 oz (47 nuts)	28	2	161	6	13	1.6	6.8	3.9
722	Pumpkin and squash kernels, roasted, with salt	1 oz (142 seeds)	28	7	148	9	12	2.3	3.7	5.4
723	Refried beans, canned	1 cup	252	76	237	14	3	1.2	1.4	0.4
724	Sesame seeds	1 tbsp	8	5	47	2	4	0.6	1.7	1.9
725	Soybeans, dry, cooked	1 cup	172	63	298	29	15	2.2	3.4	8.7
	Soy products									
726	Miso	1 cup	275	41	567	32	17	2.4	3.7	9.4
727	Soy milk	1 cup	245	93	81	7	5	0.5	0.8	2.0

Choles-terol (mg)	Carbo-hydrate (g)	Total dietary fiber (g)	Calcium (mg)	Iron (mg)	Potas-sium (mg)	Sodium (mg)	Vitamin A (IU)	Vitamin A (RE)	Thiamin (mg)	Ribo-flavin (mg)	Niacin (mg)	Ascor-bic acid (mg)	Food No.
0	36	11.2	41	4.3	478	7	26	3	0.35	0.09	0.9	1	687
0	33	7.9	48	2.3	413	718	31	2	0.18	0.18	0.8	6	688
0	4	1.5	50	1.0	170	1	0	0	0.28	0.03	0.5	Tr	689
0	92	41.0	358	3.0	852	36	14	1	0.05	0.47	2.0	Tr	690
0	9	0.9	13	1.7	160	181	0	0	0.06	0.06	0.4	0	691
0	37	4.9	53	5.3	689	814	0	0	0.55	0.23	2.3	0	692
0	8	1.1	12	1.2	150	177	0	0	0.12	0.05	0.5	0	693
0	76	7.3	41	1.3	847	3	34	3	0.35	0.25	1.9	37	694
0	45	12.5	80	4.7	477	11	44	5	0.19	0.10	0.9	2	695
0	54	10.6	77	3.2	413	718	58	5	0.07	0.08	0.3	9	696
0	7	4.1	6	1.1	160	9	0	0	0.03	0.01	0.2	1	697
0	12	7.2	11	1.9	285	16	0	0	0.05	0.02	0.4	3	698
0	44	4.2	14	1.8	313	244	0	0	0.03	0.02	0.4	1	699
0	19	11.2	131	5.4	782	0	46	5	0.74	0.13	2.1	7	700
0	5	2.7	32	1.3	193	0	11	1	0.18	0.03	0.5	2	701
0	2	0.8	5	0.3	32	53	4	Tr	0.03	0.01	0.1	0	702
0	40	15.6	38	6.6	731	4	16	2	0.33	0.14	2.1	3	703
0	17	10.7	94	3.6	486	355	0	0	0.95	0.12	3.0	1	704
0	4	2.3	20	0.8	103	75	0	0	0.20	0.02	0.6	Tr	705
0	7	2.6	20	1.0	169	190	4	Tr	0.06	0.06	1.3	Tr	706
0	6	2.6	31	0.9	165	185	5	1	0.14	0.06	1.4	Tr	707
0	6	2.3	15	0.6	187	230	0	0	0.12	0.03	3.8	0	708
0	31	11.7	79	3.3	961	9	0	0	0.64	0.14	19.7	0	709
0	6	2.3	15	0.6	187	2	0	0	0.12	0.03	3.8	0	710
0	27	13.2	127	2.6	982	624	0	0	0.36	0.16	20.6	0	711
0	5	2.6	25	0.5	193	123	0	0	0.07	0.03	4.0	0	712
0	3	0.9	6	0.3	107	75	0	0	0.01	0.02	2.1	0	713
0	3	1.1	7	0.3	120	78	0	0	0.02	0.02	2.2	0	714
0	6	0.9	6	0.3	120	97	0	0	0.05	0.01	2.6	0	715
0	41	16.3	27	2.5	710	4	14	2	0.37	0.11	1.7	1	716
0	15	10.4	76	2.7	443	0	83	9	0.71	0.14	1.3	1	717
0	4	2.7	20	0.7	116	0	22	2	0.19	0.04	0.3	Tr	718
0	4	1.3	7	2.6	170	1	8	1	0.23	0.05	1.0	1	719
0	1	0.4	2	0.8	52	Tr	2	Tr	0.07	0.02	0.3	Tr	720
0	8	2.9	31	1.2	293	121	151	15	0.24	0.04	0.4	1	721
0	4	1.1	12	4.2	229	163	108	11	0.06	0.09	0.5	1	722
20	39	13.4	88	4.2	673	753	0	0	0.07	0.04	0.8	15	723
0	1	0.9	10	0.6	33	3	5	1	0.06	0.01	0.4	0	724
0	17	10.3	175	8.8	886	2	15	2	0.27	0.49	0.7	3	725
0	77	14.9	182	7.5	451	10,029	239	25	0.27	0.69	2.4	0	726
0	4	3.2	10	1.4	345	29	78	7	0.39	0.17	0.4	0	727

Food No.	Food Description	Measure of edible portion	Weight (g)	Water (%)	Calories (kcal)	Protein (g)	Total fat (g)	Fatty acids Saturated (g)	Fatty acids Monounsaturated (g)	Fatty acids Polyunsaturated (g)

Legumes, Nuts, and Seeds (continued)

Soy products (continued)
Tofu

728	Firm	¼ block 81		84	62	7	4	0.5	0.8	2.0
729	Soft, piece 2½" x 2¾" x 1"	1 piece 120		87	73	8	4	0.6	1.0	2.5
730	Sunflower seed kernels, dry roasted, with salt	¼ cup 32		1	186	6	16	1.7	3.0	10.5
731		1 oz 28		1	165	5	14	1.5	2.7	9.3
732	Tahini	1 tbsp 15		3	89	3	8	1.1	3.0	3.5
733	Walnuts, English	1 cup, chopped ... 120		4	785	18	78	7.4	10.7	56.6
734		1 oz (14 halves) ... 28		4	185	4	18	1.7	2.5	13.4

Meat and Meat Products

Beef, cooked
Cuts braised, simmered, or pot roasted
Relatively fat, such as chuck blade, piece, 2½" x 2½" x ¾"

| 735 | Lean and fat | 3 oz ...... 85 | | 47 | 293 | 23 | 22 | 8.7 | 9.4 | 0.8 |
| 736 | Lean only | 3 oz ...... 85 | | 55 | 213 | 26 | 11 | 4.3 | 4.8 | 0.4 |

Relatively lean, such as bottom round, piece, 4⅛" x 2¼" x ½"

| 737 | Lean and fat | 3 oz ...... 85 | | 52 | 234 | 24 | 14 | 5.4 | 6.2 | 0.5 |
| 738 | Lean only | 3 oz ...... 85 | | 58 | 178 | 27 | 7 | 2.4 | 3.1 | 0.3 |

Ground beef, broiled

739	83% lean	3 oz 85		57	218	22	14	5.5	6.1	0.5
740	79% lean	3 oz 85		56	231	21	16	6.2	6.9	0.6
741	73% lean	3 oz 85		54	246	20	18	6.9	7.7	0.7
742	Liver, fried, slice, 6½" x 2⅜" x ⅜"	3 oz 85		56	184	23	7	2.3	1.4	1.5

Roast, oven cooked, no liquid added
Relatively fat, such as rib, 2 pieces, 4⅛" x 2¼" x ¼"

| 743 | Lean and fat | 3 oz ...... 85 | | 47 | 304 | 19 | 25 | 9.9 | 10.6 | 0.9 |
| 744 | Lean only | 3 oz ...... 85 | | 59 | 195 | 23 | 11 | 4.2 | 4.5 | 0.3 |

Relatively lean, such as eye of round, 2 pieces, 2½" x 2½" x ⅜"

| 745 | Lean and fat | 3 oz ...... 85 | | 59 | 195 | 23 | 11 | 4.2 | 4.7 | 0.4 |
| 746 | Lean only | 3 oz ...... 85 | | 65 | 143 | 25 | 4 | 1.5 | 1.8 | 0.1 |

Steak, sirloin, broiled, piece, 2½" x 2½" x ¾"

747	Lean and fat	3 oz 85		57	219	24	13	5.2	5.6	0.5
748	Lean only	3 oz 85		62	166	26	6	2.4	2.6	0.2
749	Beef, canned, corned	3 oz 85		58	213	23	13	5.3	5.1	0.5
750	Beef, dried, chipped	1 oz 28		57	47	8	1	0.5	0.5	0.1

Lamb, cooked
Chops
Arm, braised

| 751 | Lean and fat | 3 oz ...... 85 | | 44 | 294 | 26 | 20 | 8.4 | 8.7 | 1.5 |
| 752 | Lean only | 3 oz ...... 85 | | 49 | 237 | 30 | 12 | 4.3 | 5.2 | 0.8 |

Loin, broiled

| 753 | Lean and fat | 3 oz ...... 85 | | 52 | 269 | 21 | 20 | 8.4 | 8.2 | 1.4 |
| 754 | Lean only | 3 oz ...... 85 | | 61 | 184 | 25 | 8 | 3.0 | 3.6 | 0.5 |

Choles-terol (mg)	Carbo-hydrate (g)	Total dietary fiber (g)	Calcium (mg)	Iron (mg)	Potas-sium (mg)	Sodium (mg)	Vitamin A (IU)	(RE)	Thiamin (mg)	Ribo-flavin (mg)	Niacin (mg)	Ascor-bic acid (mg)	Food No.
0	2	0.3	131	1.2	143	6	6	1	0.08	0.08	Tr	Tr	728
0	2	0.2	133	1.3	144	10	8	1	0.06	0.04	0.6	Tr	729
0	8	2.9	22	1.2	272	250	0	0	0.03	0.08	2.3	Tr	730
0	7	2.6	20	1.1	241	221	0	0	0.03	0.07	2.0	Tr	731
0	3	1.4	64	1.3	62	17	10	1	0.18	0.07	0.8	0	732
0	16	8.0	125	3.5	529	2	49	5	0.41	0.18	2.3	2	733
0	4	1.9	29	0.8	125	1	12	1	0.10	0.04	0.5	Tr	734
88	0	0.0	11	2.6	196	54	0	0	0.06	0.20	2.1	0	735
90	0	0.0	11	3.1	224	60	0	0	0.07	0.24	2.3	0	736
82	0	0.0	5	2.7	240	43	0	0	0.06	0.20	3.2	0	737
82	0	0.0	4	2.9	262	43	0	0	0.06	0.22	3.5	0	738
71	0	0.0	6	2.0	266	60	0	0	0.05	0.23	4.2	0	739
74	0	0.0	9	1.8	256	65	0	0	0.04	0.18	4.4	0	740
77	0	0.0	9	2.1	248	71	0	0	0.03	0.16	4.9	0	741
410	7	0.0	9	5.3	309	90	30,689	9,120	0.18	3.52	12.3	20	742
71	0	0.0	9	2.0	256	54	0	0	0.06	0.14	2.9	0	743
68	0	0.0	9	2.4	318	61	0	0	0.07	0.18	3.5	0	744
61	0	0.0	5	1.6	308	50	0	0	0.07	0.14	3.0	0	745
59	0	0.0	4	1.7	336	53	0	0	0.08	0.14	3.2	0	746
77	0	0.0	9	2.6	311	54	0	0	0.09	0.23	3.3	0	747
76	0	0.0	9	2.9	343	56	0	0	0.11	0.25	3.6	0	748
73	0	0.0	10	1.8	116	855	0	0	0.02	0.12	2.1	0	749
12	Tr	0.0	2	1.3	126	984	0	0	0.02	0.06	1.5	0	750
102	0	0.0	21	2.0	260	61	0	0	0.06	0.21	5.7	0	751
103	0	0.0	22	2.3	287	65	0	0	0.06	0.23	5.4	0	752
85	0	0.0	17	1.5	278	65	0	0	0.09	0.21	6.0	0	753
81	0	0.0	16	1.7	320	71	0	0	0.09	0.24	5.8	0	754

Food No.	Food Description	Measure of edible portion	Weight (g)	Water (%)	Calories (kcal)	Protein (g)	Total fat (g)	Fatty acids		
								Saturated (g)	Monounsaturated (g)	Polyunsaturated (g)

Meat and Meat Products (continued)

Lamb (continued)
Leg, roasted, 2 pieces, 4⅛" x 2¼" x ¼"

Food No.	Food Description	Measure of edible portion	Weight (g)	Water (%)	Calories (kcal)	Protein (g)	Total fat (g)	Saturated (g)	Monounsaturated (g)	Polyunsaturated (g)
755	Lean and fat	3 oz	85	57	219	22	14	5.9	5.9	1.0
756	Lean only	3 oz	85	64	162	24	7	2.3	2.9	0.4
	Rib, roasted, 3 pieces, 2½" x 2½" x ¼"									
757	Lean and fat	3 oz	85	48	305	18	25	10.9	10.6	1.8
758	Lean only	3 oz	85	60	197	22	11	4.0	5.0	0.7
	Pork, cured, cooked									
	Bacon									
759	Regular	3 medium slices	19	13	109	6	9	3.3	4.5	1.1
760	Canadian style (6 slices per 6-oz pkg)	2 slices	47	62	86	11	4	1.3	1.9	0.4
	Ham, light cure, roasted, 2 pieces, 4⅛" x 2¼" x ¼"									
761	Lean and fat	3 oz	85	58	207	18	14	5.1	6.7	1.5
762	Lean only	3 oz	85	66	133	21	5	1.6	2.2	0.5
763	Ham, canned, roasted, 2 pieces, 4⅛" x 2¼" x ¼"	3 oz	85	67	142	18	7	2.4	3.5	0.8
	Pork, fresh, cooked									
	Chop, loin (cut 3 per lb with bone)									
	Broiled									
764	Lean and fat	3 oz	85	58	204	24	11	4.1	5.0	0.8
765	Lean only	3 oz	85	61	172	26	7	2.5	3.1	0.5
	Pan fried									
766	Lean and fat	3 oz	85	53	235	25	14	5.1	6.0	1.6
767	Lean only	3 oz	85	57	197	27	9	3.1	3.8	1.1
	Ham (leg), roasted, piece, 2½" x 2½" x ¾"									
768	Lean and fat	3 oz	85	55	232	23	15	5.5	6.7	1.4
769	Lean only	3 oz	85	61	179	25	8	2.8	3.8	0.7
	Rib roast, piece, 2½" x 2½" x ¾"									
770	Lean and fat	3 oz	85	56	217	23	13	5.0	5.9	1.1
771	Lean only	3 oz	85	59	190	24	9	3.7	4.5	0.7
	Ribs, lean and fat, cooked									
772	Backribs, roasted	3 oz	85	45	315	21	25	9.3	11.4	2.0
773	Country style, braised	3 oz	85	54	252	20	18	6.8	7.9	1.6
774	Spareribs, braised	3 oz	85	40	337	25	26	9.5	11.5	2.3
	Shoulder cut, braised, 3 pieces, 2½" x 2½" x ¼"									
775	Lean and fat	3 oz	85	48	280	24	20	7.2	8.8	1.9
776	Lean only	3 oz	85	54	211	27	10	3.5	4.9	1.0
	Sausages and luncheon meats									
777	Bologna, beef and pork (8 slices per 8-oz pkg)	2 slices	57	54	180	7	16	6.1	7.6	1.4
778	Braunschweiger (6 slices per 6-oz pkg)	2 slices	57	48	205	8	18	6.2	8.5	2.1
779	Brown and serve, cooked, link, 4" x ⅞" raw	2 links	26	45	103	4	9	3.4	4.5	1.0
	Canned, minced luncheon meat									
780	Pork, ham, and chicken, reduced sodium (7 slices per 7-oz can)	2 slices	57	56	172	7	15	5.1	7.1	1.5
781	Pork with ham (12 slices per 12-oz can)	2 slices	57	52	188	8	17	5.7	7.7	1.2
782	Pork and chicken (12 slices per 12-oz can)	2 slices	57	64	117	9	8	2.7	3.8	0.8

Choles-terol (mg)	Carbo-hydrate (g)	Total dietary fiber (g)	Calcium (mg)	Iron (mg)	Potas-sium (mg)	Sodium (mg)	Vitamin A (IU)	(RE)	Thiamin (mg)	Ribo-flavin (mg)	Niacin (mg)	Ascor-bic acid (mg)	Food No.
79	0	0.0	9	1.7	266	56	0	0	0.09	0.23	5.6	0	755
76	0	0.0	7	1.8	287	58	0	0	0.09	0.25	5.4	0	756
82	0	0.0	19	1.4	230	62	0	0	0.08	0.18	5.7	0	757
75	0	0.0	18	1.5	268	69	0	0	0.08	0.20	5.2	0	758
16	Tr	0.0	2	0.3	92	303	0	0	0.13	0.05	1.4	0	759
27	1	0.0	5	0.4	181	719	0	0	0.38	0.09	3.2	0	760
53	0	0.0	6	0.7	243	1,009	0	0	0.51	0.19	3.8	0	761
47	0	0.0	6	0.8	269	1,128	0	0	0.58	0.22	4.3	0	762
35	Tr	0.0	6	0.9	298	908	0	0	0.82	0.21	4.3	0	763
70	0	0.0	28	0.7	304	49	8	3	0.91	0.24	4.5	Tr	764
70	0	0.0	26	0.7	319	51	7	2	0.98	0.26	4.7	Tr	765
78	0	0.0	23	0.8	361	68	7	2	0.97	0.26	4.8	1	766
78	0	0.0	20	0.8	382	73	7	2	1.06	0.28	5.1	1	767
80	0	0.0	12	0.9	299	51	9	3	0.54	0.27	3.9	Tr	768
80	0	0.0	6	1.0	317	54	8	3	0.59	0.30	4.2	Tr	769
62	0	0.0	24	0.8	358	39	5	2	0.62	0.26	5.2	Tr	770
60	0	0.0	22	0.8	371	40	5	2	0.64	0.27	5.5	Tr	771
100	0	0.0	38	1.2	268	86	8	3	0.36	0.17	3.0	Tr	772
74	0	0.0	25	1.0	279	50	7	2	0.43	0.22	3.3	1	773
103	0	0.0	40	1.6	272	79	9	3	0.35	0.32	4.7	0	774
93	0	0.0	15	1.4	314	75	8	3	0.46	0.26	4.4	Tr	775
97	0	0.0	7	1.7	344	87	7	2	0.51	0.31	5.0	Tr	776
31	2	0.0	7	0.9	103	581	0	0	0.10	0.08	1.5	0	777
89	2	0.0	5	5.3	113	652	8,009	2,405	0.14	0.87	4.8	0	778
18	1	0.0	3	0.3	49	209	0	0	0.09	0.04	0.9	0	779
43	1	0.0	0	0.4	321	539	0	0	0.15	0.10	1.8	18	780
40	1	0.0	0	0.4	233	758	0	0	0.18	0.10	2.0	0	781
43	1	0.0	0	0.7	352	539	0	0	0.10	0.12	2.0	18	782

Food No.	Food Description	Measure of edible portion	Weight (g)	Water (%)	Calories (kcal)	Protein (g)	Total fat (g)	Saturated (g)	Mono-unsaturated (g)	Poly-unsaturated (g)

Fatty acids

Meat and Meat Products (continued)

Sausages and luncheon meats (continued)

Food No.	Food Description	Measure	Weight	Water	Calories	Protein	Total fat	Satu-rated	Mono-unsat.	Poly-unsat.
783	Chopped ham (8 slices per 6-oz pkg)	2 slices....21		64	48	4	4	1.2	1.7	0.4
	Cooked ham (8 slices per 8-oz pkg)									
784	Regular	2 slices....57		65	104	10	6	1.9	2.8	0.7
785	Extra lean	2 slices....57		71	75	11	3	0.9	1.3	0.3
	Frankfurter (10 per 1-lb pkg), heated									
786	Beef and pork	1 frank....45		54	144	5	13	4.8	6.2	1.2
787	Beef	1 frank....45		55	142	5	13	5.4	6.1	0.6
	Pork sausage, fresh, cooked									
788	Link (4" x ⅞" raw)	2 links....26		45	96	5	8	2.8	3.6	1.0
789	Patty (3⅞" x ¼" raw)	1 patty....27		45	100	5	8	2.9	3.8	1.0
	Salami, beef and pork									
790	Cooked type (8 slices per 8-oz pkg)	2 slices....57		60	143	8	11	4.6	5.2	1.2
791	Dry type, slice, 3⅛" x 1/16"	2 slices....20		35	84	5	7	2.4	3.4	0.6
792	Sandwich spread (pork, beef)	1 tbsp....15		60	35	1	3	0.9	1.1	0.4
793	Vienna sausage (7 per 4-oz can)	1 sausage....16		60	45	2	4	1.5	2.0	0.3
	Veal, lean and fat, cooked									
794	Cutlet, braised, 4⅛" x 2¼" x ½"	3 oz....85		55	179	31	5	2.2	2.0	0.4
795	Rib, roasted, 2 pieces, 4⅛" x 2¼" x ¼"	3 oz....85		60	194	20	12	4.6	4.6	0.8

Mixed Dishes and Fast Foods

Mixed dishes

Food No.	Food Description	Measure	Weight	Water	Calories	Protein	Total fat	Satu-rated	Mono-unsat.	Poly-unsat.
796	Beef macaroni, frozen, HEALTHY CHOICE	1 package....240		78	211	14	2	0.7	1.2	0.3
797	Beef stew, canned	1 cup....232		82	218	11	12	5.2	5.5	0.5
798	Chicken pot pie, frozen	1 small pie....217		60	484	13	29	9.7	12.5	4.5
799	Chili con carne with beans, canned	1 cup....222		74	255	20	8	2.1	2.2	1.4
800	Macaroni and cheese, canned, made with corn oil	1 cup....252		82	199	8	6	3.0	NA	1.3
801	Meatless burger crumbles, MORNINGSTAR FARMS	1 cup....110		60	231	22	13	3.3	4.6	4.9
802	Meatless burger patty, frozen, MORNINGSTAR FARMS	1 patty....85		71	91	14	1	0.1	0.3	0.2
803	Pasta with meatballs in tomato sauce, canned	1 cup....252		78	260	11	10	4.0	4.2	0.6
804	Spaghetti bolognese (meat sauce), frozen, HEALTHY CHOICE	1 package....283		78	255	14	3	1.0	0.9	0.9
805	Spaghetti in tomato sauce with cheese, canned	1 cup....252		80	192	6	2	0.7	0.3	0.3
806	Spinach souffle, home-prepared	1 cup....136		74	219	11	18	7.1	6.8	3.1
807	Tortellini, pasta with cheese filling, frozen	¾ cup (yields 1 cup cooked)....81		31	249	11	6	2.9	1.7	0.4

Choles-terol (mg)	Carbo-hydrate (g)	Total dietary fiber (g)	Calcium (mg)	Iron (mg)	Potas-sium (mg)	Sodium (mg)	Vitamin A (IU)	Vitamin A (RE)	Thiamin (mg)	Ribo-flavin (mg)	Niacin (mg)	Ascor-bic acid (mg)	Food No.
11	0	0.0	1	0.2	67	288	0	0	0.13	0.04	0.8	0	783
32	2	0.0	4	0.6	189	751	0	0	0.49	0.14	3.0	0	784
27	1	0.0	4	0.4	200	815	0	0	0.53	0.13	2.8	0	785
23	1	0.0	5	0.5	75	504	0	0	0.09	0.05	1.2	0	786
27	1	0.0	9	0.6	75	462	0	0	0.02	0.05	1.1	0	787
22	Tr	0.0	8	0.3	94	336	0	0	0.19	0.07	1.2	1	788
22	Tr	0.0	9	0.3	97	349	0	0	0.20	0.07	1.2	1	789
37	1	0.0	7	1.5	113	607	0	0	0.14	0.21	2.0	0	790
16	1	0.0	2	0.3	76	372	0	0	0.12	0.06	1.0	0	791
6	2	Tr	2	0.1	17	152	13	1	0.03	0.02	0.3	0	792
8	Tr	0.0	2	0.1	16	152	0	0	0.01	0.02	0.3	0	793
114	0	0.0	7	1.1	326	57	0	0	0.05	0.30	9.0	0	794
94	0	0.0	9	0.8	251	78	0	0	0.04	0.23	5.9	0	795
14	33	4.6	46	2.7	365	444	514	50	0.28	0.16	3.1	58	796
37	16	3.5	28	1.6	404	947	3,860	494	0.17	0.14	2.9	10	797
41	43	1.7	33	2.1	256	857	2,285	343	0.25	0.36	4.1	2	798
24	24	8.2	67	3.3	608	1,032	884	93	0.15	0.15	2.1	1	799
8	29	3.0	113	2.0	123	1,058	713	NA	0.28	0.25	2.5	0	800
0	7	5.1	79	6.4	178	476	0	0	9.92	0.35	3.0	0	801
0	8	4.3	87	2.9	434	383	0	0	0.26	0.55	4.1	0	802
20	31	6.8	28	2.3	416	1,053	920	93	0.19	0.16	3.3	8	803
17	43	5.1	51	3.5	408	473	492	48	0.35	3.77	0.5	15	804
8	39	7.8	40	2.8	305	963	932	58	0.35	0.28	4.5	10	805
184	3	NA	230	1.3	201	763	3,461	675	0.09	0.30	0.5	3	806
34	38	1.5	123	1.2	72	279	50	13	0.25	0.25	2.2	0	807

Food No.	Food Description	Measure of edible portion	Weight (g)	Water (%)	Calories (kcal)	Pro-tein (g)	Total fat (g)	Fatty acids Satu-rated (g)	Mono-unsatu-rated (g)	Poly-unsatu-rated (g)
	Mixed Dishes and Fast Foods (continued)									
	Fast foods									
	Breakfast items									
808	Biscuit with egg and sausage	1 biscuit	180	43	581	19	39	15.0	16.4	4.4
809	Croissant with egg, cheese, bacon	1 croissant	129	44	413	16	28	15.4	9.2	1.8
	Danish pastry									
810	Cheese filled	1 pastry	91	34	353	6	25	5.1	15.6	2.4
811	Fruit filled	1 pastry	94	29	335	5	16	3.3	10.1	1.6
812	English muffin with egg, cheese, Canadian bacon	1 muffin	137	57	289	17	13	4.7	4.7	1.6
813	French toast with butter	2 slices	135	51	356	10	19	7.7	7.1	2.4
814	French toast sticks	5 sticks	141	30	513	8	29	4.7	12.6	9.9
815	Hashed brown potatoes	½ cup	72	60	151	2	9	4.3	3.9	0.5
816	Pancakes with butter, syrup	2 pancakes	232	50	520	8	14	5.9	5.3	2.0
	Burrito									
817	With beans and cheese	1 burrito	93	54	189	8	6	3.4	1.2	0.9
818	With beans and meat	1 burrito	116	52	255	11	9	4.2	3.5	0.6
	Cheeseburger									
	Regular size, with condiments									
819	Double patty with mayo type dressing, vegetables	1 sandwich	166	51	417	21	21	8.7	7.8	2.7
820	Single patty	1 sandwich	113	48	295	16	14	6.3	5.3	1.1
	Regular size, plain									
821	Double patty	1 sandwich	155	42	457	28	28	13.0	11.0	1.9
822	Double patty with 3-piece bun	1 sandwich	160	43	461	22	22	9.5	8.3	1.8
823	Single patty	1 sandwich	102	37	319	15	15	6.5	5.8	1.5
	Large, with condiments									
824	Single patty with mayo type dressing, vegetables	1 sandwich	219	53	563	28	33	15.0	12.6	2.0
825	Single patty with bacon	1 sandwich	195	44	608	32	37	16.2	14.5	2.7
826	Chicken fillet (breaded and fried) sandwich, plain	1 sandwich	182	47	515	24	29	8.5	10.4	8.4
	Chicken, fried. See Poultry and Poultry Products.									
827	Chicken pieces, boneless, breaded and fried, plain	6 pieces	106	47	319	18	21	4.7	10.5	4.6
828	Chili con carne	1 cup	253	77	256	25	8	3.4	3.4	0.5
829	Chimichanga with beef	1 chimichanga	174	51	425	20	20	8.5	8.1	1.1
830	Coleslaw	¾ cup	99	74	147	1	11	1.6	2.4	6.4
	Desserts									
831	Ice milk, soft, vanilla, in cone	1 cone	103	65	164	4	6	3.5	1.8	0.4
832	Pie, fried, with fruit filling (5" x 3¾")	1 pie	128	38	404	4	21	3.1	9.5	6.9
833	Sundae, hot fudge	1 sundae	158	60	284	6	9	5.0	2.3	0.8
834	Enchilada with cheese	1 enchilada	163	63	319	10	19	10.6	6.3	0.8
835	Fish sandwich, with tartar sauce and cheese	1 sandwich	183	45	523	21	29	8.1	8.9	9.4
836	French fries	1 small	85	35	291	4	16	3.3	9.0	2.7
837		1 medium	134	35	458	6	25	5.2	14.3	4.2
838		1 large	169	35	578	7	31	6.5	18.0	5.3
839	Frijoles (refried beans, chili sauce, cheese)	1 cup	167	69	225	11	8	4.1	2.6	0.7

Choles-terol (mg)	Carbo-hydrate (g)	Total dietary fiber (g)	Calcium (mg)	Iron (mg)	Potas-sium (mg)	Sodium (mg)	Vitamin A (IU)	Vitamin A (RE)	Thiamin (mg)	Ribo-flavin (mg)	Niacin (mg)	Ascor-bic acid (mg)	Food No.
302	41	0.9	155	4.0	320	1,141	635	164	0.50	0.45	3.6	0	808
215	24	NA	151	2.2	201	889	472	120	0.35	0.34	2.2	2	809
20	29	NA	70	1.8	116	319	155	43	0.26	0.21	2.5	3	810
19	45	NA	22	1.4	110	333	86	24	0.29	0.21	1.8	2	811
234	27	1.5	151	2.4	199	729	586	156	0.49	0.45	3.3	2	812
116	36	NA	73	1.9	177	513	473	146	0.58	0.50	3.9	Tr	813
75	58	2.7	78	3.0	127	499	45	13	0.23	0.25	3.0	0	814
9	16	NA	7	0.5	267	290	18	3	0.08	0.01	1.1	5	815
58	91	NA	128	2.6	251	1,104	281	70	0.39	0.56	3.4	3	816
14	27	NA	107	1.1	248	583	625	119	0.11	0.35	1.8	1	817
24	33	NA	53	2.5	329	670	319	32	0.27	0.42	2.7	1	818
60	35	NA	171	3.4	335	1,051	398	65	0.35	0.28	8.1	2	819
37	27	NA	111	2.4	223	616	462	94	0.25	0.23	3.7	2	820
110	22	NA	233	3.4	308	636	332	79	0.25	0.37	6.0	0	821
80	44	NA	224	3.7	285	891	277	66	0.34	0.38	6.0	0	822
50	32	NA	141	2.4	164	500	153	37	0.40	0.40	3.7	0	823
88	38	NA	206	4.7	445	1,108	613	129	0.39	0.46	7.4	8	824
111	37	NA	162	4.7	332	1,043	406	80	0.31	0.41	6.6	2	825
60	39	NA	60	4.7	353	957	100	31	0.33	0.24	6.8	9	826
61	15	0.0	14	0.9	305	513	0	0	0.12	0.16	7.5	0	827
134	22	NA	68	5.2	691	1,007	1,662	167	0.13	1.14	2.5	2	828
9	43	NA	63	4.5	586	910	146	16	0.49	0.64	5.8	5	829
5	13	NA	34	0.7	177	267	338	50	0.04	0.03	0.1	8	830
28	24	0.1	153	0.2	169	92	211	52	0.05	0.26	0.3	1	831
0	55	3.3	28	1.6	83	479	35	4	0.18	0.14	1.8	2	832
21	48	0.0	207	0.6	395	182	221	57	0.06	0.30	1.1	2	833
44	29	NA	324	1.3	240	784	1,161	186	0.08	0.42	1.9	1	834
68	48	NA	185	3.5	353	939	432	97	0.46	0.42	4.2	3	835
0	34	3.0	12	0.7	586	168	0	0	0.07	0.03	2.4	10	836
0	53	4.7	19	1.0	923	265	0	0	0.11	0.05	3.8	16	837
0	67	5.9	24	1.3	1,164	335	0	0	0.14	0.07	4.8	20	838
37	29	NA	189	2.2	605	882	456	70	0.13	0.33	1.5	2	839

Food No.	Food Description	Measure of edible portion	Weight (g)	Water (%)	Calories (kcal)	Pro-tein (g)	Total fat (g)	Fatty acids Satu-rated (g)	Mono-unsatu-rated (g)	Poly-unsatu-rated (g)

Mixed Dishes and Fast Foods (continued)

Fast foods (continued)
  Hamburger
    Regular size, with condiments

Food No.	Food Description	Measure	Weight	Water	Cal	Pro	Fat	Sat	Mono	Poly
840	Double patty .... 1 sandwich........215			51	576	32	32	12.0	14.1	2.8
841	Single patty....................... 1 sandwich.........106			45	272	12	10	3.6	3.4	1.0
	Large, with condiments, mayo type dressing, and vegetables									
842	Double patty .... 1 sandwich.........226			54	540	34	27	10.5	10.3	2.8
843	Single patty....................... 1 sandwich.........218			56	512	26	27	10.4	11.4	2.2
	Hot dog									
844	Plain.................................. 1 sandwich...........98			54	242	10	15	5.1	6.9	1.7
845	With chili .... 1 sandwich.........114			48	296	14	13	4.9	6.6	1.2
846	With corn flour coating (corndog)......................... 1 corndog.........175			47	460	17	19	5.2	9.1	3.5
847	Hush puppies .... 5 pieces...............78			32	257	5	12	2.7	7.8	0.4
848	Mashed potatoes .... ⅓ cup..................80			79	66	2	1	0.4	0.3	0.2
849	Nachos, with cheese sauce ..... 6-8 nachos 113			40	346	9	19	7.8	8.0	2.2
850	Onion rings, breaded and fried.............................. 8-9 rings 83			37	276	4	16	7.0	6.7	0.7
	Pizza (slice = ⅛ of 12" pizza)									
851	Cheese .... 1 slice 63			48	140	8	3	1.5	1.0	0.5
852	Meat and vegetables .... 1 slice 79			48	184	13	5	1.5	2.5	0.9
853	Pepperoni .... 1 slice 71			47	181	10	7	2.2	3.1	1.2
854	Roast beef sandwich, plain .... 1 sandwich.........139			49	346	22	14	3.6	6.8	1.7
855	Salad, tossed, with chicken, no dressing......................... 1½ cups 218			87	105	17	2	0.6	0.7	0.6
856	Salad, tossed, with egg, cheese, no dressing............ 1½ cups 217			90	102	9	6	3.0	1.8	0.5
	Shake									
857	Chocolate .... 16 fl oz 333			72	423	11	12	7.7	3.6	0.5
858	Vanilla .... 16 fl oz 333			75	370	12	10	6.2	2.9	0.4
859	Shrimp, breaded and fried...... 6-8 shrimp 164			48	454	19	25	5.4	17.4	0.6
	Submarine sandwich (6" long), with oil and vinegar									
860	Cold cuts (with lettuce, cheese, salami, ham, tomato, onion) .... 1 sandwich.........228			58	456	22	19	6.8	8.2	2.3
861	Roast beef (with tomato, lettuce, mayo) .... 1 sandwich.........216			59	410	29	13	7.1	1.8	2.6
862	Tuna salad (with mayo, lettuce) .... 1 sandwich.........256			54	584	30	28	5.3	13.4	7.3
863	Taco, beef........................... 1 small...............171			58	369	21	21	11.4	6.6	1.0
864	1 large................263			58	568	32	32	17.5	10.1	1.5
865	Taco salad (with ground beef, cheese, taco shell) .... 1½ cups 198			72	279	13	15	6.8	5.2	1.7
	Tostada (with cheese, tomato, lettuce)									
866	With beans and beef............. 1 tostada 225			70	333	16	17	11.5	3.5	0.6
867	With guacamole .... 1 tostada 131			73	181	6	12	5.0	4.3	1.5

Choles-terol (mg)	Carbo-hydrate (g)	Total dietary fiber (g)	Calcium (mg)	Iron (mg)	Potas-sium (mg)	Sodium (mg)	Vitamin A (IU)	(RE)	Thiamin (mg)	Ribo-flavin (mg)	Niacin (mg)	Ascor-bic acid (mg)	Food No.
103	39	NA	92	5.5	527	742	54	4	0.34	0.41	6.7	1	840
30	34	2.3	126	2.7	251	534	74	10	0.29	0.24	3.9	2	841
122	40	NA	102	5.9	570	791	102	11	0.36	0.38	7.6	1	842
87	40	NA	96	4.9	480	824	312	33	0.41	0.37	7.3	3	843
44	18	NA	24	2.3	143	670	0	0	0.24	0.27	3.6	Tr	844
51	31	NA	19	3.3	166	480	58	6	0.22	0.40	3.7	3	845
79	56	NA	102	6.2	263	973	207	37	0.28	0.70	4.2	0	846
135	35	NA	69	1.4	188	965	94	27	0.00	0.02	2.0	0	847
2	13	NA	17	0.4	235	182	33	8	0.07	0.04	1.0	Tr	848
18	36	NA	272	1.3	172	816	559	92	0.19	0.37	1.5	1	849
14	31	NA	73	0.8	129	430	8	1	0.08	0.10	0.9	1	850
9	21	NA	117	0.6	110	336	382	74	0.18	0.16	2.5	1	851
21	21	NA	101	1.5	179	382	524	101	0.21	0.17	2.0	2	852
14	20	NA	65	0.9	153	267	282	55	0.13	0.23	3.0	2	853
51	33	NA	54	4.2	316	792	210	21	0.38	0.31	5.9	2	854
72	4	NA	37	1.1	447	209	935	96	0.11	0.13	5.9	17	855
98	5	NA	100	0.7	371	119	822	115	0.09	0.17	1.0	10	856
43	68	2.7	376	1.0	666	323	310	77	0.19	0.82	0.5	1	857
37	60	1.3	406	0.3	579	273	433	107	0.15	0.61	0.6	3	858
200	40	NA	84	3.0	184	1,446	120	36	0.21	0.90	0.0	0	859
36	51	NA	189	2.5	394	1,651	424	80	1.00	0.80	5.5	12	860
73	44	NA	41	2.8	330	845	413	50	0.41	0.41	6.0	6	861
49	55	NA	74	2.6	335	1,293	187	41	0.46	0.33	11.3	4	862
56	27	NA	221	2.4	474	802	855	147	0.15	0.44	3.2	2	863
87	41	NA	339	3.7	729	1,233	1,315	226	0.24	0.68	4.9	3	864
44	24	NA	192	2.3	416	762	588	77	0.10	0.36	2.5	4	865
74	30	NA	189	2.5	491	871	1,276	173	0.09	0.50	2.9	4	866
20	16	NA	212	0.8	326	401	879	109	0.07	0.29	1.0	2	867

Poultry and Poultry Products

Food No.	Food Description	Measure of edible portion	Weight (g)	Water (%)	Calories (kcal)	Protein (g)	Total fat (g)	Fatty acids Saturated (g)	Mono-unsaturated (g)	Poly-unsaturated (g)
	Chicken									
	Fried in vegetable shortening, meat with skin									
	Batter dipped									
868	Breast, ½ breast (5.6 oz with bones)	½ breast	140	52	364	35	18	4.9	7.6	4.3
869	Drumstick (3.4 oz with bones)	1 drumstick	72	53	193	16	11	3.0	4.6	2.7
870	Thigh	1 thigh	86	52	238	19	14	3.8	5.8	3.4
871	Wing	1 wing	49	46	159	10	11	2.9	4.4	2.5
	Flour coated									
872	Breast, ½ breast (4.2 oz with bones)	½ breast	98	57	218	31	9	2.4	3.4	1.9
873	Drumstick (2.6 oz with bones)	1 drumstick	49	57	120	13	7	1.8	2.7	1.6
	Fried, meat only									
874	Dark meat	3 oz	85	56	203	25	10	2.7	3.7	2.4
875	Light meat	3 oz	85	60	163	28	5	1.3	1.7	1.1
	Roasted, meat only									
876	Breast, ½ breast (4.2 oz with bone and skin)	½ breast	86	65	142	27	3	0.9	1.1	0.7
877	Drumstick (2.9 oz with bone and skin)	1 drumstick	44	67	76	12	2	0.7	0.8	0.6
878	Thigh	1 thigh	52	63	109	13	6	1.6	2.2	1.3
879	Stewed, meat only, light and dark meat, chopped or diced	1 cup	140	56	332	43	17	4.3	5.7	4.0
880	Chicken giblets, simmered, chopped	1 cup	145	68	228	37	7	2.2	1.7	1.6
881	Chicken liver, simmered	1 liver	20	68	31	5	1	0.4	0.3	0.2
882	Chicken neck, meat only, simmered	1 neck	18	67	32	4	1	0.4	0.5	0.4
883	Duck, roasted, flesh only	½ duck	221	64	444	52	25	9.2	8.2	3.2
	Turkey									
	Roasted, meat only									
884	Dark meat	3 oz	85	63	159	24	6	2.1	1.4	1.8
885	Light meat	3 oz	85	66	133	25	3	0.9	0.5	0.7
886	Light and dark meat, chopped or diced	1 cup	140	65	238	41	7	2.3	1.4	2.0
	Ground, cooked									
887	Patty, from 4 oz raw	1 patty	82	59	193	22	11	2.8	4.0	2.6
888	Crumbled	1 cup	127	59	298	35	17	4.3	6.2	4.1
889	Turkey giblets, simmered, chopped	1 cup	145	65	242	39	7	2.2	1.7	1.7
890	Turkey neck, meat only, simmered	1 neck	152	65	274	41	11	3.7	2.5	3.3
	Poultry food products									
	Chicken									
891	Canned, boneless	5 oz	142	69	234	31	11	3.1	4.5	2.5
892	Frankfurter (10 per 1 lb pkg)	1 frank	45	58	116	6	9	2.5	3.8	1.8
893	Roll, light meat (6 slices per 6-oz pkg)	2 slices	57	69	90	11	4	1.1	1.7	0.9

Choles-terol (mg)	Carbo-hydrate (g)	Total dietary fiber (g)	Calcium (mg)	Iron (mg)	Potas-sium (mg)	Sodium (mg)	Vitamin A (IU)	Vitamin A (RE)	Thiamin (mg)	Ribo-flavin (mg)	Niacin (mg)	Ascor-bic acid (mg)	Food No.
119	13	0.4	28	1.8	281	385	94	28	0.16	0.20	14.7	0	868
62	6	0.2	12	1.0	134	194	62	19	0.08	0.15	3.7	0	869
80	8	0.3	15	1.2	165	248	82	25	0.10	0.20	4.9	0	870
39	5	0.1	10	0.6	68	157	55	17	0.05	0.07	2.6	0	871
87	2	0.1	16	1.2	254	74	49	15	0.08	0.13	13.5	0	872
44	1	Tr	6	0.7	112	44	41	12	0.04	0.11	3.0	0	873
82	2	0.0	15	1.3	215	82	67	20	0.08	0.21	6.0	0	874
77	Tr	0.0	14	1.0	224	69	26	8	0.06	0.11	11.4	0	875
73	0	0.0	13	0.9	220	64	18	5	0.06	0.10	11.8	0	876
41	0	0.0	5	0.6	108	42	26	8	0.03	0.10	2.7	0	877
49	0	0.0	6	0.7	124	46	34	10	0.04	0.12	3.4	0	878
116	0	0.0	18	2.0	283	109	157	46	0.16	0.39	9.0	0	879
570	1	0.0	17	9.3	229	84	10,775	3,232	0.13	1.38	5.9	12	880
126	Tr	0.0	3	1.7	28	10	3,275	983	0.03	0.35	0.9	3	881
14	0	0.0	8	0.5	25	12	22	6	0.01	0.05	0.7	0	882
197	0	0.0	27	6.0	557	144	170	51	0.57	1.04	11.3	0	883
72	0	0.0	27	2.0	247	67	0	0	0.05	0.21	3.1	0	884
59	0	0.0	16	1.1	259	54	0	0	0.05	0.11	5.8	0	885
106	0	0.0	35	2.5	417	98	0	0	0.09	0.25	7.6	0	886
84	0	0.0	21	1.6	221	88	0	0	0.04	0.14	4.0	0	887
130	0	0.0	32	2.5	343	136	0	0	0.07	0.21	6.1	0	888
606	3	0.0	19	9.7	290	86	8,752	2,603	0.07	1.31	6.5	2	889
185	0	0.0	56	3.5	226	85	0	0	0.05	0.29	2.6	0	890
88	0	0.0	20	2.2	196	714	166	48	0.02	0.18	9.0	3	891
45	3	0.0	43	0.9	38	617	59	17	0.03	0.05	1.4	0	892
28	1	0.0	24	0.5	129	331	46	14	0.04	0.07	3.0	0	893

Food No.	Food Description	Measure of edible portion	Weight (g)	Water (%)	Calories (kcal)	Protein (g)	Total fat (g)	Fatty acids Saturated (g)	Monounsaturated (g)	Polyunsaturated (g)

Poultry and Poultry Products (continued)

Poultry food products (continued)
Turkey

894	Gravy and turkey, frozen ..... 5-oz package 142		142	85	95	8	4	1.2	1.4	0.7
895	Patties, breaded or battered, fried (2.25 oz) .... 1 patty..................64		64	50	181	9	12	3.0	4.8	3.0
896	Roast, boneless, frozen, seasoned, light and dark meat, cooked.................. 3 oz 85		85	68	132	18	5	1.6	1.0	1.4

Soups, Sauces, and Gravies

Soups
Canned, condensed
Prepared with equal volume
of whole milk

897	Clam chowder, New England................ 1 cup 248		248	85	164	9	7	3.0	2.3	1.1
898	Cream of chicken .... 1 cup 248		248	85	191	7	11	4.6	4.5	1.6
899	Cream of mushroom .... 1 cup 248		248	85	203	6	14	5.1	3.0	4.6
900	Tomato .... 1 cup 248		248	85	161	6	6	2.9	1.6	1.1

Prepared with equal volume
of water

901	Bean with pork .... 1 cup 253		253	84	172	8	6	1.5	2.2	1.8
902	Beef broth, bouillon, consomme .... 1 cup 241		241	96	29	5	0	0.0	0.0	0.0
903	Beef noodle .... 1 cup 244		244	92	83	5	3	1.1	1.2	0.5
904	Chicken noodle................ 1 cup 241		241	92	75	4	2	0.7	1.1	0.6
905	Chicken and rice............... 1 cup 241		241	94	60	4	2	0.5	0.9	0.4
906	Clam chowder, Manhattan .... 1 cup 244		244	92	78	2	2	0.4	0.4	1.3
907	Cream of chicken .... 1 cup 244		244	91	117	3	7	2.1	3.3	1.5
908	Cream of mushroom .... 1 cup 244		244	90	129	2	9	2.4	1.7	4.2
909	Minestrone........................ 1 cup 241		241	91	82	4	3	0.6	0.7	1.1
910	Pea, green .... 1 cup 250		250	83	165	9	3	1.4	1.0	0.4
911	Tomato .... 1 cup 244		244	90	85	2	2	0.4	0.4	1.0
912	Vegetable beef .... 1 cup 244		244	92	78	6	2	0.9	0.8	0.1
913	Vegetarian vegetable .... 1 cup 241		241	92	72	2	2	0.3	0.8	0.7

Canned, ready to serve,
chunky

914	Bean with ham .... 1 cup 243		243	79	231	13	9	3.3	3.8	0.9
915	Chicken noodle .... 1 cup 240		240	84	175	13	6	1.4	2.7	1.5
916	Chicken and vegetable .... 1 cup 240		240	83	166	12	5	1.4	2.2	1.0
917	Vegetable .... 1 cup 240		240	88	122	4	4	0.6	1.6	1.4

Canned, ready to serve, low
fat, reduced sodium

918	Chicken broth..................... 1 cup 240		240	97	17	3	0	0.0	0.0	0.0
919	Chicken noodle .... 1 cup 237		237	92	76	6	2	0.4	0.6	0.4
920	Chicken and rice .... 1 cup 241		241	88	116	7	3	0.9	1.3	0.7
921	Chicken and rice with vegetables .... 1 cup 239		239	91	88	6	1	0.4	0.5	0.5
922	Clam chowder, New England................... 1 cup 244		244	89	117	5	2	0.5	0.7	0.4
923	Lentil .... 1 cup 242		242	88	126	8	2	0.3	0.8	0.2
924	Minestrone .... 1 cup 241		241	87	123	5	3	0.4	0.9	1.0
925	Vegetable .... 1 cup 238		238	91	81	4	1	0.3	0.4	0.3

Choles-terol (mg)	Carbo-hydrate (g)	Total dietary fiber (g)	Calcium (mg)	Iron (mg)	Potas-sium (mg)	Sodium (mg)	Vitamin A (IU)	Vitamin A (RE)	Thiamin (mg)	Ribo-flavin (mg)	Niacin (mg)	Ascor-bic acid (mg)	Food No.
26	7	0.0	20	1.3	87	787	60	18	0.03	0.18	2.6	0	894
40	10	0.3	9	1.4	176	512	24	7	0.06	0.12	1.5	0	895
45	3	0.0	4	1.4	253	578	0	0	0.04	0.14	5.3	0	896
22	17	1.5	186	1.5	300	992	164	40	0.07	0.24	1.0	3	897
27	15	0.2	181	0.7	273	1,047	714	94	0.07	0.26	0.9	1	898
20	15	0.5	179	0.6	270	918	154	37	0.08	0.28	0.9	2	899
17	22	2.7	159	1.8	449	744	848	109	0.13	0.25	1.5	68	900
3	23	8.6	81	2.0	402	951	888	89	0.09	0.03	0.6	2	901
0	2	0.0	10	0.5	154	636	0	0	0.02	0.03	0.7	1	902
5	9	0.7	15	1.1	100	952	630	63	0.07	0.06	1.1	Tr	903
7	9	0.7	17	0.8	55	1,106	711	72	0.05	0.06	1.4	Tr	904
7	7	0.7	17	0.7	101	815	660	65	0.02	0.02	1.1	Tr	905
2	12	1.5	27	1.6	188	578	964	98	0.03	0.04	0.8	4	906
10	9	0.2	34	0.6	88	986	561	56	0.03	0.06	0.8	Tr	907
2	9	0.5	46	0.5	100	881	0	0	0.05	0.09	0.7	1	908
2	11	1.0	34	0.9	313	911	2,338	234	0.05	0.04	0.9	1	909
0	27	2.8	28	2.0	190	918	203	20	0.11	0.07	1.2	2	910
0	17	0.5	12	1.8	264	695	688	68	0.09	0.05	1.4	66	911
5	10	0.5	17	1.1	173	791	1,891	190	0.04	0.05	1.0	2	912
0	12	0.5	22	1.1	210	822	3,005	301	0.05	0.05	0.9	1	913
22	27	11.2	78	3.2	425	972	3,951	396	0.15	0.15	1.7	4	914
19	17	3.8	24	1.4	108	850	1,222	122	0.07	0.17	4.3	0	915
17	19	NA	26	1.5	367	1,068	5,990	600	0.04	0.17	3.3	6	916
0	19	1.2	55	1.6	396	1,010	5,878	588	0.07	0.06	1.2	6	917
0	1	0.0	19	0.6	204	554	0	0	Tr	0.03	1.6	1	918
19	9	1.2	19	1.1	209	460	920	95	0.11	0.11	3.4	1	919
14	14	0.7	22	1.0	422	482	2,010	202	0.05	0.13	5.0	2	920
17	12	0.7	24	1.2	275	459	1,644	165	0.12	0.07	2.6	1	921
5	20	1.2	17	0.9	283	529	244	59	0.05	0.09	0.9	5	922
0	20	5.6	41	2.7	336	443	951	94	0.11	0.09	0.7	1	923
0	20	1.2	39	1.7	306	470	1,357	135	0.15	0.08	1.0	1	924
5	13	1.4	31	1.5	290	466	3,196	319	0.08	0.07	1.8	1	925

Soups, Sauces, and Gravies (continued)

Food No.	Food Description	Measure of edible portion	Weight (g)	Water (%)	Calories (kcal)	Protein (g)	Total fat (g)	Saturated (g)	Monounsaturated (g)	Polyunsaturated (g)
	Soups (continued)									
	Dehydrated									
	Unprepared									
926	Beef bouillon	1 packet	6	3	14	1	1	0.3	0.2	Tr
927	Onion	1 packet	39	4	115	5	2	0.5	1.4	0.3
	Prepared with water									
928	Chicken noodle	1 cup	252	94	58	2	1	0.3	0.5	0.4
929	Onion	1 cup	246	96	27	1	1	0.1	0.3	0.1
	Home prepared, stock									
930	Beef	1 cup	240	96	31	5	Tr	0.1	0.1	Tr
931	Chicken	1 cup	240	92	86	6	3	0.8	1.4	0.5
932	Fish	1 cup	233	97	40	5	2	0.5	0.5	0.3
	Sauces									
	Home recipe									
933	Cheese	1 cup	243	67	479	25	36	19.5	11.5	3.4
934	White, medium, made with whole milk	1 cup	250	75	368	10	27	7.1	11.1	7.2
	Ready to serve									
935	Barbecue	1 tbsp	16	81	12	Tr	Tr	Tr	0.1	0.1
936	Cheese	¼ cup	63	71	110	4	8	3.8	2.4	1.6
937	Hoisin	1 tbsp	16	44	35	1	1	0.1	0.2	0.3
938	Nacho cheese	¼ cup	63	70	119	5	10	4.2	3.1	2.1
939	Pepper or hot	1 tsp	5	90	1	Tr	Tr	Tr	Tr	Tr
940	Salsa	1 tbsp	16	90	4	Tr	Tr	Tr	Tr	Tr
941	Soy	1 tbsp	16	69	9	1	Tr	Tr	Tr	Tr
942	Spaghetti/marinara/pasta	1 cup	250	87	143	4	5	0.7	2.2	1.8
943	Teriyaki	1 tbsp	18	68	15	1	0	0.0	0.0	0.0
944	Tomato chili	¼ cup	68	68	71	2	Tr	Tr	Tr	0.1
945	Worcestershire	1 tbsp	17	70	11	0	0	0.0	0.0	0.0
	Gravies, canned									
946	Beef	¼ cup	58	87	31	2	1	0.7	0.6	Tr
947	Chicken	¼ cup	60	85	47	1	3	0.8	1.5	0.9
948	Country sausage	¼ cup	62	75	96	3	8	2.0	2.9	2.2
949	Mushroom	¼ cup	60	89	30	1	2	0.2	0.7	0.6
950	Turkey	¼ cup	60	89	31	2	1	0.4	0.5	0.3

Sugars and Sweets

Food No.	Food Description	Measure of edible portion	Weight (g)	Water (%)	Calories (kcal)	Protein (g)	Total fat (g)	Saturated (g)	Monounsaturated (g)	Polyunsaturated (g)
	Candy									
951	BUTTERFINGER (NESTLE)	1 fun size bar	7	2	34	1	1	0.7	0.4	0.2
	Caramel									
952	Plain	1 piece	10	9	39	Tr	1	0.7	0.1	Tr
953	Chocolate flavored roll	1 piece	7	7	25	Tr	Tr	Tr	0.1	0.1
954	Carob	1 oz	28	2	153	2	9	8.2	0.1	0.1
	Chocolate, milk									
955	Plain	1 bar (1.55 oz)	44	1	226	3	14	8.1	4.4	0.5
956	With almonds	1 bar (1.45 oz)	41	2	216	4	14	7.0	5.5	0.9
957	With peanuts, MR. GOODBAR (HERSHEY)	1 bar (1.75 oz)	49	1	267	5	17	7.3	5.7	2.4
958	With rice cereal, NESTLE CRUNCH	1 bar (1.55 oz)	44	1	230	3	12	6.7	3.8	0.4
	Chocolate chips									
959	Milk	1 cup	168	1	862	12	52	31.0	16.7	1.8
960	Semisweet	1 cup	168	1	805	7	50	29.8	16.7	1.6
961	White	1 cup	170	1	916	10	55	33.0	15.5	1.7
962	Chocolate coated peanuts	10 pieces	40	2	208	5	13	5.8	5.2	1.7
963	Chocolate coated raisins	10 pieces	10	11	39	Tr	1	0.9	0.5	0.1
964	Fruit leather, pieces	1 oz	28	12	97	Tr	2	0.3	0.9	0.8

Choles-terol (mg)	Carbo-hydrate (g)	Total dietary fiber (g)	Calcium (mg)	Iron (mg)	Potas-sium (mg)	Sodium (mg)	Vitamin A (IU)	Vitamin A (RE)	Thiamin (mg)	Ribo-flavin (mg)	Niacin (mg)	Ascor-bic acid (mg)	Food No.
1	1	0.0	4	0.1	27	1,019	3	Tr	Tr	0.01	0.3	0	926
2	21	4.1	55	0.6	260	3,493	8	1	0.11	0.24	2.0	1	927
10	9	0.3	5	0.5	33	578	15	5	0.20	0.08	1.1	0	928
0	5	1.0	12	0.1	64	849	2	0	0.03	0.06	0.5	Tr	929
0	3	0.0	19	0.6	444	475	0	0	0.08	0.22	2.1	0	930
7	8	0.0	7	0.5	252	343	0	0	0.08	0.20	3.8	Tr	931
2	0	0.0	7	Tr	336	363	0	0	0.08	0.18	2.8	Tr	932
92	13	0.2	756	0.9	345	1,198	1,473	389	0.11	0.59	0.5	1	933
18	23	0.5	295	0.8	390	885	1,383	138	0.17	0.46	1.0	2	934
0	2	0.2	3	0.1	28	130	139	14	Tr	Tr	0.1	1	935
18	4	0.3	116	0.1	19	522	199	40	Tr	0.07	Tr	Tr	936
Tr	7	0.4	5	0.2	19	258	2	Tr	Tr	0.03	0.2	Tr	937
20	3	0.5	118	0.2	20	492	128	32	Tr	0.08	Tr	Tr	938
0	Tr	0.1	Tr	Tr	7	124	14	1	Tr	Tr	Tr	4	939
0	1	0.3	5	0.2	34	69	96	10	0.01	0.01	0.1	2	940
0	1	0.1	3	0.3	64	871	0	0	0.01	0.03	0.4	0	941
0	21	4.0	55	1.8	738	1,030	938	95	0.14	0.10	2.7	20	942
0	3	Tr	5	0.3	41	690	0	0	0.01	0.01	0.2	0	943
0	17	4.0	14	0.5	252	910	462	46	0.06	0.05	1.1	11	944
0	3	0.0	18	0.9	136	167	18	2	0.01	0.02	0.1	2	945
2	3	0.2	3	0.4	47	325	0	0	0.02	0.02	0.4	0	946
1	3	0.2	12	0.3	65	346	221	67	0.01	0.03	0.3	0	947
13	4	0.4	4	0.3	48	236	0	0	0.10	0.04	0.7	Tr	948
0	3	0.2	4	0.4	64	342	0	0	0.02	0.04	0.4	0	949
1	3	0.2	2	0.4	65	346	0	0	0.01	0.05	0.8	0	950
Tr	5	0.2	2	0.1	27	14	0	0	0.01	Tr	0.2	0	951
1	8	0.1	14	Tr	22	25	3	1	Tr	0.02	Tr	Tr	952
0	6	Tr	2	Tr	7	6	1	Tr	Tr	0.01	Tr	Tr	953
1	16	1.1	86	0.4	179	30	7	2	0.03	0.05	0.3	Tr	954
10	26	1.5	84	0.6	169	36	81	24	0.03	0.13	0.1	Tr	955
8	22	2.5	92	0.7	182	30	30	6	0.02	0.18	0.3	Tr	956
4	25	1.7	53	0.6	219	73	70	18	0.08	0.12	1.6	Tr	957
6	29	1.1	74	0.2	151	59	30	9	0.15	0.25	1.7	Tr	958
37	99	5.7	321	2.3	647	138	311	92	0.13	0.51	0.5	1	959
0	106	9.9	54	5.3	613	18	35	3	0.09	0.15	0.7	0	960
36	101	0.0	338	0.4	486	153	60	2	0.11	0.48	1.3	1	961
4	20	1.9	42	0.5	201	16	0	0	0.05	0.07	1.7	0	962
Tr	7	0.4	9	0.2	51	4	4	1	0.01	0.02	Tr	Tr	963
0	22	1.0	5	0.2	46	114	33	3	0.01	0.03	Tr	16	964

Food No.	Food Description	Measure of edible portion	Weight (g)	Water (%)	Calories (kcal)	Pro-tein (g)	Total fat (g)	Satu-rated (g)	Mono-unsatu-rated (g)	Poly-unsatu-rated (g)

Sugars and Sweets (continued)

Candy (continued)

| 965 | Fruit leather, rolls .... | 1 large.................21 | | 11 | 74 | Tr | 1 | 0.1 | 0.3 | 0.1 |
| 966 | | 1 small................14 | | 11 | 49 | Tr | Tr | 0.1 | 0.2 | 0.1 |

Fudge, prepared from recipe
Chocolate

| 967 | Plain.................... | 1 piece ....17 | | 10 | 65 | Tr | 1 | 0.9 | 0.4 | 0.1 |
| 968 | With nuts .... | 1 piece ....19 | | 7 | 81 | 1 | 3 | 1.1 | 0.8 | 1.0 |

Vanilla

| 969 | Plain.................... | 1 piece ....16 | | 11 | 59 | Tr | 1 | 0.5 | 0.2 | Tr |
| 970 | With nuts .... | 1 piece ....15 | | 8 | 62 | Tr | 2 | 0.6 | 0.5 | 0.8 |

Gumdrops/gummy candies

971	Gumdrops (¾" dia)	1 cup 182		1	703	0	0	0.0	0.0	0.0
972		1 medium 4		1	16	0	0	0.0	0.0	0.0
973	Gummy bears	10 bears 22		1	85	0	0	0.0	0.0	0.0
974	Gummy worms....................	10 worms...........74		1	286	0	0	0.0	0.0	0.0
975	Hard candy...........................	1 piece 6		1	24	0	Tr	0.0	0.0	0.0
976		1 small piece 3		1	12	0	Tr	0.0	0.0	0.0
977	Jelly beans...........................	10 large..............28		6	104	0	Tr	Tr	0.1	Tr
978		10 small...............11		6	40	0	Tr	Tr	Tr	Tr
979	KIT KAT (HERSHEY)	1 bar (1.5 oz).......42		2	216	3	11	6.8	3.1	0.3

Marshmallows

| 980 | Miniature...................... | 1 cup ....50 | | 16 | 159 | 1 | Tr | Tr | Tr | Tr |
| 981 | Regular .... | 1 regular ....7 | | 16 | 23 | Tr | Tr | Tr | Tr | Tr |

M&M's (M&M MARS)

982	Peanut...........................	¼ cup..................43		2	222	4	11	4.4	4.7	1.8
983		10 pieces..............20		2	103	2	5	2.1	2.2	0.8
984	Plain...........................	¼ cup..................52		2	256	2	11	6.8	3.6	0.3
985		10 pieces................7		2	34	Tr	1	0.9	0.5	Tr
986	MILKY WAY (M&M MARS)................	1 fun size bar.......18		6	76	1	3	1.4	1.1	0.1
987		1 bar (2.15 oz).....61		6	258	3	10	4.8	3.7	0.4
988	REESE'S Peanut butter cup (HERSHEY)	1 miniature cup 7		2	38	1	2	0.8	0.9	0.4
989		1 package (contains 2) 45		2	243	5	14	5.0	5.9	2.5
990	SNICKERS bar (M&M MARS)................	1 fun size bar.......15		5	72	1	4	1.3	1.6	0.7
991		1 king size bar (4 oz) 113		5	541	9	28	10.2	11.8	5.6
992		1 bar (2 oz) 57		5	273	5	14	5.1	6.0	2.8
993	SPECIAL DARK sweet chocolate (HERSHEY)	1 miniature 8		1	46	Tr	3	1.7	0.9	0.1
994	STARBURST fruit chews (M&M MARS)................	1 piece 5		7	20	Tr	Tr	0.1	0.2	0.2
995		1 package (2.07 oz) 59		7	234	Tr	5	0.7	2.1	1.8

Frosting, ready to eat

| 996 | Chocolate .... | 1/12 package..........38 | | 17 | 151 | Tr | 7 | 2.1 | 3.4 | 0.8 |
| 997 | Vanilla .... | 1/12 package..........38 | | 13 | 159 | Tr | 6 | 1.9 | 3.3 | 0.9 |

Frozen desserts (nondairy)

998	Fruit and juice bar	1 bar (2.5 fl oz) ...77		78	63	1	Tr	0.0	0.0	Tr
999	Ice pop...................................	1 bar (2 fl oz) 59		80	42	0	0	0.0	0.0	0.0
1000	Italian ices.............................	½ cup..................116		86	61	Tr	Tr	0.0	0.0	0.0
1001	Fruit butter, apple	1 tbsp 17		56	29	Tr	0	0.0	0.0	0.0

Gelatin dessert, prepared with gelatin dessert powder and water

| 1002 | Regular .... | ½ cup..................135 | | 85 | 80 | 2 | 0 | 0.0 | 0.0 | 0.0 |
| 1003 | Reduced calorie (with aspartame)................ | ½ cup..................117 | | 98 | 8 | 1 | 0 | 0.0 | 0.0 | 0.0 |

Choles-terol (mg)	Carbo-hydrate (g)	Total dietary fiber (g)	Calcium (mg)	Iron (mg)	Potas-sium (mg)	Sodium (mg)	Vitamin A (IU)	Vitamin A (RE)	Thiamin (mg)	Ribo-flavin (mg)	Niacin (mg)	Ascor-bic acid (mg)	Food No.
0	18	0.8	7	0.2	62	13	24	3	0.01	Tr	Tr	1	965
0	12	0.5	4	0.1	41	9	16	2	0.01	Tr	Tr	1	966
2	14	0.1	7	0.1	18	11	32	8	Tr	0.01	Tr	Tr	967
3	14	0.2	10	0.1	30	11	38	9	0.01	0.02	Tr	Tr	968
3	13	0.0	6	Tr	8	11	33	8	Tr	0.01	Tr	Tr	969
2	11	0.1	7	0.1	17	9	30	7	0.01	0.01	Tr	Tr	970
0	180	0.0	5	0.7	9	80	0	0	0.00	Tr	Tr	0	971
0	4	0.0	Tr	Tr	Tr	2	0	0	0.00	Tr	Tr	0	972
0	22	0.0	1	0.1	1	10	0	0	0.00	Tr	Tr	0	973
0	73	0.0	2	0.3	4	33	0	0	0.00	Tr	Tr	0	974
0	6	0.0	Tr	Tr	Tr	2	0	0	Tr	Tr	Tr	0	975
0	3	0.0	Tr	Tr	Tr	1	0	0	Tr	Tr	Tr	0	976
0	26	0.0	1	0.3	10	7	0	0	0.00	0.00	0.0	0	977
0	10	0.0	Tr	0.1	4	3	0	0	0.00	0.00	0.0	0	978
3	27	0.8	69	0.4	122	32	68	20	0.07	0.23	1.1	Tr	979
0	41	0.1	2	0.1	3	24	1	0	Tr	Tr	Tr	0	980
0	6	Tr	Tr	Tr	Tr	3	Tr	0	Tr	Tr	Tr	0	981
4	26	1.5	43	0.5	149	21	40	10	0.04	0.07	1.6	Tr	982
2	12	0.7	20	0.2	69	10	19	5	0.02	0.03	0.7	Tr	983
7	37	1.3	55	0.6	138	32	106	28	0.03	0.11	0.1	Tr	984
1	5	0.2	7	0.1	19	4	14	4	Tr	0.01	Tr	Tr	985
3	13	0.3	23	0.1	43	43	19	6	0.01	0.04	0.1	Tr	986
9	44	1.0	79	0.5	147	146	66	20	0.02	0.14	0.2	1	987
Tr	4	0.2	5	0.1	25	22	5	1	0.02	0.01	0.3	Tr	988
2	25	1.4	35	0.5	158	143	33	9	0.11	0.08	2.1	Tr	989
2	9	0.4	14	0.1	49	40	23	6	0.01	0.02	0.6	Tr	990
15	67	2.8	106	0.9	366	301	172	44	0.11	0.17	4.7	1	991
7	34	1.4	54	0.4	185	152	87	22	0.06	0.09	2.4	Tr	992
Tr	5	0.4	2	0.2	25	1	3	Tr	Tr	0.01	Tr	0	993
0	4	0.0	Tr	Tr	Tr	3	0	0	Tr	Tr	Tr	3	994
0	50	0.0	2	0.1	1	33	0	0	Tr	Tr	Tr	31	995
0	24	0.2	3	0.5	74	70	249	75	Tr	0.01	Tr	0	996
0	26	Tr	1	Tr	14	34	283	86	0.00	Tr	Tr	0	997
0	16	0.0	4	0.1	41	3	22	2	0.01	0.01	0.1	7	998
0	11	0.0	0	0.0	2	7	0	0	0.00	0.00	0.0	0	999
0	16	0.0	1	0.1	7	5	194	0	0.01	0.01	0.8	1	1000
0	7	0.3	2	0.1	15	1	20	2	Tr	Tr	Tr	Tr	1001
0	19	0.0	3	Tr	1	57	0	0	0.00	Tr	Tr	0	1002
0	1	0.0	2	Tr	0	56	0	0	0.00	Tr	Tr	0	1003

Sugars and Sweets (continued)

Food No.	Food Description	Measure of edible portion	Weight (g)	Water (%)	Calories (kcal)	Pro-tein (g)	Total fat (g)	Satu-rated (g)	Mono-unsatu-rated (g)	Poly-unsatu-rated (g)
1004	Honey, strained or extracted	1 tbsp 21		17	64	Tr	0	0.0	0.0	0.0
1005		1 cup 339		17	1,031	1	0	0.0	0.0	0.0
1006	Jams and preserves...................	1 tbsp 20		30	56	Tr	Tr	Tr	Tr	0.0
1007		1 packet (0.5 oz) 14		30	39	Tr	Tr	Tr	Tr	0.0
1008	Jellies	1 tbsp 19		29	54	Tr	Tr	Tr	Tr	Tr
1009		1 packet (0.5 oz) 14		29	40	Tr	Tr	Tr	Tr	Tr
	Puddings									
	Prepared with dry mix and 2% milk									
	Chocolate									
1010	Instant	½ cup.................147		75	150	5	3	1.6	0.9	0.2
1011	Regular (cooked)	½ cup.................142		74	151	5	3	1.8	0.8	0.1
	Vanilla									
1012	Instant	½ cup.................142		75	148	4	2	1.4	0.7	0.1
1013	Regular (cooked)	½ cup.................140		76	141	4	2	1.5	0.7	0.1
	Ready to eat									
	Regular									
1014	Chocolate...........................	4 oz 113		69	150	3	5	0.8	1.9	1.6
1015	Rice	4 oz 113		68	184	2	8	1.3	3.6	3.2
1016	Tapioca...............................	4 oz 113		74	134	2	4	0.7	1.8	1.5
1017	Vanilla................................	4 oz 113		71	147	3	4	0.6	1.7	1.5
	Fat free									
1018	Chocolate...........................	4 oz 113		76	107	3	Tr	0.3	0.1	Tr
1019	Tapioca...............................	4 oz 113		77	98	2	Tr	0.1	Tr	Tr
1020	Vanilla................................	4 oz 113		76	105	2	Tr	0.1	Tr	Tr
	Sugar									
	Brown									
1021	Packed	1 cup 220		2	827	0	0	0.0	0.0	0.0
1022	Unpacked.............................	1 cup 145		2	545	0	0	0.0	0.0	0.0
1023		1 tbsp 9		2	34	0	0	0.0	0.0	0.0
	White									
1024	Granulated...........................	1 packet 6		0	23	0	0	0.0	0.0	0.0
1025		1 tsp 4		0	16	0	0	0.0	0.0	0.0
1026		1 cup 200		0	774	0	0	0.0	0.0	0.0
1027	Powdered, unsifted..............	1 tbsp 8		Tr	31	0	Tr	Tr	Tr	Tr
1028		1 cup 120		Tr	467	0	Tr	Tr	Tr	0.1
	Syrup									
	Chocolate flavored syrup or topping									
1029	Thin type	1 tbsp 19		31	53	Tr	Tr	0.1	0.1	Tr
1030	Fudge type...........................	1 tbsp 19		22	67	1	2	0.8	0.7	0.1
1031	Corn, light............................	1 tbsp 20		23	56	0	0	0.0	0.0	0.0
1032	Maple...................................	1 tbsp 20		32	52	0	Tr	Tr	Tr	Tr
1033	Molasses, blackstrap	1 tbsp 20		29	47	0	0	0.0	0.0	0.0
1034		1 cup 328		29	771	0	0	0.0	0.0	0.0
	Table blend, pancake									
1035	Regular	1 tbsp 20		24	57	0	0	0.0	0.0	0.0
1036	Reduced calorie...................	1 tbsp 15		55	25	0	0	0.0	0.0	0.0

Choles-terol (mg)	Carbo-hydrate (g)	Total dietary fiber (g)	Calcium (mg)	Iron (mg)	Potas-sium (mg)	Sodium (mg)	Vitamin A (IU)	Vitamin A (RE)	Thiamin (mg)	Ribo-flavin (mg)	Niacin (mg)	Ascor-bic acid (mg)	Food No.
0	17	Tr	1	0.1	11	1	0	0	0.00	0.01	Tr	Tr	1004
0	279	0.7	20	1.4	176	14	0	0	0.00	0.13	0.4	2	1005
0	14	0.2	4	0.1	15	6	2	Tr	0.00	Tr	Tr	2	1006
0	10	0.2	3	0.1	11	4	2	Tr	0.00	Tr	Tr	1	1007
0	13	0.2	2	Tr	12	5	3	Tr	Tr	Tr	Tr	Tr	1008
0	10	0.1	1	Tr	9	4	2	Tr	Tr	Tr	Tr	Tr	1009
9	28	0.6	153	0.4	247	417	253	56	0.05	0.21	0.1	1	1010
10	28	0.4	160	0.5	240	149	253	68	0.05	0.21	0.2	1	1011
9	28	0.0	146	0.1	185	406	241	64	0.05	0.20	0.1	1	1012
10	26	0.0	153	0.1	193	224	252	70	0.04	0.20	0.1	1	1013
3	26	1.1	102	0.6	203	146	41	12	0.03	0.18	0.4	2	1014
1	25	0.1	59	0.3	68	96	129	40	0.02	0.08	0.2	1	1015
1	22	0.1	95	0.3	110	180	0	0	0.02	0.11	0.4	1	1016
8	25	0.1	99	0.1	128	153	24	7	0.02	0.16	0.3	0	1017
2	23	0.9	89	0.6	235	192	174	52	0.02	0.12	0.1	Tr	1018
1	23	0.1	76	0.2	99	251	121	36	0.02	0.09	0.1	Tr	1019
1	24	0.1	86	Tr	123	241	174	52	0.02	0.10	0.1	Tr	1020
0	214	0.0	187	4.2	761	86	0	0	0.02	0.02	0.2	0	1021
0	141	0.0	123	2.8	502	57	0	0	0.01	0.01	0.1	0	1022
0	9	0.0	8	0.2	31	4	0	0	Tr	Tr	Tr	0	1023
0	6	0.0	Tr	Tr	Tr	Tr	0	0	0.00	Tr	0.0	0	1024
0	4	0.0	Tr	Tr	Tr	Tr	0	0	0.00	Tr	0.0	0	1025
0	200	0.0	2	0.1	4	2	0	0	0.00	0.04	0.0	0	1026
0	8	0.0	Tr	Tr	Tr	Tr	0	0	0.00	0.00	0.0	0	1027
0	119	0.0	1	0.1	2	1	0	0	0.00	0.00	0.0	0	1028
0	12	0.3	3	0.4	43	14	6	1	Tr	0.01	0.1	Tr	1029
Tr	12	0.5	15	0.2	69	66	3	1	0.01	0.04	0.1	Tr	1030
0	15	0.0	1	Tr	1	24	0	0	Tr	Tr	Tr	0	1031
0	13	0.0	13	0.2	41	2	0	0	Tr	Tr	Tr	0	1032
0	12	0.0	172	3.5	498	11	0	0	0.01	0.01	0.2	0	1033
0	199	0.0	2,821	57.4	8,174	180	0	0	0.11	0.17	3.5	0	1034
0	15	0.0	Tr	Tr	Tr	17	0	0	Tr	Tr	Tr	0	1035
0	7	0.0	Tr	Tr	Tr	30	0	0	Tr	Tr	Tr	0	1036

Food No.	Food Description	Measure of edible portion	Weight (g)	Water (%)	Calories (kcal)	Pro-tein (g)	Total fat (g)	Fatty acids Satu-rated (g)	Mono-unsatu-rated (g)	Poly-unsatu-rated (g)

Vegetables and Vegetable Products

Food No.	Food Description	Measure	Weight	Water	Calories	Protein	Total fat	Saturated	Mono	Poly
1037	Alfalfa sprouts, raw	1 cup ...33		91	10	1	Tr	Tr	Tr	0.1
1038	Artichokes, globe or French, cooked, drained	1 cup ...168		84	84	6	Tr	0.1	Tr	0.1
1039		1 medium ...120		84	60	4	Tr	Tr	Tr	0.1
	Asparagus, green Cooked, drained									
1040	From raw	1 cup ...180		92	43	5	1	0.1	Tr	0.2
1041		4 spears...60		92	14	2	Tr	Tr	Tr	0.1
1042	From frozen	1 cup ...180		91	50	5	1	0.2	Tr	0.3
1043		4 spears...60		91	17	2	Tr	0.1	Tr	0.1
1044	Canned, spears, about 5" long, drained	1 cup ...242		94	46	5	2	0.4	0.1	0.7
1045		4 spears...72		94	14	2	Tr	0.1	Tr	0.2
1046	Bamboo shoots, canned, drained	1 cup ...131		94	25	2	1	0.1	Tr	0.2
	Beans Lima, immature seeds, frozen, cooked, drained									
1047	Ford hooks	1 cup ...170		74	170	10	1	0.1	Tr	0.3
1048	Baby limas	1 cup ...180		72	189	12	1	0.1	Tr	0.3
	Snap, cut Cooked, drained From raw									
1049	Green	1 cup ...125		89	44	2	Tr	0.1	Tr	0.2
1050	Yellow	1 cup ...125		89	44	2	Tr	0.1	Tr	0.2
	From frozen									
1051	Green	1 cup ...135		91	38	2	Tr	0.1	Tr	0.1
1052	Yellow	1 cup ...135		91	38	2	Tr	0.1	Tr	0.1
	Canned, drained									
1053	Green	1 cup ...135		93	27	2	Tr	Tr	Tr	0.1
1054	Yellow	1 cup ...135		93	27	2	Tr	Tr	Tr	0.1
	Beans, dry. See Legumes. Bean sprouts (mung)									
1055	Raw	1 cup ...104		90	31	3	Tr	Tr	Tr	0.1
1056	Cooked, drained	1 cup ...124		93	26	3	Tr	Tr	Tr	Tr
	Beets Cooked, drained									
1057	Slices	1 cup ...170		87	75	3	Tr	Tr	0.1	0.1
1058	Whole beet, 2" dia	1 beet ...50		87	22	1	Tr	Tr	Tr	Tr
	Canned, drained									
1059	Slices	1 cup ...170		91	53	2	Tr	Tr	Tr	0.1
1060	Whole beet	1 beet ...24		91	7	Tr	Tr	Tr	Tr	Tr
1061	Beet greens, leaves and stems, cooked, drained, 1" pieces	1 cup ...144		89	39	4	Tr	Tr	0.1	0.1
	Black eyed peas, immature seeds, cooked, drained									
1062	From raw	1 cup ...165		75	160	5	1	0.2	0.1	0.3
1063	From frozen	1 cup ...170		66	224	14	1	0.3	0.1	0.5
	Broccoli Raw									
1064	Chopped or diced	1 cup ...88		91	25	3	Tr	Tr	Tr	0.1
1065	Spear, about 5" long	1 spear ...31		91	9	1	Tr	Tr	Tr	0.1
1066	Flower cluster	1 floweret ...11		91	3	Tr	Tr	Tr	Tr	Tr
	Cooked, drained From raw									
1067	Chopped	1 cup ...156		91	44	5	1	0.1	Tr	0.3
1068	Spear, about 5" long	1 spear ...37		91	10	1	Tr	Tr	Tr	0.1
1069	From frozen, chopped	1 cup ...184		91	52	6	Tr	Tr	Tr	0.1

Choles-terol (mg)	Carbo-hydrate (g)	Total dietary fiber (g)	Calcium (mg)	Iron (mg)	Potas-sium (mg)	Sodium (mg)	Vitamin A		Thiamin (mg)	Ribo-flavin (mg)	Niacin (mg)	Ascor-bic acid (mg)	Food No.
							(IU)	(RE)					
0	1	0.8	11	0.3	26	2	51	5	0.03	0.04	0.2	3	1037
0	19	9.1	ˋ76	2.2	595	160	297	30	0.11	0.11	1.7	17	1038
0	13	6.5	54	1.5	425	114	212	22	0.08	0.08	1.2	12	1039
0	8	2.9	36	1.3	288	20	970	97	0.22	0.23	1.9	19	1040
0	3	1.0	12	0.4	96	7	323	32	0.07	0.08	0.6	6	1041
0	9	2.9	41	1.2	392	7	1,472	148	0.12	0.19	1.9	44	1042
0	3	1.0	14	0.4	131	2	491	49	0.04	0.06	0.6	15	1043
0	6	3.9	39	4.4	416	695	1,285	128	0.15	0.24	2.3	45	1044
0	2	1.2	12	1.3	124	207	382	38	0.04	0.07	0.7	13	1045
0	4	1.8	10	0.4	105	9	10	1	0.03	0.03	0.2	1	1046
0	32	9.9	37	2.3	694	90	323	32	0.13	0.10	1.8	22	1047
0	35	10.8	50	3.5	740	52	301	31	0.13	0.10	1.4	10	1048
0	10	4.0	58	1.6	374	4	833	84	0.09	0.12	0.8	12	1049
0	10	4.1	58	1.6	374	4	101	10	0.09	0.12	0.8	12	1050
0	9	4.1	66	1.2	170	12	541	54	0.05	0.12	0.5	6	1051
0	9	4.1	66	1.2	170	12	151	15	0.05	0.12	0.5	6	1052
0	6	2.6	35	1.2	147	354	471	47	0.02	0.08	0.3	6	1053
0	6	1.8	35	1.2	147	339	142	15	0.02	0.08	0.3	6	1054
0	6	1.9	14	0.9	155	6	22	2	0.09	0.13	0.8	14	1055
0	5	1.5	15	0.8	125	12	17	1	0.06	0.13	1.0	14	1056
0	17	3.4	27	1.3	519	131	60	7	0.05	0.07	0.6	6	1057
0	5	1.0	8	0.4	153	39	18	2	0.01	0.02	0.2	2	1058
0	12	2.9	26	3.1	252	330	19	2	0.02	0.07	0.3	7	1059
0	2	0.4	4	0.4	36	47	3	Tr	Tr	0.01	Tr	1	1060
0	8	4.2	164	2.7	1,309	347	7,344	734	0.17	0.42	0.7	36	1061
0	34	8.3	211	1.8	690	7	1,305	130	0.17	0.24	2.3	4	1062
0	40	10.9	39	3.6	638	9	128	14	0.44	0.11	1.2	4	1063
0	5	2.6	42	0.8	286	24	1,357	136	0.06	0.10	0.6	82	1064
0	2	0.9	15	0.3	101	8	478	48	0.02	0.04	0.2	29	1065
0	1	0.3	5	0.1	36	3	330	33	0.01	0.01	0.1	10	1066
0	8	4.5	72	1.3	456	41	2,165	217	0.09	0.18	0.9	116	1067
0	2	1.1	17	0.3	108	10	514	51	0.02	0.04	0.2	28	1068
0	10	5.5	94	1.1	331	44	3,481	348	0.10	0.15	0.8	74	1069

Food No.	Food Description	Measure of edible portion	Weight (g)	Water (%)	Calories (kcal)	Protein (g)	Total fat (g)	Fatty acids Saturated (g)	Monounsaturated (g)	Polyunsaturated (g)

Vegetables and Vegetable Products (continued)

Food No.	Food Description	Measure	Weight	Water	Calories	Protein	Total fat	Satu-rated	Mono-unsat	Poly-unsat
	Brussels sprouts, cooked, drained									
1070	From raw	1 cup	156	87	61	4	1	0.2	0.1	0.4
1071	From frozen	1 cup	155	87	65	6	1	0.1	Tr	0.3
	Cabbage, common varieties, shredded									
1072	Raw	1 cup	70	92	18	1	Tr	Tr	Tr	0.1
1073	Cooked, drained	1 cup	150	94	33	2	1	0.1	Tr	0.3
	Cabbage, Chinese, shredded, cooked, drained									
1074	Pak choi or bok choy	1 cup	170	96	20	3	Tr	Tr	Tr	0.1
1075	Pe tsai	1 cup	119	95	17	2	Tr	Tr	Tr	0.1
1076	Cabbage, red, raw, shredded	1 cup	70	92	19	1	Tr	Tr	Tr	0.1
1077	Cabbage, savoy, raw, shredded	1 cup	70	91	19	1	Tr	Tr	Tr	Tr
1078	Carrot juice, canned	1 cup	236	89	94	2	Tr	0.1	Tr	0.2
	Carrots Raw									
1079	Whole, 7½" long	1 carrot	72	88	31	1	Tr	Tr	Tr	0.1
1080	Grated	1 cup	110	88	47	1	Tr	Tr	Tr	0.1
1081	Baby	1 medium	10	90	4	Tr	Tr	Tr	Tr	Tr
	Cooked, sliced, drained									
1082	From raw	1 cup	156	87	70	2	Tr	0.1	Tr	0.1
1083	From frozen	1 cup	146	90	53	2	Tr	Tr	Tr	0.1
1084	Canned, sliced, drained	1 cup	146	93	37	1	Tr	0.1	Tr	0.1
	Cauliflower									
1085	Raw	1 floweret	13	92	3	Tr	Tr	Tr	Tr	Tr
1086		1 cup	100	92	25	2	Tr	Tr	Tr	0.1
	Cooked, drained, 1" pieces									
1087	From raw	1 cup	124	93	29	2	1	0.1	Tr	0.3
1088		3 flowerets	54	93	12	1	Tr	Tr	Tr	0.1
1089	From frozen	1 cup	180	94	34	3	Tr	0.1	Tr	0.2
	Celery Raw									
1090	Stalk, 7½ to 8" long	1 stalk	40	95	6	Tr	Tr	Tr	Tr	Tr
1091	Pieces, diced	1 cup	120	95	19	1	Tr	Tr	Tr	0.1
	Cooked, drained									
1092	Stalk, medium	1 stalk	38	94	7	Tr	Tr	Tr	Tr	Tr
1093	Pieces, diced	1 cup	150	94	27	1	Tr	0.1	Tr	0.1
1094	Chives, raw, chopped	1 tbsp	3	91	1	Tr	Tr	Tr	Tr	Tr
1095	Cilantro, raw	1 tsp	2	92	Tr	Tr	Tr	Tr	Tr	Tr
1096	Coleslaw, home prepared	1 cup	120	82	83	2	3	0.5	0.8	1.6
	Collards, cooked, drained, chopped									
1097	From raw	1 cup	190	92	49	4	1	0.1	Tr	0.3
1098	From frozen	1 cup	170	88	61	5	1	0.1	Tr	0.4
	Corn, sweet, yellow Cooked, drained									
1099	From raw, kernels on cob	1 ear	77	70	83	3	1	0.2	0.3	0.5
	From frozen									
1100	Kernels on cob	1 ear	63	73	59	2	Tr	0.1	0.1	0.2
1101	Kernels	1 cup	164	77	131	5	1	0.1	0.2	0.3
	Canned									
1102	Cream style	1 cup	256	79	184	4	1	0.2	0.3	0.5
1103	Whole kernel, vacuum pack	1 cup	210	77	166	5	1	0.2	0.3	0.5
1104	Corn, sweet, white, cooked, drained	1 ear	77	70	83	3	1	0.2	0.3	0.5

*White varieties contain only a trace amount of vitamin A; other nutrients are the same.

Choles-terol (mg)	Carbo-hydrate (g)	Total dietary fiber (g)	Calcium (mg)	Iron (mg)	Potas-sium (mg)	Sodium (mg)	Vitamin A (IU)	Vitamin A (RE)	Thiamin (mg)	Ribo-flavin (mg)	Niacin (mg)	Ascor-bic acid (mg)	Food No.
0	14	4.1	56	1.9	495	33	1,122	112	0.17	0.12	0.9	97	1070
0	13	6.4	37	1.1	504	36	913	91	0.16	0.18	0.8	71	1071
0	4	1.6	33	0.4	172	13	93	9	0.04	0.03	0.2	23	1072
0	7	3.5	47	0.3	146	12	198	20	0.09	0.08	0.4	30	1073
0	3	2.7	158	1.8	631	58	4,366	437	0.05	0.11	0.7	44	1074
0	3	3.2	38	0.4	268	11	1,151	115	0.05	0.05	0.6	19	1075
0	4	1.4	36	0.3	144	8	28	3	0.04	0.02	0.2	40	1076
0	4	2.2	25	0.3	161	20	700	70	0.05	0.02	0.2	22	1077
0	22	1.9	57	1.1	689	68	25,833	2,584	0.22	0.13	0.9	20	1078
0	7	2.2	19	0.4	233	25	20,253	2,025	0.07	0.04	0.7	7	1079
0	11	3.3	30	0.6	355	39	30,942	3,094	0.11	0.06	1.0	10	1080
0	1	0.2	2	0.1	28	4	1,501	150	Tr	0.01	0.1	1	1081
0	16	5.1	48	1.0	354	103	38,304	3,830	0.05	0.09	0.8	4	1082
0	12	5.1	41	0.7	231	86	25,845	2,584	0.04	0.05	0.6	4	1083
0	8	2.2	37	0.9	261	353	20,110	2,010	0.03	0.04	0.8	4	1084
0	1	0.3	3	0.1	39	4	2	Tr	0.01	0.01	0.1	6	1085
0	5	2.5	22	0.4	303	30	19	2	0.06	0.06	0.5	46	1086
0	5	3.3	20	0.4	176	19	21	2	0.05	0.06	0.5	55	1087
0	2	1.5	9	0.2	77	8	9	1	0.02	0.03	0.2	24	1088
0	7	4.9	31	0.7	250	32	40	4	0.07	0.10	0.6	56	1089
0	1	0.7	16	0.2	115	35	54	5	0.02	0.02	0.1	3	1090
0	4	2.0	48	0.5	344	104	161	16	0.06	0.05	0.4	8	1091
0	2	0.6	16	0.2	108	35	50	5	0.02	0.02	0.1	2	1092
0	6	2.4	63	0.6	426	137	198	20	0.06	0.07	0.5	9	1093
0	Tr	0.1	3	Tr	9	Tr	131	13	Tr	Tr	Tr	2	1094
0	Tr	Tr	1	Tr	8	1	98	10	Tr	Tr	Tr	1	1095
10	15	1.8	54	0.7	217	28	762	98	0.08	0.07	0.3	39	1096
0	9	5.3	226	0.9	494	17	5,945	595	0.08	0.20	1.1	35	1097
0	12	4.8	357	1.9	427	85	10,168	1,017	0.08	0.20	1.1	45	1098
0	19	2.2	2	0.5	192	13	167	17	0.17	0.06	1.2	5	1099
0	14	1.8	2	0.4	158	3	133*	13*	0.11	0.04	1.0	3	1100
0	32	3.9	7	0.6	241	8	361*	36*	0.14	0.12	2.1	5	1101
0	46	3.1	8	1.0	343	730	248*	26*	0.06	0.14	2.5	12	1102
0	41	4.2	11	0.9	391	571	506*	50*	0.09	0.15	2.5	17	1103
0	19	2.1	2	0.5	192	13	0	0	0.17	0.06	1.2	5	1104

Vegetables and Vegetable Products (continued)

Food No.	Food Description	Measure of edible portion	Weight (g)	Water (%)	Calories (kcal)	Pro-tein (g)	Total fat (g)	Satu-rated (g)	Mono-unsatu-rated (g)	Poly-unsatu-rated (g)
	Cucumber									
	Peeled									
1105	Sliced	1 cup	119	96	14	1	Tr	Tr	Tr	0.1
1106	Whole, 8¼" long	1 large	280	96	34	2	Tr	0.1	Tr	0.2
	Unpeeled									
1107	Sliced	1 cup	104	96	14	1	Tr	Tr	Tr	0.1
1108	Whole, 8¼" long	1 large	301	96	39	2	Tr	0.1	Tr	0.2
1109	Dandelion greens, cooked, drained	1 cup	105	90	35	2	1	0.2	Tr	0.3
1110	Dill weed, raw	5 sprigs	1	86	Tr	Tr	Tr	Tr	Tr	Tr
1111	Eggplant, cooked, drained	1 cup	99	92	28	1	Tr	Tr	Tr	0.1
1112	Endive, curly (including escarole), raw, small pieces	1 cup	50	94	9	1	Tr	Tr	Tr	Tr
1113	Garlic, raw	1 clove	3	59	4	Tr	Tr	Tr	Tr	Tr
1114	Hearts of palm, canned	1 piece	33	90	9	1	Tr	Tr	Tr	0.1
1115	Jerusalem artichoke, raw, sliced	1 cup	150	78	114	3	Tr	0.0	Tr	Tr
	Kale, cooked, drained, chopped									
1116	From raw	1 cup	130	91	36	2	1	0.1	Tr	0.3
1117	From frozen	1 cup	130	91	39	4	1	0.1	Tr	0.3
1118	Kohlrabi, cooked, drained, slices	1 cup	165	90	48	3	Tr	Tr	Tr	0.1
1119	Leeks, bulb and lower leaf portion, chopped or diced, cooked, drained	1 cup	104	91	32	1	Tr	Tr	Tr	0.1
	Lettuce, raw									
	Butterhead, as Boston types									
1120	Leaf	1 medium leaf	8	96	1	Tr	Tr	Tr	Tr	Tr
1121	Head, 5" dia	1 head	163	96	21	2	Tr	Tr	Tr	0.2
	Crisphead, as iceberg									
1122	Leaf	1 medium	8	96	1	Tr	Tr	Tr	Tr	Tr
1123	Head, 6" dia	1 head	539	96	65	5	1	0.1	Tr	0.5
1124	Pieces, shredded or chopped	1 cup	55	96	7	1	Tr	Tr	Tr	0.1
	Looseleaf									
1125	Leaf	1 leaf	10	94	2	Tr	Tr	Tr	Tr	Tr
1126	Pieces, shredded	1 cup	56	94	10	1	Tr	Tr	Tr	0.1
	Romaine or cos									
1127	Innerleaf	1 leaf	10	95	1	Tr	Tr	Tr	Tr	Tr
1128	Pieces, shredded	1 cup	56	95	8	1	Tr	Tr	Tr	0.1
	Mushrooms									
1129	Raw, pieces or slices	1 cup	70	92	18	2	Tr	Tr	Tr	0.1
1130	Cooked, drained, pieces	1 cup	156	91	42	3	1	0.1	Tr	0.3
1131	Canned, drained, pieces	1 cup	156	91	37	3	Tr	0.1	Tr	0.2
	Mushrooms, shiitake									
1132	Cooked pieces	1 cup	145	83	80	2	Tr	0.1	0.1	Tr
1133	Dried	1 mushroom	4	10	11	Tr	Tr	Tr	Tr	Tr
1134	Mustard greens, cooked, drained	1 cup	140	94	21	3	Tr	Tr	0.2	0.1
	Okra, sliced, cooked, drained									
1135	From raw	1 cup	160	90	51	3	Tr	0.1	Tr	0.1
1136	From frozen	1 cup	184	91	52	4	1	0.1	0.1	0.1
	Onions									
	Raw									
1137	Chopped	1 cup	160	90	61	2	Tr	Tr	Tr	0.1
1138	Whole, medium, 2½" dia	1 whole	110	90	42	1	Tr	Tr	Tr	0.1
1139	Slice, ⅛" thick	1 slice	14	90	5	Tr	Tr	Tr	Tr	Tr

Choles-terol (mg)	Carbo-hydrate (g)	Total dietary fiber (g)	Calcium (mg)	Iron (mg)	Potas-sium (mg)	Sodium (mg)	Vitamin A (IU)	Vitamin A (RE)	Thiamin (mg)	Ribo-flavin (mg)	Niacin (mg)	Ascor-bic acid (mg)	Food No.
0	3	0.8	17	0.2	176	2	88	8	0.02	0.01	0.1	3	1105
0	7	2.0	39	0.4	414	6	207	20	0.06	0.03	0.3	8	1106
0	3	0.8	15	0.3	150	2	224	22	0.02	0.02	0.2	6	1107
0	8	2.4	42	0.8	433	6	647	63	0.07	0.07	0.7	16	1108
0	7	3.0	147	1.9	244	46	12,285	1,229	0.14	0.18	0.5	19	1109
0	Tr	Tr	2	0.1	7	1	77	8	Tr	Tr	Tr	1	1110
0	7	2.5	6	0.3	246	3	63	6	0.08	0.02	0.6	1	1111
0	2	1.6	26	0.4	157	11	1,025	103	0.04	0.04	0.2	3	1112
0	1	0.1	5	0.1	12	1	0	0	0.01	Tr	Tr	1	1113
0	2	0.8	19	1.0	58	141	0	0	Tr	0.02	0.1	3	1114
0	26	2.4	21	5.1	644	6	30	3	0.30	0.09	2.0	6	1115
0	7	2.6	94	1.2	296	30	9,620	962	0.07	0.09	0.7	53	1116
0	7	2.6	179	1.2	417	20	8,260	826	0.06	0.15	0.9	33	1117
0	11	1.8	41	0.7	561	35	58	7	0.07	0.03	0.6	89	1118
0	8	1.0	31	1.1	90	10	48	5	0.03	0.02	0.2	4	1119
0	Tr	0.1	2	Tr	19	Tr	73	7	Tr	Tr	Tr	1	1120
0	4	1.6	52	0.5	419	8	1,581	158	0.10	0.10	0.5	13	1121
0	Tr	0.1	2	Tr	13	1	26	3	Tr	Tr	Tr	Tr	1122
0	11	7.5	102	2.7	852	49	1,779	178	0.25	0.16	1.0	21	1123
0	1	0.8	10	0.3	87	5	182	18	0.03	0.02	0.1	2	1124
0	Tr	0.2	7	0.1	26	1	190	19	0.01	0.01	Tr	2	1125
0	2	1.1	38	0.8	148	5	1,064	106	0.03	0.04	0.2	10	1126
0	Tr	0.2	4	0.1	29	1	260	26	0.01	0.01	0.1	2	1127
0	1	1.0	20	0.6	162	4	1,456	146	0.06	0.06	0.3	13	1128
0	3	0.8	4	0.7	259	3	0	0	0.06	0.30	2.8	2	1129
0	8	3.4	9	2.7	555	3	0	0	0.11	0.47	7.0	6	1130
0	8	3.7	17	1.2	201	663	0	0	0.13	0.03	2.5	0	1131
0	21	3.0	4	0.6	170	6	0	0	0.05	0.25	2.2	Tr	1132
0	3	0.4	Tr	0.1	55	Tr	0	0	0.01	0.05	0.5	Tr	1133
0	3	2.8	104	1.0	283	22	4,243	424	0.06	0.09	0.6	35	1134
0	12	4.0	101	0.7	515	8	920	93	0.21	0.09	1.4	26	1135
0	11	5.2	177	1.2	431	6	946	94	0.18	0.23	1.4	22	1136
0	14	2.9	32	0.4	251	5	0	0	0.07	0.03	0.2	10	1137
0	9	2.0	22	0.2	173	3	0	0	0.05	0.02	0.2	7	1138
0	1	0.3	3	Tr	22	Tr	0	0	0.01	Tr	Tr	1	1139

Food No.	Food Description	Measure of edible portion	Weight (g)	Water (%)	Calories (kcal)	Pro-tein (g)	Total fat (g)	Fatty acids Satu-rated (g)	Mono-unsatu-rated (g)	Poly-unsatu-rated (g)

Vegetables and Vegetable Products (continued)

Food No.	Food Description	Measure	Weight	Water	Calories	Protein	Total fat	Saturated	Mono	Poly
1140	Cooked (whole or sliced), drained	1 cup 210		88	92	3	Tr	0.1	0.1	0.2
1141		1 medium 94		88	41	1	Tr	Tr	Tr	0.1
1142	Dehydrated flakes	1 tbsp 5		4	17	Tr	Tr	Tr	Tr	Tr
	Onions, spring, raw, top and bulb									
1143	Chopped	1 cup 100		90	32	2	Tr	Tr	Tr	0.1
1144	Whole, medium, 4⅛" long	1 whole 15		90	5	Tr	Tr	Tr	Tr	Tr
1145	Onion rings, 2"-3" dia, breaded, par fried, frozen, oven heated	10 rings 60		29	244	3	16	5.2	6.5	3.1
1146	Parsley, raw	10 sprigs 10		88	4	Tr	Tr	Tr	Tr	Tr
1147	Parsnips, sliced, cooked, drained	1 cup 156		78	126	2	Tr	0.1	0.2	0.1
	Peas, edible pod, cooked, drained									
1148	From raw	1 cup 160		89	67	5	Tr	0.1	Tr	0.2
1149	From frozen	1 cup 160		87	83	6	1	0.1	0.1	0.3
	Peas, green									
1150	Canned, drained	1 cup 170		82	117	8	1	0.1	0.1	0.3
1151	Frozen, boiled, drained	1 cup 160		80	125	8	Tr	0.1	Tr	0.2
	Peppers									
	Hot chili, raw									
1152	Green	1 pepper 45		88	18	1	Tr	Tr	Tr	Tr
1153	Red	1 pepper 45		88	18	1	Tr	Tr	Tr	Tr
1154	Jalapeno, canned, sliced, solids and liquids	¼ cup 26		89	7	Tr	Tr	Tr	Tr	0.1
	Sweet (2¾" long, 2½" dia)									
	Raw									
	Green									
1155	Chopped	1 cup 149		92	40	1	Tr	Tr	Tr	0.2
1156	Ring (¼" thick)	1 ring 10		92	3	Tr	Tr	Tr	Tr	Tr
1157	Whole (2¾" x 2½")	1 pepper 119		92	32	1	Tr	Tr	Tr	0.1
	Red									
1158	Chopped	1 cup 149		92	40	1	Tr	Tr	Tr	0.2
1159	Whole (2¾" x 2½")	1 pepper 119		92	32	1	Tr	Tr	Tr	0.1
	Cooked, drained, chopped									
1160	Green	1 cup 136		92	38	1	Tr	Tr	Tr	0.1
1161	Red	1 cup 136		92	38	1	Tr	Tr	Tr	0.1
1162	Pimento, canned	1 tbsp 12		93	3	Tr	Tr	Tr	Tr	Tr
	Potatoes									
	Baked (2⅓" x 4¾")									
1163	With skin	1 potato 202		71	220	5	Tr	0.1	Tr	0.1
1164	Flesh only	1 potato 156		75	145	3	Tr	Tr	Tr	0.1
1165	Skin only	1 skin 58		47	115	2	Tr	Tr	Tr	Tr
	Boiled (2½" dia)									
1166	Peeled after boiling	1 potato 136		77	118	3	Tr	Tr	Tr	0.1
1167	Peeled before boiling	1 potato 135		77	116	2	Tr	Tr	Tr	0.1
1168		1 cup 156		77	134	3	Tr	Tr	Tr	0.1
	Potato products, prepared									
	Au gratin									
1169	From dry mix, with whole milk, butter	1 cup 245		79	228	6	10	6.3	2.9	0.3
1170	From home recipe, with butter	1 cup 245		74	323	12	19	11.6	5.3	0.7
1171	French fried, frozen, oven heated	10 strips 50		57	100	2	4	0.6	2.4	0.4

Choles-terol (mg)	Carbo-hydrate (g)	Total dietary fiber (g)	Calcium (mg)	Iron (mg)	Potas-sium (mg)	Sodium (mg)	Vitamin A (IU)	Vitamin A (RE)	Thiamin (mg)	Ribo-flavin (mg)	Niacin (mg)	Ascor-bic acid (mg)	Food No.
0	21	2.9	46	0.5	349	6	0	0	0.09	0.05	0.3	11	1140
0	10	1.3	21	0.2	156	3	0	0	0.04	0.02	0.2	5	1141
0	4	0.5	13	0.1	81	1	0	0	0.03	0.01	Tr	4	1142
0	7	2.6	72	1.5	276	16	385	39	0.06	0.08	0.5	19	1143
0	1	0.4	11	0.2	41	2	58	6	0.01	0.01	0.1	3	1144
0	23	0.8	19	1.0	77	225	135	14	0.17	0.08	2.2	1	1145
0	1	0.3	14	0.6	55	6	520	52	0.01	0.01	0.1	13	1146
0	30	6.2	58	0.9	573	16	0	0	0.13	0.08	1.1	20	1147
0	11	4.5	67	3.2	384	6	210	21	0.20	0.12	0.9	77	1148
0	14	5.0	94	3.8	347	8	267	27	0.10	0.19	0.9	35	1149
0	21	7.0	34	1.6	294	428	1,306	131	0.21	0.13	1.2	16	1150
0	23	8.8	38	2.5	269	139	1,069	107	0.45	0.16	2.4	16	1151
0	4	0.7	8	0.5	153	3	347	35	0.04	0.04	0.4	109	1152
0	4	0.7	8	0.5	153	3	4,838	484	0.04	0.04	0.4	109	1153
0	1	0.7	6	0.5	50	434	442	44	0.01	0.01	0.1	3	1154
0	10	2.7	13	0.7	264	3	942	94	0.10	0.04	0.8	133	1155
0	1	0.2	1	Tr	18	Tr	63	6	0.01	Tr	0.1	9	1156
0	8	2.1	11	0.5	211	2	752	75	0.08	0.04	0.6	106	1157
0	10	3.0	13	0.7	264	3	8,493	849	0.10	0.04	0.8	283	1158
0	8	2.4	11	0.5	211	2	6,783	678	0.08	0.04	0.6	226	1159
0	9	1.6	12	0.6	226	3	805	80	0.08	0.04	0.6	101	1160
0	9	1.6	12	0.6	226	3	5,114	511	0.08	0.04	0.6	233	1161
0	1	0.2	1	0.2	19	2	319	32	Tr	0.01	0.1	10	1162
0	51	4.8	20	2.7	844	16	0	0	0.22	0.07	3.3	26	1163
0	34	2.3	8	0.5	610	8	0	0	0.16	0.03	2.2	20	1164
0	27	4.6	20	4.1	332	12	0	0	0.07	0.06	1.8	8	1165
0	27	2.4	7	0.4	515	5	0	0	0.14	0.03	2.0	18	1166
0	27	2.4	11	0.4	443	7	0	0	0.13	0.03	1.8	10	1167
0	31	2.8	12	0.5	512	8	0	0	0.15	0.03	2.0	12	2268
37	31	2.2	203	0.8	537	1,076	522	76	0.05	0.20	2.3	8	1169
56	28	4.4	292	1.6	970	1,061	647	93	0.16	0.28	2.4	24	1170
0	16	1.6	4	0.6	209	15	0	0	0.06	0.01	1.0	5	1171

Food No.	Food Description	Measure of edible portion	Weight (g)	Water (%)	Calories (kcal)	Pro-tein (g)	Total fat (g)	Fatty acids Satu-rated (g)	Mono-unsatu-rated (g)	Poly-unsatu-rated (g)

Vegetables and Vegetable Products (continued)

Potato products, prepared (continued)
Hashed brown

Food No.	Food Description	Measure	Weight	Water	Calories	Protein	Total fat	Saturated	Mono	Poly
1172	From frozen (about 3" x 1½" x ½")	1 patty	29	56	63	1	3	1.3	1.5	0.4
1173	From home recipe	1 cup	156	62	326	4	22	8.5	9.7	2.5
	Mashed									
1174	From dehydrated flakes (without milk); whole milk, butter, and salt added	1 cup	210	76	237	4	12	7.2	3.3	0.5
	From home recipe									
1175	With whole milk	1 cup	210	78	162	4	1	0.7	0.3	0.1
1176	With whole milk and margarine	1 cup	210	76	223	4	9	2.2	3.7	2.5
1177	Potato pancakes, home prepared	1 pancake	76	47	207	5	12	2.3	3.5	5.0
1178	Potato puffs, from frozen	10 puffs	79	53	175	3	8	4.0	3.4	0.6
1179	Potato salad, home prepared	1 cup	250	76	358	7	21	3.6	6.2	9.3
	Scalloped									
1180	From dry mix, with whole milk, butter	1 cup	245	79	228	5	11	6.5	3.0	0.5
1181	From home recipe, with butter	1 cup	245	81	211	7	9	5.5	2.5	0.4
	Pumpkin									
1182	Cooked, mashed	1 cup	245	94	49	2	Tr	0.1	Tr	Tr
1183	Canned	1 cup	245	90	83	3	1	0.4	0.1	Tr
1184	Radishes, raw (¾" to 1" dia)	1 radish	5	95	1	Tr	Tr	Tr	Tr	Tr
1185	Rutabagas, cooked, drained, cubes	1 cup	170	89	66	2	Tr	Tr	Tr	0.2
1186	Sauerkraut, canned, solids and liquid	1 cup	236	93	45	2	Tr	0.1	Tr	0.1
	Seaweed									
1187	Kelp, raw	2 tbsp	10	82	4	Tr	Tr	Tr	Tr	Tr
1188	Spirulina, dried	1 tbsp	1	5	3	1	Tr	Tr	Tr	Tr
1189	Shallots, raw, chopped	1 tbsp	10	80	7	Tr	Tr	Tr	Tr	Tr
1190	Soybeans, green, cooked, drained	1 cup	180	69	254	22	12	1.3	2.2	5.4
	Spinach									
	Raw									
1191	Chopped	1 cup	30	92	7	1	Tr	Tr	Tr	Tr
1192	Leaf	1 leaf	10	92	2	Tr	Tr	Tr	Tr	Tr
	Cooked, drained									
1193	From raw	1 cup	180	91	41	5	Tr	0.1	Tr	0.2
1194	From frozen (chopped or leaf)	1 cup	190	90	53	6	Tr	0.1	Tr	0.2
1195	Canned, drained	1 cup	214	92	49	6	1	0.2	Tr	0.4
	Squash									
	Summer (all varieties), sliced									
1196	Raw	1 cup	113	94	23	1	Tr	Tr	Tr	0.1
1197	Cooked, drained	1 cup	180	94	36	2	1	0.1	Tr	0.2
1198	Winter (all varieties), baked, cubes	1 cup	205	89	80	2	1	0.3	0.1	0.5
1199	Winter, butternut, frozen, cooked, mashed	1 cup	240	88	94	3	Tr	Tr	Tr	0.1
	Sweetpotatoes									
	Cooked (2" dia, 5" long raw)									
1200	Baked, with skin	1 potato	146	73	150	3	Tr	Tr	Tr	0.1
1201	Boiled, without skin	1 potato	156	73	164	3	Tr	0.1	Tr	0.2

Choles-terol (mg)	Carbo-hydrate (g)	Total dietary fiber (g)	Calcium (mg)	Iron (mg)	Potas-sium (mg)	Sodium (mg)	Vitamin A (IU)	Vitamin A (RE)	Thiamin (mg)	Ribo-flavin (mg)	Niacin (mg)	Ascor-bic acid (mg)	Food No.
0	8	0.6	4	0.4	126	10	0	0	0.03	0.01	0.7	2	1172
0	33	3.1	12	1.3	501	37	0	0	0.12	0.03	3.1	9	1173
29	32	4.8	103	0.5	489	697	378	44	0.23	0.11	1.4	20	1174
4	37	4.2	55	0.6	628	636	40	13	0.18	0.08	2.3	14	1175
4	35	4.2	55	0.5	607	620	355	42	0.18	0.08	2.3	13	1176
73	22	1.5	18	1.2	597	386	109	11	0.10	0.13	1.6	17	1177
0	24	2.5	24	1.2	300	589	13	2	0.15	0.06	1.7	5	1178
170	28	3.3	48	1.6	635	1,323	523	83	0.19	0.15	2.2	25	1179
27	31	2.7	88	0.9	497	835	363	51	0.05	0.14	2.5	8	1180
29	26	4.7	140	1.4	926	821	331	47	0.17	0.23	2.6	26	1181
0	12	2.7	37	1.4	564	2	2,651	265	0.08	0.19	1.0	12	1182
0	20	7.1	64	3.4	505	12	54,037	5,405	0.06	0.13	0.9	10	1183
0	Tr	0.1	1	Tr	10	1	Tr	Tr	Tr	Tr	Tr	1	1184
0	15	3.1	82	0.9	554	34	954	95	0.14	0.07	1.2	32	1185
0	10	5.9	71	3.5	401	1,560	42	5	0.05	0.05	0.3	35	1186
0	1	0.1	17	0.3	9	23	12	1	0.01	0.02	Tr	Tr	1187
0	Tr	Tr	1	0.3	14	10	6	1	0.02	0.04	0.1	Tr	1188
0	2	0.2	4	0.1	33	1	119	12	0.01	Tr	Tr	1	1189
0	20	7.6	261	4.5	970	25	281	29	0.47	0.28	2.3	31	1190
0	1	0.8	30	0.8	167	24	2,015	202	0.02	0.06	0.2	8	1191
0	Tr	0.3	10	0.3	56	8	672	67	0.01	0.02	0.1	3	1192
0	7	4.3	245	6.4	839	126	14,742	1,474	0.17	0.42	0.9	18	1193
0	10	5.7	277	2.9	566	163	14,790	1,478	0.11	0.32	0.8	23	1194
0	7	5.1	272	4.9	740	58	18,781	1,879	0.03	0.30	0.8	31	1195
0	5	2.1	23	0.5	220	2	221	23	0.07	0.04	0.6	17	1196
0	8	2.5	49	0.6	346	2	517	52	0.08	0.07	0.9	10	1197
0	18	5.7	29	0.7	896	2	7,292	730	0.17	0.05	1.4	20	1198
0	24	2.2	46	1.4	319	5	8,014	802	0.12	0.09	1.1	8	1199
0	35	4.4	41	0.7	508	15	31,860	3,186	0.11	0.19	0.9	36	1200
0	38	2.8	33	0.9	287	20	26,604	2,660	0.08	0.22	1.0	27	1201

Food No.	Food Description	Measure of edible portion	Weight (g)	Water (%)	Calories (kcal)	Protein (g)	Total fat (g)	Saturated (g)	Mono-unsaturated (g)	Poly-unsaturated (g)

Vegetables and Vegetable Products (continued)

Sweet potatoes (continued)

| 1202 | Candied (2½" x 2" piece) | 1 piece | 105 | 67 | 144 | 1 | 3 | 1.4 | 0.7 | 0.2 |

Canned

1203	Syrup pack, drained	1 cup	196	72	212	3	1	0.1	Tr	0.3
1204	Vacuum pack, mashed	1 cup	255	76	232	4	1	0.1	Tr	0.2
1205	Tomatillos, raw	1 medium	34	92	11	Tr	Tr	Tr	0.1	0.1

Tomatoes
Raw, year round average

| 1206 | Chopped or sliced | 1 cup | 180 | 94 | 38 | 2 | 1 | 0.1 | 0.1 | 0.2 |
| 1207 | Slice, medium, ¼" thick | 1 slice | 20 | 94 | 4 | Tr | Tr | Tr | Tr | Tr |

Whole

1208	Cherry	1 cherry	17	94	4	Tr	Tr	Tr	Tr	Tr
1209	Medium, 2⅜" dia	1 tomato	123	94	26	1	Tr	0.1	0.1	0.2
1210	Canned, solids and liquid	1 cup	240	94	46	2	Tr	Tr	Tr	0.1

Sun dried

1211	Plain	1 piece	2	15	5	Tr	Tr	Tr	Tr	Tr
1212	Packed in oil, drained	1 piece	3	54	6	Tr	Tr	0.1	0.3	0.1
1213	Tomato juice, canned, with salt added	1 cup	243	94	41	2	Tr	Tr	Tr	0.1

Tomato products, canned

1214	Paste	1 cup	262	74	215	10	1	0.2	0.2	0.6
1215	Puree	1 cup	250	87	100	4	Tr	0.1	0.1	0.2
1216	Sauce	1 cup	245	89	74	3	Tr	0.1	0.1	0.2

Spaghetti/marinara/pasta sauce. See Soups, Sauces, and Gravies.

| 1217 | Stewed | 1 cup | 255 | 91 | 71 | 2 | Tr | Tr | 0.1 | 0.1 |
| 1218 | Turnips, cooked, cubes | 1 cup | 156 | 94 | 33 | 1 | Tr | Tr | Tr | 0.1 |

Turnip greens, cooked, drained

1219	From raw (leaves and stems)	1 cup	144	93	29	2	Tr	0.1	Tr	0.1
1220	From frozen (chopped)	1 cup	164	90	49	5	1	0.2	Tr	0.3
1221	Vegetable juice cocktail, canned	1 cup	242	94	46	2	Tr	Tr	Tr	0.1

Vegetables, mixed

1222	Canned, drained	1 cup	163	87	77	4	Tr	0.1	Tr	0.2
1223	Frozen, cooked, drained	1 cup	182	83	107	5	Tr	0.1	Tr	0.1
1224	Waterchestnuts, canned, slices, solids and liquids	1 cup	140	86	70	1	Tr	Tr	Tr	Tr

Miscellaneous Items

| 1225 | Bacon bits, meatless | 1 tbsp | 7 | 8 | 31 | 2 | 2 | 0.3 | 0.4 | 0.9 |

Baking powders for home use
Double acting

1226	Sodium aluminum sulfate	1 tsp	5	5	2	0	0	0.0	0.0	0.0
1227	Straight phosphate	1 tsp	5	4	2	Tr	0	0.0	0.0	0.0
1228	Low sodium	1 tsp	5	6	5	Tr	Tr	Tr	Tr	Tr
1229	Baking soda	1 tsp	5	Tr	0	0	0	0.0	0.0	0.0
1230	Beef jerky	1 large piece	20	23	81	7	5	2.1	2.2	0.2
1231	Catsup	1 cup	240	67	250	4	1	0.1	0.1	0.4
1232		1 tbsp	15	67	16	Tr	Tr	Tr	Tr	Tr
1233		1 packet	6	67	6	Tr	Tr	Tr	Tr	Tr
1234	Celery seed	1 tsp	2	6	8	Tr	1	Tr	0.3	0.1
1235	Chili powder	1 tsp	3	8	8	Tr	Tr	0.1	0.1	0.2

Chocolate, unsweetened, baking

| 1236 | Solid | 1 square | 28 | 1 | 148 | 3 | 16 | 9.2 | 5.2 | 0.5 |
| 1237 | Liquid | 1 oz | 28 | 1 | 134 | 3 | 14 | 7.2 | 2.6 | 3.0 |

*For product with no salt added: If salt added, consult the nutrition label for sodium value.

Choles-terol (mg)	Carbo-hydrate (g)	Total dietary fiber (g)	Calcium (mg)	Iron (mg)	Potas-sium (mg)	Sodium (mg)	Vitamin A (IU)	Vitamin A (RE)	Thiamin (mg)	Ribo-flavin (mg)	Niacin (mg)	Ascor-bic acid (mg)	Food No.
8	29	2.5	27	1.2	198	74	4,398	440	0.02	0.04	0.4	7	1202
0	50	5.9	33	1.9	378	76	14,028	1,403	0.05	0.07	0.7	21	1203
0	54	4.6	56	2.3	796	135	20,357	2,035	0.09	0.15	1.9	67	1204
0	2	0.6	2	0.2	91	Tr	39	4	0.01	0.01	0.6	4	1205
0	8	2.0	9	0.8	400	16	1,121	112	0.11	0.09	1.1	34	1206
0	1	0.2	1	0.1	44	2	125	12	0.01	0.01	0.1	4	1207
0	1	0.2	1	0.1	38	2	106	11	0.01	0.01	0.1	3	1208
0	6	1.4	6	0.6	273	11	766	76	0.07	0.06	0.8	23	1209
0	10	2.4	72	1.3	530	355	1,428	144	0.11	0.07	1.8	34	1210
0	1	0.2	2	0.2	69	42	17	2	0.01	0.01	0.2	1	1211
0	1	0.2	1	0.1	47	8	39	4	0.01	0.01	0.1	3	1212
0	10	1.0	22	1.4	535	877	1,351	136	0.11	0.08	1.6	44	1213
0	51	10.7	92	5.1	2,455	231	6,406	639	0.41	0.50	8.4	111	1214
0	24	5.0	43	3.1	1,065	85*	3,188	320	0.18	0.14	4.3	26	1215
0	18	3.4	34	1.9	909	1,482	2,399	240	0.16	0.14	2.8	32	1216
0	17	2.6	84	1.9	607	564	1,380	138	0.12	0.09	1.8	29	1217
0	8	3.1	34	0.3	211	78	0	0	0.04	0.04	0.5	18	1218
0	6	5.0	197	1.2	292	42	7,917	792	0.06	0.10	0.6	39	1219
0	8	5.6	249	3.2	367	25	13,079	1,309	0.09	0.12	0.8	36	1220
0	11	1.9	27	1.0	467	653	2,831	283	0.10	0.07	1.8	67	1221
0	15	4.9	44	1.7	474	243	18,985	1,899	0.07	0.08	0.9	8	1222
0	24	8.0	46	1.5	308	64	7,784	779	0.13	0.22	1.5	6	1223
0	17	3.5	6	1.2	165	11	6	0	0.02	0.03	0.5	2	1224
0	2	0.7	7	0.1	10	124	0	0	0.04	Tr	0.1	Tr	1225
0	1	Tr	270	0.5	1	488	0	0	0.00	0.00	0.0	0	1226
0	1	Tr	339	0.5	Tr	363	0	0	0.00	0.00	0.0	0	1227
0	2	0.1	217	0.4	505	5	0	0	0.00	0.00	0.0	0	1228
0	0	0.0	0	0.0	0	1,259	0	0	0.00	0.00	0.0	0	1229
10	2	0.4	4	1.1	118	438	0	0	0.03	0.03	0.3	0	1230
0	65	3.1	46	1.7	1,154	2,846	2,438	245	0.21	0.18	3.3	36	1231
0	4	0.2	3	0.1	72	178	152	15	0.01	0.01	0.2	2	1232
0	2	0.1	1	Tr	29	71	61	6	0.01	Tr	0.1	1	1233
0	1	0.2	35	0.9	28	3	1	Tr	0.01	0.01	0.1	Tr	1234
0	1	0.9	7	0.4	50	26	908	91	0.01	0.02	0.2	2	1235
0	8	4.4	21	1.8	236	4	28	3	0.02	0.05	0.3	0	1236
0	10	5.1	15	1.2	331	3	3	Tr	0.01	0.08	0.6	0	1237

Food No.	Food Description	Measure of edible portion	Weight (g)	Water (%)	Calories (kcal)	Protein (g)	Total fat (g)	Fatty acids Saturated (g)	Mono-unsaturated (g)	Poly-unsaturated (g)

Miscellaneous Items (continued)

Food No.	Food Description	Measure	Weight	Water	Calories	Protein	Total fat	Saturated	Mono	Poly
1238	Cinnamon	1 tsp	2	10	6	Tr	Tr	Tr	Tr	Tr
1239	Cocoa powder, unsweetened	1 cup	86	3	197	17	12	6.9	3.9	0.4
1240		1 tbsp	5	3	12	1	1	0.4	0.2	Tr
1241	Cream of tartar	1 tsp	3	2	8	0	0	0.0	0.0	0.0
1242	Curry powder	1 tsp	2	10	7	Tr	Tr	Tr	0.1	0.1
1243	Garlic powder	1 tsp	3	6	9	Tr	Tr	Tr	Tr	Tr
1244	Horseradish, prepared	1 tsp	5	85	2	Tr	Tr	Tr	Tr	Tr
1245	Mustard, prepared, yellow	1 tsp or 1 packet	5	82	3	Tr	Tr	Tr	0.1	Tr
	Olives, canned									
1246	Pickled, green	5 medium	17	78	20	Tr	2	0.3	1.6	0.2
1247	Ripe, black	5 large	22	80	25	Tr	2	0.3	1.7	0.2
1248	Onion powder	1 tsp	2	5	7	Tr	Tr	Tr	Tr	Tr
1249	Oregano, ground	1 tsp	2	7	5	Tr	Tr	Tr	Tr	0.1
1250	Paprika	1 tsp	2	10	6	Tr	Tr	Tr	Tr	0.2
1251	Parsley, dried	1 tbsp	1	9	4	Tr	Tr	Tr	Tr	Tr
1252	Pepper, black	1 tsp	2	11	5	Tr	Tr	Tr	Tr	Tr
	Pickles, cucumber									
1253	Dill, whole, medium (3¾" long)	1 pickle	65	92	12	Tr	Tr	Tr	Tr	0.1
1254	Fresh (bread and butter pickles), slices 1½" dia, ¼" thick	3 slices	24	79	18	Tr	Tr	Tr	Tr	Tr
1255	Pickle relish, sweet	1 tbsp	15	62	20	Tr	Tr	Tr	Tr	Tr
1256	Pork skins/rinds, plain	1 oz	28	2	155	17	9	3.2	4.2	1.0
	Potato chips									
	Regular									
	Plain									
1257	Salted	1 oz	28	2	152	2	10	3.1	2.8	3.5
1258	Unsalted	1 oz	28	2	152	2	10	3.1	2.8	3.5
1259	Barbecue flavor	1 oz	28	2	139	2	9	2.3	1.9	4.6
1260	Sour cream and onion flavor	1 oz	28	2	151	2	10	2.5	1.7	4.9
1261	Reduced fat	1 oz	28	1	134	2	6	1.2	1.4	3.1
1262	Fat free, made with olestra	1 oz	28	2	75	2	Tr	Tr	0.1	0.1
	Made from dried potatoes									
1263	Plain	1 oz	28	1	158	2	11	2.7	2.1	5.7
1264	Sour cream and onion flavor	1 oz	28	2	155	2	10	2.7	2.0	5.3
1265	Reduced fat	1 oz	28	1	142	2	7	1.5	1.7	3.8
1266	Salt	1 tsp	6	Tr	0	0	0	0.0	0.0	0.0
	Trail mix									
1267	Regular, with raisins, chocolate chips, salted nuts and seeds	1 cup	146	7	707	21	47	8.9	19.8	16.5
1268	Tropical	1 cup	140	9	570	9	24	11.9	3.5	7.2
1269	Vanilla extract	1 tsp	4	53	12	Tr	Tr	Tr	Tr	Tr
	Vinegar									
1270	Cider	1 tbsp	15	94	2	0	0	0.0	0.0	0.0
1271	Distilled	1 tbsp	17	95	2	0	0	0.0	0.0	0.0
	Yeast, baker's									
1272	Dry, active	1 pkg	7	8	21	3	Tr	Tr	0.2	Tr
1273		1 tsp	4	8	12	2	Tr	Tr	0.1	Tr
1274	Compressed	1 cake	17	69	18	1	Tr	Tr	0.2	Tr

Choles-terol (mg)	Carbo-hydrate (g)	Total dietary fiber (g)	Calcium (mg)	Iron (mg)	Potas-sium (mg)	Sodium (mg)	Vitamin A (IU)	Vitamin A (RE)	Thiamin (mg)	Ribo-flavin (mg)	Niacin (mg)	Ascor-bic acid (mg)	Food No.
0	2	1.2	28	0.9	11	1	6	1	Tr	Tr	Tr	1	1238
0	47	28.6	110	11.9	1,311	18	17	2	0.07	0.21	1.9	0	1239
0	3	1.8	7	0.7	82	1	1	Tr	Tr	0.01	0.1	0	1240
0	2	Tr	Tr	0.1	495	2	0	0	0.00	0.00	0.0	0	1241
0	1	0.7	10	0.6	31	1	20	2	0.01	0.01	0.1	Tr	1242
0	2	0.3	2	0.1	31	1	0	0	0.01	Tr	Tr	1	1243
0	1	0.2	3	Tr	12	16	Tr	0	Tr	Tr	Tr	1	1244
0	Tr	0.2	4	0.1	8	56	7	1	Tr	Tr	Tr	Tr	1245
0	Tr	0.2	10	0.3	9	408	51	5	0.00	0.00	Tr	0	1246
0	1	0.7	19	0.7	2	192	89	9	Tr	0.00	Tr	Tr	1247
0	2	0.1	8	0.1	20	1	0	0	0.01	Tr	Tr	Tr	1248
0	1	0.6	24	0.7	25	Tr	104	10	0.01	Tr	0.1	1	1249
0	1	0.4	4	0.5	49	1	1,273	127	0.01	0.04	0.3	1	1250
0	1	0.4	19	1.3	49	6	303	30	Tr	0.02	0.1	2	1251
0	1	0.6	9	0.6	26	1	4	Tr	Tr	0.01	Tr	Tr	1252
0	3	0.8	6	0.3	75	833	214	21	0.01	0.02	Tr	1	1253
0	4	0.4	8	0.1	48	162	34	3	0.00	0.01	0.0	2	1254
0	5	0.2	Tr	0.1	4	122	23	2	0.00	Tr	Tr	Tr	1255
27	0	0.0	9	0.2	36	521	37	11	0.03	0.08	0.4	Tr	1256
0	15	1.3	7	0.5	361	168	0	0	0.05	0.06	1.1	9	1257
0	15	1.4	7	0.5	361	2	0	0	0.05	0.06	1.1	9	1258
0	15	1.2	14	0.5	357	213	62	6	0.06	0.06	1.3	10	1259
2	15	1.5	20	0.5	377	177	48	6	0.05	0.06	1.1	11	1260
0	19	1.7	6	0.4	494	139	0	0	0.06	0.08	2.0	7	1261
0	17	1.1	10	0.4	366	185	1,469	441	0.10	0.02	1.3	8	1262
0	14	1.0	7	0.4	286	186	0	0	0.06	0.03	0.9	2	1263
1	15	0.3	18	0.4	141	204	214	28	0.05	0.03	0.7	3	1264
0	18	1.0	10	0.4	285	121	0	0	0.05	0.02	1.2	3	1265
0	0	0.0	1	Tr	Tr	2,325	0	0	0.00	0.00	0.0	0	1266
6	66	8.8	159	4.9	946	177	64	7	0.60	0.33	6.4	2	1267
0	92	10.6	80	3.7	993	14	69	7	0.63	0.16	2.1	11	1268
0	1	0.0	Tr	Tr	6	Tr	0	0	Tr	Tr	Tr	0	1269
0	1	0.0	1	0.1	15	Tr	0	0	0.00	0.00	0.0	0	1270
0	1	0.0	0	0.0	2	Tr	0	0	0.00	0.00	0.0	0	1271
0	3	1.5	4	1.2	140	4	Tr	0	0.17	0.38	2.8	Tr	1272
0	2	0.8	3	0.7	80	2	Tr	0	0.09	0.22	1.6	Tr	1273
0	3	1.4	3	0.6	102	5	0	0	0.32	0.19	2.1	Tr	1274

APPENDIX B
Dietary Reference Intakes

Age (yr)	Reference BMI (kg/m)	Reference Height cm (in)	Reference Weight kg (lb)	Water AI (L/day)	Energy EER (kcal/day)	Carbohydrate (g/day)	Total Fiber AI (g/day)	Total Fat AI (g/day)	Linoleic Acid AI (g/day)	Linolenic Acid AI (g/day)	Protein (g/day)	Protein (g/kg/day)
MALES												
0.05	—	62 (24)	6 (13)	0.7	570	60	—	31	4.4	0.5	9.1	1.52
0.5–1	—	71 (28)	9 (20)	0.8	743	95	—	30	4.6	0.5	11	1.2
1–3	—	86 (34)	12 (27)	1.3	1046	130	19	—	7	0.7	13	1.05
4–8	15.3	115 (45)	20 (44)	1.7	1742	130	25	—	10	0.9	19	0.95
9–13	17.2	144 (57)	36 (79)	2.4	2279	130	31	—	12	1.2	34	0.95
14–18	20.5	174 (68)	61 (134)	3.3	3152	130	38	—	16	1.6	52	0.85
19–30	22.5	177 (70)	70 (154)	3.7	3067	130	38	—	17	1.6	56	0.8
31–50				3.7	3067	130	38	—	17	1.6	56	0.8
>50				3.7	3067	130	30	—	14	1.6	56	0.8
FEMALES												
0–0.5	—	62 (24)	6 (13)	0.7	520	60	—	31	4.4	0.5	9.1	1.52
0.5–1	—	71 (28)	9 (20)	0.8	676	95	—	30	4.6	0.5	11	1.2
1–3	—	86 (34)	12 (27)	1.3	992	130	19	—	7	0.7	13	1.05
4–8	15.3	115 (45)	20 (44)	1.7	1642	130	25	—	10	0.9	19	0.95
9–13	17.4	144 (57)	37 (81)	2.1	2071	130	26	—	10	1.0	34	0.95
14–18	20.4	163 (64)	54 (119)	2.3	2368	130	26	—	11	1.1	46	0.85
19–30	21.5	163 (24)	57 (126)	2.7	2403	130	25	—	12	1.1	46	0.8
31–50				2.7	2403	130	21	—	12	1.1	46	0.8
>50				2.7	2403	130	21	—	11	1.1	46	0.8

(continued)

Age (yr)	Reference BMI (kg/m)	Reference Height cm (in)	Reference Weight kg (lb)	Water AI (L/day)	Energy EER (kcal/day)	Carbohydrate (g/day)	Total Fiber AI (g/day)	Total Fat AI (g/day)	Linoleic Acid AI (g/day)	Linolenic Acid AI (g/day)	Protein (g/day)	Protein (g/kg/day)
PREGNANCY												
1st trimester				3.0	10	175	28	—	13	1.4	125	1.1
2nd trimester				3.0	1340	175	28	—	13	1.4	125	1.1
3rd trimester				3.0	1452	175	28	—	13	1.4	125	1.1
LACTATION												
1st 6 months				3.8	1330	210	29	—	13	1.3	125	1.1
2nd 6 months				3.8	1400	210	29	—	13	1.3	125	1.1

Notes:

1. The values for infants (up to 1 year old) are all Adequate Intakes.

2. The values under "Energy-EER" are based on an active person at the reference height and weight. For males over 19, subtract 10 kcalories from the EER given for each year above 19. For females over 19, subtract 7 kcalories from the EER given for each year above 19.

Source: Adapted with permission from the *Dietary References Intakes* series, © 1997, 1998, 2000, 2001, 2002, 2004, by the National Academy of Sciences. Courtesy of the National Academies Press, Washington, D.C.

Recommended Dietary Allowance (RDA) and Adequate Intakes (AI) for Vitamins

Age (yr)	Thiamin RDA (mg/day)	Riboflavin RDA (mg/day)	Niacin RDA (mg/day)	Biotin AI (µg/day)	Pantothenic Acid AI (mg/day)	Vitamin B$_6$ (mg/day)	Folate RDA (µg/day)	Vitamin B$_{12}$ RDA (µg/day)	Choline AI (mg/day)	Vitamin C RDA (mg/day)	Vitamin A RDA (µg/day)	Vitamin D AI (µg/day)	Vitamin E RDA (mg/day)	Vitamin K AI (µg/day)
INFANTS														
0–0.5	0.2	0.3	2	5	1.7	0.1	65	0.4	125	40	400	5	4	2.0
0.5–1	0.3	0.4	4	6	1.8	0.3	80	0.5	150	50	500	5	5	2.5
CHILDREN														
1–3	0.5	0.5	6	8	2	0.5	150	0.9	200	15	300	5	6	30
4–8	0.6	0.6	8	12	3	0.6	200	1.2	250	25	400	5	7	55

(continued)

Age (yr)	Thiamin RDA (mg/day)	Riboflavin RDA (mg/day)	Niacin RDA (mg/day)	Biotin AI (µg/day)	Pantothenic Acid AI (mg/day)	Vitamin B_6 (mg/day)	Folate RDA (µg/day)	Vitamin B_{12} RDA (µg/day)	Choline AI (mg/day)	Vitamin C RDA (mg/day)	Vitamin A RDA (µg/day)	Vitamin D AI (µg/day)	Vitamin E RDA (mg/day)	Vitamin K AI (µg/day)
MALES														
9–13	0.9	0.9	12	20	4	1.0	300	1.8	375	45	600	5	11	60
14–18	1.2	1.3	16	25	5	1.3	400	2.4	550	75	900	5	15	75
19–30	1.2	1.3	16	30	5	1.3	400	2.4	550	90	900	5	15	120
31–50	1.2	1.3	16	30	5	1.3	400	2.4	550	90	900	5	15	120
51–70	1.2	1.3	16	30	5	1.7	400	2.4	550	90	900	10	15	120
>70	1.2	1.3	16	30	5	1.7	400	2.4	550	90	900	15	15	120
FEMALES														
9–13	0.9	0.9	12	20	4	1.0	300	1.8	375	45	600	5	11	60
14–18	1.0	1.0	14	25	5	1.2	400	2.4	400	65	700	5	15	75
19–30	1.1	1.1	14	30	5	1.3	400	2.4	425	75	700	5	15	90
31–50	1.1	1.1	14	30	5	1.3	400	2.4	425	75	700	5	15	90
51–70	1.1	1.1	14	30	5	1.5	400	2.4	425	75	700	10	15	90
>70	1.1	1.1	14	30	5	1.5	400	2.4	425	75	700	15	15	90
PREGNANCY														
≤18	1.4	1.4	18	30	6	1.9	600	2.6	450	80	750	5	15	75
19–30	1.4	1.4	18	30	6	1.9	600	2.6	450	85	770	5	15	90
31–50	1.4	1.4	18	30	6	1.9	600	2.6	450	85	770	5	15	90
LACTATION														
≤18	1.4	1.6	17	35	7	2.0	500	2.8	550	115	1200	5	19	75
19–30	1.4	1.6	17	35	7	2.0	500	2.8	550	120	1300	5	19	90
31–50	1.4	1.6	17	35	7	2.0	500	2.8	550	120	1300	5	19	90

Notes:

1. Niacin requirements are expressed as niacin equivalents (NE), except for infants younger than 6 months, which are expressed as preformed niacin.

2. Folate requirements are expressed as dietary folate equivalents (DFE).

3. Vitamin A requirements are expressed as retinol activity equivalents (RAE).

4. Vitamin D requirements are expressed as cholecalciferol and assume no vitamin D is derived from sunlight.

5. Vitamin E requirements are expressed as alpha-tocopherol.

6. The values for infants (up to 1 year old) are all Adequate Intakes.

Source: Adapted with permission from the *Dietary References Intakes* series, © 1997, 1998, 2000, 2001, 2002, 2004, by the National Academy of Sciences. Courtesy of the National Academies Press, Washington, D.C.

Tolerable Upper Intake Levels (UL) for Vitamins

Age (yr)	Niacin (mg/day)	Vitamin B$_6$ (mg/day)	Folate (μg/day)	Choline (mg/day)	Vitamin C (mg/day)	Vitamin A (μg/day)	Vitamin D (μg/day)	Vitamin E (mg/day)
INFANTS								
0–0.5	—	—	—	—	—	600	25	—
0.5–1	—	—	—	—	—	600	25	—
CHILDREN								
1–3	10	30	300	1000	400	600	50	200
4–8	15	40	400	1000	650	900	50	300
9–13	20	60	600	2000	1200	1700	50	600
ADOLESCENTS								
14–18	30	80	800	3000	1800	2800	50	800
ADULTS								
19–70	35	100	1000	3500	2000	3000	50	1000
>70	35	100	1000	3500	2000	3000	50	1000
PREGNANCY								
≤18	30	80	800	3000	1800	2800	50	800
19–50	35	100	1000	3500	2000	3000	50	1000
LACTATION								
≤18	30	80	800	3000	1800	2800	50	800
19–50	35	100	1000	3500	2000	3000	50	1000

Notes:

1. The UL for niacin and folate includes intake only from supplements and fortified foods.

2. The UL for vitamin A includes only the preformed vitamin found, for example, in vitamin A–fortified milk. It does not include food sources of beta-carotene such as carrots.

3. The UL for vitamin E only includes supplemental alpha-tocopherol and fortified foods.

Source: Adapted with permission from the *Dietary References Intakes* series, © 1997, 1998, 2000, 2001, 2002, 2004, by the National Academy of Sciences. Courtesy of the National Academies Press, Washington, D.C.

Tolerable Upper Intake Levels (UL) for Minerals

Age (yr)	Sodium (mg/day)	Chloride (mg/day)	Calcium (mg/day)	Phosphorus (mg/day)	Magnesium (mg/day)	Iron (mg/day)
INFANT						
0–0.5	—	—	—	—	—	40
0.5–1	—	—	—	—	—	40
CHILDREN						
1–3	1500	2300	2500	3000	65	40
4–8	1900	2900	2500	3000	110	40
9–13	2200	3400	2500	4000	350	40
ADOLESCENTS						
14–18	2300	3600	2500	4000	350	45
ADULTS						
19–70	2300	3600	2500	4000	350	45
>70	2300	3600	2500	3000	350	45
PREGNANCY						
<18	2300	3600	2500	3500	350	45
19–50	2300	3600	2500	3500	350	45
LACTATION						
<18	2300	3600	2500	4000	350	45
19–50	2300	3600	2500	4000	350	45

Note:

1. The UL for magnesium includes only intake from supplements or drugs.

Source: Adapted with permission from the *Dietary References Intakes* series, © 1997, 1998, 2000, 2001, 2002, 2004, by the National Academy of Sciences. Courtesy of the National Academies Press, Washington, D.C.

Recommended Dietary Allowance (RDA) and Adequate Intakes (AI) for Minerals

Age (yr)	Sodium AI (mg/day)	Chloride AI (mg/day)	Potassium AI (mg/day)	Calcium AI (mg/day)	Phosphorus RDA (mg/day)	Magnesium RDA (mg/day)	Iron RDA (mg/day)	Zinc RDA (mg/day)	Iodine RDA (mg/day)	Selenium RDA (µg/day)	Copper RDA (µg/day)	Manganese AI (mg/day)	Fluoride AI (mg/day)	Chromium AI (µg/day)	Molybdenum RDA (µg/day)
INFANTS															
0–0.5	120	180	400	210	100	30	0.27	2	110	15	200	0.003	0.01	0.2	2
0.5–1	370	570	700	270	275	75	11	3	130	20	220	0.6	0.5	5.5	3
CHILDREN															
1–3	1000	1500	3000	500	460	80	7	3	90	20	340	1.2	0.7	11	17
4–8	1200	1900	3800	800	500	130	10	5	90	30	440	1.5	1	15	22
MALES															
9–13	1500	2300	4500	1300	1250	240	8	8	120	40	700	1.9	2	25	34
14–18	1500	2300	4700	1300	1250	410	11	11	150	55	890	2.2	3	35	43
19–30	1500	2300	4700	1000	700	400	8	11	150	55	900	2.3	4	35	45
31–50	1500	2300	4700	1000	700	420	8	11	150	55	900	2.3	4	35	45
51–70	1300	2000	4700	1200	700	420	8	11	150	55	900	2.3	4	30	45
>70	1200	1800	4700	1200	700	420	8	11	150	55	900	2.3	4	30	45
FEMALES															
9–13	1500	2300	4500	1300	1250	240	8	8	120	40	700	1.6	2	21	34
14–18	1500	2300	4700	1300	1250	360	15	9	150	55	890	1.6	3	24	43
19–30	1500	2300	4700	1000	700	310	18	8	150	55	900	1.8	3	25	45
31–50	1500	2300	4700	1000	700	320	18	8	150	55	900	1.8	3	25	45
51–70	1300	2000	4700	1200	700	320	8	8	150	55	900	1.8	3	20	45
>70	1200	1800	4700	1200	700	320	8	8	150	55	900	1.8	3	20	45
PREGNANCY															
≤18	1500	2300	4700	1300	1250	400	27	12	220	60	1000	2.0	3	29	50
19–30	1500	2300	4700	1000	700	350	27	11	220	60	1000	2.0	3	30	50
31–50	1500	2300	4700	1000	700	360	27	11	220	60	1000	2.0	3	30	50
LACTATION															
≤18	1500	2300	5100	1300	1250	360	10	14	290	70	1300	2.6	3	44	50
19–30	1500	2300	5100	1000	700	310	9	12	290	70	1300	2.6	3	45	50
31–50	1500	2300	5100	1000	700	320	9	12	290	70	1300	2.6	3	45	50

Note:

1. The values for infants (up to 1 year old) are all Adequate Intakes.

Source: Adapted with permission from the *Dietary References Intakes* series, © 1997, 1998, 2000, 2001, 2002, 2004, by the National Academy of Sciences. Courtesy of the National Academies Press, Washington, D.C.

APPENDIX C
Expanded Serving Sizes
for MyPyramid

WHAT COUNTS AS AN OUNCE-EQUIVALENT OF GRAINS?

In general, 1 slice of bread, 1 cup of ready-to-eat cereal, or $\frac{1}{2}$ cup of cooked rice, cooked pasta, or cooked cereal can be considered as 1 ounce-equivalent from the grains group. The following chart lists specific amounts that count as 1 ounce-equivalent of grains toward the daily recommended intake. In some cases the number of ounce-equivalents for common portions is also shown.

		Amount That Counts as 1 Ounce-Equivalent	Common Portions and Ounce-Equivalents
Bagels	WG: whole wheat RG: plain, egg	$\frac{1}{2}$ "mini" bagel	1 large bagel = 4 ounce-equivalents
Biscuits	RG: baking powder, buttermilk	1 small (2-inch diameter)	1 large (3-inch diameter) = - 2 ounce-equivalents
Breads	WG: 100 percent whole wheat RG: white, wheat, French, sourdough	1 regular slice 1 small slice French 4 snack-size slices rye bread	2 regular slices = 2 ounce-equivalents
Bulgur	Cracked wheat (WG)	$\frac{1}{2}$ cup cooked	
Cornbread	RG	1 small piece ($2\frac{1}{2}$ × $1\frac{1}{4}$ × $1\frac{1}{4}$ inches)	1 medium piece ($2\frac{1}{2}$ × $2\frac{1}{2}$ × $1\frac{1}{4}$ inches) = 2 ounce-equivalents
Crackers	WG: 100 percent whole-wheat, rye RG: saltines, snack crackers	5 whole-wheat crackers 2 rye crispbreads 7 square or round crackers	
English muffins	WG: whole wheat RG: plain, raisin	$\frac{1}{2}$ muffin	1 muffin = 2 ounce-equivalents
Muffins	WG: whole wheat RG: bran, corn, plain	1 small ($2\frac{1}{2}$-inch diameter)	1 large ($3\frac{1}{2}$-inch diameter) = 3 ounce-equivalents
Oatmeal	(WG)	$\frac{1}{2}$ cup cooked 1 packet instant 1 ounce dry (regular or quick)	

(continued)

		Amount That Counts as 1 Ounce-Equivalent	Common Portions and Ounce-Equivalents
Pancakes	WG: whole wheat, buckwheat	1 pancake (4½-inch diameter)	3 pancakes (4½-inch diameter) = 3 ounce-equivalents
	RG: plain, buttermilk	2 small pancakes (3-inch diameter)	
Popcorn	(WG)	3 cups, popped	1 microwave bag, popped = 4 ounce-equivalents
Ready-to-eat breakfast cereal	WG: whole-wheat flakes, toasted oats	1 cup flakes or rounds 1¼ cups puffed	
	RG: corn flakes, puffed rice		
Rice	WG: brown, wild	½ cup cooked	1 cup cooked = 2 ounce-equivalents
	RG: white	1 ounce dry	
Pasta—spaghetti, macaroni, noodles	WG: whole wheat	½ cup cooked	1 cup cooked = 2 ounce-equivalents
	RG: durum, enriched	1 ounce dry	
Tortillas	WG: whole wheat, whole-grain corn	1 small flour tortilla (6-inch diameter)	1 large tortilla (12-inch diameter) = 4 ounce-equivalents
	RG: flour, corn	1 corn tortilla (6-inch diameter)	

WG = whole grains, RG = refined grains. This is shown when products are available both in whole-grain and refined-grain forms.

WHAT COUNTS AS A CUP OF VEGETABLES?

In general, 1 cup of raw or cooked vegetables or vegetable juice, or 2 cups of raw leafy greens can be considered as 1 cup from the vegetable group. The chart lists specific amounts that count as 1 cup of vegetables (in some cases equivalents for ½ cup are also shown) toward your recommended intake.

	Amount That Counts as 1 Cup of Vegetables	Amount That Counts as ½ Cup of Vegetables
DARK GREEN VEGETABLES		
Broccoli	1 cup chopped or florets 3 spears, 5 inches long, raw or cooked	
Greens: Collards, mustard greens, turnip greens, kale	1 cup cooked	
Spinach	1 cup cooked 2 cups raw is equivalent to 1 cup of vegetables	1 cup raw is equivalent to ½ cup of vegetables
Raw leafy greens: Spinach, romaine, watercress, dark green leafy lettuce, endive, escarole	2 cups raw is equivalent to 1 cup of vegetables	1 cup raw is equivalent to ½ cup of vegetables

(continued)

	Amount That Counts as 1 Cup of Vegetables	Amount That Counts as ½ Cup of Vegetables
ORANGE VEGETABLES		
Carrots	1 cup strips, slices, or chopped, raw or cooked	
	2 medium	1 medium carrot
	1 cup baby carrots (about 12)	About 6 baby carrots
Pumpkin	1 cup mashed, cooked	
Sweet potato	1 large baked (2¼ inches or more diameter)	
Winter squash (acorn, butternut, Hubbard)	1 cup cubed, cooked	½ acorn squash, baked = ¾ cup
DRY BEANS AND PEAS		
Dry beans and peas, such as black, garbanzo, kidney, pinto, soybeans, black-eyed peas, or split peas	1 cup whole or mashed, cooked	
Tofu	1 cup ½-inch cubes (about 8 ounces)	1 piece, 2½ × 2¾ × 1 inch, about 4 ounces
STARCHY VEGETABLES		
Corn, yellow or white	1 cup	
	1 large ear (8–9 inches long)	
Green peas	1 cup	
White potatoes	1 cup diced, mashed	
	1 medium, boiled or baked potato 2½–3 inches diameter)	
	French fried: 20 medium to long strips (2½–4 inches long) (Contains discretionary kcalories)	
OTHER VEGETABLES		
Beans sprouts	1 cup cooked	
Cabbage, green	1 cup, chopped or shredded, raw or cooked	
Cauliflower	1 cup, pieces or florets, raw or cooked	
Celery	1 cup, diced or sliced, raw or cooked	
	2 large stalks (11–12 inches long)	1 large stalk (11–12 inches long)
Cucumbers	1 cup, raw, sliced or chopped	
Green or wax beans	1 cup cooked	
Green or red peppers	1 cup chopped, raw or cooked	
	1 large pepper (3-inch diameter, 3¾ inches long)	1 small pepper
Lettuce, iceberg or head	2 cups raw, shredded, or chopped = 1 cup of vegetables	1 cup raw, shredded, or chopped = ½ cup of vegetables
Mushrooms	1 cup raw or cooked	
Onions	1 cup chopped, raw, or cooked	
Tomatoes	1 large raw, whole (3 inches)	1 small raw, whole (2¼ inches)
	1 cup chopped or sliced, raw, canned, or cooked	1 medium canned
Tomato or mixed vegetable juice	1 cup	½ cup
Summer squash (zucchini)	1 cup cooked, sliced or diced	

WHAT COUNTS AS A CUP OF FRUIT?

In general, 1 cup of fruit or 100 percent fruit juice or ½ cup of dried fruit can be considered as 1 cup from the fruit group. The following specific amounts count as 1 cup of fruit (in some cases equivalents for ½ cup are also shown) toward the daily recommended intake.

	Amount That Counts as 1 Cup of Fruit	Amount That Counts as ½ Cup of Fruit
Apple	½ large (3¼-inch diameter)	
	1 small (2½-inch diameter)	
	1 cup sliced or chopped, raw or cooked	½ cup sliced or chopped, raw or cooked
Applesauce	1 cup	1 snack container (4 oz.)
Banana	1 large (8–9 inches long)	1 small (less than 6 inches long)
	1 cup sliced	
Cantaloupe	1 cup diced or melon balls	1 medium wedge (⅛ of a medium melon)
Grapes	1 cup whole or cut up	
	32 seedless grapes	16 seedless grapes
Grapefruit	1 medium (4-inch diameter)	½ medium (4-inch diameter)
	1 cup sections	
Mixed fruit (fruit cocktail)	1 cup diced or sliced, raw or canned, drained	1 snack container (4 oz.), drained = ⅜ cup
Orange	1 large (3¹⁄₁₆-inch diameter)	1 small (2⅜-inch diameter)
	1 cup sections	
Orange, mandarin	1 cup canned, drained	
Peach	1 large (2¾-inch diameter)	1 small (2⅜-inch diameter)
	1 cup sliced or diced, raw, cooked, or canned, drained	1 snack container (4 oz.), drained = ⅜ cup
Pear	1 medium pear (2.5 per pound)	
	1 cup sliced or diced, raw, cooked, or canned, drained	1 snack container (4 oz.), drained = ⅜ cup
Pineapple	1 cup chunks, sliced, or crushed, raw, cooked, or canned, drained	1 snack container (4 oz.), drained = ⅜ cup
Plum	1 cup sliced raw or cooked	
	3 medium or 2 large plums	1 large plum
Strawberries	About 8 large berries	
	1 cup whole, halved, or sliced, fresh or frozen	½ cup whole, halved, or sliced, fresh or frozen
Watermelon	1 small wedge (1 inch thick)	6 melon balls
	1 cup diced or balls	
Dried fruit (raisins, dried apricots, etc.)	½ cup dried fruit is equivalent to 1 cup of fruit	¼ cup dried fruit is equivalent to ½ cup of fruit
100% fruit juice (such as orange, apple, grape)	1 cup	½ cup

WHAT COUNTS AS 1 CUP IN THE MILK GROUP?

In general, 1 cup of milk or yogurt, $1\frac{1}{2}$ ounces of natural cheese, or 2 ounces of processed cheese can be considered as 1 cup from the milk group.

	Amount That Counts as 1 Cup in the Milk Group	Common Portions and Cup Equivalents
Milk (choose fat-free or low-fat milk most often)	1 cup 1 half-pint container $\frac{1}{2}$ cup evaporated milk	
Yogurt (choose fat-free or low-fat most often)	1 regular container (8 oz.) 1 cup	1 small container (6 oz.) = $\frac{3}{4}$ cup 1 snack-size container (4 oz.) = $\frac{1}{2}$ cup
Cheese (choose low-fat cheeses most often)	$1\frac{1}{2}$ oz. hard cheese (cheddar, mozzarella, Swiss, Parmesan) $\frac{1}{3}$ cup shredded cheese 2 oz. processed cheese (American) $\frac{1}{2}$ cup ricotta cheese 2 cups cottage cheese	1 slice of hard cheese is equivalent to $\frac{1}{2}$ cup milk 1 slice of processed cheese is equivalent to $\frac{1}{3}$ cup milk
Milk-based desserts (choose fat-free or low-fat most often)	1 cup pudding made with milk 1 cup frozen yogurt $1\frac{1}{2}$ cups ice cream	1 scoop ice cream is equivalent to $\frac{1}{3}$ cup milk

WHAT COUNTS AS AN OUNCE-EQUIVALENT IN THE MEAT AND BEANS GROUP?

In general, 1 ounce of meat, poultry, or fish, $\frac{1}{4}$ cup cooked dry beans, 1 egg, 1 tablespoon of peanut butter, or $\frac{1}{2}$ ounce of nuts or seeds can be considered as 1 ounce-equivalent from the meat and beans group.

	Amount That Counts as 1 Ounce-Equivalent in the Meat and Beans Group	Common Portions and Ounce-Equivalents
Meats	1 oz. cooked lean beef	1 small steak (eye of round, sirloin tip) = $3\frac{1}{2}$ to 4 ounce-equivalents
	1 oz. cooked lean pork or ham	1 small lean hamburger = 2 to 3 ounce-equivalents
Poultry	1 oz. cooked chicken or turkey, without skin	1 small chicken breast half = 3 ounce-equivalents
	1 sandwich slice of turkey ($4\frac{1}{2} \times 2\frac{1}{2} \times \frac{1}{8}$ inch)	$\frac{1}{2}$ Cornish game hen = 4 ounce-equivalents
Fish	1 oz. cooked fish or shellfish	1 can tuna, drained = 3 to 4 ounce-equivalents
		1 salmon steak = 4 to 6 ounce-equivalents
		1 small trout = 3 ounce-equivalents
Eggs	1 egg	
Nuts and Seeds	$\frac{1}{2}$ oz. nuts (12 almonds, 7 walnut halves, 24 pistachios)	1 oz. nuts or seeds = 2 ounce-equivalents
	$\frac{1}{2}$ oz. seeds (pumpkin, sunflower, or squash seeds, hulled, roasted)	
	1 tablespoon of peanut butter or almond butter	
Dry Beans	$\frac{1}{4}$ cup cooked dry beans, such as kidney, pinto, black, or white beans	1 cup split-pea soup = 2 ounce-equivalents
	$\frac{1}{4}$ cup cooked dry peas, such as split peas or lentils	1 cup lentil soup = 2 ounce-equivalents
	$\frac{1}{4}$ cup baked beans or refried beans	1 cup bean soup = 2 ounce-equivalents
	$\frac{1}{4}$ cup (about 2 oz.) of tofu	
	1 oz. tempeh, cooked	
	$\frac{1}{4}$ cup roasted soybeans, 1 falafel patty (4 oz.–$2\frac{1}{4}$-inch diameter)	1 soy or bean burger patty = 2 ounce-equivalents
	2 Tbsp. Hummus	

MIXED DISHES IN MYPYRAMID

Many popular dishes don't fit neatly into one MyPyramid food group. For example, a cheese pizza counts in several groups: the crust in the grains group, the tomato sauce in the vegetable group, and the cheese in the milk group. Some other common mixed dishes and the way they count in each food group are listed in the chart. Some mixed foods also contain a lot of fat, oil, or sugar, which adds kcalories. The estimated total kcalories in each dish is also shown. The values listed are estimates based on how these foods are often prepared. The amounts in an item you eat may be more or less than these examples.

Food and Sample Portion	Grains Group (Ounce-Equivalents)	Vegetable Group (Cups)	Fruit Group (Cups)	Milk Group (Cups)	Meat and Beans Group (Ounce-Equivalents)	Estimated Total Kcalories
Cheese pizza, thin crust (1 slice from medium pizza)	1	1/8	0	1/2	0	215
Lasagna (1 piece, 3 1/2 by 4 inches)	2	1/2	0	1	1	445
Macaroni and cheese (1 cup, made from packaged mix)	2	0	0	1/2	0	260
Tuna noodle casserole (1 cup)	1 1/2	0	0	1/2	2	260
Chicken pot pie (8-oz. pie)	2 1/2	1/4	0	0	1 1/2	500
Beef taco (2 tacos)	2 1/2	1/4	0	1/4	2	370
Bean and cheese burrito (1)	2 1/2	1/8	0	1	2	445
Egg roll (1)	1/2	1/8	0	0	1/2	150
Chicken fried rice (1 cup)	1 1/2	1/4	0	0	1	270
Stuffed peppers with rice and meat (1/2 pepper)	1/2	1/2	0	0	1	190
Beef stir-fry (1 cup)	0	3/4	0	0	1 1/2	185
Clam chowder, New England (1 cup)	1/2	1/8	0	1/2	2	165
Clam chowder, Manhattan (chunky, 1 cup)	0	3/8	0	0	2	135
Cream of tomato soup (1 cup)	1/2	1/2	0	1/2	0	160
Large cheeseburger	2	0	0	1/3	3	500
Turkey sub (6-inch sub)	2	1/2	0	1/4	2	320
Peanut butter and jelly sandwich	2	0	0	0	2	375
Tuna salad sandwich	2	1/4	0	0	2	290
Chef salad (3 cups, no dressing)	0	1 1/2	0	0	3	230
Pasta salad with vegetables (1 cup)	1 1/2	1/2	0	0	0	140
Apple pie (1 slice)	2	0	1/4	0	0	280
Pumpkin pie (1 slice)	1 1/2	1/8	0	1/4	1/4	240

APPENDIX D
Growth Charts

Birth to 36 months: Boys
Length-for-age and Weight-for-age percentiles

NAME _____

RECORD # _____

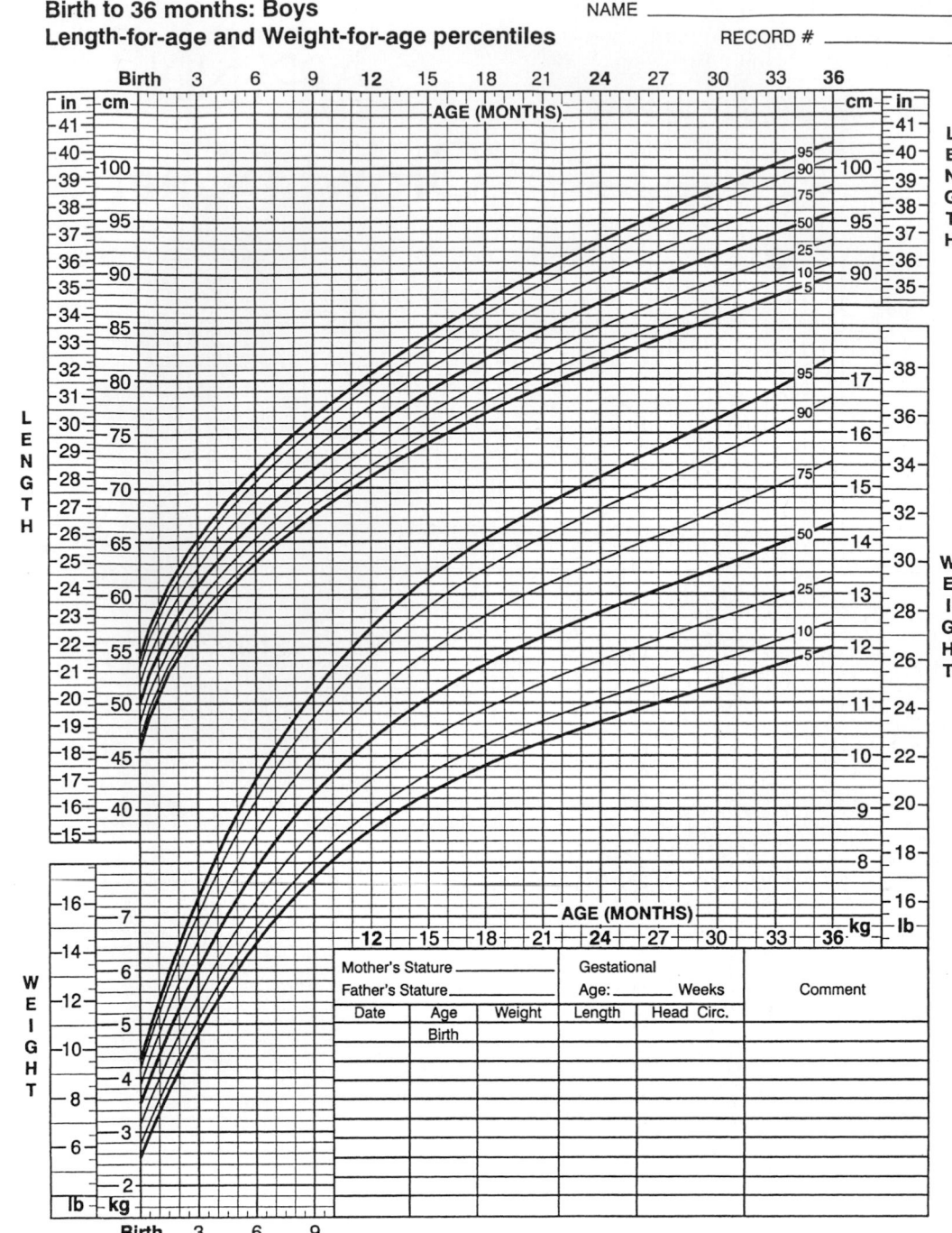

Revised November 21, 2000.

Birth to 36 months: Girls
Length-for-age and Weight-for-age percentiles

NAME _____

RECORD # _____

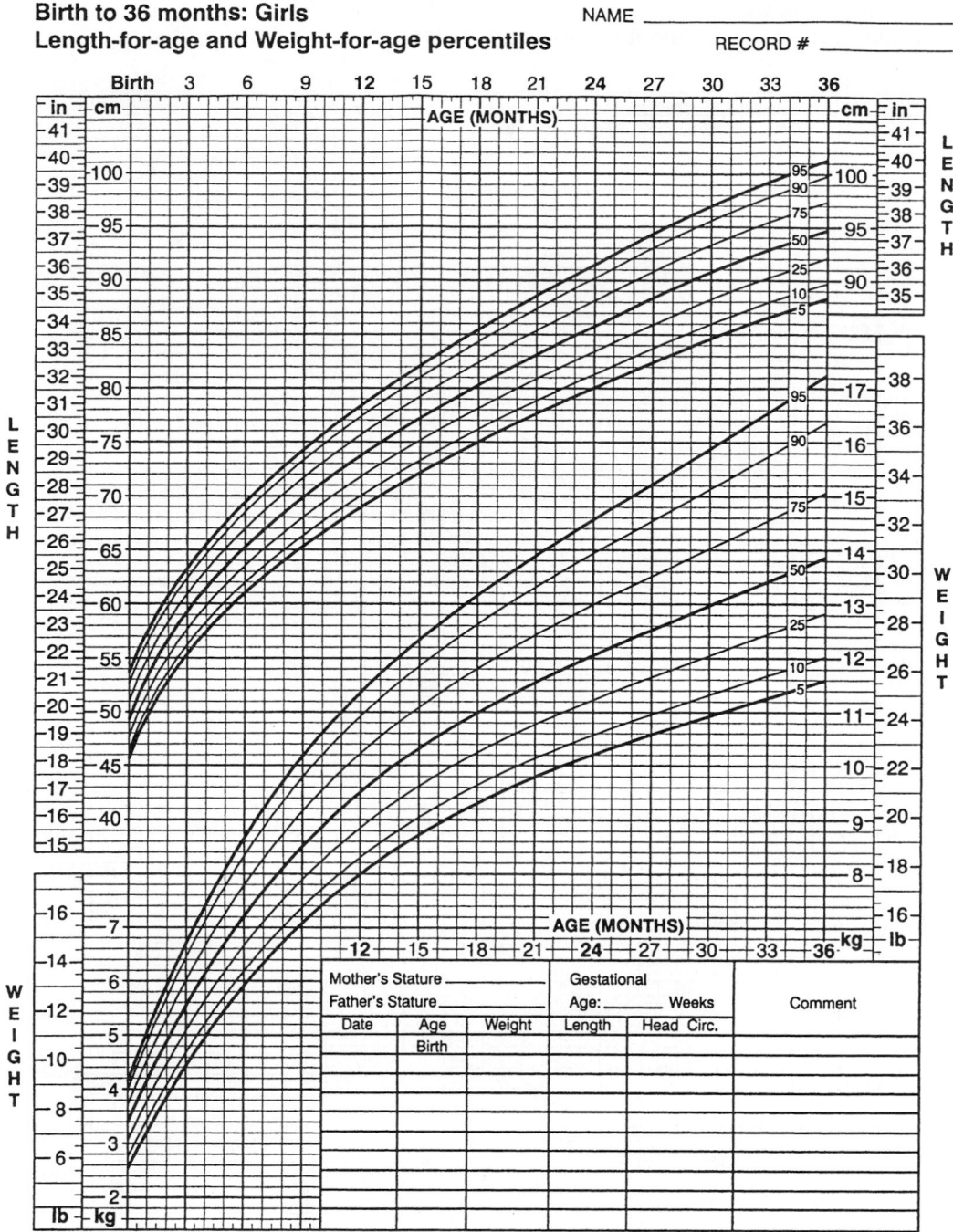

Revised November 21, 2000.

2 to 20 years: Boys
Stature-for-age and Weight-for-age percentiles

NAME _____

RECORD # _____

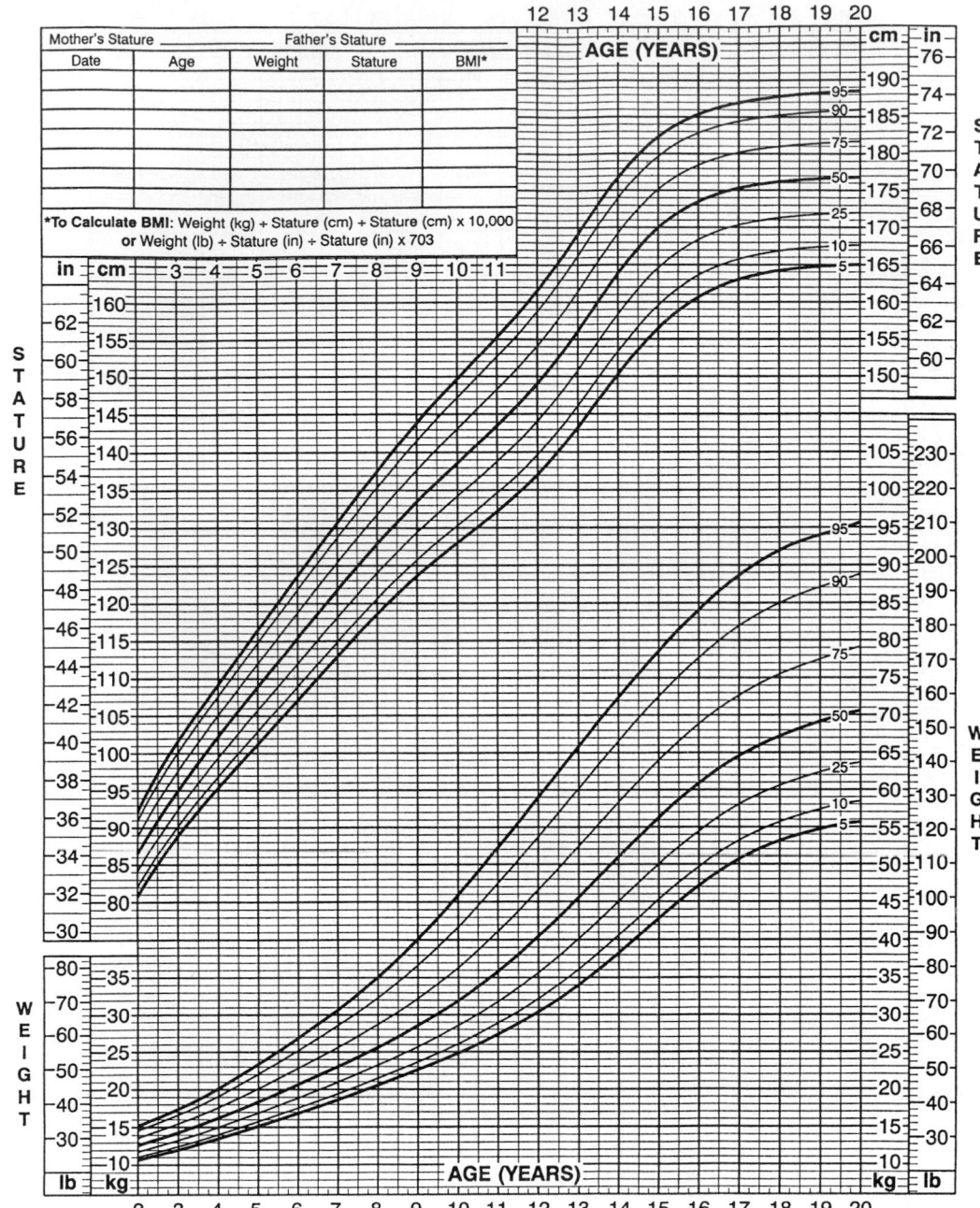

Revised and corrected November 21, 2000.

2 to 20 years: Girls
Stature-for-age and Weight-for-age percentiles

NAME _____

RECORD # _____

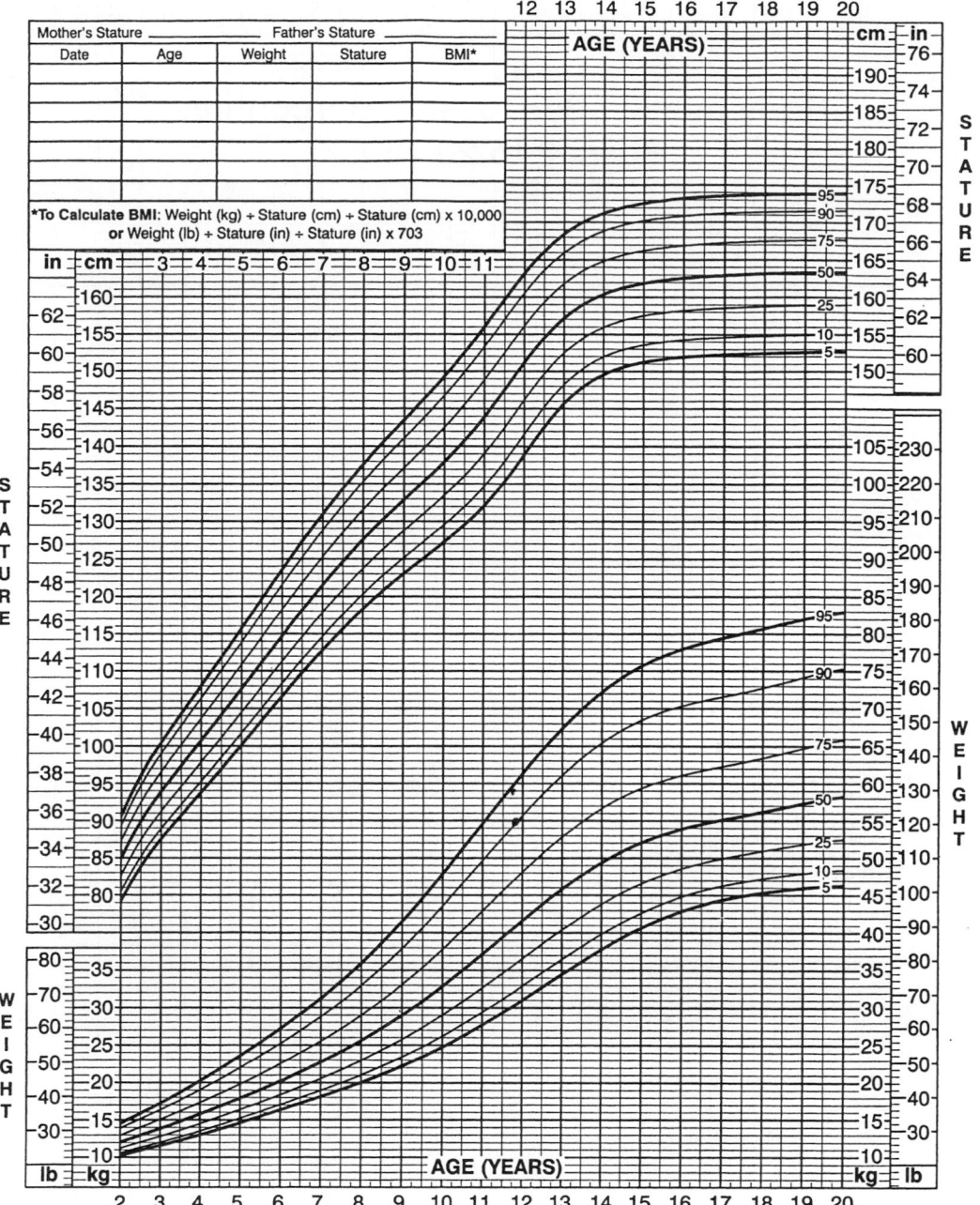

Revised and corrected November 21, 2000.

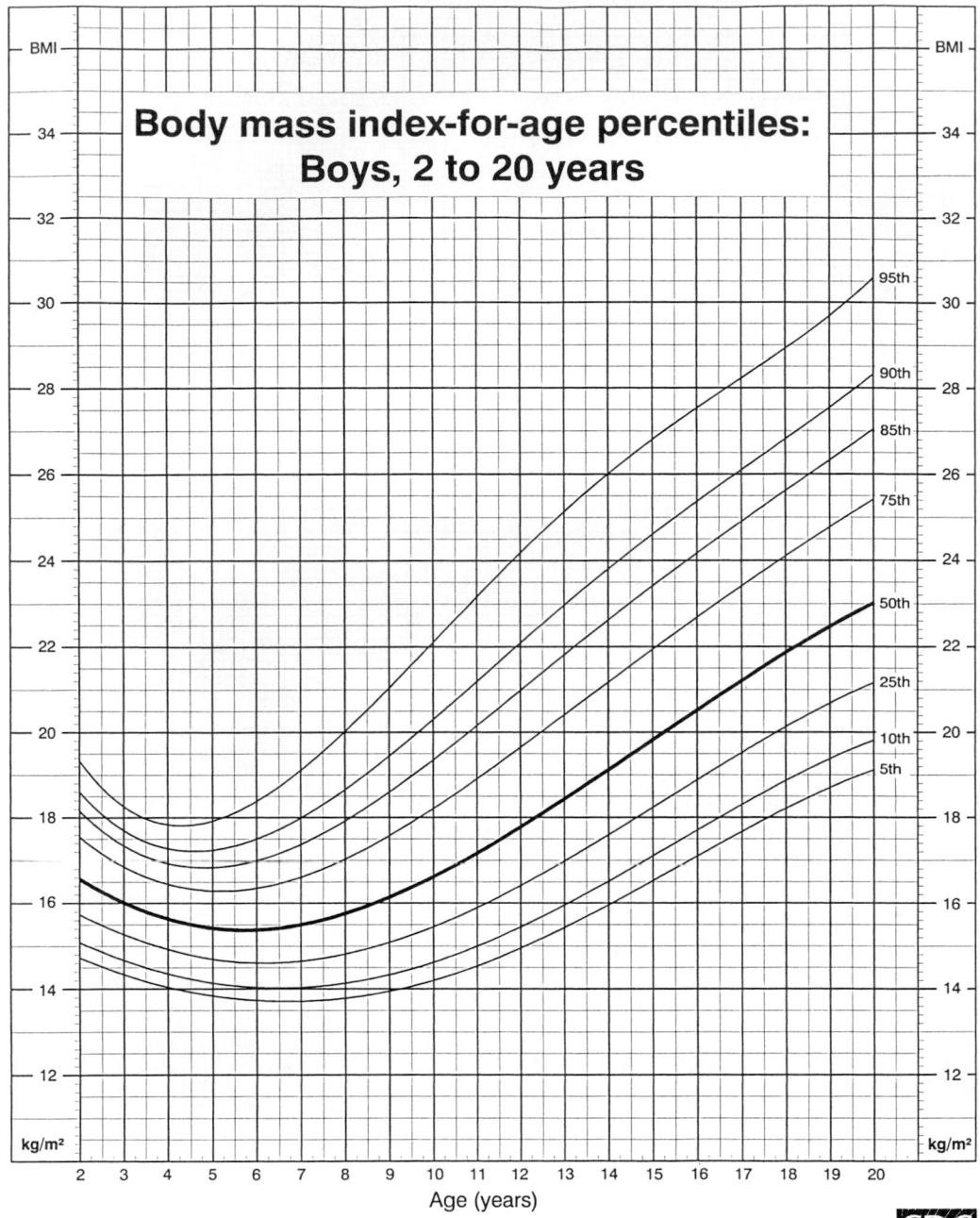

Body mass index-for-age percentiles: Boys, 2 to 20 years

Published May 30, 2000.
SOURCE: Developed by the National Center for Health Statistics in collaboration with
the National Center for Chronic Disease Prevention and Health Promotion (2000).

CDC
SAFER · HEALTHIER · PEOPLE™

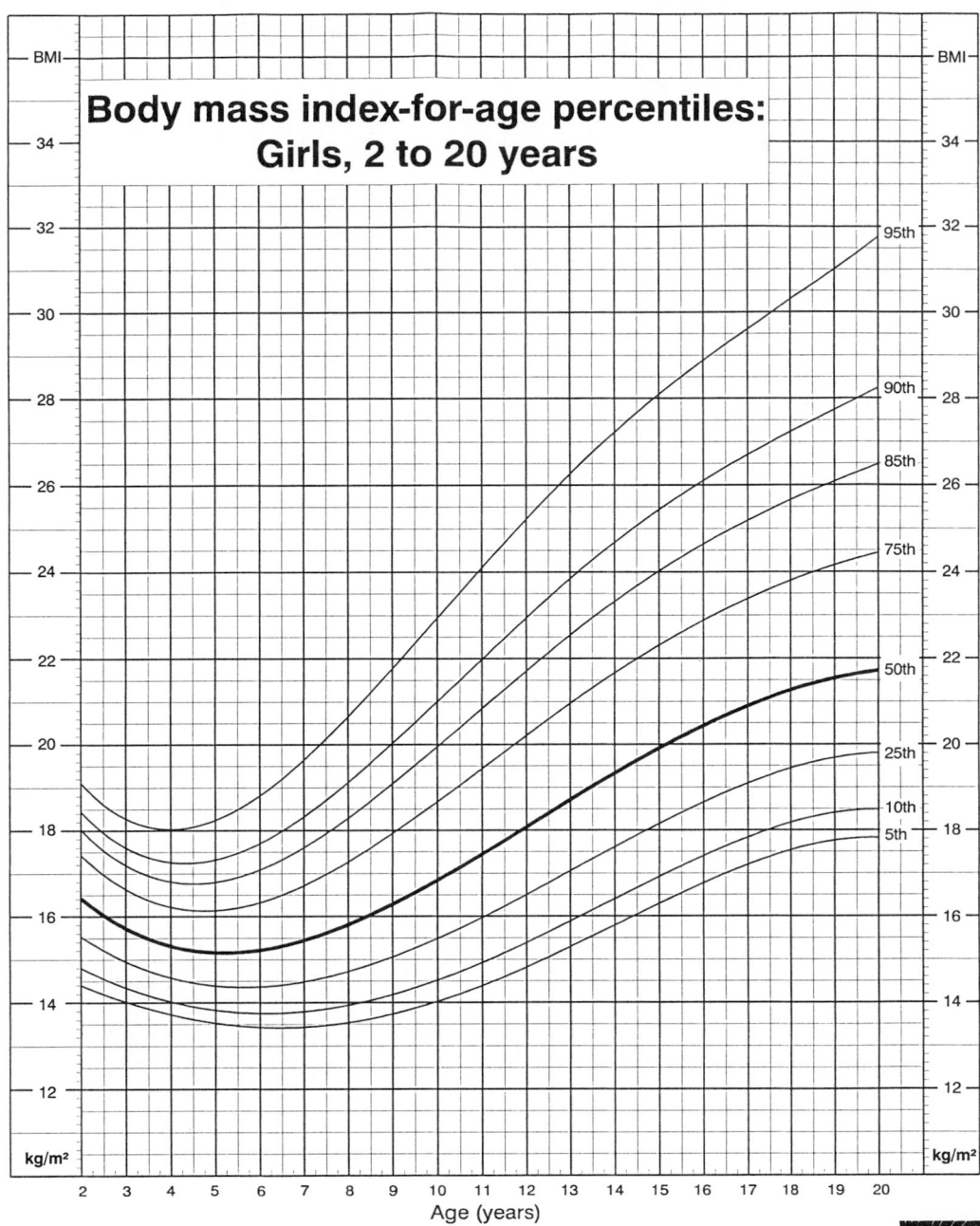

Body mass index-for-age percentiles:
Girls, 2 to 20 years

Age (years)

Published May 30, 2000.
SOURCE: Developed by the National Center for Health Statistics in collaboration with
    the National Center for Chronic Disease Prevention and Health Promotion (2000).

SAFER·HEALTHIER·PEOPLE™

APPENDIX E

Answers to
Check-Out Quizzes

CHAPTER 1

1. Carbohydrate: Provides energy
   Lipid: Provides energy, promotes growth and maintenance; regulates body processes
   Protein: Provides energy, promotes growth and maintenance; regulates body processes
   Vitamins and Minerals: Promote growth and maintenance; regulate body processes
   Water: Supplies the medium in which chemical changes of the body occur; promotes growth and maintenance; regulates body processes
2. RDA: Value that meets requirements of 97 to 98 percent of individuals
   AI: Value used when there is not enough scientific data to support an RDA
   UL: Maximum safe intake level
   EAR: Value that meets requirements of 50 percent of individuals in a group
   EER: Value for kilocalories
3. Absorption: Process of nutrients entering the tissues from the gastrointestinal tract
   Enzyme: Substance that speeds up chemical reactions
   Anabolism: Process of building substances
   Peristalsis: Involuntary muscular contraction
   Catabolism: Process of breaking down substances
4. c
5. b
6. a
7. b
8. a
9. a
10. b

CHAPTER 2

1. Energy: Don't exceed kcaloric needs
   Physical activity: At least 30 minutes most days of moderate-intensity activity
   Whole grains: Eat 3 ounce-equivalents per day
   Total fat: 20 to 35 percent of kcalories
   Saturated fat: Less than 10 percent of kcalories
   Trans fat: Keep consumption as low as possible
   Sodium: 2300 mg or less
   Cholesterol: 300 mg or less
2.

1 ounce-equivalent of grains	1 slice of bread
	1 cup ready-to-eat cereal
	½ cup cooked rice, pasta, or cooked cereal
1 cup of vegetables	2 cups leafy salad greens
	1 cup vegetable juice
	1 cup(s) cooked vegetables
1 cup milk	1 cup yogurt
1 ounce-equivalent meat	1 egg
	¼ cup cooked beans
	1 Tbsp. peanut butter

3. d
4. b
5. d
6. c
7. b
8. b
9. a
10. a

CHAPTER 3

1. White bread: starch
   Whole-wheat bread: starch, fiber
   Apple juice: natural sugars
   Baked beans: fiber, starch
   Milk: natural sugars
   Bran flakes: fiber
   Sugar-frosted whole oats: added sugars, fiber
   Cola drink: added sugars
   Broccoli: fiber
2. b
3. a
4. b
5. a
6. b
7. a
8. b
9. a
10. b

CHAPTER 4

1.
Food	Fat	Cholesterol
1. Butter	X	X
2. Margarine	X	
3. Split peas		
4. Peanut butter	X	
5. Porterhouse steak	X	X
6. Flounder		
7. Skim milk		
8. Cheddar cheese	X	X
9. Chocolate-chip cookie made with vegetable shortening	X	
10. Green beans		

2. 1. b
   2. a
   3. b
   4. a, b
   5. b
3. 1. f
   2. g
   3. d
   4. a
   5. c
   6. e
   7. b

CHAPTER 5

1. a
2. a
3. a
4. b
5. a
6. b
7. b
8. a
9. a
10. a

CHAPTER 6

1. b
2. b
3. a
4. a
5. b
6. b
7. a
8. b
9. a
10. a
11. a. Vitamin B$_{12}$
    b. Thiamin
    c. Vitamin C, folate
    d. Vitamin B$_{12}$
    e. Vitamin D
    f. Niacin
    g. Vitamin K
    h. Vitamin B$_6$
    i. Vitamin K
    j. Choline
    k. Vitamin C
    l. Vitamin A
    m. Vitamin D
    n. Vitamins A, C, E
    o. Vitamins A, C, D
    p. Vitamin A
    q. Vitamin D

CHAPTER 7

1. a
2. b
3. a

4. b
5. a
6. b
7. a
8. a
9. b
10. a. calcium, phosphorus, magnesium, zinc, fluoride, manganese
   b. calcium
   c. sodium, potassium, chloride
   d. potassium
   e. chloride
   f. potassium
   g. fluoride
   h. chloride
   i. iron
   j. iron
   k. selenium, iodine

CHAPTER 8

1. a
2. b
3. b
4. b
5. a
6. a
7. b
8. b
9. b
10. a

CHAPTER 9

1. You can define a healthy meal as one that includes whole grains, fruits, vegetables, lean protein, and small amounts of healthy oils. Another way to look at a balanced meal is to look at the nutrients it contains, such as the following:
   - 800 kcalories or less
   - 35 percent or fewer kcalories from fat, emphasizing oils high in monounsaturated and polyunsaturated fats
     - 10 percent or less of total kcalories from saturated and trans fats
     - 100 milligrams or less of cholesterol
     - no trans fat
   - 45 to 65 percent kcalories from carbohydrates
     - 10 grams or more of fiber
     - 10 percent or fewer kcalories from added sugars
   - 15 to 25 percent kcalories from protein
   - 800 milligrams or less of sodium
2. Use existing items. Make an existing item more nutritious. Create new menu items.
3. 1. Change/add healthy preparation techniques.
   2. Change/add healthy cooking techniques.
   3. Change an ingredient by reducing it, eliminating it, or replacing it.
   4. Add a new ingredient(s), particularly to build flavor.
4. Any of these answers.
   - Is the menu item tasty? Taste is the key to customer acceptance and the successful marketing of these items. If the food does not taste delicious and have a creative presentation, no matter how nutritious it is, it is not going to sell.
   - Does the menu item blend with and complement the rest of the menu?
   - Does the menu item meet the food habits and preferences of the guests?
   - Is the food cost appropriate for the price that can be charged?
   - Does each menu item require a reasonable amount of preparation time?
   - Is there a balance of color in the foods themselves and in the garnishes?
   - Is there a balance of textures, such as coarse, smooth, solid, and soft?
   - Is there a balance of shape, with different-sized pieces and shapes of food?
   - Are flavors varied?
   - Are the food combinations acceptable?
   - Are cooking methods varied?
   - Can each menu item be prepared properly by the cooking staff?

CHAPTER 10

1. a
2. c
3. b
4. b
5. d

CHAPTER 11

1. a
2. a
3. b
4. b
5. a
6. a
7. b
8. a
9. a
10. a

CHAPTER 12

1. b
2. a
3. b
4. b
5. b
6. b
7. a
8. a
9. a
10. a

CHAPTER 13

1. a
2. b
3. a
4. b
5. b
6. a
7. a
8. b
9. a
10. a
11. b
12. a
13. b
14. a
15. b
16. a
17. a
18. a
19. a
20. b

Glossary

Absorption The passage of digested nutrients through the walls of the intestines or stomach into the body's cells. Nutrients are then transported through the body via the blood or lymph system.

Acceptable Macronutrient Distribution Range (AMDR) The percent of total kilocalories coming from carbohydrate, fat, or protein that is associated with a reduced risk of chronic disease while providing adequate intake.

Acid-base balance The process by which the body buffers the acids and bases normally produced in the body so that the blood is neither too acidic nor too basic.

Acidosis A dangerous condition in which the blood is too acidic.

Added sugars Sugars added to a food for sweetening or other purposes; they do not include the naturally occurring sugars in foods such as fruit and milk.

Adequate diet A diet that provides enough kcalories, essential nutrients, and fiber to keep a person healthy.

Adequate Intake (AI) The dietary intake that is used when there is not enough scientific research to support an RDA.

Adipose cell A cell in the body that readily takes up and stores triglycerides; also called a fat cell.

Advertising Any paid form (such as radio) of calling public attention to the goods, services, or ideas of a company or sponsor.

Alkalosis A dangerous condition in which the blood is too basic.

Alpha-linolenic acid An omega-3 fatty acid found in several oils, notably canola, flaxseed, soybean, walnut, and wheat germ oils (or margarines made with canola or soybean oil); this essential fatty acid is vital to growth and development, maintenance of cell membranes, and the immune system and is inadequate in many Americans' diets.

Alpha-tocopherol The most active form of vitamin E in humans; also a powerful antioxidant.

Alternative sweeteners Sweeteners that contain either no or very few calories.

Amino acid pool The overall amount of amino acids distributed in the blood, organs, and body cells.

Amino acids The building blocks of protein.

Amniotic sac The protective bag, or sac, that cushions and protects the fetus during pregnancy.

Anabolism The metabolic process by which body tissues and substances are built.

Anaphylaxis A rare allergic reaction that is very serious and can result in death if not treated immediately.

Angina Symptoms of pressing or intense pain in the heart area, often due to stress or exertion when the heart muscle gets insufficient blood.

Anorexia Lack of appetite.

Anorexia nervosa An eating disorder most prevalent in adolescent females, who starve themselves.

Antibodies Proteins in the blood that bind with foreign bodies or invaders.

Antigens Foreign invaders in the body.

Antioxidant A compound that combines with oxygen to prevent oxygen from oxidizing or destroying important substances; antioxidants prevent the oxidation of unsaturated fatty acids in the cell membrane, DNA, and other cell parts that substances called free radicals try to destroy.

Anus The opening of the digestive tract through which feces travels out of the body.

Arterial blood pressure The pressure of blood within arteries as it is pumped through the body by the heart.

Artificial sweeteners Substitutes for sugar (such as aspartame and saccharin) that provide no, or almost no, kcalories.

Atherosclerosis The most common form of artery disease, characterized by plaque buildup along artery walls.

Attention deficit hyperactivity disorder A developmental disorder of children characterized by impulsiveness, distractibility, and hyperactivity.

Baby bottle tooth decay Serious tooth decay in babies caused by letting a baby go to bed with a bottle of juice, formula, cow's milk, or breast milk.

Balanced diet A diet in which foods are chosen to provide kcalories, essential nutrients, and fiber in the right proportions.

Basal metabolism The minimum energy needed by the body for vital functions when at rest and awake.

Beta-carotene A precursor of vitamin A that functions as an antioxidant in the body; the most abundant carotenoid.

Bile A substance made in the liver that is stored in the gallbladder and released when fat enters the small intestine because it helps digest fat.

Bile acids A component of bile that aids in the digestion of fats in the duodenum of the small intestine.

Binge eating disorder An eating disorder characterized by episodes of uncontrolled eating or bingeing.

Bioavailability The degree to which a nutrient is absorbed and available to be used in the body.

Biotechnology A collection of scientific techniques, including genetic engineering, that are used to create, improve, or modify plants, animals, and microorganisms.

Blood glucose level (blood sugar level) The amount of glucose found in the blood; glucose is vital to the proper functioning of the body.

Body mass index A method of measuring degree of obesity that is a more sensitive indicator than height-weight tables.

Bolus A ball of chewed food that travels from the mouth through the esophagus to the stomach.

Bran In cereal grains, the part that covers the grain and contains much fiber and other nutrients.

Bulimia nervosa An eating disorder characterized by a destructive pattern of excessive overeating followed by vomiting or other "purging" behaviors to control weight.

Cancer A group of diseases characterized by unrestrained cell division and growth that can disrupt the normal functioning of an organ and also spread beyond the tissue in which it started.

Carbohydrate or glycogen loading A regimen involving both decreased exercise and increased consumption of carbohydrates before an athletic event to increase the amount of glycogen stores.

Carbohydrates A large class of nutrients, including sugars, starch, and fibers, that function as the body's primary source of energy.

Carcinogen Cancer-causing substance.

Cardiovascular disease (CVD) A disease of the heart and blood vessels such as coronary artery disease, stroke, and high blood pressure.

Carotenoids A class of pigments that contribute a red, orange, or yellow color to fruits and vegetables; can be converted to retinol or retinal in the body.

Catabolism The metabolic processes by which large, complex molecules are converted to simpler ones.

Cholesterol The most abundant sterol (a category of lipids); a soft, waxy substance present only in foods of animal origin; it is present in every cell in your body.

Chutney A sauce from India that is made with fruits, vegetables, and herbs.

Chylomicron The lipoprotein responsible for carrying mostly triglycerides, and some cholesterol, from the intestines through the lymph system to the bloodstream.

Chyme A semiliquid mixture in the stomach that contains partially digested food and stomach secretions.

Clinical trials Research studies that assign similar participants randomly to two groups; one group receives the experimental treatment while the other does not.

Coenzyme A molecule that combines with an enzyme and makes the enzyme functional.

Collagen The most abundant protein in the body; a fibrous protein that is a component of skin, bone, teeth, ligaments, tendons, and other connective structures.

Colostrum A yellowish fluid that is the first secretion to come from the mother's breast a day or so after delivery of a baby; it is rich in proteins, antibodies, and other factors that protect against infectious disease.

Complementary proteins The ability of two protein foods to make up for the lack of certain amino acids in each other when eaten over the course of a day.

Complete protein Food proteins that provide all the essential amino acids in the proportions needed by the body.

Complex carbohydrates (polysaccharides) Long chains of many sugars that include starches and fibers.

Compote A dish of fruit, fresh or dried, cooked in syrup flavored with spices or liqueur; it is often served as an accompaniment or dessert.

Conditionally essential amino acids Nonessential amino acids that may, under certain circumstances, become essential.

Constipation Infrequent passage of feces.

Coronary heart disease Damage to the heart caused by narrowing or lockage of the coronary arteries.

Coulis A sauce made of a puree of vegetables or fruits.

Couscous A granular form of semolina, like a tiny pasta.

Cretinism (congenital hypothyroidism) Lack of thyroid secretion; causes mental and physical retardation during fetal and later development.

Cruciferous vegetables Members of the cabbage family; they contain phytochemicals that may help prevent cancer.

Culture The behaviors and beliefs of a certain social, ethnic, or age group.

Daily Value A set of nutrient-intake values developed by the Food and Drug Administration that are used as a reference for expressing nutrient content on nutrition labels.

Deglazing Adding liquid to the hot pan used in making sauces and meat dishes; any browned bits of food sticking to the pan are scraped up and added to the liquid.

Denaturation A process in which a protein uncoils and loses its shape, causing it to lose its ability to function; it can be caused by high temperatures, whipping, and other circumstances.

Dental caries Tooth decay.

Diabetes mellitus A disorder of carbohydrate metabolism characterized by high blood sugar levels and inadequate or ineffective insulin.

Diastolic pressure The pressure in the arteries when the heart is resting between beats—the bottom number in blood pressure.

Diet The food and beverages you normally eat and drink.

Dietary fiber Polysaccharides and lignin (a nonpolysaccharide) that are not digested and absorbed.

Dietary folate equivalents (DFEs) The unit for measuring folate; takes into account the amount of folate that is absorbed from natural and synthetic sources.

Dietary Guidelines for Americans A set of dietary recommendations for Americans that is periodically revised.

Dietary recommendations Guidelines that discuss specific foods and food groups to eat for optimal health.

Dietary Reference Intake (DRI) Nutrient standards that include four lists of values for dietary nutrient intakes of healthy Americans and Canadians.

Digestion The process by which food is broken down into its components in the mouth, stomach, and small intestine with the help of digestive enzymes.

Dipeptides A peptide with two amino acids.

Disaccharides Double sugars such as sucrose.

Discretionary kcalories The balance of kcalories you have after meeting the recommended nutrient intakes by eating foods in low-fat or no added sugar forms. Your discretionary kcalorie allowance may be used to select forms of foods that are not the most nutrient-dense (such as whole milk rather than fat-free milk) or may be additions to foods, such as sugar and butter.

Diverticulosis A disease of the large intestine in which the intestinal walls become weakened, bulge out into pockets, and at times become inflamed.

Duodenum The first segment of the small intestine, about 1 foot long.

Edema Swelling due to an abnormal accumulation of fluid in the intercellular spaces.

Electrolytes Chemical elements or compounds that ionize in solution and can carry an electric current; they include sodium, potassium, and chloride.

Embryo The name of the fertilized egg from conception to the eighth week.

Empty-kcalorie foods Foods that provide few nutrients for the number of kcalories they contain.

Endosperm In cereal grains, a large center area high in starch.

Energy-yielding nutrients Nutrients that can be burned as fuel to provide energy for the body, including carbohydrates, fats, and proteins.

Enriched food A food to which nutrients are added to replace the same nutrients that were lost in processing.

Enzymes Compounds that speed up the breaking down of food so that nutrients can be absorbed. Also perform other functions in the body.

Epidemiological research Research that looks at how disease rates vary among different populations and also factors associated with disease.

Epiglottis The flap that covers the air tubes to the lungs so that food does not enter the lungs during swallowing.

Esophagus The muscular tube that connects the pharynx to the stomach.

Essential amino acids Amino acids that either cannot be made in the body or cannot be made in the quantities needed by the body; must be obtained in foods.

Essential fatty acids Fatty acids that the body cannot produce, making them necessary in the diet: linoleic acid and linolenic acid.

Essential nutrients Nutrients that either cannot be made in the body or cannot be made in the quantities needed by the body; therefore, we must obtain them from food.

Estimated Average Requirement (EAR) The dietary intake value that is estimated to meet the requirement of half the healthy individuals in a group.

Estimated Energy Requirement (EER) The dietary energy intake measured in kcalories that is needed to maintain energy balance in a healthy adult.

Exchange system A tool to plan diets that groups foods by their nutrient and caloric content. Foods within each group have about the same amount of calories, carbohydrate, protein, and fat so that any food can be substituted for any other food in the same group.

Fasting hypoglycemia Low blood sugar that occurs after not eating for eight or more hours.

Fat A lipid that is solid at room temperature.

Fat-soluble vitamins A group of vitamins that generally occur in foods containing fats; these include vitamins A, D, E, and K.

Fat substitutes Ingredients that mimic the functions of fat in foods, and either contain fewer calories than fat or no calories.

Fatty acids Major component of most lipids. Three fatty acids are present in each triglyceride.

Female athlete triad An eating disorder found among female college athletes in which they have disordered eating, osteoporosis, and no menstruation.

Fetal alcohol spectrum disorders A variety of physical changes and/or brain damage that is associated with fetal exposure to alcohol during pregnancy.

Fetal alcohol syndrome (FAS) A set of symptoms occurring in newborn babies that are due to alcohol use of the mother during pregnancy; symptoms may include mental retardation and brain damage.

Fetus The infant in the mother's uterus from 8 weeks after conception until birth.

Fibrin Protein fibers involved in forming clots so that a cut or wound will stop bleeding.

First trimester The period during the first 13 weeks of pregnancy.

Flavor An attribute of a food that includes its appearance, smell, taste, feel in the mouth, texture, temperature, and even the sounds made when it is chewed.

Flavorings Substances used in cooking to add a new flavor or modify the original flavor.

Fluoride The form of fluorine that appears in drinking water and in the body.

Fluorosis A condition in which the teeth become mottled and discolored due to high fluoride ingestion.

Food allergens Those parts of food causing allergic reactions.

Food allergy An abnormal response of the immune system to an otherwise harmless food.

Food guides Guidelines that tell us the kinds and amounts of foods that constitute a nutritionally adequate diet; they are based on current dietary recommendations, the nutrient content of foods, and the eating habits of the targeted population.

Food intolerance Symptoms of gas, bloating, constipation, dizziness, or difficulty sleeping after eating certain foods.

Food jags A habit of young children in which they have favorite foods that they want to eat frequently.

Fortified foods A food to which nutrients are added that were not present originally, or to which nutrients are added that increase the amount already present.

Free radical An unstable compound that reacts quickly with other molecules in the body.

Fresh foods Raw foods that have not been processed (such as canned or frozen) or heated.

Fructose A monosaccharide found in fruits and honey.

Gag reflex The ability to cough or vomit up food (or anything) that can't be swallowed properly.

Galactose A monosaccharide found linked to glucose to form lactose, or milk sugar.

Gastric lipase An enzyme in the stomach that breaks down mostly short-chain fatty acids.

Gastrointestinal tract A hollow tube running down the middle of the body in which digestion of food and absorption of nutrients take place.

Gelatinization A process in which starches, when heated in liquid, absorb water and swell in size.

Germ In cereal grains, the area of the kernel rich in vitamins and minerals that sprouts when allowed to germinate.

Glucose The most significant monosaccharide; the body's primary source of energy.

Glycemic index A classification that quantifies the blood glucose response after eating carbohydrate-containing foods.

Glycemic response How quickly and how high blood glucose rises after eating.

Glycerol A derivative of carbohydrate that is part of triglycerides.

Glycogen The storage form of glucose in the body; stored in the liver and muscles.

Growth spurts Periods of rapid growth.

Health claims Claims on food labels that state that certain foods or food substances—as part of an overall healthy diet—may reduce the risk of certain diseases.

Heartburn A painful burning sensation in the esophagus caused by acidic stomach contents flowing back into the lower esophagus.

Heme iron The predominant form of iron in animal foods; it is absorbed and used more readily than iron in plant foods.

Hemoglobin A protein in red blood cells that carries oxygen to the body's cells.

Hemorrhagic stroke A stroke due to a ruptured brain artery.

Hemorrhoids Enlarged veins in the lower rectum.

Herbs The leafy parts of certain plants that grow in temperate climates; they are used to season and flavor foods.

High-density lipoproteins (HDLs) Lipoproteins that contain much protein and carry cholesterol away from body cells and tissues to the liver for excretion from the body.

High-fructose corn syrup Corn syrup that has been treated with an enzyme that converts part of the glucose it contains to fructose; found in most regular sodas as well as other sweetened foods.

Homeostasis A constant internal environment in the body.

Homogenized Milk that has had its fat particles broken up so finely that they remain uniformly dispersed throughout the milk.

Hormones Chemical messengers in the body.

Hydrochloric acid A strong acid made by the stomach that aids in protein digestion, destroys harmful bacteria, and increases the ability of calcium and iron to be absorbed.

Hydrogenation A process in which liquid vegetable oils are converted into solid fats (such as margarine) by the use of heat, hydrogen, and certain metal catalysts.

Hydroxyapatite The main structural component of bone, composed mostly of calcium phosphate crystals.

Hyperglycemia High levels of blood sugar.

Hypertension High blood pressure.

Hypervitaminosis A A disease caused by prolonged use of high doses of preformed vitamin A that can cause hair loss, bone pain and damage, soreness, and other problems.

Hypoglycemia A symptom in which blood sugar levels are low.

Hypothyroidism A condition in which there is less production of thyroid hormones; this leads to symptoms such as low metabolic rate, fatigue, and weight gain.

Ileum The final segment of the small intestine.

Immune response The body's response to a foreign substance, such as a virus, in the body.

Incomplete proteins Food proteins that contain at least one limiting amino acid.

Inorganic In chemistry, any compound that does not contain carbon.

Insoluble fiber A classification of fiber that includes cellulose, lignin, and the remaining hemicelluloses; they generally form the structural parts of plants.

Insulin A hormone that increases the movement of glucose from the bloodstream into the body's cells.

Intrinsic factor A proteinlike substance secreted by stomach cells that is necessary for the absorption of vitamin B_{12}.

Ion An atom or group of atoms carrying a positive or negative electric charge.

Iron deficiency A condition in which iron stores are used up.

Iron-deficiency anemia A condition in which the size and number of red blood cells are reduced; may result from inadequate iron intake or from blood loss; symptoms include fatigue, pallor, and irritability.

Iron overload (hemochromatosis) A common genetic disease in which individuals absorb about twice as much iron from their food and supplements as other people do.

Irradiation A process of using a measured dose of radiation on foods to reduce the number of harmful microorganisms.

Ischemic stroke The most common type of stroke, in which a blood clot blocks an artery or vessel in the brain.

Jejunum The second portion of the small intestine, between the duodenum and the ileum.

Ketone bodies A group of organic compounds that cause the blood to become too acidic as a result of fat being burned for energy without any carbohydrates present.

Ketosis Excessive level of ketone bodies in the blood and urine.

Kilocalorie A measure of the energy in food, specifically the energy-yielding nutrients.

Kwashiorkor A type of PEM associated with children who are getting inadequate amounts of protein and only marginal amounts of kcalories.

Lactase An enzyme needed to split lactose into its components in the intestines.

Lacto-ovo vegetarians Vegetarians who do not eat meat, poultry, or fish but do consume animal products in the form of eggs, milk, and milk products.

Lactose A disaccharide found in milk and milk products that is made of glucose and galactose.

Lactose intolerance A condition caused by a deficiency of the enzyme lactase, resulting in symptoms such as flatulence and diarrhea after drinking milk or eating most dairy products.

Lacto vegetarians Vegetarians who do not eat meat, poultry, or fish but do consume animal products in the form of milk and milk products.

Large intestine (colon) The part of the gastrointestinal tract between the small intestine and the rectum.

Lecithin A phospholipid and a vital component of cell membranes that acts as an emulsifier (a substance that keeps fats in solution).

Limiting amino acid An essential amino acid in lowest concentration in a protein that limits the protein's usefulness unless another food in the diet contains it.

Lingual lipase An enzyme made in the salivary glands in the mouth that plays a minor role in fat digestion in adults and an important role in fat digestion in infants.

Linoleic acid Omega-6 fatty acid found in vegetable oils such as corn, safflower, soybean, cottonseed, and sunflower oils; this essential fatty acid is vital to growth and development, maintenance of cell membranes, and the immune system.

Lipids A group of fatty substances, including triglycerides and cholesterol, that are soluble in fat, not water, and that provide a rich source of energy and structure to cells.

Lipoprotein lipase An enzyme that breaks down triglycerides from the chylomicron into fatty acids and glycerol so that they can be absorbed in the body's cells.

Lipoproteins Protein-coated packages that carry fat and cholesterol through the bloodstream; the body makes four types, classified according to their density.

Low-birth-weight baby A newborn who weighs less than $5\frac{1}{2}$ pounds; these infants are at higher risk for disease.

Low-density lipoproteins (LDLs) Lipoproteins that contain most of the cholesterol in the blood; they carry cholesterol to body tissues.

Lower esophageal (cardiac) sphincter A muscle that relaxes and contracts to move food from the esophagus into the stomach.

Lanugo Downy hair on the skin.

Macronutrients Nutrients needed by the body in large amounts, including carbohydrates, lipids, and proteins.

Major minerals Minerals needed in relatively large amounts in the diet—over 100 milligrams daily.

Maltose A disaccharide made of two glucose units bonded together.

Marasmus A type of PEM characterized by severe insufficiency of kcalories and protein that accounts for the child's gross underweight and wasting away of muscles.

Marinade A seasoned liquid used before cooking to flavor and moisten foods; usually based on an acidic ingredient.

Marketing The process of finding out what your customers need and want and then developing, promoting, and selling the products and services they desire.

Megadose A supplement intake of 10 times the RDA of a vitamin or mineral.

Megaloblastic (macrocytic) anemia A form of anemia caused by a deficiency of vitamin B_{12} or folate and characterized by large, immature red blood cells.

Metabolic syndrome A combination of risk factors (excessive abdominal fat, blood-fat disorders, insulin resistance, and high blood pressure) that greatly increase a person's risk of developing coronary heart disease.

Metabolism All the chemical processes by which nutrients are used to support life.

Metastasis The condition when a cancer spreads beyond the tissue in which it started.

Micronutrients Nutrients needed by the body in small amounts, including vitamins and minerals.

Microvilli Hairlike projections on the villi that increase the surface area for absorbing nutrients.

Milk letdown The process by which milk comes out of the mother's breast to feed the baby; sucking causes the release of a hormone that allows milk letdown.

Minerals Noncaloric, inorganic chemical substances found in a wide variety of foods; needed to regulate body processes, maintain the body, and allow growth and reproduction.

Mineral water Water from an underground source that contains at least 250 parts per million total dissolved solids. Minerals and trace elements must come from the source of the underground water.

Moderate diet A diet that avoids excessive amounts of kcalories or any particular food or nutrient.

Mojo A spicy Caribbean sauce; it is a mixture of garlic, citrus juice, oil, and fresh herbs.

Monoglycerides Triglycerides with only one fatty acid.

Monosaccharides Simple sugars, including glucose, fructose, and galactose, that consist of a single ring of atoms and are the building blocks for other carbohydrates, such as dissaccharides and starch.

Monounsaturated fat A triglyceride made of mostly monounsaturated fatty acids.

Monounsaturated fatty acid A fatty acid that contains only one double bond in the chain.

Myocardial infarction (heart attack) Occurs when the blood supply to part of the heart muscle itself—the myocardium—is severely reduced or stopped.

Myocardial ischemia A temporary injury to heart cells caused by a lack of blood flow and oxygen.

Myoglobin A muscle protein that stores and carries oxygen that the muscles will use to contract.

Negative nitrogen balance A condition in which the body excretes more protein than is taken in; this can occur during starvation and certain illnesses.

Neural tube The embryonic tissue that develops into the brain and spinal cord.

Neural tube defects Diseases in which the brain and the spinal cord form improperly in early pregnancy.

Niacin equivalents (NEs) The unit for measuring niacin. One niacin equivalent is equal to one milligram of niacin or 60 milligrams of tryptophan.

Night blindness A condition caused by insufficient vitamin A in which it takes longer to adjust to dim lights after seeing a bright light at night; this is an early sign of vitamin A deficiency.

Nitrogen balance The difference between total nitrogen intake and total nitrogen loss; a healthy person has the same nitrogen intake as loss, resulting in a zero nitrogen balance.

Nonessential amino acids Amino acids that can be made in the body.

Nonheme iron A form of iron found in all plant sources of iron and also as part of the iron in animal food sources.

Noodles Pastas made from flour, water, and egg solids.

Nutrient content claims Claims on food labels about the nutrient composition of a food; regulated by the Food and Drug Administration.

Nutrient-dense foods Foods that contain many nutrients for the kcalories they provide.

Nutrient density A measure of the nutrients provided in a food per kcalorie of that food.

Nutrients The nourishing substances in food that provide energy and promote the growth and maintenance of your body.

Nutrition A science that studies nutrients and other substances in foods and in the body and the way those nutrients relate to health and disease. Nutrition also explores why you choose particular foods and the type of diet you eat.

Obese Having a body mass index of 30 or greater.

Oil A lipid that is usually liquid at room temperature.

Oral cavity The mouth.

Organic In chemistry, any compound that contains carbon.

Organic foods Foods that have been grown without most conventional pesticides, fertilizers, herbicides, antibiotics, or hormones, and without genetic engineering or irradiation.

Osteomalacia A disease of vitamin D deficiency in adults in which the leg and spinal bones soften and may bend.

Osteoporosis The most common bone disease, characterized by loss of bone density and strength; it is associated with debilitating fractures, especially in people 45 and older, due to a tremendous loss of bone tissue in midlife.

Overweight Having a body mass index of 25 or greater.

Oxalic acid An organic acid found in spinach and other leafy green vegetables that can decrease the absorption of certain minerals, such as calcium.

Palmar grasp The ability of a baby from about six months of age to grab objects with the palm of the hand.

Pasteurized A product, such as milk, that has been treated to kill harmful germs.

Pepsin The principal digestive enzyme of the stomach.

Peptidases Enzymes that break down short peptide chains into amino acids or peptides with two or three amino acids.

Peptide bonds The bonds that form between adjoining amino acids.

Peristalsis Involuntary muscular contraction that forces food through the entire digestive system.

Pernicious anemia A type of anemia caused by a deficiency of vitamin B_{12} and characterized by macrocytic anemia and deterioration in the functioning of the nervous system.

Pesco vegetarians Vegetarians who eat fish.

Pharynx A passageway that connects the oral and nasal cavities to the esophagus and air tubes to the lungs.

Photosynthesis A process during which plants convert energy from sunlight into energy stored in carbohydrate.

Phytic acid A binder found in wheat bran and whole grains that can decrease the absorption of certain nutrients, such as calcium and iron.

Phytochemicals Minute substances in plants that may reduce the risk of cancer and heart disease when eaten often.

Pincer grasp The ability of a baby at about eight months of age to use the thumb and forefinger together to pick things up.

Placenta The organ that develops during the first month of pregnancy, which provides for exchange of nutrients and wastes between fetus and mother and secretes the hormones necessary to maintain pregnancy.

Plaque (1) Deposits of bacteria, protein, and polysaccharides found on teeth that contribute to tooth decay. (2) Deposits on arterial walls that contain cholesterol, fat, fibrous scar tissue, calcium, and other biological debris.

Point of unsaturation The location of the double bond in unsaturated fatty acids.

Polypeptides Protein fragments with 10 or more amino acids.

Polyunsaturated fat A triglyceride made of mostly polyunsaturated fatty acids.

Polyunsaturated fatty acid A fatty acid that contains two or more double bonds in the chain.

Positive nitrogen balance A condition in which the body excretes less protein than is taken in; this can occur during growth and pregnancy.

Precompetition meal The meal closest to the time of a competition or event.

Precursors Forms of vitamins that the body changes chemically to active vitamin forms.

Preeclampsia Hypertension during pregnancy that can cause serious complications.

Pregnancy-induced hypertension Hypertension during pregnancy that can cause serious complications.

Press release A printed announcement by a company about its activities, written in the form of a news article and given to the media to generate publicity.

Primary (essential) hypertension A form of hypertension whose cause is unknown.

Primary structure The number and sequence of the amino acids in the protein chain.

Processed foods Foods that have been prepared using a certain procedure such as cooking, freezing, canning, dehydrating, milling, culturing, or adding vitamins and minerals.

Promoters Substances, such as fat, that advance the development of mutated cells into a tumor.

Proportionality A concept of eating relatively more foods from the larger food groups in the USDA Food Guide and fewer foods from the smaller food groups.

Proteases Enzymes that break down protein.

Protein-energy malnutrition (PEM) A broad spectrum of malnutrition from mild to serious cases; also called protein-kcalorie malnutrition.

Proteins Major structural parts of the body's cells that are made of nitrogen-containing amino acids assembled in chains; perform other functions as well; particularly rich in animal foods.

Publicity Obtaining free space or time in various media to get public notice of a program, book, and so on.

Pureeing Mashing or straining a food to a smooth pulp.

Purified water Water produced by reverse osmosis, ozonation, or other suitable processes and that meets the definition set by the U.S. Pharmacopoeia.

Pyloric sphincter A muscle that permits passage of chyme from the stomach to the small intestine.

Qualified health claims Health claims graded B, C, or D that require a disclaimer or other qualifying language to ensure that they do not mislead consumers.

Rancidity The deterioration of fat, resulting in undesirable flavors and odors.

Recommended Dietary Allowance (RDA) The dietary intake value that is sufficient to meet the nutrient requirements of 97 to 98 percent of all healthy individuals in a group.

Rectum The last section of the large intestine, in which feces, the waste products of digestion, are stored until elimination.

Reduction Boiling or simmering a liquid down to a smaller volume.

Refined or milled grain A grain in which the bran and germ are separated (or mostly separated) from the endosperm.

Registered dietitians Professionals recognized by the medical profession as the legitimate providers of nutrition care.

Retinoids The forms of vitamin A that are in the body: retinol, retinal, and retinoic acid.

Retinol A form of vitamin A found in animal foods; it can be converted to retinal and retinoic acid in the body.

Retinol activity equivalents (RAEs) The unit for measuring vitamin A. One RAE51 microgram of retinol, 12 micrograms of beta-carotene, or 24 micrograms of other vitamin A precursor carotenoids.

Rickets A childhood disease in which bones do not grow normally, resulting in bowed legs and knock knees; it is generally caused by a vitamin D deficiency.

Risk factor A habit, trait, or condition associated with an increased chance of developing a disease.

Rub A dry marinade made of herbs and spices (and other seasonings), sometimes moistened with a little oil, that is rubbed or patted on the surface of meat, poultry, or fish (which is then refrigerated and cooked at a later time).

Sales promotion Marketing activities other than advertising and public relations that offer an extra incentive.

Saliva A fluid secreted into the mouth from the salivary glands that contains important digestive enzymes and lubricates the food so that it may readily pass down the esophagus.

Salsas Chunky mixtures of vegetables and/or fruits and flavor ingredients.

Satiety A feeling of being full after eating.

Saturated fat A triglyceride made of mostly saturated fatty acids.

Saturated fatty acid A fatty acid that is filled to capacity with hydrogens.

Scurvy A vitamin C deficiency disease marked by bleeding gums, weakness, loose teeth, and broken capillaries under the skin.

Searing Exposing meat's surfaces to a high heat before cooking at a lower temperature; this process adds color and flavor to the meat.

Seasonings Substances used in cooking to bring out a flavor that is already present.

Secondary hypertension Persistently elevated blood pressure caused by a medical problem.

Secondary structure The bending and coiling of the protein chain.

Semolina The roughly milled endosperm of a type of wheat called durum wheat.

Simple carbohydrates Sugars, including monosaccharides and disaccharides.

Simple goiter Thyroid enlargement caused by inadequate dietary intake of iodine.

Small intestine The digestive tract organ that extends from the stomach to the opening of the large intestine.

Soluble fiber A classification of fiber that includes gums, mucilages, pectin, and some hemicelluloses; they are generally found around and inside plant cells.

Spices The roots, bark, seeds, flowers, buds, and fruits of certain tropical plants; they are used to season and flavor foods.

Spina bifida A birth defect in which parts of the spinal cord are not fused together properly, thus gaps are present where the spinal cord has little or no protection.

Starch A complex carbohydrate made up of a long chain of glucoses linked together; found in grains, legumes, vegetables, and some fruits; the straight form is called amylose, and the branched form is called amylopectin.

Stomach J-shaped muscular sac that holds about 4 cups of food when full and prepares food chemically and mechanically so that it can be further digested and absorbed.

Stroke Damage to brain cells resulting from an interruption of blood flow to the brain.

Structure-function claims Claims on food labels that refer to the supplement's effect on the body's structure or function, including its overall effect on a person's well-being.

Sucrose A disaccharide commonly called table sugar, granulated sugar, or simply sugar.

Sugar replacers (polyols) A group of carbohydrates that are sweet and occur naturally in plants; are used in a wide variety of low-carbohydrate foods to provide bulk and texture; they are slowly and incompletely absorbed.

Sweating Cooking slowly in a small amount of fat over low or moderate heat without browning.

Systolic pressure The pressure of blood within arteries when the heart is pumping—the top blood pressure number.

Taste Sensations perceived by the taste buds on the tongue.

Taste buds Clusters of cells found on the tongue, cheeks, throat, and roof of the mouth. Each taste bud houses 60 to 100 receptor cells. The body regenerates taste buds about every three days. These cells bind food molecules dissolved in saliva and alert the brain to interpret them.

Tertiary structure The folding and twisting of the protein chain that makes the protein able to perform its functions in the body.

Thermic effect of food The energy needed to digest and absorb food.

Thyroid gland A gland found on either side of the trachea that produces and secretes two important hormones that regulate the level of metabolism.

TLC diet A low-saturated-fat, low-cholesterol eating plan designed to fight cardiovascular disease and lower LDL; the diet calls for less than 7 percent of kcalories from saturated fat and less than 200 milligrams of cholesterol daily and also recommends only enough kcalories to maintain a desirable weight.

Tolerable Upper Intake Level (UL) The maximum intake level above which the risk of toxicity would increase.

Tonic water A carbonated water containing lemon, lime, sweeteners, and quinine.

Trace minerals Minerals needed in smaller amounts in the diet—less than 100 milligrams daily.

Trans fats (trans fatty acids) Unsaturated fatty acids that lose a natural bend or kink so that they become straight (like saturated fatty acids) after being hydrogenated; they act like saturated fats in the body.

Transitional milk The type of breast milk produced from about the third to the tenth day after childbirth, when mature milk appears.

Triglyceride The major form of lipid in food and in the body; it is made of three fatty acids attached to a glycerol backbone.

Tripeptide A peptide with three amino acids.

Tryptophan An amino acid present in protein foods that can be converted to niacin in the body.

Type 1 diabetes A form of diabetes seen mostly in children and adolescents. These patients make no insulin and therefore require frequent injections of insulin to maintain a normal level of blood glucose.

Type 2 diabetes A form of diabetes seen most often in overweight adults but increasingly seen in adolescents and children. With type 2 diabetes, either the body does not produce enough insulin or the body's cells do not use insulin properly.

Unsaturated fatty acid A fatty acid with at least one double bond.

USDA Food Guide A food guide developed by the U.S. Department of Agriculture to help healthy Americans follow the Dietary Guidelines for Americans.

Varied diet A diet in which you eat a wide selection of foods to get necessary nutrients.

Vegans Individuals eating a type of vegetarian diet in which no eggs or dairy products are eaten; their diet relies exclusively on plant foods.

Very low density lipoproteins (VLDLs) Lipoproteins made by the liver to carry triglycerides and some cholesterol through the body.

Villi Tiny fingerlike projections in the wall of the small intestines that are involved in absorption.

Vitamin D$_3$ (cholecalciferol) The form of vitamin D found in animal foods.

Vitamins Noncaloric, organic nutrients found in a wide variety of foods that are essential in small quantities to regulate body processes, maintain the body, and allow growth and reproduction.

Water balance The process of maintaining the proper amount of water in each of three body "compartments": inside the cells, outside the cells, and in the blood vessels.

Water-soluble vitamins A group of vitamins that are soluble in water and are not stored appreciably in the body; include vitamin C, thiamin, riboflavin, niacin, vitamin B$_6$, folate, vitamin B$_{12}$, pantothenic acid, and biotin.

Whole foods Foods as we get them from nature.

Whole grain A grain that contains the endosperm, germ, and bran.

Xerophthalmia Hardening and thickening of the cornea that can lead to blindness; usually caused by a deficiency of vitamin A.

Xerosis A condition in which the cornea of the eye becomes dry and cloudy; often due to a deficiency of vitamin A.

Index

CUSTOMER NOTE: IF THIS BOOK IS ACCOMPANIED BY SOFTWARE, PLEASE READ THE FOLLOWING BEFORE OPENING THE PACKAGE.

This software contains files to help you utilize the models described in the accompanying book. By opening the package, you are agreeing to be bound by the following agreement: